General Oceanography

GENERAL OCEANOGRAPHY

an Introduction

SECOND EDITION

Günter Dietrich
Kurt Kalle
Wolfgang Krauss
Gerold Siedler

Oceanographic Institute, University of Kiel
German Hydrographic Institute, Hamburg
Federal Republic of Germany

Translated by
Susanne and Hans Ulrich Roll
Hamburg

A WILEY-INTERSCIENCE PUBLICATION

JOHN WILEY & SONS
New York • Chichester • Brisbane • Toronto

Translated from Günter Dietrich, Kurt Kalle, Wolfgang Krauss, and Gerold Siedler, *Allgemeine Meereskunde,* Third Edition, © 1975 by Gebrüder Borntraeger Verlagsbuchhandlung, Berlin-Stuttgart.

Copyright © 1980 by John Wiley & Sons, Inc.

All rights reserved. Published simultaneously in Canada.

Reproduction or translation of any part of this work beyond that permitted by Sections 107 or 108 of the 1976 United States Copyright Act without the permission of the copyright owner is unlawful. Requests for permission or further information should be addressed to the Permissions Department, John Wiley & Sons, Inc.

Library of Congress Cataloging in Publication Data:

Main entry under title:
General oceanography

   Translation of *Allgemeine Meereskunde.*
   Edition of 1963 by G. Dietrich alone.
   "A Wiley-Interscience publication."
   Bibliography: p.
   Includes index.
    1. Oceanography. I. Dietrich, Günter.
General oceanography.
GC11.2.A4413   1980   551.46   80-12919
ISBN 0-471-02102-4

Printed in the United States of America

10 9 8 7 6 5 4 3 2 1

From the Preface of 1957

The more insight we gain into the interrelation of phenomena, the easier it will be to free ourselves from the misconception that not all branches of the natural sciences, be they the observing and describing branches, the investigation of chemical components, or the exploration of the everywhere existing physical forces of matter, are of equal importance for the culture and prosperity of mankind.

Alexander von Humboldt
Kosmos I, 1845

Almost 50 years have passed since Otto Krümmel published the last comprehensive German treatise on oceanography. Since then, and especially in the last decade, the science of the oceans has received great support in numerous countries and, consequently, has expanded and deepened extraordinarily. In addition to descriptive geographical considerations quantitative solutions to the problems are being sought more and more. Thus, present-day oceanography has become a part of geophysics.

The aim of this book is to provide an insight into all branches of oceanography and thereby to point out to the reader the various results which are hidden in the extensive and widely scattered literature. At the same time, the book attempts to organize the great number of specific results and present them in such a way that general relationships become apparent. For, following the above words of Alexander von Humboldt, this book endeavors to present the problems and the general importance of oceanography not only to a group of specialists but to a larger group of people interested in natural sciences. Last but not least, this book is intended to aid in the instruction and study of geography, physics, chemistry, and biology.

With these aims in mind, this book should not be considered a handbook which claims to cover the subject completely. On the contrary, it has been kept intentionally within the framework of an introduction to the science of oceanography, which includes many scientific branches. The extensive material is presented with deliberate conciseness. Furthermore, lengthy descriptions have been made unnecessary by the presentation of numerous charts and diagrams. The list of references may also serve as a short guide.

Preface to the Third German Edition
(Translated for the Second English Edition)

The aim of the first edition of this book from the year 1957 also holds for the completely revised third edition. The authors have strived to consider the development that took place within the past 15 years. In view of the enormous expansion of oceanography during this period many a restriction was necessary to keep the book within the original frame. The guiding concept was that, more than in the first edition, Chapters 1 to 9 should help prepare a systematic base for the comprehensive description of the world ocean in Chapter 10.

Kurt Kalle contributed Chapter 6 on the chemical budget of the ocean and the other sections from the field of marine chemistry. Wolfgang Krauss revised Chapters 7 and 8 on the theory of ocean currents, on surface waves and internal waves, as well as some sections on the theory of ocean tides. Gerold Siedler rewrote much of the old Chapters 2 and 3, namely Sections 2.1.2 to 2.1.12 on the physical properties of seawater, and 3.1.1 to 3.2.13 on the instruments and methods applied in physical oceanography. Special attention was paid to supplementing the terse text by graphic information in the form of charts, diagrams, and photographs, as well as by the eight plates in the annex. These materials were specially chosen so that they, in combination with their comprehensive legends, offer an introduction to oceanography.

It is a difficult job to thank all those who contributed to the new edition of this book because there are so many of them. I wish to express my gratitude to my colleagues, Professors K. Kalle, W. Krauss, and G. Siedler, who directly contributed to parts of the book, and also to the University of Hawaii in Honolulu for granting me a one-year visiting professorship at the Department of Oceanography which enabled me to prepare this book. I wish to thank many of the colleagues and assistants at the Institut für Meereskunde an der Universität Kiel who helped to clarify various problems in many discussions. My special thanks go to my closest collaborators, Mrs. Regina Bröcker whose experience with scientific literature over many years was a great help, and the cartographer Mr. Alfred Eisele who, together with his staff, contributed to the precision and clarity of the charts and diagrams.

My sincere gratitude for the indefatigable interest in this book goes to the Publisher Borntraeger-Schweizerbart, Berlin-Stuttgart.

GÜNTER DIETRICH

Kiel
September 1972

Günter Dietrich
1911–1972

Immediately after finishing his part of this book Günter Dietrich died in Kiel on October 2, 1972. The new edition was begun on his initiative and until his very last days he devoted much of his energy and enthusiasm to this book. His contribution was slightly changed only where it was absolutely necessary in order to bring it into a line with the other chapters.

With his death this book has become the final work of a scientist who always endeavored to conceive marine science as an entity in spite of all the specialized studies that are required as well.

WOLFGANG KRAUSS
GEROLD SIEDLER

Kiel
September 1975

Translators' Preface

The first English edition of Günter Dietrich's *General Oceanography*—ably translated from German by Feodor Ostapoff—was published in 1963. It was well received by the international oceanographic community.

The third German edition was completely rewritten by Dietrich and his three coauthors in the early seventies and appeared—somewhat delayed by the sudden death of the main author—at the end of 1975.

In the light of the success of the first English edition, a second English edition was envisaged, to be based on a translation of the third German edition. Since Feodor Ostapoff was no longer available as translator, another solution had to be found. In 1975, the authors approached us with the proposal that we assume responsibility for the second English edition. Although we neither speak English as our mother tongue nor live in the United States as Feodor Ostapoff does, we dared to take upon ourselves this difficult task in order to contribute to the international dissemination of this recognized oceanographic textbook.

In preparing this English version we tried to keep, as closely as possible, to the original German text. More emphasis was laid on accurate reproduction in English of the German wording than on stylistic perfection. During our work we enjoyed encouragement and assistance freely given by many oceanographers from the United States, United Kingdom, Canada, and Australia, who never failed in giving advice on the translation of technical terms when no help could be found in any dictionary. We wish to express our sincere gratitude to our colleagues for their friendly assistance.

If this translated text turns out to be readable it is also due to the effort of the editorial staff at Wiley-Interscience, who scrutinized our manuscript very carefully and, occasionally, proposed changes. We accepted those changes that did not deviate too much from the meaning of the original German text, and rejected any changes for which we—as translators—did not feel authorized.

SUSANNE AND HANS ULRICH ROLL

Hamburg
August 1980

Contents

1. **Geomorphology of the Ocean Bottom**   1

    1.1 **Topography of the Ocean Bottom**   1

         1.1.1. Distribution of land and water, 1
         1.1.2. Boundaries and names of oceans, 1
         1.1.3. Size of the oceans, 2
         1.1.4. Development of knowledge about the relief of the sea floor, 2
         1.1.5. Statistics of depth distribution, 6
         1.1.6. Large-scale topographic features of the ocean bottom, 8
         1.1.7. Terms and definitions of ocean bottom features, 13

    1.2. **Sediments at the Ocean Bottom**   16

         1.2.1. Nature of the ocean bottom, 16
         1.2.2. Methods of investigation of marine sedimentology, 16
         1.2.3. Origin of the sediments, 19
         1.2.4. Grain size distribution of the sediments, 24
         1.2.5. Chemical composition of the sediments, 24
         1.2.6. Geographic distribution of the sediments, 25

    1.3. **Mechanism of Sedimentation**   26

         1.3.1. Physical defining quantities, 26
         1.3.2. Erosion, 27
         1.3.3. Transport, 28
         1.3.4. Sedimentation, 32

    1.4. **Formation of the Bottom Features**   33

         1.4.1. Aids for interpreting the morphology of the ocean bottom, 33
         1.4.2. Continental margins, 41
         1.4.3. Mid-Ocean Ridge, 46
         1.4.4. Deep-sea basins, 48
         1.4.5. Seamounts and oceanic islands, 49
         1.4.6. Formation of the oceans, 50

2. Physical Properties and Chemical Composition of Seawater 54

2.1. Physical Properties of Seawater 54

- 2.1.1. Unique characteristics of pure water, 54
- 2.1.2. Salinity as a thermodynamic variable, 60
- 2.1.3. Electrical conductivity, 61
- 2.1.4. Density, thermal expansion, and compressibility, 63
- 2.1.5. Specific heat, evaporative heat, and adiabatic temperature change, 66
- 2.1.6. Molecular thermal conductivity and diffusion, 68
- 2.1.7. Osmosis, vapor pressure, freezing-point depression, and boiling-point increase, 69
- 2.1.8. Turbulent exchange coefficients, 71
- 2.1.9. Viscosity and surface tension, 72
- 2.1.10. Acoustic properties, 74
- 2.1.11. Optical properties, 78
- 2.1.12. Physical properties of sea ice, 85

2.2. Chemical Composition of Seawater 88

- 2.2.1. Major constituents of salinity, 88
- 2.2.2. Trace elements, 89
- 2.2.3. Gases, 89
- 2.2.4. Organic substances, 94
- 2.2.5. Suspended matter and living organisms in seawater, 96

3. Oceanographic Instruments and Observational Methods 98

3.1. Platforms for Oceanographic Observations 98

- 3.1.1. Research vehicles, 98
  - 3.1.1.1. Research vessels, 98
  - 3.1.1.2. Submersibles, 103
  - 3.1.1.3. Position fixing, 107
  - 3.1.1.4. Computers, 109
  - 3.1.1.5. Airplanes and satellites, 110
- 3.1.2. Fixed observation platforms, 111
  - 3.1.2.1. Moored systems, 111
  - 3.1.2.2. Research towers and underwater laboratories, 112
- 3.1.3. Drifting platforms, 114

3.2. Oceanographic Instruments and Methods of Measuring 116

- 3.2.1. Measured quantities, 116
- 3.2.2. Basic requirements on oceanographic measuring technique, 116
- 3.2.3. Water sampling devices, 120
- 3.2.4. Measurement of water depth, 122
- 3.2.5. Measurement of instrument depth, 127
- 3.2.6. Measurement of water level variations, 128
- 3.2.7. Measurement of ocean surface waves, 131
- 3.2.8. Measurement of temperature, 134
- 3.2.9. Measurement of salinity, 137
- 3.2.10. Measurement of sound velocity, 140
- 3.2.11. Measurement of density, 142
- 3.2.12. Measurement of currents, 143
  - 3.2.12.1. Measurements at a fixed position, 143
  - 3.2.12.2. Drift measurements, 148
- 3.2.13. Measurement of the content of suspended material, 150
- 3.2.14. Measurement of the content of dissolved substances, 152

4. Energy and Water Budgets of the World Ocean 158

4.1. Small-Scale Transfer Processes in the Atmospheric Boundary Layer over the Ocean 158

- 4.1.1. Transfer processes between ocean and atmosphere, 158
- 4.1.2. Methods for the determination of vertical transports in the maritime friction layer, 160
- 4.1.3. Vertical fluxes of momentum, heat, and water vapor from profile measurements, 163
- 4.1.4. Parameterization of the vertical transports, 164
- 4.1.5. Transfer of gases and salts, 166

4.2. Large-Scale Heat Budget 168

- 4.2.1. General remarks on the heat budget, 168
- 4.2.2. Heat transfer by incoming radiation, 169
- 4.2.3. Heat transfer by back radiation, 175
- 4.2.4. Heat transfer by direct thermal conduction, 176
- 4.2.5. Heat transfer by evaporation, 179
- 4.2.6. Total heat balance, 182
- 4.2.7. Propagation of heat in the ocean, 186

4.3. Large-Scale Water Budget — 189

- 4.3.1. The planetary wind system at the sea surface, 189
- 4.3.2. Special regional wind fields (monsoons, tropical hurricanes, storm cyclones), 191
- 4.3.3. Evaporation, 193
- 4.3.4. Precipitation, 196
- 4.3.5. Continental runoff, 197
- 4.3.6. The hydrologic cycle on the earth, 198

5. Temperature, Salinity, Density, Characteristic Water Masses, and Ice in the World Ocean — 203

5.1. Temperature in the World Ocean — 203

- 5.1.1. Diurnal variation of temperature near the sea surface, 203
- 5.1.2. Annual variation of surface temperature, 206
- 5.1.3. Distribution of surface temperature, 209
- 5.1.4. Distribution of water temperature with depth, 212
- 5.1.5. Long-term changes of temperature, 217

5.2. Salinity in the World Ocean — 222

- 5.2.1. Distribution of surface salinity, 222
- 5.2.2. Distribution of salinity with depth, 225
- 5.2.3. Diurnal and annual variations of surface salinity, 226
- 5.2.4. Long-term changes of salinity, 228

5.3. Density in the World Ocean — 230

- 5.3.1. Distribution of density at the sea surface, 230
- 5.3.2. Distribution of density with depth, 231
- 5.3.3. Variations of density, 231

5.4. Characteristic Water Masses in the World Ocean — 232

- 5.4.1. Temperature–salinity relationship, 232
- 5.4.2. Formation of characteristic water masses, 233
- 5.4.3. Characteristic water masses in the world ocean, 236

5.5. Ice in the World Ocean — 238

- 5.5.1. Formation and types of ice, 238
- 5.5.2. Ice coverage of the world ocean, 244
- 5.5.3. Ice patrol, 246

6. Chemical Budget of the Ocean — 251

6.1. Marine Geochemistry — 251

- 6.1.1. General fundamentals of geochemistry, 251
- 6.1.2. Regulating mechanisms for the chemical elements in seawater, 253
- 6.1.3. Sedimentation budget, 255
- 6.1.4. Primary natural radioactivity, 256
- 6.1.5. Radioactivity generated by cosmic rays, 257
- 6.1.6. Radioactivity produced artificially, 259

6.2. Biochemistry of the Ocean — 261

- 6.2.1. General fundamentals of biochemistry, 261
- 6.2.2. The ocean as the source of life on earth, 264
- 6.2.3. The energy production of organisms, 266
- 6.2.4. Organic production of the ocean, 271
- 6.2.5. The mechanism of plankton metabolism, 275
- 6.2.6. The carbon dioxide–calcium carbonate system, 280
- 6.2.7. Regional distribution of nutrients and oxygen, 284

7. The Theory of Ocean Currents — 288

7.1. The System of Hydrodynamic Equations — 288

- 7.1.1. Equations of motion in the absolute coordinate system, 288
- 7.1.2. The deflecting force of the rotation of the earth, 289
- 7.1.3. Field of gravity, 290
- 7.1.4. Equations of motion in a rotating coordinate system, 291
- 7.1.5. Equation of continuity, 292
- 7.1.6. Equations of thermal conduction and diffusion, 294
- 7.1.7. Boundary conditions, 294
- 7.1.8. Friction, turbulence, and mixing, 295

7.2. Statics and Kinematics — 299

- 7.2.1. Field of mass, 299
- 7.2.2. Field of pressure, 302
- 7.2.3. Determination of the relative field of pressure, 303
- 7.2.4. Representation of the relative field of pressure, 304
- 7.2.5. Stability of water stratification, 304
- 7.2.6. Representation of the field of motion, 307

7.3. Stationary Currents — 308
- 7.3.1. Geostrophic currents in a homogeneous ocean, 308
- 7.3.2. Geostrophic currents in a two-layer ocean, 310
- 7.3.3. Geostrophic currents in a continuously stratified ocean, 314
- 7.3.4. Drift current in a homogeneous ocean, 319
- 7.3.5. Ekman's elementary current system, 321
- 7.3.6. Sverdrup regime, 321
- 7.3.7. Linear theory of the western boundary currents, 324
- 7.3.8. Nonlinear theory of the western boundary currents, 327

7.4. Nonstationary Currents — 328
- 7.4.1. Currents and waves, 328
- 7.4.2. Inertial waves, 329
- 7.4.3. Planetary waves, 331

7.5. The Influence of Bottom Topography on Ocean Currents — 332
- 7.5.1. Potential vorticity, 332
- 7.5.2. Topographic Rossby waves, 333

7.6. Thermohaline Circulation — 334
- 7.6.1. Large-scale thermohaline processes, 334
- 7.6.2. Coastal currents in higher latitudes, 338
- 7.6.3. Compensation currents in ocean straits, 338
- 7.6.4. Thermoclines, 340
- 7.6.5. Numerical solutions regarding the general circulation, 341

8. Surfaces Waves and Internal Waves — 343

8.1. Classification of Waves — 343
- 8.1.1. Progressive and standing waves, surface waves and internal waves, 343
- 8.1.2. Classification with respect to restoring forces, 343
- 8.1.3. Classification of gravity and capillary waves with respect to generating forces, 344

8.2. Kinematic Properties of Waves — 346
- 8.2.1. Harmonic oscillations and wave fields, 346
- 8.2.2. Standing waves, 349
- 8.2.3. Damped waves and forced waves, 349

8.3. Short Surface Waves or Deep-Water Waves — 352

- 8.3.1. Gravity waves, 352
- 8.3.2. Capillary waves, 353
- 8.3.3. Waves of finite amplitude, 354
- 8.3.4. Nonlinear interactions, 355
- 8.3.5. Properties of sea and swell, 357
- 8.3.6. Statistical description of the sea state, 368
- 8.3.7. Generation of the sea state, 372
- 8.3.8. Wave transformation in shallow water; surf, 375

8.4. Long Surface Waves — 378

- 8.4.1. Properties of long waves, 378
- 8.4.2. Tsunamis and storm surges, 379
- 8.4.3. Edge waves, Kelvin waves, and double Kelvin waves, 385
- 8.4.4. Seiches, 387
- 8.4.5. Amphidromic systems, 391
- 8.4.6. The influence of friction on long waves, 394

8.5. Boundary Surface Waves and Internal Waves — 398

- 8.5.1. Boundary surface waves, 398
- 8.5.2. Internal waves, 400
- 8.5.3. Internal tidal waves and inertial waves, 402
- 8.5.4. Stability oscillations, 405

9. Tidal Phenomena — 407

9.1. Definitions Concerning Tidal Phenomena — 407

- 9.1.1. Tides and tidal currents, 407
- 9.1.2. Reference levels of tides, 410

9.2. Tide-Producing Forces — 411

- 9.2.1. Description of the system of tide-producing forces, 411
- 9.2.2. Derivation of the tidal potential, 413
- 9.2.3. Harmonic expansion of the tidal potential, 415

9.3. Representation of Ocean Tides — 417

- 9.3.1. Harmonic representation of tides and tidal currents, 417
- 9.3.2. Analysis of tide observations, 421
- 9.3.3. Harmonic method of tide prediction, 425
- 9.3.4. Tide characteristics, 426

9.4. Theory of Ocean Tides — 428

    9.4.1. The scope of the theory of ocean tides, 428
    9.4.2. Equilibrium theory of tides, 428
    9.4.3. Classic hydrodynamic theories, 429
    9.4.4. Co-oscillating tides, 431
    9.4.5. Numerical integration of the tide equations, 435

9.5. Tidal Phenomena of the World Ocean — 438

    9.5.1. Oceanic tides, 438
    9.5.2. Tides of adjacent seas, 442
    9.5.3. Tidal currents, in general, 446
    9.5.4. Tidal currents of the oceans, 449
    9.5.5. Tidal currents of adjacent seas, 449
    9.5.6. Superposition of astronomic tides, 454
    9.5.7. Friction of tidal currents, 455
    9.5.8. Turbulence of tidal currents and its consequences, 457

10. Regional Oceanography — 460

10.1. Stratification and Circulation in the Deep Layers of the Three Oceans — 460

    10.1.1. Water masses of the cold-water sphere, 460
    10.1.2. Antarctic bottom water, 461
    10.1.3. Arctic bottom water, 465
    10.1.4. Subpolar intermediate water, 467
    10.1.5. Deep water, 470
    10.1.6. Antarctic water masses, 474
    10.1.7. Water masses of the warm-water sphere, 476
    10.1.8. Circulation gyres of water masses, 481
    10.1.9. Variability in the ocean, 483

10.2. Stratification and Circulation of Large, Deep Mediterranean Seas — 488

    10.2.1. Circulation scheme of Mediterranean and adjacent seas, 488
    10.2.2. The European Mediterranean Sea, 490
    10.2.3. The Red Sea, 495
    10.2.4. The Austral–Asiatic Mediterranean Sea, 497
    10.2.5. The American Mediterranean Sea, 498
    10.2.6. The Arctic Mediterranean Sea, 501

10.3. Hydrographic Regions of the World Ocean — 504

- 10.3.1. Regional classification of the oceans, 504
- 10.3.2. The regions of trade wind currents T, 507
  - 10.3.2.1. Regions of trade wind currents T_E with components of motion strongly directed toward the equator, 508
  - 10.3.2.2. Regions of trade wind currents T_w with strictly westward motion, 512
  - 10.3.2.3. Regions of trade wind currents T_p with components of motion strongly directed poleward, 518
- 10.3.3. The regions of equatorial countercurrents E, 518
- 10.3.4. The regions of monsoon currents M, 521
- 10.3.5. The regions of horse latitudes H, 525
- 10.3.6. The jet stream regions J, 528
- 10.3.7. The regions of west wind drift W_E and W_P, 536
- 10.3.8. The polar regions P_I and P_O, 544
- 10.3.9. Shelf seas, 550
- 10.3.10. Coastal waters, 558

11. Appendix — 566

11.1. Salinity as a Function of Electrical Conductivity, Temperature, and Pressure — 566

11.2. Density as a Function of Temperature, Salinity, and Pressure — 567

11.3. Potential Temperature as a Function of In Situ Temperature, Salinity, and Pressure — 568

11.4. Viscosity of Seawater as a Function of Temperature, Salinity, and Pressure — 569

11.5. Sound Velocity as a Function of Temperature, Salinity, and Pressure — 569

Bibliography — 571

Author Index — 605

Subject Index — 613

General
Oceanography

1 Geomorphology of the Ocean Bottom

1.1. Topography of the Ocean Bottom

1.1.1. Distribution of land and water

The surface of the earth exhibits great irregularities, with height differences of nearly 20 km between the deepest trough, 11,022 m observed at the Vitiaz Depth in the Mariana Trench, and the highest elevation, 8848 m at Mount Everest. Under the action of gravity, the water presently available on earth (amounting to 1350×10^6 km^3) accumulates in the troughs, covering the earth only to a certain level thus determining the most important boundary on the face of the earth, that between land and ocean. For the present configuration of the earth's crust, the amount of available water suffices to form a single, largely subdivided, but connected world ocean, which covers 70.8% of the 510.1×10^6 km^2 area of the earth's surface.

A further characteristic of the distribution of land and water is of special importance to the oceanographic conditions: the percentage of the area covered by water almost continuously increases from 70°N to 60°S. This water coverage ranges from 28.7% of the area between 65 and 70°N to 99.9% between 55 and 60°S. In the Northern Hemisphere, a total of 60.7% of the surface is covered by water, whereas in the Southern Hemisphere it is 80.9%. So both hemispheres, especially the Southern Hemisphere, have a predominantly oceanic nature. If one thinks of the Earth's surface as being divided into two hemispheres with one containing most of the land areas and the other most of the water areas, the "land hemisphere" with its pole at $47\frac{1}{4}$°N $2\frac{1}{2}$°W (just off the mouth of the Loire) would still have 53% of its surface covered by water, whereas 89% of the "water hemisphere" (with its pole near New Zealand) would be covered by water.

1.1.2. Boundaries and names of oceans

The continents and the islands off their coasts divide the whole world ocean into three parts. A natural boundary is missing only in the Southern Hemisphere where the Antarctic belt of water provides free exchange among the oceans. In scientific and nautical terminology (International Hydrographic Bureau, 1953), a common understanding has been reached not to distinguish separate polar oceans. The different oceans are considered to begin at the southern tips of the three land masses (Plate 1). The meridian of Cape Agulhas (20°E) has been chosen as the boundary between the Atlantic Ocean and the Indian Ocean, the meridian of the South East Cape of Tasmania (147°E) as the boundary between the Indian Ocean and the Pacific Ocean, and the meridian of Cape Horn (68°W) as the boundary between the Pacific and the Atlantic Oceans. The Arctic Ocean is considered to be part of the Atlantic Ocean, so that the Bering Strait forms the boundary between the Atlantic Ocean and the Pacific Ocean. In all three oceans, the equator is used to distinguish the hemispheric parts under their own names (North Atlantic Ocean, South Atlantic Ocean, etc.). Further subdivisions according to "natural regions" have been devised, which lead to different boundaries, depending on the systems used. However, none of these subdivisions has been generally accepted.

Continents and chains of islands more or less separate certain oceanic areas from the open ocean, thus forming adjacent seas. Such seas are called marginal seas when they form merely an indentation on the continental coast, or mediterranean seas when they are enclosed to a great extent by land. The latter can be subdivided into large intercontinental and small intracontinental mediterranean seas. The group of the large mediterranean seas consists of the European, the American, and the Arctic Mediterranean Seas, which are adjacent seas of the Atlantic Ocean, and the Austral–Asiatic Mediterranean Sea, which is an adjacent sea of the Pacific Ocean. Intracontinental mediterranean seas are the Baltic Sea, the Hudson Bay, the Persian Gulf, and the Red Sea. The North Sea, the Gulf of St. Lawrence, the Bering Sea, the Sea of Okhotsk, the Sea of Japan, and the East China Sea are examples of marginal seas. Plate 1 gives the boundaries and names of the oceans and their adjacent seas as based on international agreement (International Hydrographic Bureau, 1953).

1.1.3. Size of the oceans

For many oceanographic problems the size of the surface area and the water content of the sea are of great importance. Table 1.01 gives a survey thereof, with additional data on the mean and maximum depths of the respective ocean areas. According to this table, the Pacific Ocean is approximately as large as the other two oceans combined. The Atlantic Ocean is comparable in size to the area of Europe, Asia, and Africa, combined. Table 1.01 also shows that, compared to the world oceans, the adjacent seas appear small and therefore, in general, exert only a slight influence on the processes in the oceans.

1.1.4. Development of knowledge about the relief of the sea floor

The distribution of land and water can be considered as almost completely known. However, although questions regarding the coastal contour lines in the Antarctic were thoroughly treated during the International Geophysical Year (IGY) in 1957–1958, many boundaries cannot be determined unambiguously because the inland ice extends below the sea level and completely fills up large, shallow sea areas. The variable edge of the shelf ice determines the boundary of the sea. This is illustrated by Fig. 1.01, which also represents the conditions of the topography of the solid earth beneath the sea. The data are based on measurements of the thickness of the ice and the elevation of the surface of the inland ice. Thus, without inland ice, Antarctica would consist of several islands, some of them as high as 2000 m, but it would also have adjacent seas with depths of down to 2500 m. The ocean bottom is covered by inland ice up to 4000 m thick. If all the Antarctic inland ice melted, the water level of the world ocean would rise by 50 m, increased by 10 more meters if all the continental ice of the earth also melted. This represents the greatest possible eustatic rise of the sea level on the earth, as demonstrated in Fig. 1.22.

The exact course of the coast lines has been surveyed and charted with different degrees of accuracy. This can be seen from the nautical charts that are being kept up to date by the hydrographic offices of the maritime nations. Coastal charts of the scale of

Table 1.01. Area, Volume, Mean and Maximum Depths of the Oceans and their Adjacent Seas

Sea	Area[a] (10⁶ km²)	Volume[a] (10⁶ km³)	Depth Mean[a] (m)	Depth Maximum[b] (m)
Oceans without adjacent seas				
Pacific Ocean	166.24	696.19	4188	11,022[c]
Atlantic Ocean	84.11	322.98	3844	9,219[d]
Indian Ocean	73.43	284.34	3872	7,455[e]
Total	323.78	1303.51	4026	—
Intercontintental mediterranean seas				
Arctic[f]	12.26	13.70	1117	5,449
Austral-Asiatic[g]	9.08	11.37	1252	7,440
American	4.36	9.43	2164	7,680
European[h]	3.02	4.38	1450	5,092
Total	28.72	38.88	1354	—
Intracontinental mediterranean seas				
Hudson Bay	1.23	0.16	128	218
Red Sea	0.45	0.24	538	2,604
Baltic Sea	0.39	0.02	55	459
Persian Gulf	0.24	0.01	25	170
Total	2.31	0.43	184	—
Marginal seas				
Bering Sea	2.26	3.37	1491	4,096
Sea of Okhotsk	1.39	1.35	971	3,372
East China Sea	1.20	0.33	275	2,719
Sea of Japan	1.01	1.69	1673	4,225
Gulf of California	0.15	0.11	733	3,127
North Sea	0.58	0.05	93	725[i]
Gulf of St. Lawrence	0.24	0.03	125	549
Irish Sea	0.10	0.01	60	272
Remaining seas	0.30	0.15	470	—
Total	7.23	7.09	979	—
Oceans, including adjacent seas				
Pacific Ocean	181.34	714.41	3940	11,022[c]
Atlantic Ocean	106.57	350.91	3293	9,219[d]
Indian Ocean	74.12	284.61	3840	7,455[e]
World ocean	362.03	1349.93	3729	11,022[c]

[a] After Menard and Smith (1966).
[b] After Ulrich (1968).
[c] Vitiaz Depth in the Mariana Trench.
[d] Milwaukee Depth in the Puerto Rico Trench.
[e] Planet Depth in the Sunda Trench.
[f] Consisting of Arctic Ocean, Barents Sea, Canadian Archipelago, Baffin Bay, and Hudson Bay.
[g] Including Andaman Sea.
[h] Including Black Sea.
[i] In the Skagerrak area.

1:50,000, and of even greater scale, are available for many seas. However, knowledge of other coastal regions is barely sufficient for the construction of charts of general character on the scale of 1:1,000,000.

Even less is known about the vertical dimension of the oceans, namely the relief of

4 Geomorphology of the Ocean Bottom

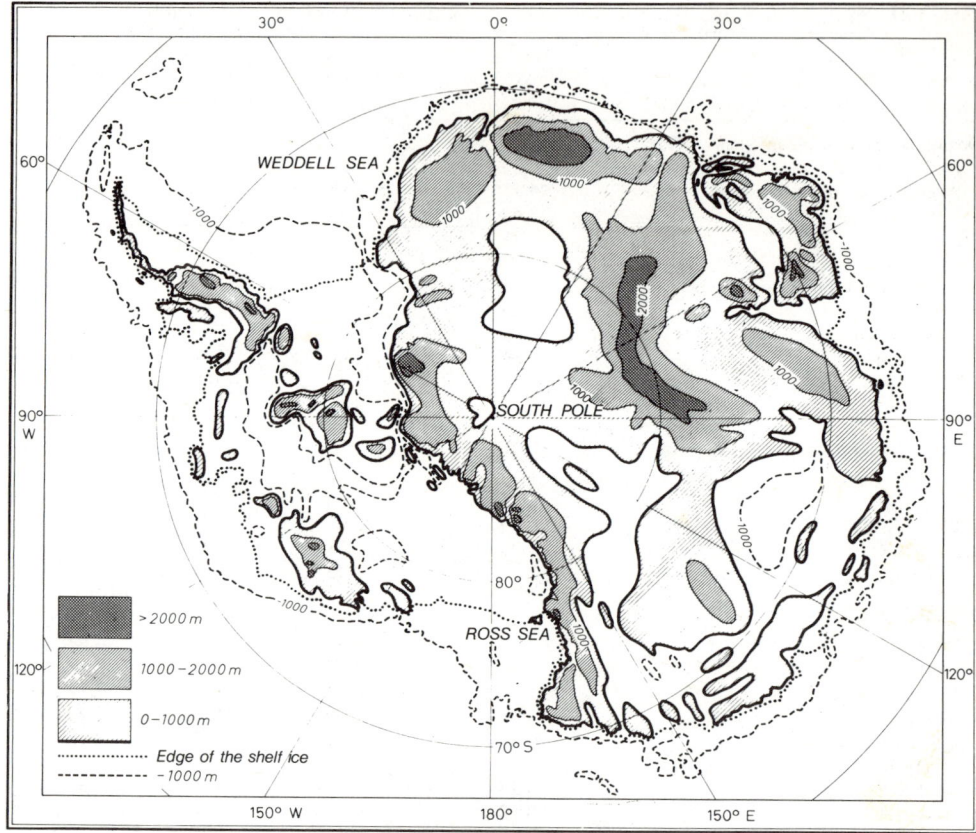

Fig. 1.01. Subglacial relief of Antarctica. (After Avsyuk et al., 1966.)

the ocean bottom. Depth determination has been one of the main tasks of oceanographers for more than 100 years. In 1845, Alexander von Humboldt still wrote in his book *Kosmos* that the depth of the ocean was almost completely unknown, a statement that may surprise many since the taking of soundings was common practice among seafarers even at that time. However, this practice was restricted to shallow waters for practical purposes of navigation. Even medieval descriptions of coasts contained information regarding soundings for navigational purposes.

The first reliable answers to questions regarding oceanic depth were obtained only after sounding methods were successfully developed. The first successful deep-sea sounding, of 4435 m, was obtained by Sir J. Clark Ross in 1840 on his cruise to Antarctic waters. This depth was later verified. The American, M. F. Maury, one of the founders of oceanography, later improved the methods and, in 1854, he published the first bathymetric chart that covered the Atlantic Ocean from 10°S to 52°N. Knowledge of the relief of the ocean bottom, at first pursued for purely scientific reasons, was soon sought after for practical reasons. In 1858, the first transatlantic cable connection was accomplished between Newfoundland and Ireland. Preliminary soundings had been taken prior to the laying of the cable to determine the most suitable path for it.

Application of the improved sounding technique was not restricted to cable laying.

After the first experimental physical-chemical and biological investigations in the deep sea, undertaken by the English vessels *Lightning* and *Porcupine* in the North Atlantic Ocean (in 1868 and 1869–1870), Sir Wyville Thomson succeeded in organizing the voyage of the *Challenger* (1872–1876) into all three oceans; even today this remains the most extensive oceanographic expedition ever made. Apart from other oceanographic investigations, numerous soundings were obtained and bottom samples were collected. At about the same time, similar work was done by the German corvette *Gazelle,* although with a less comprehensive program. Many other expeditions were organized later on. A survey of the historical development of the knowledge on the ocean bottom and its subsoil until about 1900 has been given in a study by Pfannenstiel (1970).

With the introduction of the echo sounder, new information about the topography of the ocean bottom became available. In 1919, technical problems were solved by A. Behm, making echo sounding more reliable. The first profile between the United States and Gibraltar was taken in 1922 by the American destroyer *Stewart,* and the first systematic sounding of an ocean was accomplished by the German research and survey vessel *Meteor* in the South Atlantic Ocean from 1925 to 1927. Despite the vast amount of new soundings obtained in the past decades by means of echo sounding, knowledge of the bottom relief has not increased at the same rate. Echo sounders, especially echographs, provide full detail of the depth distribution along the courses taken by ships (examples are shown in Figs. 1.04 and 1.05). However, as long as ships' courses run at large distances from each other, it is impossible to obtain a complete three-dimensional picture of the ocean bottom.

The bathymetric and nautical charts reflect the increased knowledge of the bottom relief. Besides the great number of special charts covering specific areas of the world ocean published over the years, charts covering the whole ocean are also available. Classic examples of cartographical treatment are the bathymetric charts of the oceans by Groll (1912), on the scale of 1:40,000,000, and the bathymetric chart of the Atlantic Ocean by Stocks and Wüst (1935), on the scale of 1:20,000,000. Among the best recent representations are the three ocean charts published by the Governmental Geological Committee of the USSR in 1963 and 1964. Dietrich and Ulrich (1968) used these charts as a basis for the atlas of the bottom topography of the world ocean, on the scale of 1:25,000,000. This atlas demonstrates that the topographic large-scale forms, as far as they can be represented on this scale at all, are now well known. Revolutionary discoveries such as those made in the middle of this century in the Arctic Ocean (i.e., division of the North Polar Basin, discovered by Nansen, into four basins by three deep-sea ridges) and in the Indian Ocean (i.e., the East Indian Ridge) cannot be expected in the future insofar as large-scale forms are concerned.

Most bathymetric charts are restricted to the representation of the topography by depth contours and show only scattered sounding data. The following methods were chosen to make the necessary original soundings available:

1. Frequent publication of all soundings in lists containing the position of all the stations at which they were taken, as it had already been customary at the American, British and German Hydrographic Offices. However this resulted in an impractical collection of data. Therefore the program was discontinued.
2. Registration of all soundings in charts for publication, including the bathymetric chart of the world ocean presented on 24 different sheets on the scale of 1:10,000,000 (GEBCO, General Bathymetric Chart of the Oceans),

published by the International Hydrographic Bureau in Monaco. The first edition of this compilation of charts, prepared by Thoulet and published in 1904, was due to the initiative of a patron of marine sciences, Prince Albert I of Monaco, and fulfilled its purpose at that time. The present revised fourth edition contains a valuable bathymetric chart but can, by no means, represent all soundings. A second attempt at a cartographic representation of soundings was made by Stocks (1937/61) in the *Wissenschaftliche Ergebnisse der Deutschen Atlantischen Expedition auf dem Forschungs- und Vermessungsschiff Meteor* 1925-1927. The five charts published on the scale of 1:5,000,000 will, in the long run, not be able to record the immense amount of soundings. In the third and most recent plan the burden of the work has been distributed on many shoulders. Plotting sheets, on the scale of 1:1,000,000, are to be kept up to date internationally by 17 national hydrographic offices (Ermel, 1966). The total number of sheets will amount to 603, 26 of which, covering the northern North Atlantic Ocean from 48 to 72°N, are taken care of by the German Hydrographic Institute in Hamburg. This compilation was first published in 1965. In some ocean areas the scale of 1:1,000,000 will not be sufficient to represent the existing soundings adequately. Recent improvements of the accuracy of the determination of the ship's position (see Section 3.1.1.3) permit the use of large-scale bathymetric charts up to the scale of 1:100,000, even for the open ocean. The bathymetric chart of the Great Meteor Seamount in the North Atlantic Ocean on the scale of 1:250,000 by Ulrich (1969) is an example, a reproduction of which (on a reduced scale) is given in Fig. 1.06.*

1.1.5. Statistics of depth distribution

If the frequency of the height and depth steps is determined with the aid of topographical charts as Meinardus did in 1942, two conspicuous maxima appear, the first at $+100$ m, and a second at -4950 m. These maxima indicate that the distribution of the heights and depths is not a random one, otherwise the mean level of the earth's crust (-2430 m) would occur most frequently.

The mean depth of the world ocean is 3729 m, as shown in Table 1.01. This depth will amount to 4026 m if the adjacent seas are not taken into account. In particular, the Atlantic Ocean has a mean depth of 3844 m, the Indian Ocean of 3872 m, and the Pacific Ocean of 4188 m. The total water volume of the world ocean is 1349.9×10^6 km^3. With a mean water density of 1.037 (taking the compressibility of the water into account), the mass of the world ocean amounts to 1399.9×10^{18} tonnes, which is only 0.24‰ of the total mass of the earth.

The comparison of Tables 1.01 and 1.02 proves that there is much concurrence among the three oceans. For instance, the mean depths differ from one another only by a few hundred meters, and the proportion of the depth steps is very similar. This leads to the conclusion that all the three ocean basins were formed by analogous processes. It is also remarkable that of the 362×10^6 km^2 covered by the world ocean, only 7.49% are 0 to 200 m deep (i.e., shelf). In each ocean the depth of 4000 to 5000 m is dominant, representing 32 to 37% of the total. The deep sea, with a depth of 3000 to 6000 m, covers as much as 73.83% of the world ocean. Only 0.147% of the oceans is deeper than 7000 m, a figure including part of the deep-sea trenches. The depth zone of 6000 to 7000 m, contributed mainly by the flat deep-sea basins of the three oceans, accounts for 1.232%

---

* Translators' remark: In 1973, the GEBCO program was restructured, which led to a fifth edition started in 1975.

Table 1.02. Depth Zones in the Oceans (after Menard and Smith, 1966)

Ocean Area[a]:	Depth Zone (km)											Percentage of World Ocean
	0–0.2	0.2–1	1–2	2–3	3–4	4–5	5–6	6–7	7–8	8–9	9–10	10–11	
Pacific Ocean[b]	1.631	2.583	3.250	6.856	21.796	34.987	26.884	1.742	0.188	0.063	0.019	0.001	45.919%
Austral Asiatic Mediterranean[c]	51.913	9.255	10.433	12.151	6.698	7.780	1.636	0.076	0.058	0	0	0	2.509
Bering Sea	46.443	5.975	7.623	10.330	29.629	0	0	0	0	0	0	0	0.625
Sea of Okhotsk	26.475	39.479	22.383	3.403	8.260	0	0	0	0	0	0	0	0.384
East China Sea[d]	81.305	11.427	5.974	1.239	0.055	0	0	0	0	0	0	0	0.332
Sea of Japan	23.498	15.176	19.646	20.096	21.551	0.033	0	0	0	0	0	0	0.280
Gulf of California	46.705	20.848	25.891	6.556	0	0	0	0	0	0	0	0	0.042
Atlantic Ocean[b]	7.025	5.169	4.295	8.590	19.327	32.452	22.326	0.738	0.067	0.012	0	0	23.909
Arctic Mediterranean[e]	47.083	17.427	9.317	11.153	12.834	2.195	0	0	0	0	0	0	3.386
American Mediterranean	23.443	10.674	13.518	15.313	20.796	13.440	2.572	0.193	0.051	0	0	0	1.203
European Mediterranean[f]	22.868	20.814	18.362	30.326	7.426	20.204	0	0	0	0	0	0	0.834
Baltic Sea	99.832	0.168	0	0	0	0	0	0	0	0	0	0	0.105
Indian Ocean[b]	3.570	2.685	3.580	10.029	25.259	36.643	16.991	1.241	0.001	0	0	0	20.282
Red Sea	41.454	43.058	14.920	0.568	0	0	0	0	0	0	0	0	0.125
Persian Gulf	100.000	0	0	0	0	0	0	0	0	0	0	0	0.066
World ocean	7.492	4.423	4.376	8.497	20.944	31.689	21.201	1.232	0.105	0.032	0.009	0.001	100.001%

[a] As a percentage of the surface of each ocean (cf. Table 1.01).
[b] Without adjacent seas.
[c] Including Andaman Sea.
[d] Including Yellow Sea.
[e] Consisting of Arctic Ocean, Barents Sea, Canadian Archipelago, Baffin Bay, and Hudson Bay.
[f] Including Black Sea.

of the ocean area, and about 0.25% of the world ocean covers deep-sea trenches that are deeper than 6000 m (cf. Dietrich and Ulrich, 1968).

1.1.6. Large-scale topographic features of the ocean bottom

Attempts have been made to classify the numerous submarine topographic features contained in the bathymetric charts of the world ocean. After the extraordinary increase of soundings which began by about 1950, it was possible to introduce more specific distinctions far surpassing the descriptions given in the first edition of this book in 1957. A simplified classification of the topographic features similar to that by Heezen and Wilson (1968) is given and illustrated by several examples in Fig. 1.02. International agreement was obtained regarding the terms and definitions of the ocean bottom features (International Hydrographic Bureau, 1971); these are compiled in Section 1.1.7. The concepts with respect to the formation of these topographic features are treated in Section 1.4.

Continents and oceans represent the major forms of the earth's crust. Parts of the continental shelves are covered with water and project into the world ocean as *continental margins*. The most important oceanic bottom features are the *Mid-Ocean Ridge* and the *deep-sea basins*. Each of these three large-scale features occupies about one-third of the total ocean bottom.

Continental margins form the transition from the continental shores to the deep-sea basins. They comprise the shelves, the continental slopes, the continental rises, and the deep-sea trenches. Examples of the first three regions are given in Fig. 1.03, of the latter in Fig. 1.16. In contrast to the other three regions, *continental slopes* are to be found at all continental margins. They are steep, with an inclination of more than 1:40, they extend from the shelf edge at a depth of approximately 200 m down to about 2000 m depth, they

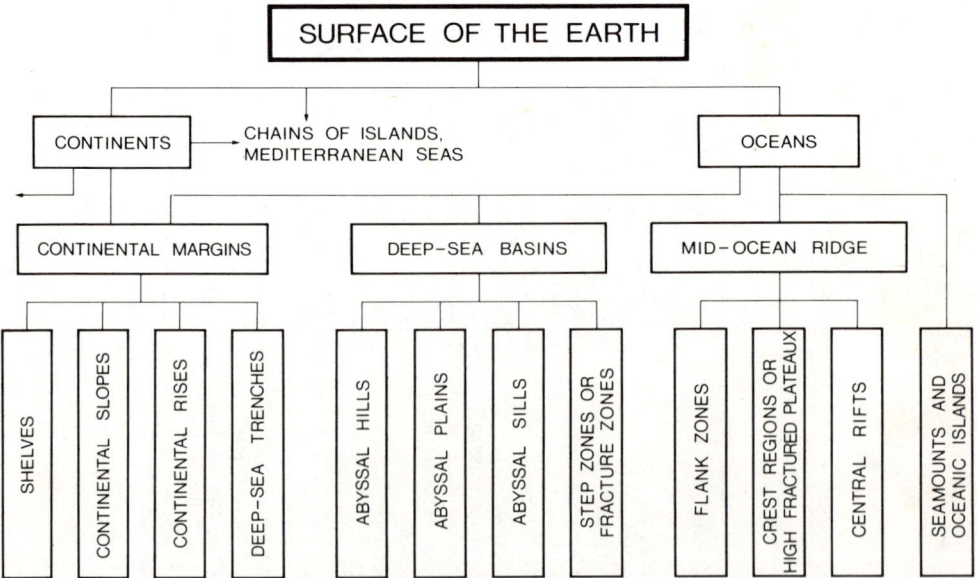

Fig. 1.02. Classification of the large-scale topographic features of the world ocean. (After Heezen and Wilson, 1968; Simplification of the scheme given by Heezen and Menard, 1963.)

quirements for the exploitation of the mineral resources of the ocean bottom resulted in the shelves generally being more densely covered by soundings and being better represented on large-scale nautical charts than the open ocean. Thus, the surface configurations of the shelf are relatively well known. Quite a number of details were discovered in areas which had a Pleistocene ice layer, such as in the North Sea and the Baltic Sea, in the Gulf of St. Lawrence, as well as off the coasts of Norway and Greenland, and, in the lower geographical latitudes, in areas with recent coral reefs.

Between these regions, highly uniform shelves are found in the middle latitudes, as shown by the example of the continental margin of North America in Fig. 1.03, where the threefold structure of the shelf is to be seen very clearly, which has been systematically investigated by Emery (1965), Uchupi (1968), and Pratt (1968).

Where there are no deep-sea trenches at the seaward side of the continental slopes, *continental rises* extend with a width of up to 300 km. They lie at depths of 2000 to 5000 m with a slight inclination of 1:700 to 1:1000 and level off into the abyssal plains. Only in very particular cases do canyons cut across the continental rise, like the Hudson Canyon (Fig. 1.03) which, like a river with a delta, ends in a sediment fan at 4700 m depth.

Over long distances, the continental slopes pass without any continental rise directly into the peripheral *deep-sea trenches*. These are elongated depressions extending from 300 to 4000 km with steep walls (gradient greater than 1:40). They run deeper than 6000 m and reach a depth of 11,000 m. They are separated from the deep-sea basins by a bottom sill. Altogether 26 peripheral deep-sea trenches are known (as shown in Plate 2). Three belong to the Atlantic Ocean (the Puerto Rico Trench with 9219 m at the Milwaukee Depth, the Cayman Trench down to 7680 m, and the South Sandwich Trench down to 8264 m at the Meteor Depth). One belongs to the Indian Ocean—the Sunda Trench, which goes down to 7455 m at the Planet Depth. The other 22 trenches belong to the Pacific Ocean, among them the Mariana Trench with the greatest depth of the world ocean (11,022 m), and the Tonga Trench with the greatest depth in the Southern Hemisphere (10,882 m).

The *Mid-Ocean Ridge* forms an interconnected mountain system which, with a length of 60,000 km, is the earth's longest. Its extent is shown in Plate 2, and in more detail in the atlas by Dietrich and Ulrich (1968). The ridge stretches across the Arctic Ocean from the Laptev Sea to the area west of Spitsbergen and is continued in the Norwegian Sea as the Mohn Ridge and the Jan Mayen Ridge. Iceland and the adjacent Reykjanes Ridge are also part of this global system. Only at three points is it connected with the continents: in the Laptev Sea with North Siberia, in the Gulf of Aden with the trench system of East Africa and the Red Sea, and in the Gulf of California. The system of ridges occasionally reaches a total width of 4000 km and a height of 1000 to 3000 m. It has a characteristic cross section, as demonstrated by the example of an echo-sounding profile through the Reykjanes Ridge (Fig. 1.04), and it is divided into a rift valley, high fractured plateaus, and flank zones.

The *central rift valley*, which is absent only in a few places, is 20 to 50 km wide and cuts 1000 to 3000 m deep into the crest region. Recent measurements have revealed that the axis of the rift valley, like the whole system of ridges, shows numerous lateral displacements (see Plate 2). Thus not only longitudinal but also cross structures make the topography of the Mid-Ocean Ridge an extraordinarily complicated one. The rugged *high fractured plateau*, in which the rift valley is embedded, is up to 1000 km wide and lies under 2000 to 4000 m of water. However, many oceanic islands also belong to this area.

The *flank zones,* adjacent to the high fractured plateau, are up to 1500 km wide and

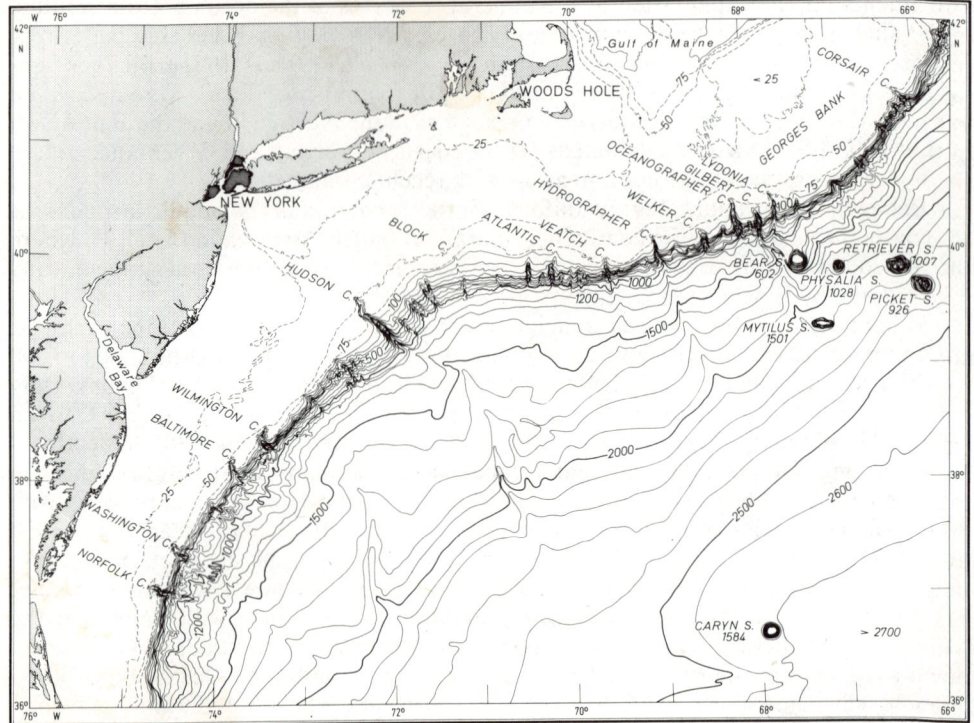

Fig. 1.03. Continental margin in the western North Atlantic Ocean: Shelf 0 to 100 fathoms, continental slope with canyons (C) 100 to 1200 fathoms, continental rise 1200 to 2600 fathoms, abyssal plain >2600 fathoms, seamounts (S) with least depth (in fathoms). (After Belding and Holland, 1970.) Depth in fathoms, 1 fathom = 1.829 m. Modified mercator projection. Scale approximately 1:5 million.

are only 20 to 100 km wide, and are considerably furrowed by *submarine canyons* (see Fig. 1.03). These canyons have various characteristics: steep walls, small width ranging approximately between 1 and 15 km, and a monotonous inclination of the axis never surpassing 1:40. Some of these bottom features, like those of the Hudson Canyon represented in Fig. 1.03, are continued in the continental rise. The canyons, first discovered in 1934 through echo soundings in connection with new methods for the determination of the precise position at sea, have proven to be worldwide bottom features of the continental slopes. Some canyons are continued in continental valleys, like those of the Indus and the Congo, but others are not, like those at the continental slope of North America, as shown in Fig. 1.03. Only in the case of the Hudson is some connection slightly indicated.

The *shelves* are part of the continental margins. They are absent where peripheral deep-sea trenches exist, which is especially true for the Pacific Ocean. Except for these cases, shelves surround the continents like flat ledges (inclination of the bottom less than 1:1000) with varying width (up to 300 km). The transition from the flat shelf to the steep continental slope is clearly marked in the world ocean. It is found at the *shelf edge* which usually lies at a depth of 150 to 200 m, but in some areas—especially in higher latitudes—not at depths less than 500 m.

The practical necessities of the shipping and fishing industries as well as the re-

Topography of the Ocean Bottom 11

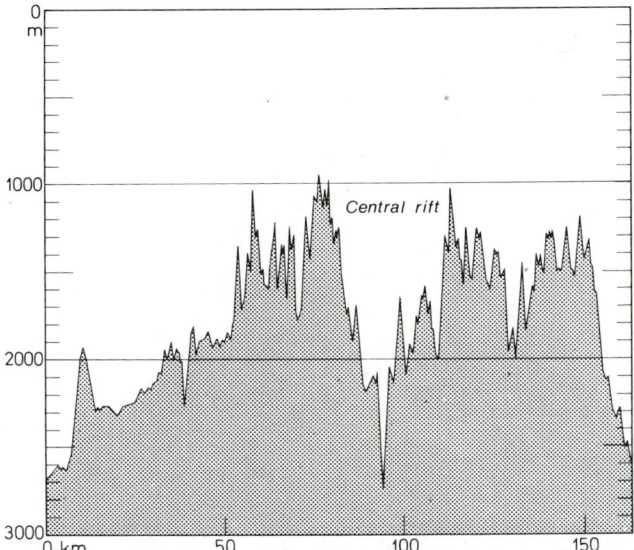

Fig. 1.04. Example of Mid-Ocean Ridge (Reykjanes Ridge at 57.5°N) with central rift (center), crest regions (on both sides of the central rift), flank regions (left and right edges) according to an echo sounding profile by fishery research vessel *Anton Dohrn* during the International Geophysical Year 1958. (After Ulrich, 1963.) Exaggeration 50 times.

mostly lie at a depth of 3000 m. They consist of several steps that decrease with increasing distance from the axis of the Mid-Ocean Ridge and pass into the abyssal hills of the deep-sea basins.

The *deep-sea basins* lie between the continental margins and the Mid-Ocean Ridge, and comprise abyssal hills, abyssal plains, deep-sea sills, and step zones. They are not identical with the oceanic basins which were described, according to Wüst (1940), in the first edition of this book in 1957 as one of the gross topographic features of the deep-sea floor. The basins in the previous sense (in contrast to the term "deep-sea basins" as it is used now) also included the continental rises of the continental margins and the flank zones of the Mid-Ocean Ridge. Wüst, however, had connected the deep-sea sills with the system of the ridges and interpreted them as boundaries between the oceanic basins, like the Greenland–Scotland Ridge, the Rio Grande Ridge, and the Walvis Ridge. The world ocean was divided into 57 basins in accordance with the soundings of 1954. Today some more basins must be added, especially due to the improved knowledge of the Arctic Ocean (4 basins instead of 1), and of the Indian Ocean (16 instead of 9), as demonstrated in the atlas by Dietrich and Ulrich (1968). The topographic division into basins remains important for the circulation of the deep-sea and bottom waters, alongside with the physiographic division of Heezen *et al.* (1959) who, when establishing a classification, also endeavored to take into account the development of the gross topographic features.

The *abyssal hills* are found to be arranged in zones, lying between the abyssal plains and the Mid-Ocean Ridge and rising up to 1000 m. Typical examples are contained in a cutting from an echogram on the right-hand side in Fig. 1.05.

The *abyssal plains* are surprisingly flat with a gradient of 1:1000 to 1:10,000. It is not uncommon for conical elevations to tower up in the plain, as shown on the left-hand

12   Geomorphology of the Ocean Bottom

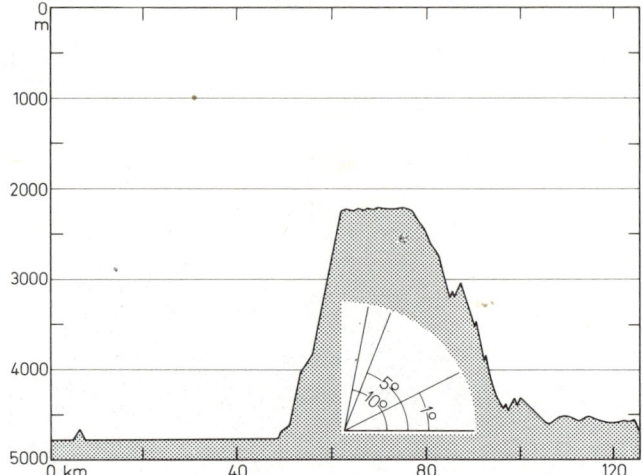

Fig. 1.05. Example from a deep-sea basin (Newfoundland Basin at 44°N and 40°W) with abyssal plain (left side), seamount (center), and abyssal hills (right side) according to an echo-sounding profile by research and survey vessel *Gauss* during the International Geophysical Year 1958. (After Dietrich, 1959.) Exaggeration 30 times, slope diagram center below.

side in Fig. 1.05. Sometimes the cones may be tremendous seamounts, as in the middle of Fig. 1.05, or like the Great Meteor Seamount shown in Fig. 1.06. The *deep-sea sills* are elongated elevations, rising above the abyssal plains by up to 4000 m. They sometimes carry islands like the Faeroe Islands on the Greenland–Scotland Ridge, or the Maldive and Laccadive Islands on the Maldive Ridge. The deep-sea sills can reach a width of 150 km and a length of 4000 km. In contrast to the Mid-Ocean Ridge, earthquakes do not occur here. The *step zones,* also called *fracture zones,* are long narrow zones (up to 100 km wide and up to 2000 km long with a step height of up to 2000 m). They consist of asymmetrical ridges and depressions and are typical for the East Pacific Ocean. The eight step zones, which are arranged in remarkably straight lines on great circles of the earth, were discovered as recently as around 1960 (see the atlas by Dietrich and Ulrich, 1968).

Seamounts—like the *oceanic islands*—are conical elevations on the sea floor which are found in each of the three gross topographic features described above (the continental margins, the Mid-Ocean Ridge, and the deep-sea basins). Their heights may lie below 100 m, but can also exceed 8000 m if the islands are also taken into account. Their total number is still unknown. If the frequency of their occurrence in the well-surveyed areas is applied to the blank spots on depth charts, a total number of about 10,000 to 20,000 for the world ocean can be estimated; 1500 were recorded as of 1969 (Ulrich, 1970). Only a few seamounts are irregularly distributed in the world ocean: most of them appear in groups of 10 to 100 and they are often arranged in chains in the weak zones of the earth's crust as, for example, the New England Seamounts in the western North Atlantic (see Fig. 1.03 in which the western part of the area concerned is shown, Uchupi et al., 1970). This is especially obvious where the seamounts rise as islands above the surface of the oceans, like the chains of the Hawaiian and Tuamotu Islands in the Pacific Ocean.

A special form of the seamounts is the *guyot,* a truncated flat-topped cone. The guyots were first described by Hess (1946), who named them guyots in memory of the Swiss-American geologist Arnold Guyot. The total number of guyots in the world ocean is es-

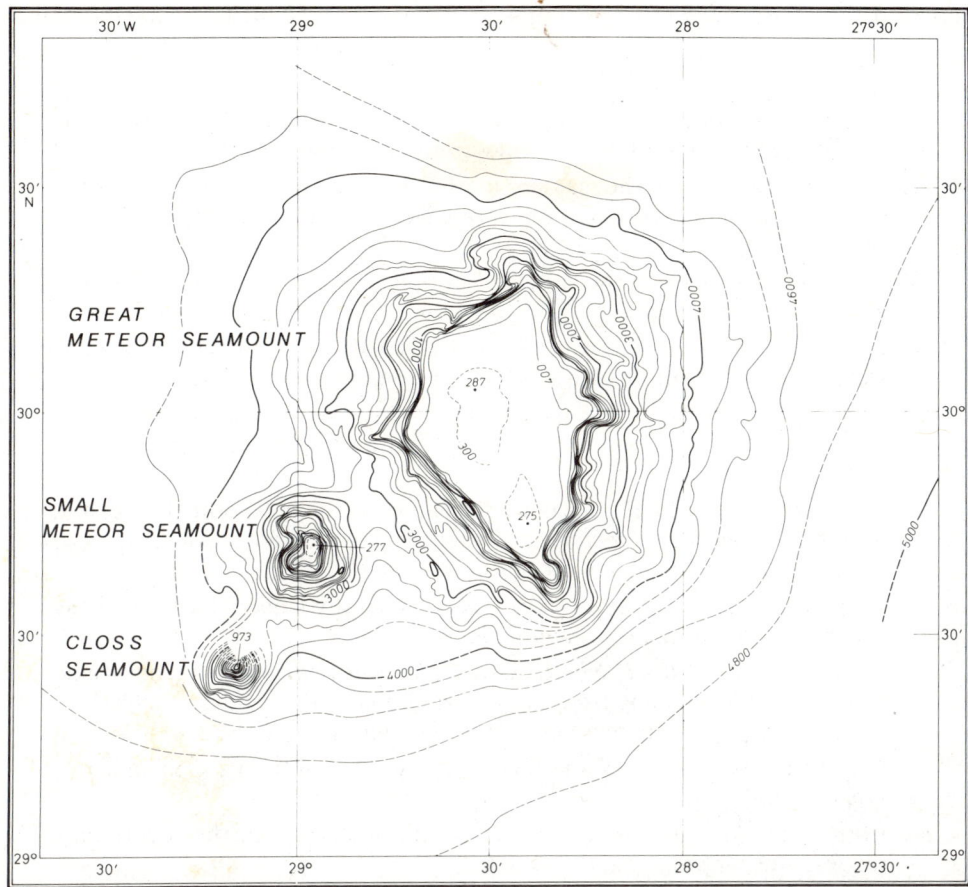

Fig. 1.06. Example of seamounts (Great Meteor Seamount, Small Meteor Seamount, and Closs Seamount) according to echo sounding profiles of research vessel *Meteor* during the "Seamounts Cruises in the Northeast Atlantic, 1967." (After Ulrich, 1969.) Depth in m, scale approximately 1:2 million.

timated at several hundred. According to Menard and Ladd (1963) more than half lie in the Central Pacific Ocean. An example of a guyot is the Great Meteor Seamount, the biggest seamount in the Atlantic Ocean (Fig. 1.06). It rises from a depth of 4800 m to 275 m below the sea surface, a height exactly equivalent to that of the top of the contiguous Small Meteor Seamount. The slope of the flanks of the cones exceeds 30°.

Some of the seamounts are active volcanoes, several of which, in historical time, occasionally or permanently emerged from the sea surface as islands. Of recent origin is the island of Surtsey, which appeared southwest of Iceland in 1963. Since that time it has been investigated under many scientific aspects. The relevant research is directed by the Surtsey Research Society, Reykjavik.

1.1.7. Terms and definitions of ocean bottom features

An international commission of GEBCO (General Bathymetric Chart of the Oceans) has adopted the nomenclature of ocean bottom features (International Hydrographic

Bureau, 1971). Evalson's (1967) terminology was the basis of the following 38 definitions:

1. *Archipelagic Apron.* A gentle slope with a generally smooth surface on the sea floor, particularly found around groups of islands or seamounts.
2. *Bank.* An elevation of the sea floor located on a *continental* (*or island*) *shelf* and over which the depth of water is relatively shallow but sufficient for safe surface navigation.
3. *Basin.* A depression of the sea floor more or less equidimensional in form and of variable extent.
4. *Canyon* (Submarine Canyon). A relatively narrow, deep depression with steep slopes, the bottom of which grades continuously downward.
5. *Channel* (Seachannel). A long, narrow U-shaped or V-shaped shallow depression of the sea floor, usually occurring on a gently sloping *plain* or *fan.*
6. *Continental Borderland.* A region adjacent to a continent, normally occupied by or bordering a *Continental Shelf,* that is highly irregular with depths well in excess of those typical of a *Continental Shelf.*
7. *Continental Rise.* A gentle slope with a generally smooth surface, rising toward the foot of the *continental slope.*
8. *Continental* (*or Island*) *Shelf.* A zone adjacent to a continent (or around an island) and extending from the low-water line to the depth at which there is usually a marked increase of slope to greater depth.
9. *Continental* (*or Island*) *Slope.* The declivity seaward from a shelf edge into greater depth.
10. *Cordillera.* An entire mountainous system, including all the subordinate ranges, interior *plateaus,* and *basins.*
11. *Escarpment* (Sea Scarp, Scarp). An elongated and comparatively steep slope of the sea floor, separating flat or gently sloping areas.
12. *Fan.* A gently sloping, fan-shaped feature normally located near the lower termination of a *canyon.*
13. *Fracture Zone.* A linear extensive zone of unusually irregular topography of the sea floor characterized by large *seamounts,* steep-sided or asymmetrical ridges, troughs, or *escarpments.*
14. *Gap.* A break in a *ridge* or *rise.*
15. *Gully.* Small valleys cut into soft sediments on the *continental shelf* or *continental slope.*
16. *Knoll* (Hill). An elevation rising less than 1000 m from the sea floor and of limited extent across the summit.
17. *Levee.* An embankment bordering the sides of a *canyon* or *channel.*
18. *Moat* (Sea-moat). An annular depression that may not be continuous, located at the base of many seamounts or islands.

19. *Plain.* A flat, gently sloping or nearly level region of the sea floor. (For example, abyssal plain.)
20. *Plateau.* A comparatively flat topped elevation of the sea floor of considerable extent across the summit and usually rising more than 200 m on all sides.
21. *Province* (morphological). A region composed of a group of similar physiographic features whose characteristics are markedly in contrast with surrounding areas (rarely used in marine cartography).
22. *Reef.* An offshore consolidated rock hazard to navigation with a least depth of 20 m or less.
23. *Ridge.* A long, narrow elevation of the sea floor with steep sides and irregular topography.
24. *Rise* (Arch, Swell). A long, broad elevation that rises gently and generally smoothly from the sea floor.
25. *Saddle.* A low part between elevations on a *ridge* or between *seamounts*.
26. *Seamount* (Peak). An isolated or comparatively isolated elevation rising 1000 m or more from the sea floor and of limited extent across the summit.
27. *Seamount Chain.* Several *seamounts* in a line with bases separated by a relatively flat sea floor.
28. *Seamount Group.* Several closely spaced *seamounts* not in a line.
29. *Seamount Range.* Several seamounts having connected bases and aligned along a *ridge* or *rise*.
30. *Shelf Edge* (Shelf Break). The line along which there is a marked increase of slope at the outer margin of a *continental* (*or island*) *shelf*.
31. *Shoal.* An offshore hazard to navigation with a least depth of 20 m or less, composed of unconsolidated material.
32. *Sill.* The low part of the *ridge* or *rise* separating ocean *basins* from one another or from the adjacent sea floor.
33. *Spur.* A subordinate elevation, *ridge,* or *rise* projecting outward from a larger feature.
34. *Strath.* A broad elongated depression with relatively steep walls located on a *continental shelf*. The longitudinal profile of the floor is gently undulating with greatest depth often found in the inshore portion.
35. *Tablemount* (Guyot). A *seamount* having a comparatively smooth flat top.
36. *Trench.* A long, narrow, and deep depression of the sea floor, with relatively steep sides.
37. *Trough.* A long depression of the sea floor, normally wider and shallower than a *trench*.
38. *Valley* (Submarine Valley). A relatively shallow, wide depression with gentle slopes, the bottom of which grades continuously downward. This term is used for features that do not have canyon-like characteristics in any significant part of their extent.

1.2. Sediments at the Ocean Bottom

1.2.1. Nature of the ocean bottom

Numerous bottom samples obtained so far from all parts of the world ocean have proven that the bottom predominantly consists of unconsolidated sediments. Deep-sea drillings, carried out by the deep-sea drilling vessel *Glomar Challenger* in all three oceans since 1968, have shown that under a sediment layer of varying thickness there is basalt, the consolidated rock of the earth's crust. The particles of the bottom sediments are of diverse origin. Until final sedimentation they are influenced more or less, depending on their grain size, by transport through water and their chemical properties are altered by contact with seawater. Consequently, close interrelations exist between seawater and sediment which determine the geographical distribution of sediment types and their stratification. Since these relations mainly concern the *origin, distribution of grain sizes,* and *chemical composition* of the sediments, the description is basically limited to these three factors, although the complexity of marine geology is much more extensive. Ever since the fundamental work by Murray and Renard in 1891 on the basis of the *Challenger* observations, marine geology has become a separate branch of geology, which is treated in special textbooks, like those by Kuenen (1950), Shepard (1963), and Seibold (1964), and which has its own scientific journals (e.g., *Marine Geology,* since 1964), as well as special methods and instruments. Lisitzin (1972) has given a summarizing representation of oceanic sediments, considering also the numerous Russian publications.

1.2.2. Methods of investigation of marine sedimentology

In marine sedimentology, in addition to the methods of modern surveying (see Sections 3.1.1.3 and 3.2.4), special procedures are used. They serve (1) for making direct observations of the bottom, (2) for taking samples from the ocean bottom, (3) for measuring the thickness and structure of sediments with geophysical methods, and (4) for analyzing these samples at the laboratory by means of physical, chemical, petrographic, and paleontologic methods.

For *direct observation,* skin diving, submersibles, bottom photography, and underwater television are used. Diving has been practiced for quite some time, in particular, systematic use of this technique was made by Wasmund of Kiel (1938). Heberlein (1968) has given a relevant historical review. Since 1946 direct observation by diving, down to a water depth of 50 m with the aid of scuba apparatus for breathing developed by Cousteau, has become very common with marine geologists.

As soon as submersibles were available (see Section 3.1.1.2) it became possible for man to reach the greatest depths. The *Bathyscaph,* developed by Piccard (1948), was used by his son Jacques (1960) when in the Mariana Trench he managed to penetrate to the greatest depth of the world ocean and reached 10,900 m. In addition to observation and photography, more modern submersibles also permit selective sampling by remote-controlled grab arms, as demonstrated by *Alvin* (see Fig. 1.07). The numerous types of submersibles, developed in the period from 1960 to 1970 in the United States, England, France, Italy, Federal Republic of Germany, the USSR, and Japan, are used not only in marine geology, but also in physical oceanography, marine biology, and marine technology. They differ primarily with regard to potential operating depth, time of diving, and sampling capacity (Table 3.03).

Underwater photography with still and motion-picture cameras has become indispensible for direct bottom observation. In addition to the equipment often used by divers,

Fig. 1.07. Deep-sea research submersible *Alvin* of Woods Hole Oceanographic Institution. Built in 1967, length 7 m, displacement 15 tonnes, maximum speed 3 knots, range under water 15 nautical mi, operating depth 2000 m (now 3600 m), crew 3. In front: A remote-controlled manipulator for sample-collecting. (Photograph courtesy of WHOI.)

special instruments have been developed for deep-sea photography [Ewing et al. (1946), Edgerton (1955), and Laughton (1957)]. With these instruments, a great many excellent photographs have been taken, giving information on the sea floor with its microrelief and population, even at the greatest depths. A comprehensive monograph of deep-sea photography and its manifold applications has been given by Hersey (1967). In Fig. 1.08, ripple marks are to be seen in the Iberian Basin at a depth of 3100 m, which can only have developed in moving water. Therefore, bottom photography has been systematically used to study the direction and velocity of bottom currents (e.g., Heezen and Hollister, 1964) that occasionally appeared in the recent geological past, giving evidence of their presence in such bottom ripples. Bottom ripples are rare; usually the deep-sea floor looks like the photo shown in Fig. 1.09. The appearance of wide areas of the Pacific deep-sea floor is quite different (Fig. 1.10); about a third of it is covered by manganese nodules.

Another device for bottom observation is the underwater-television photo camera with sampling equipment (Dietrich and Hunger, 1962). Television-eye, lights, remote-controlled photo camera, and sampling equipment are directed from a surface vessel (Fig. 1.11). The television screen and the steering gear on board permit selective photography and sampling in the deep sea. The advantages of this method are obvious: (1) there is no risk of life in contrast to skin diving; (2) much less effort is required than with the employment of submersibles; (3) photographs are taken under control and not at random, as is the case with the usual deep-sea photo camera; and (4) random sampling, as with the former sampling devices, is unnecessary because taking photographs and sample collection can be directed by means of the picture on the television screen. In Fig. 1.12,

18 Geomorphology of the Ocean Bottom

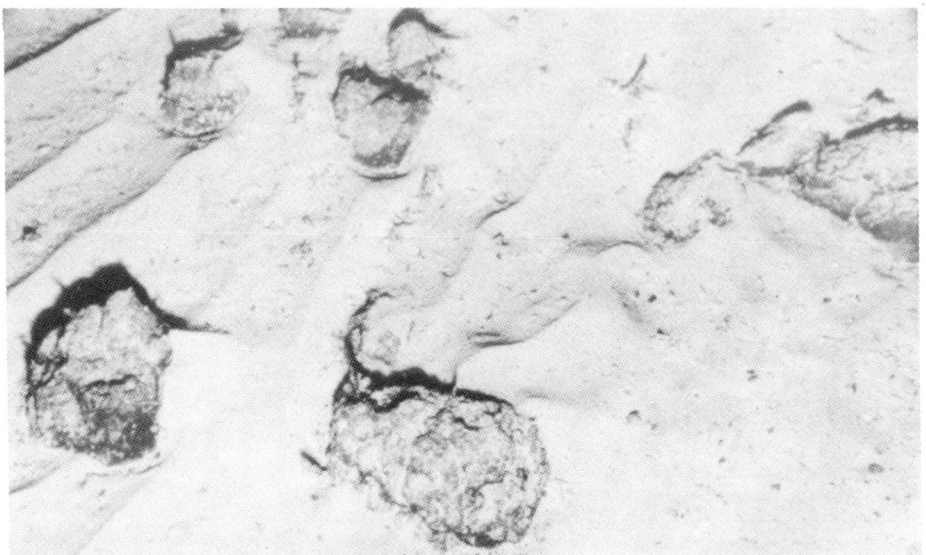

Fig. 1.08. Deep-sea floor with ripple marks in the Iberian Basin at 3127 m depth ($\varphi = 41°12'N$, $\lambda = 15°14'W$). Bottom photograph by the National Institute of Oceanographic Sciences, Wormley, England. Bottom area covered: 1.5 × 2.5 m. Bottom covered by calcareous sands and stones on a seamount. Wavelength of the sand ripples: 20 cm. Accumulations on the right-hand side of the stones, prevailing current probably from left to right. (After Laughton, 1963.)

the outcrop of basalt at the Josephine Seamount is shown. This photograph was taken with the underwater television photo camera of *Meteor* in 1967 (Schott, 1969).

Different kinds of collecting devices are in use for *taking samples*. They vary according to the demands of the relevant research program: whether samples are desired (1) from the surface of the ocean floor, (2) from the vertical distribution of the upper sediment layers (up to 20 m thick), or (3) from cores longer than 20 m. Devices for (1) have been in use for a long time and include bottom grabs or snappers, dredges, and sand traps. Devices for (2) are used to take sediment cores, either, for instance by their own weight, as the gravity corer and the box corer, or by hydrostatic pressure as the piston corer. Drilling devices are used for (3). In this case application is made of the highly developed drilling techniques of petroleum exploration, as carried out by the drilling vessel *Cuss I* during the Mohole drilling in 1961, which succeeded in penetrating 300 m into the crystalline of the earth's crust at a water depth of 3700 m.

After this successful beginning a national Oceanic Sediment Coring Program was set up in the United States in 1963, the planning and implementation of which was commissioned to a group of five great institutions, the Joint Oceanographic Institutions for Deep-Earth Sampling (JOIDES), including Lamont-Doherty Geological Observatory, Columbia University, New York; Institute of Marine Science, University of Miami; Scripps Institution of Oceanography, University of California, La Jolla; University of Washington, Seattle; and the Woods Hole Oceanographic Institution, Woods Hole, Massachusetts. Since August 1968 the deep-sea drilling vessel *Glomar Challenger* has been at the disposal of this group of institutes (see Section 3.1.1.1 and Fig. 1.13). This vessel was constructed especially for deep-sea drilling and, even at a depth of 6000 m, drillings of 750 m into the subsoil can be carried out. During the period 1968–1970, the sediment layers of the Atlantic Ocean and the Pacific Ocean were run through over 200

Fig. 1.09. Bottom of the Iberian abyssal plain at 5340 m depth ($\varphi = 41°22'$N, $\lambda = 14°24'$W). Bottom photograph by the National Institute of Oceanographic Sciences, Wormley, England. Bottom area covered: 1.5×1.5 m. Bottom covered by globigerina ooze, partly worked up by benthic fauna, which can be recognized from traces, small heaps and holes. At the upper edge either a worm or a thread (45 cm long) is visible. (After Laughton, 1963.)

times, and a rich variety of specimens was obtained. A new era in sedimentology will be opened by this highly sophisticated, but also very expensive procedure.

Even the first echo sounders recorded not only the depth of the water, but also the *thickness* of the *single sediment layers*. As early as 1935 Stocks already drew attention to this fact. In the meantime the sediment echo sounder had been developed. Higher and higher sound energies have been used which are emitted in a broad frequency band. As so-called "reflection seismology," this method has become an important aid for recording the distribution of sediments in shallow waters as well as in the deep sea from on board the moving ship (see Section 1.4.1).

A large variety of *laboratory methods* is used to analyze the collected samples. The analysis does not include only the physical properties of the sediments, such as density, porosity, sound velocity, heat conductivity, natural gamma radiation, and even paleomagnetism (current sediment-petrographic and micropaleontologic methods are used), but also the chemical composition of the stable and radioactive isotopes (e.g., for the determination of the absolute age of the sediments and of the water temperature at the time of their generation, see Section 1.3.4). Our knowledge about the paleotemperatures during the past 425,000 yr, as given in Fig. 5.09, is based on such analyses of deep-sea globigerina ooze.

1.2.3. Origin of the Sediments

The sediments are composed of four constituents which can be classified with respect to their origin as terrigenous, biogenous, halmyrogenous, and cosmogenous. The share of these components varies locally.

Terrigenous sediments originate partly from the continents, where they are made

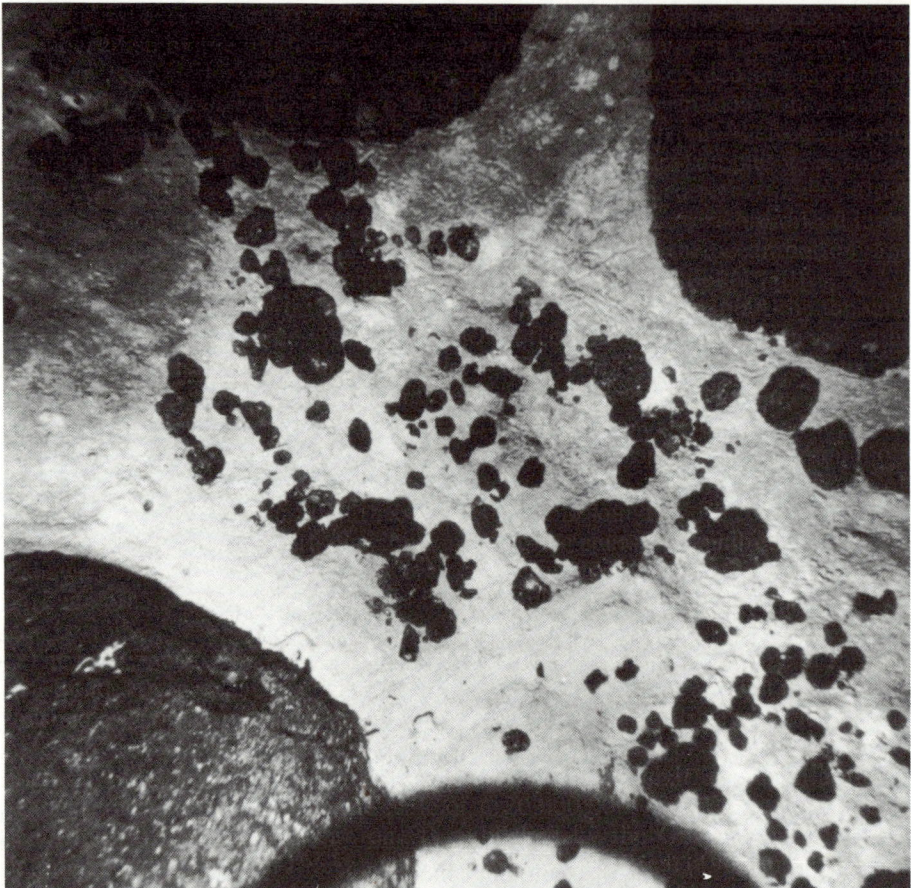

Fig. 1.10. Manganese nodules at the bottom of the South Pacific Basin at 4560 m depth ($\varphi = 42°50'S$, $\lambda = 125°32'W$). Bottom photograph taken during the "Downwind Expedition" of the Scripps Institution of Oceanography, La Jolla, California. Bottom area covered: approximately 6 m². Bottom covered with nodules: 46%, nodules without sediments. Density of the nodules: 4 g/cm³. Vertical distribution: larger nodules only at the surface of the sea floor. (After Menard and Shipek, 1958.)

available by either mechanical or chemical weathering of rocks; partly they are of volcanic origin. Their transport by rivers is by far the most important factor; the volume is estimated to be up to 12 km³ yr⁻¹. If this amount were evenly distributed on the ocean floor, it would mean an increase of 3 cm in 1000 yr. Actually the total increase of height amounts only to 1 cm per 1000 yr because a large amount of the terrigenous sediments remains on the continental margins, especially in coastal regions. Only a small part is transported as very fine turbidity by ocean currents even into parts of the ocean farthest away from land, where it forms the chief ingredient of the red deep-sea clay. However, terrigenous sediments are also transported into the sea by ice, especially by icebergs from polar regions far into the subpolar areas. Poor sorting is a characteristic of this sediment type. During the cold ages in the most recent geological past, sediments with a considerable amount of ice-transported material were carried into the present moderate and subtropical climate zones. Terrigenous sediments may also originate from dust that has been transported

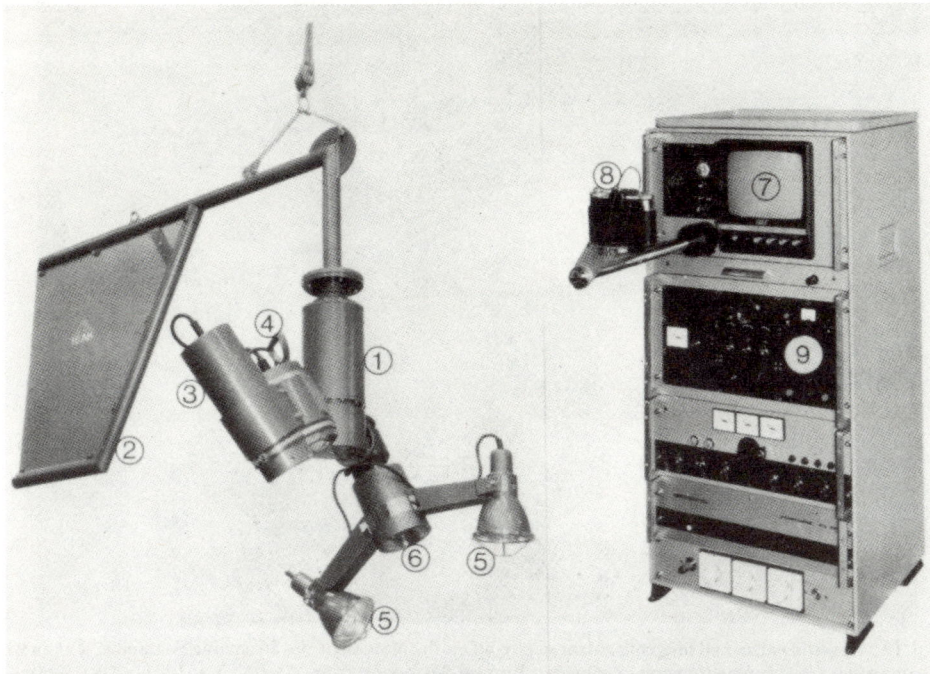

Fig. 1.11. Deep-sea television photocamera. (After Dietrich and Hunger, 1962, further development, as in 1971.) Left-hand side: Underwater unit. 1. Swiveling and tilting gear. 2. Stabilizing fin. 3. Television camera. 4. Photo camera. 5. Two search-lights. 6. Electronic flashlight. Right-hand side: Deck unit on ship-board. 7. Monitor. 8. Photo camera for television screen. 9. Operating and control unit.

by prevailing winds far out into the oceans (Griffin et al., 1968). Thus, dust from the Sahara Desert, carried by the northeasterly trades, is found in the bottom sediments west of the Cape Verde Islands, and dust from the Australian Desert, carried by the westerlies, is found as far away as east of New Zealand. Sediments of volcanic origin must be divided into those resulting from subaerial and those from submarine eruptions. Very fine ash from the former may be deposited over a considerable area as a result of transportation by wind.

The *biogenous sediments* are of organic origin. They consist of residua of shells and parts of skeletons, the remains of living benthic and pelagic organisms. Micropaleontology is the special branch of science which deals with the vast number of species of microscopically small organisms living today as well as in the geological past. A volume with contributions by 52 experts (edited by Funnell and Riedel, 1971) gives a summary of today's knowledge. The benthic sediment types consist of the residua of animals and plants living on the ocean floor. They are found almost exclusively in coastal, so-called littoral, shallow-water deposits and nearly disappear in the dark deep sea, where life on or in the sediments is very difficult for flora and fauna. There the biogenous components of the sediments are formed from the residua of zooplankton and phytoplankton, which actually inhabit the subsurface layer of the ocean where light can penetrate. However, the largest part of the dead mass of plankton is already dissolved by the seawater during the settling process or at the surface of the sediment. Only a few hard-to-dissolve residua, containing lime and silicic acid, have remained in the deep-sea sediments. In particular, the shells

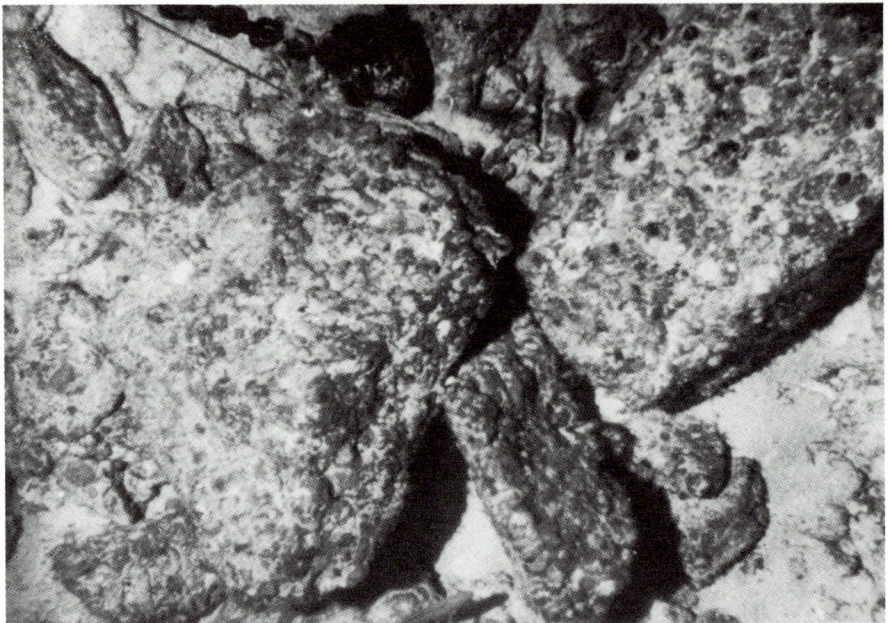

Fig. 1.12. Basaltic rocks and biogenic calcareous sand on the plateau of the Josephine Seamount. Taken with the underwater television photo camera of research vessel *Meteor* ($\varphi = 36°43.2'$N, $\lambda = 14°17.7'$W), depth 228 m, diameter of the iron ball at the upper edge of the photo is 10 cm. (After Schott, 1969.)

of the various kinds of planktonic foraminifera consist of calcite, whereas the shells of planktonic mollusks (especially of the pteropods) consist of aragonite. Also taking part in the formation of sediments are the microscopically small algae living at the sea surface. These are the coccoliths (coccolithophoridae), which have a cover of small calcareous plates. Silicic acid is contained in the sediment-forming skeletons of radiolarians and diatoms. According to the organisms which prevail in the sediments, we distinguish globigerina, pteropod, coccolith, radiolarian, and diatom oozes.

In the shelf seas, the soft parts of the organic substances are not completely decomposed by bacteria when the plankton residua reach the sea floor. This is especially the case in shallow seas rich in nutrients with high organic production, as at the estuaries of big continental rivers or in upwelling areas. In these places an explosive, abundant production of diatoms and other planktonic algae may occasionally take place. The consequence is a reddish coloring of the water which is, for instance, repeatedly observed as the "Red Tide" in the waters surrounding Florida, in the Gulf of California, and in parts of the Red Sea. The Red Sea got its name from this phenomenon. Mostly tiny dinoflagellates are involved, of which as many as 20×10^6 cells can be observed in 1 liter of seawater. They can secrete a poison leading to a catastrophic widespread dying of fish and other living organisms.

Off the coasts of Southern California, Peru, and South Africa, that is, in the main upwelling areas of the world ocean off shallow coastal waters, such disasters occur at intervals of several years. If the sunken organic material on the sea floor is covered by sands, the resultant formations might be a model for the generation of hydrocarbon deposits. At present deposits of petroleum and natural gas have frequently been found in the sediments of such marginal areas of the continents.

Fig. 1.13. Deep-sea drilling vessel *Glomar Challenger*. Since 1968, owned by Global Marine Inc., Los Angeles, operated by the Scripps Institution of Oceanography, La Jolla, Calif. for JOIDES. Length 131 m, displacement 10,500 tonnes, height of the drilling tower 59 m above the water line, operating at water depth down to 6000 m, penetrates into the sea floor as far as 750 m. Photograph courtesy of Scripps Institution of Oceanography.

The *halmyrogenous sediments* represent direct new formations of minerals which may develop when the water is oversaturated with soluble material. Lime is rather seldom directly precipitated by inorganic reactions, although the water in the warm seas of the tropics and subtropics shows an oversaturation in $CaCO_3$. In tropical shallow waters, precipitation of lime occurs locally, for example, on the Great Bahama Banks, but the complicated physicochemical system of carbon dioxide, bicarbonate, carbonate, and hydrogen ions (see Section 6.2.6) usually prevents precipitation. In warm water, however, organisms can incorporate lime into their skeletons and shells; large coral colonies can build reefs, and huge, thick-shelled mollusks and snails can survive here.

Halmyrogenous deposits in the form of concretions of ferromanganese oxides, so-called manganese nodules, are widespread in the world ocean. Their diameters vary from fractions of a millimeter to several decimeters. With the aid of deep-sea photography, it has been possible to demonstrate that the deep-sea floor can be completely covered by manganese nodules (see Fig. 1.10). Menard (1964) has written a monograph on the manganese nodules in the Pacific Ocean. These concretions contain 25% manganese, 15% iron, and 1% nickel. They are formed only at the surface of the sediment under conditions of oxidation and cover 20 to 50% of the floor of the Pacific Ocean. Their rate of growth is very small, approximately 10^{-2} mm in 10^3 yr, and a very slow sedimentation rate in the environment is the basic requirement. The stock of manganese at the sea floor and the commercial utilization of these manganese nodules was investigated by Mero (1965). The annual production of about 2×10^6 tonnes of manganese with a reserve of 10^9 tonnes on the continents is contrasted with an estimate of 2×10^{11} tonnes on the floor of the Pacific Ocean, which means a reserve for the next 500 yr under present conditions.

The *cosmogenous sediments* are tiny balls rich in iron and of a diameter of 30 to 60 μm, or rarely up to 0.2 mm. Their chemical composition is in remarkably good conformity with that of the ferric meteorites. According to investigations of cores collected during the Swedish *Albatros* Expedition, Pettersson and Frederiksson (1958) estimated the importation of this extraterrestrial material at 2500 to 5000 tonnes yr^{-1} over the entire earth. This corresponds to a rate of increase of only about 10^{-6} g cm^{-2} in 1000 yr on the floor of the world ocean, which indicates that the cosmogenic share in the marine sediments, compared with the others, is of absolutely secondary importance.

1.2.4. Grain size distribution of the sediments

The diameter of the bottom deposits varies over a wide range including stones (greater than 20 mm), gravel (20 to 2 mm), coarse sand (2 to 0.5 mm), medium sand (0.5 to 0.2 mm), fine sand (0.2 to 0.1 mm), very fine sand (0.1 to 0.02 mm), silt (0.02 to 0.002 mm), and clay (less than 0.002 mm). In shelf areas, especially in areas of Pleistocene glaciation, the distribution of grain size varies locally very much. At the deep-sea bottom, however, the distribution is more uniform, although there are great differences among the three main types of pelagic sediments: globigerina ooze, diatom ooze, and red deep-sea clay. The globigerina ooze covers a spectrum of 0.2 to 500 μm without preference for a particular grain size, the diatom ooze of 0.5 to 200 μm with a maximum at 7 μm, and the red deep-sea clay of only 0.1 to 70 μm with a maximum at 1 μm. Of the particles, 20% are smaller than the wavelength of visible light (0.4 μm). This means that these particles are no longer visually recognizable under the optical microscope.

1.2.5. Chemical composition of the sediments

In chemical composition, the marine sediments show a great variety, especially in shelf areas because there is great local variation in the terrigenous and biogenous components. Furthermore, the chemical composition of the terrigenous components shows great differences because of the variety of minerals from which they are derived. The composition of the deep-sea sediments, however, is substantially more uniform. Calcium carbonate and silicic acid form the main components of the biogenous sediments, as shown in Table 1.03. In particular, lime is the most important constituent of the globigerina and pteropod oozes, whereas silicic acid is more abundant in diatom and radiolarian oozes. Table 1.03 gives mean values obtained from a great number of samples. It should be noted that, in nature, there are transitions among the different sediments and, correspondingly, a great variety of compositions. The inorganic component, which mainly contains terrigenous material, is made very homogeneous through the natural processes of sedimentation. According to El Wakeel and Riley (1961), the red deep-sea clay, which consists predominantly of this inorganic component, contains mean values of 53.93% SiO$_2$, 17.46%

Table 1.03. Mean Chemical Composition of Pelagic Sediments in the Ocean[a]

Components	Red Deep-Sea Clay (%)	Globigerina Ooze (%)	Pteropod Ooze (%)	Diatom Ooze (%)	Radiolarian Ooze (%)
Lime (CaCO$_3$), organic	10.4	64.7	73.9	2.7	4.0
Silicic acid (SiO$_2$), organic	0.7	1.7	1.9	73.1	54.4
Other, inorganic	88.9	33.6	24.2	24.2	41.6

[a] According to Revelle (1944).

Al_2O_3, and 8.23% Fe_2O_3. The iron oxide content is responsible for the sometimes reddish, sometimes brownish coloring of the clay.

The calcium carbonate in the ocean water is subjected to special influences, as will be seen later (see Section 6.2.6), because it can be dissolved by free carbon dioxide. In the ocean, differences exist in the carbon dioxide content and, therefore, also in the solubility of calcium carbonate. The Antarctic bottom water, for example, is especially rich in carbon dioxide. Furthermore, the hydrostatic pressure, which grows with depth, causes an increase in the dissociation of carbon dioxide and, therefore, an increase in the solubility of calcium carbonate. Consequently, the calcium carbonate content of the sediments decreases with depth, especially below 4500 m.

What remains in the sediment are only the insoluble terrigenous components, which form part of the ingredients of the red deep-sea clay. The composition of red deep-sea clay, therefore, resembles that of the globigerina and pteropod oozes, but without calcium carbonate.

1.2.6. Geographic distribution of the sediments

Before the geographic distribution of sediments can be presented, a sensible classification of sediments must be found. Different criteria have been applied based on the natural characteristics of the sediments, such as origin, grain size distribution, and chemical composition (see Section 1.2.5). Sediment stratification was taken into consideration only locally, because our knowledge in this field is still insufficient. Only at the shelf is the spectrum of grain sizes so significant as to serve as a characteristic. For example, relevant data obtained in the North Sea provided an important basis for the investigation of the sediment movements in the German coastal area (Dietrich, 1969). Outside the shelf areas other criteria have been used, such as origin (Schott, 1935, 1942), or chemical composition (Arrhenius, 1963, and Chester, 1965). Still other criteria can be found in the association of clay minerals, as comprehensively described for the whole world ocean by Griffin et al. (1968). Numerous Russian papers are based on the mechanical properties of the sediments, their chemical composition and mineralogy [summary for the Indian Ocean north is 30°S is given by Bezrukov and Lisitzin (1967) and Gorbunova (1966), for the Pacific Ocean by Lisitzin (1970), and for the world ocean also by Lisitzin (1972)]. In the following, the origin of the sediments will be considered as the basis for classification. It becomes obvious that the geographic distribution will thus allow various oceanographic statements.

Deep-sea deposits of the open sea are called *pelagic*. They are divided according to their significant components into calcium carbonate sediments (globigerina and pteropod oozes), the predominantly inorganic red deep-sea clay, and the volcanic mud. Shelf deposits are collectively classified as *littoral* sediments and can vary greatly with regard to origin, grain size, and chemical composition, as already mentioned. Between littoral and pelagic sediments there exist various transitions which differ in composition and color; they are mainly found at the continental slope and are called *hemipelagic*.

Among the pelagic sediments, globigerina ooze and red deep-sea clay prevail. Their distribution is frequently reflected in the bottom topography and the hydrographic conditions. The ocean floor at great depths and the western basins of the Atlantic Ocean are covered by red deep-sea clay; the former because of the decalcification of the precipitation from the highly calcerous globigerina residua; the latter because of the additional decalcification effect caused by the carbon-dioxide-rich Antarctic bottom current, under which they lie. The radiolarian ooze is restricted to the tropical Pacific Ocean. South of approximately 50°S, globigerina ooze and red deep-sea clay are replaced by diatom ooze. The boundary coincides with the Antarctic Convergence of the surface currents which separates the cold polar surface water, carrying mainly diatoms, from the warmer surface water of the lower latitudes, containing globigerina. Likewise, in higher northern latitudes diatoms, which prefer cold water, accumulate in great masses at the surface, and at the bottom they form the diatom ooze, as, for instance, in the northern Pacific Ocean. However, if the terrigenous components in the deposits are more abundant, as in the northern Atlantic Ocean, sediments of glacial marine origin dominate. Isolated findings of diatom ooze off the coasts of Peru and Southwest Africa are related to the cold upwelling water along these coasts.

Outside the shelf regions, predominantly terrigenous deposits cover the deep-sea bottom in the high latitudes, namely in the whole of the Arctic Basin and on a broad band adjacent to Antarctica. The Arctic deposits differ from the Antarctic ones in their complete lack of coarse glacial material. The bottom of the European Mediterranean is mainly covered by littoral and hemipelagic sediments, whereas it is remarkable that at the bottom of the Black Sea the sediment is saprogenous ooze (sapropel). In the American and the Austral-Asiatic Mediterranean Seas (see Sections 10.2.5, 10.2.4) different sediments are found side by side, among them the detritus originating from coral reefs. Except for the local coral formations at the Bermuda Islands, in the Gulf of Guinea, and near the Brazilian Coast, the American Mediterranean contains the greatest amount of coral reefs in the Atlantic Ocean. In the Indian and Pacific Oceans, the importance of coral reefs and their deposits is very pronounced. But here too, their appearance is restricted to the immediate neighborhood of islands. They may, however, reach considerable growth—the Australian Barrier Reef is known all over the world.

The distribution of the various sediment types, as described above, refers only to the conditions at the present time. In the course of geological history, the processes of formation have changed or shifted. Necessarily, the sediment boundaries were displaced, and this explains why sediment cores sometimes show sediments resulting from very different settling processes. They give evidence of the development of the overlying water masses during the earth's history.

1.3. Mechanism of Sedimentation

1.3.1. Physical defining quantities

Erosion, transport, and sedimentation are collective terms for mechanical processes to which the sediments are subject. So far we have not yet been able to determine the complete quantitative relationships because too many variables are involved in these processes. Current velocity and grain size are the most important variables, but not the only ones. Vertical shear of current, water depth, solid material content, specific weight, and turbulence are further variables as far as water is concerned; bottom slope, specific weight, grain size spectrum, form and cohesion of particles are other variables as far as solid material is concerned.

Because of the great importance of erosion, transport, and sedimentation for the morphology of the coast and the offshore zone, and thus for river, harbor, and coastal engineering, several empirical formulas regarding these relations have been developed, none of which is generally applicable since not all the variables are sufficiently considered. A detailed treatment of the mechanism of sedimentation was carried out by Inman (1963) and Bagnold (1963). Since that time, remarkable progress has been made in the field of hydraulics by combining several of the variables mentioned above in the nondimensional coefficients and by experimentally determining the relations among the nondimensional parameters (the Froude and Reynolds numbers) as shown by Vollmers and Pernecker (1967). Since the two authors do not consider these results entirely satisfactory, we will only represent the critical limiting velocity at the beginning of the transport of the solid material as a function of the grain diameter (see Fig. 1.14). This simple diagram has again been confirmed by Vollmers (1969). The broad band of the critical limiting velocity between motion and rest at the bottom indicates the uncertainty of the results. The figure shows two important results. (1) There is a lower limit of erosion; its smallest values amount to approximately 20 cm sec^{-1} with grain sizes around 0.5 mm. (2) The critical limiting velocity grows with increasing and decreasing grain size. The reason for this surprising behavior of the small sediment particles lies in the increase of cohesion with decreasing grain size.

1.3.2. Erosion

In a limited sense, erosion refers to the transformation of the outcrops of rock into transportable material, which occurs mainly at rocky coasts. In a larger sense, erosion is to be understood as the transformation of existing deposits which are again made transportable. All movement of water, whether it is periodic or nonperiodic, causes erosion, at the beach as well as at the sea floor, as long as the velocity of the water exceeds a certain value as given in principle in Fig. 1.14. Apart from this, ice is the only other force causing erosion by mechanical attack on beaches and bottom.

The surface waves and the long waves, among these mainly the tidal waves, belong to the periodic water motions. The nonperiodic water motions include wind-generated currents, and gradient and density currents as well as the currents of the surf zone, especially long-shore currents. Each of the five motions will be discussed in special sections of this book. In the following, attention will be drawn only to their significance with respect to erosion.

The orbital path of the water particles in short waves quickly decreases with increasing

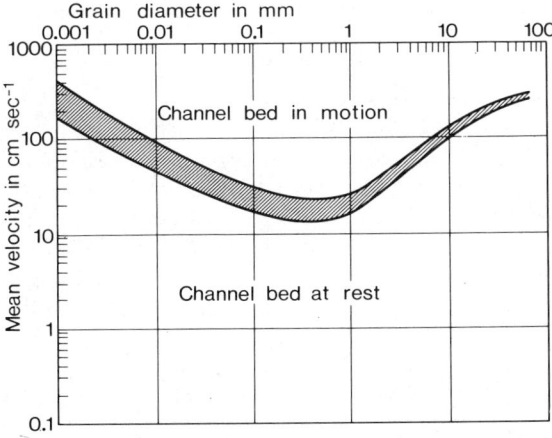

Fig. 1.14. Critical velocity at which solid material of uniform grain size begins to move at the bed of a channel. (After Hjulström, 1935 and Sundborg, 1956.)

water depth. At a depth measuring half the wavelength, the radius is reduced to $1/23$. The same is valid with regard to the velocity in the orbital path. The eroding effect of the surface waves seldom reaches deeper than to a water depth of 30 m, which means it is effective only in the shallow offshore zone. The area influenced by breakers is restricted to the still smaller strip of the surf zone, but here it is extremely effective, acting together with the long-shore current that carries off the sediments. Locally, the motion in the tidal waves is of great importance for erosion, not so much because of their vertical component—the tidal range—but because of the horizontal one—the tidal current. In contrast to the short surface waves, tidal waves, as long waves, are characterized by the fact that they reach the bottom even in oceanic depths. Therefore, seamounts, like the Great Meteor Seamount, are free of sediments because the material eroded by tidal currents is carried off into the deep water even by weak residual currents (Meincke, 1971). Other long waves, like tsunamis, storm-surge waves, and seiches may occasionally cause considerable coastal erosion.

The wind-generated surface currents rapidly decrease with depth. At the friction depth D, the undisturbed drift current has only $1/23$ of the velocity at the surface. With $D = 60$ m in the middle geographical latitudes, erosion is restricted to a thin near-surface layer of approximately 10 m depth. Density currents and slope currents, however, can be effective down to great depths, as the Gulf Stream, which was traced down to 1000 m, the outflow of the Mediterranean (1000 m), or the Antarctic bottom current. The erosion due to these deep-sea currents was demonstrated by deep-sea photography which locally revealed current ripples in the sediment (Fig. 1.08), as well as by seismic reflection surveys which, at some places, indicated a bottom free of sediments, as for example during the *Meteor* cruise in 1967 off the Spanish continental slope in the outflow of the Mediterranean (Giesel and Seibold, 1968).

Strong and highly effective erosion takes place only in a very thin upper layer of the sea, that is, in the shallow waters down to a depth of 20 to 30 m. Surface waves are effective here; above all the surf with the long-shore current and the tidal current. In the course of the earth's history, this thin layer was not fixed at a certain level. The sea surface was subject to considerable worldwide variations, estimated at 100 to 120 m within the past 20,000 yr, coming to an end only 6000 yr ago (Fig. 1.22). Thus, in the most recent geological past, a shelf zone with depths from 0 to 120 to 140 m underwent intensive erosion, which is of great importance with regard to the present-day coasts and offshore zones.

1.3.3. Transport

Sediment transport is partly effected by the same agents that cause erosion, with the exception of purely periodical processes, since these are not connected with mass transport. They are, however, indirectly involved in transport processes, for by their erosion, sediment particles that can be transported even by smallest residual currents are conveyed into the water column. Such residual currents of greatly varied strength occur in the sea as long-shore currents, as wind-generated surface currents, and as density and slope currents.

In detail, the process of transport of sediment is very complicated. At low current velocities the single grain at first is pushed and rolled, at higher velocities it will bounce, at still higher ones it will remain suspended in the water for some time until it is brought down to the bottom again by its own sinking velocity. Therefore, the sinking velocity of a particle plays an important role in the process of transport, furthermore it determines,

Mechanism of Sedimentation

Table 1.04. Dependence of Sinking Velocity w and Settling Duration s in the Ocean from 0 to 4000 m on the Particle Diameter D (after Inman, 1963)[a]

Particle Diameter D (mm)	Sinking Velocity w (cm/sec^{-1})	Settling Duration s
0.001	0.0001	127 years
0.01	0.0100	1.27 years
0.1	0.90	4.2 days
1.0	20	5.5 hr
10	80	1.4 hr

[a] w for quartz spheroids with an axis ratio of 1:4, short axis to long axis.

even without any water current, the sedimentation of the residua of the pelagic organisms.

The settling process also depends on several parameters. Here, the simplest case is given, that for motionless water and spheroidal smooth particles. According to Stokes' law in the mechanics of fluids, the sinking velocity W is determined by

$$W = \frac{1}{18} \frac{\rho_s - \rho}{\mu} g \cdot D^2$$

$\rho_s - \rho$ is the difference of density between particle and liquid, (with quartz and water = 1.65); $g = 981$ cm sec^{-2} is the acceleration of gravity, D = the diameter of the short axis of the spheroid, $\mu = 0.018$ g cm^{-1} sec^{-1} the molecular viscosity of the water (see Section 2.1.9). At $D > 0.2$ mm, the settling process is no longer laminar, and the Stokes formula needs a correction, which is considered in Table 1.04.

The settling duration of globigerina residua with the most frequent grain diameter of 0.1 mm would be a matter of days or weeks from the sea surface to a depth of 4000 m, whereas particles with a diameter $D = 0.001$ mm, as in red deep-sea clay, would need decades. Considering the small amount of motion of water in the deep sea, this means that the lateral drift of globigerina residua cannot cover more than a few kilometers and that the insoluble residua on the ocean floor will therefore reflect biotic communities of plankton species in the surface layers above. The material of the red deep-sea clay can, however, be carried off by ocean currents and horizontal exchange to places far from its area of origin because of the long duration of settlement. Therefore, even in remote oceanic areas, the sediments may be terrigenous, as can be proved by the composition of the minerals.

The *turbidity currents* are a factor of special importance in the transport processes. No one has yet seen these currents, and no one has carried out direct measurements of their velocity, their thickness, or the density of the flowing medium. But today there is no doubt that these currents which seem so mysterious really exist; there are too many indications that cannot be ignored. Turbidity currents are caused by horizontal differences of density in the water, similar to the deep-water circulation in the ocean. Here, however, it is not differences of temperature and salinity that effect a difference of density, but rather different amounts of material suspended in the water. The velocities can be estimated by applying a formula which served in the investigation of the overflow of the Iceland–Faeroe Ridge (Dietrich, 1956). Accordingly, the stationary velocity U of the current bound slope downward is

$$U = \sqrt{\frac{gZ}{C_D} \frac{\rho_2 - \rho_1}{\rho_2} \sin \alpha}$$

In this equation $g = 981$ cm sec^{-2} is the acceleration of gravity, Z is the thickness of the turbidity current, $\rho_2 - \rho_1$ is the difference of density between suspended matter and sea-water, C_D is the friction constant at the bottom (0.3 for rough bottom, 0.003 for smooth bottom), α is the inclination of the slope. Z and ρ_2 are unknowns and also C_D to some extent. At minimum conditions $Z = 10$ m, $\rho_2 = 1.05$ g cm^{-3}, $C_D = 0.3$, U becomes 0.4 m sec^{-1}. At maximum conditions, $Z = 100$ m, $\rho_2 = 1.1$ g cm^{-3}, $C_D = 0.003$, U becomes 40 m sec^{-1}, a velocity that is strong to appallingly high. The assumption here is that fine sediments on an inclined slope become suspended while sliding down. Slumping occurs if the inclination of the slope exceeds a value critical for the sediments, which, for finest deposits with high water content, is already the case at an inclination of 2°. Impulses, as for instance by earthquakes, promote slumps which may give rise to turbidity currents if the sediments are stirred up. Currents of the same physical nature are also known in the atmosphere. Relevant examples are: the suspension of snow in air forming avalanches of dry snow, and the suspension of ashes, originating from eruptions of volcanoes, in glowing clouds, like the "nuées ardentes" at the eruption of Mount Pelée on Martinique in 1885. These currents are also extremely powerful.

The mechanism of the turbidity currents was discovered by the Swiss physician Forel (1885) at the barriers of the river Rhône in the Lake of Geneva. But only in 1937 was Kuenen the first to demonstrate this mechanism by experiments at a hydraulic laboratory. Since then, heated discussions have been held about the pros and cons of these mysterious currents. In their bibliography on turbidity currents Kuenen and Humbert (1964) mentioned no less than 700 relevant papers. Today, there is no longer any doubt left, for (1) such currents are possible according to hydraulic theory; (2) they have been confirmed by laboratory experiments, and (3) their effect on the sea floor is well known. The most convincing description of the turbidity currents has been given by Heezen and Ewing (1952) with the example of the events at the continental slope of the Grand Banks off Newfoundland in November 1929. Through this catastrophe it was possible to offer an explanation of the most important morphological forces that have formed the ocean bottom over large areas. Some details about this event will illustrate the activity of these currents.

On November 18, 1929 at 8:32 p.m. an earthquake occurred south of the Grand Banks causing, simultaneously and for some time afterwards, a total of 13 breaks of overseas cables. The locations are shown in Fig. 1.15(A). An exact time check was obtained by control stations on land, so we know that the breaks followed one after the other in a very remarkable way, as can be seen in Fig. 1.15(B). On a profile perpendicular to the continental slope, this figure shows the location of the earthquake center and the position of the cable breaks, as well as the time delay of the breaks (in minutes) with reference to the initial time of the quake. Eight cables on both sides of the quake center broke almost simultaneously with the quake. With increasing distance from the base of the continental slope to over 500 km southward in the North American Basin, the breaks were delayed more and more. In the most distant case the break occurred 13 hr 17 min after the quake. Heezen and Ewing (1952) attributed this time sequence to turbidity currents that may have originated from whirling seaslides along the continental slope which, in turn, had been produced by the quake. The analysis of the relevant data leads to the determination of unusual current velocities at the bottom, as shown in Fig. 1.15(B), amounting to 28 m sec^{-1}. Even if such velocities occur for only a short time, they must be of the greatest importance for the morphology of the continental slope as well as the continental rise seaward thereof and the abyssal plains.

Turbidity currents will tear up and clear out furrows at the continental slope, the

Mechanism of Sedimentation 31

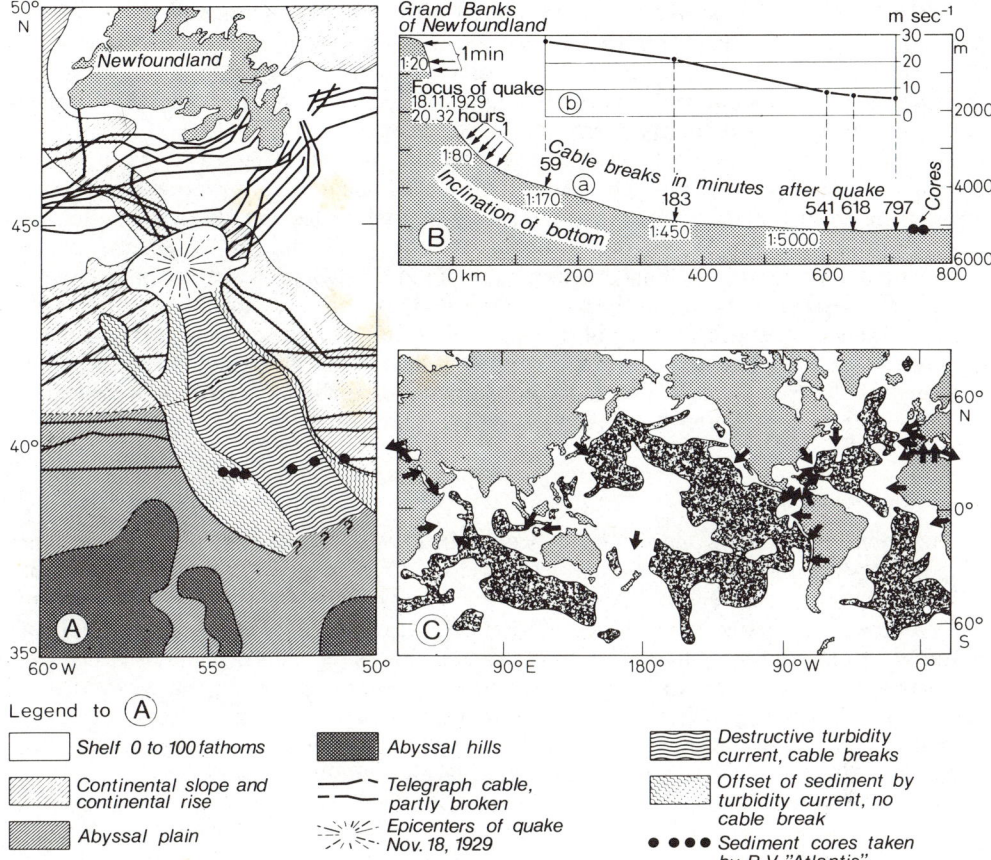

Legend to (A)

- Shelf 0 to 100 fathoms
- Continental slope and continental rise
- Abyssal plain
- Abyssal hills
- Telegraph cable, partly broken
- Epicenters of quake Nov. 18, 1929
- Destructive turbidity current, cable breaks
- Offset of sediment by turbidity current, no cable break
- ● ●●● Sediment cores taken by R.V. "Atlantis"

Fig. 1.15. A. Turbidity current in the North American Basin south of the Grand Banks of Newfoundland in November 1929. (After Heezen et al., 1954.) B. Longitudinal section through the turbidity current in Fig. A. (After Heezen and Ewing, 1952.) (a) Bottom profile in exaggeration 60:1 with position and time of the cable breaks (in minutes after the onset of the quake). (b) Velocity of the turbidity current based on the time sequence of the cable breaks. C. Areas of the world ocean inaccessible to turbidity currents from continental slopes (hatched). (After Elmendorf and Heezen, 1957.) Arrows: Turbidity currents as evidenced by cable breaks 1880–1955.

so-called submarine canyons; at the rises of the continental margins they will deposit sediment fans, and they will push forward far into the deep-sea basins, as shown in Fig. 1.15(A), smoothing the bottom there and forming abyssal plains that are flat as a table. The sediments of the same specific weight will precipitate sorted according to their grain size, beginning with the coarsest particles. The sediments will originate from far away, partly from the shelf. In fact, an analysis of the sediment cores confirms this interpretation by the special stratification as well as by the nature of the sediments, some of which originate from near-shore regions. Such local events are rare, extreme catastrophes on the earth, which fortunately do not affect human lives—apart from the breakage of valuable telegraph cables. In the course of the event of November 1929, approximately 10,000 km^2 of ocean bottom were involved and approximately 100 km^3 of sediments were displaced.

Several other cable breaks have been related with turbidity currents. Figure 1.15(C) contains 37 cases proved within the period from 1880–1955. The figure also shows the areas that have never been reached by turbidity currents, and if no other place for depositing radioactive waste material can be found other than the ocean floor, those "zones of inaccessibility" for turbidity currents should be chosen.

Up to now, it is true, no exact data could be obtained with regard to the turbidity currents in the abyssal plains, but determinations of the age composition of deposits originating from turbidity currents—the so-called turbidities—could possibly help to enlighten this problem. Catastrophes such as that at the Grand Banks certainly occur very seldom, because the amounts of sediment required for such processes are not available within short periods of time and the oceans are filled up with suspended matter only very slowly. When the eustatic lowering of the sea level as well as of the erosion level took place during the ice age (Fig. 1.22) a greater input of sediments probably appeared at the continental slope and, thus, also a greater activity of turbidity currents, which may sometimes have contributed to the formation of submarine canyons.

1.3.4. Sedimentation

The distribution of clastic and biogenous sediments in the sea is dependent on the amount of sediment-forming material and its sinking velocity, which can vary considerably. This is due not only to the difference in grain size, but also to density and the shape of the grain. Only a very rough estimate of the sedimentation rate can be derived from the supply of material and the sinking velocity. Since this value, with its regional varieties, is of decisive importance for the understanding of the history of the oceans and for the estimation of the transport of material from the continents into the world oceans, efforts have been made for direct measurements. In principle, this means determining the thickness of sediment between two stratigraphically equal levels, usually measured in cm per 1000 yr of the dry or wet deposits. As an aid for age determination, characteristic levels in a sediment core were chosen, like ash layers originating from volcanic eruptions of known date (e.g., in the European Mediterranean Sea), or certain successions of planktonic foraminifera preferring warm or cold water and associated with the warm or cold periods, respectively, of the Quaternary (Schott, 1935), or other datable faunal features. Since the natural radioactivity of the sediments was introduced for the determination of their absolute age (Arrhenius et al. 1951), and after the improvement of this method, described in a summary by Koczy (1965), rates of sedimentation have been mentioned repeatedly. Most frequently the ^{14}C method has been applied, but as the half-life of ^{14}C is about 5570 yr, reliable statements can only be dated back as far as 30,000 to 40,000 yr. Other techniques, too, have been applied: The ionium (^{230}Th)–thorium(^{232}Th) method allows sediment dating during the past 500,000 to 700,000 yr, since ionium has a half-life of 80,000 yr. With the potassium–argon method, the past 200 to 400×10^6 yr can be covered. For the Pacific Ocean Goldberg and Koide (1962) have presented a map with the distribution of the sedimentation rates, in which the values for regions far away from land lie at 1.0 cm per 1000 yr in the northern part, and at 0.4 cm per 1000 yr in the southern part. For the Atlantic Ocean, showing the highest values of all three oceans, a rate of 3 to 10 cm per 1000 yr is given. An explanation of this high rate is closely related to the large areas of inflow from rivers (see Plate 1). With the determination of age Urey (1947) combined that of the $^{16}O/^{18}O$ isotope relation, which gives information on the surface temperature at the time of sedimentation, thus opening new doors to paleoclimatology (Fig. 5.09). On continental slopes and shelves, the sedimentation rate can be much higher than on the ocean floor. Thiede (1971) found up to 20 cm per 1000 yr on the Portuguese continental slope. Hinz et al. (1971), who investigated the thickness of the late-glacial and post-glacial soft sediments in the Kiel Bight, reported 50 cm per 1000 yr; in local straths up to 500 cm per 1000 yr.

In these days the seismic reflection investigations (see Section 1.4.1) have contributed to the growing knowledge of the thickness of the sediment layers under large areas of the ocean (Fig. 1.17). On the basis of secured sedimentation rates, information can now be given with respect to the age of certain areas, which supports the theory of the continental drift (see Section 1.4.6). In the North Pacific Ocean, numerous seismic profiles of sediments have been obtained (Ewing et al., 1968), the same is true for the North Atlantic Ocean between 10 and 19°N (Collette et al., 1969), and between 30 and 70°N (Orlenok, 1968). In the North Atlantic Ocean the crest region of the Mid-Ocean Ridge in a width of 200 km have proven to be free of sediments. The flank regions have a sediment cover of around 500 m, the abyssal plains of Madeira and Guayana of 1400 and 2000 m, respectively, at the maximum. This increase of sediment thickness with the distance from the crest region may be connected with the spreading of the ocean floor (see Section 1.4.6).

Drillings carried out by the American drilling vessel *Glomar Challenger* (Fig. 1.13) in all three oceans since 1968, have run through several hundred meters of sediment as far as the floor of basalt of the earth's crust. Thus, completely new aspects have been opened to the stratigraphy of the ocean bottom as well as to the history of the oceans. Earlier, stratigraphical investigations could at best be carried out on the basis of 20 m long cores taken by the piston corer.

1.4. Formation of the Bottom Features

1.4.1. Aids for interpreting the morphology of the ocean bottom

Morphology in the sense of a genetic explanation of the topographic features, which are presently investigated with the aid of a highly developed surveying technique (see Sections 3.1.1.3 and 3.2.4), requires a well-founded concept of the geological structure of the subsoil in addition to the knowledge of the bottom coverage. On the continents such knowledge can be obtained through the study of geologic exposures and drillings. However, this approach cannot be followed in the case of oceanic regions because sediments may cover the crystalline bedrock, and with a sedimentation rate of 1 cm per 1000 yr even the longest core samples amounting to 20 m can, at best, give information only about deposits not older than 2×10^6 yr. On the shelf numerous drillings have been carried out, often in close connection with the exploitation of natural oil and gas reservoirs and therefore most of the results are company secrets. In the deep sea, drillings first had successful results in 1961 (the Mohole project of the United States, west of Lower California at 3000 m depth of water). Since 1968, new insights into the history of the oceans have been gained by drillings through the sediments down to the underlying basalt with the American special drilling vessel *Glomar Challenger* (see Fig. 1.13). Direct drilling methods were also profitable on oceanic islands. Thus it has been proven, for example, that the coral limestone of the atolls lies on the volcanic foundation of old seamounts, and that the volcanic rocks in the Pacific Ocean, except in its western part, are of basaltic nature and originate from the deeper layers of the earth's crust. All other igneous outcrops on the oceanic islands of the Atlantic and Indian Oceans, as well as on the continents, consist of basic lava with granite inclusions, originating from the upper earth crust. Besides the direct geological methods, which presently can give only very limited information with respect to the deeper oceanic subsoil, geophysical methods have increasingly become available. A detailed description including numerous results is presented in the book *The Sea,* volume 4, part 1, edited by Maxwell (1970). Consideration has been given to the interpretation of earthquakes from the point of view of tectonics (Sykes et al., 1970), and also to the results of the seismic reflection measurements (Ewing and Ewing, 1970), of seismic refraction measurements (Ludwig et al., 1970), of measurements of recent magnetism (Heirtzler, 1970), of paleomagnetism (Opdyke, 1970), and of the heat flow (Langseth and von Herzen, 1970).

Concerning *seismic investigation* the following four different methods of studying the structure of the ocean bottom were applied: (1) registration of the earthquake foci, (2) the behavior of elastic waves with refraction, (3) the behavior of elastic waves with reflection, and (4) the behavior of elastic surface waves (Love and Raleigh waves). Since Rudolph (1887) started to catalogue the geographic distribution of the foci of seaquakes, numerous investigations of the *seismicity of the ocean bottom* have supplied important information concerning the subcrustal structure. This is especially true for the systematic investigation of the entire earth by Gutenberg and Richter (1954) and the summary by Sykes et al. (1970). Four factors about earthquakes are considered as basic: (1) the geographical position of the epicenter, (2) the magnitude of the earthquake, (3) the focal depth, and (4) the travel time of seismic surface waves. Statements relating to the entire ocean can be summarized as follows: The geographical distribution of the epicenters in the world ocean (see Plate 2) clearly shows a concentration in the crest region of the Mid-Ocean Ridge, and also a concentration in those continental margins bordered by deep-sea trenches. These are the two zones of high mobility of the ocean bottom. Contrary to these, the deep-sea basins and the flank regions of the ridges are almost free of earthquakes. The earthquakes beneath the continental margins with deep-sea trenches, in contrast to those beneath the ridge system, show a characteristic distribution of the focal depth (Luvendyk, 1970), which is to be seen in the example of the Kuril Trench and the North Japanese Shelf in Fig. 1.16: shallow foci in less than 60 km depth beneath the deep-sea trench, arrangement of the earthquake foci in a relatively thin layer steeply descending beneath the continent down to a depth of 400 km. The actual slope of this layer, amounting to approximately 45°, can be taken directly from Fig. 1.16(IV), since the transverse profile is represented without exaggeration. This arrangement has been found at all marginal deep-sea trenches (Plate 2). The deepest focus observed so far, in the Sunda Trench, Indonesia, is located at a depth of 700 km. However, the most frequent earthquakes at the continental margin are shallow-focus earthquakes with foci at depths of less than 60 km; among them are also the centers of the strong earthquakes. From the 462 stronger Japanese earthquakes (greater than force 6) observed from 1926 to 1956, 74% had a focal depth of less than 60 km (Wadati, 1967). The earthquake foci beneath the Mid-Atlantic Ridge are shallow (less than 100 km) almost without exception, as demonstrated in the maps of the seismicity of the earth by Barazangi and Dorman (1969) (cf. Plate 2).

The study of seismic surface waves (Love and Raleigh waves), especially of the dispersion of these waves in the earth's crust and the earth's mantle, provides an indication of what the structure beneath the oceans and continents is like. In this way it was found that certain differences do exist (Tams, 1921), that the structures beneath the oceans are remarkably identical (Berckhemer, 1956), and that the whole Mid-Ocean Ridge shows similar, relatively low wave velocities, indicating that its material is very homogeneous (Båth, 1959).

Among all other geophysical methods, the *seismic refraction measurements* by

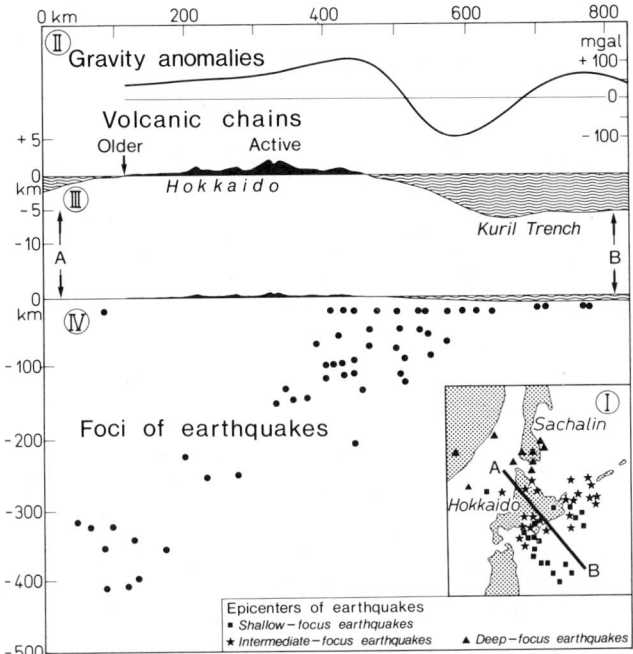

Fig. 1.16. **Earthquake foci and gravity anomalies at a deep-sea trench (Kuril Trench). (After Gutenberg and Richter, 1954.) I. Location of the earthquake foci and of the cross section A–B. II. Gravity anomalies on A–B. III. Bottom profile on A–B (exaggeration tenfold). IV. Bottom profile on A–B (not exaggerated) with the location of the foci.**

detonations have so far given the best information about the oceanic subsoil, with respect to sediment thickness and, to some extent, the material beneath the sediments and its structure (which normally can be determined only by drilling). The method requires a shooting boat and an instrument carrier equipped with hydrophones that can record the travel time of the elastic waves. Up to now such instrument carriers have been ships, helicopters, or buoys. After the first successful explorations with seismic shooting carried out on the continent by Mintrop (1923), who realized the ideas of his teacher Wiechert at Göttingen, Ewing started studying the shelf structure with this method in 1937. After 1946 the methods were rapidly developed at numerous large institutes in the U.S.A. (Lamont-Doherty Geological Observatory, Scripps Institution of Oceanography), in England (Cambridge University), in the Federal Republic of Germany (Bundesanstalt für Bodenforschung,* Hannover, and Institut für Physik des Erdkörpers, Hamburg) as well as in the USSR (Academy of Sciences), and they were applied in the entire world ocean (cf. Ludwig et al. 1970). Figures 1.17 and 1.25 contain some results of the seismic refraction and seismic reflection methods. The former represents a result of fundamental importance, namely that the Atlantic continental margin of the United States consists of sediments. The thickness of the sediment layer reaches approximately 9 km. The second figure gives a summarizing survey of the typical oceanic subsoil, which would be impossible without seismic refraction studies. The oceanic crust, with its thickness of 5 km, is considerably thinner than the continental one with an average thickness of 30 km. This had indeed been known as the large-scale mean derived from the travel times of the

---

* Now named: Bundesanstalt Für Geowissenschaften und Rohstoffe.

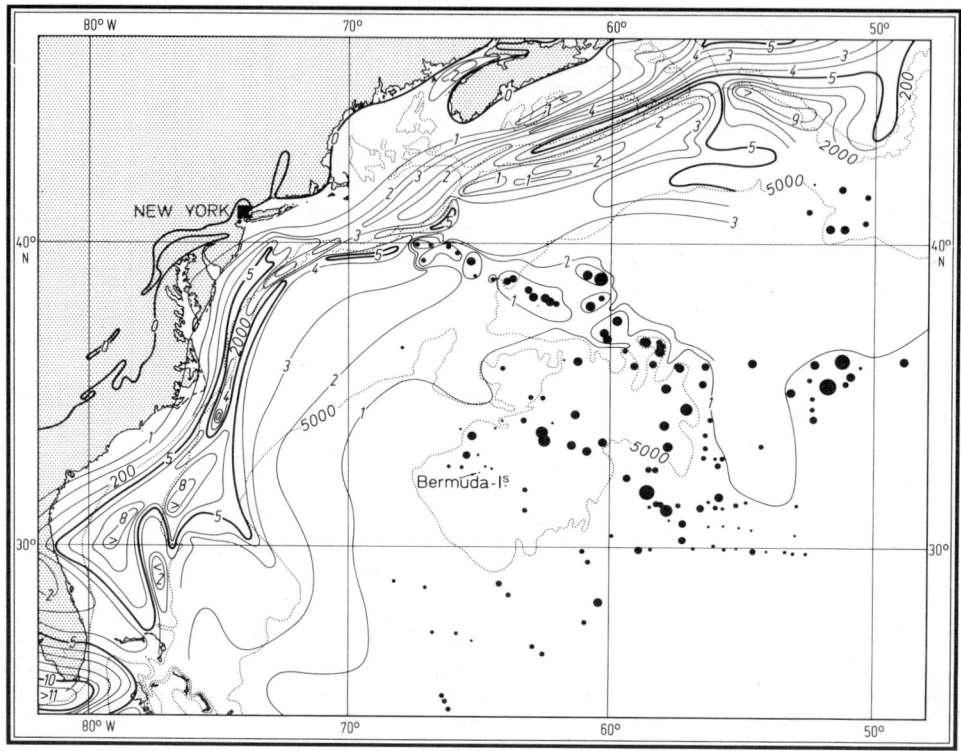

Fig. 1.17. Thickness of sediments (in km) in the western part of the North Atlantic Ocean. (After Emery, et al., 1970.) Solid circles: Seamounts. Dotted lines: Contour lines of 200, 2000 and 5000 m depth.

earthquake waves, but not in all the details, which considerably exceed the information given in Fig. 1.25.

In the *seismic reflection investigations* at sea the principle of the echosounder is applied (Section 3.2.4), except that greater sound energy is emitted on a broader frequency band ("sparker," "boomer," "air gun," and "flexor"). This method has been developed especially by Hersey since 1942 (Hersey, 1963; Ewing and Ewing, 1970). Its advantage relative to the seismic refraction method lies in the fact that, as in the case of the echograph, the reflection can be continuously recorded on the moving vessel—a simple, fast, and cheap method. But there are also grave disadvantages: (1) the sediment layer can be penetrated only down to a depth of a few hundred meters, and this has been achieved only since 1964, so that usually (2) only sediment structures can be investigated, and (3) the sound velocity in the rock is not always sufficiently known and must then be determined with the aid of seismic refraction measurements.

The various methods of seismic reflection that are applied at sea differ mainly only in the source of the sound; a more recent and efficient method makes use of the pneumatic sound generation, therefore it is called the pneuflex or air-gun method. It was applied to obtain the profile of the Great Meteor Seamount (Fig. 1.24) by *R.V. Meteor* in 1967. The simple technique of measurement and the great operational efficiency in recording distribution and thickness of sediment layers make the seismic reflection method one of the most important means in marine geology.

Gravimetric measurements give information on gravity differences on the earth and thus, to some extent, on the mass distribution in the subsoil. The first reliable gravity measurements in all three oceans were carried out by Hecker (1903), although his accuracy was 50 times less than that of the present measurements, which reach a precision of 1 to 3 mgal. He was not able at that time to detect any unusual gravity anomalies, but, after all, he recognized the important fact that, within the frame of the accuracy of measurements then possible, the oceans are isostatically compensated—which means that the obvious mass deficit of the ocean basins is compensated by denser masses in the subsoil. Important progress was achieved by Vening-Meinesz (1948) with his pendulum measurements made on board submerged Dutch submarines in all three oceans between 1923 and 1938. The evaluation of these data opened a new epoch for the understanding of the formation of oceanic bottom features, especially with the example of the East Indian area. He discovered the high negative gravity anomalies of up to -250 mgal in the deep-sea trenches, which are embedded in zones of high positive anomalies of up to $+100$ mgal. These anomalies run parallel to volcano chains and coincide with zones of high seismicity and young faults (cf. Fig. G8 in the Atlas by Dietrich and Ulrich, 1968). Figure 1.16 contains an example of the behavior of the gravity anomalies at deep-sea trenches. Vening-Meinesz developed his own hypothesis of the orogenesis of deep-sea trenches, the so-called "Buckling Theory." Since 1958 there has no longer been any need for the very burdensome process of pendulum measurements which had to be carried out from submerged submarines so that the acceleration of the pendulum instrument by the turbulent sea should be kept as small as possible. Without knowing of each other, Graf (1958) in Germany and La Coste (1959) in the United States developed sea gravimeters which can be used at the sea surface on board a ship underway. By these instruments our knowledge of the oceanic subsoil was further improved. Some of the first results concerning the Pacific deep sea have been summarized by Woollard and Strange (1962). A more comprehensive summary has been given by Talwani (1970). Here the ambiguity of gravity measurements becomes obvious: gravity anomalies depend on the thickness as well as on the composition of the earth's crust. Therefore the records of gravity must at least be related to the seismic refraction measurements of the thickness as long as we do not have any large-scale representation of the thickness of the crust. Such measurements were carried out on board the research vessel *Meteor* on the Great Meteor Seamount (Fleischer et al., 1970).

Geomagnetic measurements have given evidence of a complicated permanent magnetic field of the earth, which depends on the magnetic properties of the minerals in the upper 20 to 30 km of the subsoil. At greater depths, the Curie temperature of 650°C is reached at which ferromagnetism vanishes. A description of the nature of the subsoil can be restricted to the measurement of the total magnetic intensity. Nevertheless the records of the entire magnetic field of the earth, which were contributed by the investigations on some iron-free research vessels (American vessels *Galilee*, 1905–1908, and *Carnegie*, 1909–1929, USSR vessel *Zarya* since 1956) are not superfluous. Since 1947 attempts have been made to carry out such intensity records from shipboard and from airplanes. In 1955 the first detailed survey of an ocean area was accomplished (Mason, 1958), introducing a period of revolutionary discoveries which include the linear arrangement of anomalies in the North Pacific Basin as well as the lateral dislocation of these anomaly strips over hundreds of kilometers along the lineaments of the step or fracture zones (Vaquier, 1959), and also the arrangement of the anomalies over the Mid-Ocean Ridge, as shown in Fig. 1.18. From this, the following characteristic features can be recognized: (1) a streaky pattern of anomalies; (2) highest positive anomalies in

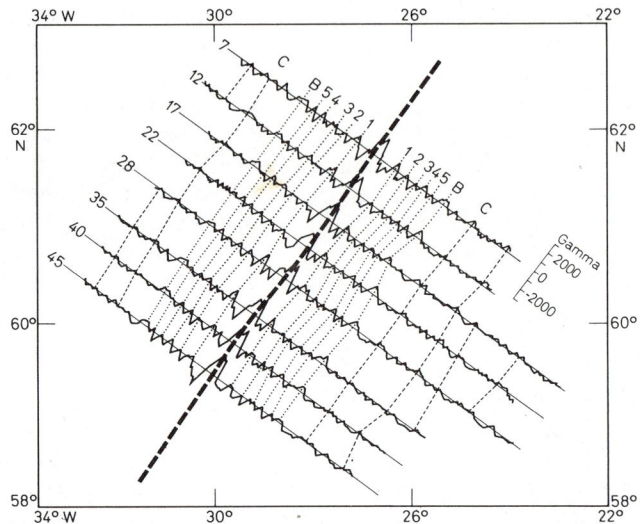

Fig. 1.18. Magnetic anomalies (in gamma) on eight profiles perpendicular to the Reykjanes Ridge. (After Heirtzler, et al., 1966.) Selected from 55 transverse profiles of an aeromagnetic survey. Figures at the side: Profile numbers. Figures and letters at the top: Characteristic anomalies, connected by broken lines. - - - Axis of the Reykjanes Ridge. (After Ulrich, 1960.)

the central axis A; (3) the central axis A in conformity with the crest of the ridge; (4) six narrow strips of anomalies, each approximately 20 km wide, on either side of the central axis A. In addition wider and more indistinct strips of anomalies occur. These results, as shown in Fig. 1.18 (which are based on an aeromagnetic special survey of the Reykjanes Ridge, Heirtzler et al., 1966), are in agreement with the results from other sections of the Mid-Ocean Ridge summarized by Mason (1967). Heirtzler (1970) has given a first monograph of recent magnetic anomalies on the sea floor.

The explanation of these patterns of strips can be based on the following three fundamental points of recent findings:

1. In the course of the history of the earth, the geomagnetic field repeatedly reversed all over the globe and very rapidly (Cox et al., 1964). The reversal is believed to have needed only a few thousand years, and the duration of the new polarity seems to have greatly varied, sometimes some thousands of years, sometimes several hundreds of thousands or millions of years.

2. Rocks containing iron and titanium are non-magnetic at high temperatures, but they become magnetic when they cool below a critical point—the Curie temperature of approximately 650°C. In each case, this temperature depends on the chemical composition of the rock. The minerals are permanently magnetized in the direction of the geomagnetic field of the period, and they retain this direction when the earth's field reverses (Cox et al., 1967).

3. The exact time of the reversals of the geomagnetic field was determined in absolute figures by means of the potassium–argon method by the same scientists in 1967. Thus, a time scale of global validity is available, in which rock magnetism finds its place. A relevant summary has been given by Opdyke

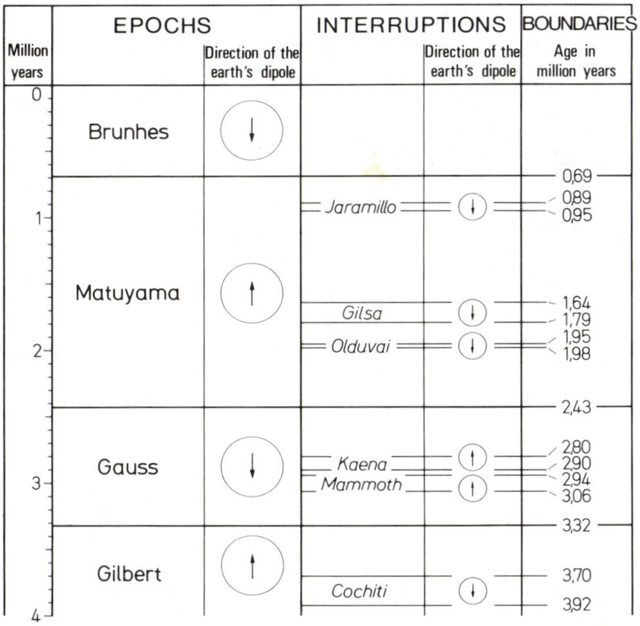

Formation of the Bottom Features   39

Million years	EPOCHS		INTERRUPTIONS		BOUNDARIES
		Direction of the earth's dipole		Direction of the earth's dipole	Age in million years
0	Brunhes	↓			
					0,69
			Jaramillo	↓	0,89
1					0,95
	Matuyama	↑			
					1,64
			Gilsa	↓	1,79
			Olduvai	↓	1,95
2					1,98
					2,43
					2,80
	Gauss	↓	Kaena	↑	2,90
3			Mammoth	↑	2,94
					3,06
					3,32
	Gilbert	↑			3,70
			Cochiti	↓	
4					3,92

Fig. 1.19. Paleomagnetic epochs of the earth's history within the past 4×10^6 years. (After Cox, 1969.)

(1970). Figure 1.19 contains a section of this time scale for the past 4×10^6 yr. Several reversals of the geomagnetic field occurred within this period; the present, the Brunhes epoch, began 690,000 yr ago. Today, the reversals of the past 70×10^6 yr are well recorded. Furthermore, Opdyke (1966) succeeded in proving that also sediments, while settling, can become magnetic in the direction of the geomagnetic field at the time of their settling, which opens quite new ways for the determination of the age of sediments and, thus, of the sedimentation rate as well.

The hypothesis of the spreading of the ocean floor, advanced by Dietz (1961) and further developed by Vine and Matthews (1963), supplied convincing arguments by investigations of paleomagnetism. Contributions were made by well-known geophysicists of the United States, the United Kingdom, and France. The interpretation of the aforementioned patterns of anomaly strips involves the following: In the oceanic crust, which is continuously pushing up at the rift of the Mid-Ocean Ridge, the direction of the respective geomagnetic field, when cooled below the Curie temperature, is "frozen" before this crust spreads sidewards. The anomaly strips reflect the reversals of the magnetic field during the horizontal spreading. From the time scale of the reversals (cf. Fig. 1.19) and the intervals between the anomaly strips we can obtain the spreading velocity of the ocean floor in one direction: 5 cm yr^{-1} in the South Pacific Ocean and 1 cm yr^{-1} in the North Atlantic Ocean (cf. Fig. 1.20 and Plate 2).

The reversals of the magnetic field are confirmed facts, and their interpretation as paleomagnetic evidence offers a unique time scale for geology, as shown for example by Pitman et al. (1971) with respect to the earth's history in the area of the North Atlantic Ocean. The mechanism of the reversals, however, is not yet clear. Thus we have the re-

40 Geomorphology of the Ocean Bottom

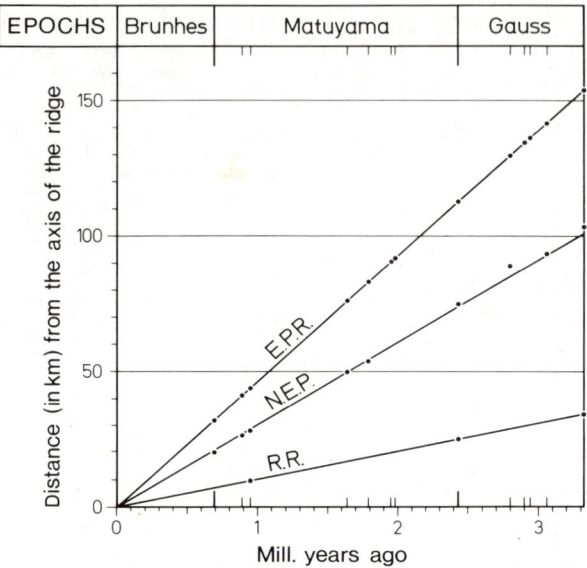

Fig. 1.20. Three examples of the spreading of the ocean floor during the past 3.32 million years. (After Vine, 1968.) They refer to the last three magnetic epochs (see Fig. 1.19). Points: Observational data. E.P.R.: Eastern Pacific Ridge. Drift 138 km per 3 million years = 4.6 cm per year. N.E.P.: North Pacific Ocean, off Juan de Fuca. Drift 90 km per 3 million years = 3.0 cm per year. R.R.: Reykjanes Ridge (see Fig. 1.18). Drift 30 km per 3 million years = 1.0 cm per year.

markable situation that the geomagnetic field in the past and in our days is one of the phenomena that are best described, but least understood.

The *heat flow* at the sea floor is a result of the thermal conditions in the crust and the upper mantle of the earth. It is the product of vertical temperature gradient and thermal conductivity, expressed in μcal cm^{-2} sec^{-1}. Except for a few drillings, such measurements at sediment cores are, at best, restricted to the uppermost 20 m beneath the sea floor. The investigations of the heat flow were initiated by Sir Edward Bullard, who started such studies in South Africa in 1939, and in 1949 began working on this problem at the ocean floor. First results were obtained in 1952 by Revelle and Maxwell. Since that time several thousand measurements of heat flow have been carried out at the floor of the world ocean. According to von Herzen and Lee (1969), as many as 5000 measurements had been taken by 1969. Summaries have been presented by Langseth and von Herzen (1970) and by Blackwell (1971). The results deserve special attention, as shown in the example of the South Pacific Ocean in Fig. 1.21. From this, the following characteristics of heat flow can be given. (1) There are considerable regional and local differences. In Fig. 1.21 they range between 0.14 and 8.09 μcal cm^{-2} sec^{-1}. (2) Very high values exist in the Mid-Ocean Ridge, especially in its central rift. (3) Very low values exist at approximately 300 to 600 km lateral distance from the crest region. (4) Very low values exist in the deep-sea trenches. It is surprising that, on the earth, the values of heat flow are the same on the continents and at the ocean floor. The most frequent value is 1.1 μcal cm^{-2} sec^{-1}, although the thickness of the crust differs considerably, amounting to 30 km under the continents and 5 km under the ocean. The interpretation of the data is difficult. The great local differences of the heat flow, as demonstrated in Fig. 1.21, induces us to believe that shallow heat sources must exist. Under the continents, the

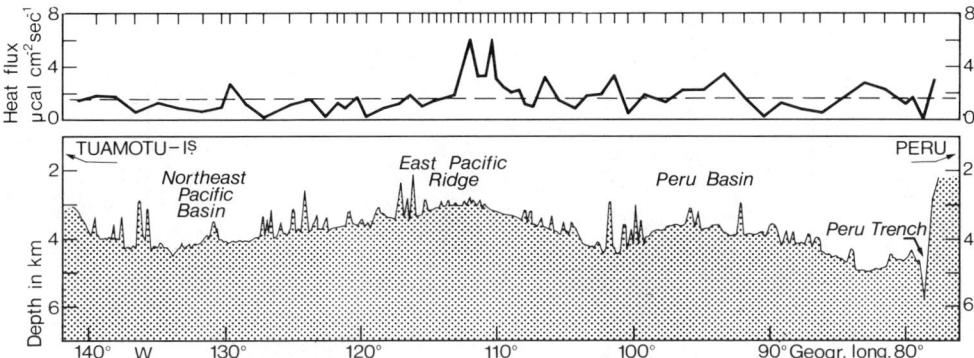

Fig. 1.21. Heat flux and bottom profile through the eastern Pacific Ocean at approximately 12°S between the Tuamotu Islands and Peru. (After von Herzen and Uyeda, 1963.) Broken line: mean value of the heat flux in the eastern Pacific Ocean equals 1.4×10^{-6} cal cm^{-2} sec^{-1}. Marks at the top: stations for heat flux measurements during the "Risepac Expedition" of the Scripps Institution of Oceanography, 1961–1962.

variable radioactivity of the earth's crust gives sufficient explanation, which is not the case under the oceans because the crust there is too thin. The source of this phenomenon must partly be sought beneath the Mohorovičič discontinuity. Interpretations that see an explanation in convective motions rising in subcrustal masses beneath the Mid-Ocean Ridge and subsiding underneath the deep-sea basins cannot be accepted without objection. Likewise, as in the case of the magnetic anomalies, the study of heat flow is a very new, not yet fully utilized aid that might help to clarify the evolution of the oceans (see Section 1.4.6). As Popova et al. (1969) showed for the Pacific Ocean, heat flow is closely related to the development of the tectonic structure of an ocean.

1.4.2. Continental margins

The continental margins, the major features of which have been mentioned in Section 1.1.6 and in Fig. 1.02 as well as in several other places, are determined by the structure of the earth's crust, especially by the considerably decreasing thickness from continental to oceanic crust—a decrease from 30 to 5 km. The profile through the North Atlantic Ocean, as presented in Fig. 1.25, is typical of this. In the course of the earth's history the original shape of the continental margin determined by the earth's crust was changed in certain areas into different forms. In the Atlantic–Indian type the following formation prevails—shelf, continental slope, and continental rise. In the Pacific type the prevalent formations are continental slope and deep-sea trench. The shelf is predominantly only rudimentary, as is the continental rise, which coincides with the landward flank of the deep-sea trench.

First of all, the *formation of the shelf* will be considered. There are two entirely different explanations. Either (1) the shelf is considered a terrace caused by abrasion, which, like an escarpment, was cut into the earth's crust by near-surface erosion (surf, tidal currents), or (2) it is interpreted to be formed by deposit of littoral sediments. In fact, all the transitional stages can be recognized, from forms developed by erosion only to those developed by accumulation only. The explanation of the shelves is made even more complicated because the continental margins were involved in flexures and in the formation of mountains, and because the surface level of the world ocean underwent

42   Geomorphology of the Ocean Bottom

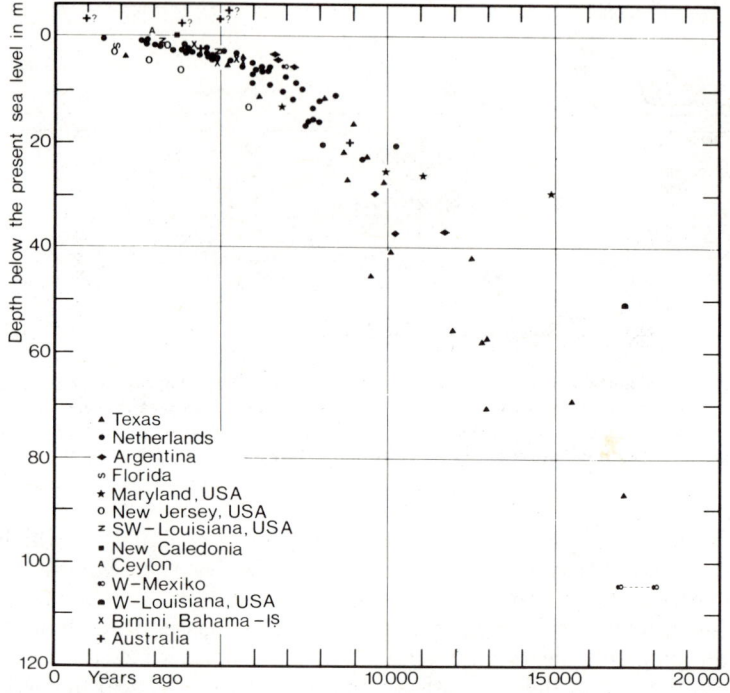

Fig. 1.22. Level of the sea surface in the world ocean during the past 20,000 years. (After Shepard, 1968.)

secular fluctuations. We have evidence that they amounted to 100 to 200 m. From the point of view of morphology, the last change was the most important one.

As late as 18,000 yr ago the sea level was 100 to 200 m lower than today, which can be proven by numerous age determinations applying the ^{14}C method on near-surface calcerous shells in areas believed to be of high geological stability (Fig. 1.22). This took place towards the end of the Würm cold age, the last of six cold ages in the past 1 to 3 $\times 10^6$ yr (Flint, 1971). Erosion, transport, and sedimentation, the effects of which strongly depend on depth, have undergone remarkable changes on the shelves within the past 18,000 yr owing to the considerable eustatic rise of the sea level. Emery (1968) has described Pleistocene relict sediments on the sea floor which can be taken as residua from times of a lower sea level; 70% of the shelves of the world ocean are of such kind.

The present high level of the sea surface was reached in historical time, approximately 5000 yr ago. So the greatest rise of the sea level amounting to more than 100 m within 13,000 yr still falls in the time of the first great epochs of human civilization. There is no definite answer to the question whether in future the sea level will continue to rise (the supply of inland ice in Antarctica (cf. Fig. 1.01), in Greenland, and in the high mountain ranges would allow a maximum rise of 60 m), or whether the earth is facing another ice age and the sea level will be lowered again. Milankovitch (see Section 4.2.2) thinks that the latter is more likely because of the correlation (good so far) of the secular course of the irradiation of the earth with water temperatures and the intermediate ice ages. But as this would take place only after several thousand to ten thousands of years, it will not be a real problem for mankind in the near future.

During the earth's history the eustatic fluctuations of the sea level have been caused

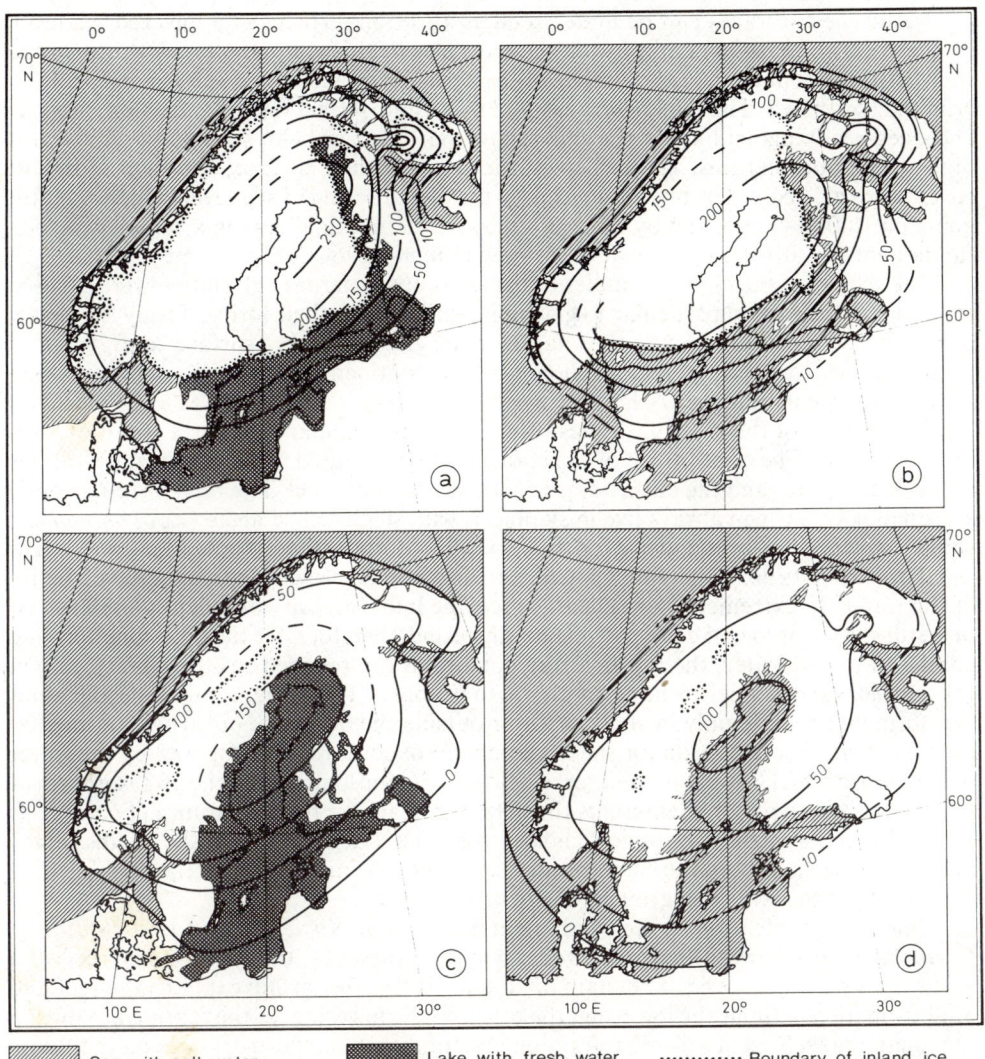

Fig. 1.23. Postglacial changes in the distribution of land and sea demonstrated by the example of the Baltic Sea. (After Sauramo, 1958.) (a) Baltic Ice Lake about 10,200 years ago. (b) Yoldia Sea about 9700 years ago. (c) Ancylus Lake about 8000 years ago. (d) Litorina Sea about 7000 years ago. The maps show the contour lines of the shore (in m) above the present sea level.

by alternating fixation of water in the form of inland ice. In the relevant regions such fixation may mean an additional weight on the earth's crust causing an "isostatic" subsidence of the continent which thus also causes a rise of the sea level. In contrast, when the ice is melting, the continent will rise and the sea level will be lowered. The history of the Baltic Sea during the past 18,000 yr has been determined by this isostatic effect (Fig. 1.23). The result was that in the Gulf of Bothnia the water level was lowered by as much as 300 m, in which is already included the 100 m worldwide eustatic rise of the sea level. Actually, the continents rose by up to 400 m within 18,000 yr.

Many shelf terraces can be understood as abrasion terraces at a lowered sea level, or as deposits from rivers at a lower water level, but this is impossible with the broad shelves off the continents, for example, off eastern North America, as shown in Fig. 1.17, or in the North Sea. The thickness measured by seismic refraction and reflection methods and deep-reaching drilling on the shelf show that these are old, subsided regions with high sedimentation rates. These shelves, the shape of their surface, and their sediment cover, were also considerably influenced by the last lowering of sea level, which was still going on 18,000 yr ago, and by the erosion by waves and tidal currents, which was then fundamentally different from what it is now (Emery, 1968).

The shelves of the higher latitudes in the areas of Quaternary glaciation were exposed to special influences, in particular to glacial modification of their forms. Today they show a pronounced relief, partly with the characteristic great variety of surface configurations that we know from the glacial morphology of the continents (North Sea and Baltic Sea as well as Greenlandic and Atlantic-Canadian waters).

The shelves of the lower latitudes often show an irregular relief as a result of reef-forming corals. The appearance of the corals is closely related to the physical conditions of the ocean water and the eustatic rise of the sea surface level. The reef-forming corals are small polyps which always live in symbiosis with single-celled algae (*Zooxanthellae*), whose presence is a prerequisite for the formation of limestone. These algae need light as all other plants do. The light penetrates to different levels, depending on the turbidity of the water. In extremely clear ocean water, the light intensity at a depth of 80 to 100 m is still sufficient to make life possible for algae and, therefore, for the limestone-forming corals. In turbid water, the life-favoring conditions are reduced to a thin layer, and in very turbid water no reef-forming corals are to be found. Therefore, corals reefs and atolls can form and remain only in shallow water outside the influence of turbid continental runoffs. A further condition for the development of coral reefs is imposed by the water temperature which should not decrease below 20°C for long. Thus, the 20° isotherm of the coldest month roughly represents the polar boundary for the appearance of coral reefs and atolls on the earth. The tropical and subtropical western margins of the oceans, which are usually warmer than the eastern sides, represent the main areas of the reef-forming corals. There the reefs can grow at a rate of $1-2$ cm yr^{-1}.

Since the publication of the fundamental work *The Structure and Distribution of Coral Reefs* by Charles Darwin in 1842, many hypotheses for the formation of reefs and atolls have been advanced. The main difference is that one group explains the reef and atoll formation without the lowering, the other by the lowering, of the sea surface during the Quaternary. Recent seismic refraction investigations and deep drillings on the Bikini and the Eniwetok Atolls (in connection with the first atom bomb experiment in the Pacific Ocean) showed calcareous corals down to a depth of 1300 m where below 200 m early Tertiary reef formation began. These and numerous other observations support Darwin's ingenious concept that the atolls were generated on a subsiding foundation. Their present shape is predominantly determined by the Quaternary subsidence of the erosion base. There are 330 atolls in the world ocean, all of which, except 9, are located in the tropical Indian–Pacific area; as many as 62 belong to the Tuamotu Islands.

The shelves also include the coasts which, because of their great practical importance for man and their easy accessibility, have been the subject of a vast number of studies. Here attention is drawn only to six summaries: by Valentin (1952), Guilcher (1958), King (1959), Shepard (1963), and Seibold (1964), as well as to a review on more than 130 recent Russian papers by Zenkovich and Schulayk (1967). From the point of view of oceanography, Shepard's classification is most obvious. Among the various coastal configurations he distinguishes primary and secondary coasts, the former determined by terrestrial, the latter by marine processes. The former include the terrestrially eroded forms which then were drowned with the rise of the sea surface level—valleys of river mouths, fjords, and karst plateaus—and also the terrestrially accumulated forms—delta coasts, dune coasts, volcanic forms, and coasts modified by active crustal movement. The secondary coasts include the coasts leveled by wave erosion as well as spits of land, coral reefs, and mangrove coasts. Considering that no agent acts alone, so that there are more mixed forms than pure forms among the aforementioned coastal types, we can readily understand the variety of the coasts of the oceans.

The *continental slopes* (see Section 1.1.6) have not developed everywhere in the same way. Like the neighboring shelf, they may be accumulations of crustal forms with a sediment cover of varying thickness. Shepard (1963) has described their development in great detail.

The most striking surface features of the continental slopes are the submarine canyons (Fig. 1.03). So far, no simple answer, which could be applied to all canyons, has been found to explain the formation of these submarine phenomena, not even in the monograph by Shepard and Dill (1966). With the exception of a few deep-cut submarine valleys, as found, for example, in the continuation of the continental valleys of the Congo and Indus which are considered to be of tectonic origin, the canyons have been explained as erosion forms. One group of scientists believes them to be subaerial formations, thus interpreting them as erosions by continental rivers. Another group considers them submarine formations, that is, developed by underwater erosion. By assuming river erosion, we can easily find a natural explanation for many details of the canyon forms. However, it would require a lowering of the sea surface by approximately 2000 m, compared with the present surface level, since canyon forms can be distinctively recognized down to this depth. Furthermore, the sinking process should have occurred in the recent geological past, because the continental slopes are not very old. There is good evidence for the global eustatic lowering of the sea surface of approximately 100 to 200 m (Fig. 1.22), but a figure 20 times as large seems to be incompatible with this evidence. The extreme theory of this unusual lowering of the sea level has been modified so that one assumes that the formation of canyons by subaerial erosion was followed by a flexure of the continental slope, which seems to be true in the Mediterranean Sea, for Corsica as well as the French and Italian rivieras.

Opinions differ with respect to the formation of canyons by submarine erosion. Some scientists would like to see them explained as submarine artesian spring sappings; others see them as caused by landslides and mudflows. Another group sought to explain them by tidal current erosion. Stronger water motions are expected from the long surface waves originating from strong seaquakes, which are called *tsunamis* in the Pacific Ocean and are feared because of their devastating force along the coast. But the worldwide distribution of the canyons cannot be explained by the Pacific tsunamis. Kuenen and Heezen suggest still another theory that uses the so-called turbidity currents as the eroding water motion (see Section 1.3.3). There is no longer any doubt about the existence of these currents (cf. Fig. 1.15), but there are very great differences of opinion as to the effect of their erosion, that is, the active cutting into the solid subsoil.

In the Atlantic and Indian Oceans the continental slopes are continued in the slightly slanted *continental rise*. This accumulated formation is characterized by an especially high thickness of sediment which can amount to several kilometers (Fig. 1.17) and even fill marginal deep-sea trenches. Some continental rises lean on the continental slopes in the form of mighty accumulated cones. They are thought to be caused by turbidity currents; an example is given in Fig. 1.15. This opinion is confirmed by the typical sorting of sediment particles in the vertical direction, which advocates that turbidities are included in the deposits.

In spite of their great depth, *deep-sea trenches* are typical phenomena of the ocean margins, especially in the Pacific Ocean, and also genetically they belong to the continental margins. Recent geophysical investigations of nearly all deep-sea trenches have shown remarkable conformities. A typical example is presented in Fig. 1.16, which is a cross section of the southern branch of the Kuril Trench in the Northwest Pacific Ocean.

1. Parallel to the axis of the deep-sea trench and offset toward the continental slope was found a narrow band of very great negative isostatic gravity anomalies, which indicates a mass deficit in the upper part of the earth's crust. Such narrow bands were first found in the area of the Sunda Trench and are called Vening-Meinesz zones after the scientist who discovered them.
2. The deep-sea trenches coincide with zones of very frequent near-surface earthquakes. As shown in Fig. 1.16, cross section IV, the foci of these earthquakes are located in a relatively thin layer that steeply slopes down toward the continent (Luvendyk, 1970). The actual slope of this layer can be readily recognized since the vertical scale in the cross section is not exaggerated.
3. Furthermore, the Kuril Trench, like the other trenches, runs parallel to chains of young volcanoes, most of which sit on geologically young folds dating from the Tertiary. Very thorough geophysical investigations of the Puerto Rico Trench (Talwani et al., 1959) and of Pacific deep-sea trenches (Hayes, 1966) confirmed three more facts concerning the deep-sea trenches.
4. They have a thick sediment cover in which the sediments are predominantly turbidities, that is, layers of deposits, sorted according to grain size, which can be explained by turbidity currents transporting sediment (Arachouchine and Hsin-Yi Ling, 1967).
5. They have low heat flux (Fig. 1.21).
6. They are magnetically undisturbed.

The geophysical results indicate that even in our day the areas in which the deep-sea trenches are located are still subject to deep-reaching and large-scale tectonic activities, concentrated at the margins of the Pacific Ocean, which must be considered as a lateral thrust and not as a pull of the continental crust towards the oceanic crust. The deep-sea trenches should, therefore, be interpreted as the result of compression and not of stretching. All the geophysical observations mentioned above can be explained by the submerging of a stiff Pacific plate below the neighboring plates (cf. Fig. 1.26). The layer of earthquake foci, which is sloping down towards the continent, is considered to be the shearing plane (cf. Fig. 1.16).

1.4.3. Mid-Ocean Ridge

The Mid-Ocean Ridge is a broad sill with a very complicated, irregular topography. Its typical profile is characterized by the crest region, the central rift imbedded therein, and the flank regions (Fig. 1.04). In the world ocean, the Mid-Ocean Ridge is the outstanding bottom feature covering an area corresponding to that of all continents together. There were numerous speculations with regard to the evolution of the Mid-Ocean Ridge, but they all have been narrowed by the results of the geophysical findings of the past decades. The following facts have been revealed.

1. In contrast to the deep-sea basins that are almost free of earthquakes, the crest region, especially the central rift valley, shows considerable seismic activity with shallow-focus earthquakes (Plate 2). Besides, there are aseismic ridges not directly connected with the Mid-Ocean Ridge, as for instance the Greenland–Scotland Ridge, the Walvis Ridge, the Rio Grande Ridge, the East Indian Ridge, and the Kerguelen Ridge.
2. The heat flow in the central rift valley is especially high, the highest on the earth outside the active volcanoes; it is concentrated in a narrow zone not exceeding 50 km in width. Fig. 1.21 gives a typical example. In the flank regions the heat flow is anomalously low.
3. Sea gravimetry (see Section 1.4.1) has shown that the ridge system is isostatically almost in balance, which means slightly inclined to positive anomalies. Talwani (1964) explained this by the anomalies of the earth's crust below the ridge.
4. The magnetic anomalies reach highest values over the central rift valley, as shown in Fig. 1.18. They are arranged in strips only around 20 km wide, abating in the flank region.
5. Seismic refraction investigations have proved that the crustal structure below the Mid-Ocean Ridge is clearly distinguished from that of the rest of the ocean. Figure 1.25 shows this in the example of a cross section through the North Atlantic Ocean.
6. Seismic reflection investigations of the entire Mid-Ocean Ridge (Ewing and Ewing, 1967), especially on profiles in the Atlantic Ocean (Collette et al., 1969) and in the Pacific Ocean (Ewing et al., 1968), have shown that the crest region with the central rift valley at a total width of about 200 km, is free of sediments, whereas on the flanks the sediment thickness is considerably increased.

By taking the geophysical observations into account, we can find an explanation of the ridge system in the worldwide extension of the earth's crust in these areas. Material from the earth's mantle rising below the ridge transports new material to the central rift, and thus the ocean floor spreads sidewards. Since this idea was presented by Dietz (1961), "Spreading of the Ocean Floor" has become a central topic of marine geophysics and marine geology. Not only the explanation of the Mid-Ocean Ridge, but also the history of the oceans (cf. Section 1.4.6) has gained new aspects from this modification of Wegener's theory of continental drift (Bullard, 1969). Detailed descriptions of the tectonic and geological structure of several oceans and some mediterranean seas are contained in *The Sea,* volume 4, part 2 (edited by Maxwell, 1970). Attention may also be drawn to Shor et al. (1970) for the Pacific Ocean, Laughton et al. (1970) for the Indian Ocean, Demintskaya and Hunkins (1970) for the Arctic Ocean, Ewing et al. (1970) for the Gulf of Mexico and the Caribbean Sea, Ryan et al. (1970) for the Mediterranean Sea, and Allan and Morelli (1970) for the Red Sea.

At present, the spreading of the ocean floor on each side of the central rift amounts to approximately 5 cm per year in the South Pacific, 3 cm per year in the North Pacific, and 1 cm per year in the North Atlantic Ocean, if the streaky character of the magnetic anomalies is taken as an indication (1) of what the magnetic field was like at the time when rising magma cooled below the Curie point, and (2) of the repeated reversals of the magnetic field within the past 7×10^6 yr. Figure 1.20 contains such spreading values referred to the time of 3.32×10^6 yr ago. It is based on geomagnetic measurements [for the Reykjanes Ridge they were obtained by Heirtzler et al. (1966), as shown in Fig. 1.18] and on the time scale of the reversals of the geomagnetic field (see Fig. 1.19). Furthermore, it is believed that there is evidence of irregularities of spreading, for instance Ewing (1967) explained thereby the sudden increase of the sediment thickness in the flank regions. Some scientists are of the opinion that the present cycle of spreading started in the world ocean 10×10^6 yr ago after a quiet period of 30 to 40×10^6 yr.

1.4.4. Deep-sea basins

The deep-sea basins represent the third large-scale topographic feature of the ocean bottom which includes the following four main types mentioned in Fig. 1.02: abyssal hills, abyssal plains, abyssal sills, and step or fracture zones. With regard to geophysics, the deep-sea basins are very uniform in contrast to the continental margins and the Mid-Ocean Ridge. They have little earthquake activity; gravity and heat flow are normal. However, the magnetic anomalies in the step or fracture zones of the eastern Pacific Ocean are especially great and arranged in bands perpendicular to the faults.

Abyssal hills and *abyssal plains* occupy the great depths of the basins. They differ only in one respect: while on the former the sediments covering the earth's crust have never been disturbed, on the latter they were redeposited and leveled probably because of turbidity currents running down the continental slopes, filling up the continental rises, and coming to a standstill on the abyssal plains. Figure 1.15 shows an example that occurred in our time. Convincing evidence for this concept can be found in sediment cores taken from these plains: the normal pelagic sediments are often interspersed with layers of well-sorted sands, so-called turbidites, the thickness of which may vary from fractions of millimeters to several meters. In the case of the turbidity current of November 1929 (Fig. 1.15), the thickness of these turbidites in the abyssal plain amounted to 100 cm. Earlier attempts to explain the extremely flat plains in the deep sea were not convincing, as shown by Laughton (1967). However, all the explanations mentioned above will help to render the worldwide occurrence of the abyssal plains understandable.

The bottom topography of the *deep-sea sills* is perfectly similar to that of the Mid-Ocean Ridge, but the two clearly differ with regard to seismicity, the deep-sea sills being completely aseismic. Further geophysical data are still lacking. These sills are considered to be fossil ridges (Schneider and Vogt, 1968) in which active processes like in the Mid-Ocean Ridge have discontinued since the Tertiary. Characteristic of such hills are high rises which play an important role in physical oceanography because they influence the distribution of the deep-sea and bottom water in the areas of the Alpha, Lomonossow, Greenland–Scotland, Rio Grande, and Walvis Ridges in the Atlantic Ocean; of the Kerguélen, Maldive, and East Indian Ridges in the Indian Ocean, and perhaps also of the Imperator and the Tuamotu Ridges in the Pacific Ocean. Further information is given in the atlas by Dietrich and Ulrich (1968). A satisfactory explanation of the nature of fossil ridges has not always been found. It is possible that they are fragments of continental crusts like the Mascarene Ridge with the Seychelles, or the Madagaskar Ridge with Madagaskar.

The *step* or *fracture zones,* which are characteristic of the eastern Pacific Ocean, include eight fracture zones, named Mendocino, Pioneer, Murray, Molokai, Clarion, Clipperton, Galapogos, and Marguesas (cf. Plate 2). They are distinguished by magnetic anomalies arranged in strips and running perpendicular to the topographic steps where they are horizontally displaced at the steps (e.g., at the Mendocino Fracture Zone by about 1140 km and at the Pioneer Fracture Zone by about 260 km). Anomalous behavior could not be recorded, by either seismic or gravimetric methods. According to the hypothesis advanced by Vine and Matthews (1963) the strips of magnetic anomalies are explained—like those of the Mid-Ocean Ridge—by a repeated reversal of the earth's magnetic field and the sideward spreading of the ocean floor. The pertinent active ridge, however, is hidden under the American continent, and only the western flank of the ridge can be recorded in the Pacific step region. The displacement at the steps are understood as blockwise drifting of the ocean floor. These are bold hypotheses which were first developed in the 1960s, but which could interpret recent geophysical observations.

1.4.5. Seamounts and oceanic islands

The seamounts, which have been treated in Section 1.1.6 are, without any exception, of volcanic origin. This is true also for the oceanic islands including atolls. Deep drillings on Pacific atolls have proved that there is always an extinct volcano hidden under the limestone reefs (see Section 1.4.2). It is remarkable that no sediments that date back further than the Cretaceous period have been found on any of the oceanic islands. Therefore, oceanic islands, and probably also the seamounts, seem to be of Mesozoic or more recent periods, as Menard (1964) has shown in his book on the geology of the Pacific Ocean.

The guyots represent a special group among the seamounts. They are believed to have been formed by surf abrasion. This means that they are considered as subsided islands, although today in the Pacific Ocean their summit plateaus usually lie at a depth of 1000 to 2000 m. The eustatic lowering of the sea level in the Pleistocene, which amounted to 100 or 120 m (Fig. 1.22), does not at all offer a satisfactory explanation with regard to guyots. A slow, but continuous subsiding process is likely to have taken place, since Mesozoic reef limestone was found on guyots at 1400 m water depth, although it is well known that corals cannot live at depths greater than 100 m. At some seamounts where the growth of the corals and the process of subsidence happened to be in balance, they became atolls. At others this was not the case; they became guyots. It is still an open question whether the subsiding process indicates a local, a regional, or a worldwide lowering of the sea floor. The latter case is unlikely but the former is partly true: volcanic seamounts represent an additional weight on the earth's crust, which will cause an isostatic subsidence of the region concerned. Relevant evidence was found for the Anton Dohrn Seamount in the Northeast Atlantic Ocean (Dietrich, 1961; also in the atlas by Dietrich and Ulrich, 1968). This seamount is surrounded by a depression 200 m deep. Certainly, regional processes will also be involved, which may be taken from the geographical distribution of the seamounts in the Northeast and Northwest Pacific Ocean.

Menard (1969) has suggested that the development of seamounts is connected with the spreading of the ocean floor. The rising of magma under the crest region of the Mid-Ocean Ridge is followed by a horizontal spreading in the form of crustal plates as well as by a downward movement of these crustal plates with increasing distance from the crest in the flank regions and the deep-sea basins. The subsiding process is estimated at 9 cm per 1000 yr for the first 10×10^6 yr, at 3.3 cm per 1000 yr for the next 30×10^6 yr, and at 2 cm per 1000 yr for the time after that. The seamounts formed on the Mid-

50 Geomorphology of the Ocean Bottom

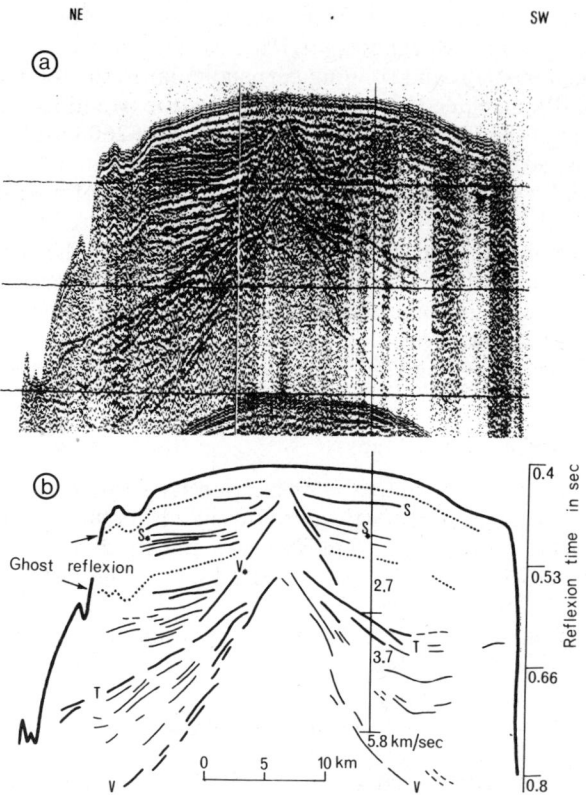

Fig. 1.24. (a) Transverse profile (NE → SW) through the Great Meteor Seamount (see Fig. 1.06), obtained by seismic reflection methods using an air gun during the "Seamounts Cruise" in the Northeast Atlantic with research vessel *Meteor* in 1967. Horizontal lines are isochrones of sound impulses at intervals of 0.133 sec, corresponding to a thickness of about 200 m. (b) Interpretation of (a) V and V_x: Upper boundary of igneous compact rock, v_p = 5.8 km sec^{-1} for compressional waves. T, S and S_x: Sediment horizons, v_p = 3.7 resp. 2.7 km sec^{-1}. Figures at vertical line: Mean velocities of compressional waves as deduced from seismic refraction measurements. Exaggeration of (a) and (b) about 30 times. (After Hinz, 1969.)

Ocean Ridge will drift 1 to 10 cm yr^{-1} and subside. They ride on the drifting crustal plates. If the volcanism is of sufficient activity, they become and remain islands (as e.g., St. Helena), but if the volcanism is only moderate, underwater seamounts are formed, and if eruptions occur only briefly and cease completely later one, the original islands will subside during the drift and form guyots. Such guyots are the Great and the Little Meteor Seamounts, which, at a subsiding velocity of 9 cm per 1000 yr, after a period of 10^6 yr, should lie at a depth of 90 m. Considering the actual depth of 275 m (see Fig. 1.06), they should still have been islands around 50×10^6 yr ago, which correlates well with the determination of the age of their limestone cover that is supposed to date from 40×10^6 yr ago (Schott, 1969). Evidence for this limestone cover on top of the volcanic cone was found by means of seismic reflection measurements (Hinz, 1969) (cf. Fig. 1.24).

1.4.6 Formation of the oceans

Theories concerning the formation of the various characteristic topographic features of the ocean bottom, as outlined above, will only be convincing when they are in good

Formation of the Bottom Features 51

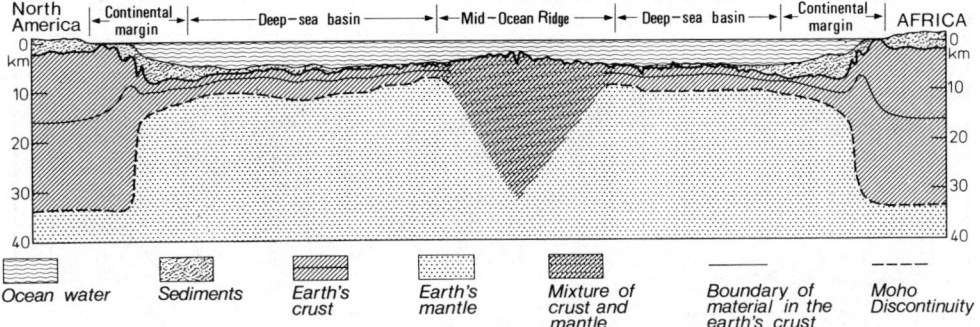

Fig. 1.25. Schematic cross section through the subsoil of the North Atlantic Ocean from New York to the Spanish Sahara. (After Heezen, 1963.) Large-scale features of the sea floor are named on top (see Fig. 1.02). The structure of the subsoil has been investigated with seismic refraction methods. The uppermost layer at the ocean floor represents the sediments; the second layer contains lithified sediments or volcanics; the third layer is part of the lower crust of the earth, and the fourth layer beneath the Moho Discontinuity belongs to the mantle of the earth.

accord with the concept regarding the formation of the oceans. How shall we imagine that the oceans came into being? This is an old question which even Alexander von Humboldt confronted. He believed in a continental drift because the shapes of the continents fit almost exactly into each other. In 1912, Alfred Wegener developed this bold idea into his large-scale theory of continental drift. For decades arguments for and against it did not come to an end, but precisely because of this, many branches of geoscience have been highly inspired by it. Even after half a century it was the topic of Runcorn's book *Continental Drift* (1962), with contributions of 12 distinguished scientists. Many of Wegener's arguments were convincing, others were not, especially those concerning the mechanism of the drift. From the viewpoint of physics it seems unrealistic to assume that continents should be able to drift in the earth's crust like ice flows through water (cf. Fig. 1.25).

Two other groups of hypotheses that tried to explain the formation of the oceans, are not in conformity with observations either. The theory of permanent oceans dates the formation of the oceans back to the time of the solidification of the earth's crust. The theory of bridge continents is a compromise between permanence and drift hypotheses. It tries to meet the demand of paleontology for land connections that might explain the different evolution of flora and fauna (e.g., the one that began between Africa and South America at the time of the Triassic).

In the 1960s, another theory was added to these hypotheses, the theory of "Sea Floor Spreading." A comprehensive representation has been given by Vine and Hess (1970) and Pitman III (1971). Properly speaking, it contains two concepts: The spreading of the ocean floor and the drifting of the primary crustal plates. The latter represents a form of the continental drift, although in contrast to Wegener's theory it is not the continents that drift but the primary crustal plates which cover the entire surface of the earth.

If the continents are included in the considerations, we get a simple picture (Le Pichon, 1968): the earth's surface is divided into six primary plates, the margins of which are marked by the epicenters of the earthquakes (see Plate 2), that is, by the Mid-Ocean Ridge and the boundaries of the Pacific Ocean. Figure 1.26 shows the primary plates and several secondary plates, all of which move as rigid bodies in the directions indicated by arrows. The central rift of the Mid-Ocean Ridge opens up; magma from the upper mantle of the earth is pressed up and solidifies; thus, after it has cooled down to below

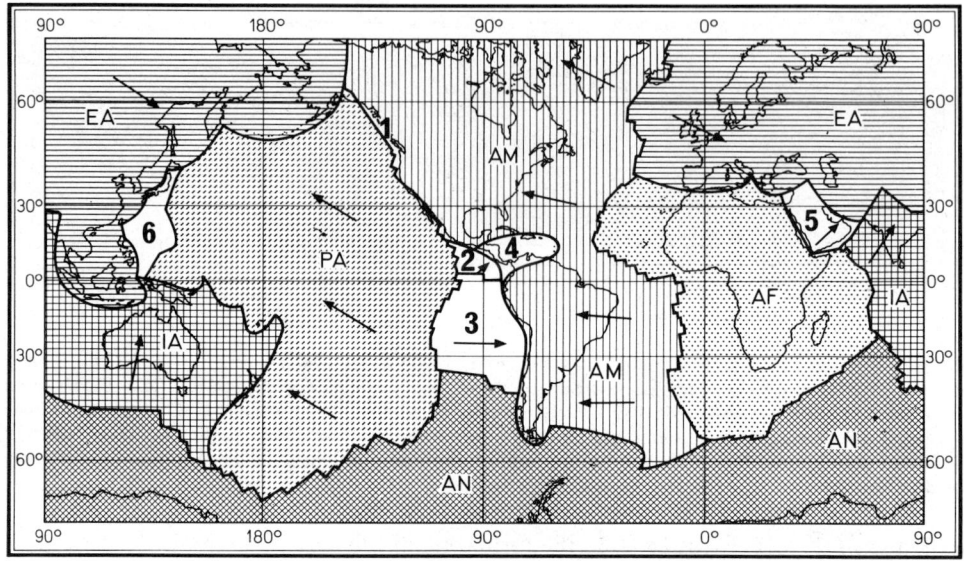

Fig. 1.26. Present position and drift of the six primary plates of the earth's surface. (After Le Pichon, 1968.) Arrows show the direction of motion relative to the African plate. There are some other small plates, six are marked by figures. EA: Eurasiatic, AF: African, IA: Indian-Australian, PA: Pacific, AM: American, and AN: Antarctic primary plates.

the Curie point, the direction of the respective magnetic field of the earth is preserved and the magma moves away sidewards together with one of the plates. Now all observations at the crest region can be explained: young topographic forms free of sediments, high seismicity, volcanism, positive gravity anomalies, the streaky structure of strong magnetic anomalies, considerable heat flow. The primary plates break into small structures which are arranged normal to the ridge. They can be determined by topographic and magnetic measurements, as shown in Plate 2. The thickness of the rigid primary plates has not yet been clarified. It is estimated at 70 to 100 km, which shows that the plates include the earth's crust and the upper mantle of the earth.

The drift velocity has been found to be 4 cm yr^{-1} in the Pacific and 1 cm yr^{-1} in the North Atlantic Oceans. Since this refers to one drift direction, the double value accounts for the total spreading. Plate 2 shows the spreading in several areas (Heirtzler et al., 1968). For instance, line 60 will say that 60×10^6 yr ago this line lay in the central rift of the Mid-Ocean Ridge. A spreading velocity of 8 cm yr^{-1} means that within 10^8 yr at a constant drift speed the ocean has been widened by 8000 km. Even if there have been periods of interrupted spreading, it must be concluded that the present oceans cannot be older than 1×10^8 to 2×10^8 yr. This is confirmed by other observations stating that up to now no sediments have been found that are older than 1.5×10^8 yr; seldom are they older than 0.8×10^8 yr. Therefore, a sedimentation rate of 1 cm per 1000 yr would correspond to a sediment thickness of 1500 or 800 m, respectively. However, since oceans have not existed on the earth only for 150×10^6 yr, sediments must have been accumulated and stored. At the moment, this takes place at those margins of the primary plates which are turned away from the Mid-Ocean Ridge. There, one plate submerges under the other at an angle of 45°. An example is given in Fig. 1.16 for the Northwest Pacific Ocean, which is characteristic of the border of the Pacific Ocean, considering the positions of the epicenters of the shallow-focus earthquakes (less than 100 km) as well as of the deep-focus earthquakes (300 to 400 km and 600 to 700 km) as given in Plate 2. The submerging of the plates is indicated by numerous observations related not only to the earthquake foci, but also to the deep-sea trenches, to the high negative gravity anomalies, to the young orogens, and the chains of volcanoes, as in the Andes, or in East or Southeast Asia. Principally, it does not make any difference if a plate submerges under a chain of islands (for instance the Tonga–Kermadec Islands, the Kurils, the Aleutians, and Antilles) and not under a continent, because the islands are part of the neighboring primary plate.

If the Atlantic Ocean started to develop out of a rift as, for example, the Red Sea does today, the question is suggested—what had taken its place during the preceding 35×10^8 yr of the earth's history? The geology of such regions gives evidence of Precambrian and Paleozoic seas, but whether they can be considered oceans or seas of marginal areas cannot be answered yet.

Although the large-scale features and the geophysical findings regarding the theory of the spreading of the oceans are in conformity with the hypothesis of plate tectonics, the mechanism is not explained thereby. Some scientists believe that the primary cause is to be found in thermal convection cells in the earth's mantle (Hess, 1965; Vening-Meinesz, 1948; and others), assuming that these cells have a rising component beneath the Mid-Ocean Ridge and a subsiding one beneath the continents. Others think that such thermal convection along lines is improbable (Bullard, 1968).

Numerous details with regard to this new hypothesis of plate tectonics are still lacking, and comprehensive investigations by geosciences are required, among which studies on the formation, motion, and disintegration of the primary crustal plates at the earth's surface. Evidence should be obtained that drifting plates contain, for instance, lava covered with sediment layers of the same age, which could be checked by drilling. The American expedition with the drilling vessel *Glomar Challenger* (Fig. 1.13) into the Atlantic, Pacific, and Indian Oceans in 1969–1971, during which the oceanic sediments were perforated at several hundred positions, represents the greatest effort so far in this respect. This expedition was the climax of investigations stimulated by the "Upper Mantle Project," which had been initiated as a great international enterprise by the IUGG (International Union of Geodesy and Geophysics). As an analogy the "Terrestrial Dynamics Project," according to a proposal by Knopoff (1969), has followed in the 1970s. At any rate, geology, previously interested more or less merely in continents, has received new stimulation from such studies concerning the development of the oceans. The fundamental change that is now under way in the concepts of the development of the earth's surface is based on the latest findings of marine geology and marine geophysics. This change means a break with the theory of the static structures and the advancement of the concept on the mobile structures of the earth.

2. Physical Properties and Chemical Composition of Seawater

2.1. Physical Properties of Seawater

2.1.1. Unique characteristics of pure water

In addition to containing small amounts of chemical constituents, seawater consists predominantly of pure water (96.5%). Therefore, its physical properties are mainly determined by those of pure water. For the natural processes on the earth, it is of special importance that not only is water the only liquid engaged in almost every process, but it also behaves most uncommonly when compared with fluids of related composition. This abnormal behavior is ultimately due to the specific structure of the water molecules.

In this chapter the following notation will be used for the description of the properties of pure water and saline water: temperature: $T(°C)$, absolute temperature: $T_{abs}(°K)$, salinity: $S(‰)$, pressure: p(dbar, or multiples).

Analogous to its chemical formula H_2O, the water molecule consists of one atom of oxygen and two atoms of hydrogen. The radius of the water molecule, which is assumed to be spherical, amounts to 1.38 Å (1 Å = 10^{-8} cm). The distance between a hydrogen atom and the oxygen atom in a water molecule is 0.99 Å. Under "normal" conditions it could be expected that the central oxygen atom is surrounded by two hydrogen atoms at an equal distance, forming, at the oxygen atom, an angle of 180°. Such a symmetrical molecule, in which the centers of the two positive charges of the hydrogen atoms and of the double negative charge of the oxygen atom coincide at the center of the water molecule, would appear as a neutral compound when seen from the outside, and there would be no remaining associative power. However, in nature, the water molecule is not a linear but rather an angled molecule, in which the two hydrogen atoms are arranged with an angle of about 105° at the central oxygen atom. This results in five effects of far-reaching importance.

1. The centers of the two electric charges do not coincide. Thus the water molecule forms an "electric dipole," * which means that the water molecule is no longer neutral towards its surroundings, but with its surplus of attracting force, it affects the neighboring water molecules. As a consequence, aggregations of water molecules are formed, the so-called polymers. These are relatively loose aggregates which, at room temperature, contain six individual water molecules on the average, corresponding to the formula $(H_2O)_6$. Such a large aggregation

---

* A charge system formed by a positive charge $+e$ and a negative charge $-e$, separated by a small distance P, is called an electric dipole. The product $e \times P$ is defined as the electrical moment of the dipole. It is measured in electrostatic units.

must react much more slowly than an individual water molecule. This is especially evident if some of the thermal properties of water are compared with the expected "normal values" obtained from data based on related hydrogen compounds showing no abnormal behavior. Then, for example, the boiling point of water would be $-80°C$ instead of $+100°C$, the freezing point $-110°C$ instead of $0°C$, and the heat of vaporization approximately 4.0 kcal mole^{-1} instead of 9.7 kcal mole^{-1}.

2. The second effect which causes water to behave abnormally is due to the large dipole moment of 1.84×10^{-18} esu (normally approximately 0.2×10^{-18} esu). Favored by the small molevolume of water, this leads to the development of the high dielectric constant of 80 (normally about 2). The immediate consequence is the great dissociative power, that is, the capability to split dissolved materials into electrically charged ions, and, thereby, the great capacity for dissolution that water has for a large number of inorganic chemical compounds. Only few compounds show a similarly high dielectric constant (DC) ($NH_3:DC = 23$, $HF:DC = 84$, $HCN:DC =$ about 95, $SO_3:DC = 13.8$), which are summarized under the term of "water-like solvent." Besides their strong dissociative power and their great capacity for dissolution, these solvents are characterized by a great capacity for associating with the dissolved substances. With water, this leads to the formation of hydrates, and, furthermore, to the formation of acids, bases, and salts with the relevant chemical processes of neutralization and hydrolysis. Here the ions formed by dissociation play an important role because their most striking property is an extraordinarily high capability for chemical reaction, with reaction times of the order of fractions of a millionth of a second. Other solvents with low dielectric constants, in general, only form molecular solutions with a reaction capacity that is much more inert. Furthermore, it is typical that pure water is only weakly ionized and, therefore, its electrical conductivity is moderately low, whereas aqueous solutions show high conductivities. Seawater has an electrical conductivity which lies midway between that of pure water and that of copper. The electric resistance of a water column 1 mm long at $18°C$ approximately equals the resistance of a seawater column 1.3 km long with a salinity of 35‰.

3. Besides the two properties discussed above, namely the association of individual water molecules forming higher polymers and the high dielectric constant, a third property is important which, to a considerable extent, contributes to the abnormal physical behavior of water. Hereby, the valence angle of $105°$, mentioned above, in which the two hydrogen atoms are connected with the central oxygen atom, plays an important role, since this angle corresponds approximately to the so-called "tetrahedral angle" which we find between the four arms if we draw the four connecting lines from the center of gravity of a tetrahedron to its four corners. The size of the tetrahedral angle is $109°28'$. Owing to this similarity, it is possible that in addition to the "spherical closest packing" arrangement, the associative power of the water molecules will permit a second arrangement in which each central oxygen atom is surrounded by four hydrogen atoms in a tetrahedral arrangement. In this arrangement, the distance of each of the two hydrogen atoms belonging to the oxygen atom, from the latter is 0.99 Å, while the distances of two other hydrogen atoms belonging to other water molecules from the central oxygen atom are 1.6 to

56 Physical Properties and Chemical Composition of Seawater

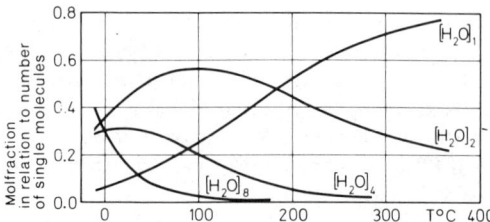

Fig. 2.01. Distribution of the different molecular aggregates of water as a function of temperature. (After Eucken, 1948.)

1.8 times larger. In contrast to the usual covalent or ionic molecular bond, this kind of linkage is called "hydrogen bond." While the bonding energy of the usual O–H bond amounts to 150 kcal mole^{-1}, the bonding energy of the hydrogen bond lies between 2 and 10 kcal mole^{-1}; that is, it is 10 to 100 times smaller. Consequently, in the range of normal temperatures, it is stable enough to permit the formation of differentiated, highly specific molecular structures; on the other hand, it is sufficiently unstable to adapt quickly to different conditions of reaction by giving up its connection. This takes only fractions of a millionth of a second. Therefore, the hydrogen bond is suitably called a chemical "zipper." A hydrogen bond may either be formed between hydrogens attached to two oxygen atoms or between those bound to an oxygen atom and a nitrogen atom.

4. Water molecules may occur not only in spatial lattice structures in a spherical closest packing arrangement, but, via the hydrogen bond, they may simultaneously appear in the form of tetrahedral crystalline lattice structures. This gives rise to the fourth peculiarity in the physical behavior of water. In contrast to normal molecular compounds, the tetrahedral molecular aggregates are of a much looser, more wide-meshed structure. Thus, in addition to other properties, the respective configurations of water can largely be distinguished by their densities. While at high temperatures only one- or two-molecule aggregates occur, the higher-molecular forms prevail as the temperature decreases (Fig. 2.01). So-called "clusters" are formed which, in their inner structure, show a wide-meshed tetrahedral crystalline lattice and are continuously rearranged. The half-life of a cluster amounts to 10^{-10} to 10^{-11} sec. The exterior appearance of all of them, however, remains that of a "spherical closest packing arrangement" (cf. Fig. 2.02). As a consequence, the changes of the physical properties, caused by cooling, are superimposed by additional changes which are characteristic of the tetrahedral water molecule. Among other characteristics there is the well-known anomaly that water at 4°C assumes an intermediate maximum of density, and thus an intermediate minimum of the specific volume (Fig. 2.03). Other physical properties of water show similar anomalies with regard to their dependence on temperature. This is expressed in the parabolic shape of the functional curves in Figs. 2.03 and 2.04. For "normal" substances, the relationships between properties and temperature are represented by nearly straight lines, as is the case for heptane in Fig. 2.04.

5. When dealing with the freezing of water, we have to pay attention to a fifth, very essential property: all water molecules join together in a large common

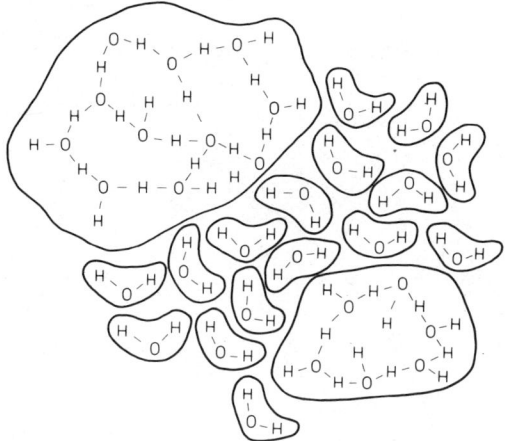

Fig. 2.02. Cluster. Model of liquid water. (From Horne, 1969.)

tetrahedral crystal-lattice structure which, in itself, is interlinked by hydrogen bonds (Fig. 2.05). Connected herewith is a large expansion of volume, and thus a decrease of density from 0.9987 to 0.9186 g cm^{-3}. Thus ice, the solid phase of water, is considerably lighter than its liquid phase. This is a property demonstrated by only very few substances. Actually, there are another seven known crystalline structures of ice that—in contrast to ordinary ice—show

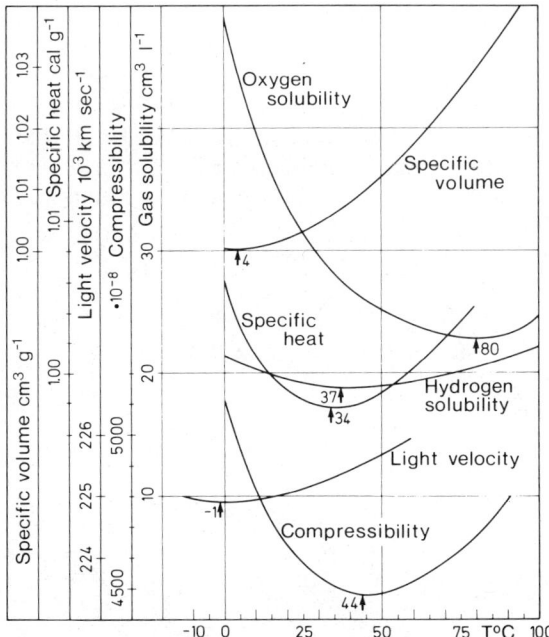

Fig. 2.03. Thermal dependence of various physical properties of water. The temperatures corresponding to the minimum of the respective curve are indicated in °C.

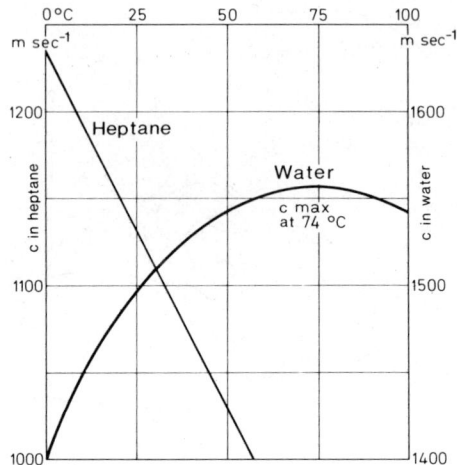

Fig. 2.04. Sound velocity c in water and in heptane.

a normal behavior and are heavier than liquid water. However, they are stable only at pressures of more than 2000 atm (Brill, 1962), and they are of no importance to oceanography. Altogether, there are four different properties of ice connected with its abnormal increase of volume during freezing. (a) Ice is very much lighter than water and, therefore, floats on it. (b) Expansion during the freezing process generates considerable pressure (up to 2000 bar). This explosive effect is an important factor in the weathering of rocks. (c) External pressure on ice results in the depression of the freezing point, possibly even in the melting of the ice. Thus, the flowing capacity of glaciers is increased. (d) The hydrogen bonds in the ice crystal give way if subjected to a relatively slight increase of pressure so that the ice becomes plastic. Consequently, the inland ice will flow, delivering icebergs to the ocean (Antarctica, Greenland). Under "normal" conditions, nearly all the water in the continental polar areas would be fixed as inland ice, and not much would remain of the world ocean.

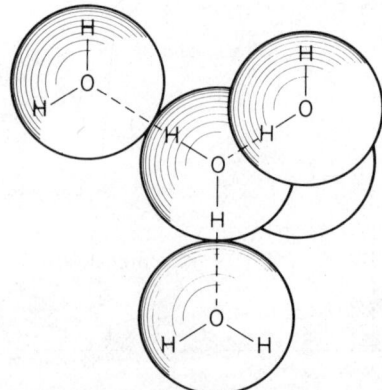

Fig. 2.05. Structure of water molecules in ice forming a tetrahedron. (From Horne, 1969.)

Physical Properties of Seawater

Table 2.01. Composition of Water with Respect to the Different Types of Water

Water Molecule	Percent of Total Water	Percent of Heavy Water	Saline Compounds of Similar Concentration in Seawater
$^1H_2^{16}O$	99.73	—	—
$^1H_2^{18}O$	0.20	73.5	Mg
$^1H_2^{17}O$	0.04	14.7	Ca
$^1H^2H^{16}O$	0.032	11.8	K
$^1H^2H^{18}O$	0.00006	0.022	N
$^1H^2H^{17}O$	0.00001	0.003	Al
$^2H_2^{16}O$	0.000003	0.001	P
$^2H_2^{18}O$	0.000000006	0.000002	Hg
$^2H_2^{17}O$	0.000000001	0.0000003	Au

Another property which has nothing to do with the anomalous structure of the water molecule but which is important for the understanding of the general behavior of water, is chemical in kind. In nature, most chemical elements are not homogeneous, but occur in various atomic forms distinguished from each other by one or several units of atomic weight; these are the isotopes. So there are three hydrogen atoms ^1H, ^2H, and ^3H. ^2H is also called *deuterium* (D) and ^3H is also known as *tritium*. Likewise, distinctions are made among three oxygen atoms ^{16}O, ^{17}O, and ^{18}O with corresponding mass relations. Combinations of the most important isotopes result in nine different forms of water:

$$^1H^{16}O^1H \quad ^1H^{16}O^2H \quad ^2H^{16}O^2H$$
$$^1H^{17}O^1H \quad ^1H^{17}O^2H \quad ^2H^{17}O^2H$$
$$^1H^{18}O^1H \quad ^1H^{18}O^2H \quad ^2H^{18}O^2H$$

Water, as found in nature, is a mixture of those nine forms, the proportions of which are given in Table 2.01.

In contrast to ordinary water ($^1H_2^{16}O$), the mixture of water compositions without $^1H_2^{16}O$ is called "heavy water" because its specific weight is greater than that of ordinary water. Although in ordinary water the constituents of heavy water are present only in very small amounts, their concentrations can still be compared to those of the important saline compounds of seawater (see Table 2.01). Therefore, it can be assumed that those water constituents play an important role not only physically but also chemically, although it is still unknown to what extent. In its physical constants, deuterium oxide $^2H_2^{16}O(D_2O)$ clearly differs from ordinary water, as shown in Table 2.02.

The differences in the vapor pressures of the various water components are of special

Table 2.02. Comparison of Some Basic Physical Properties of Natural Water and Deuterium Oxide

Property	Natural Water	Deuterium Oxide
Density at 20°C	0.99823	1.1051
Density Maximum (°C)	4	11.6
Melting Point (°C)	0	3.80
Boiling Point (°C)	100	101.42
Viscosity at 20°C (g cm^{-1} sec^{-1})	0.01005	0.0126
Surface Tension at 20°C (dyn cm^{-1})	72.53	67.8

practical importance. The vapor pressure at a temperature of 20°C decreases by 7.2% for $HD^{16}O$ and by 0.94% for $H_2^{18}O$ compared with natural water. Thus, under natural conditions, evaporation is combined with a partial fractionation of the water mixture so that the water samples of different origin show measurable differences in density compared to seawater. Lighter types of water are: snow water (by about 3.8 mg l^{-1}), natural fresh water from lakes, rivers, and springs (by about 1.5 mg l^{-1}), pure light water $^1H_2^{16}O$ (by about 260 mg l^{-1}), whereas water from the Dead Sea is heavier by approximately 1.5 mg l^{-1}.

2.1.2. Salinity as a thermodynamic variable

In addition to pure water, seawater also contains salts, dissolved gases, organic substances, and undissolved suspended particles. The anomalies of the physical properties of pure water are essentially the same in seawater. Small deviations effected by salinity can be of great importance, however. This is true, for instance, for changes of density, compressibility, the temperature of the density maximum and of the freezing point, whereas changes of viscosity, surface tension, thermal conductivity, specific heat, evaporative heat, and light absorption are of far less importance. Electrical conductivity and osmotic pressure are important only in seawater. A tabular representation of most physical properties of seawater has been given by Dietrich (1952), Joseph (1952), and Bialek (1966). The proportion and especially the composition of salts in the water of the open ocean are subject to slight fluctuation only. An exception can be seen in the very saline water found at some places at the bottom of the Red Sea (Ross et al., 1969).

    The proportion of the main constituents of sea salt was first specified with sufficient accuracy by Dittmar (1884). His results are based on the analyses of samples collected during the cruise of the *Challenger* (1873–1876). With this in mind we can state that there is a "constancy of the composition of sea salt." According to the enrichment mechanism, explained in Section 2.2.1, it can be expected that all of the chemical elements occurring in nature are present in seawater, although in concentrations that vary considerably. As a matter of fact, for many elements a direct proof has not yet been obtained (cf. Section 2.2.2).

    The true salinity of seawater, which is the sum of all dissolved salts in g per kg of seawater, shows a small difference as compared with the term salinity used in oceanography. This is due to the fact that by chemical analysis of seawater the true salinity cannot be determined with sufficient accuracy (cf. Section 3.2.9). For oceanographic purposes, however, only a very exactly reproducible determination of small differences of salinity, in space or in time, is required anyhow, whereas the total amount of salinity may be measured with less accuracy. So it is sufficient to agree on a definition of salinity which represents a good approximation instead of giving the true salinity.

    Therefore, the salinity S in parts per thousand (‰) is defined (Knudsen, 1902) as the total amount in grams of dissolved substances contained in one kilogram of seawater if all carbonates are converted into oxides and all bromides and iodides into chlorides, and if all organic substances are oxidized. Since the composition of salts is very nearly constant in seawater, an equivalent definition can be found if the relation between one component and the total salinity is known. It is obvious that for this purpose the best thing to do is to choose the component with the largest share of weight, that is, the chlorine ions. Thus, corresponding to the salinity, the chloride content $Cl(‰)$ is defined as the total amount in grams of chlorine ions contained in 1 kg of seawater, if all halogens are replaced by chlorides.

The relation $S(\text{\textperthousand}) = 0.03 + 1.805 Cl(\text{\textperthousand})$ has been found empirically. With a probability of 99% the deviations remain within ±0.024‰ of the salinity (Carrit et al., 1958). A formula giving a better description of very low salinities has been introduced later on by Wooster et al. (1969):

$$S(\text{\textperthousand}) = 1.80655 Cl(\text{\textperthousand}) \tag{2.01}$$

Besides the term "chlorinity," corresponding to the chloride content as defined above, the term "chlorosity" is occasionally used in English technical literature. It stands for the chloride content as defined above but is referred to 1 liter of seawater at 20°C. Chlorosity, therefore, is equal to chlorinity multiplied by the density of seawater at 20°C.

As mentioned above, in oceanography very precise measurements of the differences of salinity are necessary. For this purpose, a reference sample with a very well-known salinity is required as a standard. In general, the "Copenhagen Normal Water" is used (cf. Section 3.2.9), the chloride content of which has been determined very accurately. In order to avoid difficulties that might possibly arise in future because of changes of the value for the atomic weight of chlorine, chlorinity has been defined (Jacobsen et al., 1940) as follows: "the chlorinity in grams per kilogram of a seawater sample is identical with the mass in grams of atomic weight silver sufficient to precipitate the halogens in 0.3285234 kg of the seawater sample."

Recent results with respect to the specific electrical conductivity of seawater have led to a definition of salinity by means of this property (cf. Section 2.1.3).

2.1.3. Electrical conductivity

As a highly dissociated electrolyte, seawater has an electrical conductivity whose magnitude depends on the number of dissolved ions per volume (i.e., on the salinity) and on the mobility of the ions and thus on temperature and pressure. Therefore, conductivity increases with increasing salinity, increasing temperature, and increasing pressure. As to the order of magnitude, a change of the specific electrical conductivity of 0.01 mS cm^{-1} (1 mS = 1 millisiemens = mΩ^{-1}) is caused by changes of the salinity by 0.01‰, of the temperature by 0.01°C, or of the pressure by 20 dbar (1 dbar corresponds to the pressure of a seawater column 1 m long with a deviation of about 1%). Therefore, the temperature effect dominates the distribution of the electrical conductivity in the ocean.

It has been known for a long time that conductivity depends on temperature and salinity at atmospheric pressure (Thomas et al., 1934; Bein et al., 1935), but for oceanographic purposes this was only of minor importance. Today, however, due to the progress in the electrical measuring technique, salinity is predominantly determined by measuring the electrical conductivity and no longer by way of the titration of chloride. Therefore, knowledge of the electrical conductivity is of great importance now. With water samples from all oceans Cox et al. (1967) have shown that the difference between salinity data obtained by conductivity measurements and those gained by chloride determination can amount to about 0.003‰. This is probably due to the fact that at equal total salinity small deviations in the composition of the ions can cause a noticeable change in the proportion of chloride, while only a negligible change will occur in the electrical conductivity. To avoid any inconsistency of data, salinity has been redefined in terms of the specific electrical conductivity instead of the chloride content (Unesco et al., 1966; Wooster et al., 1969). Accordingly, salinity at atmospheric pressure is given as a function of the ratio

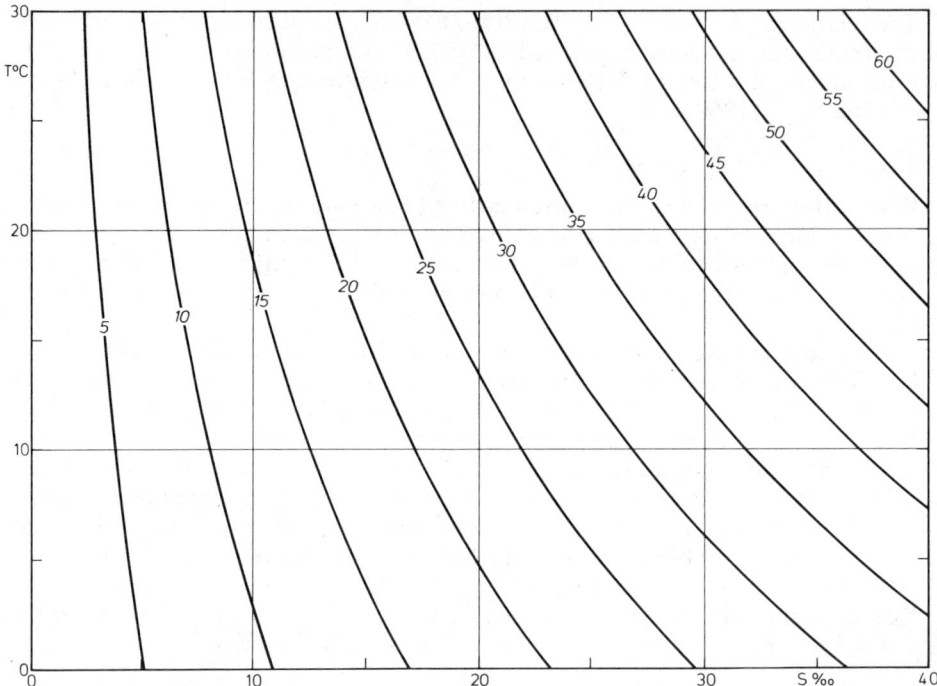

Fig. 2.06. Specific electrical conductivity C [mS cm^{-1}] at atmospheric pressure as a function of temperature T and salinity S.

between the conductivity of the sample and the conductivity of seawater with a salinity of 35‰, as well as a function of temperature.

Figure 2.06 shows the specific electrical conductivity as a function of temperature and salinity, where, according to Reeburgh (1965), for a temperature of 15°C, a salinity of 35‰ and a pressure of 0 dbar, the conductivity $C(15, 35, 0)$ has been assumed to be 42.902 mS cm^{-1}. The dependence on temperature has been described by Brown et al. (1966).

For the evaluation of the salinity from in-situ measurements of temperature, conductivity, and pressure it is necessary to know also the dependence on pressure, which has been determined by Bradshaw et al. (1965). Instructions on how to calculate the salinity from conductivity data can be found in Appendix 11.1.

Rhode (1968) has given explicit formulas for the computation of the salinity from electrical conductivity, a method that avoids steps of iteration. However, slight deviations from the results mentioned above may occur because his formulas were based on provisional data. Fedorov (1971) has developed simplified formulas for the calculation of salinity which can be used if an accuracy of ±0.02‰ is sufficient.

Deviations in the composition of seawater near the coast can lead to errors in the determination of salinity, but, in general, they do not exceed ±0.01‰ (Kremling, 1970). Biological processes may cause fluctuations of electrical conductivity. In case changes of alkalinity are chiefly caused by the carbon dioxide content, the possible variations of conductivity will correspond to a change of salinity by up to 0.01‰ (Park et al., 1964).

The maximum change of conductivity as a result of dissolved calcium carbonate corresponds to a change of salinity amounting to 0.006‰ (Park, 1964).

2.1.4. Density, thermal expansion, and compressibility

If we want to say something about the statics and dynamics of the ocean, we must know the density of seawater. Internal motions and processes of turbulent exchange are controlled by the stratification of density. Small horizontal differences of density may cause such great differences of pressure that they are related to strong geostrophic currents. Thus, density is an essential quantity for physical and biological-chemical processes in the sea.

In oceanography, density generally means a quantity which actually does not represent the mass per unit volume of seawater but the relative specific gravity that results from the relation of the specific gravity of seawater to that of pure water at 4°C. This quantity is numerically equal to the density within an accuracy range of 3×10^{-5}. The density depends on temperature, salinity, and pressure. The corresponding functional relation represents the equation of state of seawater. The density decreases with decreasing salinity. Decreasing temperature causes an increase of density if the salinity values lie above 24.7‰. For lower salinities, the density at first reaches a maximum between -1.33 and $+3.98°C$, which depends on the salinity value, and then decreases down to the freezing point.

Such dependence of the thermal expansion on salinity is of great importance for the renewal of water in the deeper layers of the ocean. When cooling off or warming up due to interaction with the atmosphere, oceanic areas with a salinity of less than 24.7‰ behave nearly like fresh-water lakes. When the surface water cools off, but not further than the temperature of the density maximum, the density increases; the cooled water particles get heavier than the particles below, and they sink. Thermal convection is started and maintained until all the water of the entire column with higher temperature has reached the temperature of the density maximum. With further cooling, the surface water becomes lighter, the convection is interrupted, and only a thin top layer, mixed by the wind, is subjected to further cooling and finally freezes when the freezing temperature is reached. In seawater with salinity values of more than 24.7‰, on the other hand, the thermal convection is not interrupted as long as the temperature lies above the freezing point. Therefore, the heat content of the entire mixed water column is utilized during the cooling process, the cooling slows down and, in addition to the lowering of the freezing point by salinity, the ocean is even better protected against freezing. The feedback effect on the atmosphere is evident, for example, in the mild winters of the zones with oceanic climate in the higher latitudes as long as the sea is free of ice.

In oceanography it is common to use the abbreviation "sigma" for the density:

$$\sigma = (\rho - 1) \cdot 10^3 \qquad (2.02)$$

Thus, the density $\rho = 1.02698$ corresponds to $\sigma = 26.98$. The term $\sigma_{S,T,p}$ denotes the density at the salinity S, temperature T and pressure p above atmospheric pressure; σ_T is the density at the salinity S, temperature T and atmospheric pressure; σ_Θ is the density at the salinity S and potential temperature Θ (cf. Section 2.1.5). Instead of the density, frequently the specific volume $\alpha = 1/\rho$ is used. Figure 2.07 gives the σ_T of seawater as a function of temperature and salinity, and Fig. 2.08 shows the coefficient

64 Physical Properties and Chemical Composition of Seawater

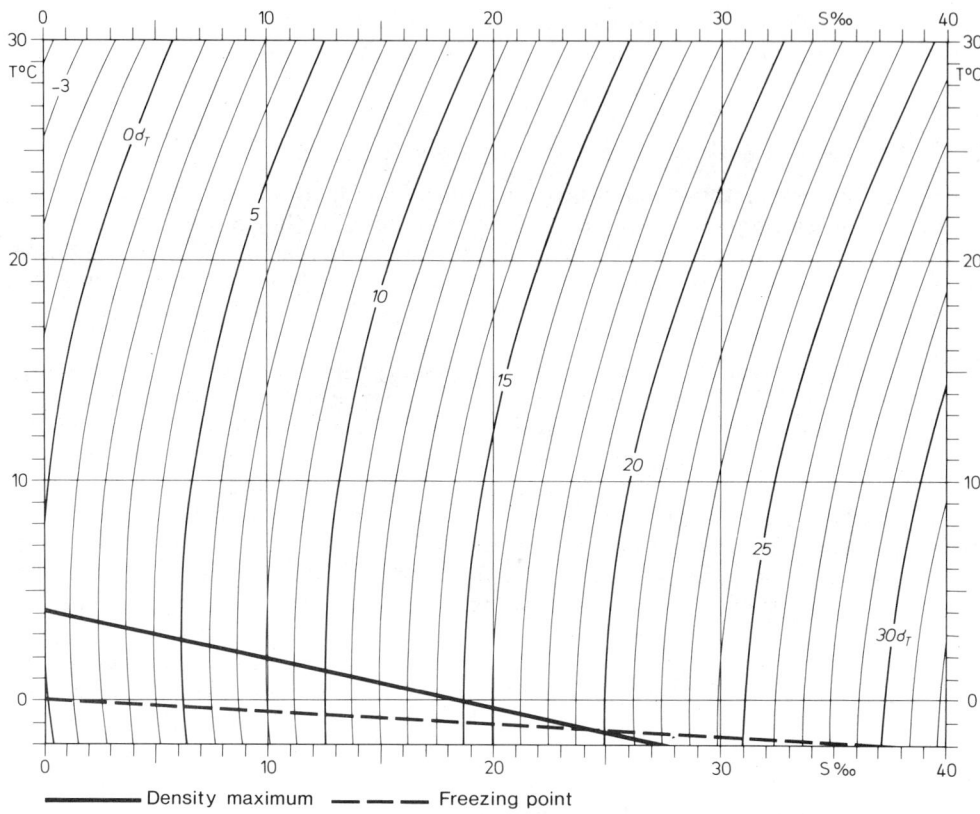

Fig. 2.07. Density σ_T of seawater as a function of temperature T and salinity S.

of thermal expansion e as a function of temperature and salinity at atmospheric pressure. Herein, e is defined by

$$e = \frac{1}{\alpha_{S,T,p}} \frac{\partial \alpha_{S,T,p}}{\partial T} \qquad (2.03)$$

Typical values for σ_T range from 26 to 29.2 in the deep ocean and down to 20 in the top layer. Near the coast, σ_T may become smaller and may be near zero in partly enclosed seas that have a strong inflow of fresh water. The possible range of the in-situ density of normal ocean water extends from $\sigma_{S,T,p} = -4.0$ to the highest value $\sigma_{S,T,p} = 75.7$ in the Challenger II Deep at 10,900 m.

The tables presently used for density and specific volume as a function of salinity, temperature, and pressure are based on the following fundamental determinations: measurements of the specific gravity at temperatures between about 0 and 30°C and at salinities between about 3 and 40‰ at atmospheric pressure taken by Forch, Knudsen, and Sørensen (Knudsen, 1902) as well as measurements of the mean isothermal compressibility for the two salinity values of 31.13 and 38.53‰ at pressures between 0 and 6000 dbar and temperatures between about 0 and 20°C made by Ekman (1908).

The mean isothermal compressibility $\tilde{\kappa}_{\text{isoth}}$ is given by

$$\alpha_{S,T,p} = \alpha_{S,T,0}(1 - \tilde{\kappa}_{\text{isoth}} p) \qquad (2.04)$$

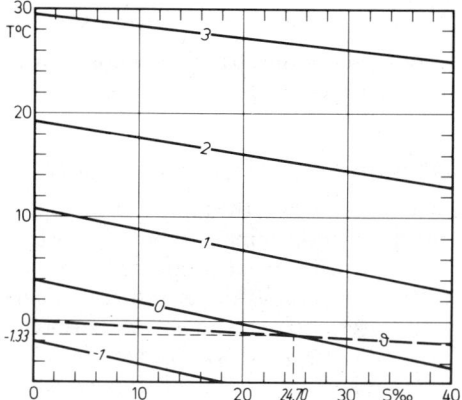

Fig. 2.08. Coefficient of thermal expansion e [10^{-4} °C^{-1}] of seawater at atmospheric pressure as a function of temperature T and salinity S. ϑ: freezing point of seawater.

At 0°C, 35‰ and atmospheric pressure, $\tilde{\kappa}_{isoth}$ amounts to approximately 4.7×10^{-5} dbar^{-1} and decreases by 15% at a pressure of 10,000 dbar. At the transition to pure water it increases by the same percentage. The dependence on temperature is smaller. The isothermal compressibility κ_{isoth} is obtained as given below:

$$\kappa_{isoth} = -\frac{1}{\alpha_{S,T,p}}\left(\frac{\partial \alpha_{S,T,p}}{\partial p}\right)_T = \frac{\tilde{\kappa}_{isoth} + p(\partial \tilde{\kappa}_{isoth}/\partial p)}{1 - \tilde{\kappa}_{isoth} p} \qquad (2.05)$$

The fundamental determinations of the specific gravity and the compressibility have been checked later by several series of measurements. The specific gravity at atmospheric pressure was determined anew by Thompson et al. (1931) and Cox et al. (1970). Cox confirmed Knudsen's results within an accuracy range of 0.01 of the σ_T for salinity values between 15 and 40‰ and, with lower salinities, he found values higher by not more than 0.08. Kremling (1972) investigated the specific gravity at salinities below 20‰ and found characteristic deviations of 0.02 to 0.03; for certain samples, σ_T was even higher by up to 0.1.

The thermal expansion at various pressures has been studied by Crease (1962), Newton et al. (1965), Wilson et al. (1968), Bradshaw et al. (1970), and Caldwell et al. (1970). The direct determinations by Bradshaw et al. agreed with those by Ekman and Knudsen within an accuracy range of 3×10^{-6} cm^3 g^{-1} °C^{-1}. The isothermal compressibility has been examined by Crease (1962) and Wilson et al. (1968). Direct measurements by Bradshaw and Schleicher showed maximum deviations corresponding to $\sigma_T = 0.05$.

According to the results obtained with fresh-water samples (Dorsey, 1940; Kritchevsky et al., 1945), we can assume that the solution of gases in seawater will not yield deviations of density—as compared to the expected values—that lie outside the range of the required accuracy. In particular, the influence of variable oxygen content should not exceed the order of magnitude of 10^{-6} of density. In general, changes of the isotope relations in seawater can also cause changes of density at the order of magnitude of 10^{-6} (Menache, 1966).

In the tables presently used for the specific volume $\alpha_{S,T,p}$ as a function of the salinity S, temperature T, and pressure p above atmospheric pressure, suitable terms of a series expansion up to the second order are summarized (according to Bjerknes et al., 1910) as follows:

66 Physical Properties and Chemical Composition of Seawater

$$\alpha_{S,T,p} = \alpha_{35,0,p} + \Delta_{S,T} + \delta_{S,p} + \delta_{T,p} = \alpha_{35,0,p} + \delta \qquad (2.06)$$

Herein δ is called the anomaly of the specific volume. Formulas for the calculation of σ, α, and δ are given in the Appendix 11.2.

2.1.5. Specific heat, evaporative heat, and adiabatic temperature change

Due to the great specific heat of seawater, the temperature in the ocean varies only little, but considerable heat transport by oceanic currents is possible in spite of the small transport velocities as compared with the atmosphere.

For seawater the specific heat at constant pressure c_p decreases with the increase of salinity, temperature, and pressure. At atmospheric pressure and at a temperature of 0°C, ocean water of 35‰ salinity contains approximately 5% less heat than an equal amount of fresh water. The tables for c_p, which had been used by oceanographers for a long time, were based on measurements of the specific heat of seawater at atmospheric pressure and at a temperature of 17.5°C (Thoulet et al., 1889) as well as on the assumption (Ekman, 1914) that the dependence of seawater on temperature corresponds to that of pure water. Later on, a complete determination of c_p at atmospheric pressure was carried out by Cox et al. (1959) which yielded values up to more than 2% higher. Figure 2.09 shows c_p values that are based on these recent results. They are given in J g^{-1} °C^{-1} (1 J = 1 Joule = 0.238846 cal). Determinations carried out more recently by Millero et al. (1973) have confirmed those results.

The dependence of c_p on pressure is obtained from the two fundamental equations of thermodynamics:

$$\left(\frac{\partial c_p}{\partial p}\right)_{T_{abs}} = -T_{abs}\left(\frac{\partial^2 \alpha}{\partial T^2}\right)_p \qquad (2.07)$$

The right-hand side of the equation is known according to earlier statements concerning density. There, α is the specific volume and T_{abs} the absolute temperature. Characteristic values of the dependence of c_p on pressure show a decrease of approximately 0.8% per 1000 dbar.

The specific heat at constant volume c_v can also be calculated:

$$c_v = c_p + T_{abs}\frac{(\partial \alpha/\partial T)_p^2}{(\partial \alpha/\partial p)_T} = c_p - \alpha T_{abs}\frac{e^2}{K_{isoth}} \qquad (2.08)$$

Here, e and K_{isoth} represent the aforementioned coefficients of thermal expansion and compressibility at constant temperature. c_v is smaller than c_p by 2% at the utmost. For seawater with a salinity of 34.85‰, the ratio $\gamma = c_p/c_v$ at 0°C has the value $\gamma = 1.0004$, and at 30°C the value $\gamma = 1.0207$.

The unusually great evaporative heat L of water, which is of great importance to the heat budget of the near-surface layer and to the atmosphere, apparently is only slightly influenced by salinity; measurements for seawater, however, are not available. For pure water between 0 and 30°C the relation is as follows:

$$L[\text{J g}^{-1}\,°\text{C}^{-1}] = 2502.9 - 2.72T[°\text{C}] \qquad (2.09)$$

In seawater, an increase of temperature may also occur without any addition of heat if the water mass is transported into greater depths and is thereby exposed to higher pressure. Such "adiabatic" change of temperature T_{ad} with increasing pressure p is given by

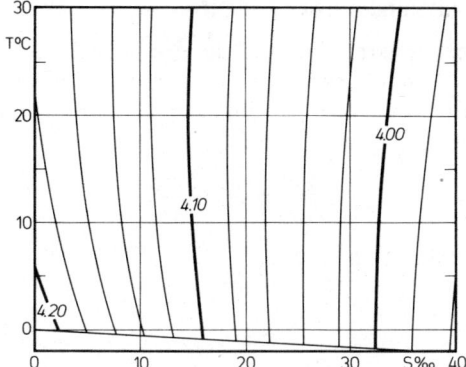

Fig. 2.09. Specific heat c_p [$J\,g^{-1}\,°C^{-1}$] of seawater at atmospheric pressure as a function of temperature T and salinity S. (After Fofonoff, 1962.)

$$\frac{\partial T_{ad}}{\partial p} = \frac{T_{abs}(\partial \alpha/\partial T)_p}{c_p} = \frac{T_{abs}\alpha e}{c_p} \qquad (2.10)$$

For a temperature of 0°C and a salinity of 35‰, one obtains 0.035°C per 1000 dbar. This value is very small as compared to the adiabatic temperature gradient in the atmosphere, which is approximately 300 times greater. Nevertheless, the adiabatic effect plays an important role in the deep ocean, because the vertical temperature gradients there are also small.

For practical purposes, when water elements from various depths are to be compared, it has proved convenient to relate their temperatures to a common reference level of pressure. The potential temperature Θ is the temperature that a water element would have if it were lifted from its in-situ depth with the pressure p above atmospheric pressure to the sea surface, without any heat exchange with its surroundings taking place:

$$\Theta = T - \Delta\Theta = T + \int_p^0 \frac{\partial T_{ad}}{\partial p^*} dp^* \qquad (2.11)$$

Characteristic values of $\Delta\Theta$ are approximately 0.1°C per 1000 dbar. The calculation of the potential temperature from salinity, pressure and in-situ temperature can be carried out according to the formula given in Appendix 11.3.

In the following three cases the potential temperature is a useful auxiliary quantity. (1) Water masses spreading in the ocean at various depths can be traced. (2) The sill depths of submarine ridges separating ocean basins from one another can be determined. Water flowing from basin A across the sill into basin B, where it sinks to the bottom, has there, at the bottom of basin B, the same potential temperature as at the level of the sill in basin A. (3) The potential temperature serves as a measure of the stability of stratification, although, strictly speaking, only in the case of constant salinity. So, in the deep water of the ocean, for different salinity values one may arrive at wrong conclusions (Veronis, 1972). At great depths, near the bottom, one always finds a temperature gradient that corresponds to the adiabatic effect. This does not mean that there is no heat flux originating from the interior of the earth. On the contrary, the adiabatic stratification indicates that the convection, maintained by the heating of the bottom water, prevents unstable stratification.

2.1.6. Molecular thermal conductivity and diffusion

Let us assume that, in seawater, there is the gradient $\partial C_M/\partial z$ of the concentration (amount per volume) C_M of a conservative property M, for instance heat or salt. Then molecular diffusion will cause that, per unit time and area, the following amount will flow in the z direction:

$$\frac{\partial M}{\partial t} = -k\frac{\partial C_M}{\partial z} \tag{2.12}$$

k is the diffusion coefficient (cm^2 sec^{-1}).

For the heat content M_T we get

$$\frac{\partial M_T}{\partial t} = -k_T\frac{\partial C_T}{\partial z} = -k_T\rho c_p\frac{\partial T}{\partial z} = -k_{DT}\frac{\partial T}{\partial z} \tag{2.13}$$

and, consequently, the temporal change of temperature is given by

$$\frac{\partial T}{\partial t} = k_T\frac{\partial^2 T}{\partial z^2} \tag{2.14}$$

with the thermal conductivity k_{DT} [cal cm^{-1} sec^{-1} °C^{-1}] and the heat diffusion coefficient $k_T = k_{DT}/(\rho c_p)$ [cm^2 sec^{-1}] (c_p is the specific heat at constant pressure).

Correspondingly, there follows for the salt content M_S

$$\frac{\partial M_S}{\partial t} = -k_S\frac{\partial C_S}{\partial z} \tag{2.15}$$

and

$$\frac{\partial S}{\partial t} = -k_S\frac{\partial^2 S}{\partial z^2} \tag{2.16}$$

So far, the coefficient k_{DT} has not yet been determined for seawater. Krümmel (1907) calculated k_{DT} from the thermal conductivities k_{DT} for pure water, assuming that, for the same volume, the ratio of the thermal conductivities approximately equals that of the heat capacities. So, for example, we obtain for k_{DT} of seawater at 17.5°C and 0‰ salinity $k_{DT} = 5.86 \times 10^{-3}$ W cm^{-1} °C^{-1}, while for 40‰ salinity and the same temperature, $k_{DT} = 5.61 \times 10^{-3}$ W cm^{-1} °C^{-1} (1 W cm^{-1} °C^{-1} = 0.238846 cal cm^{-1} sec^{-1} °C^{-1}).

The diffusion coefficient k_S for the total salinity cannot be given because k_S takes different values for the various salt compounds. The order of magnitude can be determined from the values of a sodium chloride solution, for which, at 18°C for instance, the value of $\rho \times k_S$ decreases from 1.35×10^{-5} g cm^{-1} sec^{-1} for 0‰ to 1.29×10^{-5} g cm^{-1} sec^{-1} for 40‰ salinity.

In oceanographic problems it is not so much the slight dependence of the coefficients k_T, or k_{DT} and k_{DS}, respectively, on salinity and temperature that is of importance but rather the fact that k_{DT} is about 100 times greater than k_{DS}. If, in the ocean, water masses of different combinations of temperature and salinity but of nearly equal density are arranged in layers one on top of the other, the difference in speed with which temperature and salinity will be balanced at the interface may lead to instabilities that can result in the so-called "salt fingers" and thus in steplike forms of the fine structure of temperature and salinity (Stern et al., 1969; Turner, 1973).

For the large-scale transport of heat and matter in the ocean, however, the molecular exchange processes described above are only of minor importance. If the temperature of an ocean of 0°C all over were raised by a heat source at the surface and kept at 30°C, it would take 1000 yr until, with only molecular thermal conduction acting, the temperature at 300 m depth would have reached 3°C. But the large-scale transport of heat and matter proceeds very much faster because turbulent exchange processes are dominant (cf. Section 2.1.8).

2.1.7. Osmosis, vapor pressure, freezing-point depression, and boiling-point increase

When two enclosed volumes of pure water and seawater are separated by a porous wall which is selectively permeable only for water but not for salts, diffusion of water takes place from one side to the other. This balances the different concentrations of water until a pressure gradient has been established that counteracts the diffusion. This process is called "osmosis," and the pressure p_{osm} that is developed is called "osmotic pressure." For electrolytically neutral solutions the Van't Hoff law is valid, from which it follows that the osmotic pressure in a given volume is proportional to the concentration of the dissolved substance and to the absolute temperature. This law also holds in good approximation for the electrolyte seawater. Being proportional to the absolute temperature, the osmotic pressure as a function of seawater temperature varies only slightly, but it is strongly influenced by changes of salinity. The osmotic pressure in seawater was indirectly determined by Krümmel (1907) and later by Miyake (1939). According to the results obtained by the latter we have

$$p_{osm} = 0.6955S(1 + 3.66 \times 10^{-3}\,T) \qquad (2.17)$$

where p_{osm} is given in bar (1 atm = 1.01325 bar). Figure 2.10 illustrates how the osmotic pressure is dependent on temperature and salinity.

The osmotic pressure is of special importance to the physiology of marine life, since the cell walls of the organisms are mostly semipermeable. In the interior of the cell the osmotic pressure is adjusted to the salinity of the surrounding seawater. Even small variations of salinity can impair the vitality of marine organisms because of the considerable change of the osmotic pressure in the interior of the cells. Only few species have special means by which they can quickly regulate the internal pressure of the cells. Therefore, the great sensitivity of the osmotic pressure to changes of salinity can explain two biogeographical peculiarities in the ocean: The close relation between the distribution of the individual species of fauna and flora and certain salinities, as well as the conspicuous lack of a variety of species in areas with great changes of salinity, like, above all, in zones of brackish water.

The generation of water vapor pressure at the interface between seawater and air can be understood in a way similar to the formation of osmotic pressure. The thermal motion of the water molecules causes a diffusion of water molecules through the interface into the air. A net transport upward takes place until a vapor pressure has been established that counteracts this exchange. This is the case at the saturation vapor pressure. At equal temperature, the vapor pressure of highly saline water will always be beneath that of low-salinity water because in seawater with high salinity, less water molecules per unit surface are available. Direct measurements of the vapor pressure of seawater have been carried out by Higashi et al. (1931) and Arons et al. (1954). A review of selected values of the vapor pressure difference as compared to pure water is given in Fig. 2.11 based on recent measurements.

However, a lowering of the vapor pressure also causes a depression of the freezing

70 Physical Properties and Chemical Composition of Seawater

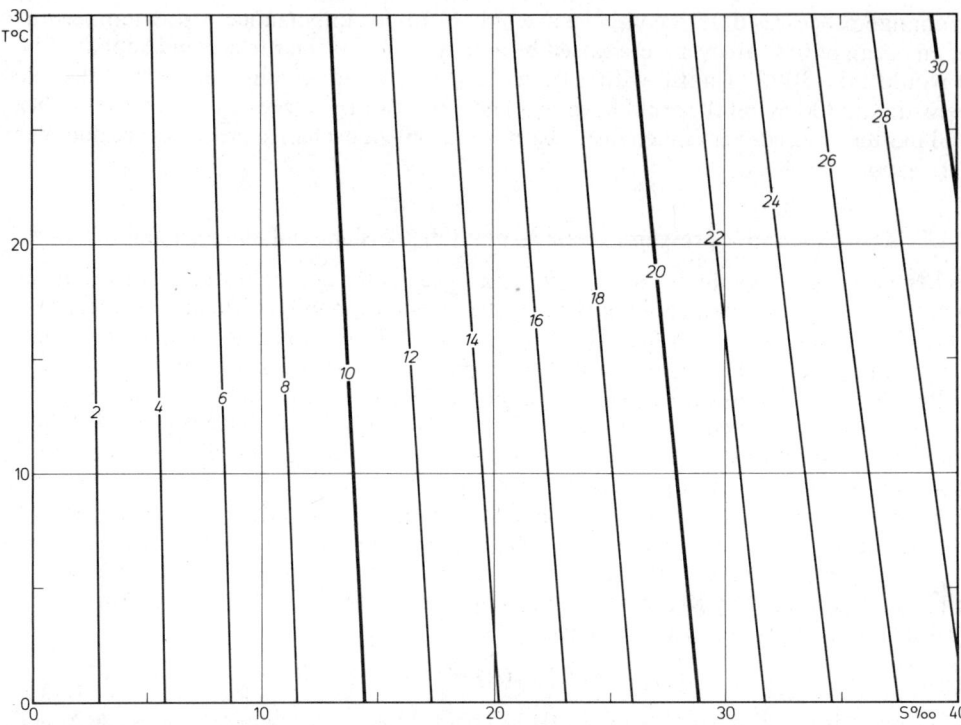

Fig. 2.10. Osmotic pressure p_{osm} [bar] of seawater, as compared to pure water, as a function of temperature T and salinity S.

point and an increase of the boiling point. In the ocean, the freezing-point depression is of special importance. On the one hand, it is important with a view to the density maximum and the convection influenced thereby (cf. Section 2.1.4) and, on the other hand, with respect to the freezing process and the structure of sea ice depending on that process

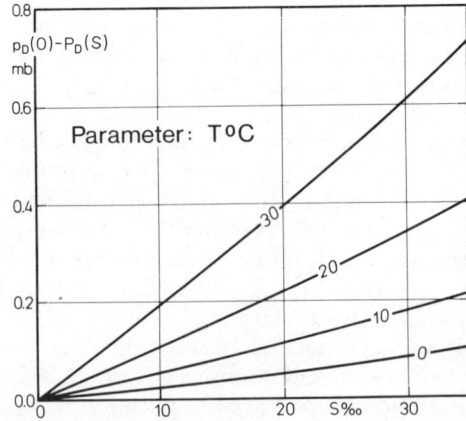

Fig. 2.11. Difference of the vapor pressure p_D (O) of pure water and the vapor pressure p_D (S) of seawater with a salinity S for selected temperatures.

(cf. Section 2.1.12). Direct measurement of the freezing-point depression were carried out by Knudsen (1903) and Miyake (1939). The dependence on salinity is shown in Fig. 2.07.

2.1.8. Turbulent exchange coefficients

There are observations (Grant et al., 1968; Woods, 1968) which show that under certain conditions laminar motion, characterized by processes of molecular exchange, prevails in the ocean. In general, however, the transfer of momentum and the exchange of heat and matter ensue from turbulent motions. These processes depend on the scale of the motions and on the stratification of currents and density. Therefore, they cannot be described by material constants as in the case of molecular processes. However, it has proved suitable to define turbulent exchange coefficients that permit us to represent the processes for certain scales and conditions of stratification in a way similar to the molecular exchange.

In analogy to the molecular exchange (cf. Section 2.1.6) the definition reads:

$$\frac{\partial M}{\partial t} = -\frac{A}{\rho}\frac{\partial C}{\partial z} \tag{2.18}$$

A/ρ [cm^2 sec^{-1}] is called the "turbulent diffusion coefficient," while A [g cm^{-1} sec^{-1}] is the "austausch coefficient." Corresponding relations hold for the turbulent transport of heat and salt:

$$\frac{\partial T}{\partial t} = \frac{A_T}{\rho}\frac{\partial^2 T}{\partial z^2} \tag{2.19}$$

$$\frac{\partial S}{\partial t} = \frac{A_S}{\rho}\frac{\partial^2 S}{\partial z^2} \tag{2.20}$$

Similarly, the transfer of momentum in z direction with a mean current U normal to z direction (cf. Section 2.1.9) is described by

$$\frac{\partial U}{\partial t} = \frac{A_M}{\rho}\frac{\partial^2 U}{\partial z^2} \tag{2.21}$$

Here, A_M/ρ is the "turbulent transfer coefficient for momentum" and A_M the austausch coefficient. Usually, the turbulent transfer coefficients exceed the corresponding molecular coefficients by several orders of magnitude.

It might be expected that with equal conditions of motion and stratification $A_S = A_T = A_M$. But this is not the case because the mechanism of momentum exchange does not completely equal that of the exchange of matter and heat. When, within a certain time interval, the momentum is transferred by processes of turbulent motion, but matter and heat are not fully mixed with the surroundings (and are therefore partly transported back), the coefficients for matter and heat will be smaller than the coefficient for momentum.

In general, we need not apply three different exchange or austausch coefficients for rectangular coordinates, but it is sufficient to have one horizontal and one vertical coefficient for heat (A_{TH}, A_{TZ}) and the same for salt (A_{SH}, A_{SZ}) and momentum (A_H, A_Z). Whereas in the ocean the horizontal exchange coefficient is mainly a function of the horizontal scale, the vertical exchange coefficient does not depend only on the vertical scale, but also on stratification. Typical orders of magnitude of the coefficients are shown

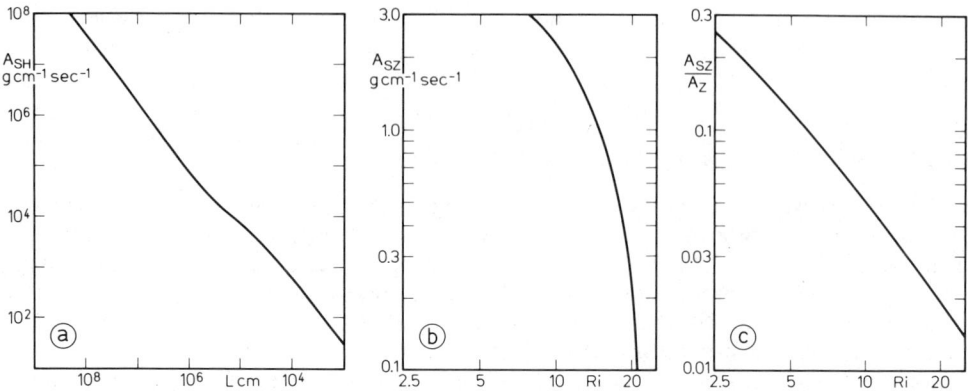

Fig. 2.12. Orders of magnitude of the austausch coefficients according to measurements in the ocean. (a) Horizontal austausch coefficient A_{SH} as a function of the horizontal scale L. (b) Vertical austausch coefficient A_{SZ} as a function of the Richardson number Ri. (c) Ratio of the vertical austausch coefficient A_{SZ} to the vertical viscosity coefficient A_Z as a function of the Richardson number Ri.

in Fig. 2.12(a)–(c). Figure 2.12(a) represents the horizontal turbulent diffusion coefficient A_{SH} as a function of the horizontal scale L (Okubo et al., 1970; Okubo, 1971).

It can be seen that, between 10 m and 1000 km, there are two extensive ranges where A_{SH} is proportional to $L^{-4/3}$. As an example for the vertical turbulent diffusion coefficient, Fig. 2.12(b) illustrates the dependence of A_{SZ} on the Richardson number

$$Ri = \frac{(g/\rho)\partial\rho/\partial z}{(\partial U/\partial z)^2} \tag{2.22}$$

according to measurements carried out in the Skagerrak (Ehricke, 1969). Data obtained in other oceanic areas show a similar pattern. However, they may deviate from the presented curve up to about one order of magnitude upward or downward because the structure of the turbulence depends on the generating mechanism that is significant for each particular case. The range of the values measured for A_{SZ} extends from approximately 10^{-2} to 3×10^2 g cm^{-1} sec^{-1}. Figure 2.12(c) shows the dependence of the ratio A_{SZ}/A_Z on the Richardson number (Jacobsen, 1913; Munk et al., 1948; Francis et al., 1953).

2.1.9. Viscosity and surface tension

When water layers are horizontally shifted with respect to each other, internal frictional forces act against such displacement because momentum is transferred through the interface. The force K, on a unit area normal to the z axis, that is the change of the momentum I, with a horizontal current U is given by

$$K = \frac{\partial I}{\partial t} = \mu \frac{\partial U}{\partial z} = T_{XZ} \tag{2.23}$$

μ [g cm^{-1} sec^{-1}] is the coefficient of molecular viscosity; T_{XZ} the tangential shear stress. For the change of the current we then obtain:

$$\frac{\partial U}{\partial t} = \frac{\mu}{\rho}\frac{\partial^2 U}{\partial z^2} \tag{2.24}$$

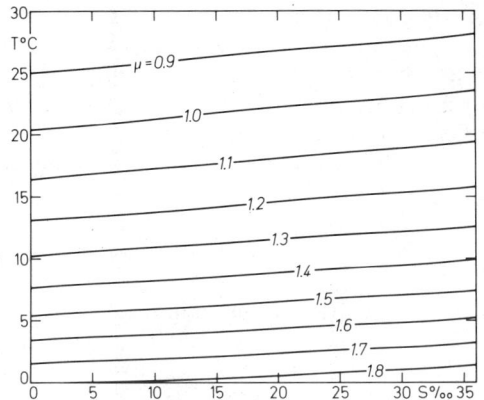

Fig. 2.13. The viscosity μ [10^{-2} g cm^{-1} sec^{-1}] of seawater at atmospheric pressure as a function of temperature and salinity.

μ/ρ is called the kinematic viscosity coefficient. For pure water at 0°C the value of μ amounts to 0.0179 g cm^{-1} sec^{-1}. For seawater this value decreases rapidly with increasing temperature, and it slightly increases with increasing salinity. Former investigations of the viscosity by Krümmel et al. (1905) have recently been complemented by measurements of Miyake et al. (1948), Horne et al. (1966), and Stanley et al. (1969). Figure 2.13 shows the results for atmospheric pressure; the formulas for the computation of the viscosity as a function of temperature, salinity, and pressure are contained in Appendix 11.4.

We see that the molecular viscosity coefficient μ is much smaller than the turbulent viscosity coefficient or eddy transfer coefficient for momentum A_M (cf. Section 2.1.8). Therefore, in the description of the large-scale transport of momentum, μ can be neglected relative to A_M. The processes of molecular friction plays an important part, however, for the description of the physical processes in the boundary layers at the sea surface and at the ocean bottom as well as for the investigation of the dissipation of the kinetic energy of ocean currents into heat at very small spatial scales.

Furthermore, molecular friction plays a significant role for marine life. Plankton has only very little motion of its own, if any at all. The sinking velocity, however, is restricted by friction, and the plankton almost hovers in the water. The larger the surface is relative to the volume, the greater is the influence of friction. In the warm water of the lower latitudes we find especially fine-structured forms of plankton which, owing to their large specific surfaces, are well adapted to the lower friction value at high temperature.

At free surfaces, the attractive forces among the molecules of seawater, by which the viscosity is caused, produce a resulting force that tries to reduce the size of the surface. If the free surface is extended by δF, the work δA that must be performed against that surface tension can be described by

$$\delta A = \alpha \cdot \delta F \qquad (2.25)$$

α is the constant of the surface tension which, for pure water at 0°C, amounts to about 75 dyn cm^{-1}. When the temperature rises to 30°C, this constant decreases by about 5%, and it increases by about 1% when the salinity increases to 40‰ (Krümmel, 1907).

The surface tension is very important for very small and short surface waves, for which the restoring force at a deflected surface is almost exclusively determined by the surface tension and not by gravity. Such waves are called "capillary waves" (cf. Section 8.3.2). Due to only slight dependence of the quantity α on salinity, capillary waves in seawater behave like capillary waves in pure water at equal temperature. Furthermore, the surface tension plays an important part when water drops are generated in breaking waves. Pollution of the sea surface exerts a considerable influence on the magnitude of the surface tension.

2.1.10. Acoustic properties

For practical purposes, the propagation of sound in the ocean is very important because acoustical techniques of sounding as well as of direction and range finding are essential aids for civil and military navigation and for fishing. However, they also represent a basis for numerous measuring methods in marine research.

The phase velocity c of sound waves is given by the Laplace equation

$$c = \sqrt{\frac{\gamma}{\rho \kappa_{\text{isoth}}}} \qquad (2.26)$$

Here, $\gamma = c_p/c_v$ is the ratio of the values of specific heat at constant pressure and constant volume, ρ is the density, and κ_{isoth} the isothermal compressibility (cf. Section 2.1.4). From the fundamental determinations mentioned before, γ, ρ and κ_{isoth} are known functions of salinity, temperature, and pressure so that c can be computed. This method was used by Heck et al. (1924), Kuwahara (1938), and Matthews (1939). For a long time Kuwahara's tables formed the basis for the determination of sound velocity. A problem, however, was that the knowledge on compressibility was still lacking sufficient accuracy. Later on, direct laboratory measurements of sound velocity in seawater were carried out by Wilson (1960) and Del Grosso (1970). Wilson's formulas represent the basis for the new tables by Bialek (1966). Anderson (1971) corrected those formulas with respect to the combinations of salinity, temperature, and pressure that actually occur in the ocean. The functional relation described in Appendix 11.5 is based on these formulas.

Figure 2.14 shows the sound velocity as a function of temperature and salinity at atmospheric pressure and of temperature and pressure at a salinity of 35‰. It is seen that the sound velocity in the open ocean with salinities roughly between 34 and 37‰ is predominantly influenced by the changes of temperature and pressure, and only slightly by those of salinity.

In lower and middle latitudes, the temperature beneath an almost homogeneous surface layer sharply drops with depth. In this thermocline the sound velocity decreases with depth. At greater depth the effect of pressure prevails, and the sound velocity increases with increasing depth. Figure 2.15 gives some examples for the vertical distribution of sound velocity in various oceanic regions.

The profiles show minima at depths between 700 and 1300 m in regions with pronounced thermal stratification. In areas with an annual variation of the distribution of temperature the sound velocity accordingly varies in an annual rhythm, too.

Numerous problems of sound propagation in the ocean can be solved by considering sound rays and neglecting the phenomena of diffraction. Here the laws of geometrical optics apply analogously: (1) the sound propagates along a straight line if the sound velocity is constant; (2) different sound rays are independent of each other; (3) the path of the ray is reversible; (4) at the sea floor, at the sea surface, and at objects and interfaces

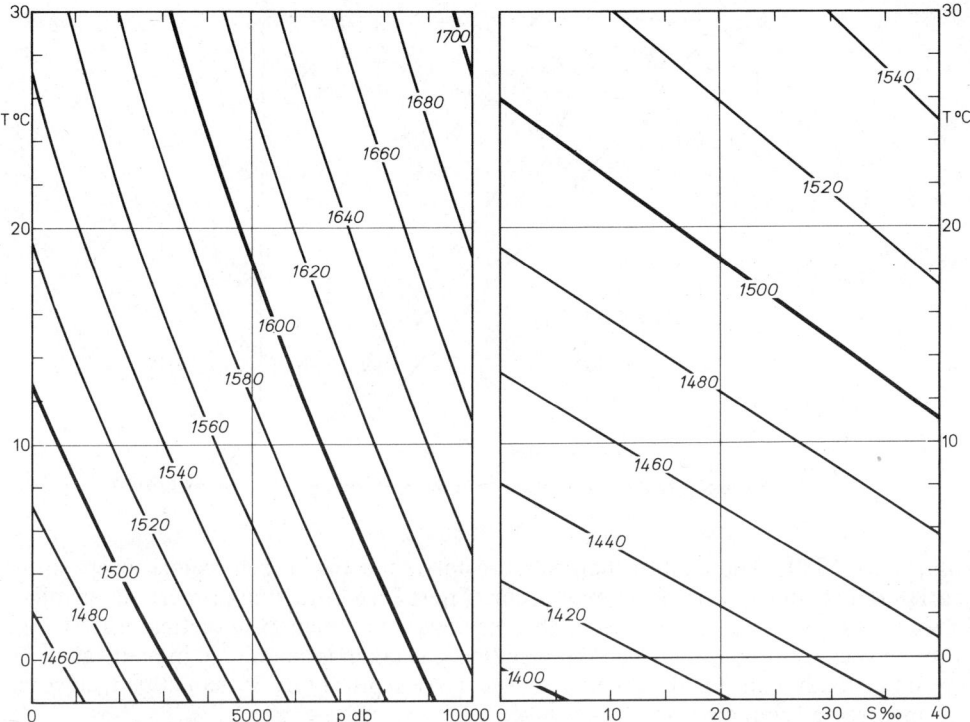

Fig. 2.14. Sound velocity c [m sec^{-1}] as a function of the temperature T and the pressure p (left) as well as the salinity S (right).

in seawater, the reflection law holds—according to which the angle of incidence relative to a line perpendicular to the interface is equal to the angle of reflection; (5) at interfaces in seawater the law of refraction applies—according to which the angle of incidence α_1 in water with the sound velocity c_1 and the angle of refraction α_2 in water with the sound velocity c_2 are interrelated by

$$\frac{\sin \alpha_1}{\sin \alpha_2} = \frac{c_1}{c_2} \qquad (2.27)$$

Since the stratification of the sound velocity in the ocean is mainly horizontal, sound propagates in the vertical along a straight line. This fact is used for echo sounding. If the depth is to be determined from the travel time of a sound signal, which is reflected at the sea floor, the mean sound velocity in the water column must be known. With the travel time t_0 from the sea surface to the sea floor at the depth h (coordinate z from the sea surface downwards) the following relation is obtained:

$$h = \bar{c}(h) \cdot t_0 \quad \text{with} \quad \bar{c}(h) = h / \int_0^h \frac{dz}{c(z)} \qquad (2.28)$$

If $c(z)$ is known, h can be determined. Since, in the deep sea, the distribution of sound velocity at a position varies only slightly, $\bar{c}(h)$ can be considered generally valid for various oceanic areas. For practical application one does not use tables of the sound velocity itself, but tables of the correction necessary for the deviation from $\bar{c}(h) = 1500$ m sec^{-1}

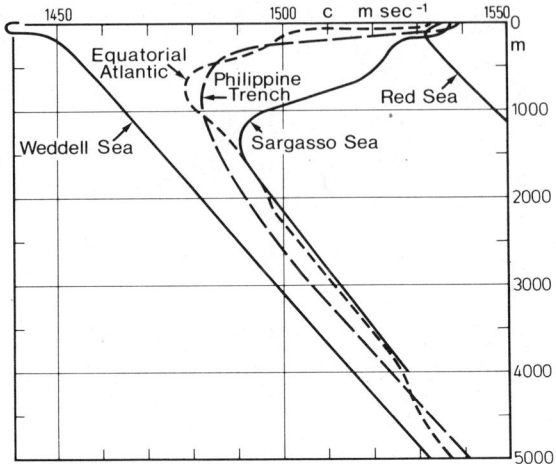

Fig. 2.15. Examples of the sound velocity c in the world ocean. (After Dietrich, 1952.)

(Matthews, 1939). Those tables, however, are not applicable in shelf regions where strong spatial and temporal variations may occur. There, $\bar{c}(h)$ must be properly determined for each case. In accordance with an international agreement, the vertical distribution of sound velocity at the time when the sounding was taken is considered in nautical charts for all the depth data that do not exceed 200 m. At depths of more than 200 m, however, the uncorrected sounding depths are given.

The propagation of inclined sound beams in the ocean may become highly complicated by refraction and reflection. Figure 2.16 gives some examples of simple profiles of sound velocity. It can be recognized that there are regions which lie in shadow zones (zones of silence) which—according to the laws of refraction and reflection—a sound ray from a given source cannot reach. Nor can, due to the reversibility of sound rays, any sound signal travel from the region of the shadow zone to that source. Submarines take advantage of such acoustic conditions when they want to escape from their pursuers navigating at the sea surface. As a matter of fact, the sound shadow is not quite complete because, to some extent, phenomena of diffraction may permit sound energy to reach the shadow zones. Of particular interest is the sound propagation in the area of the minimum of sound velocity at middle depths, as shown in Fig. 2.16. In the case of a sound source at such depth, most of the sound energy, due to refraction and reflection, remains in this intermediate layer, in the so-called SOFAR channel (Sound Fixing And Ranging), and can be traced over a distance of many thousands of kilometers.

Thus, the position of an airplane or a ship in distress, or the point of impact of a missile, can be determined even from a great distance when an explosive charge is fired in the SOFAR channel and the resulting sound signals are observed at coastal stations. In oceanography, drifting buoys with sound transmitters, floating at the depth of the SOFAR channel, are located in a similar way.

The transmission loss during sound propagation, that is, the decrease of intensity, has two causes: first, the geometry of the path of the beam, second, the absorption. The former effect depends on the stratification in the ocean, on the location of the sound source, and its directional dependence. The latter effect is a consequence of the transformation of sound energy into other forms of energy.

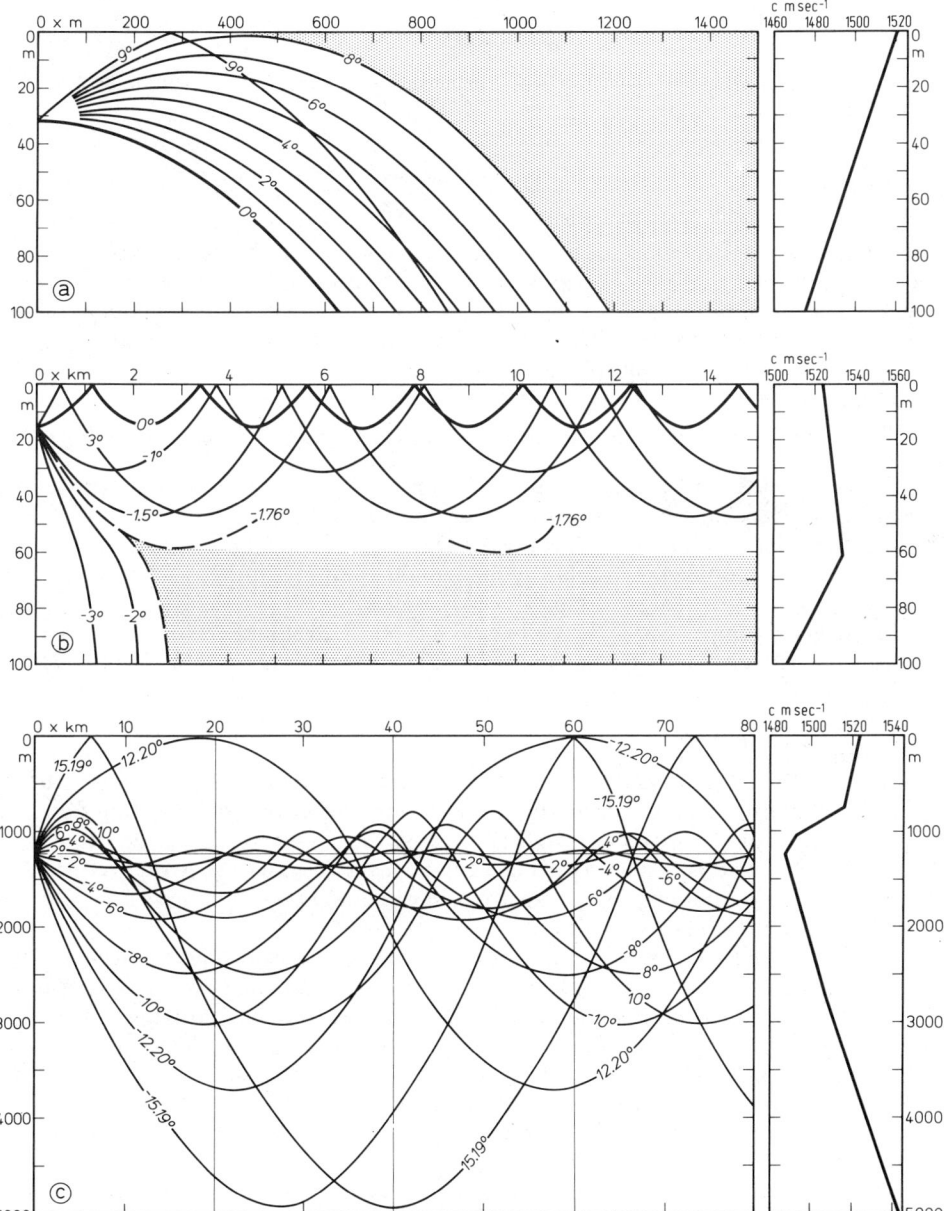

Fig. 2.16. Paths of sound beams at selected distributions of sound velocity for a sound source beneath the water surface. (a) Linear decrease of sound velocity with depth. (After Urick, 1967.) (b) Two layers, each with linear decrease of sound velocity. (After Urick, 1967.) (c) Typical distribution of sound velocity in the SOFAR channel. (After Ewing et al., 1948.)

77

The decrease of the sound intensity I per distance x from the initial intensity I_0 as a result of absorption can be described by means of the absorption coefficient ν:

$$I(x) = I_0 e^{-\nu x} \tag{2.29}$$

Due to the viscosity of seawater, the absorption coefficient for frequencies f of more than about 100 kHz is given by (Litovitz, 1965):

$$\nu = 3.1\mu \cdot \frac{16\pi^2 f^2}{3\rho c^3} \tag{2.30}$$

The absorption coefficient is thus proportional to the square of the frequency of the sound wave, which means that the absorption strongly increases with frequency. Nevertheless, the range of sound propagation is much larger in water than it is in air, because not only the density ρ but also the sound velocity c are large. Accordingly, for acoustic locating methods, the attainable range of propagation increases with decreasing frequency, but at the same time, the wavelength is increased and, due to the reduced spatial resolution as a consequence of diffraction, the applicable range of frequency is restricted in its lower part.

As the quantities ρ, c, and μ are functions of salinity, temperature, and pressure, the corresponding dependence results for ν. At low frequency, ν is greater by a factor of about 30 than the value given by Eq. (2.30) because energy is taken up by processes of dissociation and association, in particular by those of magnesium sulfate (Liebermann, 1949; Atkinson, 1971).

For selected temperatures Schulkin et al. (1963) determined the dependence of the absorption coefficient ν on frequency. Selected values are contained in Fig. 2.17.

2.1.11. Optical properties

In the ocean the intensity and the spectral composition of the electromagnetic radiation are influenced by the optical properties of seawater. The latter are important factors in the heat budget of ocean and atmosphere because they determine the proportion of the radiative energy that is taken up by the sea and converted into heat. Moreover, they are essential for the spatial limits of those areas in which the photosynthesis processes, the basis of all marine life, can take place.

The total, mostly shortwave radiation from the sun and sky which reaches the ocean (cf. Section 4.2.2) is partly reflected at the sea surface. The remaining radiation penetrates into the water and is simultaneously refracted. It is scattered and absorbed and, thus, attenuated. Part of the scattered radiation returns into the atmosphere as shortwave back-radiation. The absorbed radiation is at the disposal of the ocean in the form of thermal energy or chemical energy. Part of the thermal energy returns into the atmosphere as longwave radiation of the ocean. A discussion of the properties of reflection, refraction, scattering, and absorption by seawater follows.

The reflectivity of the sea surface depends on the angle of incidence of the light, the refractive index of seawater and the formation of foam at the interface. For a smooth surface, Fresnel's formula can be applied according to which the reflectivity at a given refractive index is a function of the angle of incidence. Since the sea surface is very seldom smooth, this formula does not suffice for describing the reflection there. But if the increase of the reflectivity due to foam formation is neglected, Fresnel's formula can be applied to small surface elements and, if the inclination of the surface is known, the reflectivity

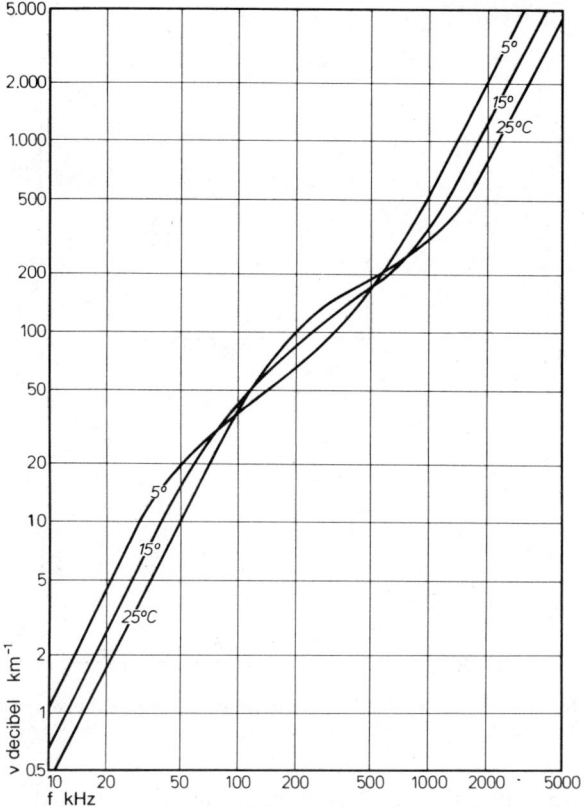

Fig. 2.17. Coefficient of sound absorption ν for selected temperatures as a function of the frequency f. (After Urick, 1967.) (10 decibel correspond to a decrease of intensity by the factor 10^1; 20 decibel to a decrease by 10^2 etc.)

can be computed from statistics of the sea state. In this connection the variability of the refractive index can be disregarded, because it varies only by about 0.2% within the range of the temperatures and salinities in the ocean and only by about 3% within the range of the wavelengths from 300 to 2000 nm. The result for which the influence of the shaded areas was also considered (Saunders, 1967) is given in Fig. 2.18. The value s = 0 corresponds to a smooth sea surface, that is, directly to Fresnel's formula.

More important than the knowledge of the reflectivity is, for practical problems, the knowledge of the ratio of outgoing to incoming radiation, the albedo. If E_a and E_b are the radiative intensities resulting from the radiation that arrives at the sea surface from above (E_a) and from below (E_b), we can write for the albedo A:

$$A = \frac{E_b}{E_a} \tag{2.31}$$

The quantity E_b contains not only the reflected radiation but also the radiation scattered back from the sea (underlight), which may amount to up to 25% of the total value. The albedo is primarily a function of the altitude of the sun as well as of the cloud cover. To a certain degree, it depends on the state of the sea and the turbidity of the seawater. In

80 Physical Properties and Chemical Composition of Seawater

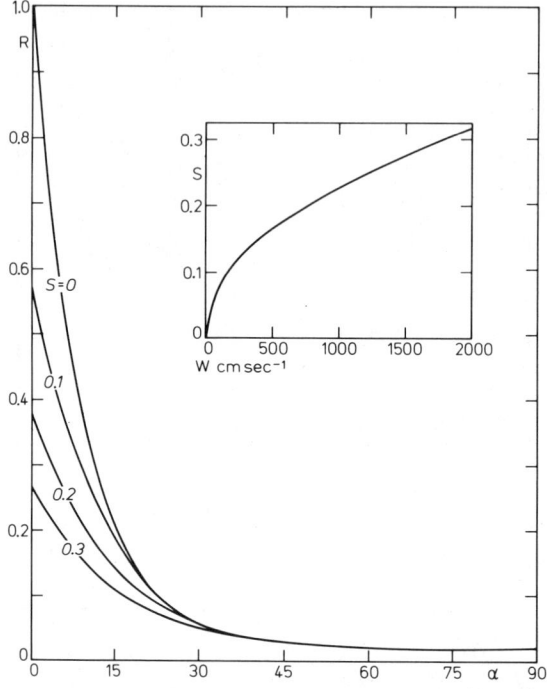

Fig. 2.18. Reflectivity R of the sea surface as a function of the elevation α of the source of radiation and of the mean square slope s of the sea surface depending on the wind velocity W. (After Payne, 1971.)

order to eliminate the difficulty connected with the quantitative description of the influence of cloudiness, Payne (1972) has determined the albedo as a function of the transparency D of the atmosphere and of the altitude α of the sun. D is defined by

$$D = \frac{E_a}{(S \sin \alpha)/r^2} \qquad (2.32)$$

where S is the solar constant (0.135 J cm^{-2} sec^{-1}; or 1.94 cal cm^{-2} min^{-1}) and r is the ratio of the true to the mean distance to the sun. The albedo as a function of D and α is shown in Fig. 2.19.

For exclusively diffuse radiation one obtains an albedo of 0.06 corresponding to theoretical predictions (Burt, 1953). With middle altitudes of the sun the albedo depends on the state of the sea; the decrease of 1%, connected with the increase of the wind of about 2 km hr^{-1}, can be taken as a variation characteristic of the open ocean. Typical deviations between coastal and ocean waters amount to 10 to 20%.

The refraction at interfaces between different types of water, or between water and air, is determined by the difference of the refractive indices. As mentioned before, the dependence of the refractive index n on temperature and salinity is so small that a constant value ($n \approx 4/3$) can be assumed when considering the radiation budget. Measurements by Krümmel (1907), Utterback et al. (1934), Bein et al. (1935), and Miyake (1939) have shown that the refractive index increases with salinity by about 2×10^{-4} per ‰ and with temperature by about 0.1×10^{-4} per °C. The exact knowledge of this relationship is important if the salinity is to be determined from measurements of the refractive index

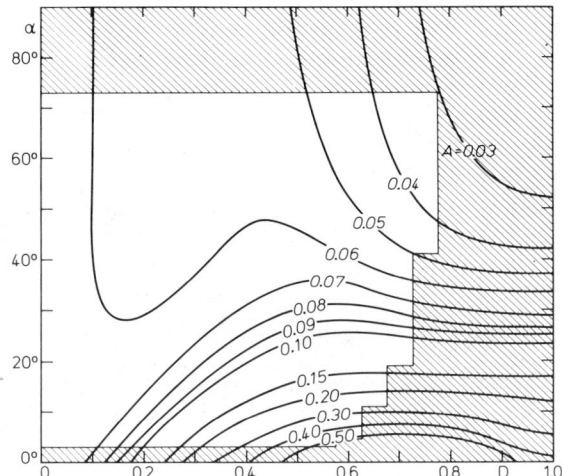

Fig. 2.19. Albedo A of the sea surface as a function of the sun's altitude α and the transparency D of the atmosphere. (After Payne, 1972.) (The curves in the hatched region have been extrapolated.)

and of temperature. According to Rusby (1967), the salinity is obtained from the difference Δn between the refractive index $n_{S,T}$ of a seawater sample and the refractive index $n_{35,T}$ of Copenhagen Normal Water with 35.000‰ salinity at the temperature T and atmospheric pressure as follows:

$$S = 35.000 + 5.3302 \times 10^{-3} \Delta n + 2.274 \times 10^5 \Delta n^2 + 3.9 \times 10^6 \Delta n^3 \\ + 10.59 \Delta n (T - 20) + 2.5 \times 10^2 \Delta n^2 (T - 20) \quad (2.33)$$

This equation holds for the ranges of $30.9‰ \leq S \leq 38.8‰$ and $17°C \leq T \leq 30°C$ for the light of the green Hg line at the wavelength of $\lambda = 546.227$ nm (1 nm = 10^{-9} m = 10 Å). The refractive index decreases by about 4% at an increase of the wavelength from 250 to 1250 nm. For further details consult Lauscher (1955).

The scattering of the electromagnetic radiation in the ocean is caused by three different physical processes, namely by the diffraction at particles, by refraction and reflection in the interior of particles, and by reflection at the outer interfaces (or surfaces) of particles. In this case, particles can be suspended organic and inorganic substances as well as inhomogeneities of the density of water. If a light beam arrives at a scattering volume element dv with the irradiation E, the following relation is obtained for the intensity of the scattered light dI within the angular element $d\omega$:

$$dI = \beta(\alpha) E d\omega \quad (2.34)$$

where α is the angle relative to the direction of propagation of the light beam and $\beta(\alpha)$ denotes the volume-scattering function.

The radiation loss of the light beam solely by scattering, when passing along the distance x, can be described by:

$$I(x) = I_0 e^{-bx} \text{ with } b = 2\pi \int_0^{2\pi} \beta(\alpha) \sin \alpha \, d\alpha \quad (2.35)$$

I_0 and $I(x)$ are the light intensities before and after passing the distance x, b is the scattering coefficient. The functions $\beta(\alpha)$ and, therewith, b depend on the type of the

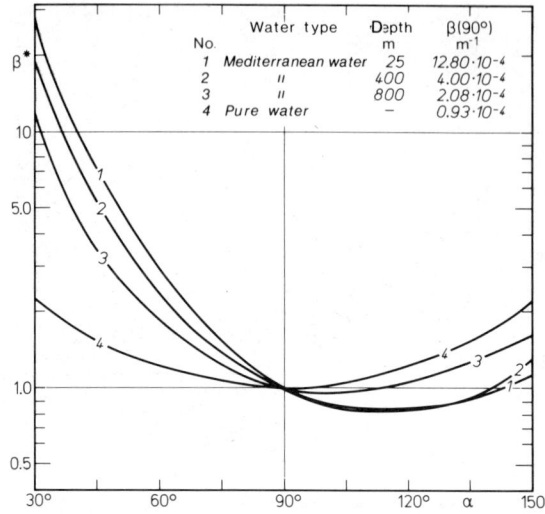

Fig. 2.20. Scattering function β^*, normalized to $\alpha = 90°$, as a function of the scattering angle α for three different types of seawater according to laboratory tests and for pure water according to theory. (After Ivanoff et al., 1964 and Morel, 1966.)

scattering particles and the wavelength λ of the radiation. The dependence of the scattering behavior of pure seawater on wavelength largely corresponds to that of pure water in which the intensity of the scattering light varies proportional to λ^{-4}. The salt content, however, causes an increase of the scattered part for small scattering angles α, which means increased forward scattering. Figure 2.20 shows the volume-scattering function $\beta^*(\alpha) = \beta(\alpha)/\beta(90°)$, normalized to $\beta(90°)$, for clear seawater in comparison to pure water.

If suspended matter is present in seawater, one obtains a similar form of the scattering function β dependent on α, but β can become very much greater. Here, the dependence on wavelength changes with the size of the scattering suspended particles. For very small particle sizes, the scattering coefficient is proportional to λ^{-m} with $m = 4$; for increasing particle sizes, however, the exponent m tends to assume smaller and smaller values and reaches $m = 0$ for particles with a diameter of more than approximately 1 μm, thus being independent on wavelength. The polarization of scattered light depending on the angle α shows an inverse behavior with respect to the scattering function, so the maximum is usually at $\alpha = 90°$ (Hinzpeter, 1962).

The total attenuation of a light beam in seawater is caused by the simultaneous effects of scattering and absorption. It can be described by the attenuation coefficient c (formerly called extinction coefficient) and the absorption coefficient a:

$$c = a + b, \qquad (2.36)$$

where a can only be determined as the difference between c and b. Figure 2.21 shows the coefficients of attenuation and scattering for pure water (Clarke et al., 1939; Le Grand, 1939; Curcio et al., 1951) as a function of wavelength. It can be seen that the scattering in pure water only plays a role in the longwave visible range. Pure seawater from the open ocean behaves similarly.

The attenuation of the radiation in natural seawater can differ essentially from that in pure seawater, as demonstrated by the two examples in Fig. 2.22.

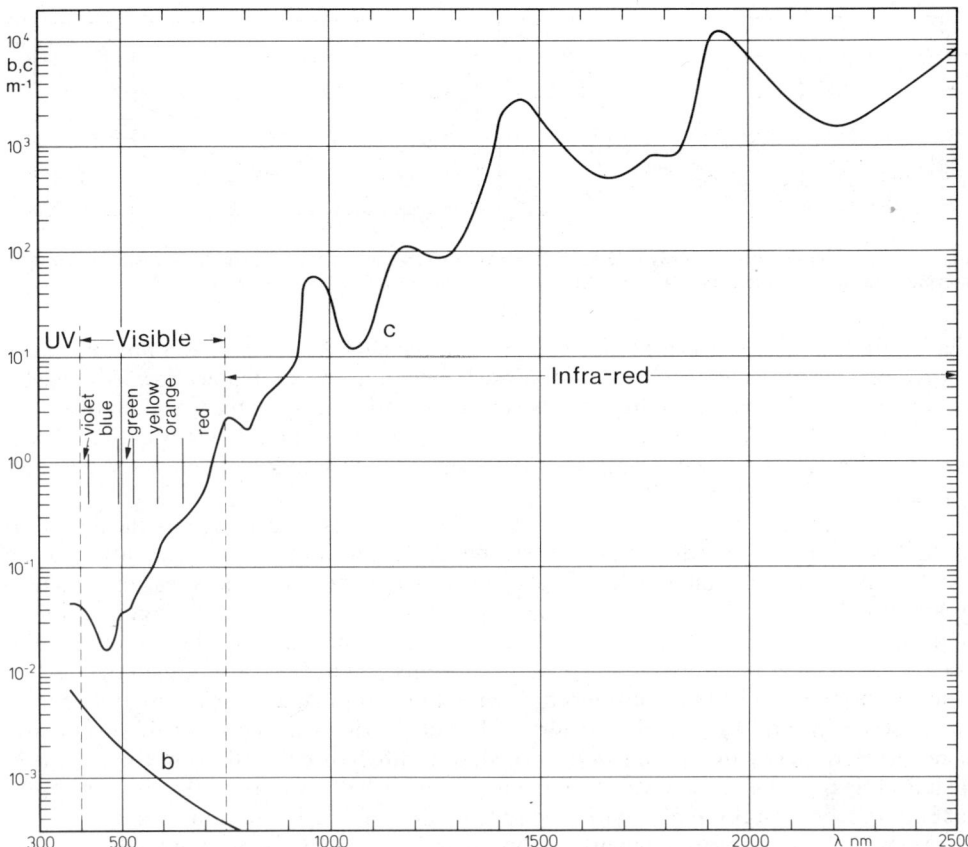

Fig. 2.21. Attenuation coefficient c and scattering coefficient b for pure water as a function of the wavelength λ of the radiation.

The curves at (a) show the ratio $I(x)/I_0$ for $x = 1$ m in percent for the extremely clear water of the Sargasso Sea northwest of the Bermuda Islands, the curves at (b) for water near the American east coast. It is obvious that the attenuation is increased when the water contains suspended particles. Furthermore, it will be noticed that, after filtration, the water from the Sargasso Sea shows almost the same behavior as pure water, whereas the coastal water does not. Therefore, in addition to the absorption in pure water and the scattering at suspended particles eliminated by filtration, there must occur some absorption, depending on wavelength, by dissolved substances. This is caused by intensely absorptive humic substances, the so-called "yellow substances" (Kalle, 1966), consisting of long-lived organic metabolic products which the ocean receives mainly from continental rivers but which are also formed in the ocean by the decomposition of plankton.

The above considerations concern the attenuation of a light ray in seawater. Marine biologists, however, are more interested in the attenuation of a flow of radiation with very large horizontal extension, when the depth is increasing. Such attenuation of radiation is smaller than that of a vertical light ray, because the scattering into the forward half space favors the irradiation of a deeper horizontal plane. At the left-hand side of Fig. 2.23, the mean attenuation of daylight depending on wavelength is shown for open-ocean and near-shore waters, and at the right-hand side we see the corresponding decrease of

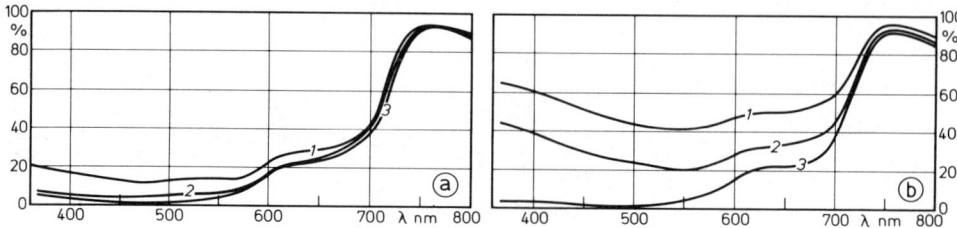

Fig. 2.22. Attenuation of a light ray in % per meter. (After Clarke et al., 1939.) (a) in the Sargasso Sea and (b) in American coastal waters, before (1) and after (2) filtration, in comparison to pure water (3).

the irradiation of a horizontal plane relative to the surface for $\lambda = 465$ nm. With the transition from the clearest areas of the world ocean in the subtropics towards coastal waters, the maximum of the light transparency is shifted to greater wavelengths.

Jerlov (1951, 1964) has made proposals for a relevant optical classification of seawater. The attenuation coefficient is often correlated with salinity as well as temperature (Dietrich, 1953; Joseph, 1959).

If the reflected light is ignored, the color of seawater is determined by the light that is back-scattered upwards. Thus it corresponds to the minimum of the attenuation coefficient and varies from the deep blue of the tropical and subtropical seas to the bluish green in higher latitudes, the green in the upwelling and shelf areas of the higher latitudes and the yellow-greenish discoloration of very turbid coastal waters. The deep blue color is characteristic of water extremely poor in nutrients and is, therefore, called the "desert color of the ocean." With a considerable content of yellow substances present, the color turns greenish, and the peculiar colors of larger particles in seawater, if sufficiently concentrated, may cause greenish, brownish or reddish discolorations of the seawater. The Red Sea, for instance, owes its name to the reddish discoloration that in some areas is temporarily brought about by the overproduction of a red plankton species.

Summarizing descriptions with a view to optical properties of seawater have been given by Joseph (1952) and Jerlov (1968).

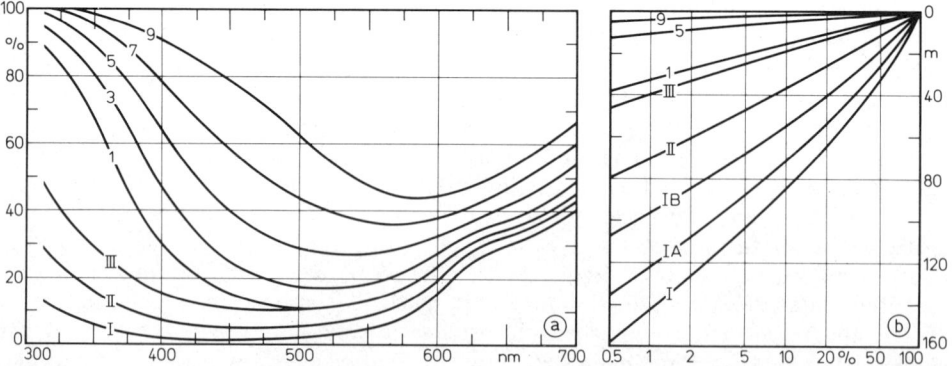

Fig. 2.23. (a) Attenuation of daylight in % per meter as a function of the wavelength λ. (After Jerlov, 1951.) Curve I: extremely pure ocean water; curve II: turbid tropical-subtropical ocean water; curve III: ocean water of moderate latitudes; curves 1 to 9: coastal waters of different degrees of turbidity. (Sun's altitude 90° for the first three cases, 45° for the other curves.) (b) Attenuation of daylight with the wavelength $\lambda = 465$ nm for the same types of water as a function of depth. (After Jerlov, 1968.)

2.1.12. Physical properties of sea ice

When studying the ice that covers the seas of the higher latitudes, we must distinguish between fresh-water ice and sea ice. The fresh-water ice is predominantly glacier ice from precipitation, which is accumulated on land and then ejected into the sea by calving glaciers. Sea ice, on the other hand, is formed from freezing seawater with some share of precipitation that has fallen on the sea ice.

When the temperature of the sea surface reaches the freezing point (cf. Section 2.1.4), the formation of pure ice begins. Seawater is trapped among the ice crystals and, on further cooling, the various salts contained in the water by and by crystallize, for instance, $Na_2SO_4 \cdot 10H_2O$ at $-8.2°C$ and $NaCl \cdot 2H_2O$ at $-22.9°C$. Below $-55°C$, the remaining brine is completely frozen, but this condition is rarely reached in situ.

During the freezing process the structure of the ice in deeper layers is such that drains are formed which are predominantly vertical. In this system the brine is gradually seeping out of the sea ice, owing to diffusion and convection (Untersteiner, 1967; Lake et al., 1970). Therefore, the salt content of sea ice is very much smaller than that of seawater. Furthermore, air dissolved in seawater is incorporated in the ice during the freezing process. In addition, air directly from the atmosphere takes the place of the brine that has seeped out. Thus, at temperatures between the freezing point and $-8.2°C$ sea ice consists of pure ice, brine, and air bubbles, and below $-8.2°C$ it also contains salt crystals with a composition that depends on temperature. In contrast to the physical properties of seawater, those of sea ice are not solely determined by salinity, temperature, and pressure, but also by the air content and partly by the special structure of the ice. Therefore, they also depend on the age and the history of the generation of sea ice.

The gradual seepage of the brine is the reason why the salt content of sea ice is smaller the slower the freezing process, the older the sea ice, and the greater its distance from the water–ice interface. When sea ice warms up, pure ice is dissolved in the brine; the ice gets more porous, and the brine escapes more quickly. Thus finally, almost pure fresh water can develop at the surface of sea ice.

Typical salinity values of sea ice are 2‰ for very old ice, 3–5‰ for ice after the first winter, 10‰ for new ice when it begins to develop, and 20‰ for ice that is formed from seawater flooding sea ice.

To a large degree the density of sea ice depends on its content of air and salt and, near the freezing temperature, also essentially on temperature. According to Schwerdtfeger (1963) the following equation holds:

$$\rho_I = (1 - \alpha)\left(1 + \frac{4.56S}{T}\right) 0.917 \qquad (2.37)$$

where ρ_I is the density in g cm^{-3}, α is the air content in volume percent, S is the salinity in ‰, and T is the temperature in °C. Figure 2.24 shows the functional relation between α and T for selected values of ρ_I and S. The rapidity of the ice formation has comparatively little influence on the density, because faster freezing leads to a higher content of brine and simultaneously to a higher content of air, and because these two effects act on density in an opposite sense.

The thermal properties of sea ice are of great importance for the heat budget of ocean and atmosphere in higher latitudes. The specific heat at temperatures just below the freezing point is very much greater than that of pure ice, because with decreasing temperature, the heat balance is almost exclusively determined by the freezing of more brine. Thus, the specific heat is practically only a function of temperature. At very low tem-

86 Physical Properties and Chemical Composition of Seawater

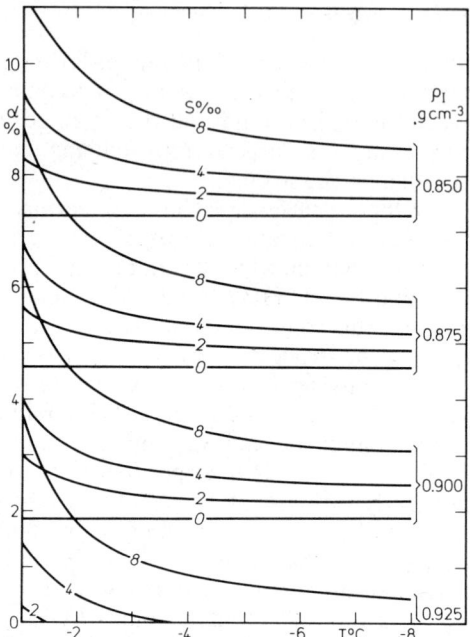

Fig. 2.24. Air content α of sea ice as a function of the temperature T for selected values of the density ρ_I and the salinity S. (After Schwerdtfeger, 1963.)

peratures its value approaches that of pure ice since the brine content becomes small. Figure 2.25 (Schwerdtfeger, 1963) shows the specific heat of sea ice and of pure ice as a function of temperature for selected salinities. In contrast to pure ice, sea ice has no definite melting point since, when being warmed, the ice will pass into the liquid phase only gradually because of the heterogeneous proportions of salt.

The thermal conductivity of sea ice increases with decreasing temperature, decreasing salinity, and decreasing air content and, in addition, it is influenced by the particular structure of the ice. Near the freezing point the dependence on temperature is especially great. At very low temperatures, however, the thermal conductivity of sea ice very closely approaches that of pure ice (Schwerdtfeger, 1963; Lewis, 1967). Characteristic values lie between 15×10^{-3} and 25×10^{-3} W cm^{-1} °C^{-1}. Thus, the thermal conductivity of sea ice is only slightly greater than the molecular thermal conductivity of seawater and therefore sea ice represents a very good natural protection of the polar seas against further cooling.

The absorption of radiative energy by sea ice depends on how far the ice is covered by water. The albedo A (cf. Section 2.1.11) is chiefly a function of the ratio R of the ice surface covered by water to the whole surface of the sea ice. Langleben (1971) found the following relation for the shortwave radiation at high altitude of the sun:

$$A = 0.59 - 0.32R \tag{2.38}$$

The mechanical properties of sea ice are important with regard to its form and distribution. Under quickly changing stresses sea ice behaves like an elastic body, but when the stress varies slowly, it is permanently deformed by forces of shear and tension, which may even result in breaking. Within the temperature range down to −22.9°C the breaking

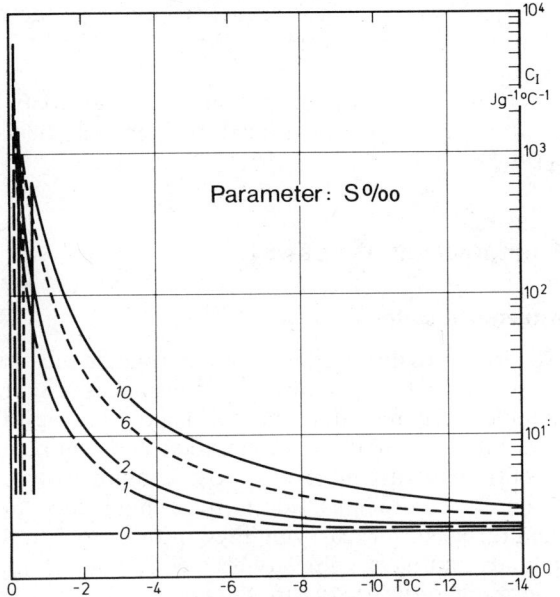

Fig. 2.25. Specific heat c_I of sea ice as a function of the temperature T for selected salinities S. (After Schwerdtfeger, 1963.)

strength, which equals the magnitude of the tensile stress at breaking, is smaller with sea ice than with pure ice. The decrease of the strength is proportional to the square root of the brine content of sea ice. Characteristic values of the bending and tensile strengths lie around 10 bar (0.981 bar = 1 kP cm^{-2}), but very much smaller values may also occur.

The elastic properties can be described by the elasticity modulus E and Poisson's elasticity number μ. With a tension P_x in the x direction, the relative change e_x of the length of a body in the x direction results from

$$P_x = E e_x \tag{2.39}$$

The ratio between the transverse contraction and the linear extension is given by μ. E and μ are determined by measurements of the phase velocity of transversal and longitudinal seismic waves in sea ice. According to Langleben (1962) the following equation holds for E as a function of the brine content ν:

$$E = 10^4(10.00 - 0.351\nu) \tag{2.40}$$

where E is the elasticity modulus in bar (1 bar = 10^6 dyn cm^{-2}) and ν the brine content in %. Very old sea ice yields somewhat smaller values. Poisson's elasticity number μ is fairly independent of the salinity, temperature, and age of the ice and has the value of $\mu = 0.3$.

The electrical properties of sea ice depend on the brine content and therefore on temperature and salinity as well as on the structure of the ice (Fujino, 1967; Addison et al., 1967). The dielectric constant increases with rising temperature and increasing salinity. Both these electrical properties change their behavior with the transition to temperatures beneath $-22.9°C$ and to frequencies of the alternating electric field that exceed 1 to 10 kHz. Typical values at low frequencies and high temperatures range between 10^4 and 10^6 for the dielectric constant and between 1 and 10^{-2} mS cm^{-1} for the specific electrical conductivity.

Comprehensive reviews on the physical properties of sea ice can be found in the following publications: Malmgren (1927), National Academy of Sciences (1958), Pounder (1965), and Oura (1967).

2.2. Chemical Composition of Seawater

2.2.1. Major constituents of salinity

In addition to atmospheric gases dissolved in seawater, organic substances, and suspended particles insoluble in seawater, the accompanying substances of seawater are composed predominantly of soluble inorganic salts. The world ocean has rightly been called "Nature's big sink": as a result of evaporation, precipitation, and continental runoff, the water masses at the surface of the earth are subject to a large continuous cycle (cf. Section 4.3.6). Hereby the runoff from the continents is constantly being loaded anew with salts originating from continental rocks, thus carrying them into the ocean. Furthermore, there are inorganic compounds that have primarily been emitted into the atmosphere by volcanism and then have reached the ocean through precipitation. Thus, during the course of the earth's history, the ocean has become enriched with more and more salts, a process that is still taking place today. As seen from the viewpoint of geological periods, the mixing of the world ocean through the effect of currents and turbulence is a perfect process. Any possible local differences in the amounts of various types of salts found in seawater vanish in the total water mass due to mixing in a relatively short time. It has been confirmed again and again by numerous analyses of seawater from different parts of the world ocean that, within very small deviations (less than 0.005‰), we are justified to speak of the "constancy of the composition of sea salt" (cf. Section 2.1.2).

According to what has been said before, it is to be expected that all chemical elements present on earth will be encountered in seawater, although in widely differing concentrations. For reasons of expedience, we distinguish between chief constituents and trace elements. The value of about 1 mg of salt per liter or kg of seawater has been agreed upon as the boundary between these two groups, although silicon (3 mg l^{-1}), as a biological minimum element, is included in the group of trace elements. While the chief constituents, due to their great mass, mainly influence the physical properties of seawater (cf. Section 2.1.2), the importance of the trace elements lies especially in their role in marine biology.

Owing to the considerable electrical dissociative power of water (cf. Section 2.1.1), the different salts are dissolved in the water in disperse ionic forms and, in a physico-chemical respect, they are subject to the law of mass action. A detailed description of the chief constituents has been given by Culkin (1965). The results of a great number of seawater analyses are represented in Table 2.03, where relative deviations of several percent occur for constituents present only in small amounts.

Including hydrogen and oxygen, seawater consists of 13 chief constituents. The other elements present in seawater as trace elements amount to less than 5 mg per kg of seawater, that is less than 0.2‰ of the total salt content.

The behavior of seawater as a physicochemical system is determined by the fact that the amount of cations expressed in milliequivalents per kg is greater than that of anions.

Therefore, seawater is alkaline, with a pH value always greater than 7; the typical value is pH = 8.2. Thus, in contrast to the solution of common salt, seawater has a strong buffering capacity towards the addition of acids or bases. The biological influence on the chief components of salinity remains within narrow limits. Variations of alkalinity mainly indicate a change of the carbon dioxide content, or the precipitation or solution of calcium. As a consequence, the specific alkalinity in the ocean ranges between 0.110 and 0.127 meq per kg and ‰ chloride content (Koczy, 1956).

2.2.2. Trace elements

Trace elements in seawater are all those elements with a concentration of less than 1 mg l^{-1}. Silicon, which as a biological minimum element is included in the group of trace elements, is an exception. Although all the elements of the periodic system must be present in seawater at some concentration, some of them have not yet been detected by analytical methods. However, including hydrogen and oxygen, the occurrence of 62 elements out of the 89 present in nature is known as far as the order of magnitude is concerned (Table 2.03).

In contrast to the chief constituents of salinity, which, due to their high concentration, greatly influence the physical behavior of seawater, the importance of the trace elements, the total concentration of which is less than 5 mg l^{-1}, that is, 0.01% of the total salt content, lies in the geochemical and biological fields. The processes of selective chemical adsorption (cf. Section 6.1.2) which control the sedimentation budget to a great extent, as well as the important phenomena of marine life, occur best under the conditions of low concentration as is the case with the trace elements.

2.2.3. Gases

Apart from the salts, the atmospheric gases, above all, play an important role in the physicochemical behavior of the ocean. At the air–sea interface the atmosphere is composed as follows: nitrogen 77.0% (by volume), oxygen 20.6%, argon 0.9%, hydrogen (mean value) 1.47%, carbon dioxide 0.03%, trace gases 0.0024% (all percentages by volume). If we disregard the content of water vapor, and if we, for the moment, include the small content of argon into that of nitrogen because of the similarity of their chemical behavior, only three constituents are left: nitrogen, oxygen, and carbon dioxide. All three of them differ with regard to their solubility in seawater. Nitrogen interacts with seawater practically solely on the basis of the physical gas laws. (For balance purposes, the process of N_2 fixation and the reverse process of nitrification, which takes place in the deep sea owing to the lack of oxygen, can be left out of consideration because of their small magnitude.) Therefore, the cycle of nitrogen in the ocean remains rather simple and clear, whereas atmospheric oxygen—once it has entered the ocean—participates very intensively in biochemical metabolic processes by interacting very actively with carbon dioxide (cf. Section 6.2.1). As for oxygen, which—like nitrogen—is found in the ocean only in a dissolved state, its conversion processes can be easily recognized, while with carbon dioxide the conversion processes are very complicated. During the process of dissolution, carbon dioxide is subject not only to physical laws but, at the same time, to the chemical law of mass action due to the fact that it is in reciprocal equilibrium with the carbonate and bicarbonate ions dissolved in seawater (cf. Section 6.2.6).

The basis for an understanding of the behavior of atmospheric gases in the ocean is

Table 2.03. Composition of Seawater [at a chlorinity of 19,000 mg l^{-1} according to Goldberg (1965) and Culkin (1966)]

Element	Symbol	(mg l^{-1})	Element	Symbol	(mg l^{-1})
Hydrogen	H	108,000.	Silver	Ag	0.00004
Helium	He	0.000005	Cadmium	Cd	0.00011
			Indium	In	0.02
Lithium	Li	0.17	Tin	Sn	0.0008
Beryllium	Be	0.0000006	Antimony	Sb	0.0005
Boron	B	4.6	Tellurium	Te	
Carbon	C	28	Iodine	J	0.06
Nitrogen	N	0.5	Xenon	Xe	0.0001
Oxygen	O	857,000			
Fluorine	F	1.3	Caesium	Cs	0.0005
Neon	Ne	0.0001	Barium	Ba	0.03
Sodium	Na	10,721	Lanthanum	La	1.2 × 10^{-5}
Magnesium	Mg	1,350	Cerium	Ce	5.2 × 10^{-6}
Aluminum	Al	0.01	Praseodymium	Pr	2.6 × 10^{-6}
Silicon	Si	3.0	Neodymium	Nd	9.2 × 10^{-4}
Phosphorus	P	0.07	Promethium	Pm	
Sulphur	S	901	Samarium	Sm	1.7 × 10^{-6}
Chlorine	Cl	19,000	Europium	Eu	4.6 × 10^{-7}
Argon	Ar	0.6	Gadolinium	Gd	2.4 × 10^{-6}
			Terbium	Tb	
Potassium	K	398	Dysprosium	Dy	2.9 × 10^{-6}
Calcium	Ca	410	Holmium	Ho	8.8 × 10^{-7}
Scandium	Sc	0.00004	Erbium	Er	2.4 × 10^{-6}

Titanium	Ti	0.001		Thulium	Tm	5.2×10^{-7}
Vanadium	V	0.002		Ytterbium	Yb	2.0×10^{-6}
Chromium	Cr	0.00005		Cassiopeium	Cp	4.8×10^{-7}
Manganese	Mn	0.002				
Iron	Fe	0.01		Hafnium	Hf	
Cobalt	Co	0.0001		Tantalum	Ta	
Nickel	Ni	0.002		Tungsten	W	0.0001
				Rhenium	Re	
Copper	Cu	0.003		Osmium	Os	
Zinc	Zn	0.01		Iridium	Ir	
Gallium	Ga	0.00003		Platinum	Pt	
Germanium	Ge	0.00006				
Arsenic	As	0.003		Gold	Au	0.000004
Selenium	Se	0.0004		Mercury	Hg	0.00003
Bromine	Br	67		Thallium	Tl	<0.00001
Krypton	Kr	0.0003		Lead	Pb	0.00003
				Bismuth	Bi	0.00002
Rubidium	Rb	0.12		Polonium	Po	
Strontium	Sr	7.7		Astatine	At	
Yttrium	Y	0.0003		Radon	Rn	0.6×10^{-15}
Zirconium	Zr					
Niobium	Nb	0.00001		Francium	Fr	
Molybdenum	Mo	0.01		Radium	Ra	1.0×10^{-10}
Technetium	Tc			Actinium	Ac	
Ruthenium	Ru			Thorium	Th	0.00005
Rhodium	Rh			Protactinium	Pa	2.0×10^{-9}
Palladium	Pd			Uranium	U	0.003

to be found in the exchange processes at the sea surface where the liquid and gaseous phases are in contact with each other. Here the state of saturation of the various gases is established, in accordance with the relevant temperature and salinity, either by the absorption of gases from the atmosphere or by the release of gases into it. As shown in Table 2.04, the saturation values* decrease when temperature and salinity increase, which, as far as the dependence on temperature is concerned, again points to some anomaly of the water. With ordinary fluids it is the other way round: the saturation values increase with increasing temperature. It is important to note that the relative content of the chemically and biologically active gases (oxygen and carbon dioxide) is higher in water than in the atmosphere. While in the atmosphere the ratio of oxygen to nitrogen is approximately 1:4, it is increased in solution to 1:2. Still more pronounced is the behavior of carbon dioxide. Its absolute solubility in seawater is even of the same order of magnitude as in the atmosphere. In this context, we have so far been dealing only with the free carbon dioxide, dissolved in the gaseous state, without taking into account the further, very considerable increase of solubility which is a result of its chemical interaction (cf. Section 6.2.6).

If the oceanic water cycle is followed further, the fundamental dissimilarities of the various kinds of gases become apparent. Whereas the gaseous nitrogen content remains at full saturation even at great depths (no significant changes indicating participation in biological processes could be shown in experiments), oxygen, on its path through the ocean, participates very actively in the biological processes of oxidation and respiration occurring everywhere. As a result, the depletion of oxygen from oceanic water masses is the greater, the longer they have been prevented from interacting with the atmosphere at the sea surface, and the more intensive marine life has been in the areas they have traversed on their path through the depths of the ocean. Under extreme conditions, as, for example, in the Black Sea, in the Norwegian fjords, and in the deep basins of the Baltic Sea complete oxygen depletion can actually occur and, as a consequence, hydrogen sulfide is released, making all organic life impossible whether floral or faunal.

If we consider the conditions existing for carbon dioxide, we can see that, on the basis of the biological equilibrium between oxygen and carbon dioxide, an abundance of carbon dioxide must be present where an oxygen deficiency exists and vice versa (see Section 6.2.1).

As has been shown by Rakestraw et al. (1951), many processes of biological conversion are accompanied by a shift in the ratio of the isotopes depending on the intensity of these processes. From the ratio $^{18}O/^{16}O$ in the calcareous shells of foraminifera, conclusions can be drawn as to the temperature conditions under which those organisms had lived.

König (1963) has attempted to determine the solubility of rare gases, all of them, except argon, belonging to the group of trace elements in seawater. With growing atomic weight a strong increase of solubility has been observed, as shown in the following short list (which refers to normal conditions: 19.12‰ chloride content, 0°C, and 760 torr). Solubilities at 10°C in cm^3 kg^{-1}: He = 7.40, Ne = 8.67, Ar = 32.8, Kr = 58.2, and Xe = 103.

---

* In oceanography the concept of "gas saturation" is always based on the mean atmospheric pressure of one atmosphere (760 mm Hg or 760 torr) acting on the sea surface. No consideration is given to the hydrostatic pressure which increases with depth and correspondingly influences the saturation of gas, since the exchange with the atmosphere under the given natural conditions can take place only at the sea surface. [For more detail see Kalle (1945), p. 52–53.]

Table 2.04. Concentration of the Most Important Atmospheric Gases in the Atmosphere as Well as their Saturation Values in Fresh Water and Seawater (reduced to dry atmosphere, 1013 mb and 0°C) (after Fox, 1907)[a]

Gas	Concentration in Atmosphere		Saturation in Fresh Water, 0‰ S					
			0°C		10°C		30°C	
	cm³ l⁻¹	vol %	cm³ l⁻¹	vol %	cm³ l⁻¹	vol %	cm³ l⁻¹	vol %
Nitrogen	780.9	78.09	18.10	61.4	14.60	62.4	10.98	63.8
Oxygen	209.5	20.95	10.29	35.0	8.02	34.3	5.57	38.2
Argon	9.3	0.93	0.54	1.8	0.42	1.8	0.30	1.8
Carbon dioxide	0.3	0.03	0.52	1.8	0.36	1.5	0.20	1.2
	1000.0	100.00	29.45	100.0	23.40	100.0	17.05	100.0
			Saturation in Seawater, 35‰ S					
Nitrogen			14.04	61.2	11.72	62.6	9.08	65.1
Oxygen			8.04	35.1	6.41	34.2	4.50	32.2
Argon			0.41	1.8	0.31	1.6	0.21	1.5
Carbon dioxide[b]			0.44	1.9	0.31	1.6	0.18	1.2
			22.93	100.0	18.75	100.0	13.97	100.0

[a] The old saturation values by Fox (1907) can still be considered valid, although recently various attempts have been made to carry out improved determinations of the saturation values of gases in seawater. So far no final agreement has been attained among oceanographers (cf. Richards, 1965).
[b] The saturation values of CO_2 refer to a NaCl solution of 35‰ salinity.

2.2.4. Organic substances

All living organisms and, to an even greater extent, all dead organisms present in seawater constantly release small amounts of metabolic and decay products into the surrounding water. While, shortly afterwards, a large part of these substances is subjected to processes of degradation by bacteria and transformed into basic inorganic substances, the remaining part may, for a very long time, resist complete degradation. This "conserving" effect of seawater is favored by the fact that the concentration of dissolved organic substances in seawater is generally relatively low and, furthermore, that in the usually rather clear water only a small amount of suspended substances is present, which, at their surfaces, can enrich the organic substances by adsorption and offer the bacteria the solid nucleus to which they can cling and which they need for their life process.

The following rough calculation conveys a picture of the conditions in the sea. As proved by numerous investigations, a content of dissolved organic carbon of an order of magnitude of 1 mg l^{-1} can be assumed for the open ocean (Fig. 2.26). In terms of the turnover of substances in the ocean, this amount represents a surprisingly high value. Assuming that all organic carbon originates from the assimilation processes of plants in the near-surface zone of the ocean and that about $9/10$ of those assimilation products are subject to biological decomposition when passing through the food chain, we may expect an annual supply of 0.8 mg C cm^{-2} yr^{-1} from which the carbon, remaining in solution, is supplemented (average rate of production in the ocean: 8 mg C cm^{-2} yr^{-1}). With a value of 1 mg C l^{-1} and a mean depth of 3800 m we obtain the value of 380 mg C cm^{-2} of the sea surface as the average stock of carbon dissolved in seawater. Thus the mean life time of the carbon compounds can be calculated, it is $380:0.8 = 475$ yr. This means that, at least in the deep sea, the degradation processes must be greatly impeded. As will be shown later, this is all the more striking, because the greater part of these organic substances does not include particularly stable compounds or compounds which are not biologically utilizable, but rather mainly structural biological material such as carbohydrates, lipids, and amino acids.

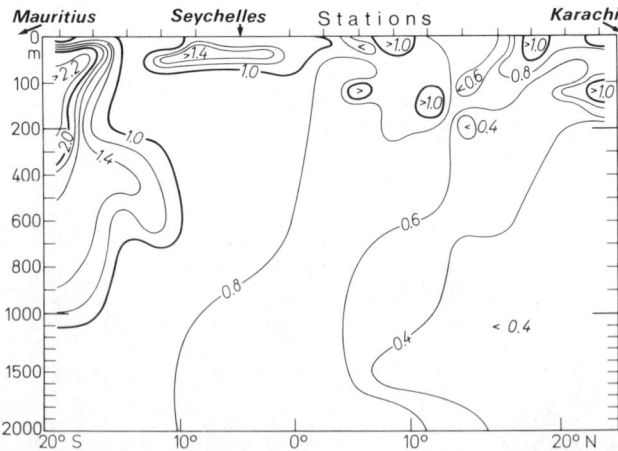

Fig. 2.26. Dissolved organic carbon in milligram C per liter on a section across the western Indian Ocean from Mauritius (20°S) to Karachi (25°N), according to the investigations carried out with research vessel *Anton Bruun* during the International Indian Ocean Expedition in October/November 1963. (After Menzel, 1964.) Pay attention to the three different depth scales: 0–200, 200–1000, and 1000–2000 m.

Table 2.05. Share of the Various Fatty Acids at Different Depths in the Gulf of Mexico[a]

Fatty acid[b]	Share in percent			
Depth (m):	10	300	900	1900
C.10	—	—	7	—
C.10 =	0	0	6	6
C.12	6.5	12	42	94
C.14	20	11	15	0
C.14 =	4	7	0	0
C.16	35.5	35	22	0
C.16 =	16	14	0	0
C.18	8.5	9	0	0
C.18 =	1.5	5	0	0
C.18 = =	2	0	0	0
C.?	6	5	8	0
Total concentration in mg l^{-1}	0.5	0.4	0.5	0.3

[a] Slowey et al., 1962.
[b] = means one carbon double bond and = = means two carbon double bonds in the molecule.

Up to the 1960s only some general results about the distribution of organic compounds dissolved in seawater had been known, but since then great progress has been made in this branch of marine chemistry after suitable microchemical methods of analysis had been developed. Some examples will illustrate this.

The vertical distribution of various fatty acids in the Gulf of Mexico is represented in Table 2.05. Accordingly, in the Gulf of Mexico the total content of fatty acids in the entire water column from the sea surface down to a depth of 1900 m showed amazingly high values between 0.5 and 0.3 mg per liter of seawater. What is remarkable is the shift in the ratio of high and low molecular fatty acids expressed as percentage. While in the surface water the maximum lies with the palmitic acid, containing 16 carbon atoms, and the monounsaturated acid corresponding to it, with increasing depth the ratio is shifted towards low molecular carbon chains and further to fatty acids with 12 carbon atoms (lauric acid). This points to the possibility that with increasing age a slow degradation of the carbon chain may take place.

Other organic substances dissolved in seawater also show a remarkably high concentration, as demonstrated by examples from the North Pacific Ocean, where the following maximum values of mass density in g cm^{-3} have been obtained: glycollic acid 140×10^{-8}, acetic acid 82×10^{-8}, formic acid 32×10^{-8}, and lactic acid 8×10^{-8} (Koyama et al., 1959). During the first deep-sea drilling (Project Mohole, 1961) off Lower California at a water depth of 3720 m, the vertical distribution of organic substances was determined (Degens et al., 1964). The examples in Fig. 2.27 show that the concentration of the total amino acids lies within mass densities of 10^{-7} and 10^{-8}, while that of the sugar species is somewhat lower. With increasing depth the concentration of the amino acids as well as of the sugars show a clear increase. In Fig. 2.27 a fact is confirmed that is well known from other oceans, namely that the content of particulate substance lies considerably below the value of the dissolved organic substance. A rule of thumb for the ratio of these contents says: the ratio of dissolved organic carbon and total particulate organic carbon and living organic carbon is 100:10:1, while in the sediments the concentration of the two compound groups is higher by three to four orders of magnitude.

Some other organic compounds detected as dissolved in seawater may be mentioned:

96 Physical Properties and Chemical Composition of Seawater

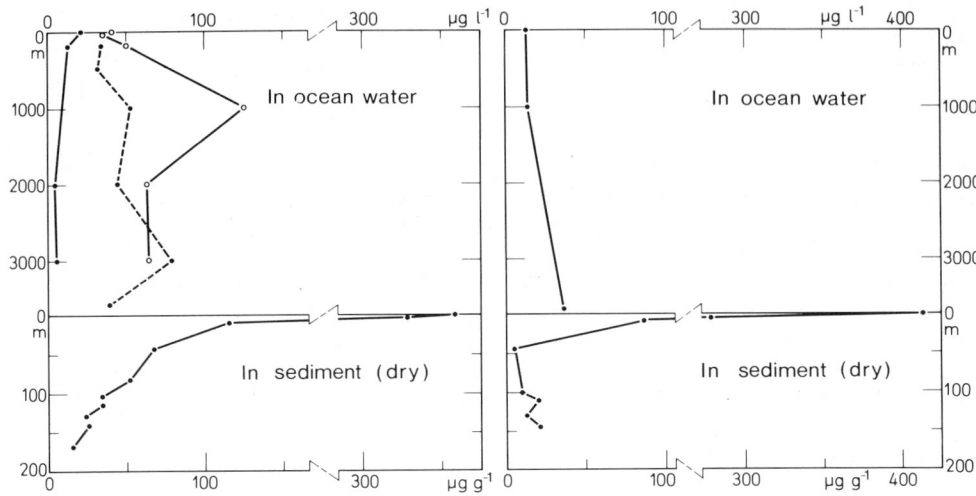

Fig. 2.27. Vertical distribution of organic substances in the East Pacific Ocean in seawater and in sediment (Mohole drilling). (After Degens et al., 1964.)
Dots: Measured values at $\varphi = 32°32'N$, $\lambda = 120°30'W$, depth 3720 m.
Circles: Measured values at $\varphi = 29°13'N$, $\lambda = 117°37'W$, depth 3990 m.
Left: Total amino acids; left-hand curve: in suspended form; the two right-hand curves: in dissolved form.
Right: Sugar in dissolved form.

urea, the three phenol acids p-hydroxybenzoic acid, syringic acid, and vanillic acid, and also humic degradation products. The latter are characterized by their yellow color and by their capability to fluoresce blue in ultraviolet light. While the types of compounds discussed up to now have mostly been investigated by chromatographic methods, whereby the lowest mass density attained was 10^{-9}, specific biological methods have permitted investigations into the vitamins in seawater at sensitivities two to three orders of magnitude higher (i.e., 10^{-12}–10^{-13}). Since in the minimum case the threshhold value, at which the receptive organs of highly specialized organisms may still respond to specific substances in their surroundings, lies at a mass density of approximately 10^{-18} (odoriferous substances, pheromones, food-indicating substances, and neurosecretions), many problems remain to be solved in the future when analytical methods have been improved.

2.2.5. Suspended matter and living organisms in seawater

Apart from dissolved additives, ocean water always contains particles in various states of suspension, ranging from the smallest microscopic order of magnitude up to the giants among the marine animal organisms, the whales—if such a subdivision can be strictly followed. What is of special importance in physicochemical respect is, above all, the turbidity which occurs in seawater as a result of the suspension of finer particles. This fraction is particularly significant in the formation of the sediments deposited by and by at the sea floor (cf. Section 6.1.2). In oceanography, the turbidity is called "seston" and is characterized by its capability to be passively suspended in seawater. It contains the following components:

1. Mineral turbidity, which is carried into the sea by rivers and sometimes transported far out into the ocean by the winds.

2. Detritus, the fine residua, organic and inorganic, originating from the decomposition of dead organisms. A special kind of detritus is the so-called "sea snow," which in the surface layers of the ocean, under the influence of small ascending air bubbles, separates from the organic excreta of plankton organisms (Riley, 1963).
3. Nanoplankton (dwarf plankton), mesoplankton, and macroplankton, which are carried along in passive suspension by the ocean currents, quite in contrast to the "necton," larger organisms which—like fish—can change their position on their own by active motion. The typical values of plankton diameters range from 1 μm to 1 mm. Here it must be noted that the smallest organisms occurring in the ocean, like bacteria, yeasts and fungi, cannot develop in seawater independently, but are mostly associated with the detritus because they need a solid base for their existence.

The total turbidity, which can easily be measured by optical means (cf. Section 3.2.13), is a valuable indicator for the identification and localization of certain water masses in the ocean. It is usually well pronounced in coastal waters, especially in the vicinity of major river mouths, and it normally decreases with increasing distance from the shore. In the lower course of tidal rivers the turbidity suddenly increases, whereas it decreases upstream. Minimum turbidity values are found in subtropical oceanic regions, poor in life and characterized by a deep blue color.

The living organisms in the oceans are dealt with by a special branch of science, that is, marine biology, which will be briefly discussed here only as far as, from the physicochemical point of view, it is important for the chemical budget of the ocean (cf. Section 6.2). In contrast to the conditions on the continents, where photosynthesis, of fundamental importance for the life processes mentioned above, is carried out by more highly organized plants, the life process in the ocean depends almost exclusively on microscopically small monocellular algae. These algae live in the ocean in great numbers, and exhibit great variation in their horizontal and vertical distribution.

Everywhere at the surface the ocean contains microscopic planktonic organisms with more than 1000 cells l^{-1}, while at 50 m depth there are rarely less than 1000 cells l^{-1}. In coastal areas the content may increase to 2×10^6, sometimes even 6×10^6 cells l^{-1}. In spite of such great numbers, the mass of the organisms is relatively small. On the average, approximately 0.0002 mg of dry weight of living substance corresponds to 1000 cells.

In order to give the reader a general idea of the abundance of planktonic forms in the ocean appreciable to the human eye when centrifugal separates and net catches are examined under the microscope, some typical major forms of North Sea plankton are reproduced in Fig. 6.13 according to their relationship and dependence.

3 Oceanographic Instruments and Observational Methods

3.1. Platforms for Oceanographic Observations

3.1.1. Research vehicles

3.1.1.1. Research vessels

In oceanography, platforms are required as instrument carriers and, to some extent, also directly as measuring devices. The former group includes research vessels and anchored measuring systems, called moorings; the latter mainly consists of drifting buoys.

The *requirements* that a research vessel must fulfill depend on the kind of tasks as well as on the sea area and its wave and climatic conditions. The tasks call for two fundamentally different types of research vehicles, namely (1) research vessels, which prevail in number and importance, and (2) submersibles, which have recently been used more frequently.

Research vessels can be of very different sizes according to their operational areas. Near the coast and in protected waters simple, small vessels will do. For work in the open ocean bigger seaworthy vessels with a larger range of action are required. But even the largest oceanographic research vessels are still so small that with rough seas the ship's motions limit the research work. In order to save costs, freighters and naval ships have often been converted for oceanographic purposes. Most of the modern oceanographic vessels, however, were designed for the specific tasks of oceanography. The total number of research vessels is about 500; around 150 of them are larger ones.

If oceanographic methods of observation are to be applied on board, the ship must have a number of fundamental qualifications: (1) it must be easy to maneuver; (2) it must be equipped with modern navigational and echo-sounding devices; (3) the working decks must be open and as close as possible to the sea surface; (4) special winches and gear for handling heavy equipment must be available. The laboratories should preferably be located in the quietest part of the ship.

Many research vessels were designed as multipurpose ships for application in all the branches of marine science. Therefore, they contain versatile laboratory facilities as well as a great variety of winches. Some vessels, however, were built to serve special purposes, for example, for the surveying of the ocean bottom, deep-sea drilling, observations in areas covered by ice, or they have auxiliary functions for the work with submersibles.

In the following, only a few *examples of research vessels* will be described because, especially in recent time, the number of research ships in the world has increased so much that even the enumeration of only the most important ones would be beyond the scope of this book. The most important German vessels and their operational areas are listed in Table 3.01, while Table 3.02 contains a selection of research ships of other nations. In the history of oceanography, we find many ships whose names are connected with great

Table 3.01. Ships of German Oceanic Research Cruises

Ship	Period of Cruise	Tonnage[a]	Operational Area
Gazelle	1874–1876	1900 t	World ocean
National	1889	835 grt	Atlantic Ocean
Valdivia	1898–1899	2176 grt	Atlantic and Indian Oceans
Gauss	1901–1903	1332 t	Atlantic and Indian Oceans
Planet	1906	650 t	Atlantic and Indian Oceans
Möwe	1911	650 t	Atlantic and Indian Oceans
Deutschland	1911–1912	598 grt	Atlantic Ocean
Meteor	1925–1938	1178 t	Atlantic Ocean
Altair	1938	4000 grt	Atlantic Ocean
Gauss	since 1949	846 grt	Atlantic Ocean
Anton Dohrn	1955–1972	999 grt	Atlantic Ocean
Walther Herwig	1963–1972	1987 grt	Atlantic Ocean
Meteor	since 1964	2615 grt	Atlantic and Indian Oceans
Planet	since 1967	1852 grt	Atlantic Ocean
Komet	since 1969	1252 grt	Atlantic Ocean
Valdivia	since 1969	1300 grt	World ocean
Anton Dohrn (former Walther Herwig)	since 1972	1944 grt	Atlantic Ocean
Walther Herwig	since 1972	2251 grt	Atlantic Ocean

[a] t = weight tonnes water displacement, grt = gross tons.

expeditions (Wüst, 1964). This may be attributed to the widely practised custom of publishing the results of an expedition in a comprehensive report under the name of the relevant ship. Thus we have the *Challenger* reports, the *Meteor* reports, and numerous other series of publications of such kind.

As a first example, a small ship will be described (cf. Dietrich et al., 1966), which was constructed for work in the North Sea and the Baltic Sea. The German research

Table 3.02. Ships of Selected Research Cruises of Other Nations

Ship	Period of Cruise	Nationality	Tonnage[a]	Operational Area
Challenger	1873–1876	United Kingdom	2306 t	World ocean
Fram	1893–1896	Norway	530 t	Arctic Ocean
Carnegie	1928–1929	U.S.A.	568 t	Pacific Ocean
Will. Snellius	1929–1930	Netherlands	1055 t	Indian Ocean
Pourquoi-Pas	1911–1936	France	500 grt	Atlantic Ocean
Dana II	1921–1935	Denmark	360 grt	World ocean
Sedow	1938–1940	USSR	1538 grt	Arctic Ocean
Albatross	1947–1948	Sweden	1450 t	World ocean
Mikhail Lomonosov	since 1957	USSR	5960 t	World ocean
Discovery III	since 1962	United Kingdom	2800 t	Atlantic and Indian Oceans
Atlantis II	since 1963	U.S.A.	2300 t	World ocean
Hudson	since 1963	Canada	4660 t	Atlantic Ocean
Jean Charcot	since 1965	France	2200 t	Atlantic Ocean
Ryofu Maru	since 1966	Japan	1598 t	Pacific Ocean
Professor Viese	since 1966	USSR	6934 t	World ocean
Glomar Challenger	since 1968	U.S.A.	10500 t	World ocean
G.O. Sars	since 1968	Norway	1500 grt	Atlantic Ocean
Knorr	since 1970	U.S.A.	2075 t	World ocean

[a] t = weight tonnes water displacement, grt = gross tons.

100 Oceanographic Instruments and Observational Methods

Fig. 3.01. Research cutter *Alkor*, a multipurpose vessel, 31 m in length, 236 gross tons, for research in the North Sea and the Baltic Sea (Photo: Institut für Meereskunde, Kiel).

cutter *Alkor* (Figs. 3.01, 3.02) is a multipurpose vessel, 31 m in length and of 236 gross tons. She has a good maneuverability because of a variable pitch propeller and a Becker single-blade fin rudder. An asymmetrical deck with a low freeboard of approximately 1 m above water level facilitates working over the side and the stern. The equipment includes two trawl and cargo winches, two wire and cable winches, one derrick and two tiltable A-frames as well as modern navigational facilities like Decca (cf. Sections 3.1.1.3) and an automatic weather station and a facsimile recorder for weather charts. A laboratory for electronic equipment and a wet laboratory are located in the quietest part of the ship, where, in addition, care was taken to prevent vibration and noise as much as possible by a special mounting of the diesel engines. The vessel has a sea endurance of 10 days and a cruising speed of 10 knots (1 knot = 1 nautical mi hr^{-1} = 1.852 km hr^{-1}). The ship's crew consists of eight men. Up to 12 scientists can be accommodated on board.

As some examples of typical deep-sea research vessels the following group will now be described: The German *Meteor* (Figs. 3.03, 3.04), the American vessels *Atlantis II* and *Knorr* (Fig. 3.05), the British *Discovery III* (Fig. 3.06), and the French *Jean Charcot* (Fig. 3.07). The ships' tonnages range between 2000 and 3000 tonnes, the lengths between 60 and 100 m. Larger vessels are often used in the USSR (cf. Table 3.02).

In addition to the common propulsion by propeller with horizontal axis, the vessels of the aforementioned group, in order to improve maneuverability, have bow thrusters and active rudders permitting the vessels to traverse. As a specialty the *Knorr* possesses two Voight–Schneider propellers with vertical axes which allow maneuvering in any direction. Apart from the usual cranes and winches, the *Jean Charcot* is equipped at the stern with two heavy-weight cranes for the handling of submersibles. Details of the vessels' properties will be illustrated by the example of the *Meteor*.

The research vessel *Meteor* has an overall length of 82 m, a beam of 13.5 m, a draft of 4.8 m, and a displacement of 3020 tonnes (Johannsen et al., 1965). She has a diesel

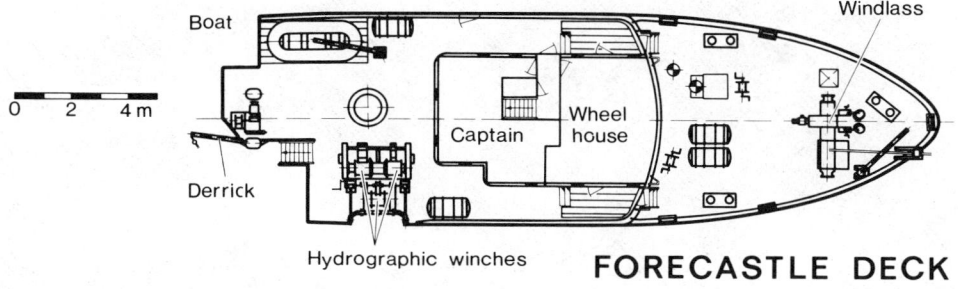

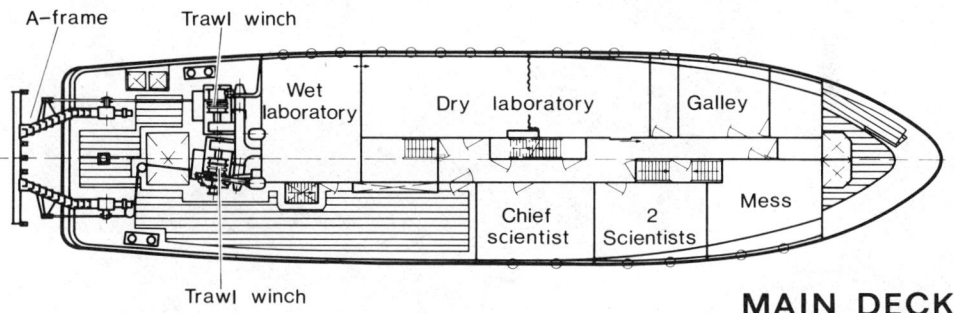

Fig. 3.02. Inboard profile of the research cutter *Alkor* (cf. Fig. 3.01).

electric engine plant producing 2000 S.H.P. (shaft horse power) and, in addition to the single-propeller propulsion, an active rudder and two bow thrusters for traversing. At a cruising speed of 12 knots, her operating range is more than 12,000 nautical mi. The main working deck, which measures 24 × 4.5 m, lies on the starboard side, only 2.5 m above water level. Another smaller working deck is at the stern. Nearly all the laboratories are located in the quietest part of the ship near the main deck. Furthermore, the ship is equipped with a meteorological station, a sounding and sonar room as well as a photo laboratory, a drafting room, an electronics workshop, a mechanical workshop, and a scientific library.

The vessel has a deep-sea winch with a tension load of 12 tonnes as well as a variety of other deep-sea winches for operation with wires and electric cables. Numerous smaller winches and lifting gear, above all a crane on the main deck, complete this equipment. There is also a hangar as well as a landing deck for a helicopter. Apart from modern navigational aids including satellite navigation, the permanent equipment among others comprises a wind-weather radar, a deep-sea echo-sounder of high accuracy, a sea gravimeter, various laboratory facilities as well as aquariums of different kinds. The crew of 54 consists of the nautical personnel, sailors, engineers and technicians, and the general service personnel; there are also a doctor and a meteorologist, a weather radio operator, surveying technicians, and technical personnel for operating and servicing the scientific measuring equipment. Accommodations are provided for 24 scientists.

As an example of a specialized vessel, the American deep-sea drilling ship *Glomar Challenger* will be briefly described here (cf. Table 3.02, Fig. 1.13). The most significant feature of the vessel, which is 131 m long, is the drilling tower with a height of 59 m above

102 Oceanographic Instruments and Observational Methods

Fig. 3.03. Research vessel *Meteor* and research cutter *Hermann Wattenberg* at the pier of the Institut für Meereskunde an der Universität Kiel (Photo: Institut für Meereskunde, Kiel, Luftbild SH-438-13).

sea level and a shaft with an opening 7×7 m wide. More than 8000 m of drill pipe can be stored on board. For deep-sea drilling, the aids for keeping the ship exactly on position over the drilling hole are of particular importance. This is achieved by means of three hydrophones that take bearings of a sound transmitter positioned at the sea floor. The signals received on board are evaluated by a computer which directly steers the twin-screw propulsion and the four bow and stern thrusters. Modern navigational equipment, like satellite navigation (cf. Section 3.1.1.3) and a weather station with satellite picture receiver, are available on the vessel.

As a second example of a highly specialized vessel, the American auxiliary ship for submersibles, named *Lulu* (Fig. 3.05), may be mentioned here. This vessel is a catamaran, 32 m in length, with a displacement of 470 tonnes; the cruising speed is 6 knots. The deep-sea submersible *Alvin* (Fig. 1.07) can be lowered from the working deck down to the water level by means of a lift located amidships. The catamaran is equipped with container laboratories for maintenance service.

Furthermore, there are other specialized vessels which fully or partly serve scientific purposes, for example, vessels for mineral resources research especially designed for applied geophysics and geology; weather ships, lying at almost the same position all the time, with the primary task of collecting weather data for safeguarding the transoceanic

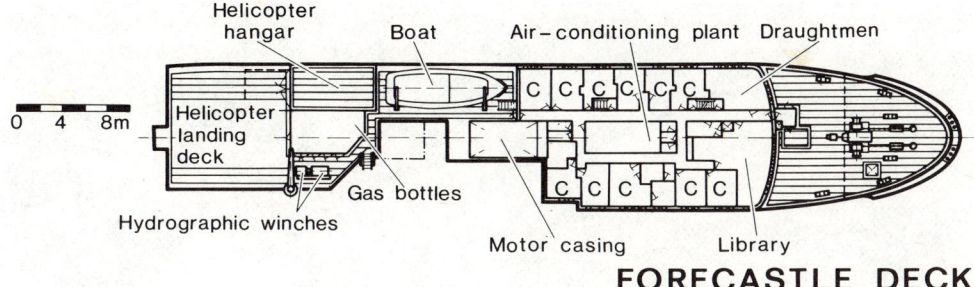

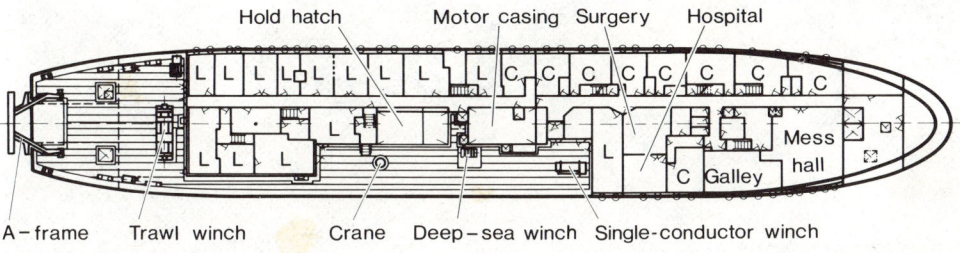

Fig. 3.04. Inboard profile of the research vessel *Meteor* (cf. Fig. 3.03). (C = cabin, L = laboratory.)

air traffic; survey vessels which carry out echo soundings for the compilation of sea charts; and, finally, light vessels which, as navigational aid for shipping, lie at selected positions and, thus, make long-time hydrographic measurements possible.

Summarizing descriptions of research vessels can be found in the collected publications of the IGY World Data Center A (1961, 1963, 1966) and in the compilations by Myers et al. (1969) and by Nelson (1971).

3.1.1.2. Submersibles

Manned submersibles and, in shallow waters, skin-divers, are important to marine sciences at those places where direct observation and sampling, aimed at a specific object, are required, and where this task cannot be fulfilled by remote-controlled, unmanned tools with television cameras. This, in particular, applies to a number of biological and geological investigations. Thus, submersibles make it possible to observe the behavior of marine organisms, to carry out population studies of benthic and pelagic forms of life, and to elucidate the causes for the formation of deep echo-scattering layers. In those layers some species of jellyfish, siphonophores, or lanternfish, have often proven to act as sound-scattering objects; but the same has been found to be true for other organisms, depending on sea area and time. Examples of the geological work with deep-sea submersibles can be seen in the observations of the local distribution of sediments and solid rocks at the ocean bottom, the precise sampling with grabbing or drilling devices, and the surveying of the microrelief of the sea floor which cannot be recorded by echo sounding from shipboard (cf. Section 3.2.4). Observations of turbidity currents, generated artificially by the influence of the vehicle, were of special interest.

The diving operations of *Trieste* in 1960, when a depth of 10,910 m was attained,

104 Oceanographic Instruments and Observational Methods

Fig. 3.05. Research vessels at the pier of the Woods Hole Oceanographic Institution, Woods Hole, Massachusetts, U.S.A.: Front left *Atlantis II*, behind *Knorr*, front right *Lulu*, right edge *Chain*, at the key pier in the background *Gosnold* (Photo: Woods Hole Oceanographic Institution).

have shown that any depth of the oceans can be reached by manned vehicles today. In accordance with the kind of problems in question, special types of submersibles have been designed for work over the shelf, at the continental slope as well as in deep-sea basins and deep-sea trenches. If we disregard the conventional submarines, which have occasionally been used for marine research, we find that work with the submersibles is rather restricted because of their relatively short operating time and their slow speed. The energy is usually supplied by lead or silver zinc batteries and the typical operative period is between 4 and 24 hr with an additional reserve for emergency cases. The typical cruising speed is only about 1 knot; the typical maximum speed is 2 to 3 knots. The crew consists of two to three, sometimes up to six men. The vehicles have portholes for visual and photographic observations. More recently constructed submersibles are equipped with manipulators and sampling devices as well as with oceanographic measuring facilities.

It has proven necessary to use auxiliary vessels at the sea surface when submersibles are in operation. Normal research vessels will serve this purpose only in exceptional cases. The precise navigation of a submersible under water represents an extremely difficult problem. New methods have been developed which use moored acoustic transponders (cf. Section 3.1.1.3) and computers.

Table 3.03 gives the features of some selected deep-sea submersibles. In the following, the American submersible *Alvin* (Figs. 1.07, 3.08) will be described in more detail as

Table 3.03. Properties of Selected Deep-Sea Submersibles

Name	Maximum Operating Depth (m)	Speed (knots)		Propulsion Endurance at Given Speed	Source of Electric Power (Batteries)	Length (m)	Mission Duration (hr)		Crew Number
		Cruising	Maximum				Normal	Maximum	
Archimedes (1961)	11,000	1	2	12 hr/1 knot	Alkali	23	12	32	3
Trieste (1964)	11,000	2	—	5 hr/2 knots	Lead	22	10	24	3
Alvin (1965)	3,600	1.2	3	8 hr/2 knots	Lead	7	10	24	3
Aluminaut (1965)	5,000	3	3.8	32 hr/3 knots	Silver Zinc	17	32	72	4-6
Deepstar 4000 (1966)	1,300	1	3	6 hr/3 knots	Lead	6	12	48	3

Fig. 3.06. Research vessel *Discovery III* (Photo: Institute of Oceanographic Sciences, Wormley).

an example of a typical research submersible. The vessel displaces 15 tonnes and has a pressure-proof steel sphere, 2 m in diameter, for the crew as well as for the equipment for scientific observation and for control and navigation. Visual observations can be made through four portholes in front and at the sides. Propulsion and control is achieved by a large propeller at the stern, which can be turned from side to side, and two lift propellers attached to either side, which can be rotated so as to direct their thrust ahead, astern,

Fig. 3.07. Research vessel *Jean Charcot* (Photo: CNEXO, Paris).

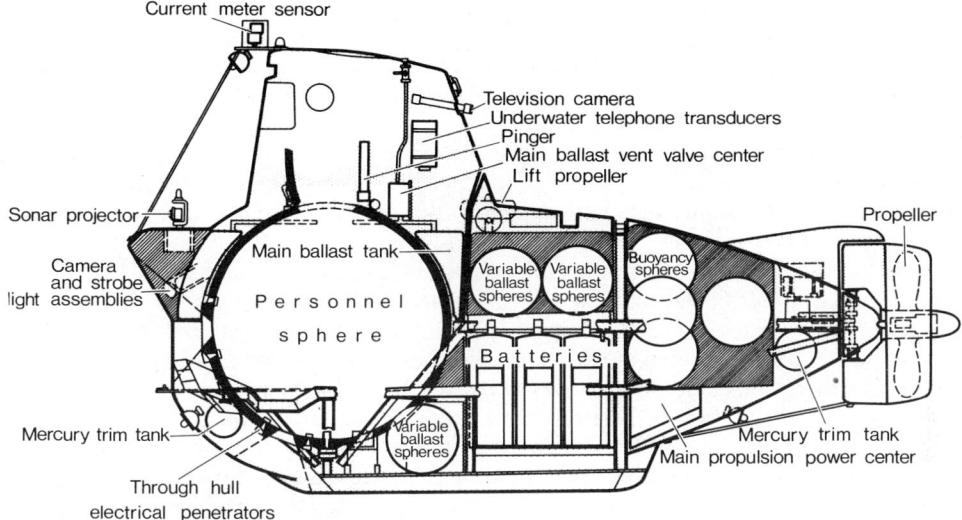

Fig. 3.08. Inboard profile of the submersible *Alvin*. According to the operational requirements additional measuring and sampling devices are installed. For bottom sampling a remote-controlled mechanical arm and a sample tray are attached to the vehicle at the lower left-hand side.

up, and down. A mercury ballast system, consisting of three tanks, permits control of the trim of the submersible. In addition, the vehicle is equipped with a variable ballast system to compensate for the varying weights of crew and instruments.

The equipment comprises a system for monitoring the oxygen supply, navigational instrumentation, an echo sounder and sonar (cf. Section 3.2.4), a closed-circuit television system, lights, an underwater telephone, and a radio transmitter for use at the sea surface. A remote-controlled mechanical arm, some sampling devices, and electronic instruments for scientific observation are made available depending on the particular operational requirements. The aforementioned catamaran *Lulu* serves as support vessel (cf. Section 3.1.1.1).

Summarizing descriptions of submersibles are given in a publication of the Interagency Committee on Oceanography (1965) and by Herring et al. (1971).

3.1.1.3. Position fixing

The *accuracy requirements* for navigational aids on board research vessels are especially high during oceanographic investigations. The classical methods include terrestrial navigation, which determines the ship's position by taking bearings of land marks or buoys, astronomical navigation, which uses the location of selected stars for position fixing, and dead reckoning by which the position in question is extrapolated from the ship's course and speed at a given starting point. In general, all these methods of position fixing are not accurate enough for oceanographic investigations. Therefore, the more accurate methods of electronic navigation play a major role in oceanography. In the following, the most important of such methods will be discussed.

With the *electronic methods* "position lines" are obtained by observing the propagation of radio waves. The point where these position lines intersect each other represents the ship's position. According to the particular method chosen, one obtains two straight lines (radio direction finding), one straight line and a circle (radar), two hyperbolas (e.g., Decca, Loran, Omega), or two circles (e.g., Hydrodist). Special problems arise for radio navigation methods because of the different paths of the direct wave and the wave reflected in the ionosphere, the varying velocities of propagation of the radio waves, and the exact timing that is necessary in most cases.

With *radio direction finding,* the transmitters ashore generally operate within the range of 250 to 420 kHz. For determination of the direction of the signal received on board, rotatable loop antennas with special radio receivers are used, or vectorially receiving goniometer systems. The method of radio direction finding can be applied only within distances of about 100 nautical mi; the standard accuracy of position fixing being a few nautical miles (1 nautical mi = 1.852 km).

With *radar* the travel time is measured that a pulse signal transmitted from a rotating antenna takes to be reflected at a land mark or buoy until it is received again by the antenna. The direction of the signal reflected at the object of the bearing is determined by comparison with a compass. The returned signal is indicated on a Braun tube serving as plan position indicator (PPI). Radar transmitters usually operate at frequencies of 3000 or 10,000 MHz. The range can amount to as much as 50 nautical mi; in general the accuracy is about ±1% of the distance and ±1° in direction. An extension of the range can be achieved by radar transponders which, as soon as a transmitted signal arrives, emit a stronger return signal.

Hyperbolic navigation methods are the most important methods for position fixing during oceanographic investigations in the open ocean. They make use of the fact that the differences of distance between a ship and two fixed points are constant on hyperbolas. Thus, if synchronized signals from two stations are transmitted, the difference measured between the times of arrival of the two signals at the ship's location yields a position line in the form of a hyperbola. With a further pair of transmitters a second line of position can be determined and, thus, the ship's position is given at the point where the two lines intersect each other. The ambiguity of hyperbolas can be eliminated by fixed shifts in time between the pulses of the pairs of transmitters or by splitting the time measurement into subranges.

When the Loran-A method is applied, short pulses of radio waves with frequencies of around 2 MHz are emitted from two or more pairs of transmitters. Each of the transmitter pairs is identified by its specific pulse repetition rate. The time differences between the pulses received are measured on shipboard, and the corresponding line of position is determined on special Loran charts. The range of transmission amounts up to 600 nautical mi during daytime (direct wave) and up to 1400 nautical mi at night (reflected wave). In general, the accuracy lies at ±0.5 to 3 nautical mi. With the Loran-C method a larger range is reached by the lower transmitting frequency of about 100 kHz, as well as a higher accuracy by an improved measurement of time, whereby not only the enveloping curves of the received signals are compared, but also the phases of the individual waves. The range of transmission can reach 1400 nautical mi during daylight hours and up to 2000 nautical mi at night. In general the accuracy is between 100 m and several nautical miles.

In near-shore navigation the Decca method plays an important role. Continuous waves are emitted at frequencies between 70 and 130 kHz, and the differences in travel time are determined by comparison of the phase differences of the incoming signals. The range of transmission amounts to about 250 nautical mi, the accuracy normally is between 40 and 300 m. A similar hyperbolic method for the region immediately near the coast is, for example, Hi-Fix which is characterized by ranges of 50 to 100 nautical mi and attains accuracies of ±2 to 10 m. In this case the main transmitter can be installed on the research vessel, as it is also the case with the Hydrodist method, so that circles are obtained as position lines.

Omega has won increasing importance as a hyperbolic method for very long distances. Long pulses are emitted by modulating radio waves with frequencies between 10 and 14 kHz and the timing is obtained by comparing the phases. The range of transmission amounts to 6000 nautical mi, the standard accuracy is ±1 nautical mi in daytime and ±2 nautical mi at night. There are intentions to develop this navigation system for worldwide use.

Today, *satellite navigation systems* using the Transit method are often installed on large research vessels. They combine a global range of application with very high accuracy. A disadvantage is seen in the considerable cost as well as in the fact that position fixing is possible only about every 2 hr according to the time interval between passages of a Transit satellite at the ship's position. In this method the fact is utilized that, due to the satellite's motion, the fixed frequency of a continuous signal emitted from the satellite is subject to a Doppler frequency shift which changes its sign at the minimum distance to the ship. The orbit data of the satellite are continuously controlled and corrected by ground stations and stored in the satellite. From there they are transmitted to the ship. The ship's position is finally determined from the Doppler shifted frequency and from the orbit and time data by a computer on board. When the satellite passes the position of a stationary ship, an accuracy of ±50 to 100 m can be obtained.

The accuracy that is practically attainable at sea with satellite navigation depends on how exactly the course and speed of the vessel with respect to the ground are known. Accuracies around ±0.5 to 1 nautical mi are typical. By combination with other measuring systems, the accuracy can be improved to approximately ±300 m. Beside hyperbolic navigation systems, the *inertial navigation* and the *Doppler sonar* can be applied for this purpose. With inertial navigation, the ship's position is obtained by determining her way since leaving a fixed starting point by means of double integration of the measured accelerations of the ship. This system is not often used in oceanography today. In contrast, increasing interest is being paid to the acoustic Doppler sonar method. Here, the ship's speed, relative to the sea floor or to water layers at some hundred meters' depth, is determined by measuring the Doppler frequency shift between an acoustic signal emitted from the moving vessel and the received signal after it has been reflected.

Acoustic transponders are used for the navigation of submersibles. Here a transponder sends an acoustic signal as a reply to a specifically coded sound signal from the submersed vehicle. The time difference between transmission and reception aboard gives the distance if the distribution of the sound velocity is known; this means that the position is approximately indicated by spherical surfaces. The combination of three or more transponders yields the position of the submersible, but electronic computers are required for quick and exact position fixing. Surface vessels occasionally make use of acoustic transponders as navigational aids in order to exactly regain selected positions or to locate moored instruments.

Summarizing descriptions of navigational methods have been given by Anderson (1966), Runcorn (1967), Müller-Kraus (1968), and Freiesleben (1968).

3.1.1.4. Computers

Although on land the computer has very quickly become a general and indispensable aid to oceanographers with a view to data evaluation and modelling studies, its application aboard research vessels, started in 1962, has developed only slowly. In the beginning, few systems technically suitable for use on shipboard were available. Furthermore, only little was known about the possibilities of their application. But today computers represent important parts of the equipment of many larger research vessels, and their application is increasing very fast.

The introduction of electronic computers aboard research vessels took place in several stages. At first, the computer was used for tasks that had already before been carried out on board without computers, although much more slowly, for instance, data recording, calibration, correction, and quality control of data. For example, the hydrographic data of in-situ salinometers (cf. Section 3.2.9) are now often processed by computers. The second stage has brought aids for making decisions when research cruises are being planned. This concerns data reduction, comparison with stored data obtained during earlier investigations, and the production of graphical presentations. In modern systems, data concerning navigation, echo sounding, and hydrography are combined by the use of a computer which prints or plots the results.

From these first two stages the further development leads to the application of the typical process computer with automatic data retrieval which can give essential aid for the optimum determination of the time and position of measurement even while the measuring series is still going on. Thus, records of in-situ salinometers are occasionally evaluated during the measurement in order to operate remote-controlled Nansen bottles at suitable depths. This shows that computers on shipboard—in spite of the increasing number of data—do not only help to reduce the time needed for the evaluation of the data but, above all, contribute to a more effective utilization of the ship's time.

The development of computer technique as well as of its application proceeds so rapidly that the discussion of special computers does not seem to fit into the framework of this book. It cannot be excluded that, before long, part of the organization of measuring programs can be taken over by computers on shipboard. Furthermore, computers can serve for the communication among ships, moored or drifting buoys, airplanes, and satellites, as well as for steering functions over radio or acoustic links. A first example, which shows that the communication between ship and satellite can be taken over by a computer, is the application of computers in satellite navigation (cf. Section 3.1.1.3). Another example is the automatic positioning of the drilling vessel *Glomar Challenger* (cf. Section 3.1.1.1) where a computer monitors the position of the vessel relative to an acoustic transmitter at the sea floor and controls the ship's propulsion.

3.1.1.5. Airplanes and satellites

The application of airplanes and satellites as research craft for oceanography opens new avenues for quick spatial surveying of the physical properties of the sea surface (Ewing, 1965). Airplanes are used only as carrier platforms of instruments for remote sensing, whereas satellites can also operate as relay stations for data transmission from buoys and ships (cf. Section 3.2.12.2) and as aids to navigation (cf. Section 3.1.1.3). In future, satellites will perhaps also be applied to the determination of the sea level.

The following quantities can be recorded from aboard an *airplane:* the surface temperature (cf. Section 3.2.8) by observation of the infrared radiation of the sea; the state of the sea (cf. Section 3.2.7) by measuring the scattering of microwaves, the reflection of radar or laser beams, or by simultaneous observation of the infrared and microwave radiation of the sea; the ice cover by observation of the visible radiation; the chlorophyll content by observing the spectral distribution of scattered visible radiation;

the salinity by observation of the microwave radiation of the sea with simultaneous measurements of the temperature and the state of the sea. The two latter possibilities cannot yet be utilized at present because the errors are too large, and the sources of error are not known exactly. But already these days, the methods of measuring temperature and ocean surface waves play an important role.

Satellites can be equipped with sensors, according to the measuring program. The stable position of a satellite and its high speed are advantageous when oceanic areas are to be surveyed rapidly and repeatedly. The disadvantages are the rather small horizontal resolution, because of the great distance from the sea surface, and the fact that many observations are not accurate enough, or that they are even made impossible due to absorption, scattering, and emission of radiation in the atmosphere—for instance when clouds are present. In cloudless areas, methods for measuring sea surface temperature, ice cover, and the color of seawater have become important.

3.1.2. Fixed observation platforms

3.1.2.1. Moored systems

Whereas until a few years ago the research vessel was by far the most important platform for measurements at selected positions in the ocean, today moored carriers of measuring instruments with automatic data recording devices are of equal significance. The advantage of the ship's flexibility, with regard to the operational area and the kind of measurements, is counterbalanced by the possibility of moored systems supplying long-time series of measurements that are only little disturbed by the movements of the platform itself, and that are less expensive than those obtained from shipboard.

Fixed measuring towers are the ideal aids when the question is how the motion of the carrier of instruments can be reduced to extremely small values. But they can be employed only in shallow water within the shelf region. Flexible moored systems with cables, on the other hand, can now be assembled for any water depth. Figure 3.09 shows two examples of deep-sea mooring systems. According to the relevant kind of problem, there are variable arrays of steel cables or synthetic ropes and buoyancy elements with numerous measuring instruments (cf. Section 3.2.12.1) which contain devices for automatic data recording and, occasionally, also for telemetering. Surface mooring systems [Fig. 3.09(a)] offer the possibility of obtaining oceanographic data not only in deep water but also near the surface and, simultaneously, meteorological data on a buoy at the same position. In contrast, underwater mooring systems [Fig. 3.09(b)], have the important advantage that they are not disturbed by motions due to ocean surface waves, and that they are less endangered by shipping. In order to recover the underwater moorings, some release mechanism is required which is controlled acoustically from the ship or mechanically by a clock. At the desired time the release mechanism separates the measuring array from the anchor at the sea floor so that the array can rise to the surface. Such release gear is usually also applied to surface moorings in order to minimize the stress on the system during the recovering.

In addition to those moorings of measuring instruments which are particularly used in the open ocean, there are several other systems, for example, moorings with a bottom line (Dietrich et al., 1963), underwater winches for measuring vertical profiles by means of instruments moved up and down (Siedler et al., 1964), and measuring masts with mobile joints (Krauss, 1960). Another system, which has been developed especially for near-bottom measurements in the deep sea, is shown in Fig. 3.10.

112 Oceanographic Instruments and Observational Methods

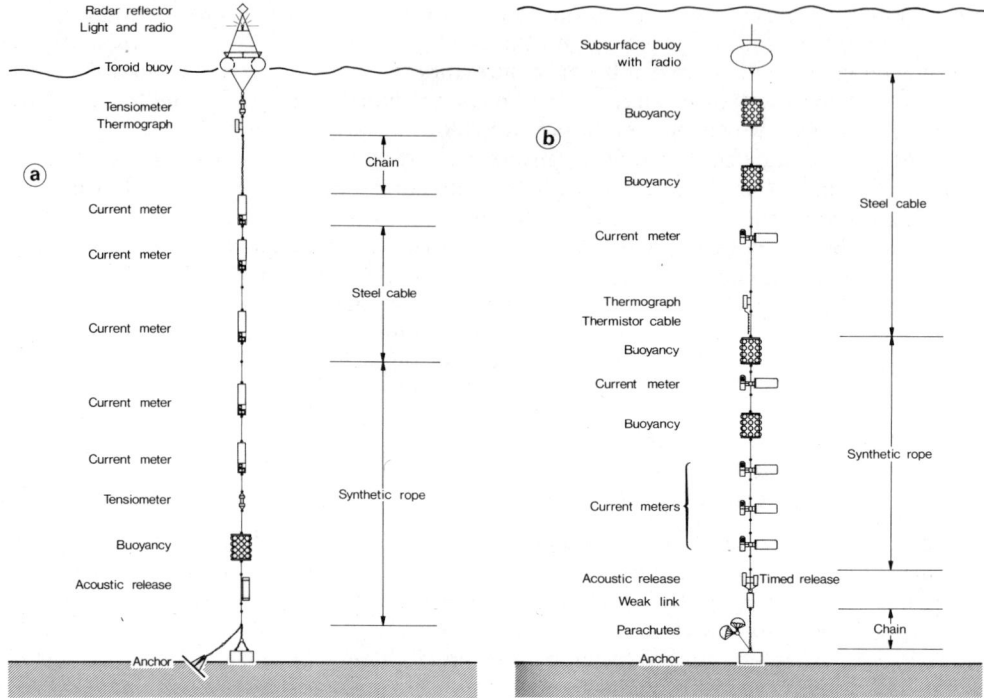

Fig. 3.09. Typical moorings: (a) surface mooring, (b) underwater mooring.

For measurements near the sea surface, surface buoys are the only devices available. Two groups of such buoys can be distinguished: the "surface-following" buoys and the "stable" buoys. Variations of the sea surface affect the buoyant forces acting on the stable buoy only slightly so that it is almost independent of the state of the sea. This is mainly achieved by having the height of the buoy large as compared to its diameter near the sea surface. With long wave periods, however, dangerous resonance oscillations might be incited. To prevent this, stable buoys for the open ocean have very large dimensions.

In addition to unmanned buoys, manned surface buoys have also been developed, like the French buoy *Borha II* (Fig. 3.11). The largest buoy of this kind is the American buoy *Flip* (Fig. 3.12), which can be operated in both ways, drifting or anchored. In horizontal position she can be towed like a ship without propulsion to the desired station where she may be brought into vertical position by flooding of the ballast tanks. *Flip* is 116 m long, displaces 1700 tonnes in the horizontal position and 2500 tonnes in the vertical position, and has a crew of up to 21. When used as a buoy, she has an immersion depth of 98 m. Surface-following buoys show stronger motions in the seaway, but even those of small dimensions are not subject to resonance oscillations. An example of this type is the Toroid buoy, shown in Fig. 3.13. The largest buoy of this kind is the American Monster Buoy with a diameter of 13 m (Fig. 3.14).

3.1.2.2. Research towers and underwater laboratories

Fixed research towers can be erected on the shelf and used as platforms for measuring instruments which are mounted so that they are rigid or movable. As an example, Fig. 3.15 shows a tower installed for oceanographic investigations off the American West

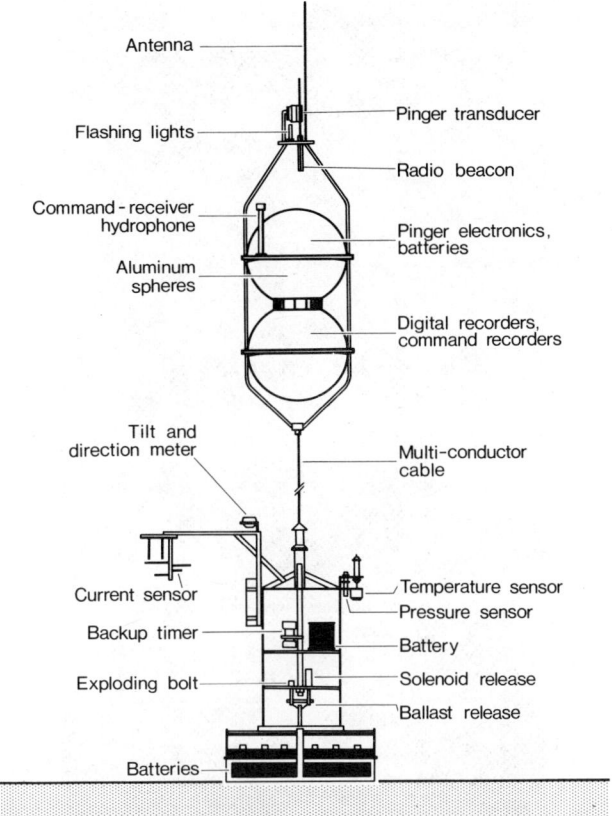

Fig. 3.10. System for near-bottom measurements. The array falls freely from shipboard down to the ocean bottom, where it automatically records the physical parameters, and then returns to the sea surface after the anchor has been released by acoustic remote control. (After Caldwell et al., 1969.)

Coast. Often, however, towers constructed for military purposes or oil and gas drilling are employed as carriers for oceanographic measuring devices. Occasionally isolated lighthouses, which had been built as navigational aids, are also used for purposes of marine research. For example, the Kiel lighthouse contains automatic oceanographic measuring equipment with data transmission by radio.

Underwater laboratories standing on the sea floor have been constructed in particular for studies in the field of marine biology. The "underwater houses," developed by J. Y. Cousteau since 1960, have been followed by many others, in the United States by *Sealab I, II,* and *III* and others, as well as in Germany by the *Underwater Laboratory (UWL) Heligoland* from 1968 onwards (Kinne et al., 1973). The *UWL Heligoland,* which was especially designed for operation in the North Sea, will be briefly described in the following as an example of such a manned research facility at the sea floor. The *UWL* consists of a horizontal steel cylinder with a diameter of 2.5 m. Inside, it is divided into living quarters, 2.5 m in length, a room for instruments and machinery, 6.5 m in length, and a 4 m long wet room. Normally, air under high pressure, oxygen, nitrogen, helium, fresh water, and electric power are provided by a moored supply buoy at the sea surface. In addition, the laboratory contains supplies for about 10 days to last out periods of bad

Fig. 3.11. Moorable manned surface buoy *Borha II*. Total height 79.3 m, 60 m of which are immersed. The diameter at the head of the buoy is 10 m, at the sea surface 2.8 m, at the foot of the buoy 7 m. The displacement is approximately 910 tonnes. Natural period: pitching motion 24 sec, rolling motion 28 sec (Photo: CNEXO, Paris).

weather. The *UWL* is part of a comprehensive system also including decompression chambers, transfer cases, by-labs, and underwater storage facilities. The contact with the control station by radio and television is effected via the supply buoy as a relay station.

3.1.3. Drifting platforms

Oceanographers use floats if they want to follow the trajectory of a water mass. This method of current measurements will be explained in more detail in Section 3.2.12.2. For a task with still higher demands, namely the determination of the variation in time of certain physical and chemical quantities in a drifting water body, floats have been applied as measuring platforms. An example is the buoy *Flip* (Fig. 3.12) mentioned above, from which such measurements can be carried out. Smaller drifting buoys with long-

Fig. 3.12. The giant buoy *Flip* (Floating Instrument Platform) on her way to the operational area in horizontal position (a), while turning into the vertical position during the flooding of the ballast tanks (b), and in vertical position at the measuring site (c) (Photo: Scripps Institution of Oceanography, La Jolla).

(c)

Fig. 3.12. *Continued*

distance telemetering devices have also been developed. In polar regions even drifting ice floes have been used as oceanographic measuring platforms.

3.2. Oceanographic Instruments and Methods of Measuring

3.2.1. Measured quantities

The physical and chemical behavior of seawater can be described by the (1) quantities of temperature, pressure, and salinity as well as by its content of dissolved substances as far as it deviates from that of normal seawater, (2) by its content of suspended particles, and (3) by the state of motion. But such properties cannot be represented by simple scalar or vectorial quantities for all problems in oceanography; often particulars of the composition must be added when the content of dissolved substances and suspended particles is given. Numerous properties of seawater, however, can be determined satisfactorily from data concerning temperature, pressure, and salinity alone, as discussed in Chapter 2. Furthermore, the shape of the boundaries is also of great importance for the course of physical and chemical processes in the ocean. Therefore, besides the measurement of the parameters mentioned before, the tasks of oceanographic measuring technique also include the determination of the position of the upper and lower boundaries, that is, the sea surface and the sea floor.

3.2.2. Basic requirements on oceanographic measuring technique

The demands on the measuring technique vary in accordance with the specific kind of oceanographic problem to be solved. With regard to oceanographic measurements, however, a number of requirements appears particularly frequently.

Fig. 3.13. Toroid buoy operating within a surface mooring system. A radio beacon, an anemometer, and a quick-flashing light unit are mounted on the instrument platform (Photo: Woods Hole Oceanographic Institution, Woods Hole).

1. There is a need for high sensitiveness and good reproducibility so that even small fluctuations of the thermodynamic variables can be identified; for example, for temperature measurements in the deep sea, a resolution and reproducibility of about 10^{-2} °C is necessary, and for salinity, the respective figure is about 10^{-3} ‰.

2. The instruments that are exposed to seawater or to air containing seawater must be resistant against this, because seawater is not chemically neutral and, in particular, as an electrolyte, together with metals, it forms galvanic cells so that it can be corrosive to metals.

3. Instruments used in water must be able to withstand hydrostatic pressure and to supply data that have not been falsified by pressure effects. Otherwise, if appropriate countermeasures are not taken, the pressure could deform an in-situ sonde for measuring electrical conductivity, even at a few hundred meters depth, so much that the data obtained would be useless.

Fig. 3.14. Monster Buoy with measuring mast, arrays of sensors and antenna. Energy supply and electronic units are installed in the disk-shaped body of the buoy (Photo: General Dynamics, Convair Division).

4. The instruments must be robust to cope with the difficult conditions at sea caused by the motion of the ship in the seaway. The strongest mechanical stress mostly occurs when the measuring gear is lowered from the ship into the sea or taken on board again.

Oceanographic Instruments and Methods of Measuring 119

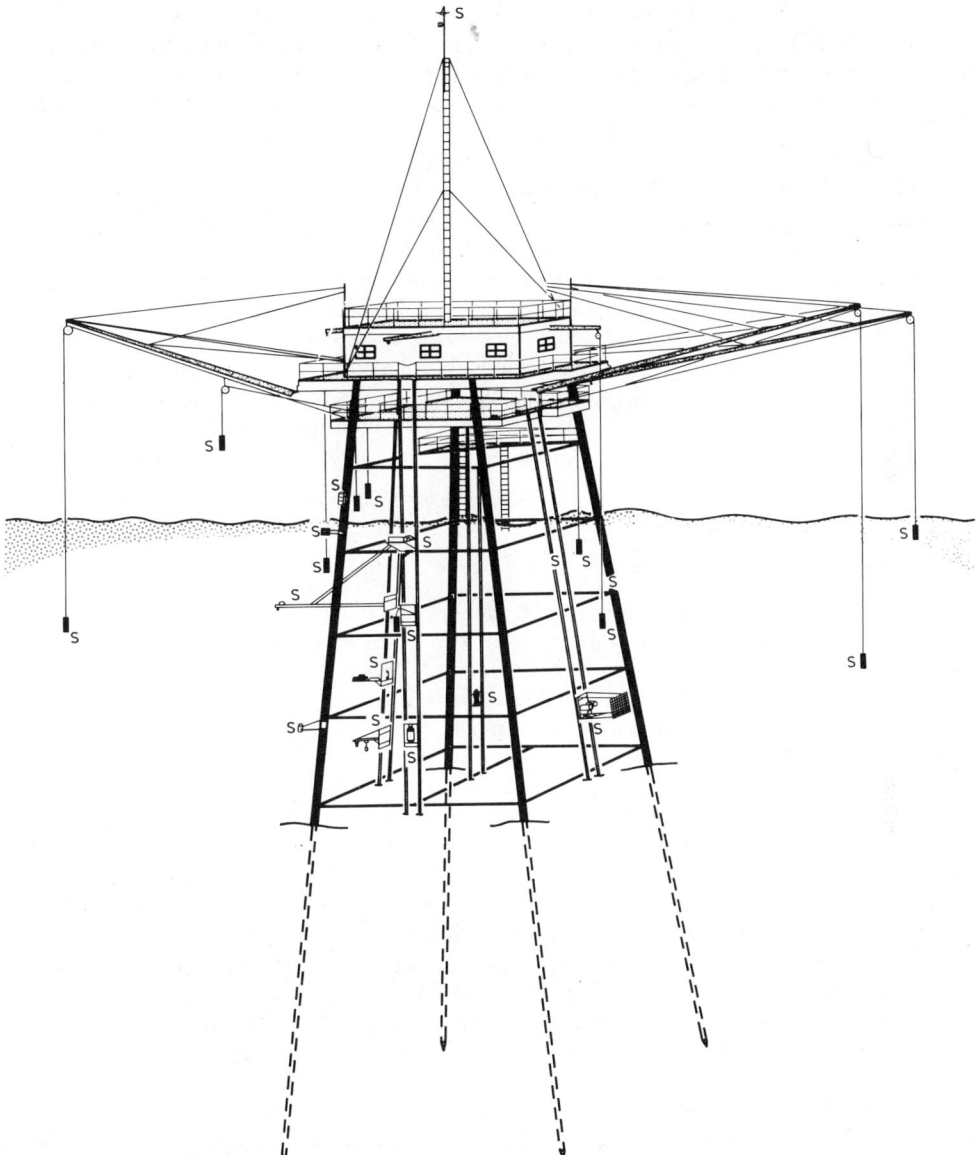

Fig. 3.15. Construction scheme of an oceanographic research tower of the U.S. Navy Electronics Laboratory off the Californian coast. According to the actual measuring program, sensors (S) of different types are arranged at variable positions. The upper part of the tower contains the living quarters, supplies, and recording facilities. (After La Fond, 1965.)

As in other geosciences, in oceanography we are often faced with the difficulty of recording changes of the measured quantities with sufficient resolution in time and space. For instance, features of the sea floor require a very great amount of the ship's time if they are to be observed well enough for the spatial interpolation of the data to permit a representation of all important bottom features.

Measurements of processes that vary only slowly in space or time are often so much falsified by rapid fluctuations that special care must be taken to avoid such disturbances, for instance, by the application of high scanning rates or of inert sensors. By way of example, tidal currents, or currents that change even more slowly, can only be determined with high accuracy in the immediate vicinity of the sea surface if the influence of the sea state on the measurement is eliminated by scanning rates in the range of seconds and vectorial averaging afterwards (cf. Section 3.2.12.1). It is still more difficult to measure currents that change rapidly with sufficient spatial resolution, since here the application of a very large number of measuring instruments is required.

Owing to the progress in electronics, in particular, the oceanographic measuring technique is developing very fast. Many a measuring instrument is already technically outdated after a short time. Therefore, it seems appropriate within the framework of this book to discuss only the most important technical problems of measurement and to present only some selected examples of their solution. Summaries of the development of oceanographic instruments can be found, for example, in the volumes of the Instrument Society of America (1962, 1963, 1965, 1968), the Marine Technology Society (1964, 1967) and with Kruppa et al. (1973).

3.2.3. Water sampling devices

Physicochemical investigations of seawater can only be partly carried out in situ. In many cases one depends on laboratory analyses of the water samples taken. To obtain such samples from the sea surface, a bucket will suffice. Only on vessels with high freeboards, or on those that are fast, are well-flushed streamlined samplers preferred that can be towed. Obtaining water samples from deeper layers is more difficult. For this purpose sampling devices must comply with two conditions: (1) the sample must be taken at a chosen depth which can be determined precisely; (2) the sample must be protected from any falsification while the instrument is hoisted.

Several types of water samplers have been developed which differ in the release mechanism of the closing valve at the desired depth as well as in the type of the closing valve itself. The release of the closing valve is effected: (1) by a messenger that is released on board and travels down along a wire to which the sampler is attached, thus activating the closing valve after contact with the sampler, or (2) by a propeller that is coasting while the sampler is being lowered and then, after some resolutions, activates the closing valve as soon as the sampler is hoisted, or (3) by hydrostatic pressure that activates the closing valve when a certain pressure is reached, that is, at a certain depth, or (4) by electric remote control through a cable.

The closing is achieved either by flap valves or by plug valves. Sometimes samplers with flap valves are also designed as insulating samplers. At the samplers with plug valves, the closing is usually activated when the sampler reverses, whereby, at the same time, the temperature in situ is fixed in the reversing thermometer attached to the sampler. Of all the samplers, the reversing sampler with a plug valve, as developed by Nansen and improved repeatedly later on, is the one used most widely. A number of these samplers can be attached to the wire of a hydrographic winch and will thus supply a series of water samples from the desired levels of depth.

In Fig. 3.16 such a reversing bottle is shown with explanations on how it works. At the top and at the bottom, the metal cylinder a is fitted with the plug valves b_1 and b_2 which ensure good flushing of the metal cylinder while it is lowered. The bottle is attached to the wire d of the hydrographic winch by the binding clamp c. The release mechanism

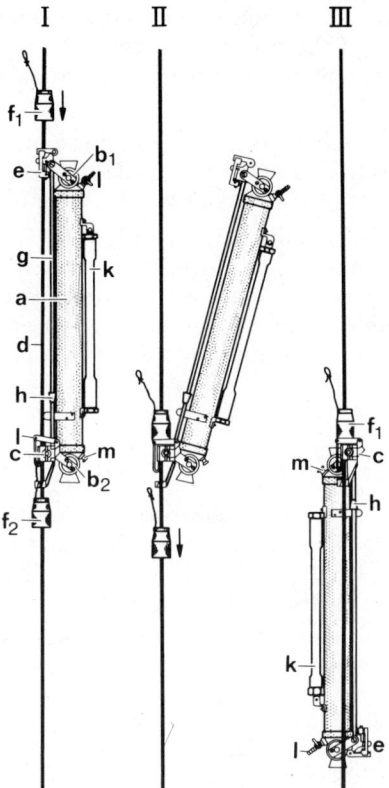

Fig. 3.16. Nansen water bottles before (I), during (II), and after (III) the reversing process.

e keeps the bottle attached to the wire. When the sampler has arrived at the desired depth, the messenger f_1 is sent down along the wire. It hits the release mechanism e, thus setting the upper part of the bottle free so that the bottle reverses with the lower plug valve b_2 as the center of rotation (stage II in Fig. 3.16). The guide rod g between the upper and lower plug valves turns both cones of the valves at the same time. After the reversal (stage III) the valves are firmly closed, and the cone h of the guide rod prevents any accidental opening of the sampler even when swinging wildly. When the bottle is reversing, the messenger, after having activated the release mechanism e, proceeds further down to the clamping device c where it hits the release lever i which, in turn, releases a second messenger f_2. This messenger now continues to slide down along the wire to the next bottle where it starts the same process all over again. To the outside wall of the bottle, a removable metallic frame k is attached which serves as a casing for two or three reversing thermometers. The drain valve l at the lower end of the bottle and the air valve m at the upper end permit the water sample to be poured into storage containers when back on board.

To obtain representative water samples from below the sea surface, the ship must be stopped, and the wire must be as vertical as possible. This is a requirement for all water samplers with the exception of the Spilhaus sampler with which water sampling in the uppermost 300 m can be carried out from a moving vessel. New designs of water samplers have three aims.

1. The investigation of certain trace elements in seawater requires samplers that prevent any contact of the seawater sample with metal. This is attained by the plastic coating of common samplers or by samplers wholly made of plastics.
2. The analysis of trace elements in very low concentrations is possible only if large amounts of water samples are available. As an example, a 270 liter sampler for investigations of ^{3}H and ^{14}C is shown in Fig. 3.17.
3. For the comparison of salinity determinations by an in-situ salinometer and by a laboratory salinometer, an in-situ measurement and a water sample obtained at the same time and place are required. For this purpose, single or multiple water samplers (Rosette samplers) are used which are placed in the immediate vicinity of the in-situ salinometer and released by electric remote control.

3.2.4. Measurement of water depth

To determine the topography of the ocean bottom, the time is measured that a sound signal needs to travel from the ship to the sea floor and back again. This measuring principle was already suggested by Bonnycastle (1838) in America. But when, after a detonation at the sea surface, he tried to hear an echo from the sea floor at a depth of more than 1000 m with a simple listening tube, his attempts failed. For a long time the opinion prevailed that the absorption at the sea floor was too great for obtaining audible echoes, until in 1912 Behm proved the contrary. He also solved the problem of measuring short time intervals and, in 1919, he manufactured the first reliable acoustic sounding device which he called "echo sounder," a name which is now generally adopted.

The echo sounder consists of a sound transmitter with a transmitter transducer, a sound receiver with a receiver transducer, a time-interval measuring device and, usually, a recording mechanism. Today, in general, transmitter and receiver transducers are identical since one transducer is used successively for both functions. Characteristic sound frequencies within the audible or the ultrasonic range lie between 10 and 30 kHz. A short sound signal is produced in the transmitter, and through the transducer it is radiated downward. At the same moment the time measurement starts, and it ends as soon as the signal that has been reflected at the sea floor is received by the transducer. If the mean sound velocity c is known (cf. Eq. 2.28), the water depth h is obtained from the total travel time t:

$$h = \tfrac{1}{2} ct \qquad (3.01)$$

Packages of nickel sheets which, by means of coils, are exposed to an alternating magnetic field, serve as transducers. Owing to the magnetostrictive effect, ferromagnetic substances, like nickel, undergo changes in their form as soon as there is a magnetic field; such changes lead to analogous variations of pressure in the adjacent water, that is, to the generation of a pressure wave. A good sound beam can be produced by a suitable combination of several transducers. Here, high frequency, due to the associated small wavelength, has a favorable effect on the beam width and thus on the spatial resolution of the depth measurement. In water, the frequency of 10 kHz corresponds to a wavelength of about 15 cm; 30 kHz to about 5 cm. A very great increase of the frequency is not possible, however, because the sound absorption (cf. Section 2.1.10) increases with the square of the frequency, whereas the range of sound propagation decreases correspondingly.

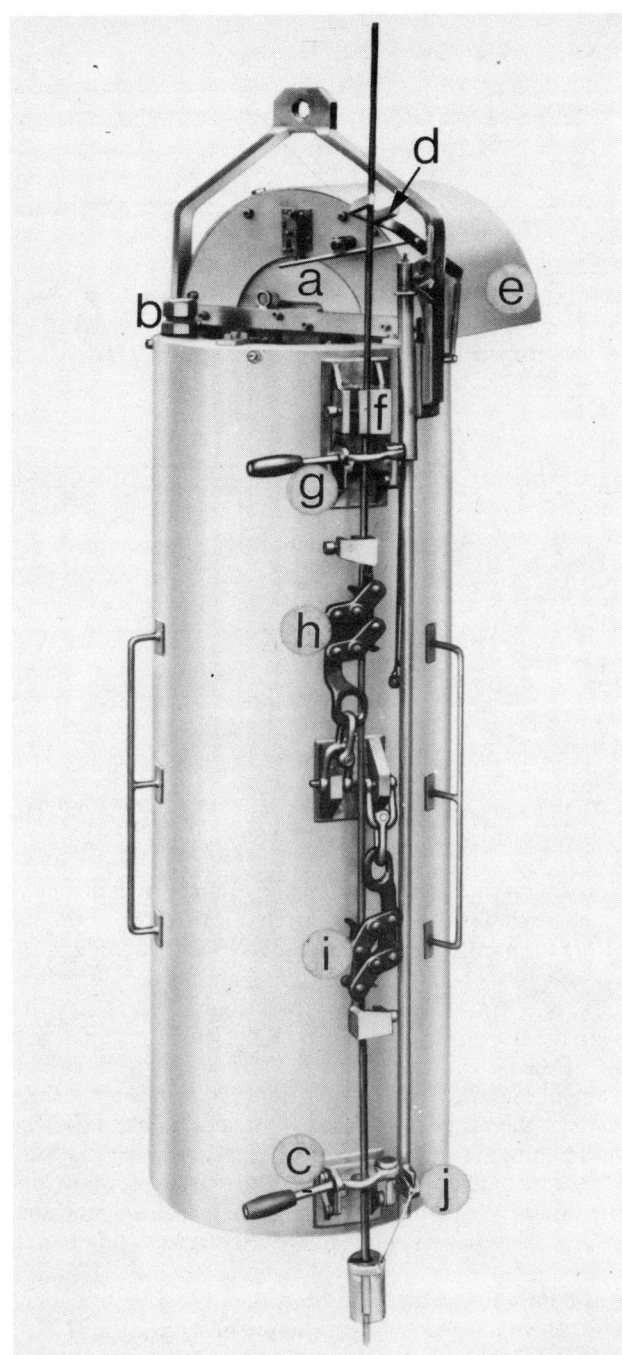

Fig. 3.17. 270-liter water bottle for sampling large amounts of water. During the lowering of the bottle, the water flows in through the cap, then along a separating wall to the bottom of the bottle from where it rises upwards and leaves the bottle through an opening which can be seen in front on top. When released by a messenger, the cover tips over and closes the bottle. In addition, the messenger is released to repeat the same process on the bottle following below. A high-pressure relief valve takes care of the pressure balance during the heaving of the instrument. Designations: (a) = cover, (b) = high-pressure relief valve, (c) = safety device for the wire, (d) = release lever, (e) = cap, (f) = clamping device for the wire, (g) = safety device for the wire, (h) = (i) = clamping device for the wire, (j) = release for series messenger (Studio photo: Hydro-Bios, Kiel).

124 Oceanographic Instruments and Observational Methods

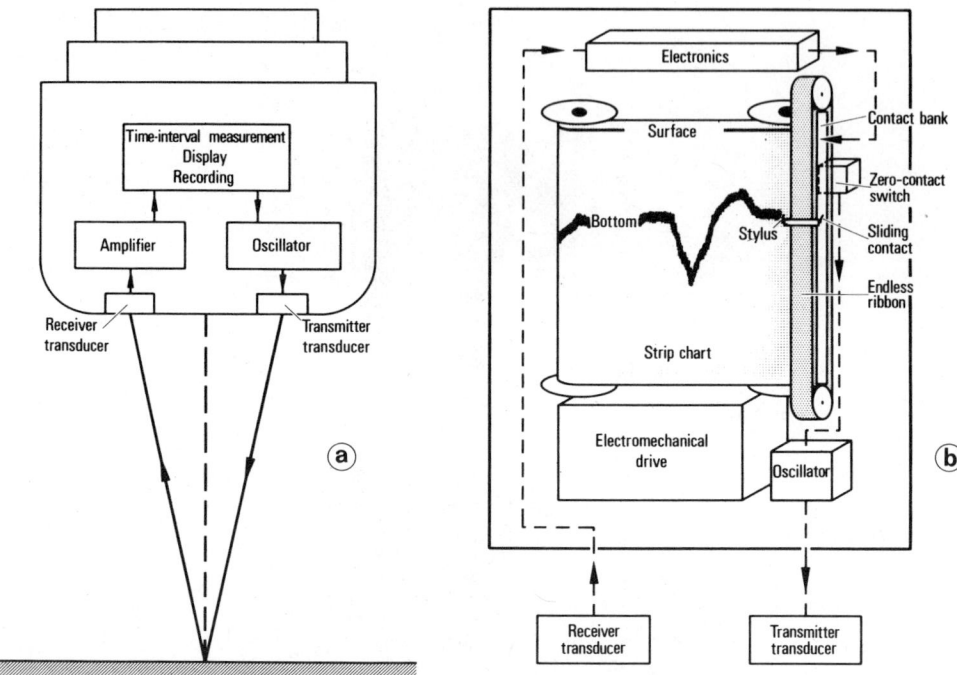

Fig. 3.18. Schematic representation of an echo sounder: (a) measuring principle, (b) recording and control mechanism.

In echo sounders for navigation purposes, the microtiming device often is a disk rotating within a circular scale with a neon tube that flashes up at the moments of transmission or receipt, or an oscillograph. For research echo sounders and also for many navigational echo sounders, a combination of timing and recording device is used, in which styluses on a ribbon, or spiral wires on a cylinder, slide along a strip chart lying on a contact bank. During the transmission and reception electrical signals are produced which cause a blackening of the paper at certain points. If the strip chart is shifted a little between one measurement and the next, two traces are obtained from the sequence of dots, one for the sea surface, the other for the sea floor. For work in the ocean the scales of German echo sounders are arranged so that they yield correct data if the mean sound velocity is $c = 1500$ m sec^{-1}. Additional corrections are necessary for exact records of depth (cf. Section 2.1.10). A schematic representation of an echo sounder is found in Fig. 3.18, and Fig. 3.19 shows an example of a deep-sea sounding system with recording gear.

The accuracy of depth determination is dependent on the time measurement, on the knowledge of the mean sound velocity, on the beam width, the ship's motion, and the bottom topography. While the exact timing does not involve any serious problem, and the variations of sound velocity can usually be corrected without great difficulty, the three latter items require some more attention. Little beaming of the sound yields echoes not only from the sea floor directly beneath the ship but also from other points at the bottom (cf. Fig. 3.20). It may occur that such additional echoes take a shorter way and indicate a depth that is too small, as compared to the real value. Such disturbing effects are ag-

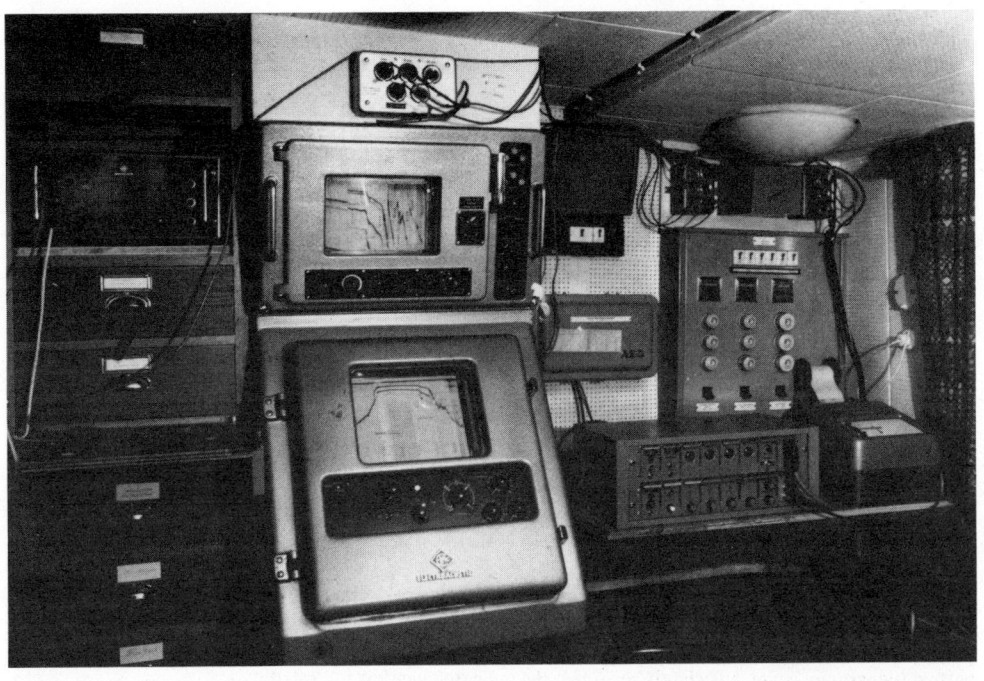

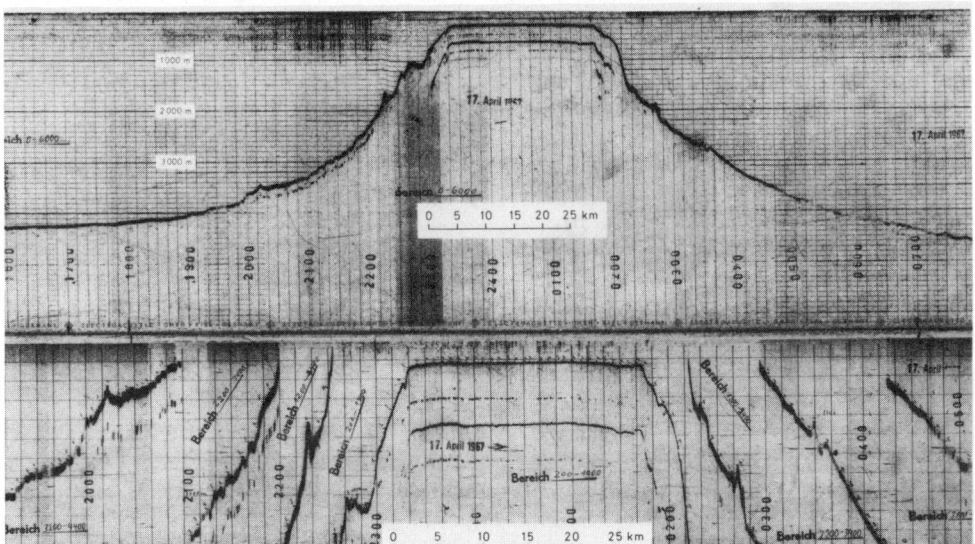

Fig. 3.19. Deep-sea sounding equipment on board the research vessel *Meteor* (above) and an example of a record (below). Above left: the recording instruments for the total depth range and for a selectable spreaded subrange, above right: the digital display unit and a printer. The record shows a profile over the Great Meteor Seamount in the North Atlantic Ocean, above in the total depth range, and below in various subranges, where details of the bottom topography can be recognized (Photo: Institut für Meereskunde, Kiel).

126 Oceanographic Instruments and Observational Methods

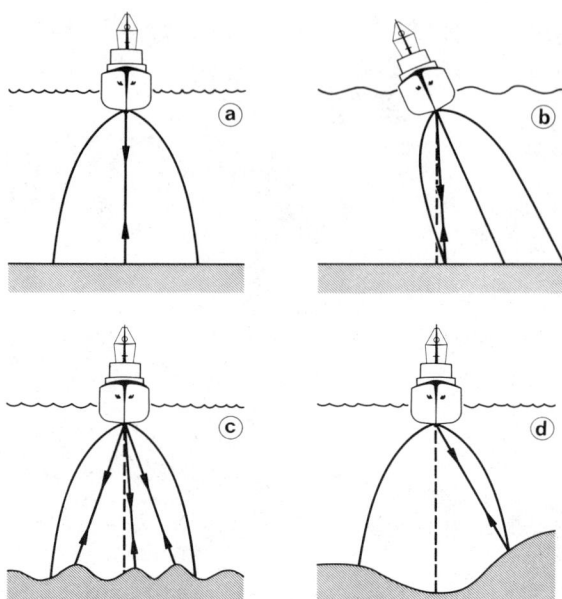

Fig. 3.20. Sources of errors in echo sounding: (a) correct measurement, (b) ship's motion causes too large depth record, (c) instead of the bottom topography the depth of the highest elevation below the ship is recorded, (d) bottom topography is not sufficiently recorded, the indicated depth is too small.

gravated by the rolling and pitching motions of the ship, which enlarge the reactive area, that is, the water space from which echoes are received. Furthermore, the upward and downward motions of the ship alter the distance between the ship and the sea floor. With such ship's motions, very narrow beaming, however, causes the sound signals to be radiated in various directions, in accordance with the position of the transducer, so that they are received only occasionally. Such records are useless. Therefore, the radiation cones that are usually applied are between 10 and 20°. But the region of the sea floor thus scanned would be too large for deep-sea research echo sounders. That is why some modern echo sounders (cf. Fig. 3.19) have a very narrow beam width of about 3° with simultaneous stabilization of the transducer which is always kept in a horizontal position by a gyro-control system.

Disturbing signals result from repeated reflections at the bottom and at the surface of the sea. In general, they are easy to identify. Signals are also received from scattering layers at medium depths. They may originate from fish, from accumulations of plankton or detritus, or from inhomogeneities of seawater. The reflection from fishes has led to the application of echo sounders for fish finding. A special measuring device is the fishing-net sonde, for which a transducer is fixed to the trawl. To a certain extent, the structure of the sediment can also be recognized by means of echo sounders. To penetrate deeper into the subsoil, however, seismic reflection systems are applied which use much lower frequencies, that is, of around 20 Hz to 1 kHz.

Echo sounding can also be carried out in horizontal direction, or any other, when underwater vehicles, fishing nets, or moored instruments are to be located. The evaluation of such "sonar" signals is difficult since the propagation of a sound beam seldom takes place along a straight line (cf. Section 2.1.10), and, furthermore, because the echoes are of very different qualities. An instrument that represents a particular development is the side-scan sonar where the sound is radiated in a broad band abeam to the ship's course from the ship itself or from some towed gear. Thus it is possible to record the topographic features of a certain area of the bottom.

3.2.5. Measurement of instrument depth

For the determination of the depth of an instrument, various methods are at our disposal, for instance, the measurement of the wire length, the measurement of the hydrostatic pressure, the measurement of the sinking time, and special measurements by means of echo sounders.

The *length of the wire* that is paid out could be considered a direct measure for the distance between the lowered instrument and the sea surface if the wire were not inclined owing to ocean currents and the ship's drift. At shallow depths it can often be assumed that the wire is hanging in a straight line, and that measuring the angle at the sea surface will suffice. At a greater depth, the deviation of the wire from a straight line must be taken into account. This can be done by measuring the hydrostatic pressure or, occasionally, also by measuring the inclination of the wire at selected points with interpolation following for the instruments that hang in between. Series of measurements with water samplers have long proved to be satisfactory if the depth is determined by the thermometric method (cf. Section 3.2.8).

From the hydrostatic pressure p of a water column, the mean density ρ of which is known, the depth z is obtained according to the hydrostatic basic equation:

$$z = \frac{p}{\rho g} \tag{3.02}$$

Here g is the acceleration of gravity. For normal seawater density, the pressure reading in dbar numerically corresponds—within an accuracy of about 1%—to the depth in meters.

For *pressure measurements* the elastic deformation of selected measuring bodies is determined. For example, when applying the thermometric depth measurement, one compares the temperature readings of two reversing thermometers, one being pressure-protected, the other unprotected (cf. Section 3.2.8). In the latter, the pressure causes a compression of the capillary and, thus, an extension of the mercury column, that is, an apparently higher temperature. The difference between the temperature readings from protected and unprotected thermometers is proportional to the hydrostatic pressure. The thermometer constants, which are required for the calculation of the pressure in the ocean, can be determined by laboratory calibrations with regard to pressure and temperature. At 5000 m the accuracy of depth measurement amounts to ±20 m.

Especially simple methods of pressure measurement are provided when Bourdon tubes or Well tubes are employed. The Bourdon tube, which is a sickle-, horseshoe-, or spiral-like tube closed at one end, bends up when the pressure within the tube is increased. The change of position at the open end can be shown by a mechanical indicator, or converted into a turn of a potentiometer or a variation of the vent hole in a coil core array. From the resulting changes of resistance or inductivity, electrical signals can be obtained through suitable transformers. Such signals represent a measure for the pressure. Depending on the type of tube, the standard accuracy of the metallic Bourdon tube lies between ±1 and 3% of the total measuring range. Errors can arise especially because variations of the surrounding temperature may result in an additional deformation of the tube, and because of hysteresis, that is, if the deformation is not fully elastic and, with the return to a certain pressure value, the reading does not exactly equal the previous one. This effect is even greater with the Well tube, which is a bellow-like metallic tube; so it can only be applied in case of low demands on accuracy, that is, between ±3 and 5% of the total range. In oceanography, the Well tube is widely used as pressure sensor in bathythermographs (cf. Section 3.2.8).

For the exact measurement of pressure with electronic methods, pressure boxes are used in which the bending of a membrane is determined by means of strain gauges: thin, metallic wires or films are deformed, which changes their electrical resistance. Inductive and capacitive methods have also been applied. The accuracy of the sensor in the membrane box lies between ±0.2 and 1% of the measuring range. In oceanography, another method has also attained great importance in which the bending of the membrane is measured through the change in the tension of a string attached to the membrane ("vibrotron"). The resonance frequency of the oscillation of the string, which is generated and scanned electromagnetically, decreases with increasing bending of the membrane and, thus, is a measure of the pressure. The accuracy lies at ±0.1% of the measuring range, the resolution can be better than this value by 2 to 3 orders of magnitude.

The *time measurement at constant sinking velocity* of a free-falling probe offers a very simple method for the determination of the depth. It is employed with the XBT (Expendable Bathythermograph), which will be described in Section 3.2.8.

Acoustic methods have proved to be very useful for the determination of the distance between the measuring gear and the sea floor, above all for water samplers, photo cameras, and television cameras as well as for geological samplers. This particular form of echo sounding makes use of a sound transmitter at the instrument, the so-called pinger, which emits a short signal approximately once per second. The signal returns to the ship via two different paths: first, directly and, second, after reflection at the sea floor. The time interval between the two resulting signals represents a measure of the distance from the sea floor. Thus, even with depths of many thousand meters, the distance between the instrument and the sea floor can be determined with an accuracy in the range of decimeters.

3.2.6. Measurement of water level variations

It has long been understood that knowledge on the sea level and its variations with time is of great importance to the practical concerns of coastal engineering, of shipping, of land and sea surveying, and of hydrology. Therefore, numerous devices, adapted to the particular measuring problems, are available now. According to the purpose they serve, they may be divided into two groups: coastal gauges and high-sea gauges.

As coastal gauges, three different types are used: tide staffs, float gauges with direct mechanical recording or with electrical telemetering of the data, and compressed-air gauges.

The *tide staff* is a measuring staff with a centimeter scale from which the water level can be read directly. For practical reasons, the zero mark should be placed as low as possible on the staff; at German gauges it is generally at -5.00 m relative to the mean sea level of the North Sea. In this way only positive values of the water level are obtained. For any gauge, the zero level must be fixed by triangulation with respect to a geodetic bench mark on shore.

Float gauges, with direct mechanical registration of the water level, are among the oldest recording instruments used in oceanography. In 1831 the first recording tide gauge

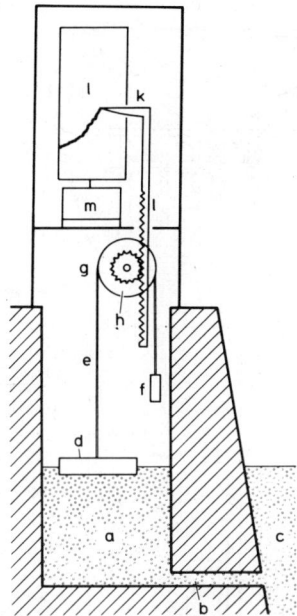

Fig. 3.21. Schematic structure of a float gauge with mechanical recording: shaft (a), connecting tube (b), open sea (c), float (d), wire (e), counterweight (f), pulley (g), gear rim (h), rack (i), pen (k), recording drum (l), and clock-work (m).

was installed in England. The German standard gauge for precision measurement was developed by Seibt–Fuess in 1891. Its functional principle, which it has in common with numerous modern float gauges, is shown in Fig. 3.21.

In a shaft, which is connected with the open sea by a thin tube, a float moves according to the water level, thus controlling a pen. The diameter of the tube must be around 1000 times smaller than that of the shaft to provide for sufficient damping of wave action. In addition to the direct recording of the water level at the measuring site, the telemetering of data by cable and radio is also possible. Such transmission is necessary for the short-term forecasting of storm surges.

The float gauges, which can also be adjusted to different requirements regarding time and height scales, have proved reliable in many ways. The sole disadvantage is that it is necessary to construct an expensive gauge shaft which must be protected not only from waves and ice but also from sand deposits. Therefore, even at an early stage, attempts were made to find other principles of measurement that did not need such shafts.

The *pneumatic gauge* meets this demand. A diving bell lies on the sea floor at such a great depth that pressure variations due to surface waves can be neglected. By means of an air-filled tube, the hydrostatic pressure is connected with a recording manometer on shore or on a measuring pole. Although this principle seems simple enough, it has not always functioned satisfactorily when used in continuous operation. In particular, the tube can be clogged by condensed water, which may falsify the accuracy of the measurement.

High-sea gauges are employed to measure water level variations in the open sea, especially to obtain tidal records. All of these gauges are pressure gauges with automatic registration, recording fluctuations of the pressure at the bottom of the sea. Older in-

130 Oceanographic Instruments and Observational Methods

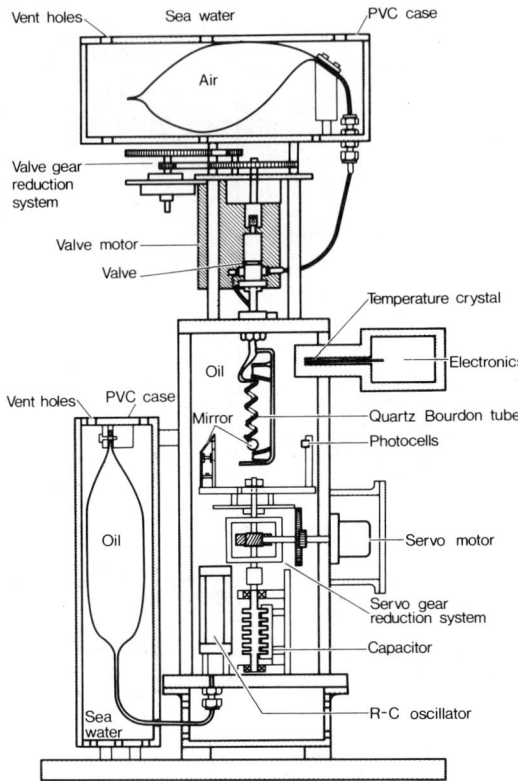

Fig. 3.22. Schematic structure of the Harvard deep-sea gauge. After the gauge is lowered to the sea floor, the valve closes the inner part of the quartz Bourdon tube, and the sensor reacts on pressure variations in its surroundings. The resulting turning of the mirror is recorded via photo cells, then converted by a servo system into capacity changes of the capacitor and used for the generation of an electrical alternating voltage signal with variable frequency. (After Baker, 1971.)

struments, for example, the high-sea gauge after Rauschelbach, measure the volume of an enclosed quantity of air which, according to the law of Boyle–Mariotte, is proportional to the temperature and inversely proportional to the pressure. Since the variation of the volume depends on temperature, a simultaneous temperature measurement with a mercury thermometer is necessary. The recording is done by photography. The maximum depth of operation lies at approximately 300 m. Later developments after Graafen permit an automatic fixing of the measuring range after the instrument has been laid out.

In recent deep-sea gauges, the vibrotron (Eyries, 1968; Caldwell et al., 1969), or membrane boxes with capacitive scanning, or quartz Bourdon tubes are used as sensors. As an example, the Harvard deep-sea gauge (Baker, 1971) may be briefly described here (Fig. 3.22 shows the schematic structure). The sensor is a quartz Bourdon tube of exceptionally great stability; it is free of hysteresis. What is measured is not the absolute pressure but the pressure relative to a particular reference pressure so that a high accuracy may be achieved when variations of pressure are measured. During the lowering of the instrument, the valve between air bag and Bourdon tube is open, and the pressure inside the measuring tube is equal to that outside. After the instrument has arrived at the sea floor, the valve is closed as soon as the temperature inside the instrument is adjusted to the temperature of the water. Then the pressure in the Bourdon tube is only dependent on temperature, while the pressure in the oil-filled case outside the measuring tube equals the hydrostatic pressure.

Deformations of the quartz tube by pressure variations make the mirror, which is fixed to the free end of the tube, turn. A photoelectric servosystem follows the turning of the mirror and thereby shifts a capacitor which, as part of an oscillator, changes the frequency of an alternating voltage. The frequency of this signal, as well as that of a corresponding signal from a temperature measurement with a quartz thermometer (cf. Section 3.2.8), is measured electronically and stored in digital form on a magnetic tape together with the data of a quartz clock. The deep-sea gauge can be employed at more than 6000 m water depth up to 6 months, and it shows a stability of the pressure registration corresponding to a few millimeters of the water column.

3.2.7. Measurement of ocean surface waves

For shipping, wind waves and swell imply severe impediment which designers try to reduce to a minimum by suitable construction of the vessels. A necessary prerequisite is the knowledge of the characteristic properties of the state of the sea. The same holds true if the effect of the waves at the coast is to be determined with regard to sand transport, beach formation, and coastal structures. In oceanographic research, special significance is given to the sea state as an important process in the energy transfer between ocean and atmosphere. Wave measurements can be carried out from the sea floor, at the sea surface, or from an airplane. Most of the methods yield information on wave height and wave period only.

The most important method for *measurement at the sea floor* is the observation of the bottom pressure fluctuations that occur in connection with surface waves and the following calculation of the surface displacements in accordance with the linear wave theory (cf. Section 8.2.1). At any rate, the waves must be long enough for noticeable variations of pressure to be sensed at the bottom of the sea. Such calculation yields correct values with an accuracy of approximately 20%. Comparative measurements show that, in general, the fluctuations of pressure are smaller than predicted by theory. The agreement is fairly good at depths where the fluctuations of pressure are already very small. Pressure sensors which can be applied at the sea floor or moored at medium depth are described in Section 3.2.5. The recording takes place, according to the type of the instrument, either in the underwater casing of the instrument or on shore after the data have been transmitted by cable.

Another method, though of less importance up to date, is the measurement with a reversed echo sounder (cf. Section 3.2.4). Here, an acoustic transducer, which is located at the sea floor or moored at medium depth, emits narrow-beam signals in upward direction. The travel time of the sound from the transducer to the reflecting sea surface and back is determined by short-time measurement. The main difficulty with this method is that, in rough weather, the water beneath the sea surface contains numerous air bubbles which, owing to scattering and absorption of sound, prevent the generation of any usable reflected signal. With long swell waves, however, this method has proved practicable.

A greater number of methods is available for *measurement at the sea surface*. Float gauges or electrical measuring devices are suited for measurement from fixed platforms, like bridges at the coast, poles, and research towers. The float gauges for the measurement of surface waves correspond in their structure to those for the measurement of tides, as discussed in Section 3.2.6. There is, however, some difference: the surrounding water can enter the shaft through a number of vent holes without being hindered, that is, there is no hydraulic filtering, and the recording speed is much greater.

With electrical methods, variations of the water level are converted into changes of electrical resistance or capacitance. In the former method, several pairs of electrodes, mounted vertically one upon the other, are short-circuited by the seawater, so that contacts are actuated which change an arrangement for measuring resistance. Thus, the total resistance is dependent on the water level and can be determined by electrical bridge arrangements. Pairs of resistance wires, suspended vertically and parallel to each other, are also used, so that the change of resistance between the two upper connecting points represents a direct measure of the displacement of the water surface.

With the capacitive method, an insulated wire is used which is stretched and clamped in vertical position. In the range where it is wetted by water, it is a cylindrical capacitor. Then the variations in length of the wetted cylinder, or the variations in size of the capacitor, respectively, correspond to the variations of the water level. The water level can be determined by a capacitance measurement, for instance by the resonance oscillation of an oscillatory circuit built up with this capacitor. However, this method is to be employed only with some limitation because, with the water level decreasing, a film of water will remain, for some time, on the cable above the water surface, falsifying the record.

Wave observations from a ship can be carried out visually, photographically, or by means of built-in wave-measuring devices. The most complete representation of the form of surface waves is obtained by stereophotogrammetric pictures. Here, a spatial picture of the sea surface is taken by simultaneous photographic shots by two cameras at either end of a fixed base on shipboard. The complicated evaluation and the restricted spatial extension of the range of the picture have prevented wider application of this method.

On the other hand, the "Shipborne Wave Recorder" (Tucker, 1956) has considerably contributed to the knowledge of the sea state in the world ocean, although it has been employed on few ships only. With this recorder, the pressure is measured by a sensor mounted on the ship below the water line. Thus, waves can be observed whose lengths are short compared to the length of the ship. An accelerometer records the vertical displacement of the ship and, after double integration, it yields a measure of the long waves. The motion of the ship in the water can be eliminated because the motion of the accelerometer relative to the mean water level is accompanied by an apparent displacement, in the opposite direction, of the mean water level at the pressure gauge.

In the open sea, those methods of wave measurement employing buoys are the most important ones by far. These are surface-following buoys of disk-like or spherical shape with accelerometers installed within. In the "Waverider" buoy (Fig. 3.23), which is moored, the accelerometer is mounted in gimbals so that it is affected by vertical acceleration only. The twice-integrated data provide a measure of the water level. They are transmitted by radio to a receiver station on land or on a ship where they are recorded. In case information is desired about the dependence of waves on direction in addition to wave height and period, data must also be obtained with regard to the slope and, better still, also to the curvature of the sea surface. This demand is met by a combination of accelerometer and slope meter within the buoy, the so-called "Pitch and Roll Buoy," and, finally, by a combination of at least three such buoys in a measuring array. These accelerometer buoys can be employed when moored or when connected with a ship by cable.

Oceanographic Instruments and Methods of Measuring   133

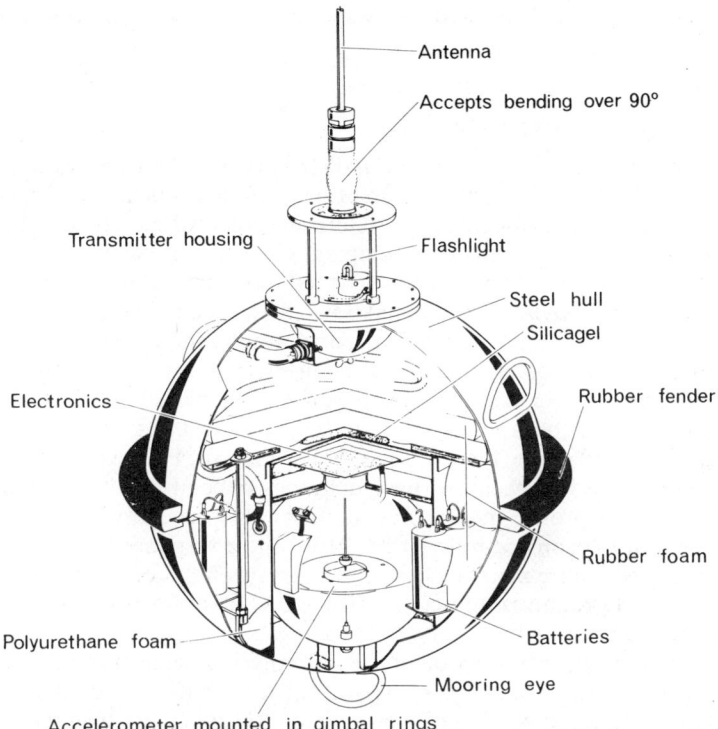

Fig. 3.23. "Waverider" buoy for measurement of the sea state. The vertical motions of the elastically moored buoy are recorded through acceleration measurements and the data are transmitted by radio to a receiver station on a ship or on land (Drawing: Datawell, Haarlem).

Other methods have also been in use for which a pressure gauge is attached to a surface-following buoy at a depth where the variations of pressure caused by the sea state are so small that they may be neglected. Therefore, the pressure sensor reacts exclusively on any variations of the hydrostatic pressure that are consequences of its vertical displacement.

From low-level *aircraft* and, to a certain extent, also from *satellites* (cf. Section 3.1.1.5) wave data can be obtained by several methods (Hasselmann, 1972). Hereby the back scattering of electromagnetic microwaves in the centimeter to decimeter range of wavelength can be observed. The motion of the sea surface leads to Doppler frequency shifts in the back-scattered signal which represent a measure of the frequency spectrum of ocean surface waves. On the other hand, it is also possible to obtain information on the sea state from the change of the increase of impulse in the radar signals reflected at the sea surface.

Finally, it should be mentioned that attempts are being made to draw conclusions with respect to the sea state from simultaneous measurements of the sea surface temperature by infrared and microwave radiation. This method is based on the fact that, as a consequence of surface roughness, the apparent temperature of a nonideal black body, as represented by the ocean, varies depending on the wavelength of the radiation. However, the formation of white caps in the seaway may also cause a variation of the apparent temperature of the sea surface.

3.2.8. Measurement of temperature

In practice, the determination of the spatial distribution of the temperature of the ocean and its variations with time involves the solution of three problems: (1) the temperature measurement at the sea surface, (2) the determination of the vertical temperature distribution, and (3) the observation of the temporal variations at particular points of certain depth levels. The instruments have been adapted to these requirements.

Temperature-measuring sensors make use of the dependence on temperature of various properties of solid or liquid bodies, for example, the thermal expansion of solid or liquid substances, the behavior of crystals in their characteristic oscillation, or the electrical conductivity of conductor or semiconductor material. The selection of the sensor that is most suitable for oceanographic investigations depends, above all, upon the demands on accuracy and on the response speed of the sensor. However, rules that are valid generally cannot be given because those demands depend on the specific measuring problem. For routine profile measurements in the deep ocean with instruments from shipboard, for example, an accuracy of $\pm 0.01°C$ and a response time of approximately 0.1 sec are required, while near the coast, the measurement of temperature variations with time often only requires an accuracy of $\pm 0.1°C$ and a response time of several minutes (cf. Section 3.2.2).

From the time of the introduction of the fluid thermometer in the 17th century until the middle of this century, our knowledge on the distribution of oceanic *surface temperature* was mainly based on the ordinary mercury thermometer giving the temperature of water samples taken by buckets. By this method an accuracy of $\pm 0.05°C$ can be attained, provided that changes by external influences like transfer of sensible heat from the ambient air or from the bucket, radiation, and evaporation are kept negligible. Occasionally, special insulated water samplers have been used for this purpose. More recently, the electrical measurement of sea surface temperature has become more important. In general with this method, the temperature at the intake of the cooling water into the vessel is measured.

The surface temperature of the ocean can also be measured from airplanes or satellites (cf. Section 3.1.1.5) by observing the intensity of the infrared radiation of seawater within the wavelength range of 8 to 12 μm. Since at these wavelengths the absorption in seawater is rather high, only the radiation from a very thin surface layer—less than one-tenth of a millimeter thick—gets directly into the atmosphere. Special problems concerning remote sensing with this method arise from the additional attenuation and emission in the atmospheric layer between the water surface and the measuring platform, from the occurrence of reflected radiation, and from the necessity to receive radiation from a very narrowly limited area in order to achieve a good horizontal resolution (Saunders, 1970). Measurements carried out from an airplane, flying at an altitude of a few hundred meters, can reach an accuracy of $\pm 0.2°C$ and a horizontal resolution of approximately 20 m. The accuracy of data obtained from satellites is smaller by one to two orders of magnitude. Methods of this kind are especially suitable for the observation of large-scale variations in the surface temperature. So, for instance, the position and form of the Gulf Stream have repeatedly been observed over a number of years.

The measurement of the *water temperature in deeper layers* is impossible with water samples taken by buckets, since the time interval between sampling and reading would

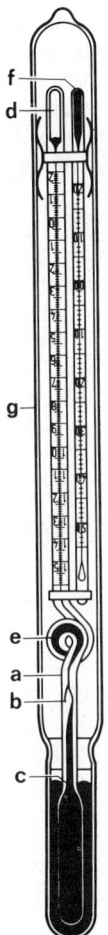

Fig. 3.24. Protected reversing thermometer by Richter and Wiese.

be too long. Thus, the temperature measurement must be carried out in situ, that is, at the respective depth, and the data must be immediately fixed or telemetered. After first experiments with insulated water samplers, maximum–minimum thermometers were employed until the 19th century. Those thermometers had two significant deficiencies: a maximum–minimum thermometer can indicate the correct temperature only if the temperature either decreases or increases with depth. Temperature inversions, however, remain undetected. Furthermore, unavoidable shocks can easily result in uncontrollable displacement of the index particles and can therefore indicate incorrect values.

The problem of how to obtain exact and reliable temperature measurements at arbitrary depths was solved with the discovery of the breaking principle, which is used in the reversing thermometers. This principle was introduced by Negretti and Zambra in 1878. Around 1900, Richter of Berlin, in collaboration with Nansen, developed the reversing thermometer to its present degree of perfection. With a maximum accuracy of ±0.01°C, this instrument meets stringent requirements. Figure 3.24 represents a reversing thermometer. At the point where the appendix b of the capillary joins the capillary a of the main instrument, the mercury column breaks when the thermometer is reversed by 180°, and the mercury supply vessel c turns to the top. The separated portion of the mercury runs into the extension d of the capillary, and the height in the capillary thus corresponds to the temperature. The loop e of the capillary catches the mercury which, in the reversed state of the thermometer, must necessarily flow out of the mercury vessel c when the temperature becomes higher than it was at the time of the breaking. With changing temperatures, the separated mercury in the capillary is also subjected to volume changes which influence its length in the capillary. These volume changes depend on the temperature T' of the main thermometer, on the temperature t of the auxiliary thermometer at the time of reading, on the volume of the amount of separated mercury $V_0 + T'$, and on the constant K which is determined by the thermal expansion of the glass and the mercury (for Jena glass $K = 6100$). The volume V_0 is an instrumental constant. It is defined as the volume of the vessel d in Fig. 3.24 up to the zero point of the thermometer scale expressed in units that correspond to the volume of the capillary per 1°C. The temperature t is obtained by the auxiliary thermometer f in Fig. 3.24. The deviation C of the temperature can be calculated from:

$$C = \frac{(T' - t)(V_0 + T')}{K - [\frac{1}{2}(T' - t) + (V_0 + T')]} \tag{3.03}$$

Besides this correction, which takes into account the deviations of the recording that result from changes of the ambient temperature when the thermometer is taken aboard, an index or scale correction is necessary. Such procedure eliminates errors that are caused by insufficient congruence of the length of the mercury column and the scale.

To guarantee the required accuracy of the temperature, apart from the calibration of the scale, checks of the zero point of the thermometer are necessary and must be repeated from time to time since the zero point is likely to change with the aging of the glass. For protection against water pressure, the primary and auxiliary thermometers are surrounded by a strong glass tube g. To provide good thermal conduction between the seawater and the mercury vessel of the main thermometer, the space between the vessel c and the glass tube is filled with mercury.

If the protecting tube is kept open at one end, the interior of the thermometer is subjected to hydrostatic pressure, and, because of elastic deformation of the thin glass vessel, the mercury height in the capillary will increase at a rate proportional to the pressure at that particular depth. In this way, an unprotected reversing thermometer in conjunction with a protected thermometer serves as a tool for thermometric depth determinations (cf. Section 3.2.5).

In marine research, fluid thermometers—apart from the surface and reversing thermometers with mercury—have been successfully employed only in the bathythermograph (Spilhaus, 1937). Here the fluid thermometer is a spiral tube filled with an organic fluid, and at its end there is a Bourdon tube with a stylus. A temperature increase is followed by a volume increase and, thus, by a pressure increase in the Bourdon tube (cf. Section 3.2.5), the top of which gets deflected. The stylus scratches a trace on a sooted or gilded glass plate which is shifted, vertically to the plate, by a Well tube at a rate proportional to pressure. This instrument is a simple one, and it can be applied down to 270 m depth from a ship underway. The measuring accuracy of ±0.1 to 0.2°C is considerably inferior to that of the reversing thermometer, but with strong vertical temperature gradients in subsurface layers it will often meet the demands for measurement.

More recently, electrical methods for measuring temperature have gained great importance. For this purpose, platinum resistance thermometers or thermistors are used almost exclusively as sensors. Such platinum resistance thermometers permit high accuracy and reproducibility of the measured quantity with nearly linear dependence of the electrical resistance on temperature. There is one disadvantage, however: a temperature increase of 20% causes an increase of resistance of only 7%, and so a considerable amount of electronics is needed if high accuracy and resolution are to be attained. Thermistors consisting of semiconductor material, which are also called NTC resistors (negative temperature coefficient), have the disadvantage of an exponential decrease of resistance with increasing temperature. Furthermore, their stability is usually less than that of platinum resistance thermometers. Several more recent thermistors, however, have shown very good stability values of less than 0.001°C over a period of some months. A great advantage of the thermistors is their pronounced change of resistance with temperature. Typical values show a decrease of resistance between 3 and 5% with a temperature increase of 1°C. Therefore, the necessary amount of electronics is smaller, but the record is more or less nonlinear.

For quick spatial recording of temperature stratification, towed instruments are employed. "Thermistor chains" (Richardson et al., 1952) are towed cables with several thermistors, where the data are simultaneously recorded in the form of isotherm graphs. Instruments that periodically move up and down in the mixed layer, like the "Dolphin" (Joseph, 1962) or the "Batfish," provide a resolution that is higher in the vertical but smaller in the horizontal direction. Towed instruments for measuring isopycnic surfaces in deep water were applied by Katz (1973).

There is a variety of instruments for temperature measurement with platinum thermometers or thermistors which usually utilize Wheatstone Bridge methods. Measurement of temperature is especially important when in-situ salinometers are used, which will be discussed in Section 3.2.9. Furthermore, the "Expendable Bathythermograph" (XBT) should be mentioned, which is a free-falling temperature sonde to be employed only once. It is connected with the ship by a thin cable that breaks off as soon as a maximum depth is passed. This instrument can be applied from a ship underway and yields an accuracy in temperature of ±0.2°C.

An electrical sensor with particularly high accuracy is the oscillating quartz. In the electronic measuring technique, the piezoelectrical effect with quartz crystals is often used for generating signals of alternating voltage with very stable frequency. This is achieved by means of the mechanical resonance of the crystal. Disturbances, caused by temperature variations, may shift this frequency, but that can be prevented by thermostatizing the quartz. Conversely, for oceanographic purposes, the influence of temperature on the frequency of oscillation can be drawn upon for temperature measurements. Such an effect is particularly great for specific directions of oscillation in the crystal. Advantages of this sensor are seen in the measuring accuracy and stability that can be attained by it (0.01°C month^{-1} up to 0.003°C within 2 yr) as well as in its suitability for telemetering. The size of this sensor, however, which implies a slow response to temperature variations, is a disadvantage. Quartz thermometers have been successfully employed in measuring the vertical temperature change in deep-sea trenches and thus have proved that the temperature increase with depth very closely corresponds to the adiabatic temperature gradient (cf. Section 2.1.5).

3.2.9. Measurement of salinity

As discussed in Chapter 2, the physical properties of seawater can be represented as a function of salinity, temperature, and pressure. In principle, therefore, salinity can be obtained by determining physical properties like density, optical refractive index, electrical conductivity, or sound velocity, and, in addition, temperature and pressure, and then by calculating the salinity on the basis of the well-known functional relation. Some difficulty arises from the strong dependence of all those physical quantities on temperature as well as from the high measuring accuracy that is required. At a salinity (S) of 35‰, a temperature (T) of 15°C, and atmospheric pressure, a change in the salinity by $\Delta S = 0.01$‰ is caused by the following deviations:

Measurement of density σ_T: $\Delta \sigma_T = 0.0077$ $\Delta T = 0.035°C$

Measurement of the
optical refractive index n: $\Delta n = 0.0000019$ $\Delta T = 0.021°C$

Measurement of the specific
electrical conductivity L: $\Delta L = 0.011$ mS cm^{-1} $\Delta T = 0.011°C$

Measurement of the
sound velocity c: $\Delta c = 0.012$ m sec^{-1} $\Delta T = 0.004°C$

Because of such difficulties, oceanographers at first used a *chemical method of determination*. At the suggestion of the International Council for the Exploration of the Sea in Copenhagen, Knudsen, Forch, and Sørensen (1902) developed a volumetric method for the determination of the chloride content. Due to the constancy of the composition of seawater, as discussed in Section 2.1.2, the salinity can be calculated, in very good approximation, from the chloride content. The accuracy required for titration is obtained by a special type of glass apparatus—the Knudsen burette and the Knudsen pipette.

The process of operation is as follows: the Knudsen burette is filled with a solution of silver nitrate. Then the seawater sample that is to be examined is put with a Knudsen pipette into a beaker, and a bit of potassium chromate solution is added. While stirring, one lets the silver nitrate solution run as a thin stream into the yellow-colored seawater sample, whereupon silver chloride precipitates in the seawater sample forming a white caseous deposit. When all the chlorine ions that were contained in the seawater sample have precipitated as silver chloride, silver chromate is formed, which is to be recognized by some red color appearing in the deposit. From the amount of the silver nitrate needed until the change of color takes place, and from the amount of seawater used, the chloride content of the seawater sample can be calculated. The process of the Cl titration, according to Mohr, is based on the following chemical formulas:

$$\text{NaCl} + \text{AgNO}_3 = \text{AgCl (white)} + \text{NaNO}_3$$
$$58.454 \text{ g} \quad 169.888 \text{ g} \quad 143.337 \text{ g} \quad 85.005 \text{ g}$$
$$2\text{AgNO}_3 + \text{K}_2\text{CrO}_4 = \text{Ag}_2\text{CrO}_4 \text{ (red)} + 2\text{KNO}_3$$

An accuracy of ±0.02‰ in the determination of salinity is attained by comparative measurements with Copenhagen Normal Water or with a substandard water sample prepared from it. Such comparative measurements are always repeated after a certain number of titrations.

Progress in electronical measuring technique, however, has now given greater importance to a physical method, namely the determination of salinity by way of measuring *electrical conductivity*. In laboratories, it is possible to compensate the temperature influence on conductivity in water samples to such an extent that an accuracy of $\Delta S = \pm 0.002$‰ is achieved as long as calibration with Copenhagen normal water is carried

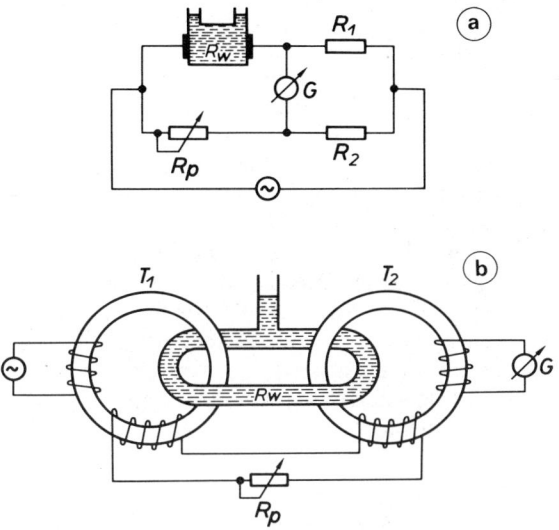

Fig. 3.25. Principles of measurement of the electrical conductivity. (a) After the galvanic method with electrodes the electrical resistance R_w of the electrolyte seawater is determined. In the arrangement illustrated here, the alternating current in the galvanometer G is set to zero if the potentiometer R_p is adjusted to $R_w/R_1 = R_p/R_2$. (b) After the inductive method the transformers T_1 and T_2 are coupled by a seawater loop with the resistance R_w and by a compensation loop. The degree of the coupling is a function of R_w. In the system shown here, the alternating current in the galvanometer G is set to zero if the potentiometer R_p is adjusted in such a way that the magnetic field induced over the seawater loop in T_2 is just compensated by the magnetic field induced by the compensation loop in T_2. (After Siedler, 1968b.)

out frequently. With continuous in-situ measurements of conductivity, temperature, and pressure, accuracies of approximately $\Delta S = \pm 0.02‰$ can still be attained. Instruments for salinity determination via conductivity are called "salinometers."

In oceanography, the determination of electrical conductivity is carried out by means of inductive or galvanic methods, the principle of which is explained in Fig. 3.25. In the simplest case of a galvanic measurement, the resistance between two electrodes is determined by a Wheatstone Bridge. Alternating current in the low-frequency range is used in order to avoid polarization effects (cf. e.g., Wenner et al., 1930; Schleicher et al., 1956). As demonstrated in Fig. 3.25(a), the seawater resistance R_w can be calculated from $R_w = R_p \cdot R_1/R_2$, if the potentiometer R_p is adjusted in such a way that the current passing through the galvanometer G becomes zero. Usually the electrodes are ring shaped and fixed to the inside of a glass tube. When in-situ measurements are performed with an open glass tube through which water is flowing, arrays of three electrodes, with the two outer rings on the same electrical potential, are employed to avoid any conduction that is susceptible to disturbance outside the measuring cell. For very precise in-situ measurements it is best to use arrays of four electrodes: a constant electric current is sent over the two outer electrodes, and the voltage drop occurring over the distance between those two electrodes is measured by the two inner electrodes and a voltmeter of high input resistance.

For inductive conductivity measurement with alternating current of the low-frequency range, two transformers are used which are coupled over a seawater loop (Hinkelmann, 1958, Brown et al., 1961). In an arrangement as shown in Fig. 3.25(b), the resistance R_w of the seawater loop can be drawn from the adjustment of the potentiometer R_p, if the current induced in the transformer T_2 over the seawater loop is just compensated by the compensation current of the R_p loop. The galvanometer, then, reads zero. The two ring-shaped transformer cores are arranged one on top of the other so that water can flow through in vertical direction. Customary measuring instruments have different compensation circuits and presentation devices. As to laboratory salinometers, the stepwise compensation by changeover switching the transformer winding has received the greatest attention. In addition, a temperature compensation is carried out over a special circuit by a thermistor.

In-situ salinometers, which have become known as "Bathysonde," "STD," "CTD," and "Multisonde," permit the simultaneous measurement of electrical conductivity, temperature, and pressure (Hinkelmann, 1957; Siedler, 1963; Kroebel, 1973). They consist of an underwater instrument, a cable connection to the ship as well as a control and recording device on board which, according to the relevant model, provides for visual presentation, analog recording and registration on computer-adapted formats. Occasionally, self-contained measuring devices with automatic recorders are also applied. Figure 3.26 gives an example of an in-situ salinometer with data transmission by cable.

Methods using *measurement of density or refractive index* play but a minor role next to titration and conductivity determination. Sometimes, however, it may be useful to employ them when the demands on accuracy and applicability are limited. For the routine determination of salinity via density measurement, areometers are used in the form of calibrated glass floats, mostly stalk areometers: the water sample is filled into a cylinder, care is taken that the temperatures are balanced, and the aerometer is put in. From a calibrated scale (graded in ‰ salinity) in the stalk of the aerometer the distance that the instrument is immersed in the seawater can be read. Simultaneous measurement of temperature and the corresponding correction of the salinity reading are necessary. This method is a simple one and yields accuracies of about $\Delta S = \pm 0.05$‰. For salinity determination by measurement of the optical refractive index (cf. Section 2.1.11), the immersion refractometer after Pulfrich is employed in a thermostatized seawater sample. The accuracy that is attainable lies at $\Delta S = \pm 0.1$‰.

Since great progress has been made in measuring technique with a view to the determination of salinity by measurement of sound velocity, the application of this method might become practical within a reasonable space of time.

3.2.10. Measurement of sound velocity

The propagation of sound in the sea is greatly dependent upon the stratification of sound velocity. The distribution of sound velocity in space is determined either indirectly from measurements of temperature, salinity, and pressure (cf. Section 2.1.10) or directly by measuring the travel time or wavelength.

For the determination of travel time, the time interval Δt is measured that a sound signal needs to pass over a certain distance Δx from the transmitter transducer to the receiver transducer. Then the sound velocity c results from:

$$c = \frac{\Delta x}{\Delta t} = \Delta x \cdot f \quad \text{with} \quad f = \frac{1}{\Delta t} \qquad (3.04)$$

The time measurement can be performed very easily if the signal arriving at the receiver transducer immediately triggers a new pulse for transmission. If small errors, caused by the time the signal travels through the electrical circuits, are disregarded, the repetition

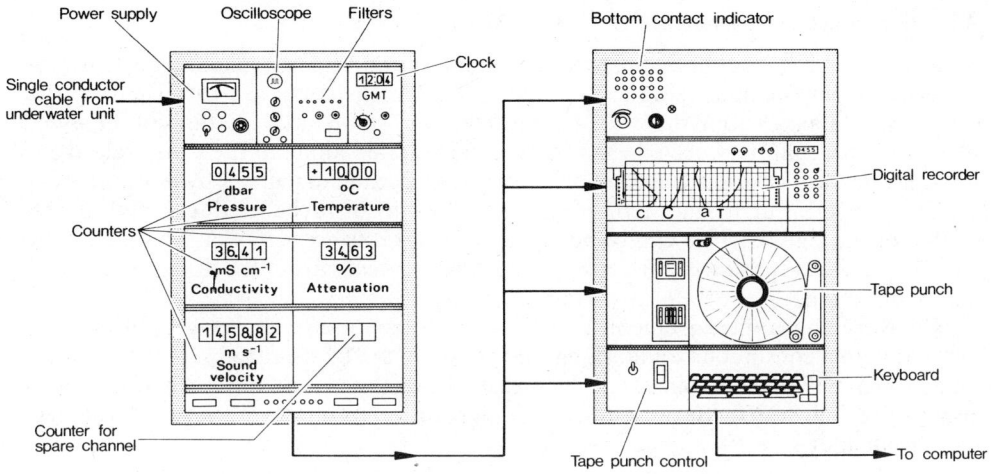
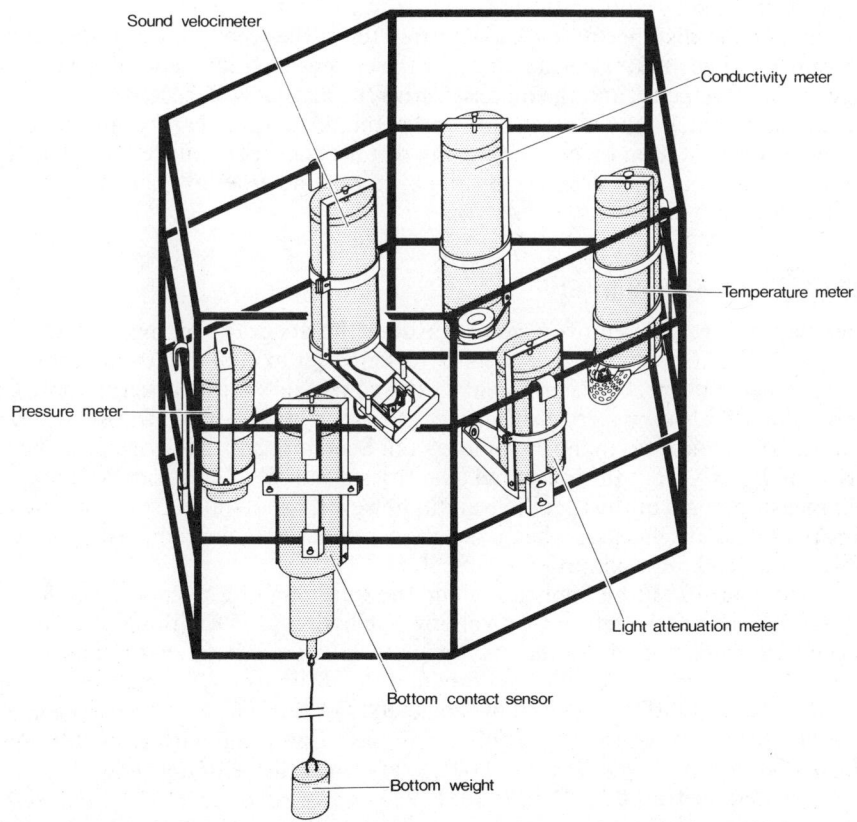

Fig. 3.26. Schematic construction of the Kiel Multisonde. Below: Underwater unit with casings for sensors measuring various quantities and for the bottom contact sensor. Above left: Shipboard equipment with control and presentation devices. Above right: Shipboard equipment with bottom contact indicator and data recorder (Design: Institut für Angewandte Physik, Kiel).

frequency of the revolution is proportional to the sound velocity (Eq. 3.04). This type of measurement is called "sing-around method" (Greenspan et al., 1957). For in-situ measurements, precaution must be taken to keep very small any variations of the sound velocity record that may occur when the sound is carried along in the water relative to the transducers. These variations may be caused by ocean currents or by motions of the instrument itself. Furthermore, there should not be any disturbance due to multiple reflection of the signals. This can be achieved by a suitable geometry of the sound path, that is, by the installation of reflectors so that the greatest part of the distance is passed in both directions. The attainable accuracy is about some 10 cm sec^{-1}.

The wavelength measurement, which, however, is not often employed, utilizes the fact that with a continuous sound signal the phases of the transmitted signal and the received signal will always coincide if the measuring distance is a multiple n of the wavelength (Kroebel, 1970). Because of the relation between sound velocity c, wavelength λ and frequency f:

$$c = \lambda \cdot f = \frac{\Delta x}{n} \cdot f \qquad (3.05)$$

This can be reached by shifting the frequency of the signal until the phases coincide. With a fixed measuring distance and a known value for n, the frequency is a measure of the sound velocity. The greater n is, the greater the change of frequency will be at a certain change of sound velocity, and the more sensitive the method will become. The ambiguity of the measurement, due to n being only approximately known, can be eliminated by the simultaneous application of two measuring distances of different length. The method is more expensive but also more accurate—by about one order of magnitude—than the "sing-around method."

3.2.11. Measurement of density

In the open ocean, the requirements regarding the accuracy of the measurement of density, or of specific weight (cf. Section 2.1.4) amount to $\Delta \sigma_T < 0.01$, and thus are too great as to permit direct measurement in the course of routine observations. Only in coastal waters with very strong differences in the hydrographic stratification may it sometimes be practical to employ aerometers (cf. Section 3.2.9). So, in general, the density is derived indirectly from the measured quantities of salinity, temperature, and pressure. Direct measurements of the specific weight, however, are required for fundamental determinations, and also in cases when it is not guaranteed that the composition of seawater is constant in good approximation (cf. Section 2.1.2).

Methods that might be applied include the weighing of a pycnometer, that is, of a glass vessel with exactly determined volume, the hydrostatic weighing of a float, or the frequency determination of the characteristic oscillation of a body that is dependent upon the density of seawater. For the fundamental determination of the specific weight of seawater Knudsen (1902) employed pycnometers. For this purpose, the pycnometer must be weighed when empty and when filled with pure water and with seawater. Knudsen attained a reproducibility of $\Delta \sigma_T = \pm 0.003$. Hydrostatic weighing showed less sources of error, as demonstrated by Bein et al. (1935) and Cox et al. (1970). Here, too, the weighing must be carried out several times, namely the weight of the float must be determined in air, then fully immersed in seawater and in pure water. The measurements of Cox et al. yielded a reproducibility of $\Delta \sigma_T = \pm 0.001$.

For the direct measurement of the specific weight, Kremling (1972) made use of the fact that the resonance frequency of a small, V-shaped glass tube, filled with seawater, is dependent on the density of seawater. By thermostatizing the measuring arrangement, by inducing oscillation electronically, and by measuring the frequency of oscillation electronically, he attained a reproducibility of $\Delta\sigma_T = \pm 0.005$.

3.2.12. Measurement of currents

3.2.12.1. Measurements at a fixed position

Currents are measured for the purpose of determining either stationary current vectors or stream lines corresponding to the Eulerian form of the equations of motion, or the drift of water particles, that is, their trajectories, according to the equations of motion after Lagrange (cf. Chapter 7). Two groups of measuring methods result from this difference.

Stationary current vectors can be determined not only directly but, in the case of geostrophic currents, also indirectly from the mass field and the resulting pressure field. The method will be described in Chapter 7. The prerequisites of direct stationary measurements are fixed points in the ocean, that is, measuring platforms. For this purpose, it is usual that moored systems as described in Section 3.1.2.1 are employed. Only occasionally are moored ships or research towers used.

Two quantities are to be determined by *sensors for current measurement:* either the absolute value and the direction of the velocity or its components in a right-angled coordinate system. The absolute value is usually obtained by means of mechanical sensors like propellers, rotors, paddle wheels, or turnstiles with hemispherical bowls, the rotation rates of which are measured. A standing or hanging pendulum-shaped body with great current resistance can be used likewise because the deflection from its position of rest is a measure of the current velocity. The direction is determined by means of a current vane relative to the northern direction (permanent magnetic compass, induction compass) or to the bearing of a fixed measuring stand. The current direction is always the direction in which the current flows, in contrast to the atmosphere where the wind direction is the direction from which the wind blows. With the pendulum-shaped current meter the direct measurement of the components can also be achieved (Krause, 1973).

Recently attempts have been made to develop sensors without moving parts. Three types of devices seem to be promising. The electromagnetic current meter utilizes the fact that in a conductor moved perpendicular to a magnetic field a current is induced. Two pairs of electrodes are employed which are mounted at a right angle to each other in a horizontal plane. In their range, an alternating magnetic field is generated. When the electrolyte seawater flows over the arrangement, the induced voltage can be measured by the electrodes (Olson, 1972; Thorpe et al., 1973).

The acoustic current meter, which makes use of the fact that the sound is carried along with the moving water (Suellentrop et al., 1962), consists of two sound paths for each coordinate direction through which the sound propagates in opposite directions. The difference in the travel times is a measure of the carrier velocity. It is also possible to use only one measuring distance which is passed by the sound in both directions either simultaneously or in rapid succession. In the acoustic Doppler current meter, the frequency shift of a sound signal is measured which returns to the sound source after being reflected at a particle moving in the water (Chalupnik et al., 1962). Experiments have also been made in order to introduce a corresponding optical method by means of lasers and interferometers.

144 Oceanographic Instruments and Observational Methods

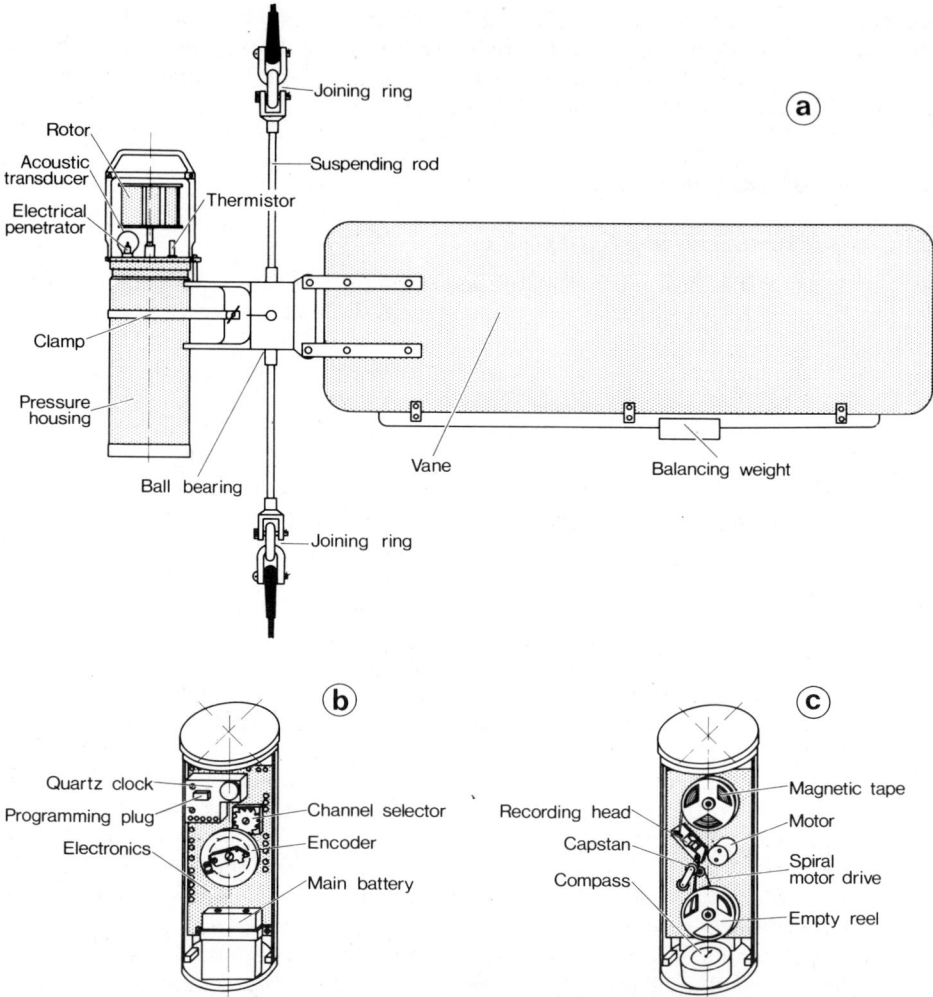

Fig. 3.27. Bergen current meter for the application in moored measuring systems (cf. Fig. 3.09): (a) total outside view with sensors and recorder casing left and current vane right, (b) inside view with control and encoder units, (c) inside view with recording magnetic tape equipment and compass.

Although in *current meters* the rapid measurement of the components with subsequent averaging is desirable in order to eliminate any disturbance by the motions of the sea or the moorings (cf. Section 3.2.2), most instruments, because of the considerably smaller expenditure, operate in such a way that the speed averaged over a certain time interval and the momentary direction of the current at the time of the data collection are determined. Current meters usually do the recording automatically; only in exceptional cases is transmission by cable or radio required. In the following, the Bergen current meter (Anderaa, 1964) and a vector-averaging current meter (VACM) will be discussed in more detail as examples of moored, automatically recording current meters.

The Bergen current meter (Fig. 3.27) has a rotor the revolutions of which are transferred to the motion of a potentiometer axis by means of a gear unit. A current vane

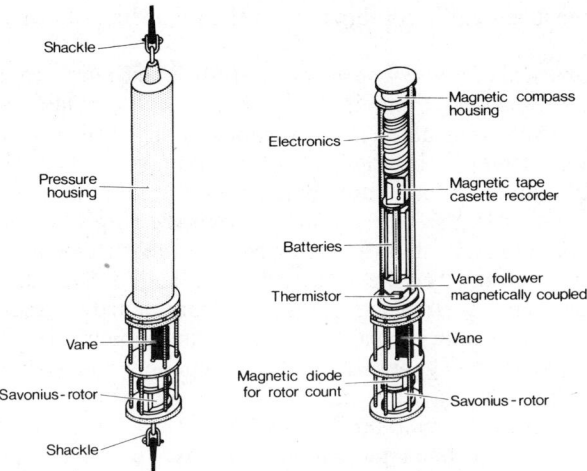

Fig. 3.28. Current meter VACM (vector averaging current meter) for the application in moored measuring systems (cf. Fig. 3.09). Outside view left, inside view right (Design: Woods Hole Oceanographic Institution, Woods Hole).

is rigidly clamped to the pressure-proof housing for the measuring instruments. The device can move freely in a ball bearing. Thus tilts of the anchor cable cannot influence the orientation of the current meter. Inside the instrument, a magnetic compass moves over an electrical conductor and, at the moment of interrogation, it closes the contacts to the conductor by means of a lifting magnet so that a resistance dependent on direction can be obtained. Furthermore, the instrument also has other, optional sensors, for example, a thermistor for measuring temperature (cf. Section 3.2.8), a Bourdon tube with a potentiometer for measuring the depth (cf. Section 3.2.5), and an inductive conductivity sensor (cf. Section 3.2.9) for measuring salinity. All the quantities are available as transformed into variable electrical resistances or voltages and are recorded by a bridge circuit which is, step by step, balanced in such a way that binary coded signals are formed. These signals are stored on magnetic tape. Batteries are used as a power supply for the instrument. The control is effected by a quartz clock which starts the cycle of interrogation at intervals adjustable from several minutes up to one day. The rotor averages the absolute value of the velocity over the measuring interval. After the moored instrument has been lifted on board again, the magnetic tape is taken out and, with a special magnetic tape reader, the data are converted into a form (punched tape, magnetic tape) that can be processed by electronic computers.

The disadvantage of the Bergen current meter, namely the sensitivity of the scanning with regard to disturbances by the motions of the sea and the mooring, is eliminated to a large degree in the VACM (Fig. 3.28). As already used in the Richardson current meter (Richardson et al., 1963), a rotor and a small current vane, freely movable at the instrument, are used here. Like the built-in compass, both sensors react within a few seconds to any change in the current velocity, that is, also to motions due to waves and shifting of the mooring. The interrogation as to the absolute value of the velocity, the direction of the current vane, and the bearing of the compass is carried out by a quartz clock control system at selected intervals within a group of measurements rapidly following each other. The data are electronically converted into the form of components and then averaged. It is only the result of the averaging that is stored in digital form on magnetic tape. Here, too, the current measurement is combined with the recording of other physical parameters.

Current measurements from a moored vessel fulfill their purpose only at high current velocities, since the motion of the ship at the anchor cable would cause considerable disturbance. From current measurements at anchor stations of the research vessel *Meteor*, Defant (1932) concluded that the yawing and swinging motions of the ship amounted to about 10 cm sec^{-1}. Such velocities are greater than the characteristic current velocities in the deep ocean. It should also be remembered that the ship's body itself represents a disturbance in the field of flow, thus distorting current measurements at levels above the draft of the ship. In ocean straits (Lacombe, 1969; Siedler, 1968) and in near-shore areas with strong tidal currents, current measurements from board an anchored ship can be carried out with good success. In the past, mechanical instruments like the Ekman current meter were in use. Today, one generally deploys propeller or rotor current meters with electrical data transmission by cable and continuous recording.

The recording of vertical current profiles, without any disturbance by the ship's motion that is caused by the sea, has been made possible by the "profiler" (Düing et al., 1972). In its design and recording system the instrument resembles the Bergen current meter described above, but it is tared by buoyancy elements in such a way that it has only very little weight in seawater. Attached to a roller fastening, the instrument slowly slides downward along a wire without reacting on the vertical motion of the wire. After the measurement the instrument can be raised to the surface by a winch, or it can float up and down due to automatic changes of buoyancy by means of variable gas volumes.

Free-falling instruments have also been employed successfully for recording vertical current profiles. In the free-falling electromagnetic current meter (Drever et al., 1970) a sonde is used (Fig. 3.29) for which the horizontal velocity is given by the motion of the surrounding water and the vertical velocity by the sinking speed. The sonde falls freely down to the bottom. At the moment of bottom contact, a weight is released, and the instrument returns to the sea surface. An acoustic transmitter installed in the sonde permits the instrument to be located and the data to be telemetered. As current sensor, an array of electrodes is applied measuring the voltage induced by the motion in the earth's magnetic field. The principle of measurement will be discussed later together with the GEK.

For the direct *measurement of the water transport* in areas with very good electronic navigation systems a special method has been developed by Richardson et al. (1965): a sonde falls from the sea surface to a given depth where a weight is released so that the sonde can ascend back to the surface. When the falling depth is known, the water transport in the layer that has been passed will result from the time interval and the distance between the starting point and the terminal point at the sea surface.

For the measurement of the surface current to be carried out from a ship underway, von Arx (1950) has developed the "geomagnetic electrokinetograph" (GEK). As early as 1832, Faraday advanced the principle of this measuring method, that is, that a voltage must be induced by the horizontal motion of the electrolyte seawater in the vertical magnetic field of the earth. Later on, this effect was experimentally proven with submarine cables and used for current measurements between two fixed points.

The electromotive force (EMF) E induced in water can be calculated quantitatively. For a line of water l cm long, in a horizontal current with the velocity of v cm sec^{-1} and with a flow density of the earth's vertical magnetic field of Z G, E (in volts) is given by

$$E = vlZ \cdot 10^8 \tag{3.06}$$

Fig. 3.29. Free-falling electro-magnetic current meter: outsides of the salt-bridge electrode system (a), conductivity sensor (b), temperature sensor (c), sound transmitter (d), release mechanism for the launching (e), for the disconnection (f) of the balloon line (g) after the ship has gone off. The radio transmitter and the quick-flashing light as well as the weights are already below the water line (Photo: Woods Hole Oceanographic Institution, Woods Hole).

It is directed perpendicular to the current direction. If, underneath the flowing water with a specific electrical resistance r, an electrically conducting sublayer is present with a specific resistance r' which closes the circuit, the current density i is

$$i = \frac{E}{(r + r')l} \qquad (3.07)$$

The potential difference e_1 between two fixed electrodes separated by the distance l is then given by

$$e_1 = E - rli \qquad (3.08)$$

If the electrodes are towed through the water by a ship, an EMF of the amount E will also be induced in the connecting cables, and the potential difference e_2 that has actually been measured will be

$$e_2 = e_1 - E = -rli \tag{3.09}$$

Elimination of i and E with the aid of the first two equations results in

$$v = -\frac{r+r'}{rZl} e_2 \cdot 10^{-8} = Ke_2 \tag{3.10}$$

with

$$K = -\frac{r+r'}{rZl} \cdot 10^{-8}$$

where the units of v are cm sec^{-1}. From this equation, it is apparent that the method is most sensitive when $r' \ll r$. The value of r' will be small if the vertical extent of the lower layer is great compared to that of the moving surface layer, that is, in very deep water. In shallow water, the conditions are not as favorable for the application of this method because the locally variable electrical resistance of the bottom is included in r' to a great extent. Therefore, the factor K becomes a function of space and must be determined by direct current measurements.

The measuring process of the GEK consists of registering, aboard ship, the potential difference between two electrodes that are towed by the ship at a distance of about 100 m. The electrodes need special attention because, in seawater, electrochemical effects may occur on their surfaces and thus falsify the measurement. Therefore, to check the zero point of the potential difference, the ship must, from time to time, change its course into the opposite direction. If the current vector is desired in addition to the current velocity perpendicular to the line connecting the two electrodes, some legs in a direction transverse to the general heading must be included.

3.2.12.2. Drift measurements

Drift measurements in surface water can be carried out with quite a variety of floats. For instance, the data concerning the drift of ships, as collected by Hydrographic Offices, represent the basis of the present charts of surface currents. The ship's drift is the difference between the predicted and the actual positions of the ship which results from the motion of the vessel as a body floating in the surface current, and also from the effect of the wind on the ship's superstructure. For the determination of surface currents, the influence of the wind must be estimated and eliminated from the data.

Systematic drift measurements in the near-surface layer are carried out with drift bottles or with drift cards in plastic envelopes. One simply has to wait until such floats, set out in great numbers, drift ashore where they can be collected. This method gives only the starting and final points and a rough estimate of the time span of the drift.

New methods of drift measurements in surface water will permit a repeated determination of the position of drifting buoys and thus the computation of the trajectories of the ambient water with the aid of satellite observations (Masterson, 1972). The drifting buoy contains a radio transmitter which continuously emits a signal of a fixed frequency. Satellites on polar orbits pass over the area of the buoy several times per day, in the middle latitudes four to six times. When the satellite is approaching, the Doppler frequency shift of the signal from the buoy, which is received by the satellite, is determined 5 to 15 times,

Oceanographic Instruments and Methods of Measuring   149

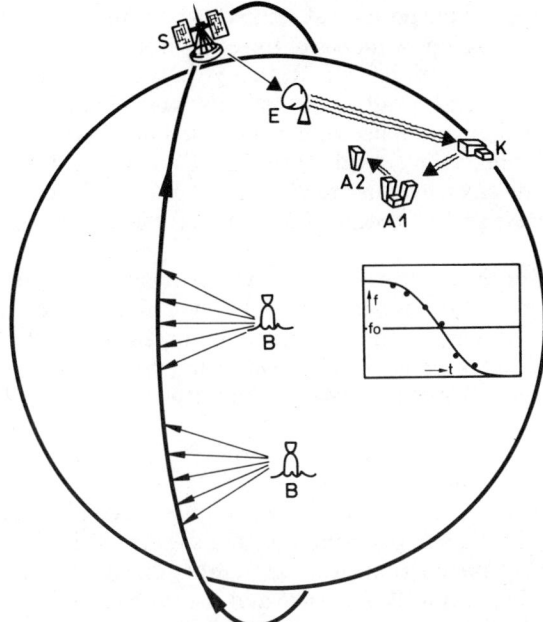

Fig. 3.30. Drifting buoy measurement with satellites. At the approach of the satellite S towards the drifting buoy B the satellite records a signal emitted from the buoy with the fixed frequency f_0 which, due to the motion of the satellite, is received with a Doppler frequency shift $f - f_0$ as a function of the time t. With $f = f_0$ the greatest approach is reached. Together with the orbital data of the satellite the position of the buoy can be derived therefrom. The data are stored and later transmitted to the receiver station E on the ground. From there they are relayed to the data control station K and to the data processing centers A. (After Masterson, 1972.)

and stored. It is a measure of the relative motion of satellite and buoy. With known orbital data of the satellite, the position of the buoy can be computed. Simultaneously, other data (for example, the temperature at the location of the buoy) can be transmitted. Besides numerous buoys and a satellite, the measuring system (Fig. 3.30) includes a receiver station on the ground to collect the stored data from the space craft, a data control station, and data processing centers.

For *drift measurements in near-surface water,* parachute buoys are employed. Such buoys are surface buoys with the smallest possible flow resistance. They carry a mark, a radar reflector or a radio transmitter, and are connected by a thin rope with a floating body at the selected depth. A parachute or another structure with high flow resistance is fixed to the float. Therefore, the entire array almost completely follows the motion of the water at that depth. Its trajectory can be tracked by taking the bearings of the surface buoy by means of optical, radar, or radio direction finding from aboard a ship or from a coastal station. In addition to measurements with floating bodies, other methods are occasionally also used, in which a fluorescent dye is mixed with the seawater, and its propagation traced by fluorometer measurements from a ship.

Drift measurements at arbitrary depth can be executed with Swallow Floats (Swallow, 1955). They are tared in such a way that, after being launched from a ship, they sink to the desired depth and remain there in a stable position. This can also be achieved in the case of small vertical differences of density as long as the compressibility of the float is smaller than that of seawater and, thus, with vertical deflection, repelling forces act in the direction of the position at rest. A sound transmitter emits acoustic signals that can be located from a ship by means of towed hydrophone arrays. Multiple bearings result in position lines (cf. Section 3.1.1.3), that is, in the position of the instrument. Simultaneous locating of several Swallow Floats is also possible by interrogation of acoustic transponders with different response signals installed in those floats.

Floats which have been developed more recently have such powerful sound transmitters that the signals can travel in the SOFAR channel (cf. Section 2.1.10) for many thousands of kilometers and be recorded in stationary acoustic receiving systems near the coast.

A special variety of the Swallow Float is the vertical current meter (Voorhis, 1968) which has obliquely mounted fins and is put into revolution by the vertical motion of the water. The rate of revolution can be determined by comparison with a magnetic compass installed in the float, and is transmitted acoustically or recorded in the float. With this instrument vertical velocities of the order of magnitude of 1 cm sec^{-1} can be measured which may occur in connection with internal waves (cf. Section 8.5) or with vertical convection (cf. Section 7.6.1). The total measuring period, however, is restricted to a few days.

Drift measurements in the bottom current have been performed with simple, mushroom-shaped floats which touch the sea floor slightly with their foot, and, thus, are carried along by the bottom current. A great number of such floats can be set out in regions exploited by fishery, and the distance and the time of drifting are obtained when they are retrieved in bottom trawls.

3.2.13. Measurement of the content of suspended material

Seawater contains suspended particles of different sizes and shapes, of organic as well as inorganic origin. Their distribution and their variations with time can be determined by direct and indirect methods of measurement. Direct measurements are obtained by filtration of water samples, as developed by Krey (1950), where the residuum is determined by weighing, and its composition is subjected to microscopic analysis. When living and dead microorganisms—plankton and detritus—form the residuum, which is almost exclusively the case in the open ocean, this method falls into the hands of marine biologists. Near beaches and in estuaries, where the residuum is predominantly of terrigenous origin, this method is employed by geologists and mineralogists.

Indirect methods for the determination of the content of suspended particles are based on the measurement of the attenuation and the scattering of light, as well as of the reflection of sound by the particles suspended in seawater. The total weakening of a light beam by absorption and scattering is determined by measurement of attenuation; but the scattering must be determined separately by measurement of the scattered light. Since absorption and scattering of pure seawater are known (cf. Section 2.1.11), the optical effect of the suspended particles can be ascertained in this way with the limitation, however, that dissolved substances, mainly humic acids called "gelbstoff," also absorb radiation. Since the suspended substances in ocean water are not of uniform grain size, and since scattering depends to a great extent on grain size, it is difficult to provide absolute figures for the mass of suspended material merely on the basis of optical measurement.

If possible, attempts are made to carry out in-situ measurements of attenuation and scattered light because the optical properties can be considerably influenced even by slight pollution in the sampler or by heavy particles deposited in the measuring device at the

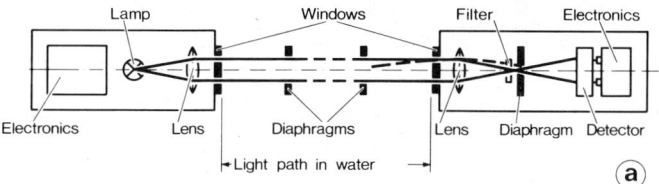

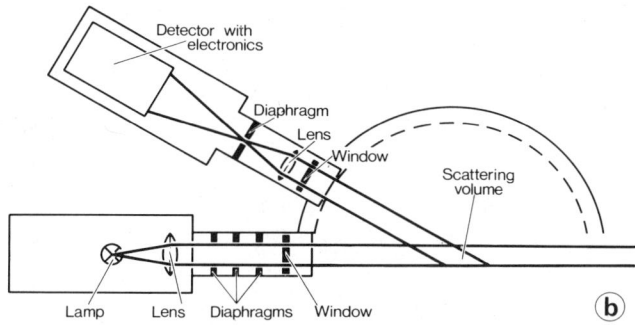

Fig. 3.31. Construction scheme of an optical attenuation meter (a) and a scattered light meter (b).

laboratory. Furthermore, measuring errors may occur owing to reflection in the measuring vessel. In-situ methods are feasible for the measurement of attenuation, whereas for the measurement of scattered light, considerable technical difficulties have to be overcome.

Attenuation measuring devices, according to the prototype of Joseph (1948), which are also called "turbidity meters" or "transparency meters," record the attenuation of a light beam over a certain measuring distance [Fig. 3.31(a)]. Such devices consist of a light source, a lens-diaphragm installation for the generation of a parallel light beam, a fixed light path in water, a lens-diaphragm-filter array on the receiver side, and a light detector. In older instruments, the power supply for the light source is on the ship, and connected by a cable, but newer instruments have a constant-power source installed in the underwater device. Usually, a photocell or a photodiode is used as the detector. The data are transmitted to the ship by cable, in more recent systems by modulating a low-frequency signal (Krause, 1963; Kroebel, 1973). Because of the great forward scattering of seawater, it is especially important that the arrangement for establishing the light path should prevent scattered light from getting into the receiver. The smaller the diameter of the light beam is the better this is achieved, and the more the incidence of scattered light is impeded by additional diaphragms.

In the laboratory, the measurement of attenuation in collected water samples is appropriate in those cases when exact information is desired with regard to the spectral shares of the light that is received. For such measurements spectral photometers are employed.

In order to measure scattered light, an installation is required where a selected volume of water is subjected to irradiation by a parallel light beam, and the scattered light is measured at various angles relative to the light beam. Because of the low intensity of the

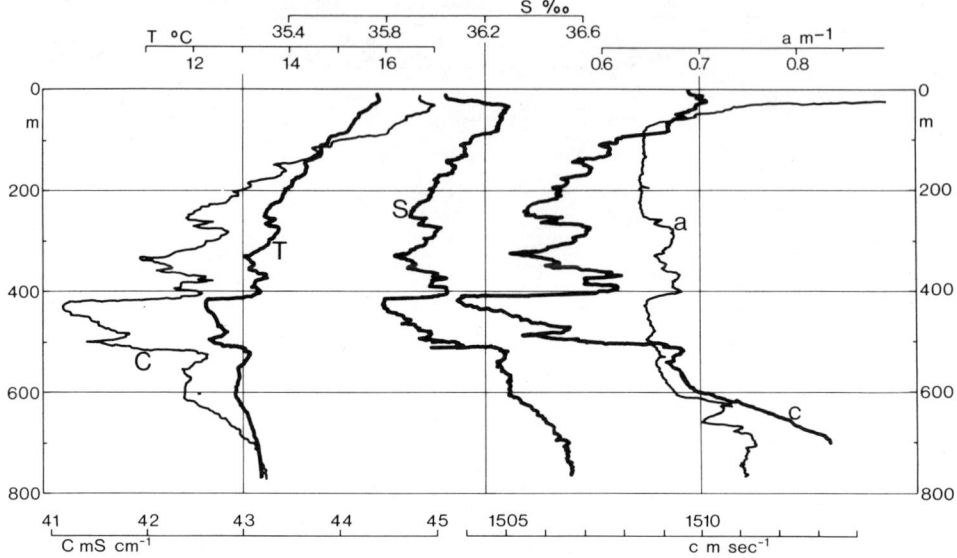

Fig. 3.32. Vertical structure of specific electrical conductivity C, temperature T, salinity S, optical attenuation a, and sound velocity c at 36°47'N and 7°58'W west of the Strait of Gilbraltar. At depths of more than 400 m the Mediterranean water flowing into the Atlantic Ocean can be recognized by its high temperature and salinity values. (After Kroebel, 1973.)

scattered light, highly sensitive detectors such as photomultipliers or photodiodes are needed. The principle of this system, as it was used by Ivanoff (1960) for laboratory measurements and by Tyler et al. (1958) and Jerlov (1961) for in-situ measurements, is demonstrated in Fig. 3.31(b). Some improvement of the measuring technique with a view to a satisfactory angular resolution also in the range of the forward scattering has recently been achieved by use of a laser beam. For the power supply and the data transmission of in-situ measuring systems the same considerations are valid as in the case of the instruments for the measurement of attenuation.

Measurements of attenuation and scattered light can be used for the characterization of seawater. Water masses of different origin often differ in their content of suspended substances and humic material (gelbstoff) so that optical measurements can serve as an aid in the investigation of the propagation of the water masses. An example of the connection among attenuation, salinity, and temperature is given in Fig. 3.32 for the area west of Gilbraltar where water from the Mediterranean Sea is found near the sea floor. Strong vertical differences of density, increased concentration of plankton, echo-scattering layers, and areas of great attenuation and scattering are often observed to occur simultaneously.

A review of optical measuring methods as employed in oceanography has been given by Jerlov (1968).

3.2.14. Measurement of the content of dissolved substances

The measurement of the substances dissolved in seawater has two aims: first, to determine the basic composition of those components which are not at all, or only very little, influenced by biological processes. Here it is the point to obtain, once, a few but very precise results where neither the time that is needed nor the difficulties of the method play a decisive role. Second, to record whole ocean areas in space and time and to observe biological–chemical processes in special limited regions. For this aim the evaluation of a vast amount of observational data within a short time is required. Here, only such methods are suitable that, with little effort, permit series of accurate measurements with rugged gear and instruments even on a rolling research vessel. These special requirements are met only by few out of the large number of chemical-analytical methods of investigation.

At the beginning of the experimental marine chemistry, about a hundred years ago, and in accordance with the state of natural science at that time, gravimetric precipitation analyses were the first to be applied, followed later by volumetric methods. Decisive progress was made when colorimetric methods and, later, spectral photometric methods were introduced for the direct determination of nutrients and for the measurement of several trace elements after they had been enriched and extracted. In the past decades the measurement techniques used in marine chemistry have been considerably improved. These include atomic absorption spectroscopy, neutron activation analysis, gas chromatography, potentiometric methods with ion-specific electrodes, inverse polarography, and microbiological assays. Volumetric methods have been developed further by the application of new specific complex-forming components. A short description of these techniques as well as a review of their present application in marine science follows.

1. *Atomic absorption spectroscopy.* Detailed discussions of this method have been given by L'Vov (1970) and Welz (1972). This method is distinguished by great selectivity and sensitivity, which is particularly useful when trace elements are investigated, provided they have an emission spectrum. Special advantages of this method are that less than 1 ml of seawater is required and that the analysis can be achieved within a short time. If flameless atomization is applied, the absolute proof limits for some elements reach values of less than 10^{-13} g. Principally, the process is as follows: a hollow cathode lamp, containing the element to be determined in its vaporous phase, serves as the source of radiation. These rays penetrate an "absorption cell"—usually a flame or, for higher sensitivity, a graphite tube that is heated electrically (flameless atomization)—in which the compounds or ions present in seawater are atomized.

   As in the Fraunhofer lines the emitted light is weakened by the atoms present in the "absorption cell" of the element that is to be investigated. The degree of the weakening is a measure of the amount of the particular element occurring in seawater. If the characteristic spectral line is filtered out by means of a monochromator, and if modern amplifying techniques are applied, the sensitivity can be increased until it reaches the value mentioned before.

2. *Neutron activation analysis* (Schütz and Turekian, 1965). This highly sensitive method is also suitable for the determination of trace elements in seawater. It is, however, much more expensive than the atomic absorption analysis since the facilities of a radioactive laboratory are prerequisites. The principle of operation is the following: the sea salt sample under investigation, which has been obtained by freeze-drying seawater, or the seawater itself, or the particulate substance collected on a filter, is exposed to the radiation from a neutron source. Here, the individual chemical elements are converted into

characteristic radioactive isotopes, and, by means of the radioactive technique of determination, they can be directly identified and measured with multi-channel gamma spectrometers. Isotopes with a very short half-life period can also be recorded in this way if the generator of neutrons and the measuring gear are connected by a pneumatic tube device so that the time between the exposure to radiation and the measurement can be controlled.

3. *Gas chromatography and paper chromatography.* The chromatographic methods, too, have a high sensitivity down to about 10^{-10} g. They are based on the great selective capacity of separation when gas mixtures or solution mixtures are passed through strongly absorbing layers such as are formed by silica gel or similar substances and by filter-paper strips with specially prepared surfaces. Since these methods require only fractions of a milligram of the initial matter and can be carried out within a few minutes, they are suitable in particular for the determination of organic substances. The great success that has been achieved within the past ten years above all in the field of organic additives in seawater and in oceanic sediments can be attributed to these methods. After the mixture of substances has been separated, the individual constituents can be identified by a combination of gas chromatography and mass spectrometry.

4. *Biochemical "microbiological" method.* In contrast to the purely chemical or physicochemical methods discussed above, this is a biochemical procedure. The multiplication rate of certain species of microorganisms, which have been deprived by selective breeding of a vital substance, serves as an indicator for the presence of this biologically important substance. A disadvantage of this method is the rather limited number of substances that can be investigated; it is mainly restricted to vitamin-like or similar organic compounds. But since the sensitivity surpasses those of chromatographic methods by two orders of magnitude, it represents a good supplement to them.

The possibilities of work are much more limited when, on board a research ship under the hard conditions at sea, large series of observational material have to be processed. Here, in general, only the volumetric methods, as already mentioned in the case of chloride titration (cf. Section 3.2.9), and the colorimetric method can be employed. Recently, however, electrochemical methods have also become more and more important, especially for the determination of trace elements (Kremling, 1973).

5. *Volumetric analysis* is convenient in those cases where the substance that is to be determined is available in sufficiently strong concentration, while the speciality of the colorimetric method can be seen in the fact that substances dissolved in seawater can be determined even if present only in traces. The boundary at which the two methods overlap in their range of application lies at the concentration of approximately 1 mg l^{-1} of the solution to be investigated. The principle of volumetric analysis is this: the substance under investigation is mixed, in a solution, with a reactant of known concentration until the equivalence point is reached. On the basis of the relevant reaction formula the amount of the initial substance can be calculated from the amount of volumetric solution needed for the reaction. The demands with regard to instruments are very modest. They include some glass vessels as used in chemistry (burettes, pipettes, flasks, and beakers). A special advantage of this

Fig. 3.33. Autoanalyzer operating on board *Alkor* (after Grasshoff). Instrument for simultaneous automatic analysis of phosphate, nitrate, ammonia, silicate and carbon dioxide in seawater samples and for continuous recording with digital data logging: (a) sampler (36 samples of 25 ml each), (b) chemical agents, (c) proportioning pumps, (d) units containing chemical reaction gear and photometers, (e) logarithmic amplification modul with analog output proportional to concentration, (f) channel selector, (g) digital voltmeter, (h) teleprinter with punched tape output.

method is that it is adequate also for work on board ship. In oceanography, the principle of the volumetric analysis is, in particular, employed for the determination of "dissolved gaseous oxygen" as well as of "alkalinity" ($CaCO_3$). Working instructions regarding the volumetric method and the colorimetric method, which will be discussed in the following, can be found in the compilations by Barnes (1959), Riley (1965), and Grasshoff (1968).

6. *Colorimetric methods* are based on the conversion of the substance under investigation into a dye by means of a suitable chemical reaction, and on the optical comparison of this colored substance with a standard solution. Since

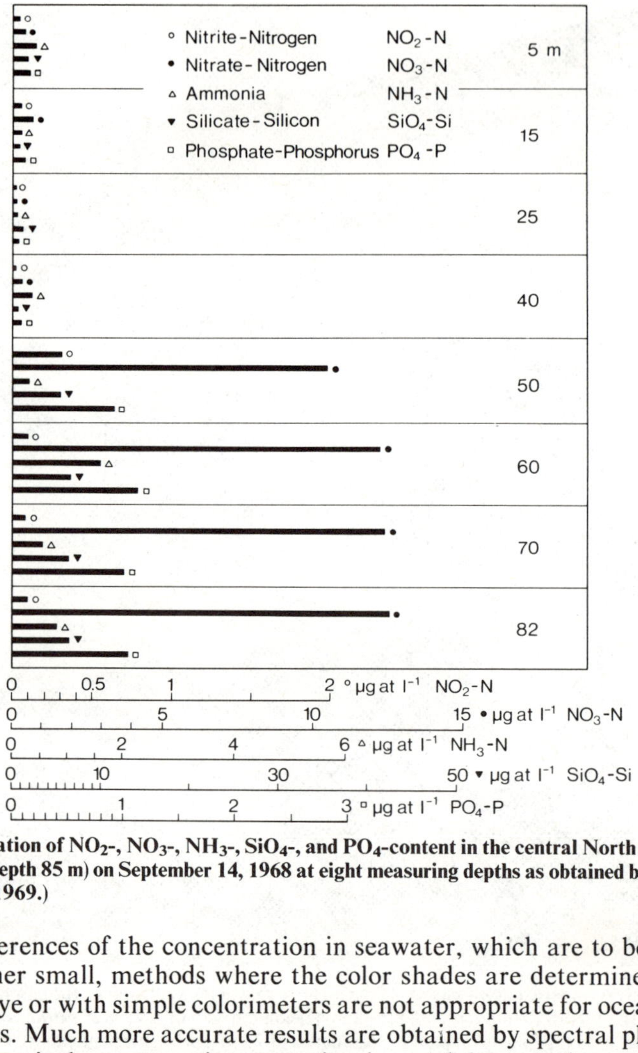

Fig. 3.34. Registration of NO_2-, NO_3-, NH_3-, SiO_4-, and PO_4-content in the central North Sea ($\varphi = 56°21'N$, $\lambda = 1°3'E$, water depth 85 m) on September 14, 1968 at eight measuring depths as obtained by the autoanalyzer. (After Grasshoff, 1969.)

the differences of the concentration in seawater, which are to be observed, are rather small, methods where the color shades are determined with the naked eye or with simple colorimeters are not appropriate for oceanographic purposes. Much more accurate results are obtained by spectral photometers with electrical compensation operating in special spectral ranges. A large number of such instruments is on the market. As to marine chemical investigations after the colorimetric method in laboratories on land as well as on shipboard, the following compounds are to be determined in particular: (a) minimum substances dissolved in seawater that are important for productive processes, such as phosphate, total phosphorus, nitrate, nitride, ammonia, total nitrogen, and silicate; (b) trace elements after previous selective extraction and concentration; (c) humus-like, yellow-colored organic substances in seawater (gelbstoff); (d) organic substances that fluoresce when exposed to longwave ultraviolet light; (e) chlorophyll and protein contained in organisms in seawater.

Recently, efforts to automate the colorimetric methods have been successful (Fig. 3.33). Thus, the number of individual determinations can be

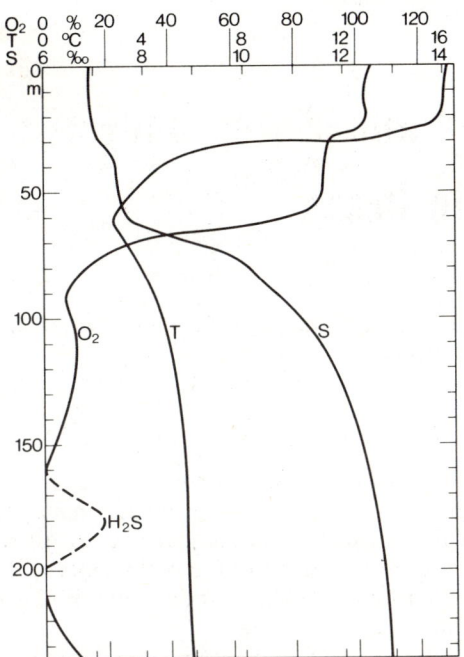

Fig. 3.35. Records of the vertical distribution of oxygen with the oxygen sonde and of temperature and salinity with the membrane salinity sensor in the Baltic Sea ($\varphi = 56°N$, $\lambda = 19°E$) on September 9, 1967. (After Gieskes and Grasshoff, 1969.)

considerably increased, and the intervals in series measurements can be shortened. The application of automatic analysis in a system of continuous flow (Technicon principle) also permits vertical as well as horizontal profile measurements of important nutrients. Likewise, temporal variations in rapid succession can be tracked by continuous measurements (Fig. 3.34) at various depths (Grasshoff, 1967; Brewer and Riley, 1967; Grasshoff, 1969a; Gieskes and Grasshoff, 1969).

Continuous recordings are also among the aims of marine chemical investigations. There are, however, basic difficulties because many chemical quantities cannot be replaced by electrical ones, as is possible with salinity which can be obtained from the electrical conductivity. Only with respect to oxygen and the hydrogen ion concentration has such development been successful. With the oxygen sonde Grasshoff (1962, 1969a) has designed a shipboard instrument for the continuous recording of the vertical distribution of oxygen. Figure 3.35 gives an example of a salinity record obtained with the membrane salinity sonde. Methods for the continuous registration of salinity by means of ion-exchange membrane electrodes have been proposed by Koske (1964) and Gieskes (1968).

4 Energy and Water Budgets of the World Ocean

4.1. Small-Scale Transfer Processes in the Atmospheric Boundary Layer over the Ocean

4.1.1. Transfer processes between ocean and atmosphere

For practical reasons, the interaction processes between ocean and atmosphere are divided into large-scale and small-scale ones. The former, predominantly the climatically effective processes, determine the lifeless environment of man and are, therefore, of special interest. The latter are physical processes which are the real agents of the interaction and must definitely be known, if the former are to be understood. Following Roll (1965), Kraus (1972) has given a summary on this subject.

There are two different groups of transfer processes at the sea–air interface, namely the transfer of energy and momentum, and the transfer of matter: (a) transfer of energy and momentum: radiation, momentum, heat; (b) transfer of matter: first, of water (evaporation or condensation, precipitation, continental runoff); second, of chemical components (minerals and gases).

Furthermore, there are processes that transport energy and matter simultaneously. The continental runoff, for instance, supplies not only water to the ocean, but also energy and minerals. With the transfer of water vapor, evaporation also achieves a transfer of heat.

The energy transfer by radiation will be treated, together with the large-scale exchange of radiation, in Sections 4.2.2 and 4.2.3, also taking into account the optical properties of seawater (cf. Section 2.1.11). Within the framework of this consideration, precipitation and runoff are not small-scale processes (cf. Section 4.3.4 and 4.3.5). Accordingly, five vertical processes are left for the discussion in Section 4.1, namely (1) momentum, (2) sensible heat, (3) latent heat of evaporation and, at the same time, loss of water in the form of water vapor, (4) gases, (5) particles, especially salts.

The nature of the sea–air boundary layer implies that quantitative estimates of the transfer of energy and matter are difficult to give. It must be assumed that immediately at the interface, in air as well as in water, the transfer takes place only through purely molecular processes, and that so-called 'laminar' boundary layers exist there. These are constantly being renewed. The molecular diffusion coefficients in these layers are known: the kinematic viscosity ν, the thermal conductivity k_{DT}, and the diffusivity of water vapor D (cf. Sections 2.1.6 and 2.1.9). With them the transports normal to the interface could be calculated from the corresponding gradients. However, in those very thin boundary layers (thickness on the order of millimeters), where a decisive part of the transfer mechanism occurs, the vertical gradients of the five parameters mentioned above are difficult to measure.

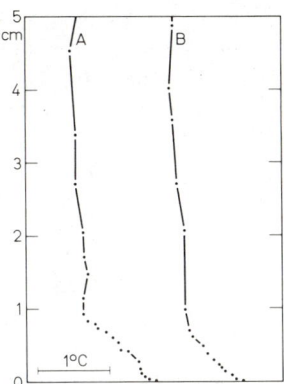

Fig. 4.01. Two temperature profiles taken in the lowest 5 cm above the sea surface. Measured with the boundary layer sonde by Hinzpeter at an anchor station of the research vessel *Meteor* ($\varphi = 30°18'N$, $\lambda = 29°25'W$) in April 1967. (After Clauss et al., 1969.)

According to Hasse (1971), there is indirect evidence of the molecular boundary layer in water. The temperature profile in the lowest layer of the atmosphere over the ocean was first measured by Hinzpeter (1969), with a resolution of less than 1 mm, on the research vessel *Meteor* during the "Seamount Cruises" in the Northeast Atlantic Ocean, 1967. He employed a special boundary layer probe. Two examples of the records taken are shown in Fig. 4.01. Basically, the instrument is a resistance thermometer with a very small time constant (less than 5×10^{-5} sec) for a resistance wire of 1 mm length and a diameter of 10^{-3} mm. One centimeter of the profile is passed within $1/100$ sec. The temperature increase in the lowest centimeter amounts to nearly 1°C.

Above this "molecular boundary layer" there is the "maritime friction layer" (up to about 300 to 500 m in height). While in the "free atmosphere," which lies on top of the others, the wind is almost geostrophic, that is, the pressure gradient and the Coriolis force are the determining elements in the equations of motion, the importance of friction increases with the approach toward the interface. This is expressed by a turn of the wind direction and a decrease of the wind velocity. It is a characteristic of the maritime friction layer that vertical transports of momentum, heat, water vapor, gases, and particles are caused by turbulence. Within the atmospheric friction layer we can also define an "atmospheric layer close to the water," the so-called "maritime Prandtl layer," up to 10 to 20 m high, in which the vertical fluxes are nearly constant with height. This layer contains the maximum of the atmospheric turbulence. It is characterized by the direct influence of the sea state.

In the maritime Prandtl layer, vertical transports are achieved by turbulence:

$$T_{xz} = -\rho \overline{u'w'} \quad \text{momentum flux}$$
$$Q_K = c_p \rho \overline{w'T'} \quad \text{sensible heat flux} \quad (4.01a)$$
$$E = \rho \overline{w'q'} \quad \text{water vapor flux}$$

Here, the primed quantities u', w', T', and q' indicate deviations from the mean values of the horizontal and vertical components of the wind u and w, of the air temperature T, and of the specific humidity q, where u is the horizontal component of the wind in the wind direction which, at the earth's surface, is also the direction of the shear stress vector.

The vertical transports within the maritime Prandtl layer can also be defined as the product of an eddy diffusion coefficient and the vertical gradient of the relevant property. This concept reflects the experience that the transports of momentum T_{XZ}, heat Q_K, and water vapor E occur in the direction of the gradient and are proportional to the absolute value of the gradient

$$T_{XZ} = \rho K_M \frac{\partial W}{\partial z}$$

$$Q_K = -c_p \rho K_H \frac{\partial \Theta}{\partial z} \qquad (4.01\text{b})$$

$$E = -\rho K_E \frac{\partial q}{\partial z}$$

Here K_M, K_H, and K_E represent the coefficients of the eddy viscosity, the eddy thermal conductivity, and the eddy diffusivity. ρ is the density of the air, W the wind velocity, Θ the potential air temperature, q the specific humidity, and c_p the specific heat of the air at constant pressure. These formulas are valid for mean values over periods of 10 min up to 1 hr.

The stochastic nature of the transport process is summarized in the eddy diffusion coefficients. They depend on the state of turbulence in the boundary layer and are therefore primarily proportional to the mean wind velocity and proportional to the distance from the interface. Furthermore, they are dependent on the stability of the density stratification in the atmospheric boundary layer close to the sea surface (maritime Prandtl layer), since stable stratification (air warmer than water) damps the turbulent vertical motions, whereas they are intensified by unstable stratification. As the vertical fluxes in the maritime Prandtl layer are nearly constant, the vertical gradients, with suitable normalization, partly behave inversely to the eddy diffusion coefficients, that is, they are inversely proportional to the distance from the interface, and they are smaller with unstable stratification than with stable stratification. For further details of the flux–gradient relation, attention may be drawn to the summarizing reviews mentioned above.

4.1.2. Methods for the determination of vertical transports in the maritime friction layer

Equations (4.01a) and (4.01b) are the basis for the determination of the vertical transports. A summarizing review of such methods and others has been given by Hasse (1968). In one of the methods computation starts from the measurement of instantaneous values of the wind components u and w, the temperature T, and the humidity q. This is the so-called cross-correlation method or direct method. Another method begins with measuring the vertical distribution of the mean values of the wind velocity W, the potential air temperature Θ, including the temperature difference air-water $\Delta\Theta$, and the specific humidity q. This is the aerodynamic profile method for which, apart from the profiles, the eddy diffusion coefficients K_M, K_H, and K_E must also be known.

With the former (the direct method), measurements of the fluctuations are needed; with the latter, those of the vertical differences between the meteorological parameters. Both methods are concerned with the lowest 10 m above the undulating sea surface. Difficulties for the measuring technique result from the rapid sequence of the fluctuations,

Fig. 4.02. Sensor-equipped buoy, stabilized by weight, for measurement of vertical profiles of wind, air temperature and humidity. Height of the mast 10 m, immersion depth of the buoy 4.5 m. In the background the research vessel *Meteor* in the tropical Atlantic during the Atlantic Trade Wind Experiment, 1969. (From Brocks, 1970.)

which requires a high resolution in time for the measurement and, in addition, from the fact that the vertical differences are very small (namely a few tenths of a degree, or fractions of meters per second), so that they can easily be falsified by external influences, on shipboard by the superstructure of the ship, for example. However, such a great number of highly sensitive instruments is necessary that a research vessel with modern laboratories is required for such measurements. The application of a sensor carrier connected with the research ship by cable or radio is the way chosen to overcome such difficulties (Brocks and Hasse, 1969). Various buoys of different shapes have been developed. Figure 4.02 shows a buoy as it was used in the equatorial Atlantic Ocean about 300 m windward from the research vessel *Meteor,* with which it was connected by a multicore floating cable (Brocks, 1970). This short reference may indicate that the determination of the vertical fluxes of energy, momentum, and matter by means of profile measurements and direct measurements involves considerable effort in the measuring technique.

Therefore, marine meteorologists were forced to develop methods that permit the worldwide determination of the vertical fluxes by means of shipboard measurements of simple meteorological quantities (temperature of air and water, wind speed, and humidity at a given height). This task is paraphrased by the word "parameterization" and has partly been solved for moderate wind velocities up to 12 m $\sec^{-1}$ (cf. Section 4.1.4). From among the numerous other methods which have been used, more or less modified, in order to determine the turbulent vertical fluxes, attention may be drawn in particular to a method by which these are obtained with the help of the heat balance. A detailed description of this will be given in Section 4.2 (cf. Sections 4.2.4 and 4.2.5).

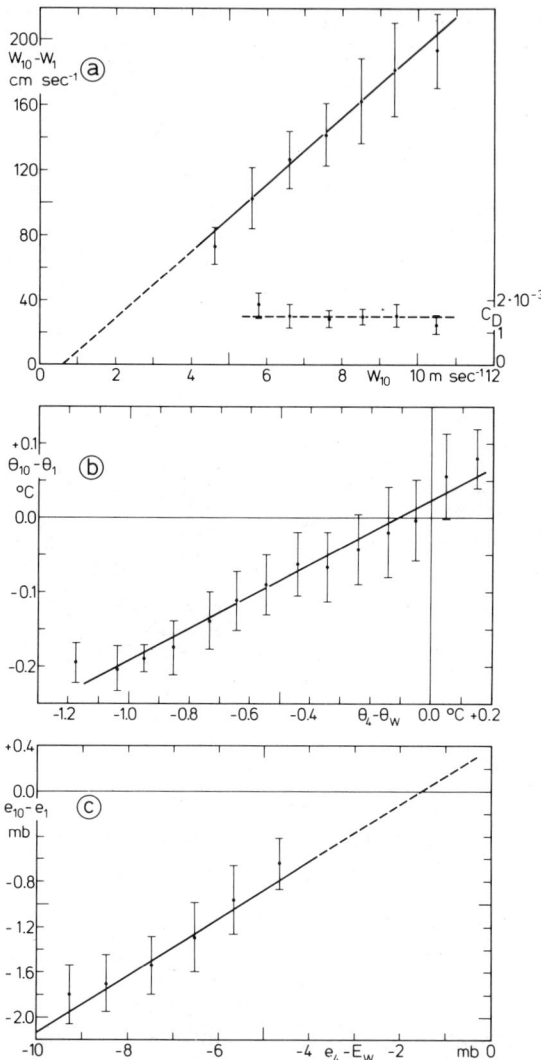

Fig. 4.03. Vertical stratification of wind (a), temperature (b), and humidity (c) in the lowermost 10 m of the atmosphere over the ocean according to records taken at the anchor station ($\varphi = 0°1'N$, $\lambda = 29°33'W$) of the research vessel *Meteor* during the Atlantic Expedition 1965 (IQSY) within the period from September 2 to October 10, 1965. (After Hoeber, 1969.) The sensors were mounted on a stabilized mast (cf. Fig. 4.02) at the distance of 300 m from the ship. Group averages with standard deviations are given. Total number of profiles for (a) 1798, (b) 1758, (c) 1409. (a) Difference of the wind velocities between 10 m and 1 m height ($W_{10} - W_1$) in cm/sec (left ordinate) and friction coefficient C_D (right-hand ordinate) at neutral stratification as a function of W_{10} (in cm/sec). (b) Difference of the potential temperatures between 10 m and 1 m height ($\theta_{10} - \theta_1$) as a function of the difference between the potential air temperature at 4 m height and the sea surface temperature ($\theta_4 - \theta_W$). (c) Difference of the water vapor pressures between 10 and 1 m height ($e_{10} - e_1$) as a function of the difference between the vapor pressure at 4 m height and the saturation vapor pressure E_W at the sea surface temperature ($e_4 - E_W$).

4.1.3. Vertical fluxes of momentum, heat, and water vapor from profile measurements

As described in Section 4.1.2, the vertical gradients are essential for the vertical exchange. In Figure 4.03 the results of numerous profile measurements are represented as the vertical differences of the wind speed ($W_{10} - W_1$), the potential temperature ($\Theta_{10} - \Theta_1$), and the water vapor pressure ($e_{10} - e_1$) between 10 and 1 m height (Hoeber, 1969). The best-fitting straight lines are given together with the group averages and standard deviations. Figure 4.03 is based on measurements in the tradewind region at moderate wind velocities (4 to 11 m sec^{-1}), with temperature differences of -1.2 to $+0.2$°C, and differences of water vapor pressure of -10 to -4 mbar.

From the diagram (a) in Fig. 4.03 the friction coefficient C_D, as defined by Eq. (4.03) (see later), can also be obtained:

$$C_D = (1.23 \pm 0.25) \times 10^{-3} \tag{4.02}$$

This means that the friction coefficient is independent of the wind speed, which holds true for continuous records taken at one site, namely the anchor station of *Meteor* at the equator.

Since the friction coefficient is of such great importance to ocean surface waves and wind-induced surface currents, Brocks' last summarizing paper (Brocks and Krügermeyer, 1972)—after his many previous ones (Brocks, 1955)—is very noteworthy. On the basis of numerous profile measurements in the North Sea and the Baltic Sea as well as in the Atlantic Ocean, the authors arrived at the result given in Fig. 4.04. At wind speeds between 4 and 12 m sec^{-1} and within the standard deviations given, evidence has been found of only a slight dependence of C_D on the wind velocity.

Time series of the vertical fluxes of sensible heat (Q_K) and latent heat of evaporation (Q_V) can be determined by means of profile measurements. A detailed example is contained in Fig. 4.05. It is based on measurements on board *Meteor* in the Atlantic Ocean at the equator in September–October 1965. This diagram demonstrates the extraordinary variability of the heat loss of the ocean. Evaporation, which had hitherto been believed to be more or less constant, showed variations by the factor of five. The amount of water vapor taken up by the trade wind, which is directed into the Intertropical Convergence Zone, should vary accordingly. This will affect the variability of the heat content and the precipitation in this most active zone of the global atmospheric circulation. Furthermore, it should be noted that the flux of sensible heat Q_K is of little significance, as compared to the flux of latent heat with evaporation Q_V; it amounts to only 5 to 10% of

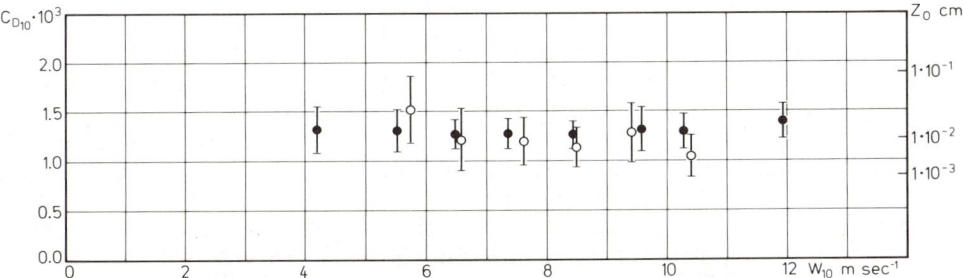

Fig. 4.04. Mean values and standard deviations of the friction coefficient C_D, resp. the roughness parameter Z_0 as a function of the mean wind speed W_{10} at 10 m height with neutral density stratification. (After Brocks and Krügermeyer, 1972.) The values are based on profile measurements. Dots: Baltic and North Seas, circles: equatorial Atlantic Ocean (cf. Fig. 4.03).

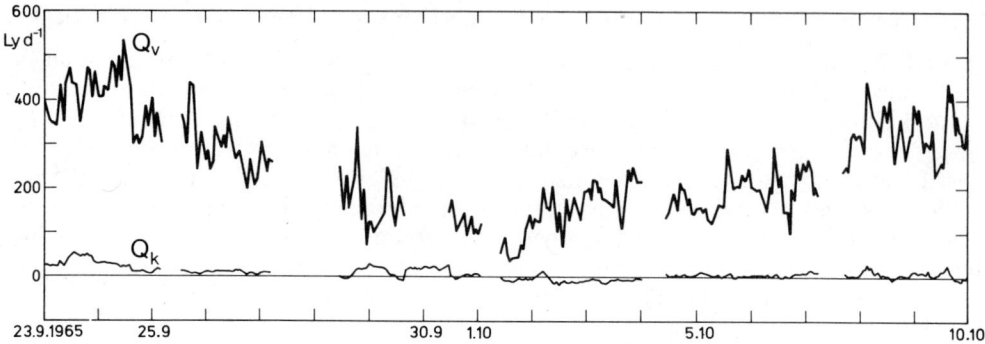

Fig. 4.05. Example of the flux (in cal cm^{-2} day $^{-1}$) of sensible heat Q_K and evaporative heat Q_V from the ocean into the atmosphere for the period from September 23 to October 10, 1965 at the anchor station of the research vessel *Meteor* ($\varphi = 0°1'N$, $\lambda = 29°33'W$) during the Atlantic Expedition 1965 (IQSY). The values are calculated on the basis of hourly means of the measured quantities. (After Hoeber, 1969.)

the latent heat flux. We must, however, take into account that the sensible heat is immediately at the disposal of the atmosphere in the near-surface friction layer, whereas the latent heat is released only after the water vapor has condensed in the clouds, which means, referred to the trade wind area, not before reaching the region of the Intertropical Convergence above the condensation level with a maximum at about 4000 m height. Moreover, if Q_V is compared with the change of the wind speed in time, we see that the considerable variations with periods of several weeks as well as the strong diurnal oscillations of Q_V are predominantly determined by the wind.

4.1.4. Parameterization of the vertical transports

Two facts are obvious:

1. The vertical transports of momentum, energy, and water vapor in the atmospheric boundary layer over the ocean hold a key position in the energy budget of the ocean–atmosphere system.

2. The measurement of such vertical transports requires sophisticated technical equipment. This can occasionally be achieved at certain locations, but not worldwide for synoptic purposes.

From such reasoning the demand results that meteorological quantities, which can be measured with simple instruments on shipboard, should be used for the determination of the vertical transports and that the decisive physical processes must be parameterized. In such a way the knowledge gained could be utilized for other investigations, especially for numerical computations of the atmospheric circulation as well as for other applications.

Suitable variables for parameterization are the mean wind speed W, the mean difference between the (potential) temperatures of air and sea surface $\Delta\Theta$, the mean difference between the specific humidity of the air and the specific saturation humidity at sea surface temperature Δq. By applying the relation $q = 0.623e/P_L$ (where P_L is air pressure), we may also use the corresponding vapor pressure e instead of the specific humidity. Then we get the following relations for the parameterization:

$$\tau = C_D \rho W^2$$
$$Q_K = -c_K c_p \rho W \Delta \theta \qquad (4.03)$$
$$E = -c_E \rho W \Delta q$$

τ stands for the vertical transport of momentum, that is, for the wind stress. These equations show an evident relationship with Eqs. (4.01a) and (4.01b). The transport coefficients C_D, c_K, and c_E are much less variable than the eddy diffusion coefficients and, for many purposes, they can therefore be considered as constants. This also includes the influence of the density stratification which is principally present and which has been proven by experiment (Hasse, 1968), but which vanishes with increasing wind speed.

The results of profile measurements after Brocks and Krügermeyer (1972) with respect to the friction (or drag) coefficient C_D (usually as C_{10} referred to wind speed measurements at a height of 10 m above the sea surface) have been quoted already in Section 4.1.3. As the mean value there has been obtained: $C_{10} = 1.3 \times 10^{-3} \pm 16\%$ (Fig. 4.04). This value has also been confirmed by the results of direct measurements in various regions of the ocean, although with some major scatter in the measurements themselves and also among each other (for summaries see Hasse, 1968, and Kraus, 1972).

Figures 4.06 and 4.07 show recent results of investigations with regard to the vertical transports of sensible heat, of water vapor and thus of latent heat (Kruspe, 1972). The parameterization products $W \Delta \theta$, or $W \Delta q$, are plotted on the abscissa and the vertical fluxes that have been measured on the ordinate.

Both Figs. 4.06 and 4.07 show close linear relationships although they are based on different methods for the determination of the vertical fluxes, on investigations in various oceanic areas, and on results obtained by several authors. In Fig. 4.06 a positive Q_K stands for the heating of the atmosphere by the ocean while a negative sign indicates cooling. The slopes of the straight lines correspond to transport coefficients of $c_K = 0.9 \times 10^{-3}$ and $c_E = 1.35 \times 10^{-3}$.

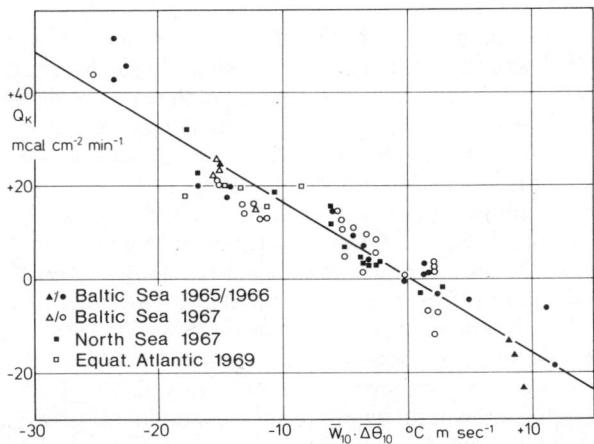

Fig. 4.06. Vertical flux of sensible heat Q_K over the sea as a function of the product $W_{10} \Delta \theta_{10}$. W_{10}: Wind speed in m/sec at 10 m height; $\Delta \theta_{10}$: Difference between the potential air temperature at 10 m height and the sea surface temperature in °C; $Q_K > 0$: Heat flux upward; solid line: best linear fit. (After Brocks et al., from Kruspe, 1972.)

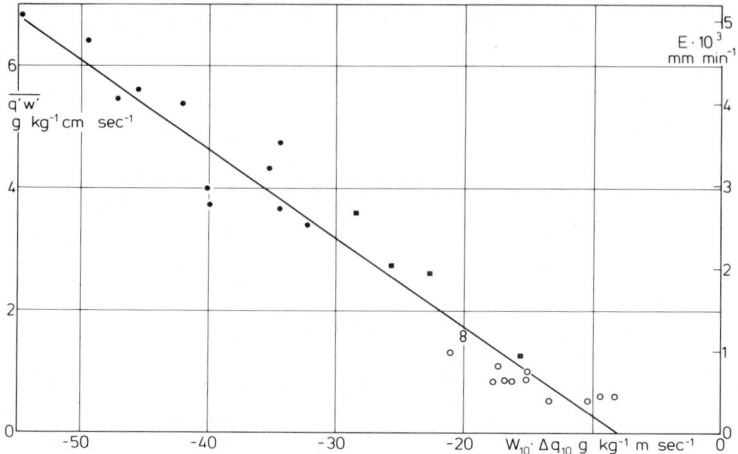

Fig. 4.07. Vertical flux of humidity (evaporation from the sea) $\overline{q'W'}$ or E as a function of the product $W_{10}\Delta q_{10}$. W_{10}: Wind speed in m/sec at 10 m height; Δq_{10}: Humidity difference air–water. (After Kruspe, 1972.) Circles: North Sea (from Kruspe); dots: tropical Atlantic (from Phelps); squares: California waters (from Phelps); solid line: best linear fit.

4.1.5. Transfer of gases and salts

As a result of the continuous interaction between the lower layers of the atmosphere and the sea surface all of the atmospheric gases are represented in the near-surface water layer.

In case of ideal behavior, the solubility of a gas in equilibrium with a fluid is described by Henry's law

$$p_i = K_T c_i$$

Here p_i denotes the partial pressure of the gas i in the gaseous phase; c_i stands for the concentration of the gas i in solution; K_T is the solubility constant depending on temperature. The dimensions of this constant depend on the units of p_i and c_i; its numerical value is determined by the kind of gas or fluid. When in a two-phase system, consisting of gas and fluid, the solution equilibrium has been established according to Henry's law, we can also speak of gas saturation in the solution phase.

For seawater in interaction with the atmospheric gases, the above relation (which—strictly speaking—is valid only for ideal solutions) can be applied as good approximation because the gases in question normally do not exceed the partial pressure of 1 atm, and the deviations from the law of ideal gases are negligibly small.

From observations it follows that, with regard to the various atmospheric gases, the near-surface layer of the ocean is saturated in general, frequently even slightly oversaturated. This holds true for nitrogen and rare gases also in deep water, but not for oxygen and carbon dioxide which, owing to their biological sources, and their activity, hold a special position among the gases dissolved in seawater.

The following may be added with a view to the dynamic processes determining the velocity of the gas exchange at the ocean–atmosphere phase interface (Davies and Rideal, 1963; Kanwisher, 1963).

A gas molecule that penetrates the sea surface must overcome a resistance R, which is composed of three partial resistances: the resistance in the gaseous boundary layer, the resistance in the mono- or multi-molecular interface proper, and the resistance in the liquid boundary layer beneath that phase interface. The latter is a fluid boundary layer which normally is not mixed by turbulence. Its thickness is determined by the force of the turbulent motion in the deeper layers of the fluid and must be overcome by molecular diffusion. Therefore, this partial resistance usually is the strongest, and the process of diffusion taking place in this zone determines the speed of reaction within the gas exchange between ocean and atmosphere.

Such conditions, though basically simple, become complicated because, in the sea, the nature of the ocean–atmosphere phase interface varies considerably, since this interface is covered very differently—in qualitative as well as in quantitative respects—by dissolved, surface-active organic material. Such surface films, more or less pronounced, render the gas transport through the phase boundary more difficult in an undefined way, and thus change the percentage of those among all the gas molecules hitting the sea surface which do not return into the gaseous phase.

A complete picture is obtained if, in air as well as in water, a boundary layer with molecular transport is assumed, and adjacent to it, on either side, a layer with increasing turbulent mixing. If the phase discontinuity and the surface films are taken into account, we get six resistances which add up. Which of these resistances will be decisive depends on the nature of the gas and of the kind of surface film.

Quantitatively, the following formula holds for the transport of substance through any plane surface:

$$\frac{dq}{dt} = A \cdot k \Delta c$$

Here, q stands for the amount of substance, t for the time, A for the surface concerned through which the transport takes place, k for the permeability coefficient (dimension: length/time) of the zone considered, and Δc for the difference of concentration of the substance to be transported within that zone in a direction normal to the surface A. The reciprocal value of k is the transport resistance R (dimension: time/length).

While in the boundary of the gaseous phase above the phase interface numerical values of the resistance R_G for gas between 5 and 80 sec cm^{-1} have been measured, the characteristic values of the resistance R_L in the liquid boundary layer, below that interface, lie between 10^2 and 10^3 sec cm^{-1}.

From comparative experiments in pure water without any surface contamination and with water containing surface-active additives differences of a factor of 10^2 to 10^4 resulted for the resistance R_I of the phase interface. This points to a considerable influence of organic contamination of the sea surface on the ocean–atmosphere gas exchange. Influences exerted by the state of turbulence of both media and by the surface film are included in k. Consequently, k and R are extremely variable.

In addition to the exchange of gases, an extensive interchange of nongaseous substances also takes place at the sea surface, from the atmosphere into the ocean as well as in the opposite direction from the ocean into the atmosphere. In absolute figures, the amount of the substances thus transferred is considerable, but as compared to the salt content of the world ocean it is negligibly small. According to Woodcock (1962) the total "fallout" of salts from the atmosphere reaches 2×10^9 tonnes yr^{-1}, which corresponds to the amount of salt in the uppermost $\frac{1}{10}$ mm of the world ocean.

In the case of the exchange of nongaseous substances between ocean and atmosphere, the most important processes are the occurrence of precipitation (in the form of dust and of rain or snow) on the one hand, and on the other hand, the formation of wind-generated foam in connection with the bursting of air bubbles at the sea surface ("bubble process"). In this way a direct transfer of water dust or spray takes place from the sea surface into the atmosphere. Here, the salts, dissolved in finest droplets, form an important constituent of the aerosol in the lower layer of the atmosphere.

In accordance with the mechanism of their formation, the concentration of the sea salts in the air directly depends on the actual wind speed. The aerosols thus formed are the subject of investigations in the field of air chemistry. In cloud physics they play a role as condensation nuclei. In his book on air chemistry, Junge (1963) has shown that in the marine atmosphere of Hawaii, sulfates and chlorides occur in a very similar ratio to sodium as in the ocean, which indicates the marine origin of those atmospheric salts.

On the other hand, when investigating precipitation of marine origin, oceanographers have found deviations, partly considerable ones, from the ion ratio in seawater. Sugawara (1965) studied more than 300 samples of rain and snow originating from the ocean and found that all essential cations present in seawater, were enriched as compared to sodium. Bloch et al. (1966) reported on similar shifts of the ion ratio in rain-water samples from the region of the Mediterranean Sea, and so did Bruyevich and Korzh (1969) on rain-water samples from the central Indian Ocean.

Possibly, such deviating conditions in the composition of salts are already present in the dissolved state at the ocean–atmosphere phase interface, which shows physico-chemical properties that differ considerably from those of seawater. On account of wind acting on the sea surface covered by bubbles, parts of this phase interface are transported into the atmosphere when foam and spray are formed. By this means, the different ion ratios are fixed in the aerosol particles.

At present, the results of experiments, particularly with newly formed spray droplets, do not yet suffice to permit a definite answer to this problem.

4.2. Large-Scale Heat Budget

4.2.1. General remarks on the heat budget

The temperature changes observed in the ocean can be understood as oscillations about a stationary state. On the average, the gain of heat must equal the loss of heat for the entire ocean. In detail, this heat balance Q_Σ consists of a number of components that can be summarized in the heat budget equation. Its complete form reads:

$$Q_\Sigma = (Q_S - Q_A) - Q_K - Q_V - Q_T + Q_C + Q_E + Q_F + Q_R \quad (4.04)$$

The meaning of the symbols is as follows:

$Q_S - Q_A$: Heat gained by transfer of radiation, where Q_S stands for the thermal energy of the solar and sky radiation absorbed in the ocean, and Q_A for the thermal energy of the effective back radiation.

Q_K: Heat lost by direct (convective) thermal conduction to the atmosphere. This loss becomes a gain of heat if, conversely, heat is conducted from the atmosphere to the water.

Q_V: Heat lost through evaporation. Gain of heat takes its place if condensation occurs instead of evaporation.

Q_T: Heat lost through heat transport by ocean currents, vertical convection, and mixing.

Q_C: Heat gained from chemical and biological processes.

Q_E: Heat gained by heat influx from the earth's interior.

Q_F: Heat gained from frictional heat.

Q_R: Heat gained from the decay of radioactive substances in seawater.

The last four terms of the heat budget equation can be disregarded in the ocean. The amount of heat gained from chemical–biological processes (Q_C) is very small. The radiation energy, which, in the ocean, is used up by plants during assimilation processes, is estimated to be, at most, 0.8% of the incoming radiation; on the average, it is thought to be less than 0.1%. It remains in the ocean as heat of oxidation and can be removed from the water only by the detritus, which is deposited permanently as bottom sediment.

The heat influx from the earth's interior Q_E, which has a mean value of 1.1 μcal cm^{-2} sec^{-1} and varies locally between 0.1 and 10 μcal cm^{-2} sec^{-1} (cf. Section 1.4.1), may offer significant indications as to physical processes in the subsoil (cf. Section 1.4.6). However, for the heat budget of the world ocean it is only of very minor importance. It corresponds to the heat flux from the earth's interior determined on the continents. In the ocean, primarily only the water near the bottom benefits from this heat. If there is no other heat transport, the water of the bottom layer, due to the temperature increase, becomes less dense than the water at higher levels. Such an unstable structure necessarily leads to vertical convection, which results in the neutral stratification that has been actually observed near the bottom in some parts of the Pacific Ocean (cf. Table 7.04).

The frictional heat Q_F, which remains in the ocean as the result of the dissipation of the kinetic energy of the water motion, is of no importance either. With wind-induced ocean currents, Q_F only amounts to about 1/10,000 of the incoming radiative energy. Q_F, however, can be larger in shallow coastal waters with tidal currents (cf. Section 9.5.7). For the Irish Sea there is given $Q_F = 0.7$ cal cm^{-2} day^{-1}. The total amount of merely 0.01 to 0.02 cal cm^{-2} day^{-1} is all the benefit the world ocean gets out of the frictional heat. Radioactive substances are present in seawater in such small quantities that their decay supplies no more heat than $Q_R = 5 \times 10^{-5}$ cal cm^{-2} day^{-1}.

The first four terms in the heat budget equation are the decisive ones. One of these, the heat transport Q_T, can be neglected when we first consider the mean state of the entire world ocean or of an individual ocean whose current system is approximately closed. Thus the heat budget equation reduces to the following terms:

$$Q_\Sigma = (Q_S - Q_A) - Q_K - Q_V \qquad (4.04)$$

In the following Sections 4.2.2 to 4.2.5, the four terms will be discussed separately.

4.2.2. Heat transfer by incoming radiation

The only heat source of importance is the radiation from the sun. It reaches the sea surface as direct solar radiation and as diffuse sky radiation, the latter resulting from the scattering of solar radiation in the atmosphere. The incoming radiation differs very much locally. It depends on the sun's altitude, that is, on the hour of the day and on the season, and, in particular, on the geographical latitude. Furthermore, solar radiation is subject to variable attenuation in the atmosphere, depending on the turbidity of the air and on the degree of cloudiness. Finally, the incoming radiation is reflected at the sea surface in different ways according to the sun's altitude and the state of the undulating sea surface, as shown by Cox and Munk (1956) (cf. Section 2.1.11). This reflection is especially pronounced when the altitude of the sun is low (35% with calm sea and a sun's altitude of 10°) and very small at a high altitude of the sun, independent of the sea state (2.0 to 2.5%). The reflection of the diffuse sky radiation, however, is approximately constant: 6.6% (Burt, 1954). In his book *Optical Oceanography,* Jerlov (1968) has described the behavior of the radiation in water.

Before starting to discuss the radiation transfer, let us consider the fundamental laws of radiation. Planck's radiation law from 1901, one of the basic laws of nature, is so comprehensive that it includes the other radiation laws (by Stefan–Boltzmann and Wien). However, it is not easy to understand, and so it will be useful to state separately what those laws say.

1. Planck's law of radiation gives the distribution of the spectral energy of a radiating body which is only a function of two universal constants and the absolute temperature. Thus, the spectrum of the radiation emitted by black bodies at equal temperature is identical.

2. The Stefan–Boltzmann law states that the total emission E_A is proportional to the fourth power of the absolute temperature T_{abs}.

$$E_A = \sigma T_{abs}^4 \qquad (4.05)$$

$\sigma = 5.673 \times 10^{-5}$ erg cm^{-2} degree^{-4} sec^{-1} is the Stefan–Boltzmann constant.

3. Wien's law of displacement states that the maximum of the emission occurs at a certain wavelength which depends on the absolute temperature of the radiating body (λ in micrometers)

$$\lambda_{max} = \frac{2897}{T_{abs}} \qquad (4.06)$$

4. The Kirchhoff law states that, at a certain wavelength, absorption and emission of a body are equal. A good absorber is also a good emitter, and vice versa.

5. The Rayleigh law deals with the influence of scattering on radiation. It says that in a medium, where the particles are small as compared to the wavelength of the radiation, the radiation is scattered inversely to the fourth power of the wavelength:

$$i = a\lambda^{-4} \qquad (4.07)$$

where a depends on the number of particles and on the degree of polarization. So we see that the radiation is much more scattered toward the short wavelengths, which—with a view to the atmosphere—explains why the sky is blue as long as water droplets do not prevent the selective scattering, in which case the sky would appear milky white.

With perpendicular incidence at the upper limit of the atmosphere, the radiant energy of the sun amounts to $S = 2.00$ cal cm^{-2} min^{-1}. This is called the solar constant, although

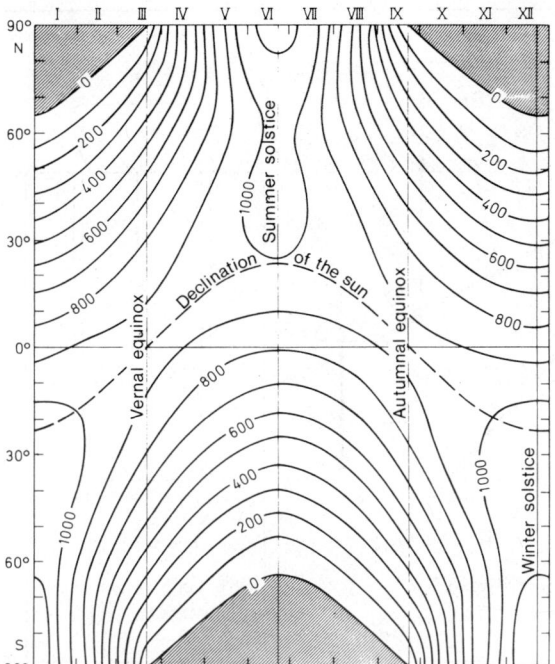

Fig. 4.08. Annual variation of the diurnal range of incoming solar radiation at the outer border of the atmosphere as a function of latitude. (After Sellers, 1965.) The curves represent the irradiation of a horizontal surface in cal cm^{-2} day^{-1} as based on the solar constant 2.0 cal cm^{-2} min^{-1}. Shaded areas show regions of continuous darkness (polar night).

it is not absolutely constant but varies irregularly as well as regularly. The first variation reaches ±1.5% of S; it is a function of certain events on the sun and cannot be predicted. The second reaches ±3.34% of S and varies with the distance between sun and earth, that is, it depends on the orbit elements of the earth and is predictable. An example is given by the curves of the irradiation of the earth as plotted by Milankovitch (1930) for the past 600,000 yr. They may explain the climatic variations caused by "astronomical" influences (cf. Fig. 5.09, covering the past 450,000 yr).

The diurnal variation of the incoming solar radiation on a horizontal plane at the outer border of the atmosphere, as represented in Fig. 4.08, is based on the mean solar constant $S = 2.00$ cal cm^{-2} min^{-1}. The figure shows that the incoming radiation is strongly dependent on the geographical latitude and on the season. With increasing latitude, the difference in the diurnal range of the irradiation between summer and winter grows until, at the poles, the irradiation becomes zero in winter. Furthermore, the figure shows that there are small deviations from the symmetry of irradiation of the hemispheres of the earth in winter and in summer. Since the earth's orbit around the sun is elliptic, the earth is at perihelion in January, that is, closest to the sun, and, thus the irradiation of the whole earth shows higher values in January than in July at the time of the aphelion. The maximum of 1185 cal cm^{-2} day^{-1} occurs at the South Pole on December 22.

On its way from the outer border of the atmosphere to the sea surface, the radiation is subject to various influences, as illustrated schematically in Fig. 4.09. For the Northern Hemisphere, the single components are given as mean values in percent of the incoming

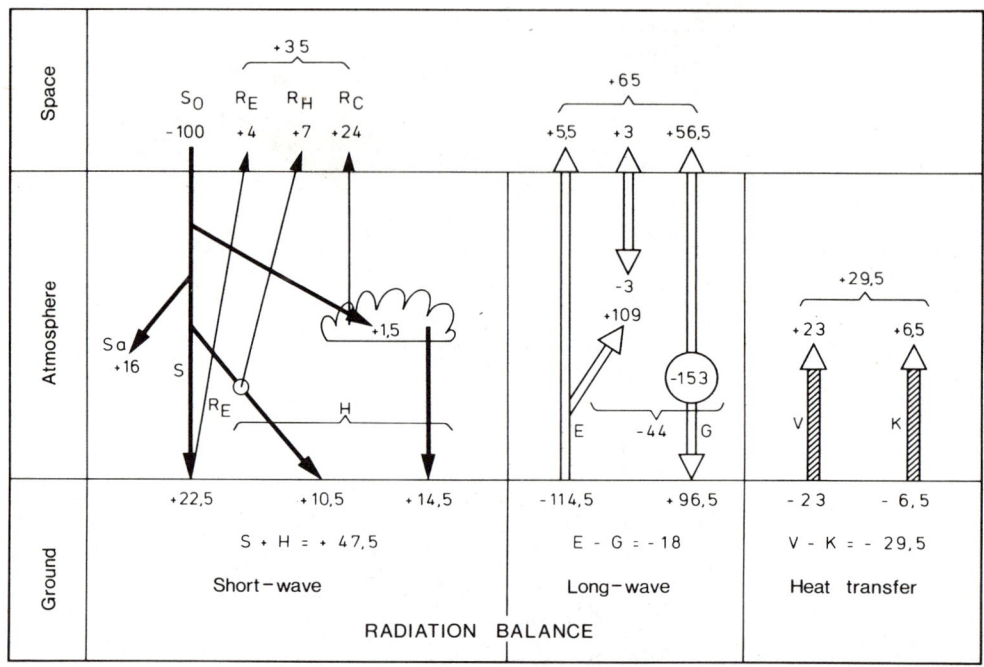

Fig. 4.09. Mean values of the energy budget for the Northern Hemisphere in percent of the incoming radiation S_0 at the outer border of the atmosphere (from Brocks, 1967, after London, 1957, and Sellers, 1965.) $S + H$ = global radiation; R_E = reflection from the sea surface; R_H = back scattering from the atmosphere; R_C = reflection from clouds; $R_0 = R_E + R_H + R_C$; S_0/R_0 = albedo; E = back radiation from the earth's surface; G = re-radiation from the atmosphere; V = heat loss of the sea by evaporation; K = heat loss by convective heat transfer; $S_0 = 100\% = 720$ cal cm^{-2} day^{-1}.

radiation. 16% of S_0 is absorbed in the atmosphere; 24% is reflected at the clouds (R_C), for which a cloud cover of 52% is assumed as the hemispheric average; 7% is lost to space as back scattering (R_H) in the atmosphere; 4% is reflected by the earth's surface (R_E); the latter value is larger at the sea surface, especially at low altitudes of the sun (cf. Section 2.1.11). Thus, the loss of shortwave radiant energy $R_0 = R_E + R_H + R_C$ amounts to 35% of S_0, which corresponds to an albedo of the earth of 0.35. Figure 4.09 also comprises the longwave part of the radiation balance (cf. Section 4.2.3) and the heat transport by convection (cf. Section 4.2.4) and by evaporation (cf. Section 4.2.5).

At the outer border of the atmosphere, the total energy of the incoming radiation S_0 is distributed to certain wavelengths, as shown in Fig. 4.10. This representation is valid for perpendicular incidence of the radiation and illustrates the spectral energy distribution of solar radiation at various depth levels of the ocean [curves (c) to (g)]. The ordinate gives the energy in W m^{-2} Å^{-1}; the abscissa the wavelength in μm. The visible part of the spectrum ranges from $\lambda = 0.4$ to 0.7 μm; the range of $\lambda < 0.4$ μm belongs to the ultraviolet (UV) and that of $\lambda > 0.75$ μm to the infrared (IR).

In Fig. 4.10, curve (b) gives the energy distribution of a black radiator of 5900°K; here it can be seen that the sun is an ideal black radiator only for wavelengths of $\lambda > 1.2$ μm; for shorter wavelengths, its emission is reduced by the absorption in the sun's atmosphere. Curve (a) represents the actual extraterrestrial incoming radiation. It shows that the maximum of the incoming solar radiation occurs at $\lambda = 0.475$ μm. As much as

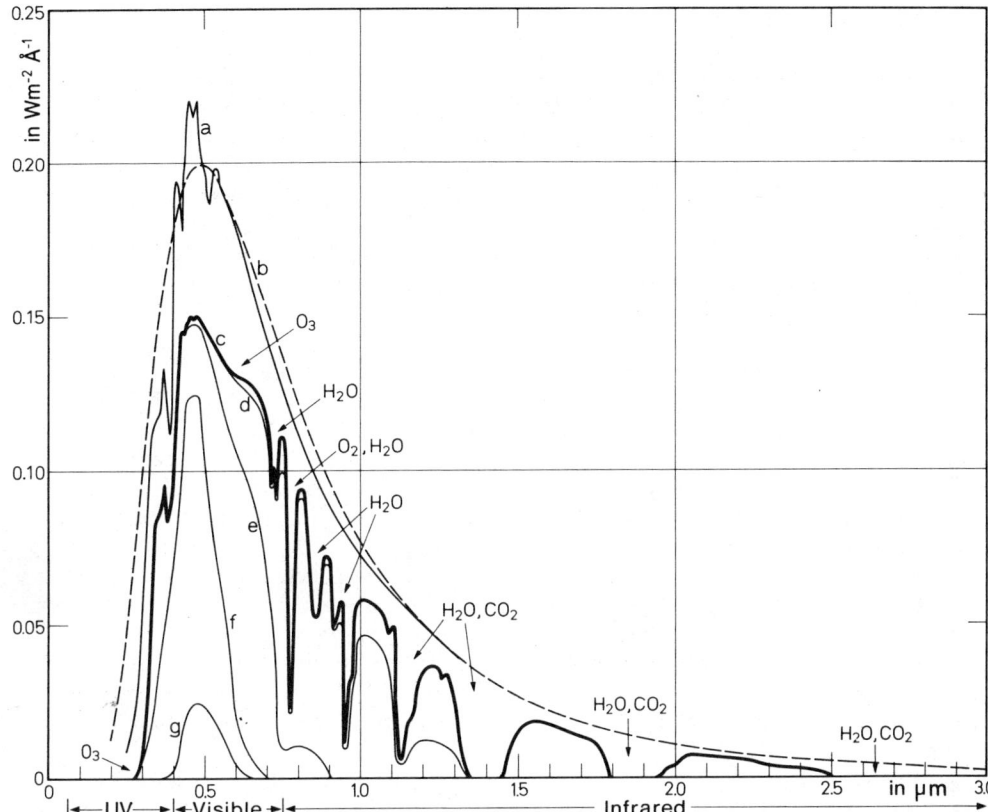

Fig. 4.10. Spectral energy distribution between wavelengths of 0.4 to 3.0 µm at perpendicular incidence of solar radiation (in W m^{-2} Å^{-1}), (solar constant: 1396 W m^{-2} = 2.00 cal cm^{-2} min^{-1}): (a) at the outer border of the earth's atmosphere; (b) for a black radiator of 5900°K; (c) at sea level with reference to the absorption of atmospheric constituents: O_3, O_2, H_2O and CO_2; (d) at a water depth of 1 cm, (e) 1 m, (f) 10 m, and (g) 100 m. (a)–(c): After Gast (1965). (d)–(g): Calculated from curve (c) with attenuation values for pure seawater after Jerlov (1968).

99% of the incoming radiation falls to the share of the electromagnetic short waves between 0.15 and 4.0 µm, 9% of which belongs to the UV range, 40% to the visible range, and 51% to the IR range.

Curve (c) in Fig. 4.10 shows how the spectral energy of the incoming radiation is distributed at the sea surface when the altitude of the sun is 90°. This curve greatly deviates from the even course of a radiation curve because, in the atmosphere, absorption and scattering have a selective effect. The absorption by oxygen (O_2) and ozone (O_3) cuts off the ultraviolet part of the radiation below $\lambda < 0.3$ µm. Without this absorption of the ultraviolet radiation, organic life would not be possible at all on the earth, because the shortwave ultraviolet radiation destroys the molecules of organic compounds. The atmosphere has a particular property: it protects the earth from destructive radiation but lets the constructive radiation pass that is necessary for photosynthesis and heating. In the infrared range, the absorption bands of water vapor and carbon dioxide become very effective. The scattering, too, is selective; according to Rayleigh, it is particularly strong for $\lambda < 0.5$ µm, which is evident when curves (a) and (c) are compared.

On account of the great number of factors determining its intensity (hour of the day, season, latitude, atmospheric turbidity), the complete quantitative determination of the incoming radiation can be achieved only in approximation, and even this only for special cases. An alternative is offered by Mosby's empirical formula (1936) which has proven to be applicable if mean monthly or annual values of the incoming radiation are concerned (units of Q_S are cal cm^{-2} min^{-1}):

$$Q_S = Kh(1 - 0.71\ C) \qquad (4.08)$$

K depends on the turbidity of the atmosphere; at the equator it amounts to 0.023 and at 70° latitude to 0.027; h is the sun's altitude averaged over the whole day, and C is the average cloudiness, expressed in fractions of the cloud cover ($C = 1$ stands for completely overcast sky). Details regarding this climatological method of computing the incoming radiation and the radiation balance have been given by Budyko (1956).

What happens to the radiation that penetrates into the sea? After being scattered in the water, a small part leaves the sea in the form of diffuse underlight. This is mainly what causes the impression of color that an observer gets when looking at the ocean (cf. Section 2.1.11). The greatest portion of the incoming radiation is absorbed. The electromagnetic energy of the radiation is transformed into molecular kinetic energy which can be sensed as heat and observed as a change of temperature. If we want to find where the incoming radiant energy remains, independent of the influences of thermal conduction, mixing, and back radiation, it is easier and more accurate to compute the temperature increase from the incoming radiation and its attenuation than to measure it directly. For that purpose we must know the spectral intensity of the radiation at the sea surface, that is, the distribution of the radiant energy to the different wavelengths, as well as the spectral attenuation in seawater. The necessary data for both of them can be supplied by observation.

In Fig. 4.10 the distribution of the spectral energy, valid for the sea surface, has been calculated for four depths by means of the well-known attenuation coefficient that depends on the wavelength. It becomes obvious that the total weakening or attenuation of the radiation in water is very strong as compared to that in the atmosphere; besides, it is selective, namely great in the ultraviolet range due to Rayleigh scattering and also in the infrared range due to absorption. The minimum of the total attenuation occurs at the $\lambda = 0.48$ μm wavelength which exactly corresponds with the wavelength for the maximum of the radiation coming from the sun. The coincidence of the maximum of incoming radiation with the minimum of attenuation has been arranged by nature very wisely but cannot be explained by any causal relation. The former depends on the sun's surface temperature, the latter on the structure of the water molecules. Nevertheless, this fact is of primary importance to all life in the ocean. In spite of the great attenuation, the uppermost layers of the ocean receive an optimum share of the incoming radiation. The best conditions imaginable for photosynthesis and therefore for organic production, and also for the direct heating of the uppermost water layers are provided by the absorption of the incoming radiation.

The spectral energy distribution in pure seawater is given in Fig. 4.10 for the depths of 1 cm, 1 m, 10 m and 100 m. When the sun is exactly at the zenith, only 73% penetrates as far as the depth of 1 cm, 44.5% to 1 m, 22.2% to 10 m and 0.53% to 100 m and merely 0.0062% to 200 m depth. If it is taken into account that the so-called compensation depth (i.e., the water depth at which the photosynthetic production is balanced by the consumption for respiration) for phytoplankton is reached with the radiant energy of 0.003 cal cm^{-2} min^{-1}, the greatest depth possible for plant production lies at approximately

Table 4.01. Total Incoming Radiation from Sun and Sky in the Wavelength Range 0.3 to 2.5 μm for Some Oceanic and Coastal Water Types of Different Transparency[c]

Depth (m)	Ocean water			Coastal water				
	I	II	III	1	3	5	7	9
0	100	100	100	100	100	100	100	100
1	44.5	42.0	39.4	36.9	33.0	27.8	22.6	17.6
5	30.2	23.4	16.8	14.2	9.3	4.6	2.1	1.0
10	22.2	14.2	7.6	5.9	2.7	0.69	0.17	0.052
25	13.2	4.2	0.97	1.3[b]	0.29[b]	0.020[b]		
50	5.3	0.70	0.041	0.022				
100	0.53	0.0228						
150	0.056	0.00080						

[a] After Jerlov, 1968.
[b] At 20 m.

220 m, assuming pure seawater and the sun at the zenith. At 300 m depth, even in the clearest seawater, the human eye is always met with complete darkness; at 600 m bioluminescence becomes greater than the remainders of the incoming radiation, and even in layers as deep as 1000 m, marine organisms daily migrate in vertical direction, which means that they must have highly sensitive receptors which can sense the tiniest bit of daylight.

Figure 4.10 indicates the greatest possible penetration depth of the incoming radiation if pure seawater and the sun at the zenith are assumed. Actually, the transparency of seawater may be very much smaller, owing to scattering at particulate substance, which acts as an opacifying agent, and owing to selective absorption by yellow substances ("gelbstoffe") which are contained in seawater, especially in coastal waters, as dissolved humic compounds (cf. Section 2.2.4). In order to give a survey of the greatly varying penetration of the radiation (cf. Table 4.01), Jerlov (1964) has optically classified ten types of water, five of them concerning the ocean and five the coastal waters (cf. Fig. 48 in Jerlov, 1968).

Since the greatest part by far of the incoming radiant energy is transformed into heat, the temperature increase in particular water layers with different turbidity can be computed, on condition that no other processes are involved. Under the assumption that 1000 cal cm^{-2} penetrate ocean water of type I, this would mean a temperature increase of 5.5°C for the layer between 0 and 1 m, whereas in the coastal water of type 9, the increase would be 8.2°C. For the layer from 5 to 6 m the corresponding values are 0.20 and 0.08°C, and for the layer from 10 to 11 m, 0.05 and 0.005°C. These values show that the heating due to the absorption of the incoming radiation is effective only in a very thin top layer, predominantly in turbid coastal water. The value of 1000 cal cm^{-2} corresponds to an optimum of diurnal incoming radiation. In spite of the fact that, in the uppermost meter of the oceans and lakes of the earth, such a temperature increase has never been observed in the course of the diurnal variation, this result must not be wrong. For the thermal effect of the absorption is concealed by the turbulent transport of heat in downward direction, by the consumption of heat for evaporation and also for back radiation as well as by the direct transfer of heat into the atmosphere.

4.2.3. Heat transfer by back radiation

For the incoming radiation Q_S, the maximum of the radiant energy—corresponding to the surface temperature of the sun of $T = 6000°K$—lies at 0.48 μm, that is, within the blue range of the visible part of the spectrum. The back radiation from the ocean at $T = 283°K$—according to the Wien law—has its maximum at 10 μm, that is, in the range of the longwave heat radiation. In contrast to the incoming radiation, the back radiation is effective all the time, and is subject to only small fluctuations, for—according to the Stefan–Boltzmann law—the energy of the back radiation depends on the fourth power of the absolute temperature T_{abs} which changes but little in the course of the diurnal and annual rhythm.

Simultaneously with the back radiation from the sea surface, the atmosphere also emits radiation, part of which is directed upward and lost to space; another part is directed downward, counteracting the back radiation from the sea surface. Therefore, it is also called re-radiation of the atmosphere. As radiation of a gas, it has a complicated spectrum with numerous emission bands in contrast to the emission spectrum of water, which is continuous like that for all solids and liquids. Both longwave radiations, namely the total back radiation and the re-radiation, are usually combined. The difference between them is the effective back radiation Q_A. Since detailed measurements of the effective back radiation as of the incoming radiation are available only very rarely, an empirical formula by Ångström has often been used. After recent corrections by Möller (1953) it has taken the following form:

$$Q_A = \sigma T_{abs}^4 [1 - (0.210 + 0.174 \cdot 10^{-0.055} e_0)(1 - 0.765C)] \quad (4.09)$$

where σ represents the Stefan–Boltzmann constant, T_{abs} the absolute temperature, e_0 the vapor pressure over the sea surface, and C, as in Eq. (4.08), is the mean cloudiness in tenths of the sky coverage. The units of Q_A are cal cm^{-2} min^{-1}. A review of the formulas for back radiation has been given by Sellers (1965) among others.

As shown by investigations of the radiative transfers Q_S and Q_A in the world ocean, there is, on the annual average, an excess of incoming shortwave radiation Q_S from the sun and sky at all latitudes when compared to the effective back radiation Q_A. Figure 4.11 indicates the regional distribution of the average annual heat gain by the radiative transfer $Q_S - Q_A$ for the North Atlantic Ocean. According to the figure, the heat excess is not distributed purely zonally, as would correspond to the mean altitude of the sun. This is caused by the regional differences of the meteorological parameters and of the water temperature. After part of it has been displaced by ocean currents, this heat excess will ultimately be given back to the atmosphere, primarily in the form of latent heat of evaporation and, to a lesser degree, in the form of direct heat transfer. The atlas published by Budyko (1963) contains maps representing the radiative transfer as well as the heat transfer by direct thermal conduction and by evaporation based on monthly and annual averages for the entire world ocean.

4.2.4. Heat transfer by direct thermal conduction

In accordance with the results of the radiative transfer, the surface temperature of the ocean is, on the average, higher than the temperature of the lowest layers of the atmosphere. Careful measurements by Kuhlbrodt (1938) have resulted in an average difference of 0.8°C. In 1965 records with the meteorological buoy after Brocks, taken under undisturbed conditions at the equator, have approximately confirmed this value: five-week continuous recording yielded 0.47°C (cf. Fig. 5.01). So we see that, in general, heat is transferred directly from the ocean to the overlying atmosphere. Here, the physical heat

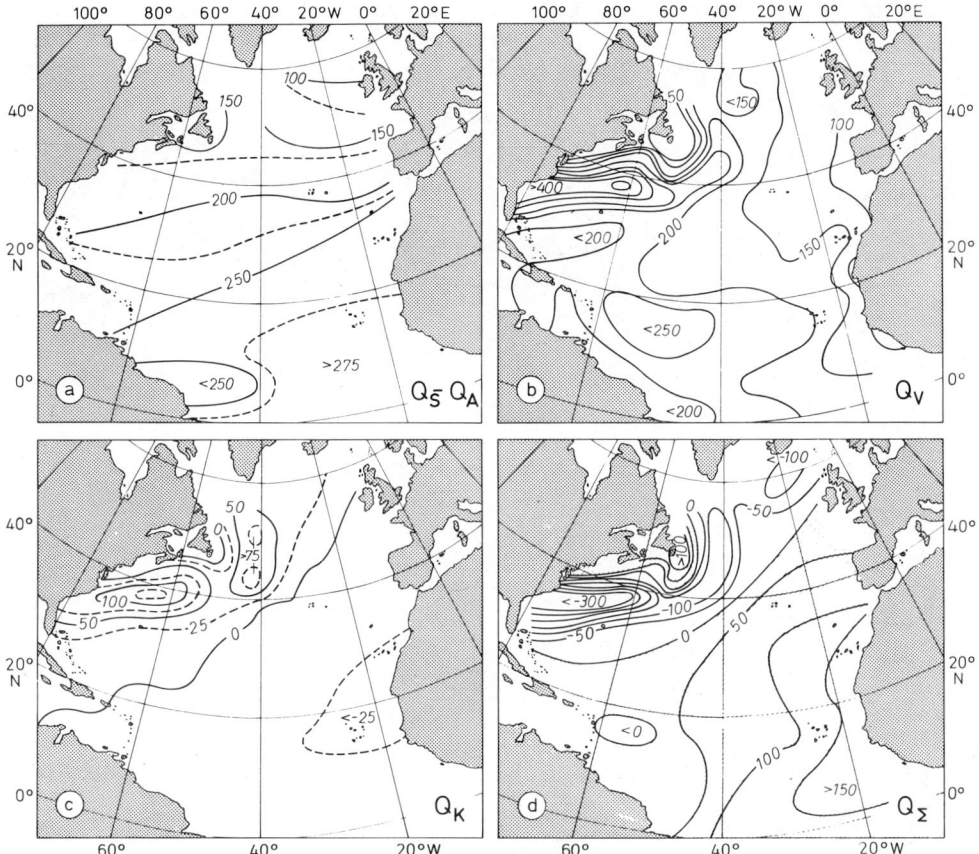

Fig. 4.11. Mean annual heat exchange at the surface of the North Atlantic Ocean in cal cm^{-2} day^{-1}. (a) Heat gain from the radiative transfer Q_S-Q_A. (After Sverdrup, 1943.) (b) Heat loss Q_V by evaporation. (After Jacobs, 1951.) (c) Heat loss Q_K by direct thermal conduction to the atmosphere. (After Jacobs, 1951.) (d) Total heat balance $Q_\Sigma = (Q_S - Q_A) - Q_V - Q_K$. Areas with positive heat balance Q_Σ are hatched.

conduction is not so much responsible as the exchange processes resulting from the turbulence of the air. The heat transport Q_K is given in Eqs. (4.01a) and (4.01b). Since the exchange is predominantly controlled by the wind velocity, it is evident that the wind and temperature gradients essentially influence thermal conduction. In this connection, only the results of Figs. 4.05, 4.11, and 4.12 may be recalled which show that direct thermal conduction from ocean to atmosphere is small, and that, regarding the world ocean, it amounts to only 10% of the heat transfer that goes along with evaporation. In detail, the direct thermal conduction varies regionally and seasonally, which can be recognized from the examples given in Figs. 4.11 and 4.12 when compared with the other components of the heat budget.

Compared with the process of heat transfer from air to water, the reverse process of direct heat transfer from water to air is definitely favored. An explanation can be found on the basis of two physical facts. First, at equal temperatures, the heat contents of air and seawater differ significantly. To heat 1 cm^3 of dry air at 0°C by 1°C, 0.00031 cal

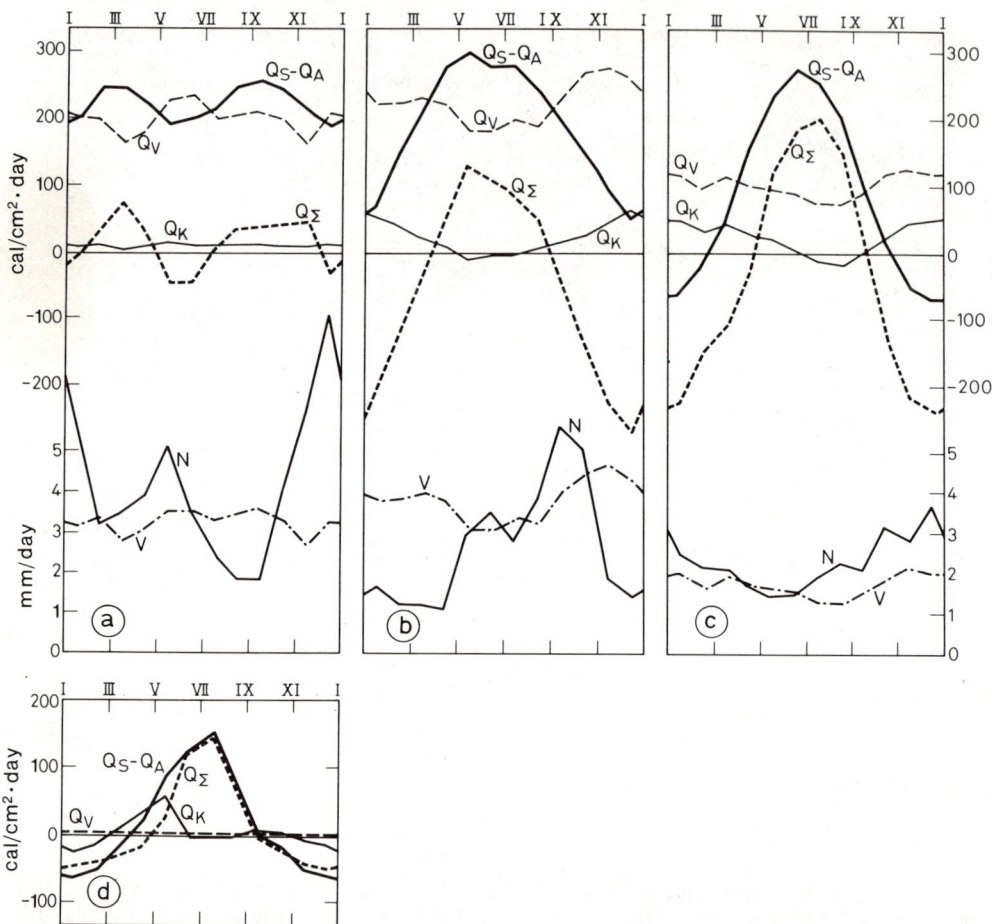

Fig. 4.12. Mean annual variation of the components of the oceanic heat budget in four different regions. (After Albrecht, 1940.) (a) Tropical: Java Sea (Discovery East Bank); (b) Subtropical: Florida Straits (Key West); (c) Temperate: Bay of Biscay (Scilly Islands); (d) Polar: Arctic Ocean (*Maud* drift 1923/24). For explanation of symbols see Section 4.2.1.

will suffice, whereas 0.97 cal are necessary to achieve the same effect for the same volume of seawater with a salinity of 35‰, that is, approximately 3100 times as much heat is needed. Therefore, if, at the air–water interface, a water layer 1 cm thick should be warmed up by 1°C by direct thermal conduction from the atmosphere, an air layer 31 m thick must cool by 1°C. In addition, there is a second process that favors the heat transfer from the ocean to the atmosphere. When heat is conducted from the ocean into the air, the lowest atmospheric layers become unstable, and vertical convection is initiated in the atmosphere, replacing air masses warmed up at the sea surface by cooler ones. Conversely, when heat is transferred from the atmosphere to the sea surface, the lower layers of the atmosphere become stable, turbulence is reduced, and near the sea surface a heavy atmospheric layer is formed which greatly inhibits the turbulent heat transfer from the overlying layers of the atmosphere.

4.2.5. Heat transfer by evaporation

In the ocean, evaporation has a dual function. In the first place, it is the most important term of the heat loss, since 51% of the incoming radiant energy is lost by the world ocean as evaporative heat through the process of evaporation (cf. Fig. 4.09). Secondly, evaporation, together with precipitation, determines the water transfer and, thus, the salinity budget at the sea surface (cf. Section 4.3.6). If we further consider that the heat of evaporation that the ocean is deprived of is gained by the atmosphere in the form of heat of condensation when water vapor is condensed, and that it represents one of its most important energy sources for the general circulation of the atmosphere, it is understandable that problems of evaporation have repeatedly been investigated from different aspects. Summaries and critical discussions of this subject have been given by Sverdrup (1951), Reichel (1952), and Swinbank (1967) who themselves have contributed significantly to the solution of these problems.

The *evaporation process* begins as soon as the air immediately above the water surface is unsaturated with water vapor. It should be noted, however, that the saturation vapor pressure, which is about 2% lower over seawater of 35‰ salinity than over pure water, increases strongly with temperature: at 0°C it is 5.99 mbar; at 10°C, 12.05 mbar; at 20°C, 22.96 mbar; and at 30°C it is as much as 41.68 mbar. Primarily, this means that the higher the air temperature, the greater the readiness of air to absorb water vapor, and, furthermore, that, when air saturated with water vapor is heated by contact with the sea surface, which generally is warmer, the air becomes unsaturated with water vapor. Therefore, over oceanic areas, evaporation is the rule, whereas condensation is the exception.

Condensation mainly occurs when warm air flows over cold water. Due to cooling, the air gets supersaturated with water vapor, and fog is formed. Most sea fogs originate in this way. Turbulence is reduced because, with the cooling of the air, the stratification of the air mass over the water becomes stable. Therefore, the heat gained by the ocean from condensation is very small.

The process of evaporation is connected with three phenomena: (1) the ocean loses water; (2) the ocean loses energy because heat of evaporation is consumed; (3) there is a vertical flux of water vapor in the atmospheric layer near the sea surface. Each of these phenomena has been drawn upon for a quantitative determination of evaporation. The first phenomenon is the basis for direct measurements in evaporation pans on shipboard; the second and third are used for computing evaporation from the heat budget equation and from the vertical transport of water vapor in the lowest atmospheric layers (cf. Section 4.1.4). Each of the three methods contains quantities that are difficult to measure and that therefore hamper the exact determination of evaporation over the ocean.

When evaporation is directly measured in evaporation pans, difficulties arise not so much from the measurement itself as from the reduction of the measured values. Evaporation is determined by the salinity increase with time of a seawater sample contained in an evaporation pan which is suspended on board ship. However, the water in the pan is subjected to external influences differing from those at the sea surface. Wüst (1920) extensively investigated the consequences for such values measured on board and, after a re-examination, he arrived at the conclusion (1954) that the actual evaporation at the sea surface, on the average, amounts to only 55% of the evaporation measured on shipboard.

The method of determining the evaporation by using the heat budget, which goes back to Schmidt (1915), is based on Eq. (4.04). If $R = Q_K/Q_V$ is introduced in this equation, and it is noted that the evaporation height measured in cm is given by $V = Q_V/L$, where $L = 585$ cal g^{-1} is the heat of vaporization for water, the evaporation is then given by

$$V = \frac{Q_S - Q_A}{L(1 + R)} \qquad (4.10)$$

To evaluate this expression for evaporation data, the radiation transfer $Q_S - Q_A$ must be determined, which, in detail, is very difficult to accomplish, as mentioned above. With sufficient accuracy, however, the calculation can be carried out for climatological averages by using Eqs. (4.08) and (4.09). Besides, it is necessary to determine the value of R, the so-called Bowen ratio. Bowen (1926) showed that this ratio can be derived from normal shipboard observations:

$$R = \frac{Q_K}{Q_V} = 0.64 \frac{T_w - T_a}{e_w - e_a} \qquad (4.11)$$

Here, T_w and T_a are the temperatures of water and air, respectively, e_w is the maximum vapor pressure at the temperature T_w, and e_a represents the actual vapor pressure in the air. If Eq. (4.11) is solved with respect to Q_K, an expression is found for the direct heat conduction:

$$Q_K = R L V \qquad (4.12)$$

where the units of Q_K are cal cm^{-2} day^{-1}.

Although the procedure for the determination of evaporation according to Eqs. (4.10) and (4.04) seems easy enough, considerable difficulties will arise in its practical application. Only the extensive material of earlier observations on shipboard is at our disposal, which is summarized in climatic tables and atlases. The systematic errors occurring in this material can exceed 0.5°C with air and water temperatures, which has been confirmed by Saur (1963) also with regard to more recent observations. These errors carry great weight, because the differences of the measured values enter into the formulas. Errors may reach 50% of the evaporation values. Dietrich (1950), however, has established correcting formulas in order to make the vast earlier observational material of the ships suitable for computations of heat transfer.

In the third method of determining evaporation, that is, if the vertical flux of water vapor is calculated, difficulties arise because several assumptions must be made about the process of turbulent exchange in the boundary layer over the water. This becomes obvious when Eqs. (4.03), following Montgomery (1940) and Jacobs (1942), are transformed so that they are more convenient for evaluation:

$$V = K_a(e_w - e_a)W_a \qquad (4.13)$$

Here, V is the evaporation in mm day^{-1}, $K_a = 0.142$ at an observation height of $a = 6$ m, and W_a is the wind speed at the height a. For the practical determination of evaporation over the ocean, this greatly simplified expression yields results of limited validity only, similar to Eq. (4.10), owing to systematic errors included in the earlier observations of water and air temperatures on shipboard.

Each of the three procedures mentioned above has been applied to the determination of oceanic evaporation. According to Wüst (1954), reduced direct measurements indicate that a water volume of 351,200 km^3 yr^{-1} evaporates from the world ocean. This corresponds to a water layer of 97.3 cm, as referred to the whole area of the oceans. The amount of heat needed to accomplish this, that is, the amount lost by the ocean, is $Q_V = 156$ cal cm^{-2} day^{-1}.

However, evaporation is not uniformly distributed over the ocean but shows great regional differences and characteristic annual variations. Figure 4.17 illustrates the zonal distribution of the total oceanic evaporation, based on reduced direct measurements. Because of the low air temperature, the atmosphere at high latitudes is less capable of taking up water vapor. Therefore, evaporation remains small. The secondary minimum, near the equator, can be explained by the fact that the air is already almost saturated with water vapor there. Furthermore, the wind speed is small.

The regional distribution of evaporation and its annual variation have been given by Jacobs (1942, 1951) for the North Atlantic and North Pacific Oceans, by Albrecht (1951) for the Indian and Pacific Oceans, and by Budyko (1956) for the entire world ocean. Jacobs' results are based on the simplified equation for water vapor flux [Eq. (4.13)]; Albrecht and Budyko used the heat budget equation, Eq. (4.04). Malkus (1962) has discussed the results obtained by Budyko (1956) in detail. In the first two cases, the climatic atlas of the oceans after McDonald (1938) served as observational basis, whereas in the third case the Soviet ocean atlas (Morskoi Atlas, 1953) was used. Some inaccuracies in the representation must be borne since ships' observations are insufficient for such purposes, as mentioned before. The basic regional and seasonal differences in evaporation, demonstrated by these investigations for the first time, should be correct, however.

Figure 4.11(b) shows, as an example, how the energy loss resulting from evaporation Q_V is distributed over the North Atlantic Ocean. If it is remembered that 100 cal cm^{-2} day^{-1} corresponds to an evaporation of 62 cm yr^{-1}, this chart will also indicate the regional distribution of the water loss caused by evaporation. For comparison, Fig. 4.11(c) gives the distribution of the energy loss by direct heat transfer Q_K, which has been determined according to Eq. (4.12). The sum of the two components of the heat budget $Q_V + Q_K$ represents the total amount of oceanic heat energy lost to the atmosphere. The maximum Q_K can reach is around 25% of Q_V. In the eastern North Atlantic Ocean off the North African coast, Q_K is slightly negative; here, the atmosphere supplies heat to the ocean by direct conduction. In the chart, the area of weak direct heat conduction from air to water appears too large because in the calculation of Q_K, based on Eqs. (4.12) and (4.11), the errors in the observation of air temperature are included. Furthermore, it should be pointed out that the western side of the ocean is characterized by a much higher loss of energy than the eastern side. This also holds true for the other oceans and affects Q_V as well as Q_K, especially in regions with pronounced poleward water transport as, for example, in the Gulf Stream between 30 and 40°N. Similar conditions have been found in the regions of the Kuroshio, the East Australian Current, the Agulhas Current, and the Brazil Current, even though the energy loss is not so extreme there as in the North Atlantic Ocean.

The energy losses presented in Figs. 4.11(b) and (c) as annual averages are subject to a distinct annual variation with great regional differences. In general, evaporation and direct heat conduction reach their maxima in winter [cf. examples (b) and (c) in Fig. 4.12], in contrast to the conditions on the continent. An explanation is seen in the fact that it is not the temperature of the water but the temperature difference between air and water—with its maximum over the ocean occurring in winter—which is the decisive factor. Besides, the wind also has its maximum in winter, thus promoting evaporation all the more. The annual variation of the energy loss is particularly pronounced on the western sides of the oceans, with the greatest amplitude found in the Gulf Stream region between 30 and 40°N.

4.2.6. Total heat balance

Only in rare cases do the necessary meteorological and oceanographic observations suffice for calculating the mean annual variation of the various components of the heat budget separately, which would permit some insight into their interaction. One of the few complete investigations of this kind has been carried out by Albrecht (1940). Examples from the tropical, subtropical, temperate, and polar regions of the world ocean, one from each, are represented in Fig. 4.12. For the former three areas, the precipitation N and the evaporation V, expressed in mm day^{-1}, are also given. We shall return to these later in connection with the water budget in Section 4.3. These data as well as the calculations are based on monthly mean values of observations made over long periods at meteorological stations located well exposed on small islands. Only the observations made during the *Maud* drift in the Arctic Ocean were obtained within one single year, from May, 1924 to May, 1925; they have been compiled here to show the annual variation.

Figure 4.12(a) shows a typical example of the annual march of the components of the heat budget in the equatorial oceanic climate: the radiation transfer $Q_S - Q_A$ varies only slightly in the course of the year, owing to radiation coming in steadily (cf. Fig. 4.08).

The vernal and autumnal maxima can still be recognized, but they have shifted from the equinoctial months of March and September to April and October. The heat transfer by evaporation Q_V roughly equals the radiation transfer. Q_K is very small, indeed, but it is directed from the ocean towards the atmosphere during the entire year. Except for one month in winter and two months in summer, the sum of the components Q_Σ is positive. At the end of those positive periods, the surface temperature consequently reaches its maximum with the result that the annual march in the tropics is characterized by two maxima, which occur one month or two after the equinoxes. The heat excess is stored in the ocean and transported toward higher latitudes by surface currents. Further details about the relations in Indonesian waters, from where this example originates, are given by Wyrtki (1957).

The annual march in the subtropics [cf. Fig. 4.12(b)] is dominated by the radiation transfer $Q_S - Q_A$. Like the direct thermal conduction Q_K, the annual variation of evaporation Q_V reaches a maximum in winter, as an effect of the warm Florida Current. In this season, the total balance Q_Σ is strongly negative; it is positive only during a period of five months. Referred to the entire year, there is a heat deficit which must be compensated by heat transported by ocean currents from lower latitudes. Jacobs (1951) has given a detailed description of the relations in space.

The characteristic annual variations of the components of the heat budget in temperate latitudes are represented in Fig. 4.12(c). The radiation transfer $Q_S - Q_A$ has a large amplitude with negative values in winter. The heat loss by evaporation Q_V is smaller here than in the subtropics, but it shows a distinct annual march with a maximum in winter. Since the direct heat conduction to the atmosphere Q_K is also relatively large with a maximum in winter, the total heat balance Q_Σ is positive only in summer. The annual average is negative so that heat must be brought along by ocean currents if the balance is to be kept up. In the above example, this is accomplished by the branches of the Gulf Stream in the North Atlantic Current, the "warm-water heating" of Europe. With regard to the area off the English Channel, further details have been described by Dietrich (1951).

The last example [Fig. 4.12(d)] is characteristic of polar conditions, where the ocean is covered by ice all year long. The radiation transfer $Q_S - Q_A$ dominates the annual variation of the heat balance. From observations during the *Maud* drift, Sverdrup (1933) deduced that there is no vertical vapor pressure gradient in the atmospheric layer just above the ice, neither in summer nor in winter, and that therefore, no noticeable evaporation can exist.

The examples show that the contribution of advective heat transported by ocean currents can be isolated from such considerations of the heat budget. Q_T cannot always be recognized as easily as in Figs. 4.12b and 4.12c with the effects of the Gulf Stream. Fofonoff and Tabata (1966) have used the observations obtained at the Canadian ocean weather station *Papa* in the North Pacific Ocean for demonstrating that all changes that have been observed with a view to the heat budget can be explained by the heat exchange with the atmosphere. On the other hand, there is a distinct advection of warmer water masses in the trade wind zone of the North Pacific Ocean, near the Hawaiian Islands, with strong trade winds blowing during the first half of the year, as shown by Seckel (1968).

The heat that remains in the ocean and influences the water temperature, is given by the total heat balance (Fig. 4.12). A positive Q_Σ contributes to an increase of the water temperature and, conversely, a negative one to a decrease. The processes concurring in the distribution of those amounts of heat in the ocean will be described in Section 4.2.7.

The important role played by ocean currents in the oceanic heat budget is demonstrated in Fig. 4.12 by the fact that Q_Σ may considerably deviate from zero in special seasons and in the annual average, as it clearly does in the examples of Fig. 4.12(b) and (c). This significant function will become still more evident if one tries to represent the regional distribution of the annual averages of Q_Σ. For this purpose, in the example of the North Atlantic Ocean in Fig. 4.11, a fourth chart, showing the total annual average of the heat balance Q_Σ, has been added to the other charts illustrating the radiation transfer $Q_S - Q_A$, the heat transfer by evaporation Q_V, and the heat transfer by direct thermal conduction Q_K. The central and eastern parts of the North Atlantic Ocean receive more heat by radiation transfer than they lose by evaporation and direct thermal conduction. In these areas, the ocean would be continuously heated if ocean currents did not carry the heat away. In contrast, the western and northern parts of the North Atlantic Ocean, restricted to the narrow Gulf Stream region, lose more heat than they gain. Continuous cooling would prevail there, if ocean currents did not supplement the deficit in this region from the areas of excess mentioned above. For the entire ocean, the total heat budget is well balanced and equals zero. A corresponding complete representation of the components of the heat budget for the North Pacific Ocean has been given by Wyrtki (1965).

The large regional differences in the heat budget of the ocean must be considered as the effect of heat transport by ocean currents. Energy transfer to the atmosphere is dominant, especially where currents flow from lower to higher latitudes. This holds true for the annual average but is particularly pronounced in winter. It is not accidental that the main frontal zone of the atmosphere is located in such an area where a maximum of energy is lost by the ocean. Very active cyclones with high wind velocities develop there, especially in winter, which take care of the transport and redistribution of the energy. A function such as assumed by the Gulf Stream region for the energy transfer between ocean and atmosphere in the North Atlantic Ocean is held by the Kuroshio region in the North Pacific Ocean.

The heat budget of the entire world ocean is summarized in Table 4.02 which is mainly based on the atlas published by Budyko (1963). The values may slightly differ from those

Table 4.02. Zonal Distribution of the Annual Averages of the Heat Budget at the Surface of the World Ocean[a,b]

Latitude (degrees)	$Q_S - Q_A$	Heat component (cal cm^{-1} day^{-1}) Q_V	Q_K	Q_T
70–60 N	63	90	44	−71
60–50	80	107	44	−71
50–40	140	145	38	−41
40–30	228	226	36	−41
30–20	309	288	25	−3
20–10	326	272	16	38
10–0	315	219	11	85
0–10 S	315	230	11	74
10–20	309	285	14	11
20–30	276	274	19	−16
30–40	225	219	22	−16
40–50	156	151	25	−19
50–60	77	85	27	−36
World ocean	225	203	22	0

[a] After Budyko, 1967.
[b] For explanation of symbols see Section 4.2.1.

of other authors, owing to the different observational material, but the determination is homogeneous in itself. The table demonstrates the following: (1) the energy transfer of latent heat Q_V to the atmosphere reaches its maximum in the subtropical high-pressure belts; (2) the direct thermal conduction Q_K is smallest in the tropics and increases toward higher latitudes; (3) the sum of the heat losses $Q_V + Q_K$ in the tropics does not suffice to balance the radiation transfer $Q_S - Q_A$. A considerable amount of heat Q_T is left. By way of the oceanic circulation, the subtropics and, in particular, the higher geographical latitudes in the world ocean benefit from it.

New insight into the radiation budget of ocean and atmosphere has been opened up by radiation measurements carried out from weather satellites. On the annual average, the radiation budget of the entire earth is approximately balanced, that is, equal to zero. The Southern Hemisphere, however, has a radiation transfer that is 10% higher than that of the Northern Hemisphere. Three reasons can be made jointly responsible: (1) areas covered by water—and 80.9% of the Southern Hemisphere is covered by water—have a lower albedo than continents. Therefore, in the Southern Hemisphere, the gain of radiation is larger by 4% than that in the Northern Hemisphere; (2) the ellipticity of the earth's orbit around the sun and the inclination of the earth's axis favor the incoming radiation on the Southern Hemisphere in summer (cf. Fig. 4.08); (3) as the total back radiation varies only slightly with latitude and even less with the seasons, the favored position of the Southern Hemisphere regarding the radiation budget is maintained.

If we separate the terms $Q_S - Q_A$ and Q_V, as has been done by Rasool and Prabharka (1966), the meridional heat flux Q_T on the earth within the ocean–atmosphere system can be calculated. Such values obtained from radiation measurements by weather satellites should enable us in future to compute the heat transport by ocean currents, provided the transport of latent heat in the atmosphere is known.

The accuracy reached up to now with the radiation measurements by satellites does not invalidate Rossby's hypothesis (1959) that especially the oceanic areas of the Southern Hemisphere serve as secular heat reservoirs. With the deep ocean circulation the deeper layers may also be involved in the transport of heat so that the heat can be kept isolated there for decades and even centuries. If 1% of the total incoming radiation per year were stored, a layer 1000 m thick would warm up per year by only 0.015°C. Suppose that the atmosphere benefits by these stored amounts of heat (with the ^{14}C method half-life periods of renewal of 400 yr have been measured). This would be an indication of how to explain the small post-glacial climatic variations in the order of magnitude of decades and centuries.

If we want to come to a better understanding of the heat budget of the earth, an outstanding aid is provided by satellites measuring the total incoming radiation, the albedo of the earth, and the total longwave back radiation. It is very difficult, however, to measure the difference between incoming radiation and back radiation $Q_S - Q_A$ of the earth toward space with sufficient accuracy (less than 0.5%). It is absolutely necessary to determine the quasisynoptic temperature distribution in the entire world ocean, as was begun during the International Geophysical Year 1957–1958 (Dietrich, 1957).

The development of the technique for measuring longwave back radiation from the sea surface by means of weather satellites has made very good progress. Synoptic records of the distribution of the radiative temperature of the sea surface have been obtained for small sea areas. The areas covered by such records can only be small ones because cloudless atmosphere and constant stratification of water vapor are prerequisites for only little disturbed back radiation from the sea surface. Only then will the radiative temperature approximately equal the actual water temperature. Figure 4.13 gives an example of the measurement of surface temperature from space. On July 3, 1966, within 8 min, the weather satellite *Nimbus II* supplied the data for the temperature chart of the Indian Ocean east of Somalia (Szekielda, 1970). For the first time, this chart shows a synoptic survey of the Somali Current which, during the southwest monsoon, is characterized by strong upwelling and great horizontal temperature differences. Due to the good resolution with regard to the distribution of the radiative temperature, even locally limited upwelling areas can be recognized which permit novel insight into the structure of ocean currents, as Düing (1970) has shown with such satellite records. In addition, it should be emphasized that comparisons between the distribution of the radiation temperature and of the "real" water temperature, measured conventionally from shipboard, are only in approximate agreement. There are several reasons for this.

1. The data for the chart of the radiation temperature were, in fact, recorded synoptically within a few minutes; for ships' records, for example, during the International Indian Ocean Expedition, weeks and months were needed.
2. The back radiation is influenced by clouds and its absorption by water vapor in the atmosphere.
3. The back radiation from the sea surface is determined by the uppermost layer with a thickness of microns. The temperature of this very thin layer can clearly differ from that of the water in the uppermost centimeters (Hasse, 1971). The radiation temperature need not correspond exactly to that of the sea surface, but the actual horizontal temperature gradients and, thus, the sharp temperature boundaries in the areas of upwelling and in the large ocean currents can be recorded. Regarding the Gulf Stream, the Brazil Current, and the Agulhas Current, relevant evidence has been found in analyses of radiation measurements obtained by means of satellites (Warnecke et al., 1971).

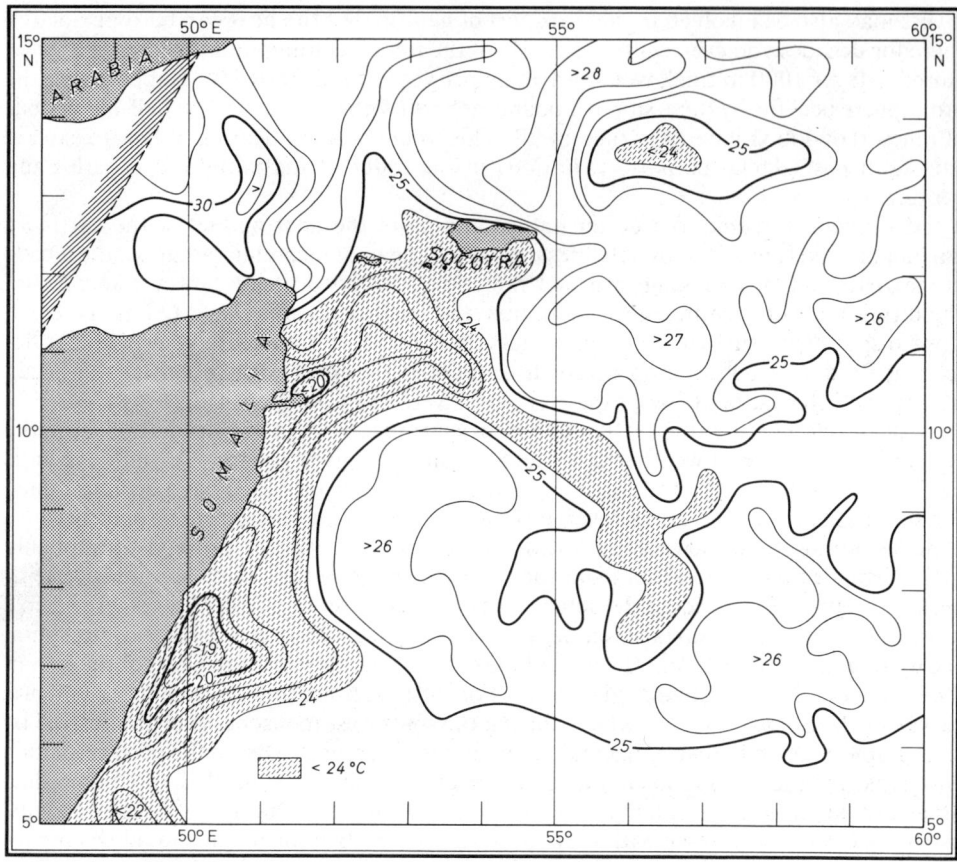

Fig. 4.13. Radiation temperature (°C) of the sea surface in the western Arabian Sea on July 3, 1966 (after Szekielda, 1970) based on infrared radiation measurements (3.6–4.2 μm) with the weather satellite *Nimbus II*. Light screen: Surface temperature below 24°C; Dark green: No observations evaluated.

4.2.7. Propagation of heat in the ocean

The effects of the four fundamental processes of heat transfer discussed in Sections 4.2.2 to 4.2.5 are (except for the first process) restricted to the sea surface. Only the incoming radiation Q_S penetrates into the ocean, but its absorption is so great that the uppermost meter of the water column is clearly warmed up during the day, but at as shallow a depth as 10 m (cf. Section 4.2.2), the effect is hardly noticeable, even under optimum conditions (high incoming radiation, sun at the zenith, pure seawater). If observations teach us, for instance, that an upper layer in the ocean, the thickness of which may vary very much locally, but which may reach down to 200 m, is involved in the annual temperature variation, then it is not the absorption of radiation that is mainly influential. Other processes of heat propagation are also effective. In the vertical direction, molecular thermal conduction, thermohaline convection, and dynamic convection are important; in the horizontal direction, advective heat transport by ocean currents and lateral mixing are involved. All these phenomena will now be discussed separately.

The effect of *molecular thermal conduction* (cf. Section 2.1.6) on the propagation of heat can be given quantitatively. The amount of heat Q, as conducted downward per unit time and unit surface, is proportional to the temperature gradient dT/dz and to the thermal conductivity k_T of water:

$$Q = -k_T \frac{dT}{dz} \qquad (4.14)$$

where $k_T = 0.0014$ cal cm^{-1} sec^{-1} °C^{-1}. The theory of thermal conduction states that a periodic temperature variation at the sea surface

$$T_0 = a_0 \cos \frac{2\pi}{\tau} t$$

takes the following form at the depth z:

$$T(z,t) = a_0 e^{-\alpha z} \cos \left(\frac{2\pi}{\tau} t - \alpha z \right) \qquad (4.15)$$

where

$$\alpha = \sqrt{\frac{c_p \rho \pi}{k_T \tau}}$$

c_p denotes the specific heat, and τ the period. The amplitude decreases and the time of occurrence of the extremes is delayed, depending on the thermal conductivity k_T, if one proceeds from the surface toward deeper layers. If the depth z, at which the amplitude a_0 is reduced to one-tenth of the surface value, is computed from k_T, which is known, the formula yields 14 cm for the diurnal variation and 2.7 m for the annual variation. The internal heat transfer should therefore be restricted to a thin surface layer, which is in contradiction to observations.

Thermohaline convection contributes much more to the propagation of heat than the molecular thermal conduction does. It is generated by an effect of gravity. When the sea surface is cooling, the water particles there get heavier and sink down, and lighter ones ascend to replace them. As long as the salinity is higher than 24.70‰, such vertical convection in the ocean is continued even at low temperatures, because the temperature of the density maximum lies below the freezing point (cf. Section 2.1.4). As the salinity in the ocean—with few exceptions and the Baltic Sea is one of them—is higher than 24.70‰, the vertical convection is not interrupted when the water is cooling, as is the case in lakes at a temperature of 4°C. This fact represents a far-reaching fundamental difference between ocean and continental waters.

The increase of density at the sea surface is mainly the consequence of cooling within the total heat exchange (cf. Section 4.2.6), to a smaller degree also the consequence of a salinity increase in the water budget (cf. Section 4.3.6), namely when evaporation exceeds precipitation. How far the thermohaline convection can penetrate into depth depends on the stability of stratification (cf. Section 7.2.5). Such convection vanishes at high stability which, in addition, varies only slightly in the course of the year, as is characteristic of tropical waters. There are some privileged areas of the world ocean where the stability of stratification disappears at least in one season. Then, deep-reaching convection will start, renewing the deep water with oxygen-enriched surface water which, besides, has a relatively high density. Such areas are the places where the deep-sea circulation (cf. Section 10.1.5) originates, and, therefore, they are also the ventilation centers of the deep ocean. In winter those areas include the Greenland Sea, as far as it is not covered by ice, down to the bottom at about 4000 m, the Weddell Sea beneath the shelf ice down to the bottom, the Labrador Sea down to 2000 to 3000 m (cf. Fig. 10.01) and others. In some smaller areas of the adjacent seas, too, thermohaline convection determines the warming of all deep water: this is true in the Mediterranean Sea, for example, for the Lion's Gulf and the area between the islands of Rhodes and Cyprus, and in the Red Sea for the Gulf of Suez (Plate 4). In the shelf seas of the temperate and higher latitudes, winter cooling mostly leads to total convection as in the North Sea (cf. Fig. 10.55), in the English Channel, the Gulf of Maine, the Gulf of St. Lawrence, but not so in the Baltic Sea, because, there, total convection is prevented by high stability, a consequence of strong salinity stratification.

Dynamic convection is a mixing process, enforced by turbulent motion in the water, which is associated with currents and waves. When dealing with the processes of heat propagation, we cannot distinguish strictly between those two forms of convection, the thermohaline one and the dynamic one.

It is clearly seen how powerful convection is, when the effective thermal conduction is determined, by means of Eq. (4.15), from the observed delay of the extremes and the decrease of the amplitude of the diurnal and annual variations with depth. For the diurnal variation, for instance, Horn (1971), using observations made at Anchor Station 9027 of the research vessel *Meteor* in the subtropical North Atlantic (Fig. 5.04), found an amplitude of 0.28°C at the sea surface and of 0.03°C at the depth of 50 m. At 50 m depth the maximum was delayed by 9 hr compared to the time when it occurred at the sea surface. According to Eq. (4.15), an effective heat conduction of A_{TZ} = 240 to 310 cm^2 sec^{-1} results from the decrease of the amplitude and the phase shift of the maximum. Here, $c_p \rho A_{TZ}$ takes the place of k_T, with the same dimensions. Therefore, this process is also called "apparent thermal conduction" or "eddy heat conduction" (cf. Section 4.1.1). In this particular case, $c_p \rho A_{TZ}$ is about 200,000 times larger than k_T; in general, it varies between 10^3 and 10^6 k_T.

Lateral mixing is effective in addition to vertically directed convection. Lateral mixing is accomplished by eddies with vertical axes and varying diameters, which, however, are always considerably larger than those of the turbulence elements in the vertical direction. Defant (1926) was the first to determine the order of magnitude of the horizontal exchange coefficient A_{TH} as amounting to 5×10^7 cm^2 sec^{-1} for large-scale oceanic processes.

Since that time numerous studies have been carried out to determine A_{TH}. This exchange coefficient depends on the diameter L of the turbulent elements, whose maximum extent equals that of the oceanic area under discussion (cf. Section 2.1.8). These elements can be interpreted by turbulence theory (cf. Section 2.1.8). The result is that A_{TH} is proportional to $L^{4/3}$ and comprises a continuous "spectrum" from 0.014 to about 10^8 cm^2 sec^{-1}.

The large value of the exchange coefficient must not give the impression that lateral mixing is dominant. In addition to dependence on the exchange coefficient, the horizontal transport also depends on the horizontal gradient of the respective property—in this case of temperature—which is very much smaller than the vertical gradient. With a view to their effects, lateral mixing and vertical mixing can be of the same order of magnitude.

The propagation of heat by advective heat transport with ocean currents can be determined in a simple way if the volume transport and the temperature distribution at a cross section are known. The actual importance of advective heat transports will be recognized only when the heat balance, namely the difference between the incoming and the outgoing heat, is calculated for an oceanic area.

The balance determined by Model (1950) for the Norwegian Sea and the Arctic Ocean may serve as an example. Whereas, on the average, water amounting to 3×10^6 m^3 sec^{-1} enters the Norwegian Sea with the Northeast Atlantic Current, the same amount of water leaves this area through the Denmark Strait with the East Greenland Current at a temperature which is lower by 5.5°C. Thus, 16.5×10^{12} cal sec^{-1} are carried into the northern seas by the extensions of the Gulf Stream. Distributed over the entire Arctic Ocean, an additional amount of 11.5 cal cm^{-2} day^{-1} or 4200 cal cm^{-2} yr^{-1} is available for melting ice and for release to the atmosphere. This heat balance is almost in agreement with the total heat balance Q_Σ which, according to the results of the *Maud* observations (cf. Fig. 4.12), amounts to 3900 cal cm^{-2} yr^{-1}.

So far, conditions have been described under which the ocean gives and the atmosphere receives. By transporting heat, the ocean currents assume the task of localizing this energy transfer. In doing so, they exert a significant influence on the prevailing winds, which, in turn, control the oceanic currents to a great extent (cf. Section 7.3). Thus, the causal chain is closed: heat transfer → wind → ocean currents → heat transfer. Any change in one link must be followed by changes in the others. Time delays may occur because the heat content of water is so much greater than that of air. Any disturbance in the oceanic currents can therefore exert its influence on air temperature and wind conditions for a relatively long period of time. The ocean does not only moderate the climate, but also prevents rapid climatic changes.

In conclusion, the following can be said about the heat budget of the ocean: the external processes of heat transfer between ocean and atmosphere as well as the internal processes of the propagation of heat in the ocean are known only in rough outline. In detail, our knowledge of the manifold and complicated processes is not sufficient for the computation of, say, the temperature distribution in the ocean. At present, we are left to inductive reasoning, that is, we have to infer the effective principles from the observed distribution. This method of investigation also applies to other branches of oceanography, a fact that oceanography has in common with other geophysical sciences. As far as the distribution of temperature and salinity in the ocean is concerned, this means that we primarily depend on observation. Some basic facts will be mentioned in Chapter 5.

4.3. Large-Scale Water Budget

4.3.1. The planetary wind system at the sea surface

The large-scale interaction between ocean and atmosphere is not restricted to the heat transfer by radiation, evaporation, and direct thermal conduction. The exchange of momentum is directly effective. Wherever the wind flows over the sea, it exerts a shear stress on the sea surface. Surface waves are generated (cf. Section 8.3.7) which grow and become major waves and, depending on the wind speed, breakers, from which water droplets containing sea salts are released into the atmosphere. Simultaneously, the integrating effect of the wind becomes the main driving force of the surface currents. It is important that we should not only understand the turbulent shear stress at the air–water interface (Section 4.1.3) but also know the main characteristics of the wind system.

It would not be appropriate to give a detailed description of the atmospheric circulation here; this can be found in textbooks on meteorology and climatology as well as in atlases. Only a few facts may be mentioned which are of special importance to oceanography. The great horizontal differences of temperature in the atmosphere, which develop as a consequence of the radiation budget depending on latitude, produce differences of air pressure which maintain a large-scale circulation of air masses. This circulation is influenced by the Coriolis force which prevents direct heat exchange between equatorial and polar regions by the shortest route. Further complications are due to the irregular distribution of land and sea as well as to the relief of the continents. The final result is the large-scale distribution of atmospheric pressure, referred to sea level, as is often represented in atlases for the months of January and July. This distribution is distinguished by a shallow trough of low pressure near the equator, by high-pressure cells in the subtropics, by a number of low-pressure regions in higher latitudes as well as by high-pressure cells in polar areas.

This distribution of air pressure causes large-scale wind belts which are especially pronounced over the oceans. From the equator toward the poles they are arranged in the following way: (1) a low-wind equatorial zone with air motion ascending vertically; (2) trade-wind zones with very steady and strong trade winds blowing from the northeast in the Northern Hemisphere and from the southeast in the Southern Hemisphere; (3) subtropical high-pressure cells with low winds, descending air motion and, therefore, with few clouds and little precipitation; (4) zones where highly variable west winds prevail which bring about a rapid transport of atmospheric properties, in particular of the most important one, water vapor; (5) zones with variable east winds. It is a general rule that direction, velocity, steadiness of the wind, etc., are the factors decisive for the effect the wind exerts on the ocean.

The singularities of the wind field are the skeleton, so to speak, into which the planetary wind system is incorporated. They include (1) the intertropical frontal zone near the equator which is the convergence of the trade winds and thus has heavy precipitation (this is also the low-wind zone of the calms or doldrums); (2) the subtropical divergence zones, also called horse latitudes, with extremely little precipitation but high evaporation; (3) the polar fronts of both hemispheres as the convergences of tropical and polar air masses (they are the zones of cyclogenesis with strong but variable winds and heavy precipitation).

Following the declination of the sun, this planetary wind system is subject to seasonal shifts. Furthermore, the wind speed within the wind belts also shows an annual variation. If we compare the areas occupied by the wind zones (cf. Table 4.03), the dominating role of the trade winds is evident. They govern as much as 31% of the world ocean; in addition, their great steadiness and high velocity must be taken into account. Therefore, the shear stress, which depends on the square of the wind speed, is very effective. So, inevitably, the surface currents, generated by the trade winds, must play an outstanding role in the world ocean.

Table 4.03. Proportions of the Wind Zones in the World Ocean

Zone	Characteristics of the Wind	Percentage of the World Ocean
Equatorial doldrums	Weak, unsteady	7
Trade winds	Strong, steady	31
Monsoons	Strong, steady, seasonal change of direction	8
Horse latitudes	Weak, unsteady	20
Westerlies	Strong, unsteady	24
Polar east winds	Strong, unsteady	10
		100

4.3.2. Special regional wind fields (monsoons, tropical hurricanes, storm cyclones)

If we want to understand the ocean–atmosphere interaction, as far as wind conditions are concerned, three exceptions from the planetary wind system, mentioned above, are of special interest: (1) the monsoons, (2) the tropical hurricanes, and (3) the storm regions of the moderate latitudes.

The *monsoon* is a wind that regularly reverses its direction between winter and summer. The directional reversal must amount to at least 120°. The monsoon can be explained by the fact that, due to the special heat budget of the continents, the singularities of the planetary wind field are greatly shifted during the annual march. The monsoon areas of the earth are represented in Fig. 4.14. According to Chromov (1957), one distinguishes three groups with regard to the frequency of the regular reversal of the wind by at least 120°: greater than 60% of the time—typical monsoon regions; 40–60% of the time—they may still be called monsoon regions; less than 40% of the time—regions with monsoonal tendency. Among the numerous monsoon zones the tropical zone over the Indian Ocean and the southeast Asiatic waters is of the greatest importance. It is characterized by the wind turning from easterly directions in winter to westerly directions in summer.

The *tropical revolving storms* are the strongest hurricanes, by far, on earth (wind speeds more than 65 knots = 116 km hr^{-1}). Besides, they are distinguished by torrential rain. In East Asia they are called typhoons, in the South Indian Ocean Mauritius cyclones, and in the West Indies hurricanes.

Among the four conditions mentioned by Palmen (1948) as necessary for the generation of tropical revolving storms, there are two that depend on the ocean:

1. Instability of the lower atmospheric layers. The water masses are warmed up along their way from east to west in each of the oceans, the lower atmosphere is heated and, due to strong evaporation, water vapor is accumulated there. Both of these processes contribute to the instability of the atmosphere.

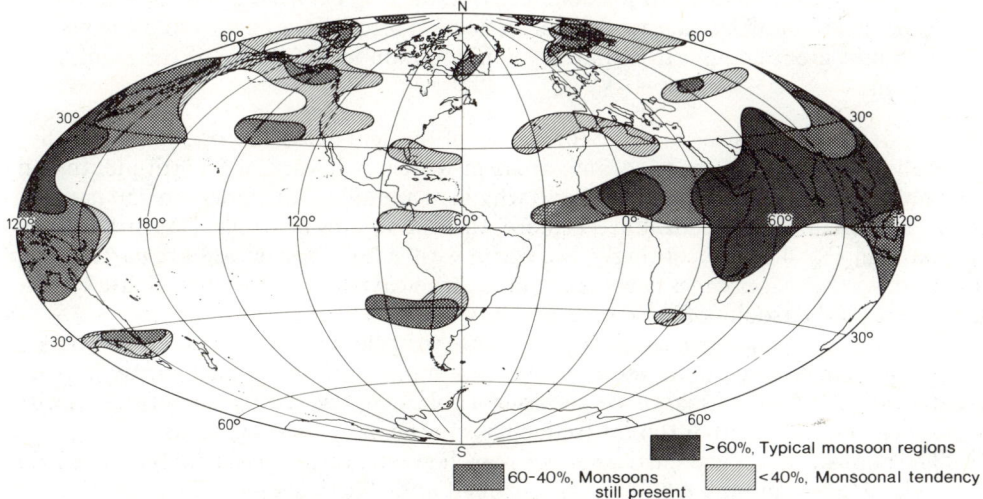

Fig. 4.14. The monsoon regions of the earth. (After Chromov, 1957.) Mean frequency of the major wind directions with changes of direction from winter to summer of at least 120°.

192 Energy and Water Budgets of the World Ocean

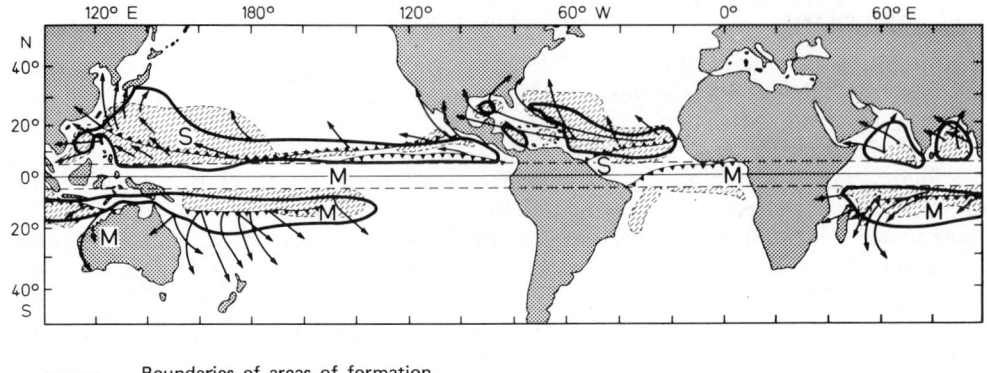

	Boundaries of areas of formation.
→	Path of tropical cyclones.
▨	Air temperature more than 27°C, in particular, in September in the Northern Hemisphere, in March in the Southern Hemisphere, outside the zone bounded by 5°N and 5°S, at a distance from shore > 250 nautical miles.
⁓⁓⁓S ⁓⁓⁓M	Tropical frontal zone of the atmosphere in September or March respectively.
----	5° geographical latitude.

Fig. 4.15. Regions with tropical cyclones. (After Bergeron, 1954.)

2. Reduced atmospheric friction over the ocean. Obviously, this is important since tropical revolving storms do not develop over the continents nor near the coasts.
3. A further condition is an atmospheric one. A convergence of air masses is a prerequisite as it is found in the tropical frontal zone separating the trade winds and the monsoons from the doldrums.
4. Finally, the fourth condition demands that the tropical frontal zone be located poleward of 5° latitude so that the deflecting force of the earth's rotation becomes great enough to initiate cyclonic motion in the waves of the frontal zone.

Figure 4.15 represents the areas of formation of tropical cyclones and shows that they generally coincide with those oceanic areas in which, poleward of 5° latitude, the air temperature exceeds 27°C in the fall and which, in addition, are more than 250 nautical mi away from any continent. Furthermore, the main paths of tropical storms are also indicated in Fig. 4.15. Accordingly, the warm west sides of the oceans are particularly afflicted, whereas hurricanes never occur in the tropical part of the eastern South Pacific Ocean nor in the tropics of the entire South Atlantic Ocean.

Tropical hurricanes are small vortices, always cyclonic, with diameters of 100 km or less, quite in contrast to the storm cyclones of higher latitudes which mostly have diameters about 10 times as large. As shown in Table 4.04, severe tropical hurricanes with wind force 12 occur with different frequencies in the individual sea areas.

The number of tropical hurricanes per year, however, varies considerably. From 1926 through 1935 the minimum number of North Atlantic hurricanes was 2 (in 1929, 1930) and the maximum 21 (in 1933).

The *storm cyclones of the moderate latitudes* of each hemisphere are characteristics

Table 4.04. Mean Annual Number of Tropical Hurricanes[a,b]

Ocean Area	Number per Year
East Asiatic waters	21
Gulf of Bengal	10
South Indian Ocean	7
Southern North Atlantic Ocean	7
Southwest Pacific Ocean	6
Northwest Australian waters	2
Arabian Sea	2
Californian waters	1

[a] After Gentilli, 1967.
[b] See Fig. 4.15 for sea area.

of the polar front zones. Like these they can occur between 30 and 75° latitude. They are large-scale cyclones that are most frequent and strongest in the winter of each hemisphere and are accompanied by heavy precipitation. The wind fields travel together with the moving depressions, which means that wind conditions are very variable locally. The storm zones of the earth (cf. Fig. 4.16) are characterized by wind forces higher than Beaufort 8 in more than 20% of the time or also more than 15 to 20% of the time. In the Southern Hemisphere there are the "roaring forties" and the "howling fifties," as well as the corresponding regions in the Northern Hemisphere, which give the sailors a great deal of trouble, above all in winter and particularly in the North Atlantic Ocean where the shortest route between Europe and North America runs across this zone. Therefore, weather routing has been introduced for shipping here.

There are many other parts or disturbances of the planetary wind system which may locally be of great importance in the ocean. For instance, in the Mediterranean Sea there are quite a number of them: the bora in the Adriatic Sea, the etesien winds in the eastern Mediterranean, the mistral in the Ligurian Sea, the scirocco in the North African waters.

4.3.3. Evaporation

In the ocean, the transfers of water and heat due to evaporation are complementary; when one of them is known, the other one can be calculated. Evaporation, as expressed by the height of the evaporated water, is $V = Q_V/L$, where $L = 585$ cal·g^{-1} stands for the heat of vaporization of water. In this way, the V values in mm day^{-1} have been computed from the Q_V values (cal cm^{-2} day^{-1}) of the four examples in Fig. 4.12. After being divided by 585, the Q_V values in Fig. 4.11(b) give the annual average of the evaporation per day, which results in an amount of evaporation of about 310 cm yr^{-1} for maximum values of $Q_V = 500$ cal cm^{-2} day^{-1}. For the world ocean Wüst (1954) found a mean value of 97.3 cm yr^{-1}, whereas the value given by Lvovitch (1971) is 124.0 cm yr^{-1} = 440,000 km^3 yr^{-1}. The considerable differences among the values obtained by various scientists are due to methodical difficulties that make the measurement of evaporation very complicated (cf. Section 4.2.5). What has been said in Section 4.2.5 about the strong regional differences and about the annual variations concerning the heat exchange by evaporation can be applied analogously to the transfer of water. The meridional distribution of evaporation in Fig. 4.17 indicates the climatic zones. Remarkable features are the high evaporation in the subtropics, caused by advection of relatively dry air over warm water, and the minimum of evaporation in the equatorial air pressure trough, reflecting the high

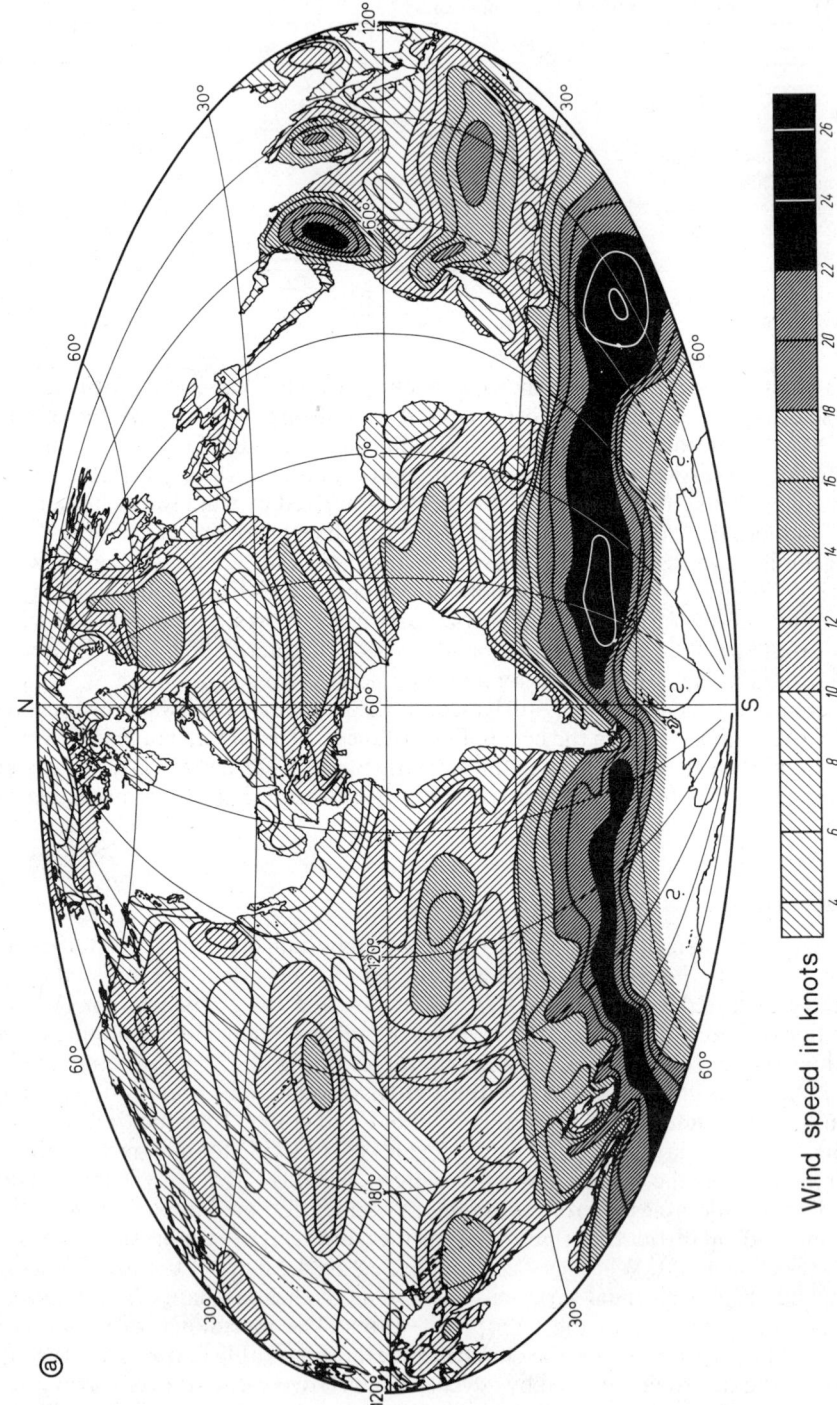

Fig. 4.16. Mean scalar wind velocity (in knots) over the world ocean after the Atlas of Climatic Charts of the Oceans by McDonald (1938), poleward of 50° latitude after the oceanographic atlases of the Arctic (1968) and Antarctic (1970). (a) June–August; (b) December–February.

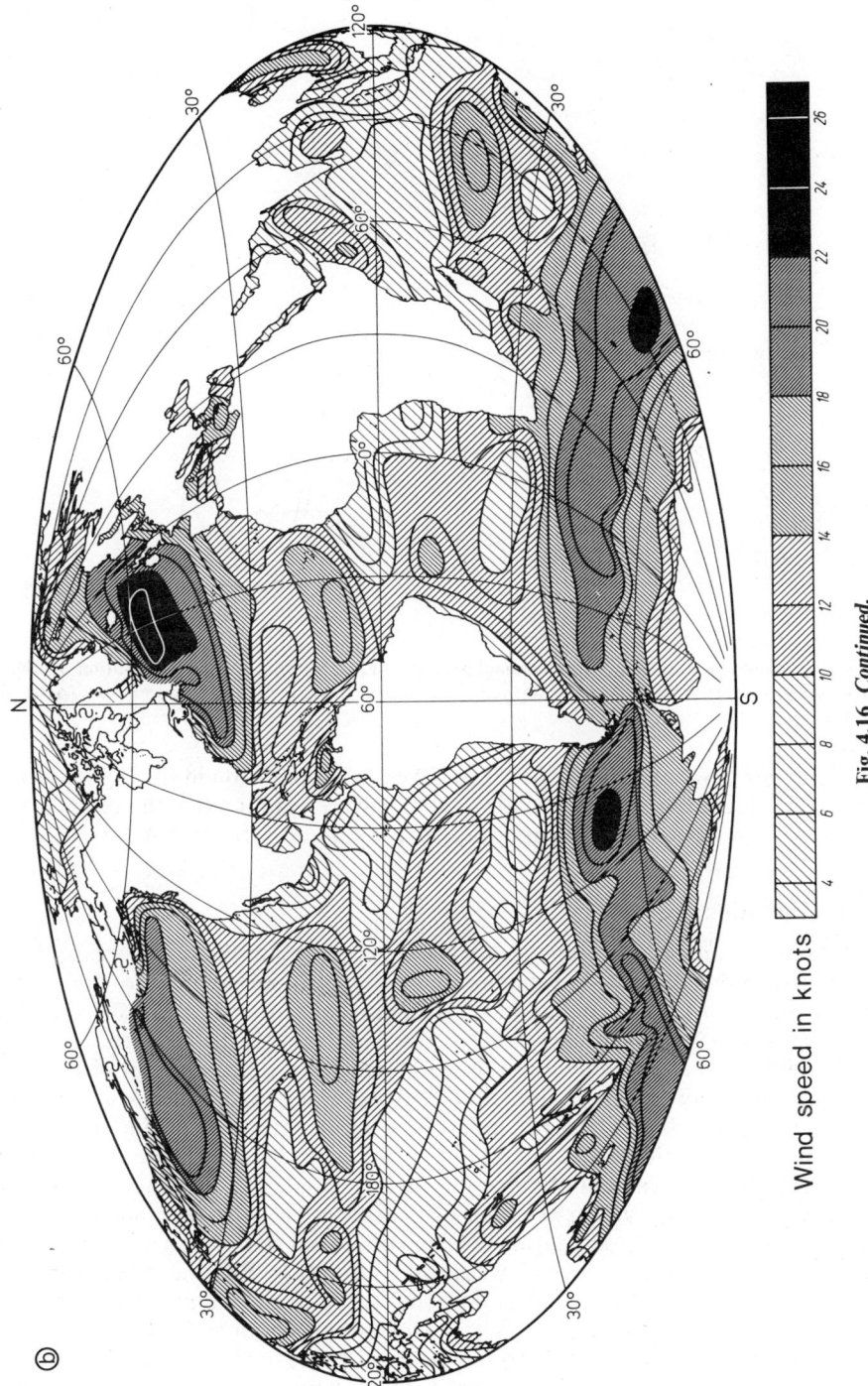

Fig. 4.16 Continued.

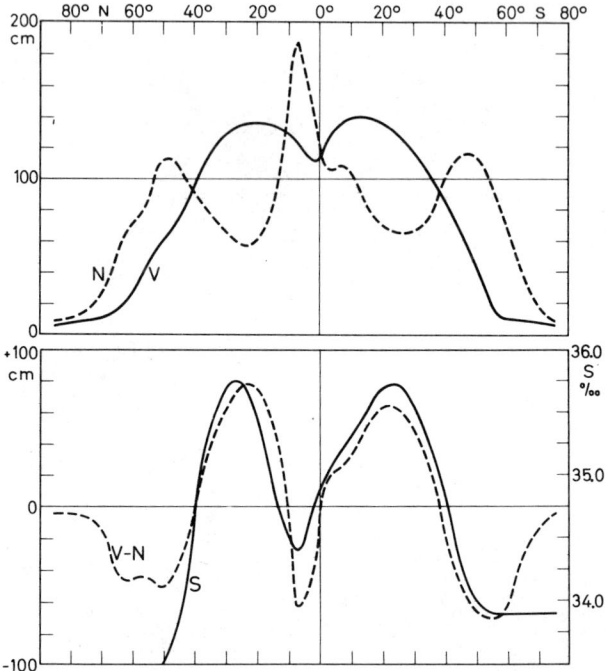

Fig. 4.17. Meridional distribution of zonal annual averages of precipitation N and evaporation V (above) as well as $V-N$ and salinity S (below) at the surface of the world ocean, including adjacent seas. (After Wüst, 1954.)

atmospheric humidity and the weak winds. The low evaporation in the higher latitudes is due to the reduced capacity of the air to absorb water vapor when the temperature is low, that is, when the saturation vapor pressure is small (cf. Section 4.2.5).

4.3.4. Precipitation

Precipitation is an important component of the water budget of the world ocean, although it cannot be recorded directly. On shipboard it is very difficult to measure precipitation; in a few cases such measurements have been carried out with some success aboard lightships (Roll, 1958). What can be determined on board is the frequency of precipitation. Precipitation charts covering the world ocean have been prepared repeatedly, among others by Möller (1951), Albrecht (1960), as well as by Kusnetsova and Sharova (1964). At first, the relation between monthly or annual averages of the frequency and of the amount of precipitation at near-shore, possibly undisturbed continental stations is determined; then, the amount of precipitation is calculated on the basis of this relation and of the frequency observed in neighboring oceanic areas. Also then, some corrections will be necessary, because precipitation at island or coastal stations is generally increased owing to orographic influences, which was proven directly by Brogmus (1952) at stations suitably situated at the Baltic Sea. An additional aid for the location of the precipitation belts is provided by synoptic photographs of cloud fields obtained by weather satellites, as used, for example, by Taylor (1970) for the Pacific Ocean. The zonal averages in Fig. 4.17 show the main belts of precipitation, which are bound to the convergence zones of the wind belts: the intertropical convergence zone near the equator and the polar fronts at about 50° latitude in both hemispheres. Together with the motion of the wind belts in an annual rhythm, the precipitation belts are shifted, too. According to Kusnetsova and Sharova (1964), the total precipitation received by the world ocean amounts to 410,500 km³ yr⁻¹ = 113.7 cm yr⁻¹.

4.3.5. Continental runoff

The continental runoff from rivers is known approximately. At the symposium "On World Water Balance," Nace (1970) gave the value of 924,000 m³ sec⁻¹ (= 29.2 × 10³ km³ yr⁻¹) as the result of most recent investigations in the framework of the International Hydrological Decade. Leaving out the areas on earth that have no runoff at all (cf. Plate 1), we may assume that 103 × 10⁶ km² of the continents are contributing to that figure for the runoff. What is largely unknown is the runoff from the vast store of ground water (cf. Table 4.06). This runoff has been estimated at 7000 m³ sec⁻¹, that is, less than 1% of the surface runoff. Here it has been taken into account that only 200,000 km of the total coast line (370,000 km in length) of the earth are free of permanent frost and thus can contribute to the runoff of ground water.

The continental runoff is very unevenly distributed among the various rivers and, also from these, the discharge to the relevant oceans is disproportionate, as illustrated in Table 4.05, where the 33 great rivers with annual means of more than 2.0 × 10³ m³

Table 4.05. Rivers with Annual Means of Continental Runoff of More Than 2.0×10^3 m³ sec⁻¹ Listed by Oceans[a,b]

River	Mean Annual Runoff (10^3 m³ sec⁻¹)	River	Mean Annual Runoff (10^3 m³ sec⁻¹)
Atlantic Ocean		**Pacific Ocean**	
Amazon	180.0	Yangtze	34.0
Congo	42.0	Mekong	15.9
Orinoco	28.0	Amur	11.0
Plate[c]	24.1	Hsiyang	11.0
Mississippi	17.5	Columbia	6.7
Yenisey	17.4	Yukon	4.3
Lena	15.5	Fraser	3.9
Ob	12.5		
St. Lawrence	10.4		
Magdalena	8.0	**Indian Ocean**	
Mackenzie	7.5		
Danube	6.5	Brahmaputra	20.0
Niger	5.8	Irrawaddi	14.0
Pechora	4.1	Indus	3.9
Kolyma	3.8	Zambesi	2.5
Dvina	3.5		
San Francisco	3.3		
Grijalva	3.3		
Neva	2.6		
Pyasina	2.6		
Nelson	2.3		
Rhine	2.2		

[a] After Marcinek, 1964.
[b] See Plate 1 for boundaries.
[c] Parana and Uruguay.

sec^{-1} are listed. They discharge 58% of the continental runoff; one of them, the Amazon, as the river with the largest abundance of water on earth, supplies as much as 19.5%. With the continental divides on the different continents in mind, we can say that the Atlantic Ocean disposes of the largest drainage region by far, and also of the largest inflow from the continents (cf. Plate 1). The European rivers are of minor importance, except for the runoff from the rivers Danube, Pechora, Dvina, Neva, and Rhine, exceeding 2.0×10^3 m^3 sec^{-1}. The river Volga discharges 8.1×10^3 m^3 sec^{-1} into a land-locked sea, the Caspian Sea. Since the Nile was dammed up in 1965, there is no river that discharges more than 2.0×10^3 sec^{-1} into the European Mediterranean Sea (with the Black Sea left out of consideration).

4.3.6. The hydrologic cycle on the earth

According to Table 4.06 the total stock of water on the earth amounts to 1434×10^6 km^3, 1350×10^6 km^3 of which are contained in the world ocean. 84×10^6 km^3 are stored in ground water as well as tied up in the inland ice and icecaps and in glaciers. If the total stock of water on our globe (1434×10^6 km^3) became water in the world ocean, the sea level would be higher by 232 m, provided that continents and oceans were distributed as at present, that is, that 362×10^6 km^2 were covered by water. If only the inland ice and the glaciers melted, and the water were discharged into the world ocean, the sea level would rise by 66 m. This is the utmost eustatic rise, if the amount of ground water remains unchanged.

The stock of water on the earth is subject to continuous cycling. Besides the water volume of the various parts of the hydrosphere, Table 4.06 contains the rates of renewal, that is, the number of years in which—on the average—a complete renewal takes place. For instance, about 3000 yr are needed to completely vaporize the water content of the world ocean of 1350×10^6 km^3 under the assumption of an annual evaporation rate of 440,000 km^3. The water of the rivers, however, amounting to 1200 km^3, is replaced once in 0.032 yr, that is, in 11.4 days.

For the present time, the total amount of water in the world ocean as well as the total amount (40×10^{18} tonnes) of salts dissolved in seawater, can be considered constant. Thus, the average salinity of seawater, which is 34.72‰, must be constant, too. In spite of this, considerable differences of salinity have been found in the world ocean, in horizontal as well as in vertical directions. These are caused by the water cycle of the ocean. When seawater evaporates or freezes, the greatest part of the salts remains in the water and contributes to an increase of salinity in the surface water. Conversely, precipitation,

Table 4.06. Stock and Exchange of Water on the Earth[a]

Parts of the Hydrosphere	Water Volume (10^6 km^3)	Water Cycle (yr)
World ocean	1350	3000
Ground water	60	5000
Inland ice and glaciers	24	8600
Lakes	0.23	10
Soil moisture	0.082	1
Rivers	0.0012	0.032
Atmosphere	0.014	0.027
Total hydrosphere	1434.327	2800

[a] After Lvovitch, 1971.

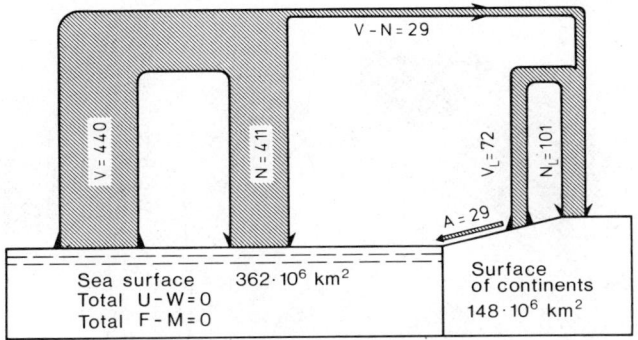

Fig. 4.18. Water balance on the earth in 10^3 km^3 yr^{-1}.

continental runoff, and water from melting ice cause a decrease of surface salinity. In this water cycle, ocean currents and mixing processes ensure that the contrasts are compensated, not only at the surface but in all layers of the ocean.

The total water exchange of the world ocean W_Σ can be summarized in a budget equation:

$$W_\Sigma = (V - N) + (F - M) + (U - W) + A \qquad (4.16)$$

The term in the first parenthesis gives the water exchange between ocean and atmosphere which takes place through evaporation V and precipitation N. The second parenthesis is concerned with the water transfer by the formation F and the melting M of ice. The third parenthesis accounts for the gain U and the loss W achieved by currents and mixing. A stands for the continental runoff. A survey of the final terms of the water exchange in the entire world ocean is given in Fig. 4.18, which is based on recent investigations concerning evaporation V, precipitation over the world ocean N, and over the continents N_L, carried out by Lvovitch (1971). It results that the total runoff from the continents is 29,000 km^3 yr^{-1}. Referred to the surface of the world ocean, the precipitation amounts to 114 cm yr^{-1}, the evaporation to 122 cm yr^{-1} and the continental runoff to 8.1 cm yr^{-1}. In the world ocean and in seas with closed current systems, $F - M$ as well as $U - W$ vanish. Since the amount of discharge from the continents is small compared to the water exchange $V - N$ and, moreover, the effect of this runoff on salinity is primarily restricted to near-shore waters, the large-scale distribution of surface salinity should essentially be determined by $V - N$. Figure 4.17 shows how well this expectation is confirmed by the zonal values, averaged over 5° of latitude, for evaporation V, for precipitation N, for $V - N$, and for surface salinity S, each quantity represented by a distribution curve from 80°N to 70°S. The curves for $V - N$ and for S practically coincide between 40°N and 60°S. Low surface salinity corresponds to a surplus of precipitation in the equatorial zone and in higher latitudes, whereas the excess of evaporation in the subtropics is reflected by higher surface salinity. The result is a linear relation for the surface salinity as given by Wüst (1954) for the Northern and Southern Hemispheres of the world ocean:

$$S = 34.47 + 0.0150(V - N) \text{ for } 70\text{--}10°\text{N}$$

$$S = 34.92 + 0.0125(V - N) \text{ for } 60°\text{S--}10°\text{N} \qquad (4.17)$$

V and N must be expressed in cm yr^{-1} so that the salinity S is obtained in ‰. This relation is of a statistical nature and says nothing about the physical processes involved.

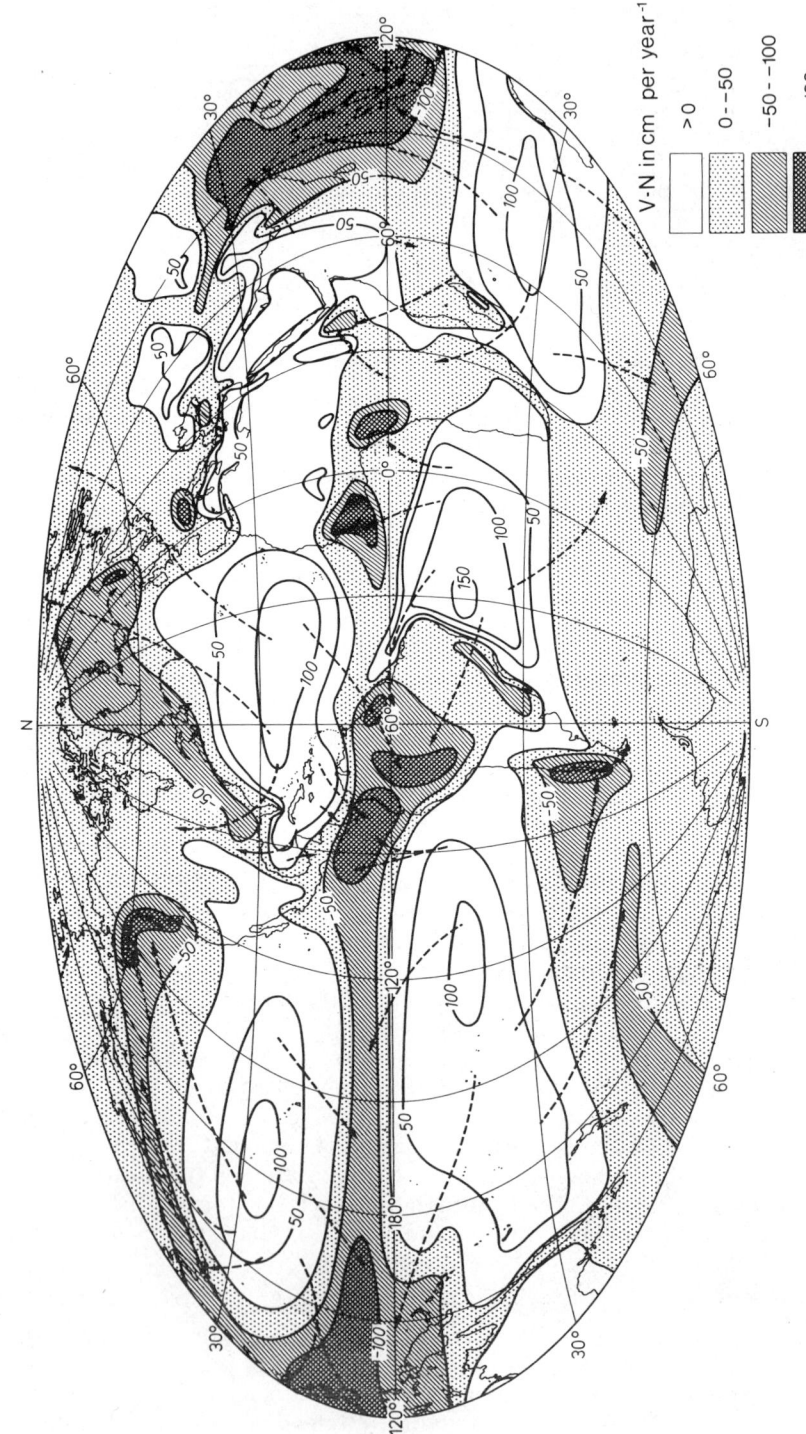

Fig. 4.19. Distribution of the difference of evaporation minus precipitation (in cm yr^{-1}) at the earth's surface, as well as main directions of the water vapor transport in the atmosphere. (Based on Albrecht, 1949, 1951 and Jacobs, 1951.)

The relations between surface salinity and $V - N$, as mentioned above, are valid only if they are based on zonal values averaged over the entire width of the individual oceans, because, in this way, the influence of the salinity transport by currents is nearly eliminated. The regional water exchange between ocean and atmosphere is better understood if we consider the geographic distribution of $V - N$ (cf. Fig. 4.19), although the data used contain a number of uncertainties, because the values for V and N need some improvement (cf. Sections 4.2.5 and 4.3.4). Therefore, the water exchange can be presented only in rough outline.

Apart from the regional distribution of the water exchange, Fig. 4.19 gives some information on the heat budget of the atmosphere and on the main directions of the water vapor transport and, thus, on the atmospheric part of the water cycle. It is obvious that the chief action areas of the water cycle lie in the subtropics, preferably on the western sides of the oceans. The evaporation surplus is transported toward the equator mainly by the trade winds and polewards by the west winds prevailing in moderate latitudes. Within the trade-wind region, evaporation greatly exceeds precipitation and the water vapor content is increased further. It is only in the intertropical convergence zone—and in the monsoon regions—that precipitation surpasses evaporation. During condensation, latent evaporative heat is released. This thermal energy heats the atmosphere and, so, maintains the circulation of the latter to a large degree. In the equatorial zone, large amounts of rain precipitate from gigantic, towering cumulus clouds, thus causing a strong decrease of sea surface salinity in the equatorial rain belt. In the moderate and higher latitudes, the excess of precipitation over evaporation also results in a large heat gain for the atmosphere, won from the heat of condensation of water vapor. The consequence is the moderation of climate in the higher latitudes. If in regions with a surplus of evaporation, or precipitation respectively, the salinity does not change over long periods, we may consider this as an indication of the equilibrium between the effects of mixing and circulation in the world ocean, on one hand, and the atmospheric water exchange $V - N$, on the other.

In Chapter 4 it has been shown that close relations exist between ocean and atmosphere. This is not surprising if we take three facts into account:

1. Ocean and atmosphere have a common interface where a mutual exchange of motion, heat, water, salts, and gases takes place.
2. Although the wind system of the atmosphere is basically driven by solar energy, the ocean serves as the necessary intermediate link, since the heat of condensation of evaporated seawater in the central tropics represents the most important driving force for the atmospheric circulation. For these condensation processes the ocean supplies additional condensation nuclei for the aerosol.

    In the shear stress the winds transfer a small part of the energy of motion to the sea surface and, thus, drive the oceanic current system. With a view to the total energy budget, this small proportion of energy would be uninteresting if it were not for the fact that small variations of the shear stress can change the whole system of oceanic circulation. In such a case, atmospheric energy sources in the ocean, originating in the heat store of the upper layers of water, will be displaced, and such displacements involve far-reaching effects in the atmosphere.
3. On this rotating planet earth the media of both spheres, air and water, are in turbulent motion that is maintained by different degrees of heating. Ocean and atmosphere are governed by the same hydrodynamic laws.

In this Chapter 4 the "interaction between ocean and atmosphere" has been described, which is a topic under which highly complicated processes are hidden. Many of them are now understood after the efforts made by numerous scientists during the past decades, but others are not. Malkus (1962), who herself has substantially contributed to the elucidation of such interaction, is very pessimistic with regard to further progress in this field: "At the risk of being reactionary in this age of scientific optimism, we must conclude that the head-on approach to the overall air–sea interaction problem is almost surely doomed to failure. Geophysicists must discipline themselves to seek critical parts of it for careful study. Thus they must try to pose questions which isolate tractable features of the complexity which can be treated under relatively controlled conditions." This confession is, at the same time, an appeal for an intensive study of the physical processes by means of model investigations. Among others, this line has been followed by leading scientists like Brocks (1972), Kitaygorodski (1970), Kraus (1972), and Stewart (1967) who, in their papers, have summarized the results that have been achieved recently.

5 Temperature, Salinity, Density, Characteristic Water Masses, and Ice in the World Ocean

5.1. Temperature in the World Ocean

5.1.1. Diurnal variation of temperature near the sea surface

The diurnal variation of the heat exchange at the sea surface as the result of incoming radiation, back radiation, evaporation, and direct heat transfer between ocean and atmosphere sets an upper limit to the diurnal variation of the water temperature. In particular, the diurnal variation also depends on the vertical heat transport in the top layer of the ocean. In the example mentioned in Section 4.2.2, the incoming radiation is assumed to 1000 cal cm^{-2}; hence, after attenuation in clearest ocean water (type I of Jerlov, 1968), the water in the uppermost meter would warm up by 5.5°C and by 0.05°C in the layer between 10 and 11 m depth. In reality, however, the average diurnal variation of the oceanic surface temperature ranges from 0.2 to 0.3°C and extends down to approximately 10–20 m. Exact and detailed measurements are difficult to make. Figure 5.01 contains a case observed with particular care on the research vessel *Meteor* (Hoeber, 1966, 1969). It is based on five-week continuous records at 30 cm depth (T_W) in the open equatorial Atlantic Ocean at equinoctial time, taken with the measuring buoy after Brocks (1959), which had been adapted to the requirements of the ocean by Dunkel (1967). There was no disturbance by the ship. In addition, the mean diurnal variation of the air temperature at 8.9 m height (T_A) is given. The characteristic quantities of the average diurnal variation of temperature are the following: diurnal range of T_W = 0.27°C, of T_A = 0.62°C; time of occurrence of the maxima: 15^h local time (T_W), 16^h (T_A), of the minima: 7^h (T_W), 5^h (T_A). The variation of temperature in water is far from being a simple harmonic oscillation, the increase of T_W lasts 8 hr, and the decrease 16 hr, whereas the corresponding values for T_A are 11 hr and 13 hr. The mean difference $T_W - T_A$ is 0.47°C. The same diurnal variation of water temperature has also been found in adjacent seas. Only in shallow waters (less than 10 m), where the normal range of vertical exchange is markedly reduced, does the variation reach higher values, as shown by Dietrich (1953). For instance, the mean maximum of daily variation of 0.24°C in the southern North Sea (water depth 50 m) in June contrasts with the value of 1.90°C in the waters around the Finnish skerries, the small coastal islands in the Gulf of Bothnia (water depth about 6 m).

The contrast between the diurnal temperature variation as observed in situ and as calculated from the attenuation of the incoming radiation indicates that processes other than radiation must also be effective; such processes are thermal conduction and turbulence. Our knowledge on the quantitative effect of these mechanisms of heat transport near the surface is still rather incomplete. There are two reasons for this insufficiency:

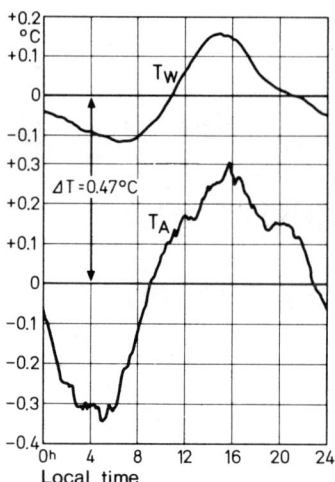

Fig. 5.01. Mean diurnal variations of water temperature (T_w at 30 cm depth) and of air temperature (T_A at 8.90 m height) at 0°1'N and 29°33'W from September 9 to October 10, 1965. Based on registrations at the anchor station of the research vessel *Meteor* during the Atlantic Expedition 1965 (IQSY). The sensors had been mounted on a stabilized mast (cf. Fig. 4.02) at 300 m distance from the ship. (After Hoeber, 1966, 1969.)

(1) technical difficulties in measuring the temperature profile in the uppermost millimeter of the sea with a resolution of at least 10 μm (0.01 mm) (the process of thermal conduction takes place in such extremely thin layers); (2) theoretical difficulties in predicting the behavior of water near the wavy sea surface that is disturbed by the wind.

An essential step toward the solution of the first (i.e., the technical problem, is illustrated in Fig. 5.02, where, on the left side, the registrations of four vertical temperature profiles are shown that were taken between 40 cm above and 40 cm below the sea surface with very calm sea in the Black Sea in the fall of 1968 (Andreyev et al., 1969). Platinum resistance thermometers with a time constant of 1.4 sec and an accuracy of ±0.05°C were used as temperature sensors. One profile was taken in 64 sec (nothing is said about the diameter of the resistance wire). The autumnal heat loss is especially pronounced in the uppermost centimeter. While the diurnal variation at the surface amounts to 5.0°C

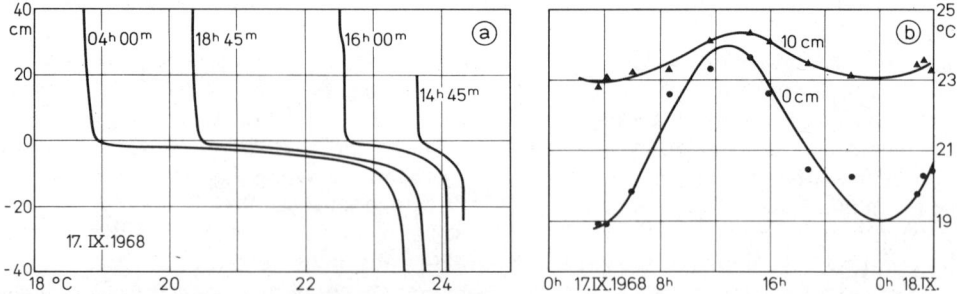

Fig. 5.02. Diurnal variation of the temperature, measured with calm sea and clear sky, in the Black Sea on 17–18 September 1968. (After Andreyev et al., 1969.) (a) Temperature profiles between 40 cm above and 40 cm below the sea surface at four different moments on 17 September 1968. (b) Diurnal variation of water temperature at the sea surface and at 10 cm depth on 17–18 September 1968.

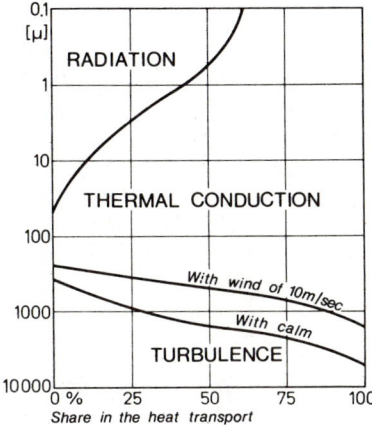

Fig. 5.03. Mechanisms of heat transport into the sea in the uppermost cm below the sea surface. (After McAlister and McLeish, 1969.)

according to Fig. 5.02(b), it is only 1.0°C at the depth of 10 cm. The strong vertical temperature gradient is nearly linear between 1 and 3 cm, which points to molecular thermal conduction.

Although the exceptional character of the sea surface is demonstrated by the experimental solution in Fig. 5.02, the resolution of the vertical gradient in the uppermost millimeter is not yet satisfactory. This is shown by the theoretical estimate given by McAlister and McLeish (1969) and partly based on empirical determinations. The results concerning the processes of vertical heat transport in the uppermost centimeter are summarized in Fig. 5.03: In the uppermost micron (0.001 mm), the temperature distribution is predominantly determined by radiation. Below this layer, in the uppermost $\frac{1}{2}$ mm with a wind of 10 m sec^{-1} and in the uppermost millimeter with calm, thermal conduction dominates the vertical heat transport. Further down, this is effected by turbulence. Apart from the uppermost millimeter, the vertical temperature distribution is determined by turbulence, in particular by wind-induced turbulence or by free turbulence which is caused by the instability of stratification when the surface is cooling. The infrared back radiation, which can be measured with radiation recorders installed in satellites (cf. Fig. 4.13), is influenced by the top water skin of the sea. It is to be expected that such measurements from satellites will differ from the real "surface temperature" observed from ships in the uppermost 1 to 2 m of water.

The depth range of the diurnal temperature variation can seldom be determined quite exactly since there often are disturbances, especially those caused by internal waves, which may conceal small daily oscillations. Therefore, longer series of measurements are necessary. Eight-day continuous records, taken by the research vessel *Meteor* in the North Atlantic at the Great Meteor Seamount in April 1967, have been analysed by Horn (1971). From the results, summarized in Fig. 5.04, it can be seen that the diurnal amplitude decreases from 0.28°C at the sea surface to 0.03°C at 50 m depth. At this depth, the temperature maximum occurs 9 hr later than at the sea surface. The vertical heat exchange, which can be calculated from the vertical distribution of amplitudes and phases, lies between 240 and 310 cm^2 sec^{-1}. From 0 to 50 m depth, the slight stability of stratification σ_T steadily increases by 0.10. So in this case, the penetration of the diurnal

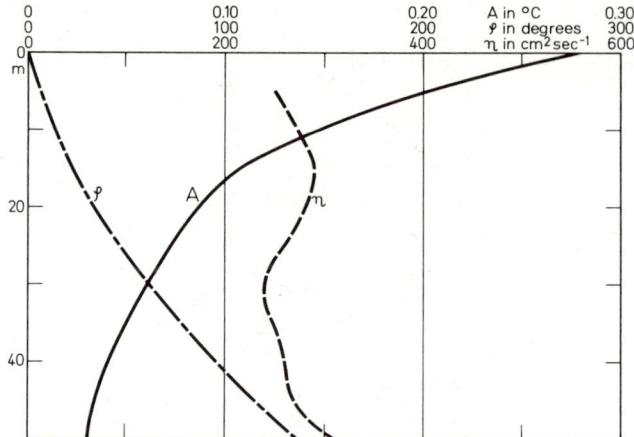

Fig. 5.04. Diurnal variation of the water temperature in the uppermost 50 m according to eight-day continuous records at the anchor station 9027 of the research vessel *Meteor* ($\varphi = 30°18'$N, $\lambda = 29°25'$W, April 19 to 27, 1967). There are represented the vertical distributions of the amplitudes A (in °C), of the phases φ (in degrees relative to the sea surface), and of the coefficient of turbulent heat exchange η, resulting from A and φ, in cm^2 sec^{-1}. (After Horn, 1971.)

temperature variation is facilitated. Altogether, the individual diurnal temperature variation is of minor importance in the ocean, but it represents the shortest rhythm with which the propagation of heat takes place in the ocean.

5.1.2. Annual variation of surface temperature

In contrast to the diurnal variation, the annual temperature variation significantly influences physical, chemical, and biological processes in the ocean. It can be represented by the amplitude T_1 and the phase ϕ_1 of a harmonic wave, which, for the sake of completeness, is supplemented by the annual mean T_0. The characteristic features of the geographical distribution of the annual variations of surface temperature in the world ocean will be demonstrated by these two examples: for the open ocean from low to higher latitudes (Fig. 5.05) by the case of the North Pacific Ocean, and for a shelf sea (Fig. 5.06) by the cases of the North Sea, the Irish Sea, and the English Channel.

The graph for the North Pacific Ocean (Fig. 5.05), evaluated by Wyrtki (1965), is based on the harmonic analysis of two-degree monthly means of surface temperature for the period from 1947 through 1960. In some peripheral regions the number of observations is insufficient as indicated in the figure. It should be mentioned that the original paper also includes the derivation of the semiannual temperature oscillation. In a similar, though less detailed investigation, Panfilova (1968) has dealt with the entire Pacific Ocean.

The graphs for the North Sea, the Irish Sea, and the English Channel in Fig. 5.06 have been taken from Dietrich (1953). They are based on the harmonic analysis of observational data of the atlas *Monthly Charts of Surface Temperature of the North Sea, the Baltic Sea, and Adjacent Waters* by Böhnecke and Dietrich (1951). This analysis covers the period from 1906 to 1938 and is based on 1° monthly means and on monthly means from observational series at numerous lightships and coastal stations. Here, too, the semiannual temperature wave has been analysed.

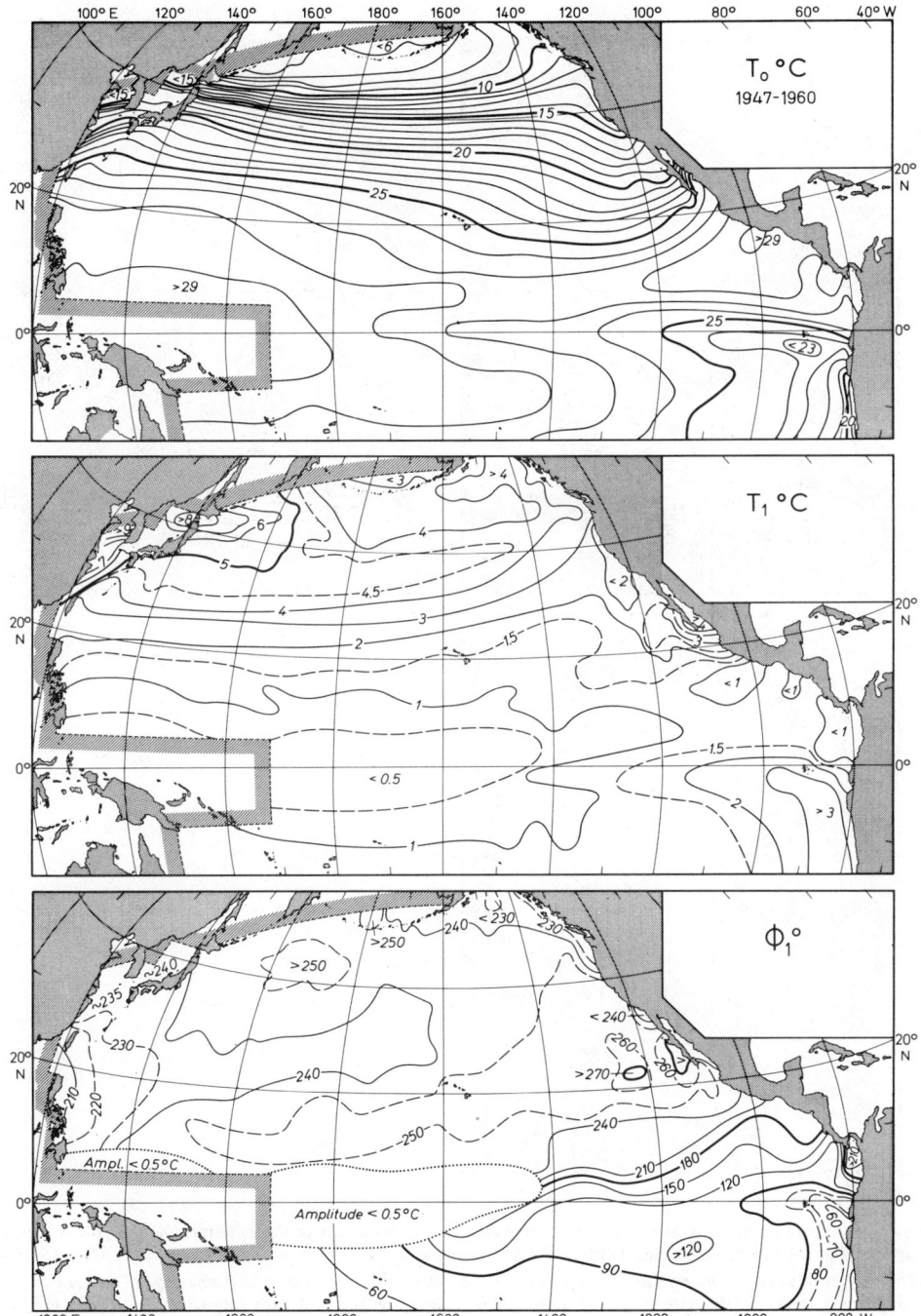

Fig. 5.05. Mean annual variation of surface temperature in the open ocean in the example of the North Pacific Ocean 1947–1960 as harmonic wave $T = T_0 + T_1 \cos(\alpha - \phi_1)$. (After Wyrtki, 1965.) Above: Annual mean T_0 in °C for 1947–1960; middle: Amplitude T_1 in °C; below: Phase ϕ_1 in degrees, referred to January 1, indicates the date of the maximum, 210° is for August 1, 230° for August 21, 250° for September 11. Dark screen: no data.

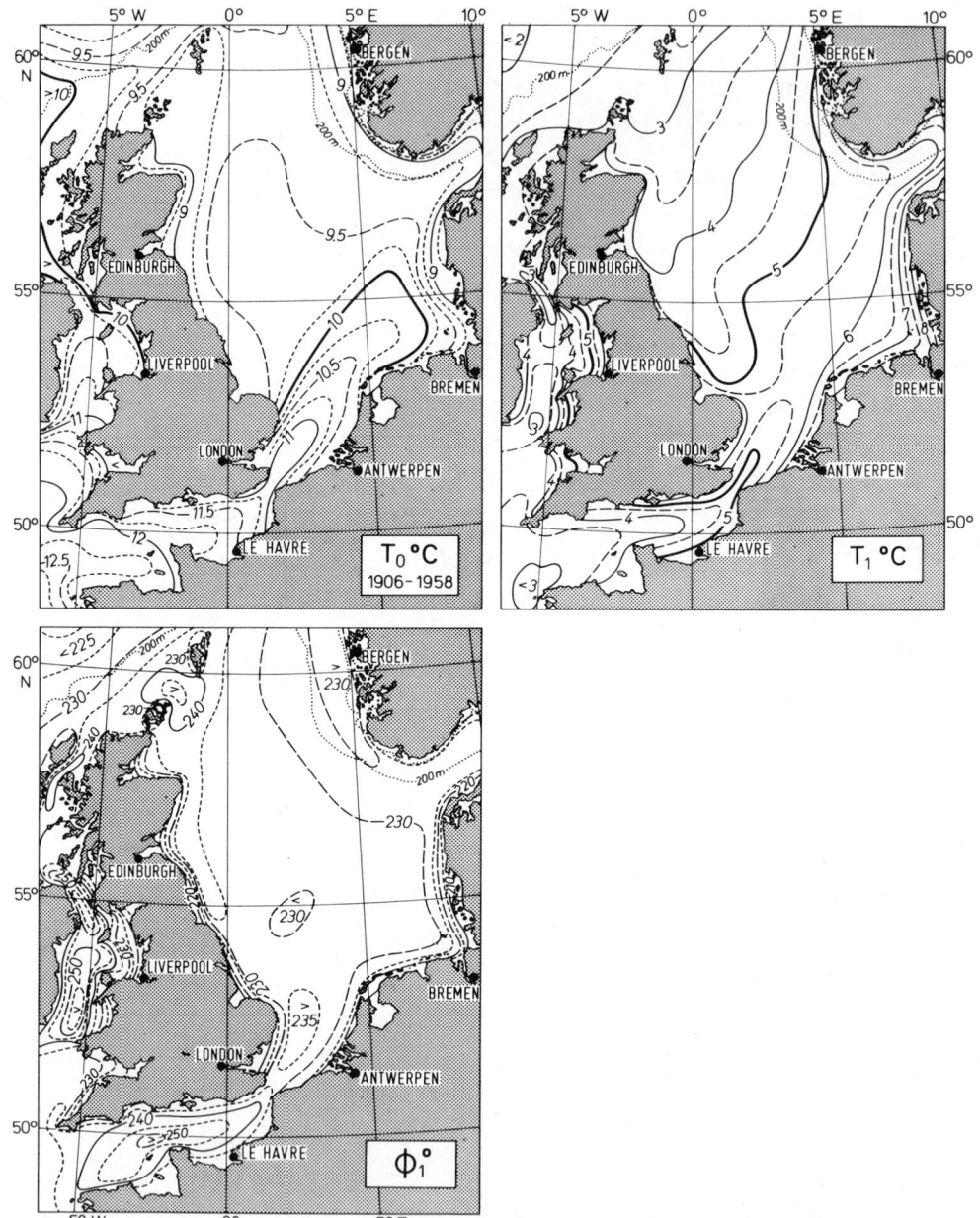

Fig. 5.06. Mean annual variation of surface temperature in shelf seas in the examples of the North Sea, the English Channel and the Irish Sea 1906–1938 as harmonic wave $T = T_0 + T_1 \cos(\alpha - \phi_1)$. (After Dietrich, 1953.) Above left: Annual mean T_0 in °C; above right: Amplitude T_1 in °C; below left: Phase ϕ_1 in degrees, referred to January 1, indicates the date of the maximum, 210° is for August 1, 230° for August 21, 250° for September 11.

According to Fig. 5.05, the annual amplitude T_1 exceeds 4.5°C in the open ocean at latitudes between 30 and 40°. It decreases toward the poles and the equator as well as when approaching the upwelling areas off California and Peru. In the equatorial region it declines below 1°C, more often even below 0.5°C, which agrees with the small annual variation of the heat exchange (cf. Fig. 4.12). Exceptions are found in the upwelling areas near the equator on the eastern side of the Pacific Ocean with amplitudes of 3°C. In the open ocean, unusually high amplitudes occur where seasonal displacement of the boundaries of water masses with characteristic thermal conditions can be expected. Figure 5.05 shows the following examples from the North Pacific Ocean: the frontal zone between the Kuroshio and the Oyashio, $T_1 > 7$°C; the frontal boundaries at the California Current, $T_1 > 3.5$°C; and at the Humboldt Current, $T_1 > 3$°C. Extreme amplitudes occur in adjacent seas, especially in shallow shelf seas with strong tidal currents. The amplitudes are relatively small where tidal currents destroy the summer stratification on deeper shelves (100 to 200 m depth) as shown in Fig. 5.08, as in the Irish Sea and the English Channel, $T_1 < 3$°C, and they are relatively large in shallow waters (depth <25 m), because, here, continental influences control the heat transfer, as in the German Bight with $T_1 > 8$°C, and in the Yellow Sea with $T_1 > 9$°C. The latter value belongs to the exceptional ones in the world ocean which can otherwise be found only in the bays of the Baltic Sea (Gulf of Riga and Gulf of Finland), in the Gulf of St. Lawrence, and in the Sea of Asov.

Phase ϕ_1 in Figs. 5.05 and 5.06 gives the average time when the maximum of surface temperature occurs within the annual cycle. Phase 210° corresponds to the first of August, 240° to the first of September, and 270° to the first of October. In the North Pacific Ocean, the maxima predominantly occur between 235 and 255°, and in extreme cases they take place at 210° (Philippines) or reach 270° (off the peninsula of Lower California). Thus, their time range, comprising 2 months from August 1 to October 1, is remarkably large. With amplitudes of $T_1 < 0.5$°C, it is not certain when the maximum occurs; therefore, they are not represented here. At the equator, we recognize the transition to the annual variation of the Southern Hemisphere with the maximum at 60°, with the Humboldt Current carrying this annual variation somewhat into the Northern Hemisphere. In the waters of Western Europe, the maximum is generally shifted to earlier dates—in the open North Sea it is found at 230° (August 21) and, in shallow coastal waters, its phase decreases further, in the inner part of the German Bight even to 210° (August 1). In deeper shelf areas with considerable mixing by strong tidal currents (cf. Fig. 9.12), the maximum is delayed to more than 245° in the Fair Isle Passage between the Shetland and Orkney Islands, to more than 250° in the English Channel and to more than 255° in the outlets of the Irish Sea. On the whole, the mixing by tidal currents has a damping effect on the annual variation by reducing the amplitudes and delaying the time of occurrence of the maxima.

5.1.3. Distribution of surface temperature

Isotherm charts of mean monthly temperature, as have been plotted repeatedly, the first time by Maury (1852) for the Atlantic Ocean, give us an idea of the distribution of surface temperatures. It is because of Maury's initiative at the first International Meteorological Conference in Brussels in 1853 that observations of surface temperature, together with other meteorological data, have been carried out on merchant ships according to uniform methods. These data have been collected by the hydrographic offices and meteorological centers of the various seafaring nations. Today they form the basis for the monthly pilot charts which are prepared for individual oceans by such institutions in Hamburg, Copenhagen, London, Tokyo, Utrecht, and Washington, D.C. Summaries are available for the entire world ocean, for the oceans and smaller oceanic areas, in particular for adjacent seas. The first group includes:

W. F. McDonald (1938): *Atlas of climatic charts of the oceans.* U.S. Weather Bureau. Washington, D.C.

U.S. Hydrographic Office (1948): *World atlas of sea surface temperatures.* Washington, D.C.

Academy of Sciences (USSR) (1950): *Ocean atlas.* Moscow. (in Russian)

To the second group belong:

G. Böhnecke (1936): *Temperatur, Saltgehalt und Dichte an der Oberfläche des Atlantischen Ozeans.* Wiss. Erg. Deutsch. Atlant. Exped. Meteor, **5**, Berlin.

Netherlands Meteorological Institute (1952): *Indian Ocean oceanographic and meteorological data.* 2nd edit. No. 135. De Bilt.

U.S. Navy (1957): *U.S. Navy marine climatic atlas of the world. III, Indian Ocean.* NAVAER 50-1C-530. Chief of Naval Operations. Washington, D.C.

U.S. Naval Oceanographic Office (1967): *Oceanographic atlas of the North Atlantic Ocean, physical properties.* Publ. No. 700. Washington, D.C.

L. E. Eber, J. F. T. Saur and O. E. Sette (1968): *Monthly mean charts. Sea surface temperature. North Pacific Ocean 1949–1962.* Bureau of Commercial Fisheries. Circular 258. Washington, D.C.

J. L. Reid (1969): *Sea-surface temperature, salinity, and density of the Pacific Ocean in summer and in winter.* Deep-Sea Res., Suppl. **16**, 215–224 (Fuglister Vol.).

The third group includes:

G. Dietrich (1963): *Mean monthly temperature and salinity of the surface layer of the North Sea and adjacent waters from 1905–1954.* Cons. Int. l'Explor. Mer. Charlottenlund.

This is not a suitable place for isotherm charts from atlases to be reproduced. Attention may be drawn, however, to a few phenomena of fundamental character which occur in the open ocean and in shelf seas, but also this only as far as the distribution of the annual means T_0 of the surface temperature is concerned. Some relevant examples are given in Figs. 5.05 and 5.06.

The highest annual means of the surface temperature T_0 with more than 28°C, in some cases with more than 29°C, are found in tropical ocean waters. As shown in Fig. 5.05 for the Pacific Ocean, such values are most pronounced west of 140°W, as well as off Central America in the region where the Equatorial Countercurrent originates. The upwelling areas off Peru and, less pronounced, off California are characterized by relatively low temperatures. The same is true for the upwelling areas along the equator from the Galapagos Islands to beyond 170°W. Corresponding values are found in the Atlantic Ocean off Northwest Africa and Southwest Africa, but hardly at all at the equator. In the Indian Ocean, due to the alternating monsoons, there are no areas where upwelling is present all year round.

In temperate latitudes, the zonal course of the isotherms shows some deviations on the eastern and western sides of the North Pacific Ocean: in the region of the subtropical anticyclone, the isotherms on the eastern side are deflected toward the equator, those on the western side toward the pole, while in the region of the subpolar cyclone it is the opposite, that is, on the eastern side toward the pole and on the western side toward the equator. The consequences of those deviations are strong meridional temperature gradients off northern Japan and weak ones off Oregon and British Columbia. So the following rule can be established: in the subtropics the western sides of the oceans are warmer than the eastern sides, whereas in temperate latitudes, it is the reverse; the western sides are colder than the eastern sides. This rule is especially valid for the North Atlantic Ocean. Here, due to the configuration of the coast, the heat transport can freely proceed towards the northeast, which, in the Pacific Ocean, is prevented by Alaska. Therefore, the eastern part of the Norwegian Sea and the western European waters are anomalously warm. In the Southern Hemisphere, the above rule applies only to the subtropics of the three oceans, which is clearly shown in the chart of the anomalies of surface temperature (Dietrich, 1950a). In higher southern latitudes, there is no subpolar cyclonic motion in the Antarctic water belt, thus, there is no contrast between the eastern and western sides. The water moves in circumpolar fashion, which, due to the eccentric position of the Antarctic continent with a view to the South Pole, results in the Indo-Atlantic sector appearing too cold as compared to the Pacific one.

In marginal seas, the distribution of the annual means of the surface temperature T_0 is greatly determined by the heat transport of oceanic currents. In the North Sea, the isotherms are arranged predominantly in meridional direction (cf. Fig. 5.06). Of course, the tongue of relatively warm water does not mean that the North Sea is dominated by water from the English Channel. This is true only for the southern part of the North Sea, as is clearly to be seen in the distribution of salinity (Fig. 5.12). The cyclonic motion prevailing in the other parts of the North Sea transports Atlantic water on the western side from north to south. But before entering the northern North Sea, this water has already lost much of its heat. What is remarkable in the distribution of T_0 in Fig. 5.06 is that the horizontal temperature gradient points toward the coast everywhere, which means that the continent derives profit from the relatively warm sea.

The statistical treatment of the vast amount of data necessarily smooths out temperature differences existing within small areas which are typical for particular regions of the higher latitudes. Such temperature differences are found on the left-hand sides of the Gulf Stream and of the Kuroshio Current, at the boundary between the Irminger Current and the East Greenland Current, and along the Antarctic Convergence in the Atlantic water belt. At these boundaries, phenomena have been observed not only with respect to temperature but also with respect to motion; these show a certain similarity with those along the polar front of the atmosphere; therefore these boundaries are also called oceanic polar fronts. Two water types of different origin abruptly adjoin here, namely cold polar–subpolar water on the one side and warm subtropical water on the other. In particular, the polar fronts are not stationary but oscillate over large areas in close connection with the course of ocean currents as shown by the example of the Gulf

Stream. When averaged over longer periods of time, these typical frontal zones can be recognized only indistinctly, as in the upper part of Fig. 5.05, or even not at all.

5.1.4. Distribution of water temperature with depth

The data gathered during oceanographic expeditions represent the source of our knowledge on the vertical distribution of temperature as well as on the other hydrographic factors. Such data are scattered throughout numerous publications, particularly in the comprehensive reports on expeditions, some of which are mentioned in this book. The first step toward achieving a survey is a list of the expeditions with the references consulted. Such complete lists have been compiled by Wüst (1935) for the Atlantic Ocean and by Vaughan (1937) for the entire world ocean. But as soon as marine research was intensified, especially during the International Geophysical Year 1957-1958, it became obvious that a fundamental change must take place with regard to archiving the observational data. National and world data centers have been established which store the data on punch cards or magnetic tape, keeping them available for retrieval. The German Oceanographic Data Center in Hamburg is operated by the German Research Society and the German Hydrographic Institute. The World Data Center A is located in Washington, D.C., and the World Data Center B in Moscow. It is their task to ensure the exchange of all data that may be wanted. Besides, data lists that have met with great interest are being published as, for instance, the list for the North Atlantic Ocean published by the International Council for the Exploration of the Sea (ICES). From 1902 to 1960 all available data were collected and published in the Bulletin Hydrographique; since then data lists have been compiled for international expeditions that are coordinated by ICES. Data lists for the Atlantic Ocean are contained in the atlas by Fuglister (1960), for the Pacific Ocean—compiled by the Pacific institutes of the United States and Canada, and covering the period of 1949-1959—they are given in 12 volumes by the Scripps Institution of Oceanography (1961-1965).

For reasons of expediency, we will deal first with the upper layer of the ocean, in which a noticeable annual variation of temperature takes place, and afterwards with the deeper layers where this is not the case. In the ocean the penetration of heat during the annual cycle is a much more complicated process than in the firm ground on land, where only the physical thermal conduction is effective. When it becomes warmer in spring, the temperature increase at first benefits a thin top layer which, thereby, becomes less dense. Thus, stratification becomes more stable and a thermocline begins to develop. In this thermocline, the vertical exchange which is caused by the turbulence of wind-induced currents and surface waves and which is responsible, almost exclusively, for the transport of heat to deeper layers, is diminished. If more heat is supplied, it will primarily be stored in the near-surface top layer, while the thermocline gets stronger, moving slightly downward. Such processes in the thermocline come to an end when the ratio between the vertical density gradient and the vertical velocity gradient reaches a limiting value, the so-called Richardson number. Then the vertical turbulence in the thermocline and, thus, the heat transport to deeper layers completely cease. The thermocline becomes an impenetrable barrier for all vertical exchange processes regardless of whether heat or chemical factors are involved. Herein lies the great importance of the summer thermocline for chemical-biological processes in the ocean.

The summer thermocline is a typical phenomenon of all areas with a great amplitude in the annual variation of the surface temperature T_1. Such areas can be recognized in Figs. 5.05 and 5.06. The annual variation of stratification is illustrated in Fig. 5.07, which

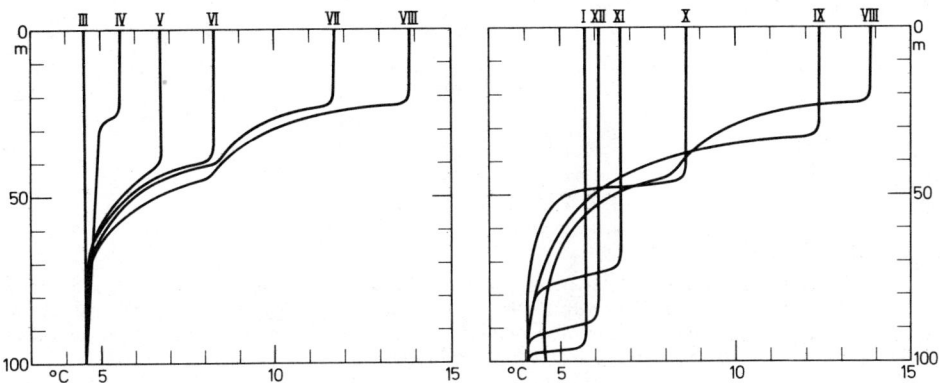

Fig. 5.07. Typical annual variations of temperature stratification and of the thermocline, based on individual vertical records taken on weather ship *Papa* in the Northeast Pacific Ocean ($\varphi = 50°N$, $\lambda = 145°W$) in 1956. (After Dodimead et al., 1963.)

is based on individual vertical registrations for each month in 1956, obtained by the Canadian weather ship *Papa* in the Northeast Pacific Ocean. Except for the change of slope in July and August at 40 to 50 m depth, caused by a halocline, the annual variation here is a typical one: in April, the winter homothermy of the month of March is turned into a thermocline at 25 to 30 m depth, in May it sinks down to 40 m, and during the summer it ascends, with the gradient intensifying, until it reaches a depth of about 25 m in August. The degradation of the stratification from September to February is effected by the decrease of temperature in the top layer and, in connection therewith, by the thermocline sinking down to about 100 m depth before the homothermy of winter is once more established.

According to Overstreet and Rattray (1969), the depth of the thermocline depends on the annual variation of the heat transfer, on the vertical velocity (i.e., on the divergence of the currents), and on the magnitude of the turbulent exchange, and, thus, on the wind. The following formula, obtained half-empirically, for the depth of the wind-mixed surface layer H_0 (in m) has been derived by Rossby and Montgomery (1935):

$$H_0 = \frac{2.38 W_0}{\sin \varphi}$$

where W_0 is the wind speed in m sec^{-1}, and φ is the geographic latitude. By means of this formula, Lumby (1955) determined the thickness of the homogeneous surface layer of the world ocean in different seasons. The results are satisfactory only in part; therefore, attempts were made to derive better relations, which can be applied more easily, from the long observational series collected by the weather ships D and K in the North Atlantic Ocean (Moiseyev, 1967). Strong and rapid heat supply in spring favors the development of a thermocline at shallow depth and leads to a seemingly paradoxical fact: the more the surface layer is warmed, the more the subsurface layer is excluded from the annual variation of temperature, thus maintaining the winter temperature. Since the speed of the spring warming necessarily increases with larger annual variation of the surface temperature, the shallowest depth of the summer thermocline must be expected in areas of high annual temperature range. This has been confirmed by observations along a meridional section in the North Pacific Ocean between the Aleutian and the Hawaiian Islands, as mentioned by Munk and Anderson (1948) in their theory on the thermocline.

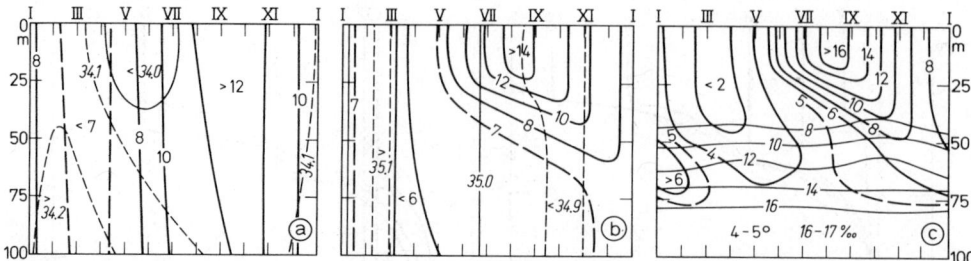

Fig. 5.08. Examples of the mean annual variations of temperature and salinity from the sea surface to the bottom at 100 m depth in shelf seas. (After Dietrich, 1950b.) (a) Irish Sea, North Channel with strong tidal currents; (b) Central North Sea with weak tidal currents; (c) Baltic Sea, Bornholm Deep without any tidal currents. (a) and (b) without, (c) with strong salinity stratification. For location of measuring stations see Fig. 10.55.

In midsummer the thermocline drops from about 20 m depth north of 35°N, where the annual range of the surface temperature amounts to more than 8°C, to a depth below 50 m near the Hawaiian Islands. Here, the annual range of surface temperature reaches merely 2°C.

In the shelf seas, conditions become more complicated when heat penetrates into the water. Beside wind-induced turbulence, originating from the surface, frictional turbulence is also active at the bottom, especially with strong tidal currents. As shown by Dietrich (1954a), this frictional turbulence does not permit any stratification to develop in the lower layer, nor even any thermocline if the tidal currents are strong. For the first case, Fig. 5.08(b) gives an example based on observations in the central part of the North Sea, and Fig. 5.08(a) represents an example from the northern Irish Sea, where the high turbulence of strong tidal currents prevents the formation of a thermocline even in summer. If, however, any noticeable tidal currents are lacking and if, in addition, the salinity increase with depth stabilizes the density stratification, in shelf seas the annual temperature variation will also not reach the bottom. The conditions in the Bornholm Basin in the Baltic Sea represent a typical example of this kind [cf. Fig. 5.08(c)].

For the northern part of the North Sea Fig. 5.08(b) indicates that the winter temperature of the surface water is maintained in the lower layer in summer, that is, that it varies from year to year according to the temperature in winter. Details can be seen in Fig. 10.55. Because this temperature determines the maturing process of herrings living in the lower layer, the beginning of the spawning migration in summer depends on the water temperature in winter. Thus, as early as in late winter, the duration of the summer herring season in the northern North Sea can be predicted on the basis of such relations, which have been detected by Dietrich, Sahrhage and Schubert (1959).

In all three examples, chosen from nearly the same latitude, the isopleth representation gives a complete picture of the annual variation. It should be noted that these diagrams show long-term mean conditions. Although, in the latter two examples, the homogeneous surface layer, the thermocline and its formation as well as the lower layer can be recognized, the actual magnitude of the vertical temperature gradient is not given because of the averaging process. In individual cases, the thermocline is limited to a few meters, as shown in Fig. 5.07, in some areas to even a few decimeters.

In the fall, when cooling begins, the surface gets denser, and, thus, vertical thermal convection becomes effective, eroding the thermocline by and by, until vertical differences of temperature have completely disappeared, which is typical of the winter stratification. In cases where salinity differences do not contribute to the density stratification, the winter homothermy in the shelf seas extends from the surface to the bottom, as shown in Fig.

5.08(b) for the central North Sea. In the open ocean of the temperate latitudes, it extends to a depth of about 200 to 300 m. In some areas of the higher latitudes, thermal convection can be traced to great depth. It leads to the formation of cold North Atlantic deep water south of Greenland (cf. Plate 4), of very cold Arctic bottom water in the Greenland Sea between Spitsbergen and Greenland, and likewise of very cold Antarctic bottom water in the Weddell Sea. Such winter circulation determines the distribution of vital gases and nutrients in the ocean and, therefore, it is of great importance biologically.

What is known best is the vertical thermal convection in the northern North Atlantic Ocean. It was one of the chief problems of the Polar Front Program in the northern North Atlantic Ocean during the International Geophysical Year 1957–1958 (Dietrich, 1957). The network of stations during the winter of 1957–1958 was repeated in the summer of 1958. The results obtained by 20 research vessels from 12 nations have been summarized in an atlas by Dietrich (1969), giving information with regard to the ranges of temperature, salinity, density, and oxygen. The vertical thermal convection in winter has been determined by Meincke (1967) with a view to geographic position and vertical extent. At some places in the Labrador Sea south of Greenland at 58°N, this convection extends over more than 2000 m, whereas at 45°N and in the middle latitudes the corresponding value is only 200 m, which is in agreement with the results derived from the heat budget by Filippov et al. (1968).

Since the thermal stratification of the upper layer is of great importance, in particular for underwater acoustic position fixing and ranging, forecasting methods have been developed. They aim at short-term forecasts in the order of 12 to 96 hr, especially with regard to surface temperature and several properties of the thermocline, namely depth of the upper boundary, maximum temperature gradient, total temperature difference, and depth of the lower boundary. Tully et al. (1963) and Wolff, Carstensen, and Laevastu (1960) have developed methods that are most commonly used. They have been improved by Laevastu and Hubert (1965) to form the SST method (Sea Surface Temperature Analysis and Forecasting) of the United States Fleet Numerical Weather Facility at Monterey, Calif.

The method developed by Tully is based on actual bathythermograph registrations and on the extrapolation of the conditions observed in accordance with the experience gained in previous years. Such a method is limited to application in a certain region. In the SST method, the heat budget (cf. Eq. 4.04) is taken into account; its fundamentals have been dealt with by Laevastu (1960). This method can be applied worldwide, but it is very demanding with regard to the basic data. It is a prerequisite that synoptic charts of several meteorological parameters are available, like wind direction and wind speed, cloud cover, vapor pressure, water temperature, air temperature, global radiation, and heat transport by ocean currents, that is, all the parameters that are included in the equation of the heat budget. The method is applicable only if a large electronic computer is available. A first step forward has been taken with the forecasting of surface temperature; the second step aims at forecasting the structure of thermal stratification in the upper layer of the ocean. Here, use is made of the well-known relations of the vertical heat transport as a function of meteorological factors, especially of the wind. According to Laevastu and Hubert (1965) a 12- to 96-hr forecast of the thermocline has been achieved by comparing the result obtained with vertical registrations and by considering corrections that might be necessary as well as by taking into account current boundaries, and convergences and divergences in the current field.

The consideration of the annual variation has shown that the heat transport into deeper layers is brought about almost exclusively by vertical convection. Combined with thermal conduction this would result, as the final stage, in a complete isothermy in the water column if other processes did not counteract it. The high net radiation in the low and middle latitudes tends to impede convection in the ocean, while the small net radiation in higher latitudes favors convection. In the atmosphere, by the way, it is just the opposite. The reason is that the hydrosphere is heated at its upper boundary, but the atmosphere at its lower boundary. Even if there were no vertical convection at all in the tropical and subtropical oceans, thermal conduction, in the course of the earth's history, would have led to uniform temperatures for the entire water column. According to observations, this definitely is not the case, merely because the heat flux from above downward is compensated by lateral advection of colder water from polar and subpolar regions. The vertical temperature distribution in the ocean must be considered as a result of oceanic circulation, which was already recognized by Alexander von Humboldt (1814) who was the first to observe the low water temperatures at great depths in the tropics. In ocean waters between 50°N and 45°S, the vertical temperature distribution shows a two-layer structure: an upper shallow warm-water sphere, on the average only 500 m thick, and an underlying mighty cold-water sphere which extends down to the bottom. As an analogy to the vertical structure of the atmosphere, Defant (1928, 1961) proposed the term "oceanic troposphere" for the upper sphere and "oceanic stratosphere" for the lower sphere. However, this definition seems useful only in the lower and middle latitudes, but even there the analogy with the atmosphere is confined to a few characteristic phenomena, as pointed out by Wüst (1949).

The large-scale vertical distribution of temperature is represented by longitudinal sections through the Pacific, the Atlantic, and the Indian Oceans (cf. Plates 3, 4, and 5). The geographic location of these sections can be taken from Fig. 10.31. In order to ensure the best possible comparability, all the three sections are reproduced in uniform scales of height and length, that is, the exaggeration of depth is the same (1:1110); the isolines shown are the same, and the density of stations is approximately the same. New comparable sections concerning the distribution of potential temperature, salinity, and density have been published by Reid and Lynn (1971).

The section through the Pacific Ocean, evaluated by Reid (1965), extends from Antarctica at 160°W in the Ross Sea as far as Alaska. Observational stations of several research vessels have been combined, which—with few exceptions—had been obtained during the summer months of the respective hemisphere from 1955 to 1958, those south of 40°S by the Russian research vessel *Ob* and, for the other part of the section, by the American research vessels *Argo, H.M. Smith, Stranger, Spencer F. Baird, Brown Bear*, and the Canadian research vessel *Whitethroat*.

The section through the Atlantic Ocean in its portion between 40°N and 60°S is based on evaluations by Wüst (1935) published in the *Meteor* Report. Observations north of 40°N have been taken from the Polar Front Survey carried out during the International Geophysical Year 1958, namely from cruises during the winter and spring of 1958 (Dietrich, 1969). For the region south of 60°S, stations of several research vessels have been used, of *General San Martin* in December 1955, of *Eltanin* in September 1963, April 1964, and February 1966, as well as of *Glacier* in February–March 1969. The section proceeds from the Antarctic continent in the Weddell Sea at 30°W along the western side of the ocean to the Labrador Sea and east of Greenland across the Greenland-Iceland Ridge into the Greenland Sea as far as 80°N.

The section through the Indian Ocean is a combination of several sections by Wyrtki (1971), contained in the Atlas of the Indian Ocean. It runs from the Antarctic continent at 57°W into the Gulf of Aden and through the Red Sea to the Gulf of Suez. In the Indian

Ocean, from Antarctica to Socotra, the section is based on observations of the research vessel *Ob* (Stations 204 to 212 in February 1957, 274 in March 1957, and 120 to 146 in May to June 1956). For the Gulf of Aden and up to the Gulf of Suez the *Atlantis II* Stations 52 to 38, taken in July to August 1963, have been used, and the *Meteor* stations of November 1964 have provided the data for the Red Sea, in particular those needed to locate the hot brines.

For the sake of completeness, the vertical distributions of salinity, oxygen, and of the deep circulation are given, which will be dealt with later in this book (cf. Section 10.1).

As can be seen in the longitudinal sections of Plates 3, 4, and 5, the transition from the warm-water sphere to the cold-water sphere takes place in a relatively thin boundary layer. In the open ocean, its center almost coincides with the depth of the isotherms for 8 to 10°C. In the tropics, this boundary layer lies at a depth of 300 to 400 m, and in the subtropics at 500 to 1000 m. From 40° latitude polewards it ascends toward the surface, intersecting it at the well-pronounced boundary between water types mentioned before, that is, at the oceanic polar front, which is mostly associated with a convergence of the surface currents. Therefore it is also called Antarctic, or in the north, Arctic Convergence. Its geographical position in the world ocean can be found in Plate 6. Poleward of the oceanic polar front, the cold-water sphere extends up to the surface.

The vertical temperature distribution in adjacent and mediterranean seas is quite different from that in the open ocean. These seas are more or less cut off from the open ocean and thus excluded from the advective propagation of polar and subpolar water masses. Here convective processes become decisive, which, however, are influenced significantly by the density stratification caused by salinity. The Red Sea (in Plate 5) is an example of this; similar conditions can be observed in the European Mediterranean Sea (cf. Section 10.2.2).

The entire world ocean, with its average temperature of 3.8°C, represents a cold sphere. Even at the equator, the temperature of the whole water column amounts to only 4.9°C. Only the near-surface layer of the lower and middle geographic latitudes serve as heat store. Because of the high specific heat of water, they can accumulate large amounts of heat. Higher latitudes benefit from this heat which is transported to them by ocean currents as well as, after transfer into the atmosphere, by atmospheric currents—above all as latent heat of evaporation gained from the transported water vapor. This is the reason why the ocean is so very important as a balancing regulator in the global heat budget.

5.1.5. Long-term changes of temperature

The annual variation and also the annual mean of surface temperature are subject to long-term changes. This is true for the recent past, as well as for historical and geological periods of time. In the following, facts, causes, and effects of temperature changes will be discussed, beginning with the recent geological past, the Pleistocene.

Up to 1947, only indicators of the respective temperature as applied in paleoclimatology, served as observations, that is, geological–morphological evidence as well as paleontological, astronomical, and physical facts were used as clues to the past. Here, reference is made to the only comprehensive textbook of paleoclimatology, written by Schwartzbach (1959, second German edition 1961, translation into English 1963), and to the all-round study of paleoclimatology, containing 52 contributions and edited by Nairn (1964).

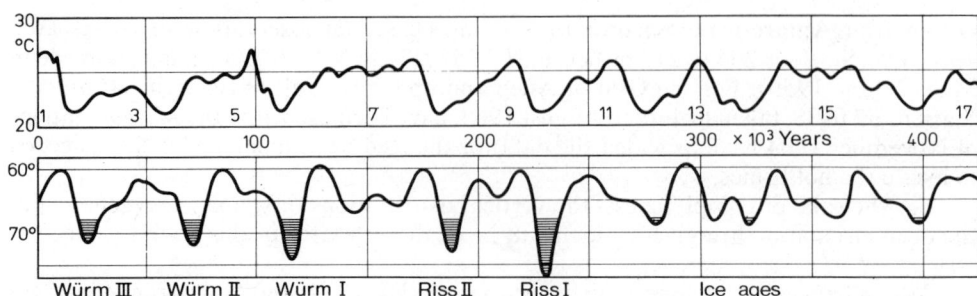

Fig. 5.09. Above: Temperature of the surface water of the central Caribbean Sea in the past 425,000 years. (After Emiliani, 1966.) (Numbers 1, 3, .. 17: interglacial warm periods.) Below: Incoming radiation in summer at 65°N in the past 425,000 years, expressed in equivalents of the radiant energy at this latitude. Ice ages are marked. (After Milankovitch, 1930.)

Since 1947, our knowledge on the water temperature of the oceans has won a new quantitative basis by isotope chemistry; for Urey (1947) discovered that the ratio of the oxygen isotopes $^{16}O/^{18}O$ in the lime $CaCO_3$ of shells and skeletons of marine organisms depends on the temperature of the water in which those animals spent their lives. Urey at once realized the consequences of his ingenious discovery: "I suddenly found myself with a geological thermometer in my hands." The "geological chronometer," as a supplement to the "geological thermometer" is available in the methods of age determination of radioactive substances (^3H, ^{14}C, ^{231}Pa, ^{230}Th, etc.). In this way, the temperature of the mixed surface layer and of the bottom water can be represented as a function of time. The prerequisite is a careful analysis of the lime residues in long sediment cores, that is, of special species of foraminifera for the surface layer temperature and of belemnites or benthonic foraminifera for the bottom temperature. The first series of measurements were executed by Urey and his collaborators Epstein, Buchsbaum, and Lowenstam (1951). More recently, further information concerning paleotemperatures in the ocean has been obtained from many oceanic areas including Arctic and Antarctic waters (Donk and Mathieu, 1969), in particular by Emiliani.

Figure 5.09, taken from Emiliani (1966, 1970), shows basic findings of this research. They resulted from the investigation of two sediment cores from the central Caribbean Sea ($\varphi = 14°59'N$, $\lambda = 69°20'W$, depth 3927 m, core length 10.54 m, and $\varphi = 14°57'N$, $\lambda = 68°55'W$, depth 4126 m, core length 14.29 m, *Globigerinoides sacculifera*). According to these investigations, the temperature of the surface layer ranged between 21 and 27°C in the past 425,000 yr; the fluctuations can be arranged in eight glacial–interglacial cycles. The present warm period (1) seems to be ending; the last glacial period (2) is dated 18,000 yr ago, which, by the way, is in agreement with the eustatic lowering of the sea level (cf. Fig. 1.22). What is remarkable in the variation of surface layer temperature with time, is that analyses from the open equatorial Atlantic Ocean have yielded the same results. This suggests that the temperature variations followed the same pattern all over the world. Another important result is the high correlation (0.997) of the time of occurrence of the temperature minima with that of the minima of incoming radiation according to the radiation curve by Milankovitch (1930). This is the full confirmation of the—repeatedly disputed—interpretation given by Köppen and Wegener (1924) in their book *Die Klimate der geologischen Vorzeit* (*The Climates of Early Geologic Time*) with regard to the first radiation curves given by Milankovitch (1920).

The radiation curve has been determined from the periods of the elements of the earth's orbit, well known to astronomers: namely the obliquity of the ecliptic, the excentricity of the earth's orbit and of the revolution of the perihelion. The ice ages Nos. 2, 4, 6, 8, 10, 12, 14, and 16 in Fig. 5.09 correspond to the subdivision applied to the glaciation of the Alps: Würm III, II, I; Riss II, I, and Mindel.

Changes in water temperature have been taking place till the present time. The first homogeneous series of temperature measurements were started as late as in the second half of the 19th century, mainly on lightships and island stations, that is, always in the vicinity of the coast. The long series of observations made by the Danish lightship *Horns-Rev* in the eastern North Sea from 1880 to 1940 has given evidence of a distinct warming (Dietrich 1954b) which, however, was not uniformly distributed throughout the year, but concentrated in particular seasons, from 1880 to 1910 mainly in winter, but from 1920 to 1940 in summer. Except for observations by the weather ships, long homogeneous temperature records are lacking from the open ocean. We are dependent on summaries of observations taken on merchant ships like they have been collected for the North Sea and the western British waters in tables covering the period from 1905 to 1954 (Dietrich, 1962). These tables have often been used for the solution of various problems (Dickson and Lee, 1969).

After weather ships were stationed in the North Atlantic and North Pacific Oceans, homogeneous series of measurements from the open ocean have been recorded ever since around 1949. They have provided deep insight into temperature changes, especially in the North Atlantic Ocean by the work done by Rodewald (latest paper 1972). Some facts are demonstrated in Fig. 5.10, where the annual means (overlapping over 5 yr) from 1951 to 1967 are given for three weather ships only, that is for A positioned on the northern side of the North Atlantic Current, for D on its southern side, and for K positioned within this current, together with the total mean of all nine weather ships. The considerable cooling at D, which between 1953 and 1958 reached a total amount of 1.56°C in the annual mean, is contrasted by a simultaneous warming of 0.5°C at A. It seems that a "heat swing" or "heat pendulum" was active during that period, which, however, was not the case from 1958 to 1965.

The facts given in the two examples of Figs. 5.09 and 5.10, as well as a great number of other direct measurements and indirect findings point to a complicated system of surface temperature changes with time. There are small-scale changes as well as oceanwide and worldwide ones. Furthermore, with a view to time there is a whole spectrum of periods: short-term fluctuations in the order of months and years (recent changes), medium-term changes in decades and centuries (historical changes) and long-term changes within thousands and millions of years (secular changes).

Scientists have repeatedly studied the *causes* of such temperature changes. Most of these changes have been discussed in the theory of the ice ages and in paleoclimatology, as for example, by Schwarzbach (1961). Oceanographers are convinced that the temperature in the ocean changes whenever the components of the heat budget equation (Eq. 4.04) change. Of the nine terms in this equation the latter four are of minor importance. The decisive ones are: incoming radiation Q_S, back radiation Q_A, evaporation Q_V, direct heat exchange with the atmosphere Q_K, and heat transport in the ocean Q_T. The quantities Q_V and Q_K can be summarized as interaction between ocean and atmosphere. In each of these five components several processes are involved so that the water temperature and its changes, after all, depend on a great number of different processes which take place on the sun, during the propagation of the radiation from the sun through space,

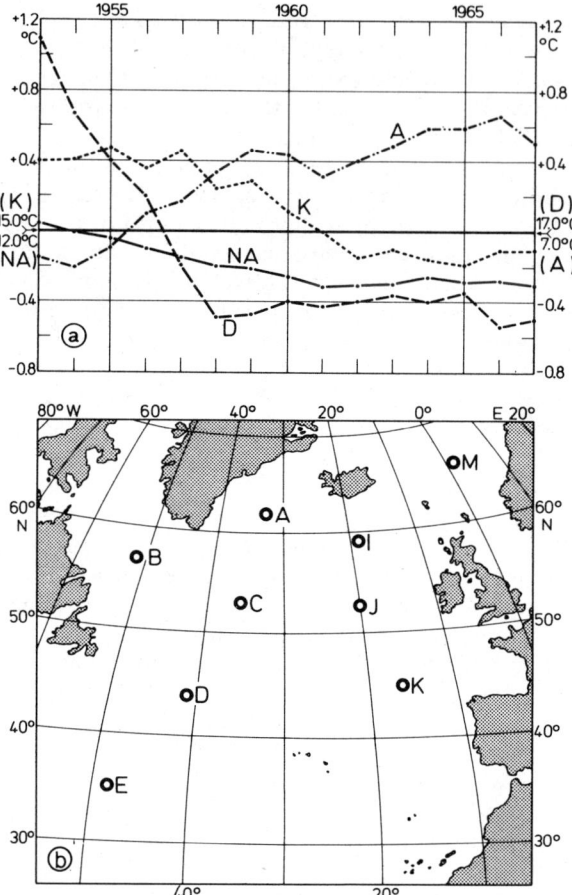

Fig. 5.10. (a) Long-term changes of surface temperature in the northern North Atlantic Ocean based on observations of the weather ships A, D, K, and of all nine ships (NA) for 1951–1969 in running 5-year averages. (After Rodewald, 1967 and according to personal communication.) Reference temperatures for A = 7.0°, D = 17.0°, K = 15.0°, NA = 12.0°C. (b) Positions of the nine North Atlantic weather ships A, B, C, D, E, I, J, K, and M.

in the atmosphere, and in the water until at last the radiation is absorbed in the ocean, and the heat, won thereby, is radiated back into space. In the following 11 groups of factors influencing temperature changes are listed:

1. Behavior of the sun. Its radiation is subject to fluctuations, for example, with the sunspot cycles of 80–90 yr and 20–24 yr (doubled 11 yr cycles) as described by Willet (1967).
2. Processes in space. Cosmic clouds, passing through the sun–earth system, influence the radiation on its path from the sun to the earth. Periods are not known.
3. The distance between sun and earth influences the incoming radiation. The periods are given by the elements of the earth's orbit: eccentricity e (period

92,000 yr), obliquity of the ecliptic ϵ (period 40,000 yr), revolution of the perihelion Π (period 21,000 yr).

Factors 1 and 3 are covered by the "solar constant" which, in fact, is not a real constant (cf. Section 4.2.2).

4. Absorption and dispersion by variable constituents in the atmosphere: water vapor, carbon dioxide, ozone, and dust. Definite periodicities are not known but only some definite tendencies caused, for example, by the increase of CO_2 in the past 100 yr as an effect of increased fossil fuel combustion. Thus, the greenhouse effect of the atmosphere has been increased, although moderated by the great absorption of CO_2 in the ocean. Volcanic dust was produced by major eruptions, like that of Krakatoa in 1883, with aftereffects in 1912–1913 (Öpik, 1967).
5. Reflective behavior of the sea surface (albedo), effective especially when the ice coverage of the ocean changes.
6. Attenuation in seawater. Variation of turbidity is effective.
7. Sea surface temperature (depending on all the other factors).
8. Re-radiation of the atmosphere (depending on Factor 4).
9. Distribution of the climatic zones and the oceans. Wandering of the poles and spreading of the oceans change the distribution of the climatic zones.
10. Atmospheric circulation and factor 9.
11. Atmospheric circulation, factor 9, and oceanic circulation.

A quantitative analysis can be carried out only with some of these factors, for example with factor 3. We are far from being able to make safe predictions for the future because an additional complication comes from the fact that a change in temperature is not the effect of one or several factors but of complex, coupled mechanisms.

The short-term temperature changes of the order of months are especially interesting. Here, Q_T is decisive. If some change in water transport occurs, for example, by the shifting of strong ocean currents or by transport fluctuations, anomalous temperatures—although locally restricted—may be caused, which will last for months or years and may be displaced by currents and mixing.

Numerous statistical relations have been discovered, mostly as correlations, by Dietrich, Sahrhage and Schubert (1959), Bjerknes (1967, 1969), Duvanin (1968), Uda (1968), Dickson and Lee (1969), Wyrtki (1969), Lamb (1969), and many others. The most comprehensive material—8×10^6 pieces of data of surface temperatures during 20 yr—from the North Pacific Ocean is the basis of Namias' work (1970). In principle, all conclusions deal with the effect that some disturbance in Q_T may have on the region concerned and on its surroundings, be it a small oceanic area or an entire ocean. Up to now, at best, information has been published on the temporal coherence of medium-term development, as by Namias (1970) who has given evidence of high autocorrelation of the surface temperature in the North Pacific Ocean. After intervals of approximately 5 yr, some large-scale disturbance will occur for which the drift current is believed to be responsible. Since the sea-surface temperature is so very important for medium-term and long-term weather forecasting, efforts are being made to establish a system of anchored measuring buoys, as is the case in the project IGOSS (Integrated Global Ocean Station System).

The *effect of the present change of temperature* is manifest in various phenomena on earth, in particular because the change is not restricted to temperature alone, but also involves wind, precipitation, and evaporation. The most obvious effects are to be seen at the glaciers in the Arctic and in high mountains. In the ocean as well as on the continents, fauna and flora are affected, which can best be recognized by the shifting of the boundaries of the habitats of various species during the past decades. Far-reaching economic consequences have partly resulted from this. Ecological studies on many species important to commercial fishery have proved that their distribution depends on temperature, that is, of fish like herring, redfish, haddock, cod, and tuna. For some species the temperature tolerance is very limited, from which it follows that, in their habitat, they are extremely sensitive to temperature changes. It is a task of fishery hydrography to monitor these temperature conditions (cf. Hela and Laevastu, 1970). Information even with a view to any possible trend towards temperature change is very valuable already. For instance, according to numerous fishery biological observations from Arctic and sub-Arctic regions collected by Jensen (1939), haddock and codfish have appeared on the banks off western Greenland, where they had not been caught before. Besides, catches of codfish suddenly increased there between 1922 and 1930, and the area turned into a new important fishing ground, remaining so until the present day.

No evidence can be found for *long-term temperature changes* in the deep sea beneath the range of the annual variation and such changes must be very small if they occur at all. Rouch (1943) has confirmed this by comparing observations from neighboring stations of different years and seasons. In the South Atlantic Ocean, the absolute temperature differences amount to 0.06°C at 2000 m depth and to only 0.04°C at a depth of 3000 m. This suggests that conditions are fairly stationary. So we may use, for instance, stations from different years and seasons when compiling the three longitudinal sections through the oceans shown in Plates 3 to 5. Within the program of the International Geophysical Year 1957–1958 the *Meteor* sections of 1925–1927 were repeated, that is, 30 yr later, but the comparison of the data does not show any appreciable difference. With respect to the comparison between the results of 1925–1927 and of 1957–1958 Fuglister (1960) said in the introduction to his atlas of the Atlantic Ocean: "Such comparison provides dramatic evidence for the steady-state character of large-scale Atlantic dynamics." Wüst (1965) has modified this statement by showing that it is true only for the characteristic salinity of the core layers of the cold-water sphere in the South Atlantic Ocean. The deviations remain within the limits of error of the measuring method, amounting to 0.03‰ salinity. These facts are valid only for short time periods; the $^{16}O/^{18}O$ method, if applied to the lime residues of bottom-dwelling species, should also give information on secular temperature changes, at least at the abyssal floor. But up to now, this is only the field of vague hypotheses.

5.2. Salinity in the World Ocean

5.2.1. Distribution of surface salinity

If completely mixed with the available seawater of 1400×10^{18} tonnes, the salts, present in the world ocean and amounting to 4861×10^{16} tonnes, would yield an average salinity of 34.72‰. It is true that, in the course of the earth's history, salinity has increased due to weathering products discharged into the ocean by continental runoff. But an increase of the average salinity would not be perceptible, even with the precise measuring methods of today, before about 10^5 yr. If, nevertheless, considerable deviations of salinity from

34.72‰ have been measured in the world ocean, in horizontal as well as in vertical directions, the cause thereof is found in the water exchange as expressed in the budget equation (cf. Eq. 4.16). Evaporation, precipitation, concentration of salts in freezing seawater and dilution with melting sea ice, continental runoff, as well as transport by ocean currents and mixing are the six factors determining the distribution of salinity. The former four are effective only at the sea surface. In the depth, the salinity distribution is governed exclusively by the latter two processes.

Charts regarding surface salinity are contained in several of the atlases mentioned in Section 5.1.3. Since salinity measurements are much more difficult to carry out than temperature measurements, the number of relevant data available is considerably smaller. Most of them were collected during research cruises. On the other hand, the individual observation of surface salinity is much more representative than is the case with surface temperature, because the annual variation of surface salinity in the ocean is quite small.

This is not the place to describe the distribution of surface salinity in detail, except that attention may be drawn to a few large-scale phenomena shown in Fig. 5.11. If we assume that a pure zonal salinity distribution, which would exist on a globe completely covered with water, is disturbed in many ways by the continents, such a continental influence can obviously be characterized by the deviations of salinity from the normal distribution. This has been done in Fig. 5.11. Since it is impossible to compute the normal data for a water-covered hemisphere on the basis of the water budget with sufficient accuracy, the salinity averages have been determined for the latitudes of the Southern Hemisphere, for there 86% of the circles of latitude between 0 and 70°S are covered by water. The normal values are given at the right-hand margin of Fig. 5.11. The chart itself shows the local deviations of the annual means from those normal values.

If, for example, a deviation of +1.5‰ from a normal value of 33.9‰ is indicated west of Scotland, the surface salinity is equal to 35.4‰. Besides the large-scale deviations from the normal state of a water-covered globe, the chart also gives the salinity distribution that has actually been observed.

Figure 5.11 clearly shows a fact in the distribution of surface salinity which should be pointed out: in the Atlantic Ocean, the surface salinity is considerably higher than in the Pacific Ocean, particularly in the Northern Hemisphere. The reason for this must be sought in the water cycle. Several processes play a distinct role and will be briefly mentioned here:

1. The water vapor taken up by the atmosphere through evaporation over the North Atlantic Ocean is partly lost to the Pacific Ocean. In Central America, as shown in Fig. 4.19, the North Atlantic trade winds extend to the Pacific Ocean and contribute to the high precipitation in the Gulf of Panama, which, with over 700 cm yr^{-1}, is among the highest on earth. Consequently, the salinity in the Gulf and the adjacent areas is lowered appreciably.
2. In the west-wind belt of the Southern Hemisphere, the southern part of the Andes acts as a rain trap that has no counterpart south of Africa. While passing over the Andes, the west winds release their water vapor in the form of precipitation, thus benefiting the waters of Chile, and replenish it from evaporation in the South Atlantic Ocean. The negative anomalies of surface salinity in the eastern South Pacific Ocean result mainly from this water transfer.
3. The area influenced by water vapor of Atlantic origin has proved to be many

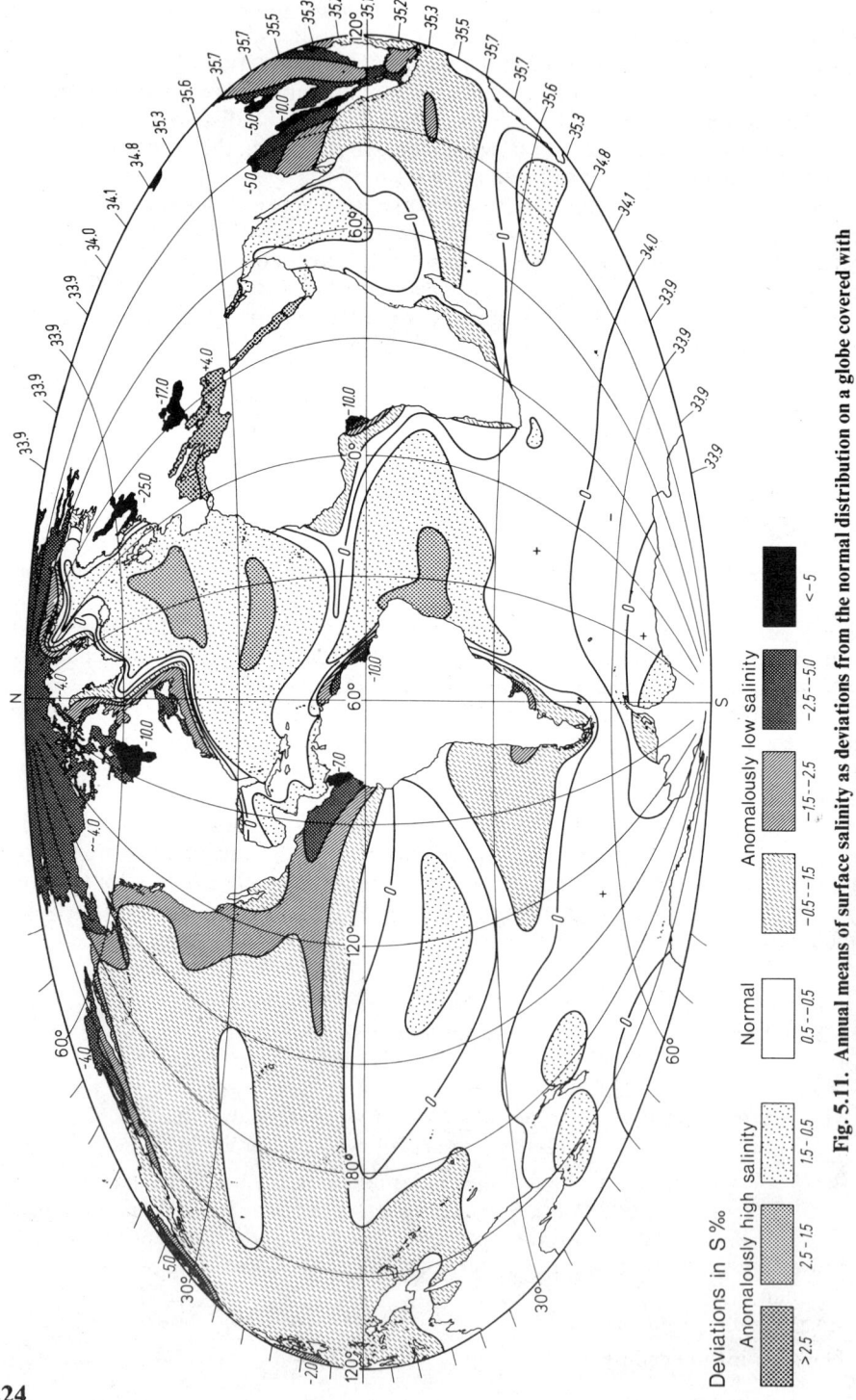

Fig. 5.11. Annual means of surface salinity as deviations from the normal distribution on a globe covered with water. (After Dietrich, 1950.) Numbers at right-hand margin: 'Normal' values for latitude circles (cf. Section 5.2.1) in S‰. Isolines of salinity anomalies (in S‰).

times larger than that influenced by water vapor from the Pacific Ocean. This can be clearly seen, if we consider the main directions of the water vapor transport as given in Fig. 4.19. The reason why the Pacific area of influence is so much smaller is found in the chain of high mountains bordering the greatest part of the coasts of the Pacific Ocean. From these orographic conditions it is understandable that, on the continents, the amounts of surface and ground water originating from the Atlantic Ocean are much larger than those of Pacific origin. Therefore, the proper balance for either ocean is a different one, resulting from the effects that oceanic circulation, mixing, and the atmospheric water cycle of evaporation minus precipitation exert on salinity.

4. In the Pacific Ocean large mediterranean seas of high salinity are lacking, whereas—like the European Mediterranean Sea—such seas contribute to the high salinity in the depth of the Atlantic Ocean. Owing to vertical mixing, the surface water of the Atlantic Ocean is influenced by that high salinity.

5. The salt transport by oceanic currents from one ocean into the other also favors the increase of salinity in the Atlantic Ocean. The Agulhas Current, for instance, transports water of high salinity from the subtropical Indian Ocean into the South Atlantic Ocean. In this way, the subtropical anticyclonic surface circulation in the South Atlantic Ocean receives a contribution of salt water, whereas the homologous area of the South Pacific Ocean has a fresh-water influx by heavy precipitation. Thus, the relatively uniform distribution of salinity in the Antarctic water belt gradually changes and, already in moderate southern latitudes, a distinct contrast between the surface salinities of the two oceans develops, which originates from the different latitudinal location of the southern tips of the South American and the African continents.

5.2.2. Distribution of salinity with depth

What has been said about the source of our knowledge on the distribution of temperature with depth is also valid for salinity (cf. Section 5.1.4). Here we will deal only with the large-scale vertical distribution of salinity together with the distributions of temperature and oxygen, as represented in longitudinal sections through the Pacific, Atlantic, and Indian Oceans (Plates 3, 4, 5). The material on which these sections are based is mentioned in Section 5.1.4.

All three oceans have in common that greater differences in salinity are restricted to the warm-water sphere between the polar fronts of the Northern and Southern Hemispheres. The accumulation of water masses with higher salinities is a characteristic feature of subtropical regions. In the cold-water sphere, the absolute values as well as the vertical and horizontal differences are largest in the Atlantic Ocean and smallest in the Pacific Ocean. The three layers, namely the low-salinity "intermediate water" (SIW and NIW), the high-salinity "deep water" (DW), and the low-salinity "bottom water" (AABW and ABW) are quite distinct in the Atlantic and Indian Oceans, where they are asymmetrically distributed with regard to the equator. In the Pacific Ocean, however, the "intermediate water" is symmetrical to the equator; farther down, the vertical differences are negligible.

Since the salinity distribution beneath the surface currents is dependent only on water transport and mixing, salinity is an important indicator of the propagation of water masses (cf. Section 10.1). The schematic representation of the oceanic circulation in the individual oceans in Plates 3, 4, and 5 is partly based on the distribution of salinity.

5.2.3. Diurnal and annual variations of surface salinity

The diurnal variation of salinity is of no importance whatsoever, even with the strong diurnal variation of evaporation in the subtropics according to Defant (1961) it remains below 0.04‰, as confirmed by observation. In general, the annual variation also plays only a minor role in the open ocean. According to the results of Smed (1943) it does not exceed 0.20‰ in the North Atlantic Ocean, if the region around Newfoundland is omitted. In areas where a thermocline is formed in summer, the extremes of the annual variation occur in the fall when thermal vertical convection begins. If, in the area concerned, a surplus of evaporation prevails, the salinity of the surface mixed layer will increase in the course of the summer. The extreme will be a maximum. Conversely, if precipitation exceeds evaporation, the extreme will be a minimum.

Unusually high annual variations of surface salinity occur where at least one of the four components in the water budget equation (Eq. 4.16) shows a strong annual variation. This applies to the annual variations of precipitation, ice coverage, transport by ocean currents, and continental runoff, that is, in the subpolar areas, in the monsoon regions, and off the estuaries of large continental rivers. In subpolar areas, salinity is decreased by the melting process in summer. The salinity maximum takes place when the melting process begins. An example is the Newfoundland region where the annual variation exceeds 0.7‰, and the surface salinity reaches its maximum in February to March. In the monsoon regions, the regular seasonal change in the direction of the monsoon winds is followed by a reversal of the oceanic currents and, therewith, by an advective shifting of the horizontal salinity distribution. In addition, the large annual variation of precipitation contributes to increasing the annual variation of salinity. So in the Bay of Bengal and in the Austral-Asiatic Mediterranean Sea the annual variations amount to 1 to 3‰. In the vicinity of great rivers the annual oscillation of the continental runoff results in annual variations that differ locally but sometimes are rather large. In the tropical western Atlantic Ocean, the water of the Amazon has a far-reaching influence (Neumann, 1969a). The North Sea with its great number of observations is a suitable object for studies regarding the annual variation of surface salinity.

Basing his studies on 200,000 individual observations from the period of 1905 through 1954 which have been published by ICES (1962) in an atlas with tables, compiled by Dietrich, Schott (1966) was the first to attempt a representation and explanation of the local dependence of the annual variations of surface salinity in the North Sea. The amplitudes and the times of occurrence of the maximum of the annual wave, both of which are obtained by Fourier analysis, are demonstrated in Fig. 5.12 together with the annual averages for 1905-1954. The isohalines of the annual means congregate in regions with considerable inflow of fresh water, for example, off the estuaries of the rivers Rhine and Elbe, at the exit of the Baltic Sea, and along the west coast of Norway. In these regions we also find a large mean annual variation: greater than 2‰ near the above estuaries, greater than 6‰ in the Skagerrak, greater than 3‰ off western Norway. The maximum occurs when the continental runoff is smallest, that is, with rivers in the fall, and in the Skagerrak in winter. In the central North Sea the annual variation is smaller than 0.2‰, which equals the values obtained in the open North Atlantic Ocean. Besides, the time of occurrence of the maximum is shifted to early summer (180° = July 1), thus showing the transition between a maximum corresponding to that of the Atlantic water (90° =

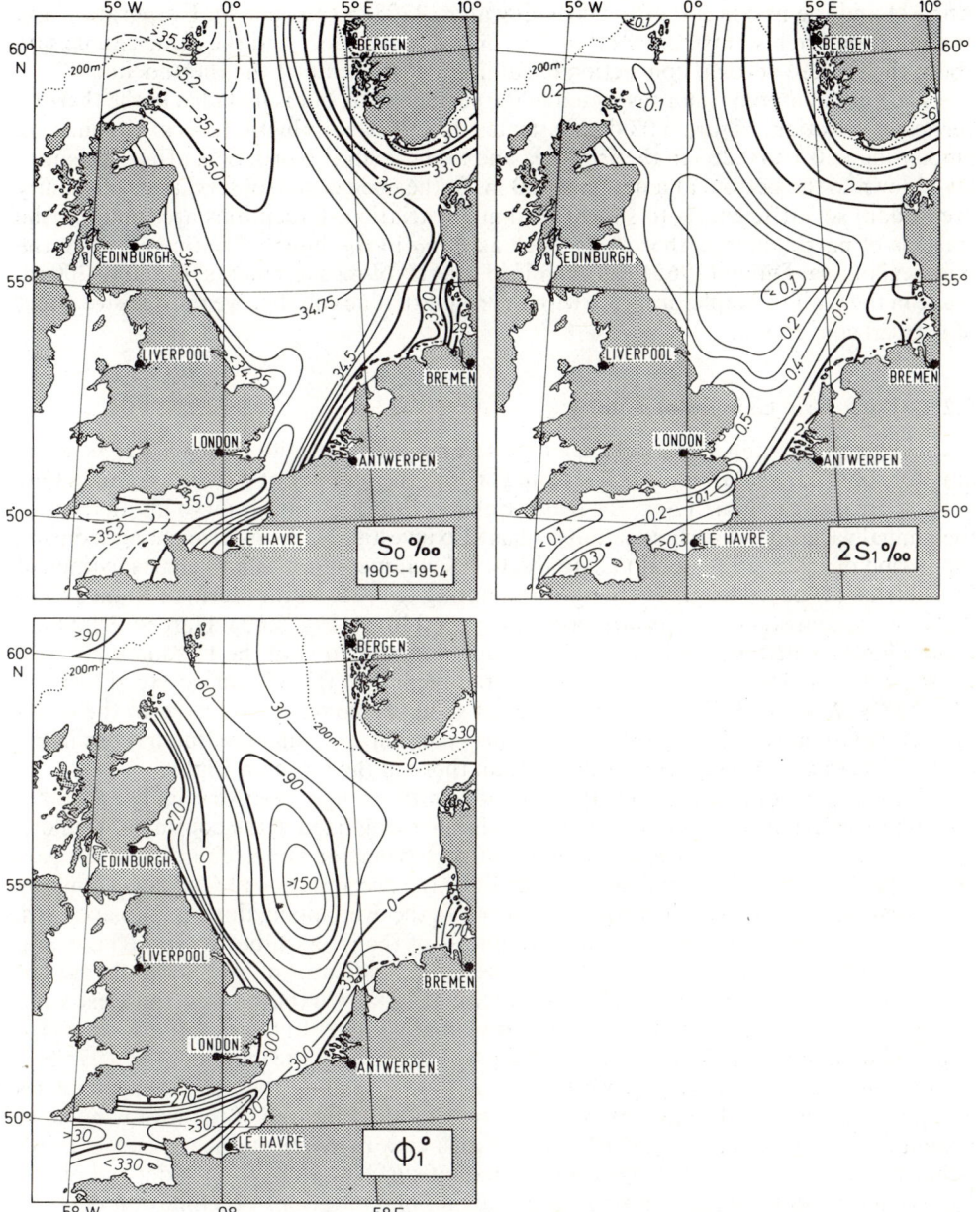

Fig. 5.12. Mean annual variation of the surface salinity in the shelf sea shown by the example of the North Sea 1905–1954 as harmonic wave $S = S_0 + S_1 \cos(\alpha - \phi)$. (After Schott, 1966.) Above left: Annual mean S_0 in ‰; above right: Annual range $2 S_1$ in ‰; below left: Phase ϕ in degrees, referred to January 1, indicates the time of occurrence of the maximum: 90° corresponds to April 1, 180° to July 1, 270° to October 1. Dotted line: 200 m depth contour line.

April 1) and the maximum of the coastal water (270° = October 1). Evaporation and precipitation are also instrumental in the central North Sea, and the same can be said about the limited vertical convection in summer, at the time of the thermocline.

The annual march is not restricted to the surface only, but is included in the thermal vertical convection. Figure 5.08 demonstrates that the small fluctuations at the surface, amounting to 0.1 to 0.2‰ in the Irish Sea and in the central North Sea, affect the entire layer down to the bottom at a depth of 100 m. In the open ocean, observations of salinity are seldom so sufficient as to show the annual variation. Exceptions are found at the stations of particular weather ships, such as *Papa* in the North Pacific and *M* in the Norwegian Sea. Düing (1965) has examined a series of measurements from 0 to 200 m, made in the Gulf of Naples in 1957, with the result that he could prove the decisive effect of vertical convection.

5.2.4. Long-term changes of salinity

Although surface salinity may change over several years, definite evidence can be given only for some particular areas because, in general, the number of data is too small. One of such areas is the North Sea, for which Schott (1966)—in addition to his analysis of the annual variation of surface salinity—has also investigated long-term changes during the period of 1922–1938, again using the ICES volume of tables and charts, compiled by Dietrich (1962). The great changes amounting to 2.0‰ in the German Bight and to 0.2‰ in the central North Sea are rather remarkable (cf. Fig. 5.13). Here, overlapping three-month averages are given, the first value referring to $1/8$ of the 1° square No. 194 ($\varphi = 54°0'-54°15'N$, $\lambda = 8°0'-8°30'E$), the second to the 1° square No. 117 ($\varphi = 56°-57°N$, $\lambda = 2°-3°E$). The two curves of salinity anomalies, referred to the mean annual variation from 1925 to 1934, run nearly parallel, although the fluctuations in the central North Sea are ten times smaller than those in the German Bight.

The conspicuous alternation between periods with strongly positive salinity anomalies and others with strongly negative ones can also be recognized in the continental runoff, the precipitation, and the wind. Examples of such behavior are given in the lower part of Fig. 5.13 showing the runoff of the river Elbe, the precipitation at Cuxhaven, and the west wind component over the English Channel. In the overlapping three-month averages, close correlations were found among the runoffs of the rivers Elbe, Weser, Rhine, and Thames. Therefore, the runoff of the Elbe is a good indicator of long-term changes of runoff quantities. The same applies to precipitation at Cuxhaven and the west wind component over the English Channel. Conspicuous changes in the runoff are most pronounced with the river Weser; therefore Schott has referred the numbers 1 to 10 in Fig. 5.13 to the river Weser. High correlation coefficients among the individual processes are quite obvious; for the anomalies of the runoff of the Elbe and of the west wind component over the English Channel r is 0.81. With a view to the North Sea, the chief result is that there are significant long-term changes of surface salinity. They depend on the continental runoff, which—in turn—depends on precipitation. Precipitation depends on the west wind component and thus on atmospheric circulation. This means that, after all, the long-term changes of surface salinity are regulated by long-term changes of the atmospheric circulation.

These relations, which are valid for the North Sea, need not necessarily apply to the open ocean, because the continental runoff is of no importance there. On the other hand, according to Smed (1943), long-term changes of up to 0.20‰ (in 1928) occurred in the North Atlantic Ocean between Iceland and the Azores during the summer months of

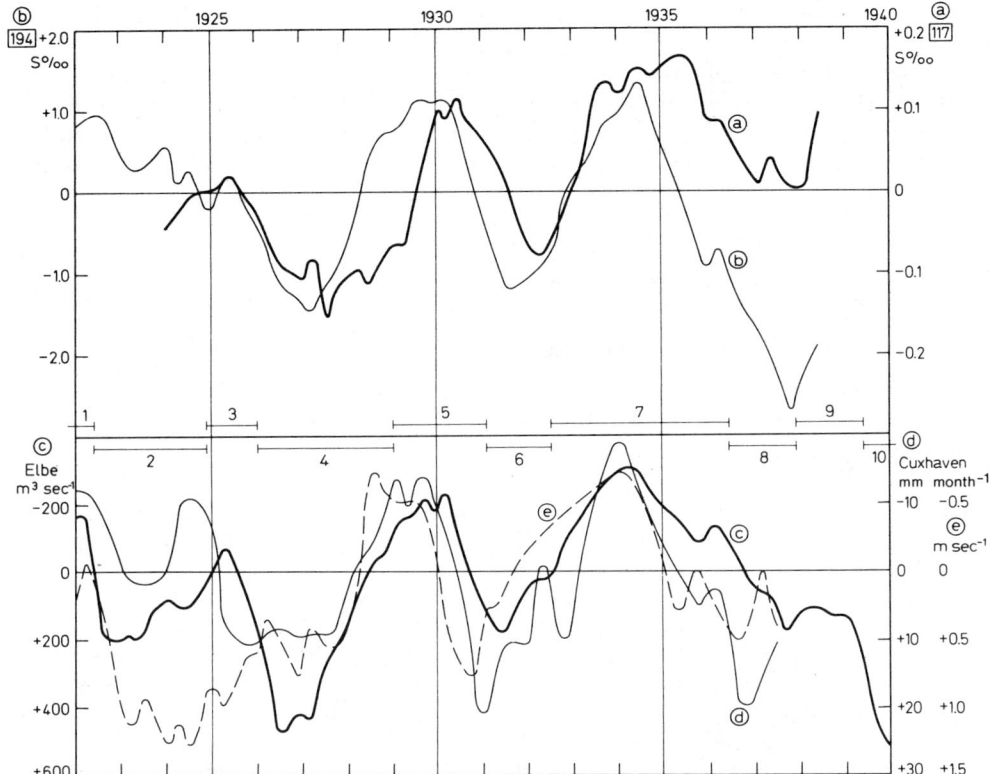

Fig. 5.13. Long-term changes 1922–1940 of (a) surface salinity in the central North Sea (square 117: 56°–57°N, 2–3°E); (b) surface salinity in the German Bight (square 194: 54°0′–54°15′N, 8°0′–8°30′E); (c) runoff of the river Elbe; (d) precipitation at Cuxhaven; (e) west wind component over the English Channel. There are represented running averages of three-month anomalies, referred to the mean annual variation 1925–1934. 2, 4, 6, 8, 10: Periods with positive anomalies of the runoff; 1, 3, 5, 7, 9: Periods with negative anomalies of the runoff. (After Schott, 1966.)

1905–1939. Probably those changes were related to major disturbances of the oceanic current system which were a consequence of disturbances in the atmospheric circulation.

At greater depths of the ocean, long-term changes of salinity have not been observed so far. Evidence of these changes has been found only in marginal areas, like in the Norwegian fjords and in the Baltic Sea, where the near-bottom water is renewed, at irregular intervals, by highly saline water of Atlantic origin. According to Segerstråle (1965a), salinity increased by about 0.6‰ at the depth of 5 m in the Gulf of Finland near Helsinki during the period of 1927 through 1964.

Since many species of marine animals and plants are highly sensitive to salinity changes, such variations can evidently cause some displacement of their habitat. This applies to the Baltic Sea into which stenohaline animals have immigrated and where other native animals have extended their spawning grounds (Segerstråle, 1965b). To some extent, commercial fisheries have profited from this.

During the International Geophysical Year 1957–1958, research vessels of the Woods Hole Oceanographic Institution repeated many of the stations of the German *Meteor*

Expedition of 1925–1927 in the South Atlantic Ocean (Fuglister, 1960), but no evidence was found of any definite systematic change after those 30 yr.

Long-term salinity changes of a specific kind have been proven in the Suez Canal (Morcos, 1972). Since the opening of the Canal in 1869, salinity decreased from the original 168‰ to 45‰ in 1967. The high values are related to the leaching of salt deposits, while the considerable decrease took place when those deposits were covered with sediments.

5.3. Density in the World Ocean

5.3.1. Distribution of density at the sea surface

In the ocean the distribution of density is determined by the distribution of temperature and salinity (cf. definition in Section 2.1.4). Density decreases not only when the water temperature increases, but also by the admixture of water from precipitation and melting as well as from continental runoff. Density increases with cooling, evaporation, and formation of ice. As soon as, by combined action of the various processes, the density is higher at the sea surface than in the lower layers, thermohaline vertical convection starts.

Charts of the density distribution at the sea surface are included in some atlases (cf. Section 5.1.3). An especially detailed representation has been given by Böhnecke (1936) for the entire Atlantic Ocean, and a similar one by Krauss (1962) for the North Atlantic Ocean. Figure 5.14 contains the zonal averages for 2° belts regarding density as well as temperature and salinity. The lowest density values have been found in the equatorial zone which is poor in salinity; high values can be met in the subtropics with their high salinity, and the highest values occur in the cold subpolar zones which have relatively high salinity.

In principle, the relationship between density distribution and geographical latitude in the Indian and Pacific Oceans is rather the same as in the Atlantic Ocean. An exception can be observed in the North Indian Ocean where the Gulf of Bengal is characterized by low values of salinity and thus of density, owing to the heavy monsoon rains and the large quantities of runoff from rivers.

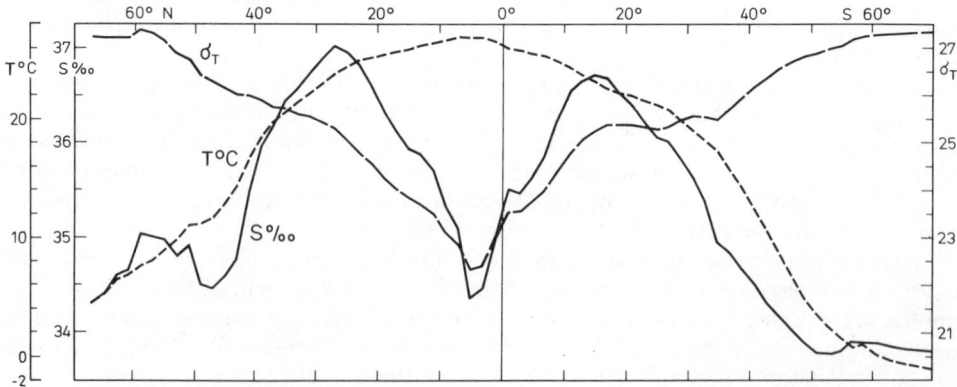

Fig. 5.14. Annual averages of temperature, salinity, and density for two-degree zones at the surface of the Atlantic Ocean. (After Krauss, 1962.)

5.3.2. Distribution of density with depth

Only the first of the three common expressions for the density of seawater will be dealt with here (cf. Section 2.1.4):

$\sigma_T = (\rho_{S,T,0} - 1) \cdot 1000$: density at atmospheric pressure, $p = 0$

$\sigma_\theta = (\rho_{S,\theta,0} - 1) \cdot 1000$: potential density at potential temperature θ (°C)

$\sigma_{S,T,p} = (\rho_{S,T,p} - 1) \cdot 1000$: density in situ, including compressibility

The density σ_T is the basic quantity which can be taken from tables, if the values of temperature T and salinity S are known. Other tabulated values permit us to calculate the potential density as well as the density in situ (cf. Appendix 11.2).

In the world ocean, the vertical distribution of density is predominantly determined by the water temperature. The very cold water of the subpolar regions in winter shows the highest density, as in the Greenland Sea ($\sigma_T = 28.1$) and in the Weddell Sea ($\sigma_T = 27.9$). Even the high surface salinity in the subtropics, as, for instance, $S = 37.3‰$ in the Sargasso Sea, cannot enhance the low density, which is determined by the high water temperature ($T = 20°C$ in winter, $\sigma_T = 26.7$), to such an extent that deep-reaching convection will take place. The stratification remains stable. Such stability is defined by means of the vertical density gradient (cf. Section 7.2.5).

Along with vertical density differences horizontal ones are also of great importance in the ocean, because water masses of different densities cannot co-exist in equilibrium side by side. The horizontal density differences, maintained at the sea surface by the climatically controlled influence of the atmosphere, cause large-scale water motions in the ocean, following the basic principle that water takes the shortest path to that layer of the ocean which has the same density, and that it spreads there horizontally. This touches on the subject of oceanic circulation which will be discussed in detail in Section 10.1. The core layer method by Wüst (1936), the isentropic analysis by Montgomery (1938), and the geostrophic total analysis have all been applied to investigate this phenomenon. All three methods require the knowledge of the vertical distribution of the density σ_T. With regard to the density distribution, a special position is held by the mediterranean seas in arid climates, for example, the European Mediterranean Sea and the Red Sea. In winter, due to high evaporation, the density regionally increases to $\sigma_T = 29.2$ in the Tyrrhenian Sea, as in the Lion's Gulf and in the Gulf of Genoa ($T = 12.5°C$, $S = 38.5‰$), and in the Ionian Sea near Rhodes ($T = 15.0°C$, $S = 39.1‰$). In the Gulf of Suez in the Red Sea ($T = 21.0°C$, $S = 42.0‰$), the density even reaches $\sigma_T = 29.9$, which is the highest value in the ocean if places with salt deposits are disregarded, like the hot brines in the central Red Sea (Krause and Ziegenbein, 1966), represented in Fig. 10.25, or the Bitter Lake in the Suez Canal with a salinity of approximately 48‰ (Morcos, 1960).

5.3.3. Variations of density

Diurnal and annual variations as well as long-term changes of the density σ_T are determined by the behavior of temperature and salinity, with temperature being the decisive factor. Like the annual variation of temperature, the annual variation of density also is a phenomenon governing the processes in the uppermost 100 m of the ocean. The summer thermocline (cf. Section 5.1.4) at the same time becomes a summer pycnocline. The

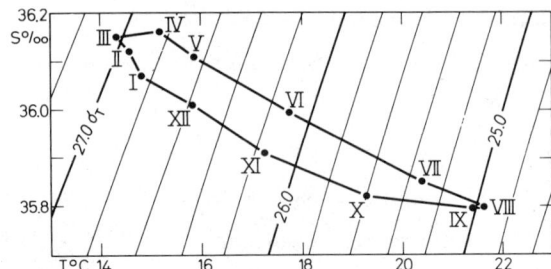

Fig. 5.15. Example of the representation of the mean annual variations of density, temperature and salinity at the sea surface. (After Neumann, 1940.) There are given 5° mean values east of the Grand Banks (40°–45°N, 40°–45°W). I–XII: months.

stability of the density stratification can increase so much that vertical turbulent exchange processes cease completely, and the pycnocline becomes an impenetrable barrier. Thus, the renewal of the water by vertical exchange processes beneath this discontinuity layer is stopped.

Since temperature, salinity, and density are closely related, it is an obvious step to represent the annual variations of all three quantities at one point in a diagram, as, for example, Neumann (1940) has done for several squares in the sea area between Newfoundland and the Azores. An example is given in Fig. 5.15. In this case the density maximum of $\sigma_T = 27.0$ occurs in March, and the density minimum with $\sigma_T = 25.0$ in August. They are mainly determined by the annual variation of surface temperature. The annual variation of surface salinity, amounting to 0.4‰, is relatively large and contributes to intensifying the density variation. The salinity maximum occurs in March to April, before the influence of the melt water of the ice in the Labrador Current is perceptible.

5.4. Characteristic Water Masses in the World Ocean

5.4.1. Temperature–salinity relationship

So far temperature, salinity, and density have been discussed separately, except for the allusion in Fig. 5.15 showing that, with regard to the sea surface, it is practical to describe the temperature T as a function of the salinity S for a certain location in the ocean, although this is without any proper physical meaning. But for depths that are beyond the range of the annual variation, conditions are different. At such depths, the TS curve shrinks to a single point. If, for a certain oceanic location, many pairs of observations obtained from various depths are plotted in a diagram, the result is not a random assembly of dots but an arrangement along a line. In Fig. 5.16 two such TS diagrams are shown together with the points of observation. They originate from two stations within the Gulf Stream region, only 50 nautical mi apart from each other. In Fig. 5.16(a) and 5.16(b) the vertical distribution of temperature and salinity at those stations is given for depths from 0 to 1200 m. Although the horizontal differences between the two stations are rather large for oceanic conditions, the TS diagram in Fig. 5.16(c) shows that the observations line up along two closely adjacent curves. In both cases, we are dealing with the same characteristic water masses; the only difference is that they lie at different depths. The numbers at some points of the curves indicate the depth at which the observation was made. So we see that the values of Station 1229 at 284 m depth are nearly identical to

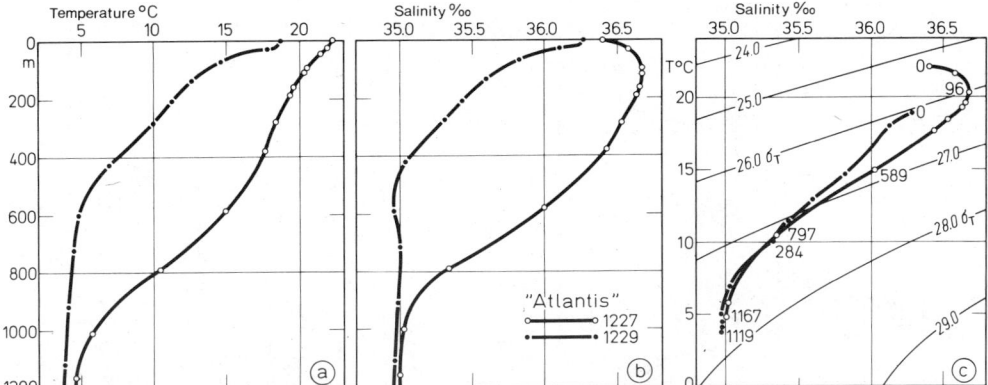

Fig. 5.16. Relation between temperature and salinity at the two neighboring stations *Atlantis* 1227 and 1229 in the Gulf Stream east of Cape Hatteras. (See Fig. 10.42 for station positions.) (a) Vertical temperature distribution; (b) vertical salinity distribution; (c) temperature-salinity relation. Figures at the points of measurement: depths in m.

the values of Station 1227 at 797 m depth. Since density is determined by temperature and salinity, the *TS* diagram can be covered with a system of σ_T lines, thus simultaneously providing information on the corresponding density values.

In this form, the *TS* diagram, introduced by Helland-Hansen (1916), has proven to be an extraordinarily valuable tool in oceanography for five reasons: (1) areas with uniform water masses are characterized by one single *TS* normal curve; (2) incorrect observations can easily be detected because they will fall outside the normal *TS* curve for that particular oceanic area; (3) if it has been demonstrated that the normal *TS* relation for an oceanic area does not show any considerable scattering when plotted but is arranged along a curve, the corresponding values of salinity or density can be obtained from the measured temperature data, which very much contributes to the rationalization of taking observations. In addition, the *TS* diagram has proven especially useful if the problems of (4) the origin and (5) the mixing of the characteristic water masses in the ocean are to be clarified. This has come into play, in particular, in the investigations of the North Atlantic Ocean by Jacobsen (1929), of the entire Atlantic Ocean by Wüst (1936), and of the world ocean by Sverdrup (1942).

5.4.2. Formation of characteristic water masses

The importance of the *TS* relationship for problems of origin and mixing of water masses is illustrated in Fig. 5.17. If two homogeneous water masses are assumed to be lying one on top of the other (the upper one with temperature T_1 and salinity S_1, the lower one with temperature T_2 and salinity S_2), the vertical distribution of temperature and salinity is given by the broken straight lines marked by A in the upper left part of Fig. 5.17. In the *TS* diagram, however, the two water masses appear as two points. If m_1 portions of the upper water body are mixed with m_2 portions of the lower body, the temperature T and the salinity S of the mixed water will obey the mixing rule as it is known from thermodynamics:

$$T = \frac{m_1 T_1 + m_2 T_2}{m_1 + m_2} \qquad S = \frac{m_1 S_1 + m_2 S_2}{m_1 + m_2}$$

In the *TS* diagram, all points of the mixed water will always lie on the straight line

234 Temperature, Salinity, Density, Characteristic Water Masses, and Ice in the World Ocean

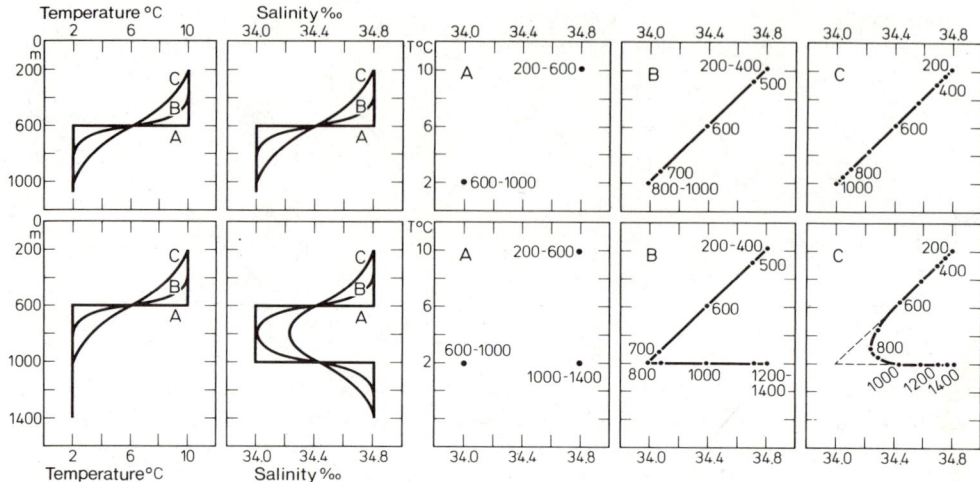

Fig. 5.17. Temperature-salinity relation for mixing of water masses. Above: mixing of two homogeneous water masses; below: mixing of three homogeneous water masses; left: vertical distribution of temperature and salinity at the initial stage A and at two stages of progressive mixing B and C; right: the three stages A, B, and C represented in a TS-diagram. Figures at the points of measurement: depths in m.

connecting the points which represent the unmixed water bodies. The ratio of their distances from these two corner points gives the mixing ratio. In the upper part of Fig. 5.17, two stages B and C illustrate progressive mixing as reflected in the vertical distributions of temperature and salinity and in the *TS* relation. The numbers at the points indicate the depth (in meters) of the corresponding *T* and *S* values. Therefore, if interrelations between temperature and salinity are clearly recognized, this applies only to those cases in which both quantities are changed by mixing alone. This is true in the world ocean if the temperature and salinity of the water masses are no longer subject to any direct influence from the sea surface, which means, at the same time, that such interrelation does not hold at the sea surface.

The *TS* relation becomes somewhat more complicated if a third water body is involved, as indicated in the lower part of Fig. 5.17. Such a water body is represented by a third point in the *TS* diagram. In the case of mixing, the *TS* diagram then consists of two broken straight lines (case B). As soon as the salinity in the core of the intermediate water body is changed by mixing from above and below, the corners in the *TS* diagram will be rounded (case C). If further water bodies are involved, the transition layers will behave correspondingly. Hence, it may be expected that, wherever in the ocean different water masses with different temperatures and salinities get mixed, the *TS* relation will appear as broken straight lines with corners rounded in accordance with the degree of mixing. This fact is the basis of two methods by which the vertical exchange can be determined, one by Jacobsen (1927), the other by Defant (1929). A detailed description of these methods has been given by Defant (1961).

Supplementing such purely schematic consideration of vertical mixing, as depicted in Fig. 5.17, Fig. 5.18 represents the *TS* relations observed on the western side of the Atlantic Ocean at five stations located between 41°S and 39°N. The stations have been taken from the longitudinal section through the Atlantic Ocean in Plate 4. The *TS* relations can be regarded as the result of the mixing of four water bodies in which the cores of the unmixed water masses have already disappeared. From this diagram the

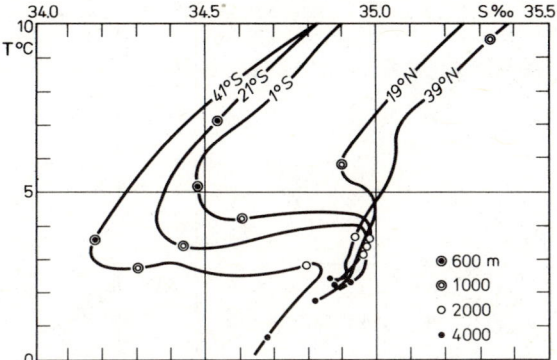

Fig. 5.18. Relation between temperature and salinity at five stations on the western side of the Atlantic Ocean between 41°S and 39°N. (After Wüst, 1936.)

portion of the water can be determined which originates from the area of formation of the respective core water. This is the fundamental idea behind the so-called core layer method by Wüst (1936), according to which the preferred spreading of the water in the particular core is derived from the geographic distribution of the mixing ratio. Such a core can be traced independently of its depth. In Fig. 5.18 three such core water masses can be recognized: cold water of low salinity, which, according to its area of origin, is called sub-Antarctic intermediate water; cold water of high salinity, the so-called North Atlantic deep water; and very cold Antarctic bottom water.

It should be emphasized that the idea of water masses in the ocean must not be too rigidly attached to the scheme presented in Fig. 5.17; for completely homogeneous water masses are very rarely formed in the ocean. Furthermore, differences in temperature and salinity occur not only in vertical direction but also along the σ_T surfaces which are almost horizontal. Lateral and vertical mixing participate in the formation of characteristic water masses, although the TS relation does not permit any distinction of the effects.

By continuously recording the vertical distribution of temperature and salinity with the bathysonde (cf. Section 3.2.9), evidence has been found of a fine structure of stratification (Siedler, 1970), especially where water bodies of different origin are mixing. In the TS diagram such structure is expressed by a characteristic irregularity or unrest. In Fig. 5.19 an example is given that represents the oceanic area west of the Strait of Gibraltar. In this case, the irregularity concerns warm, high-salinity water at 500 to 1300 m depth which has flowed out of the Mediterranean Sea. Beneath the strong salinity maximum, such stratification is the prerequisite for a peculiar type of mixing in the form of so-called "salt fingers" which creates a steplike structure (Turner, 1967). The molecular thermal conduction is around 100 times faster than the molecular diffusion of salt. As a consequence, instabilities in the density stratification develop because the water temperature changes more rapidly than the salinity when mixing takes place. The resulting vertical convection proceeds in a "fingerlike" manner, whereby the length of the fingers is limited. When this convection layer is mixed, the steplike structure of the water stratification is kept up. West of the Strait of Gibraltar, the conditions beneath the Mediterranean water, at depths from 1200 to 1500 m, are especially favorable for such structures, as shown, for example, by Zenk (1971) on the basis of *Meteor* observations (cf. Fig. 10.22).

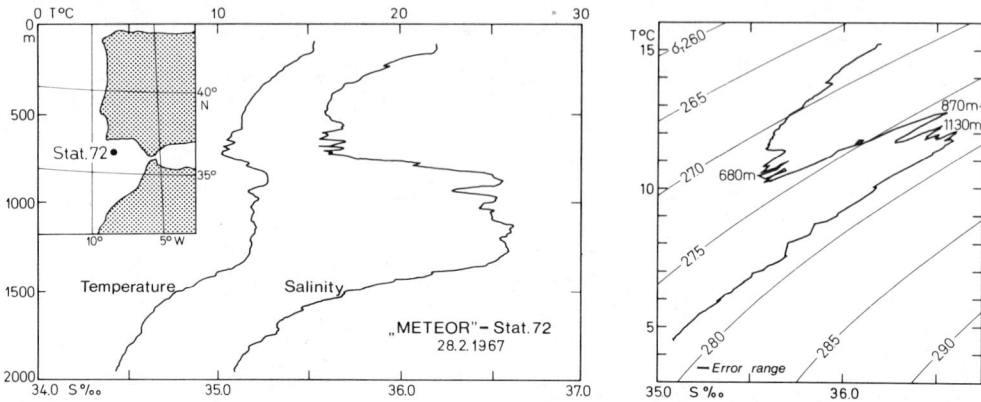

Fig. 5.19. Fine structure of water stratification based on a continuous record. Left: vertical distribution of temperature T and salinity S from the surface to 2000 m; right: record in a TS-diagram. (After Zenk, 1971.) Example *Meteor* station 72, $\varphi = 36°30'$N, $\lambda = 08°38'$W, on February 28, 1967.

5.4.3. Characteristic water masses in the world ocean

After having explained the idea and the usefulness of the *TS* relation for the characterization of ocean water in Sections 5.4.1 and 5.4.2, we will now turn to discussing this relation with a view to the water masses of the world ocean. Only the uppermost 100 to 200 m are excluded from the discussion because temperature and salinity there do not depend on spreading and mixing alone. At the same time this implies that none of the shallow shelf seas is included. If all the other pairs of temperature and salinity values, as obtained from different depths of the ocean down to the bottom, are entered into a *TS* diagram, the corresponding points of observation fall within the limits of the areas marked in Fig. 5.20. The way these areas fit into the σ_T lines also gives information on the density values that will occur.

The following deep-sea areas represent exceptions with regard to their thermohaline conditions which differ from the general oceanic relations: the Red Sea, the European Mediterranean Sea, the Arctic Ocean and the Norwegian Sea, Baffin Bay, the Sea of Japan, the Sea of Okhotsk, and the Black Sea. In these cases, the deeper parts of these seas are separated from the open ocean by submarine sills. But this alone does not suffice to explain their peculiar conditions. There are other seas which are topographically secluded in either a similar or even a stricter way, as, for example, the American Mediterranean Sea (see Section 10.2.5), the basins in the Austral-Asiatic Mediterranean Sea (see Section 10.2.4), or the Norwegian fjords, but which, nevertheless, completely remain within the limits of the *TS* relation for ocean water. The seven sea areas mentioned above show another peculiarity: under the influence of the respective climatic conditions at the sea surface, the density of seawater is sometimes increased to such an extent that deep-reaching thermohaline convection is initiated. This regularly happens in winter when water temperatures are at their lowest. In some regions this coincides with the time of highest salinity.

If the water masses of the individual oceans are considered separately, and the observations are classified according to their origin in one of the three oceans, the striking exceptional position of the Indian and Atlantic Oceans, as compared with the Pacific Ocean, will be noticed. In moderate and higher southern latitudes where the water masses belong to the Antarctic water belt, no pronounced differences in the *TS* relation have

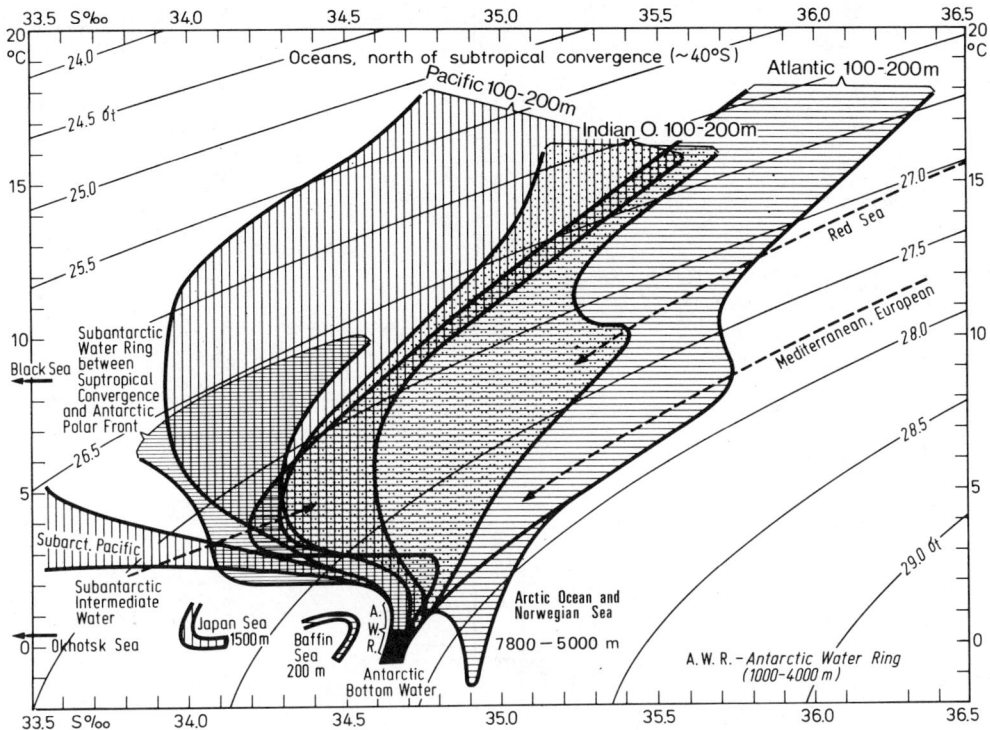

Fig. 5.20. Temperature-salinity relation in the entire world ocean below the disturbed near-surface layer. (After Dietrich, 1950a.)

been found. But north of the subtropical boundary, which—on the average—lies at approximately 40°S, a certain individuality of each ocean begins to develop, as shown in Fig. 5.20. This can be so marked that the relation for the Atlantic Ocean with temperatures of more than 8°C is almost absolutely different from that for the Pacific Ocean. From Fig. 5.20 it can already be recognized that this difference is partly caused by admixtures of high-salinity water in the deeper layers, which, for the Atlantic Ocean, originate in the European Mediterranean Sea and, for the Indian Ocean, in the Red Sea. Another part of such admixtures starts from the sea surface and is brought about by the influence of the atmospheric water cycle (cf. Section 4.3.6), as already mentioned, and by the interoceanic salt transport.

The *TS* relation can help to establish a natural division of the world ocean, as done by Sverdrup et al. (1942) in the book *The Oceans*. If the classification of the oceanic water masses is to be continued further (cf. Stepanov, 1965), generalized mixing processes in the world ocean can be traced in this way, as Mamayev (1969) has done.

In addition, the *TS* relation can be used for a statistics of water volumes with certain temperatures and salinities, as it was first accomplished by Montgomery (1958) for the entire world ocean. Figure 5.21 shows, in tabulated form, the shares of water volume that fall to the various classes. Here it becomes obvious that the combinations of temperature and salinity, as they occur in the world ocean, are restricted to a narrow range: From the total water volume of 1369×10^6 km^3 in the world ocean, 1262×10^6 km^3 (i.e., 92.2%), have salinities between 34 and 35‰ for the total temperature range from -2 to 32°C.

238 Temperature, Salinity, Density, Characteristic Water Masses, and Ice in the World Ocean

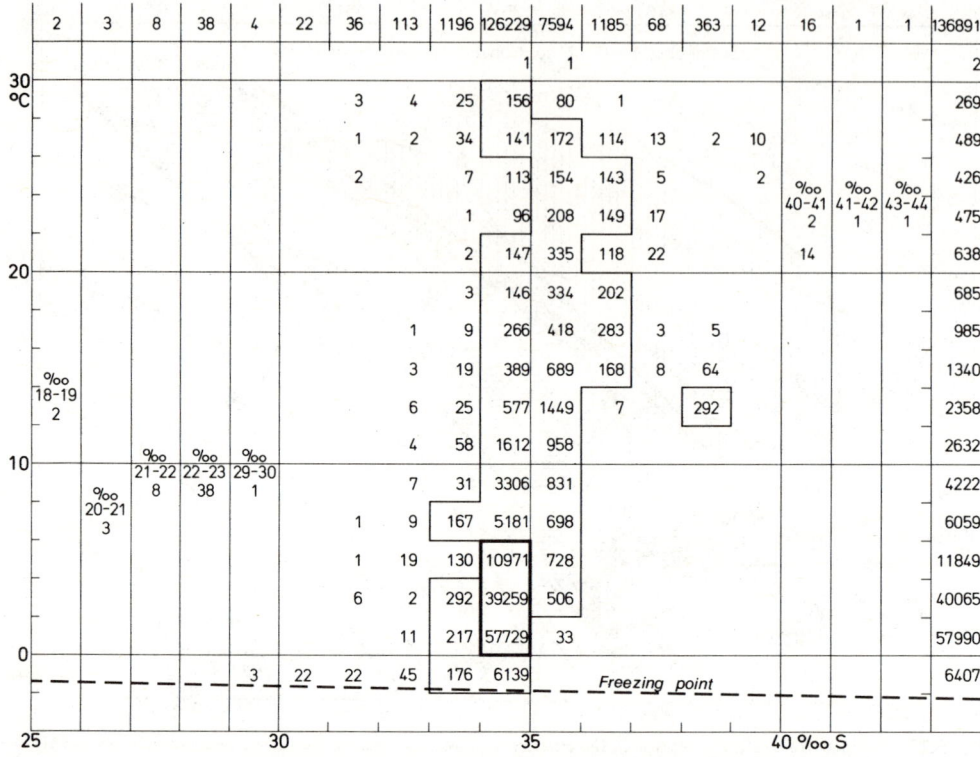

Fig. 5.21. Volumes of seawater in 10^4 km³ in the world ocean arranged for temperature classes of 2 to 2°C (potential) and for salinity classes of 1 to 1‰. (After Montgomery, 1958.) Thick framing encloses 75%, thin framing 99% of ocean water. Cumulative sums for each salinity class are given at the top of the diagram, those for each temperature class are indicated at the right-hand margin.

If the temperature range is restricted to 0 to 4°C, the relevant share still is 1080×10^6 km³, that is, 79.0%. This water is a mixture of Antarctic bottom water and North Atlantic deep water. From Fig. 5.21 further information on volumes can be taken: Antarctic and Arctic bottom waters with negative temperatures and salinities within the 34 to 35‰ class; subtropical surface water with salinities of more than 35‰ and temperatures of over 20°C; tropical surface water (T greater than 24°C, S less than 35‰); deep water and bottom water in the European Mediterranean Sea (T: 12 to 14°C, S: 38 to 39‰); water of the Red Sea (T: 20 to 22°C, S: 40 to 41‰). Such diagrams, for the first time, give a frequency distribution and mean values of the characteristic water masses. Hence, the mean potential temperature in the world ocean amounts to 3.52°C and the mean salinity to 34.72‰, which is in remarkable agreement with the values determined by Krümmel (1907), namely the mean actual temperature of 3.8°C and the mean salinity of about 34.7‰.

5.5. Ice in the World Ocean

5.5.1. Formation and types of ice

Monitoring and forecasting the distribution of ice in the ocean is of great importance with a view to the commercial utilization of many sea areas. For this purpose, knowledge on the physical properties of sea ice (cf. Section 2.1.12) is required, as well as on the processes of ice formation, on the types of ice occurring in the ocean, and on the statistics of the geographical distribution of such types of ice. Many studies have been published, particularly in those countries where navigation and shipping in areas endangered by ice play an important role. Recent summaries deal with the physics of sea ice (Pounder, 1968), with the nomenclature of ice (World Meteorological Organization, WMO, 1970; and Koslowski, 1969). Information on the geographical distribution of sea ice is contained in special atlases like that by Büdel (1950), as well as in the two atlases for the Arctic and Antarctic Oceans, published by the U.S. Oceanographic Office (1957, 1968) and in the Soviet atlas of Antarctica (1965).

The ice nomenclature mainly refers to the types of ice occurring in polar waters, but it can also be used, without any restriction, for the types of ice in the adjacent seas of the moderate latitudes. The new ice nomenclature, compiled by the World Meteorological Organization (WMO) (1970), takes into account the processes to which sea ice is subjected. Ice floating in the sea fundamentally differs depending on its origin:

Sea ice, that is ice formed by freezing seawater.

Ice of land origin. This kind is formed on the continent or on the ice shelf and then carried out into the sea. The ice shelf is the ice cover which is attached to the shore and maintained by "firn" (granulated snow from former years) or glaciers.

Lake ice is formed from fresh water in an inland lake.

River ice is formed at the surface of a river.

Apart from persistent freezing temperatures, the formation of ice depends on several other factors, namely on the surface salinity, influencing the freezing point (cf. Section 2.1.12), on the stratification of salinity as well as on the depth of the water. The two latter factors determine the thickness of the water layer that is involved in the vertical convection during the cooling process. All three factors mentioned above favor the ice formation near the coast.

Depending on the sea state and currents, different types of ice are formed. If the sea surface is calm, predominantly clear crystalline new ice is formed, whereas with a rough sea surface "grease ice" will develop which looks rather turbid because of air bubbles enclosed therein. If there is no further disturbance during the freezing process, a primary thin ice crust, called "nilas," is formed from the new ice, as occasionally encountered in the first-year ice of adjacent seas. Primary "pack ice" develops when, at the undulating sea surface, grease ice begins to grow into disks which, by constantly striking against each other, take the form of circular plates with raised rims, all of nearly the same size. This so-called "pancake ice," as seen in Fig. 5.22, represents the characteristic transition stage from grease ice to a consolidated pack ice sheet. Such transition forms may also develop when a thin sheet of new ice breaks up from increasing wave motion.

A different kind of pack ice, called "secondary pack ice" to distinguish it from the type described above, develops when a strong new-ice sheet is forceably broken up under external influences (wind, currents, waves, ships' traffic). This pack ice consists of ice floes of different sizes and irregular shapes and is prevalent in adjacent seas. During longer periods of freezing, pack ice and ice floes may be connected there by new ice and grease ice and freeze together to form a consolidated secondary fast-ice sheet.

If pack ice freezes under simultaneous pressure, thus causing the floes to pile up and override or raft each other, a last type of sea ice is formed, that is, the "compact pack ice,"

Fig. 5.22. Typical pancake ice (cf. Section 5.5.1). Initial stage of a pack ice sheet in the open sea. Photograph was taken north of Alaska in July, 1954.

which is encountered along the coasts as hummocks or ridges reaching heights of up to 20 m. Such ridges also occur in the narrow passages of the adjacent seas as, for instance, in the Baltic Sea off the Sund and the Belts, if the winter is particularly hard, as shown by the ice chart for February 20, 1963 (cf. Fig. 5.32). The main areas of compact pack ice, however, are the polar regions, especially in the Northern Hemisphere (cf. Figs. 5.23, 5.24, 5.25). There the pack ice, on the average, reaches ages of 2 to 6 yr before it is ejected through the straits between Greenland and Spitsbergen toward the south into the melting region. In the many periods of pressure, to which the ice is repeatedly subjected during its lifetime, it can form hummocks up to 8 m high. In the Antarctic Ocean which is all open to the north, the ice pressure is less severe. Besides, pack ice is rarely older than 1 to 2 yr. Therefore, pack-ice fields of flat floes prevail there.

Apart from the ice formed in the ocean itself, three other types of ice occur in the sea. They come from the continents, namely river ice, lake ice, shelf ice and ice of land origin. All of them consist of fresh water; so their physical properties deviate markedly from those of the original sea ice (cf. Section 2.1.12). Of special practical importance to ice navigation is the difference in strength which is greater for freshwater ice than for the more porous ice formed in the ocean. Although river ice and lake ice may considerably impede traffic in harbors and rivers, it does not play any role in the ocean, because it is more or less restricted to the estuaries of the large rivers. In the ocean, the occurrence

Fig. 5.23. Typical consolidated pack ice sheet in summer (cf. Section 5.5.1). Photograph was taken in September, 1954 on board the ice breaker *Westwind* of the U.S. Coast Guard in the Arctic Ocean north of Ellesmere Island. The lead in the foreground was made by the ice breaker, the puddles in the center are melt-water pools on the ice. Photo: U.S. Coast Guard.

of land and shelf ice is much more serious when it appears in the form of icebergs generated along the coasts of the polar regions.

As to icebergs, we can distinguish two different types, depending on whether they originate from valley glaciers or from shelf ice. Valley glaciers flow in narrow, but thick tongues of ice into the sea—the fastest among them advance 25 to 30 m day^{-1}, a speed 100 times faster than that of alpine glaciers. Tremendous, bizarre ice pinnacles break off at the front of the glacier tongue and immediately roll onto their sides. The calving of glaciers, as this process of formation of icebergs is called, and which usually occurs in the interior of fjords, may be accompanied by enormous water waves, which can reach several tens of meters and represent the highest surface waves recorded on earth. In the Northern Hemisphere, the main area of production of icebergs is found at the west coast of Greenland, in the general area of Disko Bay between 69 and 73°N, where the six fastest glaciers on earth reach the sea. According to Smith (1931), some 5400 icebergs are born there annually, which represent the main part of the icebergs in the Labrador Current carrying about 7500 icebergs southward every year. Figure 5.26 shows a photo of one of those dangerous colossi at the Grand Banks.

Icebergs of the shelf ice type are characteristic for the south polar regions where the Antarctic inland ice sheet does not advance in separate glacier tongues but in a compact

Fig. 5.24. Consolidated pack ice sheet (cf. Section 5.5.1) near Novaya Zemlya in summer (July, 1931). Openings (polynyas) at upper left and bottom, and numerous melt-water puddles on the ice sheet. Photograph was taken during the Arctic flight of the German airship *Graf Zeppelin* (shadow of airship, 236 m in length, in the center of the photo).

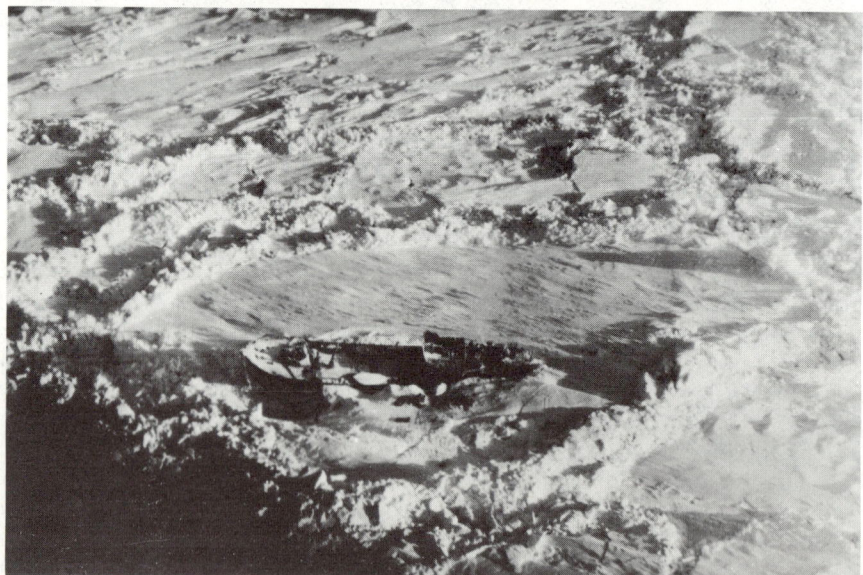

Fig. 5.25. Compact pack ice with ridges (up to 6 m in height) east of Greenland (cf. Section 5.5.1) in March, 1943 (near Shannon Island). In the center of the photograph: wreck of German weathership *Sachsen*.

Fig. 5.26. Iceberg near the Grand Banks (cf. Section 5.5.1). In background the *Acushnet* of the U.S. Coast Guard on international ice patrol service in spring 1950.

ice front towards the ocean surrounding Antarctica on all sides. Partly this ice sheet lies directly on the bottom of the shelf, partly it is already floating, as indicated by the results of seismic measurements of ice thickness summarized in Fig. 1.01. Occasionally, tabular bergs break away from this ice shelf. A typical example is given in Fig. 5.27. In some cases they reach giant dimensions. In February 1953, for example, north of the Ross Sea at 65°S 150°W, the British whaling factory ship *Balaena* observed a tabular iceberg 145 km in length and 40 km in width and with a height of 30 m above sea level.

The depth of the submerged part of an iceberg is given by the specific weight of the ice, which strongly depends on its air content (cf. Section 2.1.12). Observations on icebergs showed an air content of 1 to 10% by volume. With a density of seawater of 1.027 and an air content of 1%, as much as 88.6% of the total volume of the iceberg is submerged. With an air content of 10%, only 80.8% is submerged. For a tabular iceberg with vertical sides, the ratio between the height above sea level and the draft of the iceberg is, in the first extreme case, 1:7.8, in the second case, it is 1:4.2. Measurements of ice thickness, obtained by means of reflection-seismic methods in the South Polar Sea, showed a ratio of 1:5; that is, the height of the shelf ice above sea level amounted to 36 m, the immersion depth to 180 m.

Fig. 5.27. Characteristic tabular berg in Antarctic waters (cf. Section 5.5.1). Photograph was taken from the British research vessel *Discovery II* in the Weddell Sea. Height of the iceberg approximately 25 m.

5.5.2. Ice coverage of the world ocean

Oceanic ice conditions, in particular, depend on the type of ice, the limits of its distribution, and the variations of such boundaries with time. If the ice conditions are classified according to these viewpoints, and if we follow a suggestion by Büdel (1950), we arrive at a list of types of regional ice formation which is useful both in scientific and in practical results. Here, reference is made to this summary. Only two main types of ice will be mentioned here: the ice of the polar seas and the ice of the adjacent seas, since they represent the most significant differences in the ice coverage of the world ocean. These are elucidated in the following comparison of the main types of sea ice:

Ice of Polar Seas	Ice of Adjacent Seas
Multiyear ice always present beside first-year ice.	Only first-year ice in winter, no ice in summer.
Pack ice, partly with icebergs.	Fast ice and pack ice of flat floes.
Thickness about 2.5 to 3.5 m.	Thickness less than 0.5 to 1.0 m.
In marginal areas: ice maximum in early summer.	Ice maximum always in winter.
Navigable for ships only in leads.	Navigable for ships in channels forceably broken up.

This distinction between ice of polar seas and ice of adjacent seas has been taken into account when presenting the data on the ice coverage in Fig. 5.28. Otherwise, the representation is restricted to four especially important ice boundaries which characterize the annual rhythm of ice coverage. They are based on monthly charts from the *Atlas of*

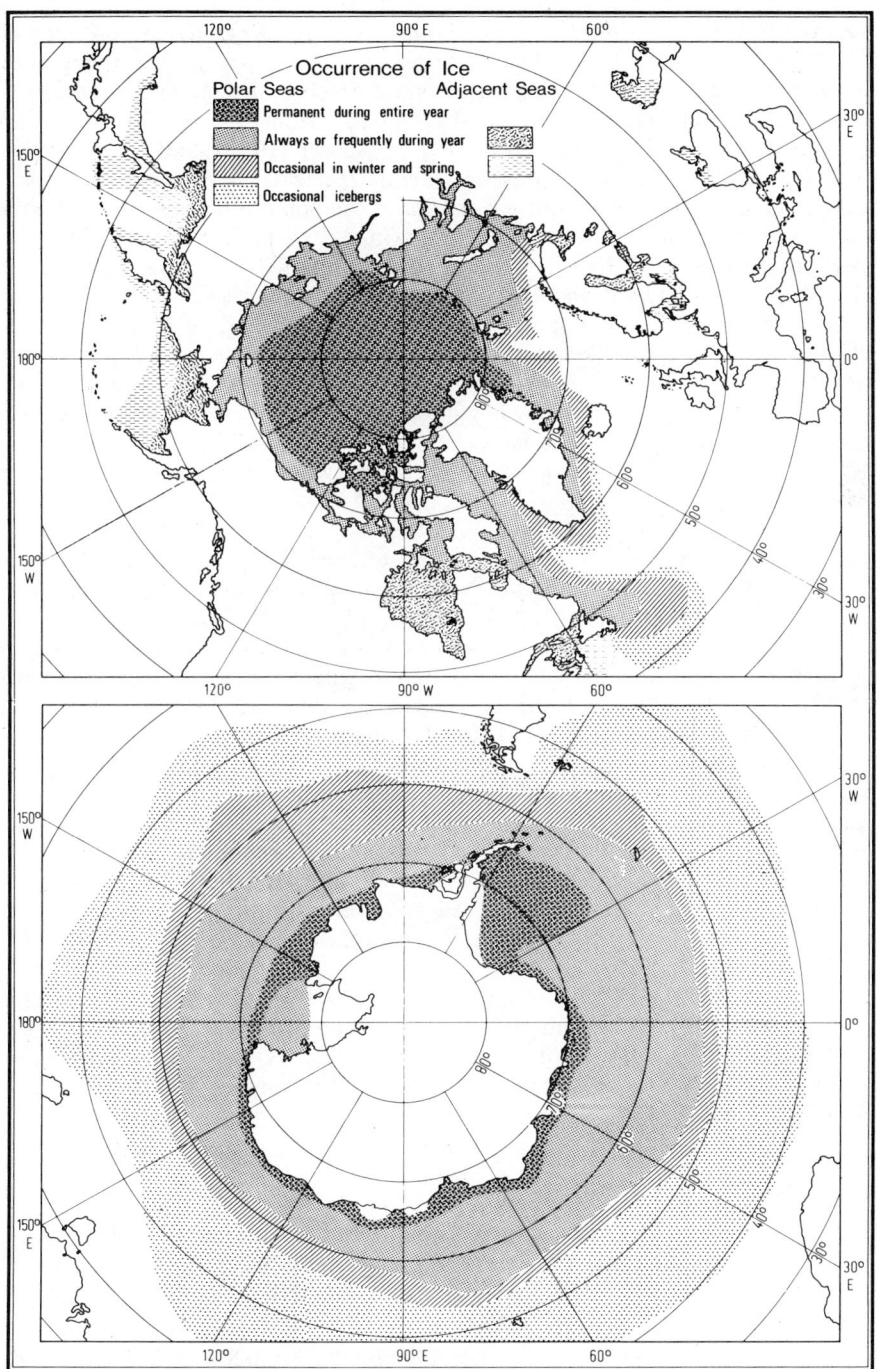

Fig. 5.28. Ice limits in the Northern and Southern Hemispheres. Scale approximately 1:100 mill. (After Büdel, 1950; Nusser, 1952 and U.S. Oceanographic Office, 1957 and 1968.)

the Ice Conditions in the Northern and Southern Polar Regions compiled by Büdel (1950) and published by Deutsches Hydrographisches Institut, Hamburg. For the Northern Hemisphere, the charts are based on ice observations made between 1919 and 1943, for the Southern Hemisphere on observations from 1929 to 1939. The summer ice limits for the South Polar Sea, as revised by Nusser (1952), and the boundaries of areas of occurrence of icebergs, based on recent American pilot charts, have also been taken into account. The inner limit for the ice of the polar seas ("permanently during the whole year"), as shown in Fig. 5.28, approximately coincides with the boundary of the minimum extension of the ice coverage (greater than 50%) in the first half of September in the Northern Hemisphere, or of February in the Southern Hemisphere, as given in the atlases of the U.S. Oceanographic Office (1957, 1968). The other limit ("always or frequently during the year") corresponds to the minimum extension of the ice coverage (greater than 50%) in March (Northern Hemisphere) or September (Southern Hemisphere). The third limit ("ocasionally in winter and spring") is approximately equivalent to the maximum extension of the ice coverage (less than 10%) in March (Northern Hemisphere). The innermost zone is always covered with ice all year round. This is the zone of permanent ice. In the second zone, one can always or very frequently count on the occurrence of ice in winter and spring, either as pack ice in the polar seas or as fast ice and flat-floe pack ice in the adjacent seas. In summer, ice conditions vary considerably in this zone: at the boundary to the inner polar region ice still occurs frequently; at the outer boundary it is observed only rarely or not at all, and the adjacent seas are ice-free during that time. In the third zone, ice of the polar and adjacent seas is met only occasionally in winter and spring, namely in about 10% of all cases. Finally, in the fourth zone, occasional occurrences of icebergs can be expected, particularly in spring. The boundaries of the polar regions are set where ice is the dominant factor that decisively determines the processes and phenomena in the uppermost water layer (cf. Section 10.2.1).

Figure 5.28 shows only the mean distribution of ice in the sea. Considering the large number of factors that influence the formation and distribution of this ice, but that vary a great deal, as, for example, the oceanic and atmospheric circulation, it is understandable that the occurrence of ice is also subject to considerable variation from one year to the next, especially with regard to its advance into lower latitudes. As an example, the occurrence of icebergs crossing 48°N in the region of the Grand Banks may be mentioned. They are the objects of particularly intensive monitoring. During the period from 1909 through 1960, the annual number of icebergs varied between less than a dozen as in the years 1924, 1940, 1941, 1951, and 1958, and more than 1000 in 1909, 1912, 1929, and 1945 (Schell, 1962), the long-term average being 400 yr^{-1}. An example concerning the variability of the pack ice boundaries has been well documented by observations north of Iceland (cf. Fig. 10.54).

5.5.3. Ice patrol

The considerable variables of the occurrence of ice in the ocean require observation and forecasting, in particular where important shipping lanes are concerned. This is the case for the Grand Banks, the Gulf of St. Lawrence, the North Canadian seaways, as well as for the Baltic Sea, the eastern North Sea, and the North Siberian seaway.

On April 15, 1912, on her maiden voyage, one of the biggest passenger liners, the *Titanic,* with a tonnage of 46,328 gross tons, collided with an iceberg south of the Grand Banks and sank (for position see Fig. 5.30), and 1490 people lost their lives. This tragedy was the direct cause for the establishment of the International Ice Patrol Service in 1913,

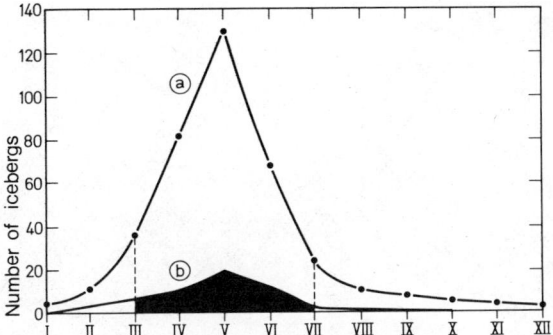

Fig. 5.29. Average number of icebergs per month. (a) Northwest Atlantic Ocean, south of 47°N; (b) south of the Grand Banks. Black: under surveillance by the International Ice Patrol Service. (After Smith, 1931.)

which is taken care of by the United States Coast Guard. In addition to ensuring the safety of the North Atlantic shipping routes between North America and Europe, the Ice Patrol Service has given great help to oceanography by elucidating the origin of the icebergs, the mechanism of their drift, and the ocean currents in the area of the Grand Banks. The cold Labrador Current, which meets here with the flank of the warm Gulf Stream, causes highly complicated current conditions.

Ice patrol is maintained from March through July. After having escaped from the embrace of the pack ice and fast ice in the Labrador Sea and the Baffin Sea, icebergs are most frequently encountered in May (cf. Fig. 5.29). Most of the 7500 icebergs annually carried southward, on the average, by the Labrador Current, run ashore and melt even before reaching the Grand Banks. In the annual mean, some 400 drift to the area north of the Island of Newfoundland, and only about 50 get to the southern tip of the Grand Banks. They do not follow the shortest path but drift along the axis of the Labrador Current over deep water east of the Banks and are included in the large current eddies south of the Banks. The observed trajectories of the icebergs, plotted in Fig. 5.30, very

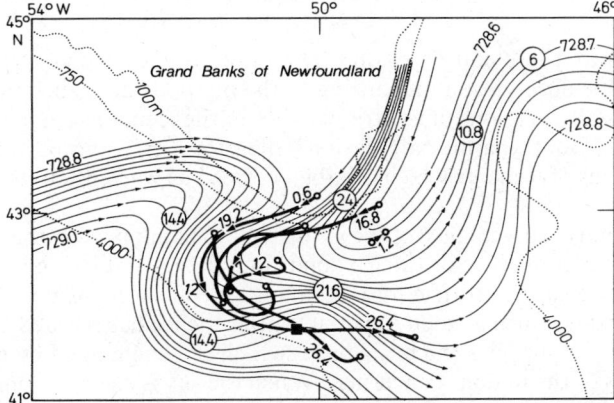

Fig. 5.30. Trajectories and velocities of five icebergs at the southern tip of the Grand Banks according to observations between June 14, and July 3, 1922. Also dynamic isobaths of the 750-dbar surface for the same period used as streamlines of the geostrophic flow (calculated velocities in circles). Velocities given in nautical miles per day. (After Smith, 1931.) ■ Position of *Titanic* shipwreck on April 15, 1912.

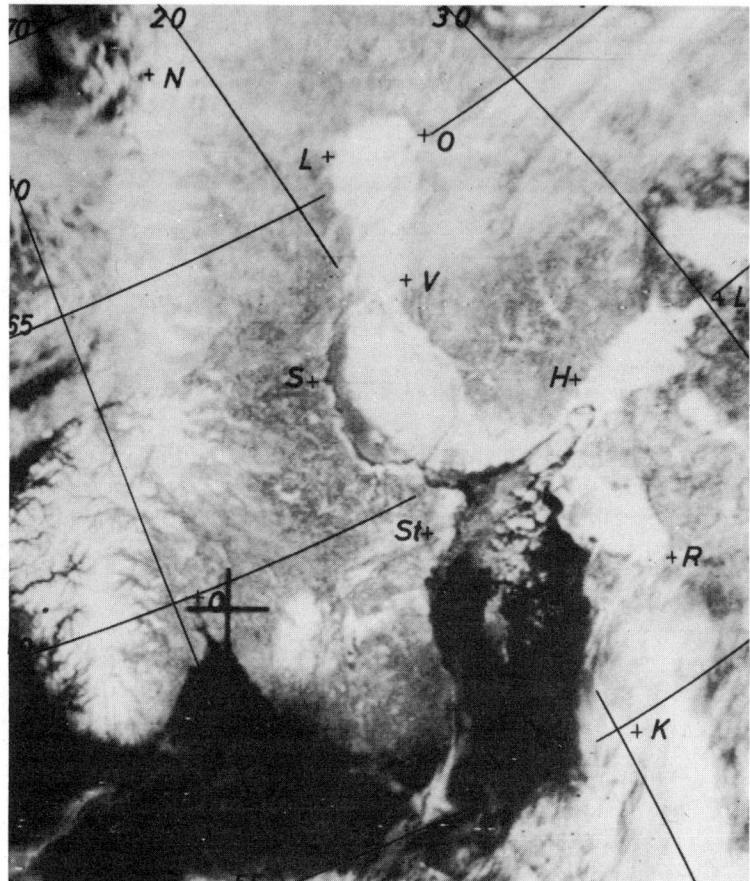

Fig. 5.31. Photograph of northern Europe taken by United States weather satellite.

clearly demonstrate these eddy motions. The internal geostrophic current, calculated from hydrographic observations according to the methods described in Chapter 7, and the simultaneous observations of the trajectories of the icebergs are in good agreement, as shown by Fig. 5.30. Dynamic calculations, therefore, are an important tool today for predicting the paths of icebergs which are then radioed as warnings to the North Atlantic ship traffic.

A different kind of surveillance of the ice distribution is applied in the Baltic and North Seas. Here, a detailed picture of the ice conditions at sea and in the harbors is of great interest to shipping so that traffic may be maintained as long as possible after the occurrence of ice and resumed as soon as possible after the ice has receded. A dense network of observing stations supplies the basic data, which are exchanged by radio among the collecting centers of the nations concerned. When the sky is clear, a valuable supplement to ground observation is provided by television pictures taken by weather satellites (Strübing, 1970), as shown in Fig. 5.31 giving information on the ice coverage of the Baltic Sea on March 26, 1969. In this way it is possible to prepare synoptic charts of ice conditions, as they are published, for instance, by the Ice Service of the Deutsches Hydro-

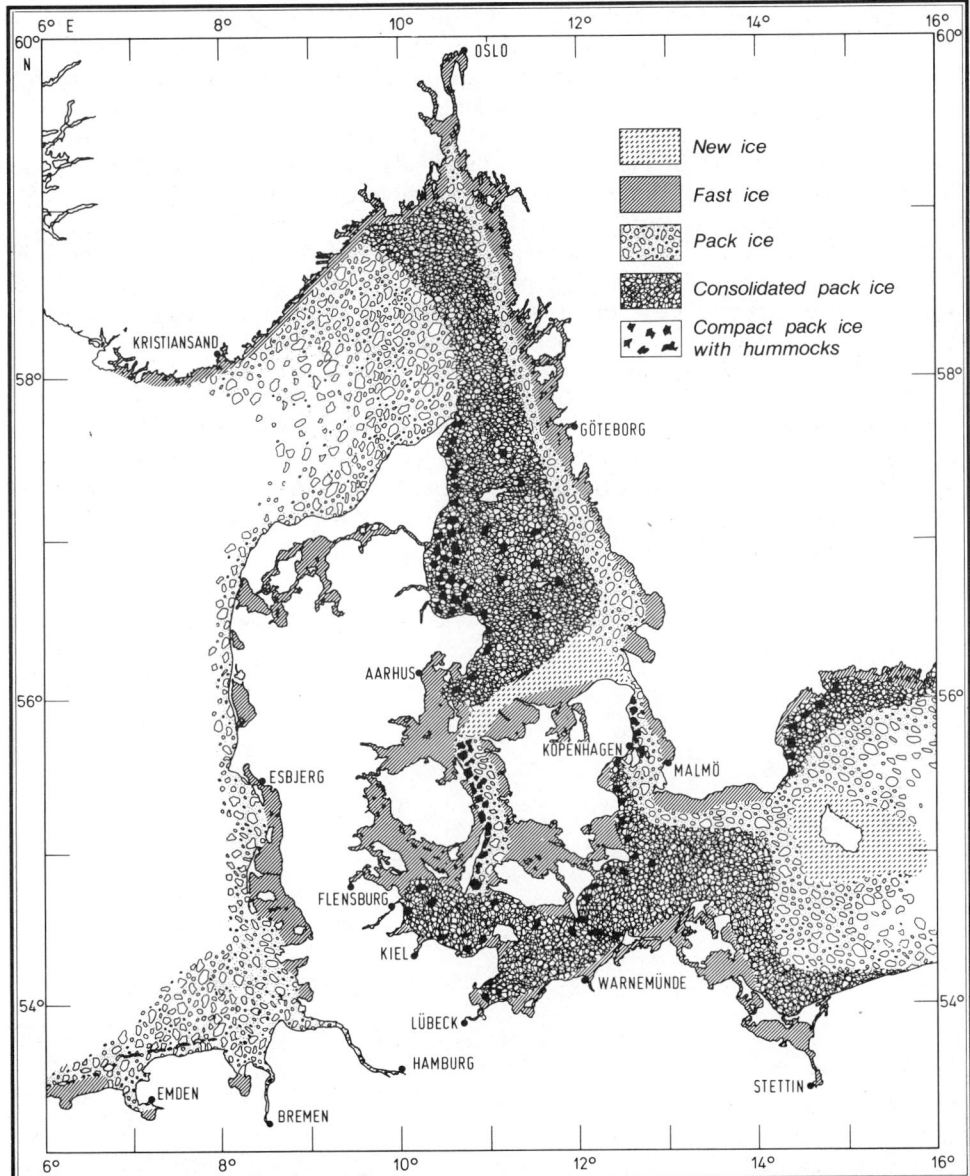

Fig. 5.32. Ice conditions in the eastern North Sea and western Baltic Sea on February 20, 1963. (After Ice Chart of Deutsches Hydrographisches Institut, Hamburg.) This represents the maximum ice coverage in the period from 1896 to 1971.

graphisches Institut in Hamburg every day whenever ice appears in the Baltic or North Seas. Figure 5.32 is a simplified cutout of such a chart representing the actual ice conditions in the passages between the North Sea and the Baltic Sea as they were on February 20, 1963. This is the ice maximum in that area reported since the beginning of regular ice observations in 1896 up to 1971. Under such extreme conditions, the entrance to the

Baltic Sea is completely blocked by ice as far out as in the Kattegat, and navigation is impossible. The German harbors along the North Sea coast, however, can always be entered, although with some difficulty. The great range of variation of the ice coverage in different winters becomes the more evident when Fig. 5.32 is compared with the maximum ice distribution in a winter with little ice. In such a case, the sea remains ice-free in the entire area covered by Fig. 5.32; only the shoals between the Frisian Islands of the North Sea and the bays of the Baltic Sea carry thin ice.

Yet another kind of ice surveillance is in use along the North Siberian seaway. It is carried out exclusively by the Soviet Union and is restricted to the polar summer. It is based on meteorological and oceanographic observations at sea and on land, completed by reconnaissance from the air. In this way, and supported by strong icebreakers, which are an additional requirement in the narrow passages between the Siberian shelf islands, convoys can be guided from the Barents Sea to the estuaries of the large Siberian rivers as well as to the Pacific Ocean. However, in spite of all efforts, this seaway is open for navigation only during a short season, that is, in August and September.

Navigation along the North Canadian seaway is even more difficult, nothwithstanding the endeavours undertaken since the opening of the oil fields in northern Alaska in 1969. A detailed summarizing study of the natural conditions of that seaway has been presented by Breslau et al., (1970).

6. Chemical Budget of the Ocean

6.1. Marine Geochemistry

6.1.1. General fundamentals of geochemistry

It is one of the tasks of geochemists to ascertain how the chemical elements are distributed on earth and to investigate the laws governing their flow. According to Goldschmidt (1933, 1937), we distinguish four stages in the geochemical development of our globe.

1. The separation of the elements according to their density into (a) Fe–Ni core, (b) oxide–sulfide shell, (c) lithosphere, (d) hydrosphere, and (e) atmosphere.
2. The separation of the elements resulting from the crystallization of mineral melts.
3. The separation of the elements from processes of weathering, sedimentation, and crystallization of aqueous solutions.
4. The separation of the elements that results from biochemical conversions associated with the phenomena of life.

The first two stages of geochemical development can be considered as completed; at present the earth is in stages 3 and 4. The special nature of those geochemical processes of development is based on the presence of water in liquid form and on the readiness of the elements for conversion which is considerably increased by the active participation of water. A comparison of the earth with other planets in our solar system demonstrates the uniqueness of this situation. On none of the other planets do conditions exist which would permit us to conclude that the structure of their surfaces, consisting of continents and oceans filled with liquid water, might be approximately similar, in the order of magnitude, to our globe. From this point of view, the earth's crust can be considered an area where geochemical processes are activated by the concurrence of a great number of favorable circumstances. The following phenomena contribute to this:

1. The addition of a liquid phase to the gaseous and solid phases and the extensive intermixing of these three phases, "solid–liquid–gaseous," along their common interfaces.
2. The existence of the "unique" water (cf. Section 2.1) as the only liquid phase, characterized by the following properties: (a) When water cools, condensation begins by 180°C earlier compared with the norm, and, thus, its liquid phase is already stable at a range of the geochemical development process at which the general physicochemical readiness for reaction is still 10^6 to 10^9 times stronger than at temperatures 180°C lower. (b) Owing to its asymmetrical

Table 6.01. Periodic Table of the Elements[a]

Period	Group I a	Group I b	Group II a	Group II b	Group III a	Group III b	Group IV a	Group IV b	Group V a	Group V b	Group VI a	Group VI b	Group VII a	Group VII b	Group VIII			Group 0
I	1 **H**																	2 He
II	3 Li		4 Be		5 **B**		6 C		7 N		8 **O**		9 **F**					10 Ne
III	11 **Na**		12 **Mg**		13 *Al*		14 **Si**		15 **P**		16 **S**		17 **Cl**					18 Ar
IV	19 **K**		20 **Ca**		21 Sc		22 *Ti*		23 V		24 Cr		25 **Mn**		26 **Fe**	27 **Co**	28 **Ni**	
		29 **Cu**		30 **Zn**		31 Ga		32 Ge		33 *As*		34 Se		35 **Br**				36 Kr
V	37 Rb		38 Sr		39 Y		40 Zr		41 Nb		42 **Mo**		43 Ma		44 Ru	45 Rh	46 Pd	
		47 Ag		48 Cd		49 In		50 *Sn*		51 Sb		52 Te		53 **I**				54 X
VI	55 Cs		56 Ba		57 La 58–71[b]		72 Hf		73 Ta		74 W		75 Re		76 Os	77 Ir	78 Pt	
		79 Au		80 Hg		81 Tl		82 *Pb*		83 Bi		84 Po		85 At				86 Rn
VII	87 Fr		88 Ra		89 Ac		90 Th		91 Pa		92 U							

[a] **Boldface** type indicates elements definitely necessary for life processes; *italic* type shows elements probably necessary for life processes; roman type indicates elements probably not necessary for life processes.
[b] 58 to 71, Rare earths: 58 Ce, 59 Pr, 60 Nd, 61 Pm, 62 Sm, 63 Eu, 64 Gd, 65 Tb, 66 Dy, 67 Ho, 68 Er, 69 Tm, 70 Yb, 71 Cp.

molecular structure, water has a variety of physicochemical properties favoring reaction (cf. Section 2.1).

3. The favorable ratio between the surfaces of the ocean and of the continents (2.4:1). With the given distribution of elevations and depressions on the earth's surface, this ratio guarantees that an almost maximum area is included in the geochemically active, humid–amphibious boundary zone.
4. The process of life as a highly developed and especially effective kind of geochemical activity.
5. The periodic action of forces based on the continuous change of the different elements of weather in annual and diurnal rhythm.

The characteristic of geochemical processes in nature is their vast dimensions in space and time. Consequently, even minute chemical effects may become important and occasionally play a dominant role in nature. The concentration of the substances involved in reactions is sometimes very small, as for example is usually the case in the processes of weathering and sedimentation. Hence, no effects at all or only very slight ones would be achieved under laboratory conditions.

For the distribution of chemical elements in the ocean, their original cosmic frequency is of fundamental importance. Owing to their increasing atomic number (number of unit positive charges of electricity in the nucleus of the atom), and, related to that, their atomic mass, beginning with H of mass 1 and ending with U of mass 238, the chemical elements can be arranged in the series of the periodic table. The following natural laws result:

1. The original cosmic frequency of the elements decreases by a factor of 10^{-8} with increasing mass number from H with mass number 1 roughly to Mo with mass number 42 (in the midst of the fifth period, Table 6.01). With further increase of the mass number it remains almost constant.
2. There are some deviations from this general law. The strong minima with Li, Be, and B and the maxima with Fe and Ni are the most conspicuous ones.
3. According to the so-called "Harkin's Rule," the elements with even mass numbers are more abundant than the elements with odd mass numbers.

If the frequency of elements that prevails in the earth's crust is compared with the cosmic data obtained by analyses of meteorites, an analogous relation is clearly recognizable, except for the conspicuous minima with the rare gases, which probably have evaporated at an early stage owing to their lack of any possibility for chemical bonding.

6.1.2. Regulating mechanisms for the chemical elements in seawater

If we compare the frequency of the elements in seawater with that in the earth's crust, it is evident that no simple relationship is found in a comparison with the cosmic values. Hence, the laws governing the distribution of the elements in seawater must be of a different nature. The differences between the logarithmic frequency values in the earth's crust and those in seawater give some indication (cf. Fig. 6.01) which elements preferably tend to pass into the ocean. Apart from H and O, such elements are B, Na, S, Cl, Br, and I. They are called "thalassophylic elements" in contrast to "thalassophobic elements," (P and Si as well as most heavy metals), only traces of which pass into the ocean.

254 Chemical Budget of the Ocean

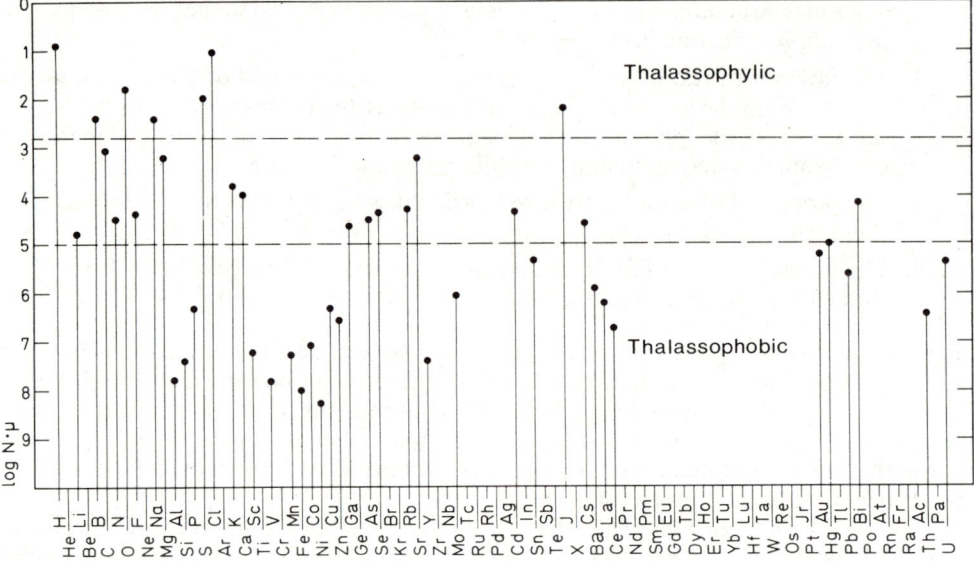

Fig. 6.01. Frequency of chemical elements in seawater and in the earth's crust, represented as the difference of the values log $(N \cdot \mu)$ earth's crust minus seawater. N = number of molecules, μ = atomic weight.

The distribution of the elements in seawater is governed by two chemical phenomena: (1) the behavior of the elements based on their ionic potential, and (2) the selective adsorption. The ionic potential is given by the ratio of the state of ionization Z (chemical valence) to the ion radius R of the elements. According to their ionic potential Z/R, the elements can be divided into three groups (Fig. 6.02):

1. Substances with a low ionic potential, below 3 (for example, Na, Ca, Mg). During the processes of weathering and subsequent transporting, they change into true ionic solutions.
2. Substances with an ionic potential between 3 and 10 (e.g., Al, Ti, Ce). In

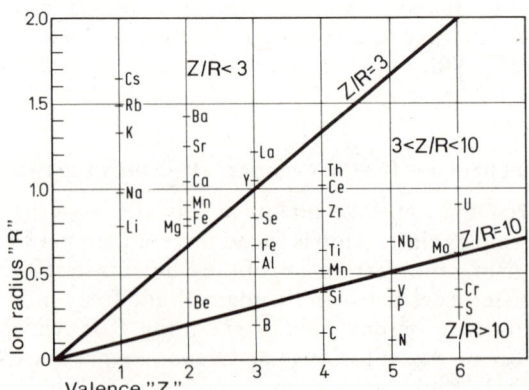

Fig. 6.02. Ionic potential Z/R of some elements. Z = valence, R = ion radius in Ångström.

aqueous solutions, hydroxyl radicals are added to them and, through hydrolysis, they are precipitated from such solutions in the form of their hydroxides.

3. Substances with a high ionic potential (above 10) (e.g., C, P, S, N). They oxidize and form anions (proton donors). In most cases they are soluble again, unless, in a secondary reaction, such behavior is prevented by the formation of sparingly soluble salts.

The second phenomenon that greatly influences the solubility properties of the elements in seawater is the "selective adsorption." In the ocean, two components of matter are present which act in this direction: first, the argillaceous material that is finely distributed everywhere in seawater. It preferably tends to combine with alkali metals and alkaline-earth metals. Second, particles of ferric hydroxide and manganese hydroxide that are suspended in seawater. They strongly adsorb heavy metals like Pb, Zn, Cu and complexing agents like As, Se, Mo, and remove them from seawater by sedimentation on the ocean floor.

6.1.3. Sedimentation budget

Selective adsorption plays an important part in the sedimentation budget. Assuming that, during the weathering process, the magmatic primeval rock disintegrated into sea salts and sedimentary rock and that, furthermore, all sodium present in seawater originated in the weathering process, Goldschmidt (1933) arrived at the following result: so far, in the course of the geological development there have disintegrated, referred to 1 cm^2 of the earth's surface,

> 157 kg of primeval rock into 132.9 kg of clay slate (80%),
> 19.1 kg of sandstone (11.5%),
> 10.0 kg of limestone (6.0%),
> 4.2 kg of dolomite (2.5%);
> a total of 166.2 kg of sediments

corresponding to a thickness of sedimentation of approximately 625.5 m. If these values are referred to the amount of seawater corresponding to 1 cm^2 (269.6 liters), a balance can be calculated for each element, as given, for instance, in the following table for Na and K:

Element	In Primeval Rock (kg)	In Sediment (kg)	In Seawater (kg)	In Seawater (%)
Na	4.55	1.55	3	66
K	4.16	4.05	0.11	2.6

Whereas 66% of Na has gone into solution in seawater, the corresponding value for K is only 2.6%. Most of the potassium, which, in fact, is easily soluble, has apparently been withdrawn from seawater through selective adsorption. In principle, this is the same process that plays an important part in agriculture in the storage of fertilizers by soil colloids. This phenomenon is of general significance. In the case of potassium, however, it has not yet reached its maximum development, which can be derived from the following comparative figures giving the percentage of the corresponding elements which has been dissolved in the course of the weathering process.

Li	Na	K	Rb	Cs	Mg	Ca	Sr
0.23	66	2.6	0.11	0.033	10.5	1.97	5.3
Ba	Ra	B	Al	Sc	Y	La	
0.023	0.017	280	0.00025	0.0014	0.0016	0.029	

Thus we see that adsorption increases considerably with increasing atomic weight. The only exceptions are Li, which is not surprising considering its position as the first element of the alkali metals, and B which, as a volatile compound, is subject to different conditions.

If the same calculations as applied to the cations are made for the proton donors, a striking result is obtained:

Cl	Br	S	I	F
6900	1900	300	29	0.96

For Cl, Br, S, and I the concentration in seawater is a multiple of the quantity that has been provided by the disintegration of primeval rock. This can only mean that there must be resources of matter other than those supplying products of weathering. This is indeed the case. The proton donors are substances that are either volatile themselves or form volatile compounds. We may therefore assume that they reach the ocean by another way, that is, through the atmosphere. Just as the weathering process and, thus, the subsequent delivery of cations to the ocean still continue today, anions produced by volcanic eruptions are also steadily carried into the ocean by way of the atmosphere. Accordingly, the increase of salinity in the ocean is still going on. It seems that, at present, the equilibrium conditions between supply and removal of substance are regulated in such a way that, if geological periods are considered, the chlorine content of the ocean increases whereas the boron content appears to decrease. For Ca, supply and removal seem to be balanced.

The selective adsorption of ferric and manganese hydroxides is very important for life processes in the ocean, since such adsorption is responsible for the "detoxicating effect" that the ocean exerts on a great number of noxious heavy metals and metalloids. Table 6.02 illustrates the actual conditions by contrasting the real concentrations to those without adsorption. Apart from such general phenomena of adsorption, some other, more individual adsorption processes also play a role in the oceanic sedimentation process. By preference, Zn, Cd, In, and Bi are found in phosphate sediments, whereas siderite (spathic iron ore) is associated with In, and the organic substances are distinguished by their content of Va, Mo, as well as of Cu and Ni.

6.1.4. Primary natural radioactivity

In the ocean three groups of radioactive elements can be distinguished:

1. those present in nature since the creation of the earth;
2. those newly generated by cosmic rays;
3. those produced artificially.

Table 6.02. Adsorption of Sea-Salt Ions on Ferric and Manganese Hydroxides

Element	Amount in Seawater (%)	Estimated Concentration without Adsorption Process (mg m^{-3})	Actual Concentration (mg m^{-3})
Adsorbed on ferric and manganese hydroxides only			
Cu	0.0086	58,000	5
Pb	0.0042	12,000	5
Mo	0.0078	9,000	0.7
Adsorbed on ferric and manganese hydroxides, and on ferrous sulfide			
As	0.52	2,900	15
Se	0.97	400	4
Hg	0.01	300	0.03
P	0.013	460,000	60

The first group includes primary elements already present at the creation of the earth with an average life period ranging at least within the order of magnitude of the earth's age, and their often relatively short-lived secondary elements. For the ocean, only the disintegration series resulting from ^{238}U, ^{235}U, and ^{232}Th, as well as from ^{40}K and, perhaps, ^{87}Rb are of any practical importance. As far as radium is concerned, seawater is extremely depleted. While the radium content in the rocks of the solid crust of the earth, on the average, amounts to 1.4×10^{-12} (2.9 to 0.9×10^{-12}), it is only of the order of magnitude of 0.2 to 3×10^{-16} (the mean being 1×10^{-16}) in seawater. This shows that the ocean contains more than 10,000 times less Ra than the rocks of the solid crust of the earth. Since, on the other hand, the sediments of the deep sea are 3 to 20 times richer in Ra than the rocks of the continents, the process of selective adsorption must be involved to a great extent. The disintegration product ^{230}Th (ionium), developing between ^{238}U and ^{226}Ra, is captured by adsorption before it can be converted into Ra.

The transfer of energy during radioactive processes in the ocean differs considerably from that in the earth's crust (cf. Table 6.03). The difference in the energy transfer between ocean and continents however, is not quite so conspicuous as the difference in the radium content. In the ocean, potassium (97.3%) plays the main role. The ratio of the total energy (per unit weight) of radioactive processes in the ocean to those in the earth's crust, amounting to 1:190, is much less pronounced than that of the Ra contents. A balance of radioactive substances in the ocean is given in Table 6.04. The disintegration products of ^{238}U and ^{235}U with disintegration periods of medium length, namely ^{230}Th (8×10^4 yr), ^{231}Pa (34,300 yr), and ^{234}U (2.5×10^5 yr), are quite useful for the determination of the age of deep-sea sediments. Helium is formed as product of the disintegration series of U, while the disintegration product of ^{40}K is ^{40}Ar. The amounts, particularly those generated in seawater, are small and practically of no importance. In this way, according to Revelle and Suess (1963), about 30 atoms of He and about 50 atoms of ^{40}Ar are produced per liter of seawater and min.

6.1.5. Radioactivity generated by cosmic rays

The sources of radioactive substances generated by cosmic rays are mainly to be found in the atmosphere. On their path through the atmosphere, the cosmic rays, which are very rich in energy and which predominantly consist of protons (91.5%), generate fast-moving neutrons, partly upon collision with gas molecules. Due to repeated collisions,

Table 6.03. Energy Transfer during Radioactive Processes in Seawater and in the Earth's Crust

Element	Radioactive Heat Production (cal g^{-1} hr^{-1})	Seawater			Earth's Crust	
		Content	cal g^{-1} hr^{-1}	%	Content	cal g^{-1} hr^{-1}	%
^{40}K	1.19 × 10^{-6}	4.6 × 10^{-6}	5.47 × 10^{-12}	97.3	2.83 × 10^{-4}	3.37 × 10^{-10}	31.7
^{235}U + ^{238}U	6.25 × 10^{-5}	2 × 10^{-9}	1.25 × 10^{-13}	2.23	4.2 × 10^{-6}	2.63 × 10^{-10}	24.8
^{226}Ra	140.00	1 × 10^{-16}	1.4 × 10^{-14}	0.25	1.4 × 10^{-12}	1.96 × 10^{-10}	18.5
^{232}Th	2.21 × 10^{-5}	4 × 10^{-10}	8.84 × 10^{-15}	0.16	1.2 × 10^{-5}	2.65 × 10^{-10}	25.0
			5.62 × 10^{-12}	99.94		10.61 × 10^{-10}	100.0

Table 6.04. Balance of Radioactive Substances[a,b]

	U	^{230}Th (Ionium)	Ra	Th (total)
Input of dissolved substance by rivers	0.810×10^{-4} g	12×10^{-10} g	5.6×10^{-12} g	1.8×10^{-6} g
Disintegration	-8×10^{-10} g			3.5×10^{-11} g
Radioactive supply		8×10^{-10} g		
Disintegration		-1.8×10^{-10} g		
Radioactive supply			1.8×10^{-10} g	
Disintegration			-1.4×10^{-10} g	
Sedimentation on ocean floor	0.8×10^{-4} g	18×10^{-10} g	0.4×10^{-10} g	1.8×10^{-6} g

[a] m² of ocean surface yr^{-1}.
[b] After Koczy, 1963.

these neutrons gradually change into slow (thermal) neutrons which, in turn, interact with other nuclei. In this process the nuclei become so highly excited that they release one or several nucleons, and radioactive nuclei, mostly short-lived ones, are produced. Such products generated by the interaction with the atmospheric gases of nitrogen, oxygen, and argon, are the following: ^{3}H (half-life: 12.3 yr), ^{14}C (5760 yr), ^{7}Be (53 days), ^{10}Be (2.5×10^{6} yr), ^{22}Na (2.6 yr), and ^{32}Si (710 yr). In marine sciences, ^{14}C is of special importance because it can be used to determine the age of dead organisms, of sediments and water masses up to a maximum age of about 40,000 yr. Until about a hundred years ago, the ratio of ^{14}C:^{12}C amounted to 1.24×10^{-12} under undisturbed conditions. Since that time, this value has noticeably decreased, because, with the development of industry, considerable amounts of ^{12}C have been returned into the natural cycle by burning coal and petroleum, both—owing to their old age—practically free of ^{14}C. Evidence has been found that the content of ^{14}C in the atmosphere has increased since 1945, which is primarily due to the neutrons produced through the explosions of hydrogen bombs (Junge, 1963).

By means of the age determination with the ^{14}C method, analyses of the CO_2 content in the Pacific bottom water, originating from the Antarctic surface water and spreading northward at the bottom of the Pacific Ocean, have resulted in a time span of around 600 yr for the water transport between the latitudes of 60°S and 30°N. This corresponds to an average current velocity of about 0.5 mm sec^{-1} at the equator.

6.1.6. Radioactivity produced artificially

Since 1945 the group of natural isotopes and radioactive isotopes induced by radiation has been enlarged by radioactive isotopes formed as waste products at the artificial fission of U and Pu (atomic bombs, reactors). As far as those products are not used for biological-medical purposes, interest, especially in marine sciences, is centered primarily on the dangers of contamination connected with the dumping of such waste material into the ocean in order to "destroy" it.

In the process of artificial nuclear fission numerous radioactive isotopes are generated, their half-life period ranging from fractions of a second up to 10^{15} yr. Table 6.05 represents a summary of the most important fission products occurring with the fission of ^{235}U by thermal neutrons. The figures are also valid, in their order of magnitude, for the nuclear

Table 6.05. Composition of the Isotope Mixture Produced at the Fission of ^{235}Uranium by slow Neutrons[a]

Isotope	Half-life	MCP$_w$[b] (10^{-8} μC/ml)	Percentage of Total Activity after 10 days	100 days	4 years
132 Te	3 days	ca. 1,000	11		
99 Mo	3 days	30,000	6.8		
133 Xe	5 days		15		
131 I	8 days	600	6.4		
147 Nd	11 days	3,000	5	0.3	
140 Ba	13 days	3,000	21	2.3	
143 Pr	14 days	5,000	10	1.5	
141 Ce	32 days	4,000	5.5	11	
95 Nb	35 days	20,000	0.6	19	
103 Ru	40 days	ca. 10,000	5.5	14	
89 Sr	51 days	700	2.4	9.5	
91 Y	58 days	3,000	3.1	13	
95 Zr	65 days	6,000	3.2	15	
144 Ce	0.8 yr	ca. 1,000	1.3	12	27
106 Ru	1.0 yr	ca. 1,000		1.0	4.2
147 Pm	3 yr	20,000		1.0	23
85 Kr	11 yr				1.9
90 Sr	28 yr	8	0.05	0.6	26
137 Cs	30 yr	20,000		0.4	17
151 Sm	80 yr	80,000			0.4
Decrease of total activity			100 %	5.9%	0.15%

[a] After Herrmann et al., 1959.
[b] MCP$_w$ = maximum concentration permissible in drinking water at continuous supply.

fission of ^{235}U by high-speed neutrons. With respect to the percentages, they differ from the above values by the factor of 2 at the most. For practical reasons, only fission products with half-lives between 3 days and 80 yr are listed, whose share of the total activity between 10 days and 4 yr is more than 0.1%. This concerns 17 elements, mainly rare ones. Their dangerousness can be recognized from their share in the total activity, greatly varying in the course of time, and from their radiobiological effect which depends on how long the respective elements remain in the organism. Hence, the following isotopes (italicized in Table 6.05) are particularly dangerous: ^{131}I, ^{89}Sr, ^{144}Ce, ^{90}Sr, and ^{137}Cs. With regard to the biological effect, the most dangerous isotope by far is ^{90}Sr, because it has the extremely low MCP value (see Table 6.05 for definition) of only 8 and a relatively long half-life. ^{137}Cs is also one of the rather dangerous isotopes, whereas ^{131}I can only occur in relatively young isotope mixtures. In addition to the radioactive fission products mentioned above, other radioactive byproducts are formed, above all ^{239}Pu and ^{3}H (in addition to ^{14}C, ^{54}Mn, diverse Co-isotopes as well as ^{55}Fe and ^{65}Zn). While tritium (^{3}H) is comparatively innocuous, ^{239}Pu is one of the isotopes that are biologically dangerous.

The understanding of the biological behavior of the various isotopes is rendered more difficult by the fact that, because of their chemical behavior, the various elements are involved in different processes in the organisms and react specifically in the biological metabolic cycle. For example, ^{90}Sr (28 yr) and ^{239}Pu (2.4 × 10^4 yr) are preferably stored in the skeleton and permanently fixed there, whereas ^{137}Cs (30 yr) is mostly deposited in the muscles, which—in contrast to the skeleton—are subject to extensive metabolic processes. Therefore, beside the physical half-life, a biological half-life must be taken into account as the period after which half of the amount of a certain substance is excreted from the organism. The pertinent figures are: ^{137}Cs = 17 days, ^{90}Sr = 7.5 yr, ^{239}Pu = 120 yr. Thus, in the cycle of the biological food chain, radioactive isotopes can accumulate considerably (up to 10^6 times) in nature, due to the specific selectivity of some organisms or because the deposition in certain organs is preferred. In this way, strong displacements in the mixing ratio of fission products are caused, and monitoring is made much more difficult.

6.2. Biochemistry of the Ocean

6.2.1. General fundamentals of biochemistry

A great part of the material transport and of the mineral deposits on the earth is based on biochemical processes. The laws governing the life processes are of such a specific kind and so well tuned to one another that they demand special consideration. It is a significant property of life processes that they join in the processes determined by inorganic nature in a "parasitical way," so to speak, utilizing the amounts of energy set free thereby. In nature and especially in the ocean, it is therefore difficult to distinguish between purely inorganic processes and purely organic ones.

Various fundamental properties characterize the phenomenon of life and the sphere where life is possible. A precondition for life is water. Hence, the sphere of life or, according to Suess (1909), the "biosphere" is therefore found in the "amphibious–humid" boundary zone of the earth's surface, the main characteristic of which is the mutual penetration of the solid, liquid, and gaseous phases. At the same time, this is the region where the development of interfaces or surfaces finds optimum conditions for evolution and where matter in the form of powder or dust is persistent. Moreover, it is one of the peculiarities of life that matter is dispersed in the finest distribution. This does not impede the virtual "geological" achievement of life which consists in the accumulation of substances present in the environment in very great dilution and in the formation of large mineral deposits that are important from the point of view of economy (cf. Section 6.1.3). Another peculiarity is the extremely high degree of dilution of the vital substances that make an optimum development of life possible and dominate the processes of life. While chemical laboratory tests generally deal with concentrations of the order of magnitude of 10^{-2} to 10^{-4} g cm^{-3}, normal life processes only demand concentrations of 10^{-6} to 10^{-10} g cm^{-3} of CO_2, O_2, P, chlorophyll, adrenaline and others. Even concentrations a thousand times smaller (down to the density of 10^{-13} g cm^{-3}) are within ranges required by the normal life process, as demonstrated by the extreme limits of sensitivity to active substances, hormones, highly specific pheromones and food-attractants, as well as to odoriferous substances and poisons. In the case of acetylcholine, even dilutions to 2×10^{-19} g cm^{-3} produce clear registrations of contractions of the smooth muscles of frogs' lungs. Here, not even one molecule of acetylcholine is contained in 1 mm^3. The dogs' sensitivity to the smell of butyric acid is of a similar order of magnitude. Even a concentration of 10^{-18} will still be noticed, which corresponds to nine molecules of butyric acid per mm^3. This pronounced sense of smell of dogs is also found with other animals. Recent investigations have shown that fish living in fresh water or in the ocean also possess a distinct sense of smell. A comparison taken from marine chemistry may elucidate these extreme conditions. A crystal of sodium phosphate the size of a pin head (2 mm^3), contains 0.265 mg of phosphorus. When it is dissolved in 26.5 liters, the result is a phosphorus concentration of 10 mg m^{-3}, an amount which, under natural conditions in the ocean, represents rather a good mean value. Even a value 100 times smaller (10^{-10}) is already recognizable in a phosphorus-free culture of plankton by the start of cell division.

The basic element of all living matter is carbon. Like water (cf. 2.1.1), carbon must also be considered as "unique" among the substances occurring in nature. Its symmetrical position in the center of the periodic table (Table 6.01) is the reason why it can form stable compounds with electropositive substances as well as with electronegative ones. In addition, several carbon atoms can combine and form chainlike and ringlike molecules, sometimes of rather a complicated structure. Thus, carbon, from which the system of "organic chemistry" has been developed, can provide the life process, which is based on it, with a multitude of different compounds that are carefully tuned with regard to their energy content and their reducing potential.

Because of its position in the periodic system, carbon can best be compared to nitrogen. "Normal" carbon, as it is to be understood due to its relationship with nitrogen, is not solid, but must be considered gaseous, which is also true for the simple forms of carbon compounds (e.g., CO_2, CO, CH_4, etc.) in analogy with the nitrogen compounds. It is of fundamental importance for the life process that the state of aggregation of carbon dioxide is gaseous, that it is present everywhere in the biosphere as the primary substance of life, and that it automatically flows in equal concentration to any place where it has been consumed. This, of course, is of utmost importance for the propagation and abundance of life.

The true biological achievement of life is the production of living substance from carbon dioxide with the aid of the energy of solar radiation. Only plants can accomplish this because of their chlorophyll content. Therefore, they are called *producers,* whereas all other organisms belong to the group of *consumers*; for, by utilizing the energy stored in the nutrients, they re-establish the original state by way of fermentation and respiration processes and, finally, by replenishing the carbon dioxide reservoir. Although the final result is unimportant from the geochemical point of view, this closed cycle is so very significant because secondary processes, which may lead to final, geologically important transports of material, are involved. Here, the transport of elementary carbon itself and that of carbon dioxide are of primary interest. Figure 6.03 gives a schematic representation of the carbon cycle in nature as far as conditions in the earth's crust are concerned. Nine different carbon deposits can be distinguished (in g cm^{-2} of the earth's surface):

(1) Carbon in the ocean	7.5
(2) Carbon in the atmosphere	0.125
(3) Carbon in plants	0.053
(4) Carbon in animals	0.00071
(5) Elementary carbon in sediments	633
(6) CO_2-carbon of carbonates and sediments	2,340
(7) Total carbon of crystalline slate	1,960
(8) Total carbon of palingenic igneous rocks	567
(9) Total carbon of juvenile igneous rocks	33
Total	5,540

There are three carbon cycles:

1. The small but intensive biological cycle: atmosphere–assimilation (8530 μg)–plants–animals–respiration (8215 μg)–atmosphere. A small shortcut leads back to the atmosphere via forest and brush fires (314 μg). Thus, the cycle is practically closed in itself.

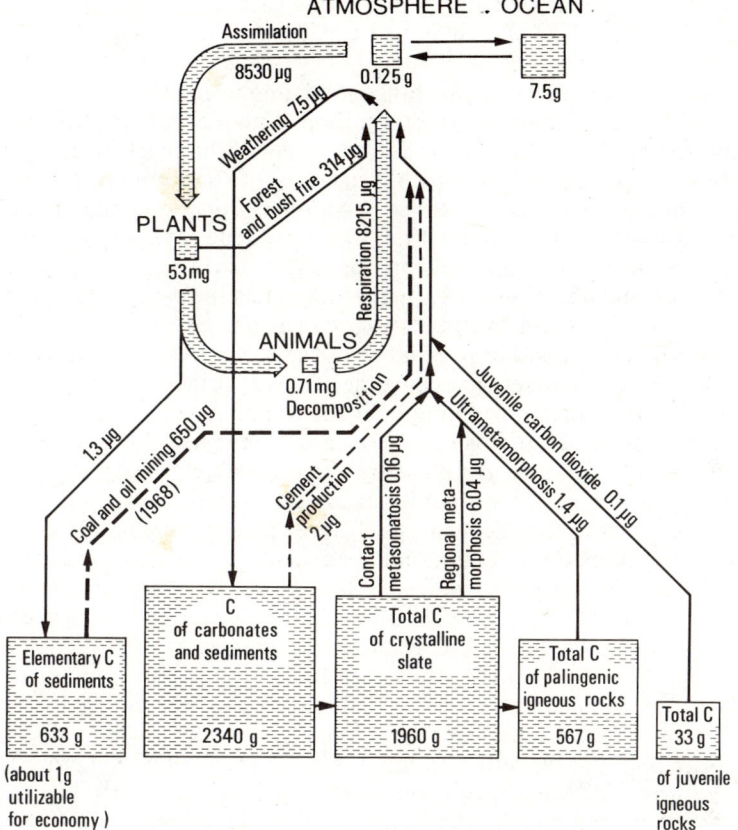

Fig. 6.03. Carbon cycle in nature. (After Kalle, 1945; improved 1970.)

2. The large geological cycle: carbon dioxide originating from respiration–weathering (7.5 µg) of carbonates and sediments and, further, by transformation in the rocks, while one portion at each stage is returned to the atmosphere (totally 7.6 µg). This natural process, too, is remarkably balanced in itself, although it should be mentioned that it is artificially unbalanced, by the production of cement, by 2 µg cm^{-2} to the debit of the rocks.

3. The biological–geological–economical cycle that is influenced artificially: atmosphere–assimilation–plants (animals) (1.3 µg)–elementary carbon of sediments–coal and oil mining (650 µg)–atmosphere. This cycle is completely unbalanced. The artificial input of about 650 µg of carbon—compared with which the juvenile carbon dioxide of about only 0.1 µg is negligible—should have led to about a doubling of the atmospheric carbon dioxide in the past decades since industrialization began. This certainly would have had far-reaching climatic consequences if another process had not been counteracting, namely the great buffer capacity of the ocean with respect to carbon dioxide. Seawater, in equilibrium with the gaseous phase, has the capacity to absorb the 60- to 100-fold excess of carbon dioxide in physical and chemical bonds

(cf. Section 6.2.6) so that any change in the gaseous phase is noticeable in the atmosphere only by as much as 1 to 1.5%.

Although carbon, linked to living substances and amounting to 54 mg cm^{-2}, represents only an extremely small portion of the total carbon content of the earth's crust, amounting to 5540 g cm^{-2}, it must be considered the actual motor of the geochemical cycles. Because of this active interference in geochemical processes, life has quite rightly been ascribed the properties of a geochemical "catalyst." Another property of the life process fits in very well with this analogy—namely that, like every true catalyst, it is in very delicate balance with the given environmental conditions and that it responds specifically to any influence from the outside. Since life, furthermore, has the capability to reproduce and multiply, the properties of an "autocatalyst" can also be ascribed to it.

As mentioned above, another peculiarity of the life process is its capability of taking part in nature's energy budget as a consumer. This is primarily achieved through the plants, which divert a certain percentage of the total solar radiation indirectly via the biological carbon cycle before it is finally released to the heat budget. Secondly, this is also brought about by animals as well as by bacteria, which are experts in utilizing, by "parasitical" intrusion, not only organic carbon compounds for their own purposes but also nearly any potential chemical energy difference occurring on the earth's surface.

Finally, a further speciality of life should be mentioned. It concerns the phenomenon of the "optical activity" of life substances, that is, the fact that a great number of substances characteristic of the life process are based on molecules occurring in configurations related spatially to each other like image and mirror image. According to their capability of turning a polarized light beam to the left or to the right out of its plane of oscillation, these two forms are denoted by L (laevus = left) and D (dexter = right). It is most peculiar that all forms of life existing on earth are built by only one of the two corresponding molecular forms and that the fermentation system, the basis of all life processes, responds only to these forms of life. Hence, the terrestrial life is but "one half." Certainly, it may be possible that, for example, the corresponding other half of the phenomenon of life has developed on another planet. There would be no possibility, however, of merging the two spheres of life although they could not be distinguished by their external appearance. Because of the strict specificity of the fermentation systems, no possibility of mutual intermixing would exist, either in the areas of nutrition and reproduction or with regard to any other phenomena of life.

6.2.2. The ocean as the source of life on earth

The discovery of the strange phenomenon of the optical activity of life substances may perhaps serve to shed some light on the origin of life. The older view by Pasteur hypothesized that circularly polarized light may have contributed to the development of life. Jordan, on the other hand, has suggested that the act of the primary creation of life may have occurred in such a way that originally, under climatically favorable conditions of past geologic periods and in accordance with the laws of probability, among others also an optically active protein molecule was created that was capable of autocatalysis, which then led to an avalanche-like increase of life processes.

It seems to be certain that the primary forms of life developed in the ocean, not in the deep sea but in the shallow offshore areas where favorable conditions with respect to light, temperature, and nutrients existed. The main reasons for this hypothesis are many. Under today's conditions, too, water is the main medium of life processes. Living

Table 6.06. Composition of Blood Serum and Seawater

Element	Seawater g kg⁻¹	Seawater %	Human Serum g kg⁻¹	Human Serum %	Mammalian Serum g kg⁻¹
Na	10.75	30.7	3.00	34.9	3.00 –3.55
K	0.39	1.1	0.20	2.3	0.19 –0.22
Ca	0.416	1.2	0.10	1.2	0.08 –0.10
Mg	1.295	3.7	0.025	0.3	0.023–0.040
Cl	19.345	55.2	3.55	41.3	3.40 –4.00
SO_4	2.701	7.7	0.02	0.2	
HPO_4	0.000185	0.0005	0.10	1.2	
HCO_3	0.145	0.4	1.60	18.6	

organisms consist predominantly of water (about 50% of trees is water, about 66% of vertebrates, and up to 99.7% of jellyfish). This includes the amounts of skeletal material, which sometimes are rather considerable. Furthermore, pure salt solutions of a single salt act in a poisonous or, at least, growth-retarding manner on animals as well as on plant life. If, however, by the addition of a second or even a third salt, the composition of the solution approximates that of seawater, the poisonous effect decreases, provided the total concentration is adjusted to that of the organism. On the other hand, the composition of the blood serum of animals, especially that of vertebrates and human beings, shows a very close relationship to the composition of seawater (cf. Table 6.06). From this table, Quinton (1912) calculated that the following relation is valid:

$$Cl:Na:K:Ca = 42 \pm 8 : 25 \pm 5 : 1.5 \pm 0.56 : 1$$

The close relations between sodium, potassium, calcium, and chlorine suggest that the animals, when passing from water to land life, internally carried part of the seawater, to which they were best adapted, with them in their cavities. The differences in the total concentration as well as in the magnesium and sulfate contents can be explained by two facts: (1) since the Cambrian period, when this change from marine to continental life took place, a considerable amount of salt has been added to the ocean; the magnesium content, in particular, has increased. (2) During the following immensely long periods the blood was adjusted to the different living conditions imposed on animals now breathing through lungs. Above all, owing to the large increase of bicarbonate ions, the buffer capacity of the serum became much stronger than that of seawater. This mainly occurred at the expense of the sulfate content which takes care of the increased carbon dioxide pressure in the alveoli of the lungs.

A simple method for examining the relationships of individual organisms with the seawater of today consists in determining the magnesium–calcium ratio of the blood serum. While for primitive marine animals the ratio approximates that of seawater (3.11 : 1), the corresponding figures are always one order of magnitude lower (0.11 : 1 to 0.34 : 1) for land and fresh-water animals, including the marine bony fish (teleostei), to which the majority of the fish belong that live today. Therefore the conclusion seems justified that, after going through a fresh-water period, these fish, in contrast to sharks and rays belonging to the class of cartilaginous fish (elasmobranchii), have repopulated the ocean secondarily in more recent geologic periods. Another relevant evidence can be seen in the fact that the cartilaginous fish, quite different from bony fish, have a blood serum that is isotonic to seawater. Because of the difference between the salinity of seawater and that of blood serum, most fish are rigidly restricted to the habitat they are found in. Only a few species of fish are able to live in seawater as well as in fresh water.

Fish that live predominantly in the ocean but ascend rivers in order to spawn (lamprey, sturgeon, shad, salmon) are called anadromous fish, whereas catadromous fish spend part of their young life in rivers but spawn in the ocean (river eel, flounder). Occasionally, also genuine sea fish like pleuronectides, sygnathides, tetrodons, as well as some sharks and rays, penetrate into rivers, and fish normally living in rivers wander into the ocean. Sticklebacks can live both in seawater and in fresh water.

6.2.3. The energy production of organisms

The terrestrial form of life has adjusted to the physical and chemical conditions on earth. Probably it is not unique, and similar possibilities may exist for nature to develop an organic life cycle also under different external conditions. The fundamental requirements on which the terrestrial life process depends are the following:

1. The life substance consists of carbon compounds.
2. Energy can primarily be supplied only by solar radiation.
3. The starting material is present everywhere in nature, either in gaseous or in liquid (or dissolved) form.
4. Because of Requirement 2, only a photochemical reaction is possible. To ensure the metabolic cycle, it must be reversible and, for reasons of expediency, it must have an energy increase as large as possible.

The life processes, as they occur almost exclusively in nature, can be essentially described by the following formula:

$$6CO_2 + 6H_2O + 674{,}000 \text{ cal} \underset{\text{respiration}}{\overset{\text{photosynthesis}}{\rightleftharpoons}} C_6H_{12}O_6 + 6O_2$$

This formula expresses the fact that six molecules of carbon dioxide and six molecules of water are rearranged in such a way that an energy increase of 674,000 cal is achieved. The products won by this reaction, that is, by photosynthesis, are glucose, as a prototype of organic substance, and oxygen. Since glucose is a highly reactive substance, it can be very easily changed back into the initial reactants by means of "respiration" if free oxygen is available. At the same time the energy taken up is set free. Thus, the life cycle is closed. In the above formula it is remarkable that it is not in agreement with Requirement 4 demanding the maximum energy increase. The process of photosynthesis doubtlessly leads to the hydration of carbon dioxide (by adding hydrogen to the molecule). Therefore, one might suspect that a life process on the basis of the following formula would create better living conditions:

$$6CO_2 + 12H_2O + 1{,}279{,}700 \text{ cal} \rightleftharpoons 6CH_4 + 12O_2$$

The energy increase resulting from such reaction would be nearly double than actually produced. That, under the conditions realized in nature, the reaction remains at a medium energy level, is apparently caused by the fact that, at this level, a great number of organic compounds are available and ready to react, a complete keyboard, so to speak, offered to nature to play on (Fig. 6.04). The farther we move in the diagram from the center to either side, the smaller is the number of possible organic compounds provided by nature. Finally, in the extreme cases of complete oxidation and complete hydration, only one

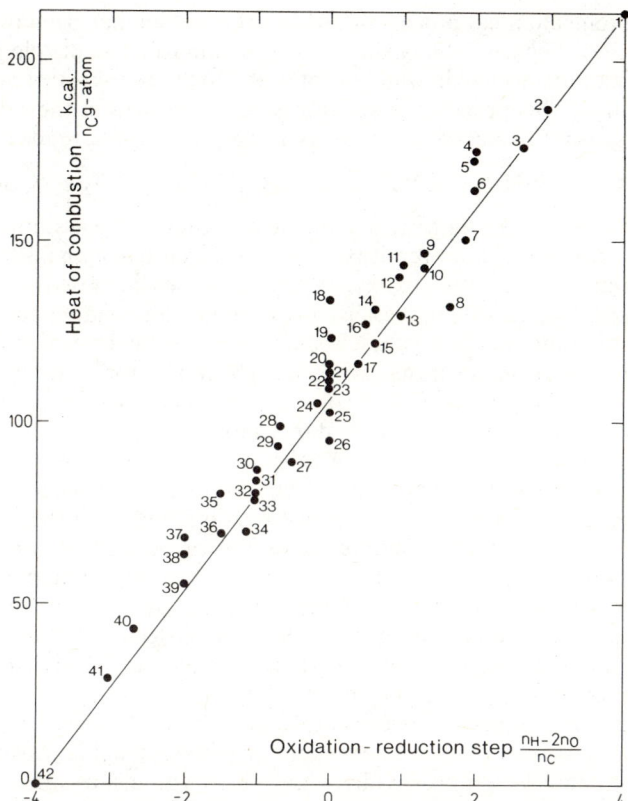

Fig. 6.04. Distribution of some physiologically important carbon-hydrogen-oxygen compounds (1–42) based on their energy content set free by oxidation. Examples: 1 Methane, 4 Methyl alcohol, 7 Stearic acid, 14 Glycerine, 18 Formaldehyde, 22 Glucose, 26 Graphite (black lead), 33 Citric acid, 37 Carbon monoxide, 42 Carbon dioxide.

single compound is available, namely carbon dioxide (CO_2) or methane (CH_4), respectively. Furthermore, the farther we go from the center toward the less hydrated compounds, the stronger the tendency to decompose, whereas strongly hydrated compounds, on the other hand, lose their capability to react chemically. In the extreme case, the completely hydrated paraffins become almost totally inactive, as expressed by their Latin name "parum affinis."

The opinion is often ventured that one of the fundamental requirements of organic life is the presence of an oxidizing atmosphere as we know it on earth. But this is not the case. In theory, a corresponding life process in a reducing atmosphere is equally conceivable as, for example, in the gas envelopes of the large planets, which hold considerable amounts of methane, ammonia, and hydrogen. The following formula clearly shows the completely analogous structure with the actual one describing terrestrial photosynthesis:

$$6CH_4 + 6H_2O + 214{,}500 \text{ cal} \rightleftharpoons C_6H_{12}O_6 + 12H_2O$$

The difference is that the pertaining energy increase amounts to only one third of that of the terrestrial life process.

Another photomechanical process of assimilation, which has been realized on earth, although to a minor degree only, is the carbon assimilation by purple bacteria. These one-celled microorganisms live in waters containing hydrogen sulfide and can assimilate CO_2 in light. During this process, free sulfur is deposited within their cells in the form of small yellow grains. The corresponding assimilation formula reads as follows:

$$6CO_2 + 6H_2S + 295{,}700 \text{ cal} \rightleftharpoons C_6H_{12}O_6 + 3O_2 + {}^{3}\!/_{4}\, S_8$$

The energy gain involved amounts to around 50% of that of the assimilation process by plants. The great difference, however, is that sulfur is deposited in an insoluble, solid form. The reverse process corresponding to oxygen respiration would close the cycle, but it does not exist in nature. Hence, this process is only of minor importance for terrestrial life.

Besides these assimilation processes based on photochemical reaction there are various other kinds of assimilation processes. Anaerobic bacteria, for instance, are capable of utilizing a chemical reaction instead of solar radiation as a source of energy by means of which they can use the carbon of carbon dioxide for producing organic living substance. All such processes have this in common: a substance offered by nature is oxidized with the aid of atmospheric oxygen. Simultaneous with this main reaction, a certain percentage of the oxygen required for this oxidizing process is obtained from the carbon dioxide of the atmosphere. Thus, the corresponding amount of carbon is released to be converted into organic substance. Nitrite bacteria act according to this principle when oxidizing ammonia to nitrite. In the same way, nitrate bacteria further oxidize the nitrite thus formed. Sulfur bacteria oxidize hydrogen sulfide, oxyhydrogen bacteria oxidize hydrogen, and iron bacteria convert bivalent ferrous carbonate into trivalent ferric hydroxide. These bacteria are also of importance in the ocean.

The second group of anaerobic bacteria can exist only in the absence of oxygen; hence, assimilation is not possible. They are dependent on a heterotrophic way of life. Their life process is a pure degradation process by which the organic substance is oxidized and, at the same time, sulfates are reduced to hydrogen sulfide and nitrates to elemental hydrogen. Whereas, under suitable conditions, the former process is of great importance in the ocean, that of denitrification seems to play a minor role there (Vaccaro, 1965).

The dependence of plants on solar radiation is the cause of several peculiarities in their habits. To begin with, their habitat in the ocean is restricted to the uppermost, fully lighted layers of water. The light intensity that is still sufficient for assimilation, the so-called "compensation light intensity," amounts to approximately 0.005 cal cm^{-2} min^{-1}, for shadow cells to approximately 0.002 cal cm^{-2} min^{-1}, as shown by experiment. Since, at the sea surface, the radiation active in photosynthesis amounts to about 0.5 cal cm^{-2} min^{-1} during the greatest part of the day in summer, the compensation light intensity is approximately reached at a depth where the surface light intensity has decreased to 1%. This corresponds to the three- or fourfold value of the "visibility depth" as measured with the Secchi disk. In the extremely clear water of the open ocean in the subtropics, this limit lies between 200 and 100 m, in very turbid coastal water at a depth of only a few decimeters. In the open North Atlantic Ocean of the moderate latitudes, the assimilation limit can be assumed to be around 40 m, in the North Sea and the Baltic Sea at a depth between 20 and 30 m. In the upper, fully lighted water layers under otherwise optimum living conditions, the assimilation activity and the rate of multiplication of phytoplankton approximately increase proportional to the exposure to light. In general, however, too strong a radiation (1 cal cm^{-2} min^{-1}) has a detrimental effect by bleaching the plant pigments, especially of chlorophyll. Therefore, the maximum production of phytoplankton is not encountered at the sea surface, but often a few m below. Under normal conditions, the cell division of diatoms will occur every 10 to 36 hr.

A further peculiarity of marine plants is a result of their dependence on daylight and the need for the best possible utilization of radiation: it is their tendency to develop the largest possible surfaces. Continental plants have solved this problem by forming assimilation organs of extremely large surfaces—the leaves—which is in accordance with their general structure requiring, above all, the development of a strong connective frame. Most marine plants, however, have chosen another way. By forming extremely small single cells or cell compounds in phytoplankton with a size from 10^{-1} to 10^{-4} cm they have enlarged their specific surfaces to the utmost. Thus they have achieved the same effect as the land plants have done through the formation of leaves.

For animal life of the ocean, conditions are quite different. With regard to their nutrition, they primarily depend on the living substance produced by phytoplankton. However, since they also feed on each other and due to their own mobility, they are not restricted with respect to their habitat, especially because food is supplied also to deeper layers of the ocean by a slow but steady "rain" of dead plankton organisms from upper layers rich in such organisms. Swarms of healthy diatoms, for example, probably sink by 1 or 2 m day^{-1}, starved ones by 10 or 20 m day^{-1}. In addition, swarms of zooplankton, by their active vertical migration, take care of the transport of nutrients into deeper layers of water. In contrast to phytoplankton, zooplankton—for physiological reasons—is limited to the development of extremely small forms. A relevant explanation is given by Bergmann–Rubner's "surface law" formulated by Rubner (1883). This law states that the metabolism of an individual becomes all the more intensive the smaller the organism. On the average, the metabolic process is nearly proportional to the surface of an organism. This means that a tenfold increase of the metabolism per unit weight of living substance is the result of a tenfold decrease in diameter, owing to the related tenfold increase of the specific surface (surface/volume). Since, in general, only a limited amount of food is available for animal organisms and, moreover, especially in the ocean, oxygen necessary for respiration is present only in relatively low concentration, it is obvious that the larger an animal the better its chance to survive in bad times. For the same reason, a lower limit must exist for the size of animal organisms below which an orderly life process is no longer possible due to lack of food and oxygen. With regard to plankton organisms in particular, the term "minimum habitat per hour" has been coined, denoting the size of the water volume needed by an organism for respiration according to its oxygen requirements calculated for 1 hr. The size of this volume is referred to the volume of the organism as a unit. For marine plankton, this value varies between 0.1 and 1000. For bacteria, it may exceed 1,000,000; but, because of their peculiar way of living, bacteria cannot physiologically be put on the same level as animals.

Another important condition which, besides the providing of energy, influences the living conditions of phytoplankton to a great extent, is their dependence on nutrient distribution. In this connection, nutrients are defined as all those saltlike substances needed by plants for building up their organisms and for maintaining their life processes. Table 6.01 presents a review of the elements necessary for life. In this table all those substances which definitely play an important physiological role in the life process, are emphasized by bold type, whereas the physiological role of the elements printed in italics is not yet clear. The remaining elements apparently are of no importance for life. Some of them, however, represent rather strong poisons. Most of the elements essential to life are found in the first periods of the periodic table.

In accordance with the "minimum law" established by von Liebig in the middle of the last century with a view to conditions in agriculture, we must expect that those elements will be the first to be removed from seawater by life processes which are present in seawater in relatively low concentration but which are badly needed by living cells, that is, for which the ratio of supply to demand reaches a minimum. Thus, the further development of life is limited even though a great excess of other elements is available in the ocean. Liebig recognized phosphorus, nitrogen, and potassium as the minimum substances in agricultural practice. In the ocean, the corresponding elements are phosphorus, nitrogen, and silicon, the latter particularly for diatoms or silicious algae, which are abundant in the ocean and need silicon for building the silicious scales of their bodies. In seawater, these three elements are present in the form of compounds as silicates, phosphates, in inorganic and organic bonds, as well as nitrates, nitrites, ammonia, and finally as organically bonded nitrogen. In seawater, nitrogen generally is the first to arrive at the minimum. Some other elements that may be exhausted under special conditions are manganese, cobalt, zinc, and molybdenum. They play a part in the cell mechanism as enzyme cofactors. Furthermore, in this context, the elements iron, copper, and vanadium should be mentioned as important cell constituents.

Apart from the trace elements just mentioned, according to our present knowledge, around 70% of the marine phytoplankton species also need some additional organic trace substances for their development, although only in extremely low concentrations. These are the following active substances: vitamin B_1 (thiamine), vitamin B_{12} (cobalamine), and vitamin H (auxine). Some species need all three vitamins, others only two or one. The concentration of these vitamins in seawater generally ranges between 10^{-10} and 10^{-13} g cm^{-3}.

Abundant flora and, thus, also fauna can be expected in oceanic areas where, with sufficient radiation from sun and sky, the water is furnished with great amounts of nutrients. This is the case in the estuaries of the continental rivers, and also in the open ocean in those areas where water from deeper layers, which is rich in nutrients due to the decomposition of dead and sinking organisms, wells up to the surface. Three different processes are known by which water is brought up from the depth to near-surface layers:

1. Vertical convection of water masses, especially that which is caused thermally. It becomes effective when the stability of stratification is diminished by loss of heat, as it occurs mainly in moderate and high latitudes during the fall and winter.
2. Wind-driven divergent surface currents. They generate a vertical circulation, which, beneath the zone of divergence, forces the water to rise in order to maintain continuity. The nutrient-rich zones of upwelling off the coasts of California, Peru, Northwest and Southwest Africa (cf. Fig. 10.32) are, in part, the result of such processes. The same is true for the Antarctic Divergence (cf. Fig. 10.46), and the Equatorial Divergences (cf. Fig. 10.34).
3. The inclination, caused hydrodynamically, of seawater stratification in all ocean currents. This allows—in the Northern Hemisphere on the left-hand flank and in the Southern Hemisphere on the right-hand flank of these currents—water rich in nutrients to be brought up from great depths to near-surface layers, as observed on the left-hand side of the Gulf Stream (cf. Fig. 10.42) or on the right-hand side of the Benguela Current (cf. Fig. 10.32). The same effect contributes to the intensification of upwelling in the four coastal areas mentioned under process 2.

From these relationships it becomes evident why—in contrast to conditions on the continents—tropical and subtropical areas, with the exception of those coinciding with the equatorial and coastal divergences, are scantily inhabited by living organisms, whereas polar and subpolar waters, as far as they are not covered by pack ice, are characterized by a great abundance of floral and faunal organisms. For the same reason it is just the largest animals, the whales, that find their habitat in the rich subpolar waters.

6.2.4. Organic production of the ocean

The magnitude of the biological performance of the various oceanic regions, namely their organic production, is particularly interesting to scientists and economists. The "production rate" is defined as the amount of organic substance produced per unit time. A year or a day is usually chosen as the time unit. Some difficulties arise from the notion "organic substance." Since we are not dealing with a uniform substance but with a multitude of components differing in matter as well as in energy (even the thermal values of the three chief nutrients show great differences: protein = 4.4 kcal g^{-1}, fat = 9.4 kcal g^{-1}, carbohydrates = 4.2 kcal g^{-1}), the clear definition of the production rate must be based on energy. Therefore the production rate equals that portion of the energy emitted by the sun and absorbed by the ocean in the course of 1 yr which is not at once converted into heat or kinetic energy but which takes a detour via the reduction of carbon dioxide to organic substance.

The schematic pattern of the life cycle, consisting of assimilation and respiration, has already been mentioned. The intensity of this cycle is identical with the production quantity. Directly, however, it is not related at all to the amount of organic substance present at a particular time. This amount is only the visible manifestation of the equilibrium conditions between production and decomposition of living matter. Therefore, we must distinguish among:

1. the amount of organic substance just present at a particular time (standing crop);
2. the increase of organic substance available as a final result after a certain period of time, the "effective production rate"; and
3. the total turnover, the "absolute production rate."

It is possible that the absolute production rate may considerably exceed the effective production rate, since, under unfavorable living conditions or during the night, part of the synthesized organic substance is subsequently consumed by respiration, and another part of this substance is immediately excreted to the surrounding seawater. The following consideration may serve to explain the principles ruling here: Let P be the production of organic substance per unit time and Z the corresponding loss due to processes of respiration and excretion. Futhermore, let the amount of substance decomposed per unit time be proportional to the amount of organic substance M present at a given time. Thus, we have $Z = kM$, where k is the coefficient of decomposition. Under equilibrium conditions, production and decomposition must just be in balance. Therefore $P = Z$ or $P = kM$, $M = P/k$. Hence, the available amount of organic substance depends likewise on the production rate as well as on the decomposition coefficient. Under equal conditions of production, thus, the amount of organic substance is inversely proportional to the decomposition coefficient that is largely dependent on temperature. For a temperature increase of 10°C, it increases by a factor of 2 to 3. If, on the average, the temperature

Table 6.07. Food Chain on the Continent and in Water

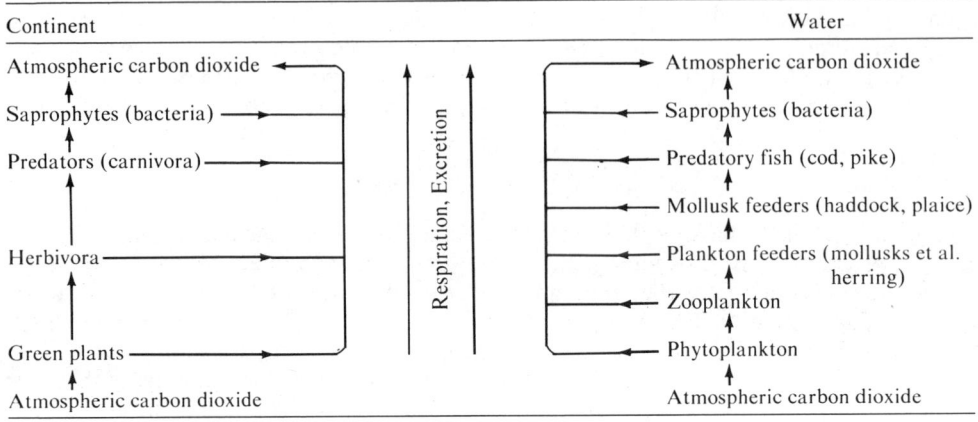

difference between the surface layers of tropical and polar regions is assumed to be 25°C, the decomposition coefficient can be expected to be ten times larger in tropical regions than in polar areas. Accordingly, we have for polar waters $M = P/1$, for tropical waters $M = P/10$; that is, under equal production conditions, we will find ten times more organic substance and, thus, living organisms in polar regions than in tropical oceanic areas. Therefore, if conclusions concerning the rate of production in one oceanic area are to be drawn from the amount of living substance in another area, utmost care will have to be taken.

The highly schematized cycle, used so far, permits only a rough insight into the real conditions. Life as a whole represents an extremely complex phenomenon combining many separate but closely interwoven life cycles to a large entity. We shall attempt to get a somewhat better understanding of the mechanism by demonstrating two examples of the production cycle, one from the continent and one from the ocean. For this purpose the concept of the *food chain* (cf. Table 6.07), commonly employed in biology, will be used here. This term implies that in the various groups of organisms, listed in this table, each higher group depends on the next lower one as the source of food. Beginning with the floral organisms which, as the producers, assimilate carbon dioxide, the organic substance is carried upward from one animal group to the next higher one, suffering considerable losses of primary organic substance at each transition as the result of respiration and excretion processes. The organic substance left after the passage through the food chain is, at last, attacked by bacteria and, thus, it returns to the great primary reservoir of atmospheric carbon dioxide from where the cycle starts once more.

The essential characteristic of the processes within the food chain is the fact that at each transition from one link of the chain to the next, the primary carbon is split into two parts, namely one part that is consumed and from which the next higher organisms benefit and another part that is used for respiration and excretion and thus returned to the reservoir of atmospheric carbon dioxide. In Fig. 6.05 the temporal course of these processes is represented in a diagram based on oceanographic measurements taken on Georges Bank for a period of one year in 1939–1940 (Riley, 1947). The organic substance is given here in the uniform unit "mg C cm^{-2}." For the sake of simplicity, the organisms are just roughly classified as phytoplankton, zooplankton, and animals of higher order. Each black band of organisms is bordered by two lines, on the left-hand side by the *growth curve* and

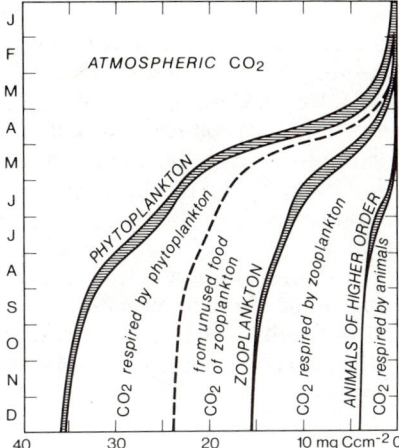

Fig. 6.05. Course of production at Georges Bank within one year. (After Kalle, 1950.)

on the right-side by the *decomposition curve*. The correlation of growth and decomposition conditions yields the amount of organisms present at a particular time in the course of the year. It is especially striking how narrow the "bands of organisms" are that extend across the diagram. It is characteristic of phytoplankton that two seasonal maxima are formed, one in mid-April and the other at the beginning of August, whereas the maximum for zooplankton occurs around the middle of May, following the first maximum of phytoplankton with a delay of 4 weeks. The great loss due to respiration and insufficient utilization of nutrients in the course of the year is also very conspicuous. The zooplankton organisms consume much more phytoplankton food than they need and return the surplus by excretion as detritus into the seawater. The difference between "absolute production" and the "effective production" is also shown very clearly in this diagram. The *absolute production* corresponds to the advance of the growth curve of phytoplankton within one year, in our case it is about 35 mg C cm^{-2}. The *effective production*, however, represents only that portion of carbon which is left when the individual loss by respiration and excretion is subtracted from the absolute production. It equals that portion of organic substance offered to the next higher group in the form of usable food. In our case, the effective production of phytoplankton, thus, amounts to around 24 mg C cm^{-2}.

It is characteristic of the metabolism in water that processes of growth and decomposition closely follow each other in such a way that the development of a large stock of living matter is not possible at all. This approximately corresponds to the concept of pasture management applied in agriculture. In contrast to the otherwise common form of storage management, here the two basic processes, that is, growth of the pasture and usage by livestock, are closely and continuously coupled with each other. For water, this procedure has the advantage that, because of the steady turnover of nutrients during the growth and the decomposition of individual organisms, a relatively high amount of "effective production" can be achieved in the course of a year even though the supply of nutrients may be small. The term *turnover* describes the phenomenon that, due to the quick growth and decomposition of individual plankton organisms, the nutrients are used by several consumers during one period of vegetation. Thus, the nutrient cycle is multiplied and the rate of production is considerably increased.

In the past decades the production problem has been studied thoroughly by oceanographers. On the whole, the available data are still rather inhomogeneous, and most of the figures represent minimum values, since, in former years, the internal mechanism of the production process was not yet clearly understood and, above all, the phenomenon of the "turnover," so important for the development of plankton, was not taken into account. This also explains why, with the production rate in the ocean lying at a level certainly comparable to that on the continent, the amount of organic substance, present at the time being and referred to unit area (standing crop), never exceeds 1% of that produced on good pasture ground (open ocean: 1 to 2 g C m^{-2}, coastal areas: 2 to 10 g C m^{-2} when plankton is blooming, and 0.5 to 1 g C m^{-2} in winter).

A big step forward in the studies on production has been achieved by Steemann Nielsen (1952, 1954) who, during the cruise of the *Galathea* round the world, determined production rates in tropical and subtropical waters by applying the radioactive ^{14}C method. With regard to production conditions, Steemann Nielsen distinguishes among four different oceanic regions:

1. Areas with high admixture of "new deep water" rising from depths beneath the zone of photosynthesis (0 to 200 m) and rich in nutrients. Rate of organic production: 0.5 to 3 g C m^{-2} day^{-1}. Example: upwelling zone off the African coast.
2. Areas with moderate admixture of "new deep water." Rate of organic production: 0.2 to 0.5 g C m^{-2} day^{-1}. Example: outer margins of upwelling areas and zones of divergence in the tropics and subtropics.
3. Areas with little admixture of "new deep water." Rate of organic production: 0.1 to 0.2 g C m^{-2} day^{-1}. Example: most of the tropical and subtropical oceanic areas.
4. Areas with minute admixture of "new deep water." Rate of organic production: less than 0.1 g C m^{-2} day^{-1}. Example: Areas of subtropical convergence. Measured value in the Sargasso Sea: 0.05 g C m^{-2} day^{-1}.

Since in tropical and subtropical waters the annual variations of light intensity and nutrient supply can be assumed to be very small, these values can be used to calculate the annual production by simple multiplication. In general, it can be stated that the deeper the photosynthetic layer extends, the smaller the production is if referred to the same unit area. The four *Galathea* stations from different regions, listed in Table 6.08, give us some idea of the great differences in the production rates in the ocean.

Table 6.08. Production Rates According to Selected *Galathea* Stations

Station Number	Date day month year	Position	Thickness of Photosynthetic Layer (m)	g C m^{-3}	Production per Day Ratio	g C m^{-2}	Ratio
1. 139	24/12/50	Walvis Bay	0.8	6.6	14,000	3.8	79
2. 693	20/ 3/52	4°23'N 164°44'W	81	0.0052	11	0.26	5.4
3. 691	17/ 3/52	6°31'S 169°42'W	113	0.0016	3.3	0.14	2.9
4. 764	7/ 6/52	31°35'N 43°31'W	121	0.00048	1	0.048	1

Since, as already mentioned, the production process in the ocean differs very much from that on land, where, in general, annual yields are regarded, any attempt to compare the two areas with respect to their production rates must remain inadequate. From the economic viewpoint, the comparison between the yields of fish catches and meat production, provided by commercial fisheries and agriculture, is justified somewhat better. Here the following figures result: cattle breeding yields 100 to 1000 kg of meat per hectar and yr, and fresh-water fish hatching yields 25 to 400 kg of fish per hectar and yr, whereas the respective catches of sea fish only amount to 3 to 30 kg. It is not surprising that the oceanic yield compares rather unfavorably. On the one hand, only part of the fish present in the ocean is caught (an estimated 50%) and, on the other hand, in agriculture and fresh-water fisheries—in contrast to fishing in the ocean—cattle breeding and fresh-water fish hatching are attentively cared for. The figures for the ocean become even more unfavorable if the comparison is restricted to the amounts of oil and fat won per unit area. Whereas in agriculture, fat yields amounting to 100 to 1000 kg hectar^{-1} yr^{-1} can be expected from vegetable cultivation and cattle breeding (the oil palm being an exception with 3000 kg hectar^{-1} yr^{-1}), the respective value from catches of herring (a proper fat fish) lies below 1 kg hectar^{-1} yr^{-1}. However, the small yield from the ocean per unit area is more than compensated by the vastness of the available fishing grounds.

How small the values for production and yield are in fact is revealed by a comparison with the amounts of energy received from the sun at the earth's surface:

Total radiative energy of solar radiation at the earth's surface per year	557 $\times 10^{21}$ cal
Total energy available in coal and crude oil	50 $\times 10^{21}$ cal
Annual energy amount of photosynthesis	0.36 $\times 10^{21}$ cal
Energy released by coal and oil combustion per year	0.014 $\times 10^{21}$ cal

The total annual fishing yield from the North Sea, in terms of energy, corresponds to only 0.004‰ of the total radiation or to the amount of radiation received by the North Sea, under clear-sky conditions, within 16 sec at noon on June 21.

6.2.5. The mechanism of plankton metabolism

Before dealing with the natural laws controlling the life process in the ocean, let us consider the chemical composition of plankton organisms (cf. Fig. 6.13). As shown in Table 6.09, life requires a number of other chemical elements in addition to carbon in order to be able to carry out the chemical conversions within the cells that are essential for life and, in part, of rather a complicated nature. So far, however, our knowledge is restricted

Table 6.09. Chemical Composition of the Copepod *Calanus finmarchicus*

Element	Content (%)	Enrichment Factor	Element	Content (%)	Enrichment Factor
O	79.99		P	0.13	43,000
H	10.21		Ca	0.04	1
C	6.10	2,100	Mg	0.03	0.2
N	1.52	50,000	Fe	0.007	1,400
Cl	1.05	0.5	Si	0.007	140
Na	0.54	0.5	Br	0.0009	0.14
K	0.29	7.4	I	0.0002	40
S	0.14	1.6			

Table 6.10. Relative Composition of Plankton Organisms

Element	Relative Content (C = 100)			Enrichment Factor (C = 2100)		
	Diatom	Peridinium	Copepod	Diatom	Peridinium	Copepod
C	100	100	100	2,100	2,100	2,100
N	18.2	13.8	25.0	34,000	28,000	50,000
P	2.7	1.7	2.1	55,000	35,000	43,000
Fe	9.6	3.4	0.11	120,000	42,000	1,400
Ca	12.5	2.7	0.66	18	4	1
Si	93.0	6.6	0.11	110,000	8,000	130

to very few conditions of the natural laws dominant in these processes and of their mutual interaction. Of course, chemical analyses of plankton organisms can give only a very rough picture of the importance of the various chemical elements for life processes because the order of magnitude of their values differs considerably for the various species. Even the ratio of dry weight to algae weight varies greatly and may lie between 20 and 50%, depending on whether we deal with organisms with a small portion of ashes or, for example, with diatoms with their strong silicious scales. But the data obtained by analyses also vary considerably if based on ash-free dry weight for evidence, as demonstrated by the following figures: carbon 46 to 56%, hydrogen 7 to 12%, oxygen 22 to 38%, nitrogen 3 to 12%, phosphorus 0.4 to 3.8%. More detailed information is given in Tables 6.09 and 6.10. Apart from the values of the contents, the tables also show the capability of enrichment that living cells have. It is not only very pronounced but also very different according to the species as illustrated by a comparison of diatoms, peridinia, and copepods. With diatoms, it is not so much their large silicon content that is conspicuous but rather their large iron content. According to Harvey (1937), they absorb iron at the surface of their bodies in the form of ferric hydroxide. Only fractions of this amount are actually needed for their growth. Harvey has found that the diatom *Nitzschia* requires iron only as much as $1/175$ of the amount of phosphorus needed at the same time.

In fact, the iron in the ocean is a very peculiar element. It is not only its absorption by organisms that is quite cryptic, but the same can be said about its solubility in seawater. Iron can pass into seawater in true solution only to a very small degree. Under normal conditions at the sea surface, its solubility is less than 10^{-7} mg Fe m^{-3}. Hence, nearly all the iron present in seawater (2 to 50 mg Fe m^{-3}) must be contained in colloidal form as ferric hydroxide, or in complex chemical linkage with organic substances or with fluorides.

The chief metabolites, which build up plankton cells, are, of course, carbohydrates, lipids, and proteins. The ratio of their relative quantities can vary greatly. The carbohydrates of phytoplankton comprise three main components: (1) the building material for the scale, (2) the dissolved carbohydrates in the cell fluid, and (3) polymeric carbohydrates, serving as a reserve for the energy budget. Glucose and galactose prevail by far among the chemical compounds. Small fatty globules of microscopic size are formed in the phytoplankton cells, especially if nitrogen is lacking, and here, too, they represent a reserve product. The proteins in phytoplankton contain a great number of various vital amino acids. Their composition differs only a little from that of zooplankton and fish. Most plant cells contain several percent of free amino acids and of polypeptides.

Other important compounds, found in phytoplankton, are nucleic acids and related compounds. Healthy cells of marine diatoms, like chlorophyceae and myxophyceae, contain 5 to 6% nucleic acids. Evidence of several percent of ascorbic acid (vitamin C) has been found also in plankton algae, while vitamin B_{12} occurs in phytoplankton cells in the order of magnitude of 10^{-8} g g^{-1} (10 molecules of vitamin B$_{12}$ per μm^3), that is, in the same order of magnitude as has been proven also in bacteria.

In the metabolism of phytoplankton, cell pigments play a particularly important role (Strickland, 1965). Four different classes of pigment can be distinguished, which do not only serve as photosensitizers (light traps) but cooperate in photosynthesis in a much more complex way. The following substances are involved: the chlorophylls a, b, and c; the carotines α, β, γ, and ϵ, and a great number of xanthophylls as well as biliproteins such as phycoerythrine and phycocyanine. The latter two, however, occur only in myxophyceae and cryptophyceae. The pigment concentration in algae differs greatly and varies very much, depending on the state of health of the cells. Referred to dry-weight basis, the following figures can be assumed:

1. Chlorophyll a: 0.2 to 1% (in chlorophyceae up to 2% at the maximum);
2. chlorophylls b and c: less than chlorophyll a, varying very much;
3. carotines: rarely more than $2/3$ of the chlorophyll concentration;
4. xanthophyll: a multiple of the chlorophyll concentration;
5. biliproteins: several parts per thousand.

Besides having the capability, common to all organisms, of accumulating nutrients by a factor of several thousand, quite a number of "specialists" are able to assimilate and to accumulate in their bodies, by means of some specific faculty, elements occurring in seawater in great dilution. This is true for the following elements: Ca, Si, Sr, Ba, Fe, Mn, S, Cu, V, Zn, Sn, Pb, I, Mo, Co, Ag. With a view to geological time periods, this capability of selective enrichment that organisms have can be of great importance. A large number of mineral deposits, exploited for economic reasons, like coal, petroleum, chalk, kieselguhr, as well as many deposits of phosphate, iron, and copper, are of biological origin. The same can be said about the extraction of iodine from the ash of seaweeds. From the viewpoint of geochemistry one may even say that this accumulation of substances represents the only lasting achievement of the terrestrial life process.

In moderate and subpolar zones a pronounced annual rhythm in the development of marine organisms and, thus, in the content of nutrients in seawater, corresponds to the seasonal variation of climatic factors. Around the end of the winter season, the living substance in the ocean reaches its minimum. In spring when, with increasing incoming radiation, a stable water stratification is formed, phytoplankton begins to bloom exuberantly, starting with the diatoms which prefer cold water. Parallel to this, the consumption of nutrients present in seawater is strongly increased, for example, of phosphates, nitrogen compounds, and silicate. Figure 6.06 shows the characteristic seasonal changes in the water chemistry occurring during that time, on the basis of the representation of the hydrogen ion concentration, of the carbon dioxide factors, and of the nutrients in the English Channel. When comparing the various curves, we very easily notice the rapid decrease of the individual nutrients, which indicates the beginning of the spring blooming from the first days of February through March. Another typical phenomenon is the summer minimum of the various nutrients, when the content of nitrogen or phosphorus often reaches a concentration of zero. The characteristic phenomenon of plankton bloom twice a year can clearly be recognized through the minima of the silicate curve in May and September. It is less obviously reflected by the phosphate curve because, here, the interplay of the nutrient factors is more complicated.

In addition to inorganic phosphate, phosphorus, organically dissolved, plays rather

278   Chemical Budget of the Ocean

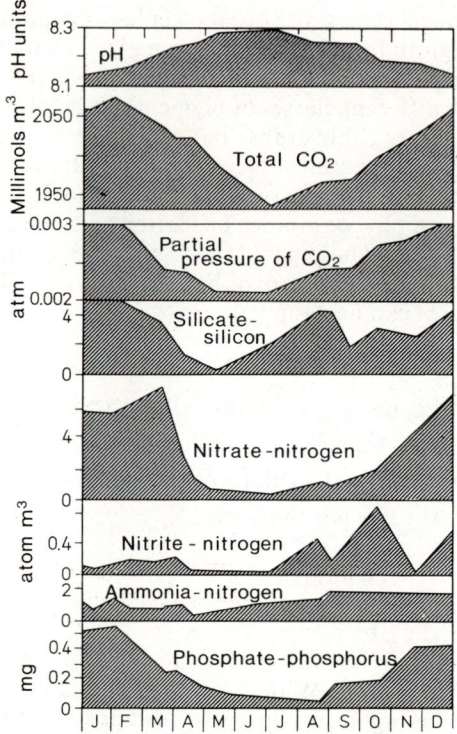

Fig. 6.06. Annual variation of nutrients at the English reference station E1 at the western exit of the English Channel at 0–25 m depth in 1931.

an important role in the development of phytoplankton. Neither for phosphorus nor for nitrogen are the assimilation processes well understood at present. Under favorable conditions, phytoplankton algae can accumulate more than the tenfold amount of phosphorus needed at the moment. It must be assumed that similar conditions are valid for nitrogen compounds. Absorption of nitrogen and cell division need not necessarily always be strictly parallel. This explains why photosynthesis and cell division will continue even when all the nitrogen available in the solution has been consumed. Visible symptoms of deficiency will appear only somewhat later. So algae, suffering from lack of phosphate, first show a relatively high content of carbohydrates and pigments until definite processes of decay become dominant, which can primarily be recognized by the increased amount of excretions into the water. The metabolism of chlorophyll a is closely connected with the nitrate metabolism. As soon as the nitrogen compounds have been consumed, the formation of chlorophyll is stopped, in contrast to the carotenoid synthesis which goes on as long as sufficient phosphate is available. As a result, the ratio of carotenoid to chlorophyll is increased. Conversely, the increasing ratio can be taken as an indicator of the growing lack of nitrogen compounds.

Conditions regarding nitrogen compounds are even more complicated than those regarding phosphorus. When organisms decay, at first amino acids are released to the water. Ammonia is formed from this, which, by and by, is oxidized first to nitrite and then to nitrate. Nitrate represents the end product of the decomposition chain of nitrogen and, thus, like phosphate, it is accumulated in especially abundant amounts in the deep sea, from where it is again brought up to the sea surface by vertical convection in polar and subpolar regions.

On a smaller scale, the process of accumulation of nutrients at the sea floor even occurs in shallow seas and, in particular, in such areas where a two-layer stratification of seawater develops in summer as the result of the heating and stabilization of the surface layer. Not until the cooling in fall and winter, which initiates vertical convection of the water masses, is the surface layer once more enriched with nutrients. So in the next spring, plankton can develop again when solar radiation increases.

Harvey (1927) was the first to point out the peculiar phenomenon that the ratio of phosphate–phosphorus to nitrate–nitrogen is nearly the same for different water masses as well as for plankton. Numerous investigations, however, have proven more recently that this rule is valid only in approximation. Calculated on atomic basis, the N/P ratio varies between 5.5 and 15 for phytoplankton, while in seawater extremes of 5, or as much as 19 have been found. The following relation, first established by Fleming (1940) and supplemented later on, is now used in oceanography as a standard rule:

$$O_2 \text{ deficit}: C \text{ (plankton)}: N: P = 212:106:16:1 \text{ atoms}$$
$$= 109:41:7.2:1 \text{ g}$$

This relation, however, cannot be applied to calcium and silicon, because the conditions of assimilation of elements needed for the formation of scales differ from those of elements necessary for the production of the protoplasm of organisms. Regarding the ratio for the transformation of oxygen into carbon dioxide, it is found that the above atomic ratio of O_2 deficit to $C = 2:1$ can be correct only as an order of magnitude approximation. In the ocean, conditions regarding the oxidation of organic substance must be quite similar to those which, in physiology, govern the respiratory metabolism. In the latter case we speak of the "respiratory coefficient" (RQ) which gives the ratio between the released carbon dioxide and the consumed oxygen by volume or by moles, respectively. With respect to the ocean, according to Schulz (1923), we speak of the "oxygen transformation factor." These values vary, depending on the composition of the organic substance. For pure carbohydrates $RQ = 1$, for lipids $RQ = 0.71$, and for protein $RQ = 0.80$, approximately. As the average value of his investigations in the Baltic Sea, Buch (1949) has found the oxygen transformation factor to amount to 0.83. Therefore, it would be more correct to write the above relation as follows:

$$O_2 \text{ deficit}: C = 256:106 \text{ atoms} = 132:41 \text{ g}$$

Here, the attempt has been made to determine some quantitative relations for plankton and plankton metabolism first in the inorganic field; but it should be understood that, with the many aspects of the life process and its adaptability in nature, it is not possible, in general, to extend such a procedure to the organic field. The table by Brandt and Raben (1919–1922) may elucidate this (Table 6.11).

Table 6.11. Composition of Dry Plankton

Main Component of Catch	Protein (%)	Fat (%)	Carbohydrates (%)	Pure ash (%)	P_2O_5 (%)	N (%)
Copepods	70.9–77.0	4.6–19.2	0–4.4	4.2–6.4	0.9–2.6	11.1–12.0
Diatoms	24.0–48.1	2.0–10.4	0–30.7	30.4–59.0	0.9–3.7	3.8–7.5
Peridinium	40.9–66.2	2.4–6.0	5.9–36.1	12.2–26.5	0.7–2.9	6.4–10.3
Sagitta	69.6	1.9	13.9	16.3	3.6	10.9

Analytical investigations of this kind are difficult to make, because, first, the portions of the individual organisms and, second, also the number of living organisms vary very much with each catch. When studying plankton in the Kiel Bight, Gillbricht (1952) came to the conclusion that, in nature, the distinction between plankton and detritus is, by far, not so easy as it would seem from the definition. Under certain circumstances, a large portion of filter catches may consist of dead plankton organisms in various stages of decomposition, widely ranging between pure plankton and inorganic detritus. Although mostly on the basis of indirect conclusions, Gillbricht has found that, on the annual average, the portion of living plankton amounts only to 4.2% of the total seston. Merely 2% of the detectable chlorophyll is associated with living plankton. Referred to dry weight, the chlorophyll content of diatoms amounts to 6.8%, that of peridinium to 3.4%. According to calculations, the production of solid substance is supposed to be 37 g C m^{-2} yr^{-1}. In addition, the diatoms apparently release rather considerable amounts (21 g C m^{-2} yr^{-1}) of dissolved organic substance into seawater.

The above example shows that also here the absolute production is very much higher than the effective production, in particular because the amount of organic substance consumed during the respiration process must also be taken into account. According to recent investigations, this amount can be assumed to run up to 5 to 10% of the assimilation process, provided that this takes place under optimum conditions. It may be left out of consideration, however, that about 50 to 70% of the respiration carbon dioxide is immediately re-assimilated. The excretive substances cannot be treated all in the same way; for among them there are substances that may be useful as nutrients for plankton, perhaps in the form of marine snow. On the other hand, it has been found that certain organisms release substances into seawater which do not only impair the growth and cell division of other organisms, but also stop their own further development. In fact, as far as we know, the process of cell division seems to be a specific and very delicate process, depending on quite a number of conditions in the surrounding seawater. Cell division and photosynthesis are absolutely independent processes. Both of them have their characteristic temperature coefficients, and they also differ in their dependence on light conditions. Cell division seems to require a critical radiation intensity, light with a wavelength of 6500 Å being the most suitable one. The rate of cell division varies a great deal. Time periods from 2 to 48 hr have been observed with phytoplankton organisms. Absolutely necessary requirements for cell division are: N, Ca, Fe, Mn, Mo, and vitamin B_{12}.

A particularly extreme form of plankton development is the mass bloom of special plankton species, most dinoflagellates, occurring in some regions of the tropics and subtropics and known under the name "red tide." Probably it is the coincidence of particularly favorable circumstances like light, water stratification, and possibly also the presence of vitamin-like growth factors, resulting in the mass development of one single species of plankton within a few days, which, however, will last only for a couple of days.

6.2.6. The carbon dioxide–calcium carbonate system

In the preceding sections, the assimilation of carbon has been described with special regard to production conditions in the ocean. The assimilation formula, mentioned above, is important not only with respect to the actual production of living matter but also to the carbon dioxide–calcium carbonate system, which represents a large chemical equilibrium system. Beside the carbon dioxide dissolved in seawater, this system includes calcium carbonate, essential for sedimentation, and also oxygen contained in seawater in its gaseous phase (and, somewhat related to and dependent on it, hydrogen sulfide). In order to be able to understand this system and its internal relationships, let us consider seawater as a solution of salts. A detailed description of the actual relationships has been given by Skirrow (1965).

Seawater differs from any ordinary salt solution, for example, a NaCl solution, in one major respect. Whereas in an ordinary salt solution the amount of cations generally is equivalent to that of anions, this relation does not hold in seawater, at least as far as the ions of strong bases and acids are concerned. There is rather an excess of 2.38 meq of cations per kg of seawater. This phenomenon is called "alkalinity." This alkalinity is responsible for the alkaline behavior of seawater which, under normal conditions, has a hydrogen ion concentration of pH = 8.2. In general, the total variation of pH may range from 7.8 to 8.3. Much more important, however, is a second factor. Since there is a free exchange between atmosphere and ocean, the latter has the opportunity to absorb as much atmospheric carbon dioxide as permitted by its alkalinity. The amount of carbon dioxide, required herefor, is considerably greater (100 to 200 times) than the amount of carbon dioxide dissolved physically in seawater. From this point of view, the ocean can be considered a bicarbonate solution that is in equilibrium with the atmospheric carbon dioxide. The most significant property of such a solution is its great buffering capacity towards additional amounts of acids and bases. The three fundamental variables of this system are (1) the alkalinity, (2) the hydrogen ion concentration, and (3) the carbon dioxide pressure, which, in turn, is a function of the free carbon dioxide as well as of temperature and salinity. Since these three basic quantities are functionally related to each other, the system is already fixed when two of these variables have been determined. In oceanographic measuring technique, the observation of alkalinity and hydrogen ion concentration has been chosen for reasons of simplicity. Figure 6.07 permits an insight into the percentage displacement of the three equilibrium components, that is, free carbon dioxide, bicarbonate ions, and carbonate ions as functions of pH, for seawater as well as for pure water. Whereas, in a strongly acidic medium below pH = 4, there is practically only free carbon dioxide, the bicarbonate component in seawater reaches its maximum at pH = 7.5. With further alkalization, carbonate ions become predominant. In the pH range from 7 to 8.5, which is normal for seawater, it so happens that all three kinds of linkage of carbon dioxide are present in measurable quantities.

For the solubility of calcium carbonate in seawater it is of fundamental importance that the CO_2 equilibrium is established. This question is most closely connected with the processes of sedimentation. When seen from the physicochemical point of view, the ocean

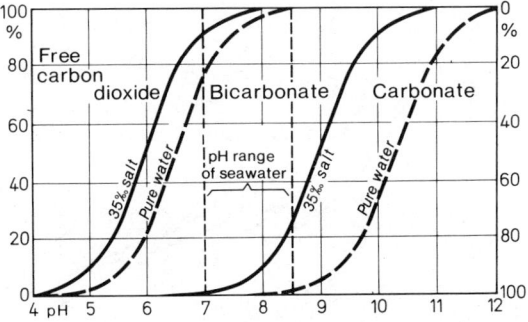

Fig. 6.07. Percentage distribution of the three kinds of bond of carbon dioxide (free carbon dioxide, bicarbonate, carbonate) in pure water and in seawater as a function of pH. (After Buch, from Wattenberg, 1943.)

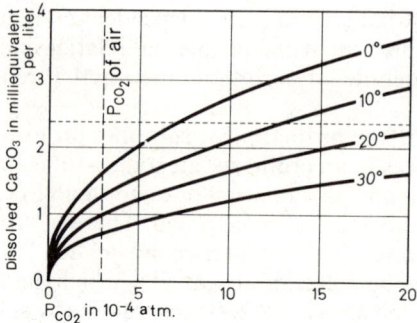

Fig. 6.08. Solubility of CaCO₃ in seawater (19‰ Cl) as a function of CO₂ pressure. (After Wattenberg and Timmermann, 1936.)

can be considered a solution in equilibrium with calcium carbonate as bottom substance. This system, too, can be treated numerically on the basis of the law of mass action. As the final result, a small decrease in the solubility of calcium carbonate is obtained for seawater as compared with pure water. Figures 6.08 and 6.09 clearly show the solubility of calcium carbonate in seawater under natural conditions. The consequence is rather an odd fact: seawater which, at the sea surface, is in equilibrium with atmospheric carbon dioxide ($P_{CO_2} = 3 \times 10^{-4}$ atm, pH = about 8.2, and alkalinity = 2.38 meq) is supersaturated with respect to calcium carbonate, and this all the more, the warmer the water is, with a maximum of up to 300% (Fig. 6.10). This, however, is not valid for deep water, which generally has a considerably higher carbon dioxide content.

The supersaturation with calcium can be understood if the fact is taken into account that calcium, when in solution, shows a strong tendency towards supersaturation. This behavior of seawater is of special biological and geological significance. The marine organisms that depend on calcium will naturally be able to satisfy their specific needs in seawater all the better and the more easily the more seawater is supersaturated with calcium carbonate. This is the reason that the typical calcareous organisms of the ocean, the corals and coccolithophoridae, are found mainly in tropical regions, whereas the cold

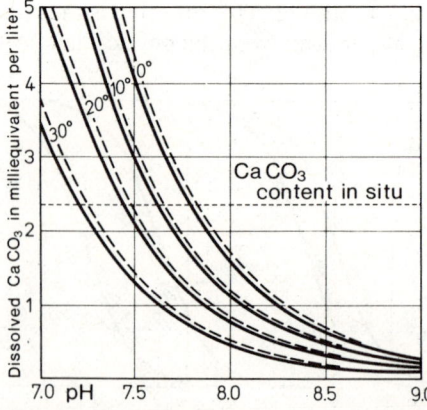

Fig. 6.09. Solubility of calcite (——) and aragonite (- - -) in seawater (19‰ Cl) as a function of pH. (After Wattenberg and Timmermann, 1936.)

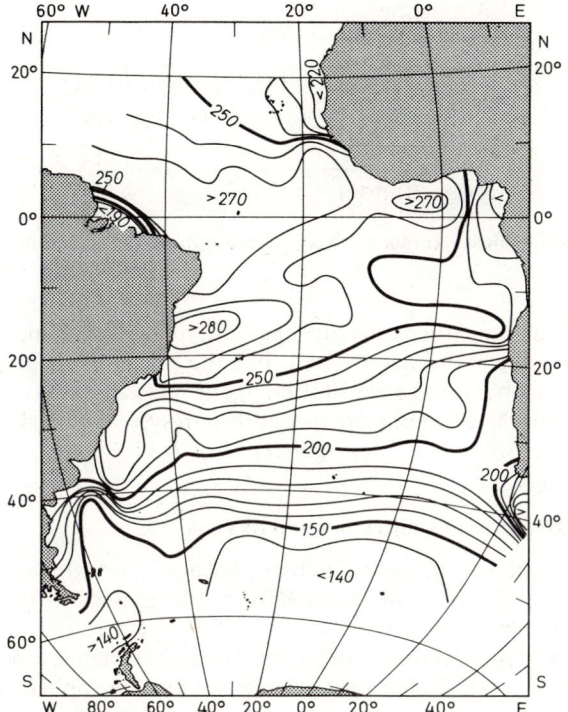

Fig. 6.10. Percentage of saturation of surface water with $CaCO_3$, based on constant CO_2 pressure of 3×10^{-4} atm. (According to Wattenberg, 1933.)

polar regions are mostly populated by diatoms, the scales of which consist of silicic acid. In the deep sea, a contrary phenomenon can be observed. The hydrostatic pressure that increases with depth becomes manifest by increased dissociation of carbon dioxide; thus, the hydrogen ion concentration is shifted toward the acid state. Consequently, the calcium solubility increases with depth. This explains why the calcium content of the sediments usually decreases with depth and why, at depths below 4000 m, the bottom sediments are nearly free of calcium.

Finally, as the last component of this system, the interplay between oxygen and hydrogen sulfide will be discussed. In enclosed oceanic basins, the oxygen may be completely consumed under certain conditions, especially when lateral advection of new, well-ventilated water is hindered by the bottom topography, and replenishment from above is impeded by the formation of a pronounced, stable discontinuity layer. The larger the supply of organic substance sinking from surface layers, the sooner this state will be reached. Under the anaerobic conditions which then develop, bacteria will appear that use the oxygen bonded in the sulfates for the oxydation of their organic food substances. At the same time, gaseous hydrogen sulfide is set free and dissolved in seawater. The same process takes place in the sediments, with the difference that hydrogen sulfide combines with heavy metals, particularly with iron, to form sulfides, which can be recognized by their dark color. Normal organic flora and fauna cannot exist in such areas poisoned by hydrogen sulfide. Areas of this kind include, first of all, the Black Sea (Fig. 10.23), several Norwegian fjords with very narrow entrances (Fig. 10.64), and the deep basins of the

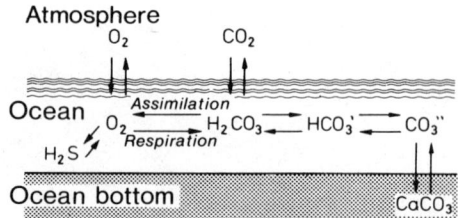

Fig. 6.11. Carbon dioxide–calcium carbonate–oxygen–hydrogen sulfide system in the ocean. (After Kalle, 1945.)

Baltic Sea (Fig. 10.58). In earlier geological periods, more of such marginal seas poisoned by hydrogen sulfide seem to have existed, but the present topography apparently is somewhat more favorable for the ventilation of the marginal seas. A schematic review of the interrelations in the great chemical equilibrium system of carbon dioxide–calcium carbonate–oxygen–hydrogen sulfide is given in Fig. 6.11.

6.2.7. Regional distribution of nutrients and oxygen

The regional distribution of nutrients such as phosphate, nitrate, and silicate is in good agreement with the circulation of the oceanic water masses in the depth. Generally speaking, the relation has been found to be valid everywhere in the ocean, which says that oxygen deficit will always correspond to an excess of nutrients, especially of phosphate and nitrate. In this context, such a relationship can also be applied to silicate, as far as purely qualitative considerations are concerned.

The ventilation of the world ocean is of fundamental importance. Conditions are very peculiar here; they are governed by climate and by the bottom topography of the ocean. There are only three major centers where, in winter, a sufficiently deep-reaching vertical convection takes care of the ventilation of the deep water. All these three areas—the "oxygen valves" of the deep sea, so to speak—are parts of the Atlantic Ocean, one is in the Labrador Sea, one in the Greenland Sea, and the third in the Weddel Sea. They are represented in Plate 4.

In the Labrador Sea, the winter cooling causes an increase of density at the sea surface so that, in the cold season, the convection of water masses can extend down to a depth of 3000 m, but not to the bottom, thereby guaranteeing the supply of oxygen to the deep water every year (Dietrich, 1961). The distribution of the oxygen content on longitudinal sections through the three oceans can be taken from Plates 3 to 5. From the Labrador Sea, water rich in oxygen spreads southward with the North Atlantic Deep Water. Along its path it mixes with water masses, also relatively rich in oxygen, which have left the European Mediterranean Sea across the sill at Gibraltar. It can be traced further down in the South Atlantic Ocean from where the Antarctic water belt finally carries it into the Indian and Pacific Oceans.

The counterpart of the area with vertical convection in the Labrador Sea is the Weddell Sea. In the cold season, large amounts of cold, well-ventilated water sink at the continental margin to the bottom at more than 5000 m depth. On the western side of the Atlantic Ocean, this water spreads northward as Antarctic Bottom Water, flowing beneath the North Atlantic Deep Water. The basins of the Indian and Pacific Oceans are similarly supplied with this highly oxygenated water from the south. Since oxygen-containing water is subject to a constant process of consumption, the oxygen content of

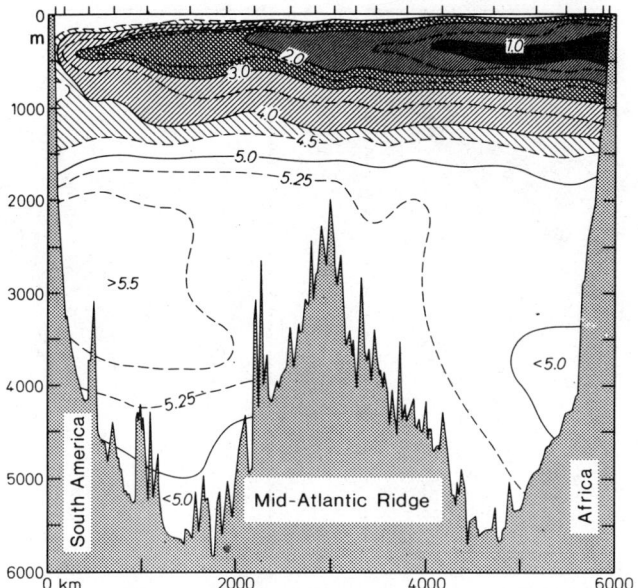

Fig. 6.12. Distribution of oxygen (ml/l) on a cross section through the South Atlantic Ocean at about 9°S obtained by research vessel *Meteor* in August–September 1926. (After Wattenberg, 1939.) Marks at the upper edge: oceanographic stations. Exaggeration of the profile: 1000-fold.

the water masses, on the whole, steadily decreases along their path from the region of origin into the other parts of the world ocean, as can be clearly seen in the oxygen longitudinal sections on Plates 3, 4, and 5. Along the path of the water masses, however, certain complications are met, since the process of consumption may greatly vary from one region to another, in correspondence with external hydrographic conditions such as temperature, supply of organic substance, and advection of water masses. The intermediate oxygen minima in the tropics and subtropics are particularly conspicuous phenomena, especially in the Atlantic and Pacific Oceans. It is a characteristic of theirs that they are extremely manifest at the eastern margins of those two oceans, whereas they become less pronounced towards the west. A typical example is given in Fig. 6.12. Furthermore, the much stronger development of the minimum areas in the Pacific Ocean, as compared to the Atlantic Ocean (cf. Plates 3 and 4), reflects the poor renewal of those water masses due to weak circulation.

The large-scale distribution of nutrients can be explained by the distribution of oxygen: wherever the oxygen content is high, the nutrient content usually is low, and vice versa. So we see that, as a whole, the deep water of the North Atlantic Ocean is the area of the world ocean which is the poorest in nutrients, while, in contrast, the deep water of the North Pacific Ocean is the richest as far as nutrients are concerned.

There are some adjacent seas where those large-scale extremes may still be surpassed. For example, because of their specific conditions as to water exchange, the upper layers of the European Mediterranean Sea and the Baltic Sea are distinguished by a particularly low content of nutrients. On the other hand, there are coastal areas characterized by an extraordinarily high nutrient content. The aforesaid relationships of mutual dependence between oxygen and nutrients can be applied with the same success also when the areas of oxygen minima in the tropics are considered. The extreme oxygen minima on the

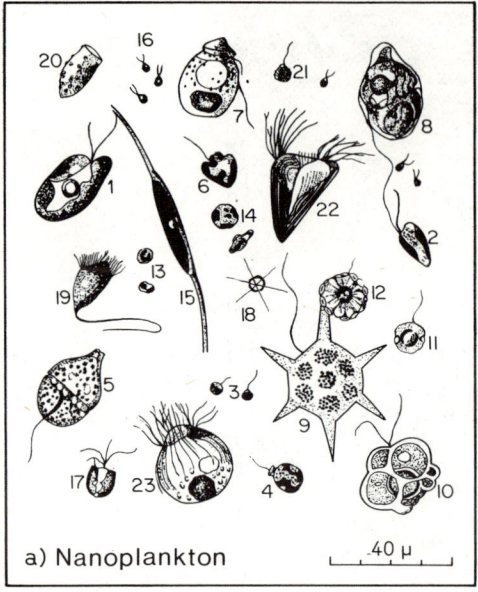

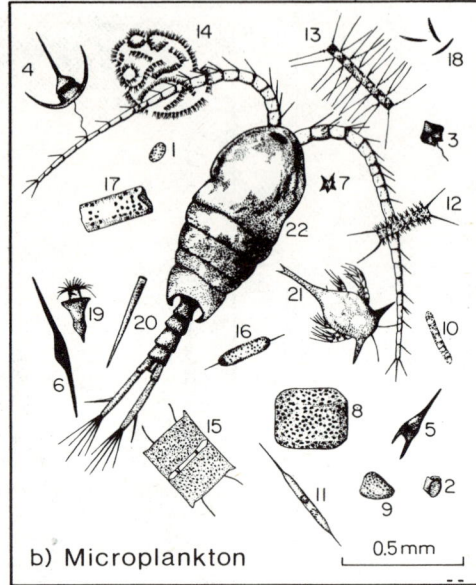

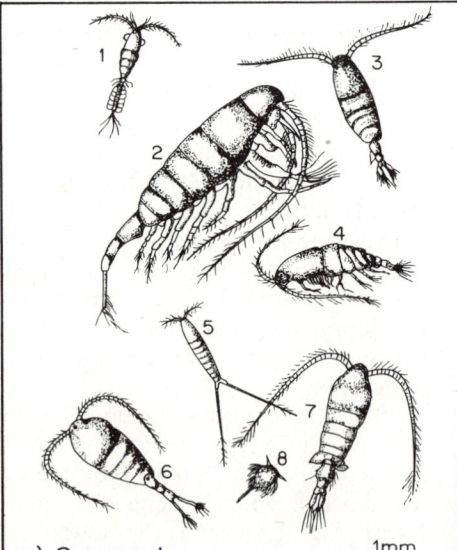

to (a)
    I. Cryptomonadinae:
        1. Rhodomonas baltica
        2. Rhodomonas pelagica
        3. Erythromonas haltericola

    II. Dinoflagellatae:
        4. Exuviella baltica
        5. Glenodinium trochoideum
        6. Amphidinium rotundatum
        7. Amphidinium crassum
        8. Pouchetia parva

    III. Silicoflagellatae:
        9. Distephanus speculum
      10. Ebria tripartita

    IV. Coccolithophoridae:
      11. Pontosphaera Huxleyi
      12. Discosphaera tubifer

    V. Diatomeae:
      13. Thalassiosira nana
      14. Thalassiosira saturni
      15. Nitzschia Closterium

    VI. Volvocales:
      16. Chlamydomonas micropland.
      17. Carteria spec.

    VII. Green unicells
      18. Meringosphaera radians

    VIII. Protozoae:
      19. Strombidium caudatum
      20. Tintinnopsis nana
      21. Calycomonas gracilis
      22. Laboea conica
      23. Mesodinium rubra

to (b)
    I. Dinoflagellatae:
        1. Prorocentrum micans
        2. Dinophysis acuta
        3. Peridinium divergens
        4. Ceratium tripos var. atlantica
        5. Ceratium furca
        6. Ceratium fusus

    II. Silicoflagellatae:
        7. Distephanus speculum

    III. Diatomeae:
        8. Coscinodiscus concinnus
        9. Coscinodiscus Granii
      10. Rhizosolenia faeroensis
      11. Rhizosolenia setigera
      12. Chaetoceras decipiens
      13. Chaetoceras boreale
      14. Chaetoceras sociale
      15 Biddulphia mobiliensis
      16. Ditylium Brightwellii
      17. Guinardia flaccida
      18. Nitzschia Closterium

    IV. Protozoae:
      19. Tintinnopsis campanula
      20. Tintinnopsis subulata

    V. Cirripediae:
      21. Nauplius v. Verruca Strömia

    VI. Copepodae:
      22. Temora longicornis

to (c)
    1. Oithona similis
    2. Limnocalanus grimaldi
    3. Acartia discaudata
    4. Acartia bifilosa
    5. Microsetella norvegica
    6. Temora longicornis
    7. Centropages hamatus
    8. Nauplius v. Verruca Strömia

Fig. 6.13. (a)–(c) Plankton forms of the ocean. (Photo: Institut für Meereskunde, Kiel.)

eastern sides of the oceans are accompanied by particularly large amounts of nutrients in those regions. But detailed investigation concerning the vertical distribution of nutrients in the area of oxygen minima have yielded the result that the nutrient maxima lie several hundred meters beneath the oxygen minima.

Another close relationship with regard to the distribution of chemical factors in the world ocean is reflected in the equilibrium between oxygen and free carbon dioxide. In deep water, corresponding to the decrease in oxygen content, the content of free carbon dioxide may increase considerably, thus causing a corresponding decrease of the pH value. So it may happen that in the Baltic Sea, for example, the value of pH of seawater drops from its normal value of 8.2 right into the acidic range beneath 7.0.

7 The Theory of Ocean Currents

7.1. The System of Hydrodynamic Equations

7.1.1. Equations of motion in the absolute coordinate system

All physical disciplines aim at establishing laws which describe the course of a process in a mathematical form. If such laws have been recognized and formulated as equations, the general solution of these equations is valid for all processes subject to this law. Formulating laws that govern the motion in the ocean is an essential task for oceanographers.

One of the fundamental axioms of mechanics is the law of the conservation of momentum applied here in the form of Newton's basic law. This law states that a force per unit mass causes an acceleration $d\mathbf{V}/dt$ acting in the same direction as the force vector $\mathbf{K}$:

$$\frac{d\mathbf{V}}{dt} = \mathbf{K} \qquad (7.01a)$$

Here, $\mathbf{V}$ is the velocity vector in an absolute coordinate system or inertial system, because the above equation is valid only in such a system. An inertial system is a coordinate system either at rest or moving with constant speed relative to the system in the state of rest. For oceanographic purposes we can imagine a coordinate system, fixed and at rest at the center of the earth, as an inertial system (cf. Fig. 7.01) around which the earth, assumed as a spherical surface, is rotating (coordinates x^a, y^a, z^a).

The forces acting on a fluid particle can be divided into two groups:

1. **Long-range forces.** They act over long distances and penetrate any matter. Due to their long range their decrease with distance is small. In oceanography, the sole force of importance is the gravitational force resulting from the mutual attraction between masses of matter. It can be divided into one part that is due to the attraction of the total mass of the earth on an individual fluid particle, and another part originating from the attracting forces of moon and sun. The former is the gravitational force of the earth's mass, $\mathbf{K}_\phi$; the latter produces the tidal forces $\mathbf{K}_G$ with the components X, Y, and Z. Both forces are proportional to the volume of the fluid particle on which they act.

2. **Short-range forces.** A rigid body is defined by the fact that it maintains its form. Deformable bodies, among them the fluids, change their forms when an external force acts on them. Thus, a state of tension is generated in the fluid. On the one hand, this is caused by the attraction exerted by the molecules of the fluid on adjacent molecules and, on the other hand, by the capability of the individual molecules of a fluid element to move freely, in contrast to the molecules in a rigid body. So some molecules may leave the volume element in question which, consequently, loses their momentum, their energy etc.; other

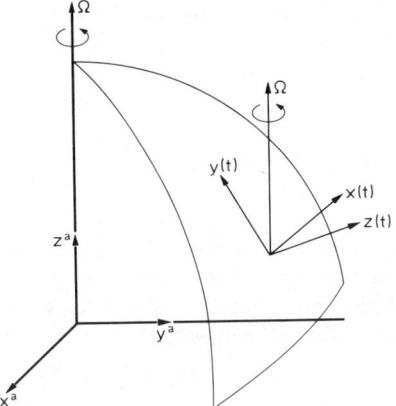

Fig. 7.01. Absolute coordinate system x^a, y^a, z^a (inertial system) and relative system $x(t)$, $y(t)$, $z(t)$ rotating with the earth.

molecules will enter it. From the state of tension, generated in this way, there result the pressure gradient force $-(1/\rho)$ grad p and the frictional force **R**.

So the equation of motion in the inertial system reads as follows:

$$\frac{d\mathbf{V}}{dt} = -\frac{1}{\rho} \operatorname{grad} p + \mathbf{K}_\phi + \mathbf{K}_G + \mathbf{R} \qquad (7.01\mathrm{b})$$

where ρ denotes the density and p the pressure. Equation (7.01b) is called the Navier–Stokes equation of motion. If the frictional force is neglected, this equation takes the form of the Eulerian equation of motion.

7.1.2. The deflecting force of the rotation of the earth

As observers we are accustomed to looking at any motion as relative to a fixed point on the surface of the earth. Due to the rotation of the earth, however, such a point is moving around the earth's axis at the angular velocity of the earth's rotation. In Fig. 7.01 this coordinate system rotating with the earth is marked by the axes x, y, z. It is not an inertial system because it rotates around the axis of the earth and therefore does not move steadily. Thus, Eq. (7.01b) is valid in this observational system only if the absolute acceleration $d\mathbf{V}/dt$ is transformed into the relative acceleration $d\mathbf{v}/dt$ of this relative system.

Such transformation yields two apparent forces:

1. The Coriolis force **F**. In a rotating system, with the positive z axis pointing downward, it has the following components as far as the Northern Hemisphere is concerned:

$$F_x = -2\Omega\,(w \cos \varphi + v \sin \varphi), \quad F_y = 2\Omega u \sin \varphi, \quad F_z = 2\Omega u \cos \varphi \qquad (7.02)$$

Here u, v, and w stand for the velocity components in the x, y, and z directions; φ is the geographical latitude and Ω the angular velocity of the earth's rotation. The value of the latter is $\Omega = 2\pi/86{,}164$ sec $= 7.29 \times 10^{-5}$ sec^{-1}; that is, one rotation of the earth takes place within one sidereal day (86,164 sec).

In general, the vertical component F_z can be neglected as compared to the gravitational force because their ratio is $10^{-7} : 1$. Furthermore, since the vertical velocity w in the ocean is very much smaller than the horizontal velocity, the term $w \cos \varphi$ in F_x is neglected against $v \sin \varphi$.

The Coriolis force was named after the French physicist Coriolis who detected it in 1835. If its horizontal components alone are considered, we also speak of the deflecting force of the earth's rotation. It always acts perpendicular to the motion—on the Northern Hemisphere toward the right-hand side, on the Southern Hemisphere toward the left-hand side, if you look in the direction of the current. In order to avoid the terms "right" and "left" we follow a suggestion by Ekman and use the terms *cum sole* and *contra solem*, that is, with or against the apparent azimuthal motion of the sun in the equatorial plane. These terms, describing the direction of rotation, correspond to the terms "anticyclonic" and "cyclonic" used in meteorology.

2. The centrifugal force **Z**. Its value is given by $\Omega^2 r \cos \varphi$. In the following it will be combined with the gravitational force and referred to as gravity.

7.1.3. Field of gravity

The centrifugal force of the rotation of the earth (**Z**) and the gravitational force of the earth ($\mathbf{K}_\phi$) are both independent of time and depend only on the geographical latitude as well as on the distance from the earth's center. Hence, both forces can be combined. The resultant is called "gravity," which corresponds to an acceleration of gravity.

In Fig. 7.02, $\phi^* = $ const represents the sea surface level of the earth at rest as derived from the gravitational force of the earth's mass. It is a spherical surface. The acceleration of gravity $-\nabla\phi^*$ would point toward the center of the earth. Due to the rotation and the centrifugal force caused thereby, the globe is oblate, and the sea surface is deformed into $\phi = $ const. The acceleration of gravity g is perpendicular to the sea surface and forms the angle φ with the equatorial plane, that is, the geographical latitude.

The acceleration of gravity—also briefly called gravity—has been determined on the earth by a great number of pendulum observations. Numerical smoothing of those results has led to the normal distribution accepted in 1930 as the "international gravity formula" valid for sea level:

$$g_0 = 9.78049(1 + 0.005288 \sin^2\varphi - 0.000006 \sin^2 2\varphi) \text{ m sec}^{-2} \quad (7.03)$$

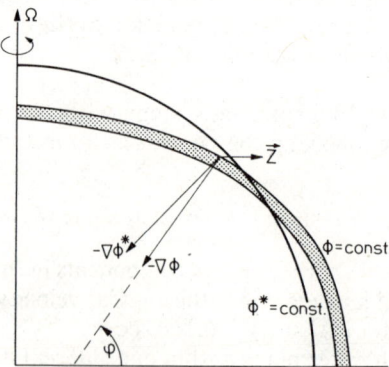

Fig. 7.02. Level surfaces of an earth at rest (ϕ^*) and a rotating earth (ϕ) in a strongly exaggerated ratio. Gravitational force $-\nabla\phi^*$ and centrifugal force **Z** result in the force of gravity $-\nabla\phi$.

Table 7.01. Corresponding Depths, Dynamic Depths, and Pressures for a Homogeneous Ocean[a]

Depth (m)	100	500	1000	5000	10,000
Dynamic depth (d.m.)	98	490	980	4903	9,811
Pressure (dbar)	101	504	1010	5098	10,310

[a] $T = 0°C$, $S = 35‰$, gravity $g_0 = 9.80$ m sec^{-2}.

Thus we see that gravity is a function of the geographic latitude φ. Because of the rotation and the oblateness of the earth, the largest values are found at the poles and the smallest at the equator. Furthermore, gravity depends on the distance from sea level and on regional inhomogeneities of the earth's crust. However, such deviations from the normal values of gravity are so small that they can be neglected as far as static and dynamic considerations regarding the ocean are concerned. This is confirmed by the following two facts: gravity increases with depth by about 0.19 mgal m^{-1}, and at 1000 m depth it has increased by not more than 0.20‰ as compared to normal gravity. Gravity anomalies resulting from inhomogeneous mass distribution in the earth's crust only very rarely exceed 100 mgal. Such a value represents a deviation of 0.11‰ from normal gravity.

Since gravity is the most important of the forces acting, it is advantageous to take this fact into account when a coordinate system is chosen. We can do so by starting not from surfaces of equal water depth but from surfaces of equal gravity, so-called level surfaces or potential surfaces. These are surfaces that are, at every point, perpendicular to the direction of a plumb line, the top one being the ideal sea surface level. They are characterized by the fact that no work is required for shifting a mass along the level surfaces as long as no other forces are acting there besides gravity. The work needed in order to displace a unit mass in the vertical is given by gh, where h denotes the distance in meters. This quantity represents a potential value. Thus, the position of any level surface is best described by its potential value. Bjerknes introduced the dynamic decimeter as the unit for the potential (Bjerknes and Sandström, 1910). It stands for the work that must be done in order to lift a unit mass by $\frac{1}{10}$ m. Hence, we can also define the level surfaces as surfaces of equal dynamic depth beneath the sea surface level, using the ideal sea level as the reference surface with the potential value zero. If the dynamic meter is chosen as the practical unit for the dynamic depth D, the relationship is $D = gh/10$, and the length of 1.02 geometric meters approximately corresponds to the unit of the gravity potential of 1 dynamic meter (d.m.). Such an approximate agreement is very convenient for practical calculations; but it is only a numerical agreement since the dynamic meter has the dimension of work (ML^2T^{-2}). Some examples of the transformation of geometric depths into dynamic depths in a homogeneous ocean are contained in Table 7.01.

7.1.4. Equations of motion in a rotating coordinate system

After the above references to the various forces occurring on the rotating earth, the equations of motion can now be given. For an x, y, z coordinate system with the positive z axis pointing downward they read as follows:

$$\frac{\partial u}{\partial t} + u\frac{\partial u}{\partial x} + v\frac{\partial u}{\partial y} + w\frac{\partial u}{\partial z} = -\frac{1}{\rho}\frac{\partial p}{\partial x} - 2\Omega v \sin \varphi + R_x + X$$

$$\frac{\partial v}{\partial t} + u\frac{\partial v}{\partial x} + v\frac{\partial v}{\partial y} + w\frac{\partial v}{\partial z} = -\frac{1}{\rho}\frac{\partial p}{\partial y} + 2\Omega u \sin \varphi + R_y + Y \quad (7.04)$$

$$\frac{\partial w}{\partial t} + u\frac{\partial w}{\partial x} + v\frac{\partial w}{\partial y} + w\frac{\partial w}{\partial z} = -\frac{1}{\rho}\frac{\partial p}{\partial z} + g + R_z + Z$$

292 Theory of Ocean Currents

Equations (7.04) contain those forces recognized as essential in the above Sections 7.1.1 to 7.1.3. The accelerations du/dt, dv/dt, and dw/dt have been transformed into the terms at the left-hand side. This is necessary because the accelerations du/dt, etc. refer to an individual water particle (Lagrange's method of description). Oceanographers, however, usually are not interested in the paths of the individual particles but in the description of the field of motion as a function of space and time (Euler's method of description). The transformation from the Lagrangian method to the Eulerian method for any property ψ of a particle is given by the following relation:

$$\frac{d\psi}{dt} = \frac{\partial \psi}{\partial t} + u\frac{\partial \psi}{\partial x} + v\frac{\partial \psi}{\partial y} + w\frac{\partial \psi}{\partial z} \tag{7.05}$$

$d\psi/dt$ denotes the individual or substantial derivative because it refers to the change in time of the property ψ of a particle.

To determine this change we must follow the motion of the particle through space. With the Eulerian method of description, the change of the property ψ at a certain location is considered. It consists of the change of ψ in time at this location, plus the change effected by the addition of other particles with a different property. Consequently, $\partial \psi / \partial t$ is the local derivative with respect to time, and the other terms are called convective (or advective) terms. In the case of $\partial \psi / \partial t = 0$, the process is called stationary.

7.1.5. Equation of continuity

Another important axiom in physics is the principle of mass conservation, called equation of continuity in hydrodynamics. It states that mass can neither be gained nor lost.

To formulate this principle by an equation, we will consider a volume element $dx\, dy\, dz$ (Fig. 7.03) in a cartesian coordinate system with the xy plane horizontal and the positive z axis pointing toward the center of the earth. Let u, v, w be the components of velocity in the directions of the x, y, z coordinates. Then $\rho u\, dy\, dz$ represents the water mass entering the volume element at one side in the x direction in unit time. At the opposite side, where ρu has been changed into

$$\rho u + \frac{\partial(\rho u)}{\partial x} dx$$

the mass

$$\left(\rho u + \frac{\partial(\rho u)}{\partial x} dx\right) dy\, dz$$

leaves the volume element. Hence, the change in mass transport along the distance dx

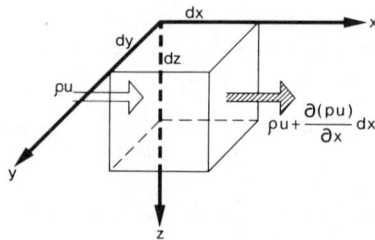

Fig. 7.03. Conditions of continuity.

is given by

$$\frac{\partial(\rho u)}{\partial x} dx\, dy\, dz.$$

Corresponding quantities result for the y and z directions. Furthermore, due to changes within the volume element, the mass may change by

$$\frac{\partial}{\partial t}(\rho\, dx\, dy\, dz)$$

per sec. For reasons of continuity the sum of the four terms must equal zero:

$$\frac{\partial \rho}{\partial t} + \frac{\partial(\rho u)}{\partial x} + \frac{\partial(\rho v)}{\partial y} + \frac{\partial(\rho w)}{\partial z} = 0 \qquad (7.06)$$

In this general equation of continuity, valid for any medium, simplifications may be made, if the physical properties of water and the particular conditions in the ocean are taken into account. Some special cases will be discussed in the following:

1. At a first approximation, water can be considered incompressible. If mixing processes and thermal conduction are neglected, the density of a particle always remains constant, $d\rho/dt = 0$, and the equation of continuity is reduced to

$$\frac{\partial u}{\partial x} + \frac{\partial v}{\partial y} + \frac{\partial w}{\partial z} = 0 \qquad (7.07)$$

because

$$\frac{\partial \rho}{\partial t} + u\frac{\partial \rho}{\partial x} + v\frac{\partial \rho}{\partial y} + w\frac{\partial \rho}{\partial z} = \frac{d\rho}{dt} = 0.$$

2. If conditions in the ocean are further simplified by $\rho = $ const (homogeneous ocean), Eq. (7.07) is valid here, too.
3. The total horizontal mass transport through a water column with a base of unit area is obtained by integration over the depth. For level sea surface and level sea floor these transports in the x direction and y direction, respectively, are given by

$$M_x = \int_0^H \rho u\, dz \quad M_y = \int_0^H \rho v\, dz. \qquad (7.08)$$

Because of the rigid surfaces at $z = 0$ and $z = H$, the integral over $\partial(\rho w)/\partial z$ vanishes. Thus, for stationary currents, we obtain from Eq. (7.06):

$$\frac{\partial M_x}{\partial x} + \frac{\partial M_y}{\partial y} = \text{div } \mathbf{M} = 0 \qquad (7.09)$$

In this case, consideration of continuity shows that the mass transport in the entire water column remains nondivergent, provided the sea surface is stationary. This, however, does not mean that the current field is nondivergent in each layer. The inflow of mass is balanced by the outflow only if the entire water column is considered. Within the water column, vertical motion of water may occur.

7.1.6. Equations of thermal conduction and diffusion

Temperature differences in the ocean will gradually become balanced because of thermal conduction unless they are maintained by other processes. This temperature balance is described by the equation of thermal conduction, which can be derived in analogy to the equation of continuity, but it should be noted that heat is transferred by thermal conduction through the boundary surfaces of the volume (cf. Fig. 7.03). The equation of thermal conduction reads as follows:

$$\frac{\partial T}{\partial t} + u\frac{\partial T}{\partial x} + v\frac{\partial T}{\partial y} + w\frac{\partial T}{\partial z} - k_T \Delta T = 0 \qquad (7.10)$$

A corresponding equation is valid for salinity. Differences in salinity are gradually balanced by diffusion, and for this case we obtain the following equation:

$$\frac{\partial S}{\partial t} + u\frac{\partial S}{\partial x} + v\frac{\partial S}{\partial y} + w\frac{\partial S}{\partial z} - k_D \Delta S = 0 \qquad (7.11)$$

The constants k_T and k_D are called coefficients of thermal conduction or diffusion, respectively. Thermal conduction and diffusion are caused by individual molecules of water or salt leaving the boundary surface of the volume element. Thus, their kinetic energy, respectively their "salinity," is lost to the volume element. Accordingly, the property of the volume element is changed by the inflow of molecules, if the entering molecules differ from those already present. This is always the case if there are local differences in temperature and salinity.

7.1.7. Boundary conditions

The equations of motion, continuity, thermal conduction, and diffusion are valid only inside a steady fluid. At the boundaries, those equations are replaced by so-called boundary conditions, which are of fundamental importance to oceanography because only gravity and tide-producing forces act as long-distance forces directly on all water particles. The entire oceanic circulation, on the other hand, is caused by forces effective at the sea surface, like the tangential shearing stress of the wind, the differences in air pressure, in precipitation and evaporation as well as the heat flux.

Each of the equations mentioned above has its corresponding boundary condition at the sea surface:

1. The equations of motion are replaced by the dynamic interface condition.
2. The kinematic interface condition takes the place of the equation of continuity.
3. The equation of thermal conduction is replaced by the equation of the heat flux through the interface.
4. Accordingly, an equation of the salt flux is valid instead of the equation of diffusion.

These boundary conditions, in their general form, are very complicated. They have been formulated by Krauss (1973). A simplification is given in the following:

1. The pressure is of equal size on either side of the sea surface, and the tangential

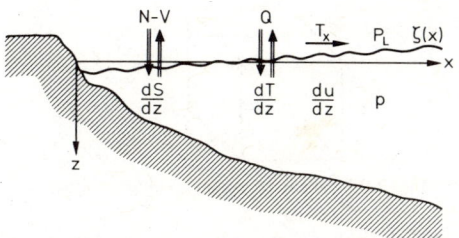

Fig. 7.04. At the sea surface $z = -\zeta(x)$, precipitation and evaporation (N–V), heat supply (Q), tangential shearing stress T_x, and variations of atmospheric pressure P_L cause differences in salinity (dS/dz), in temperature (dT/dz), in currents (du/dz), and variations of pressure (p) in the ocean.

shear stress is given by Newton's relation, that is,

$$p = P_L \qquad \mathbf{T} = -\mu \frac{d\mathbf{v}}{dz} \qquad \text{(dynamic interface condition)} \qquad (7.12)$$

where P_L is the atmospheric pressure, $\mathbf{T}$ the tangential shear stress of the wind, and μ the viscosity of the water.

2. The vertical velocity of the water particles corresponds to the motion of the sea surface, that is,

$$w = -\frac{\partial \zeta}{\partial t} \qquad \text{(kinematic interface condition)} \qquad (7.13)$$

$z = \zeta(x,y,t)$ is the sea surface.

3. The temperature gradient is proportional to the heat influx Q, that is,

$$\frac{dT}{dz} \propto Q \qquad (7.14)$$

4. The salinity gradient is proportional to precipitation minus evaporation, that is,

$$\frac{dS}{dz} \propto N - V \qquad (7.15)$$

The conditions described in the latter two equations are the cause of the thermohaline circulation. The conditions are schematically represented in Fig. 7.04.

7.1.8. Friction, turbulence, and mixing

As already shown in Section 2.1.9, water has a certain viscosity. Therefore, a water layer, gliding over another one, exerts a tangential shearing stress T_x on the underlying layer. In addition to depending on the dynamic viscosity μ, this tangential shearing stress also depends on the vertical velocity gradient $\partial u/\partial z$. Such relationship is expressed by Newton's friction law

$$T_x = \mu \frac{\partial u}{\partial z} \qquad (7.16)$$

The frictional force equals the difference of the shearing stresses acting on opposite sides

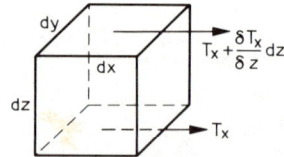

Fig. 7.05. Shear stress on a volume element.

of a volume element with the edges dx, dy, dz, as illustrated in Fig. 7.05. If the x direction coincides with the direction of motion, the frictional force R_x is

$$R_x dz = \left(T_x + \frac{\partial T_x}{\partial z} dz\right) - T_x = \frac{\partial T_x}{\partial z} dz \qquad (7.17)$$

$$R_x = \frac{\partial T_x}{\partial z} = \mu \frac{\partial^2 u}{\partial z^2}$$

The complete theory of friction, a derivation of which can be found in textbooks of hydrodynamics (e.g., Lamb, 1931), leads to complicated terms for the frictional forces which are taken into account in the Navier–Stokes equations of motion. The complete form for R_x in an incompressible medium for example, reads

$$R_x = \mu \left(\frac{\partial^2 u}{\partial x^2} + \frac{\partial^2 u}{\partial y^2} + \frac{\partial^2 u}{\partial z^2}\right) \qquad (7.18)$$

In the ocean, however, the Navier–Stokes' friction term is of minor importance, since Newton's friction law, as given in Eq. (7.16), must only be taken into consideration if the transport of momentum, which is equivalent to friction, is effected by the irregular motion of the molecules. This is the case in laminar, that is, smooth currents. But it has been found that actually nearly all motion in the ocean is irregular or turbulent, that is, the mean current is superposed by random motions of water bodies of various sizes. Because of this, great differences in velocity occur locally, thus giving considerable importance to the nonlinear terms of the equations of motion. Besides, the exchange taking place through such water bodies greatly exceeds the exchange through molecules.

Hagen (1839) was the first scientist to study the transition from regular to irregular motion. In the meantime great progress has been made in turbulence research, particularly since the fundamental experiments with currents in tubes, carried out by Reynolds (1895). His experiments have proven that, at a critical number, a sudden change takes place in the character of the motion. This number, the so-called Reynolds number, is given by $Re = \rho \bar{u} d/\mu$, where ρ is the density, $\bar{u}$ the mean velocity, and d the diameter of the tube.

For most purposes it is sufficient to consider irregular motion in the form of the so-called Reynolds friction term, which is obtained by smoothing the hydrodynamic equations of motion. If u' and w' are the turbulent velocity components in the x and z directions, and if bars denote average values, the following frictional force will occur:

$$R_x = \frac{1}{\rho} \frac{\partial}{\partial z} (\overline{\rho u' w'}) \qquad (7.19)$$

In order to determine quantitatively the frictional force of turbulent motion, let us consider a current with the mean velocity $\bar{u}$ at a level in the x direction. Water bodies that move in the vertical z direction because of turbulence lose their individuality after a certain length of path because of mixing with their surroundings. According to Prandtl, this path l is called "mixing length," and $l\partial \bar{u}/\partial z$ represents an approximate value of the velocity of the turbulent motion u'. The vertical turbulent motion w' has the same order of magnitude. Hence we have

$$\overline{\rho u'w'} = \rho l^2 \left(\frac{\partial \bar{u}}{\partial z}\right)^2 \tag{7.20}$$

This expression represents the transport of momentum per unit time, where the following symbol can be introduced:

$$\rho l^2 \left(\frac{\partial \bar{u}}{\partial z}\right) = A_Z. \tag{7.21}$$

The dimension of A_Z is the same as that of the dynamic viscosity μ, namely g cm^{-1} sec^{-1}. It is called "virtual viscosity" or also "apparent friction" or "austausch." Instead of the laminar shearing stress T_x, the effective shearing stress of turbulent motion in the xz plane results as

$$T_x = A_Z \frac{\partial \bar{u}}{\partial z} \tag{7.22}$$

and the frictional force as

$$R_x = A_Z \frac{\partial^2 \bar{u}}{\partial z^2} \tag{7.23}$$

The fundamental difference between μ and A_Z is that μ is a physical constant of matter, whereas A_Z depends on the particular state of motion and may vary within wide limits (cf. Chapter 2). In the ocean, A_Z can reach values of about 100,000 times greater than μ.

In analogy to Eq. (7.20), relations can also be given for the products $\overline{\rho u'u'}$, $\overline{\rho u'v'}$ etc. Since the mixing length is considerably longer in the horizontal direction than in the vertical, the austausch coefficient A_H resulting for the horizontal direction is different from A_Z. Then, the complete frictional force R_x is given by

$$R_x = A_H \left(\frac{\partial^2 \bar{u}}{\partial x^2} + \frac{\partial^2 \bar{u}}{\partial y^2}\right) + A_Z \frac{\partial^2 \bar{u}}{\partial z^2} \tag{7.24}$$

In stratified water turbulence decreases. The work required in order to displace water bodies in the vertical direction increases with the stability of stratification. If the work to be done per unit time becomes greater than the kinetic energy of the turbulent perturbation velocity w', turbulence ceases. From this, the Richardson number is derived:

$$\text{Ri} = \frac{g}{\rho} \frac{d\rho/dz}{(d\bar{u}/dz)^2} \tag{7.25}$$

which was introduced into the theory of flow by Richardson (1920). The approximate value for the existence of free turbulence seems to be $\text{Ri} < 0.5$. This Richardson number states that, if $\text{Ri} > 0.5$, turbulence ceases and the existing stratification of density is maintained; if $\text{Ri} < 0.5$, the water is mixed by turbulence and gradually approaches homogeneity.

If a turbulent current flows along a boundary surface, the mixing length l decreases linearly with decreasing distance z: $l = \kappa_0 z$, where $\kappa_0 = 0.4$ is the Kármán constant. In the immediate vicinity of a hydraulically smooth surface, the shearing stress becomes independent of the mixing length and is determined by viscosity. We then have a laminar boundary layer. The boundary surfaces in the ocean, however, that is, the ocean floor, sea surface, and internal discontinuity surfaces, show a certain roughness; hence, with $z = 0$, the mixing length l assumes a definite value:

$$l = \kappa_0(z + z_0), \qquad (7.26)$$

where z_0—according to Prandtl—denotes the roughness length and is related to the mean height of the small irregularities of the surface of the boundary.

If, in the equation for the effective shearing stress (Eq. 7.22), A_Z is eliminated by means of Eq. (7.21), and if it is taken into account that the mixing length depends on the roughness length [Eq. (7.26)], we obtain

$$T_x = \rho \kappa_0^2 (z + z_0)^2 \left(\frac{\partial \overline{u}}{\partial z}\right)^2 \qquad (7.27)$$

Therefore

$$\frac{\partial \overline{u}}{\partial z} = \frac{1}{\kappa_0(z + z_0)} \sqrt{\frac{T_x}{\rho}} \qquad (7.28)$$

and

$$\frac{A_Z}{\rho} = \kappa_0(z + z_0) \sqrt{\frac{T_x}{\rho}} \qquad (7.29)$$

$\sqrt{T_x/\rho}$ has the dimension of a velocity and is called shear stress velocity or friction velocity. After integration, assuming $\overline{u} = 0$ for $z = 0$, we obtain

$$\overline{u} = \frac{1}{\kappa_0} \sqrt{\frac{T_x}{\rho}} \ln \frac{z + z_0}{z_0} \qquad (7.30)$$

and

$$T_x = \rho \kappa_0^2 \overline{u}^2 \left(\ln \frac{z + z_0}{z_0}\right)^{-2} \qquad (7.31)$$

This shear stress, which is difficult to measure directly, can be determined quantitatively if we know the dimensionless quantity

$$\kappa_0^2 \left(\ln \frac{z + z_0}{z_0}\right)^{-2} = c_D$$

which plays the part of a friction coefficient.

The most important interface is the sea surface on which the wind exerts a tangential shear stress. Since, here, the friction coefficient c_D is of great importance not only for the wind-induced ocean currents but also for the generation and maintenance of surface waves as well as for evaporation, many attempts have been made to determine this coefficient c_D or $\rho \overline{u'w'}$ directly. Various oceanographic and meteorological methods have been employed. While former measurements led to $c_D = 2.6 \times 10^{-3}$, for an observational height of $z = 6$ m above sea level, more recent measurements (cf. Section 4.1.2 and Fig. 4.04) with moored or drifting measuring devices yielded $c_D = 1.2 \times 10^{-3}$ (Hasse, 1968; Miyake et al., 1970). In the following section this value will be used. Then the tangential shearing stress **T** (in g cm^{-1} sec^{-2}), exerted by the wind on the sea surface, amounts to

$$\mathbf{T} = 1.5 \times 10^{-6} |W| \mathbf{W}_{(10)} \qquad (7.32)$$

if we take $\rho = 1.25 \times 10^{-3}$ g cm^{-3} for the density of the air. Here W is the wind speed in cm sec^{-1} at a height of 10 m above the sea surface.

By the above parameterization, turbulence is understood only as a phenomenon, that is, only with regard to its effect on average conditions. In addition, turbulence is of great interest as a specific form of motion.

The spectrum of turbulence is extraordinarily wide and cannot be separated from the spectrum of the various types of waves. A comprehensive theory applicable to the ocean has been established only for very small turbulence elements. This is the theory of homogeneous turbulence. It is based on the assumption that the degree of turbulence does not show any local difference. Since the ocean is nearly always stratified, and since the horizontal and vertical extensions vary greatly, this theory can be applied only to small-scale conditions. Its most significant result is that the energy E of turbulent motion decreases with increasing wavenumber k, that is, with decreasing size of turbulence elements. The $5/3$ power law is valid:

$$E \propto k^{-5/3} \tag{7.33}$$

which has repeatedly been confirmed by measurements in the ocean.

7.2. Statics and Kinematics

7.2.1. Field of mass

The field of mass in the ocean is determined by the distribution of the density in situ $\sigma_{S,T,p}$, or by its reciprocal value, the specific volume $\alpha_{S,T,p}$. In oceanography, the latter value is preferred for the description of the field of mass. The definition of both quantities and their dependence on temperature, salinity, and pressure have already been given in Chapter 2. To simplify calculations, a standard water column with a salinity of 35‰ and a temperature $T = 0°C$ is assumed. The specific volume in situ is then given by $\alpha_{S,T,p} = \alpha_{35,0,p} + \delta$. The computation of the so-called anomaly of the specific volume δ, according to a simplified procedure suggested by Sverdrup, consists of the determination of three terms: one depending only on the density σ_T, one depending on salinity and pressure, and another one depending on temperature and pressure: $\delta = \Delta\sigma_T + \delta_{S,p} + \delta_{T,p}$. The tables needed for the determination of these terms can also be found in Sverdrup et al. (1942) and Dietrich (1952).

Strictly speaking, δ should be determined as a function of the absolute pressure. However, oceanographic observations of temperature and salinity, which are fundamental values, are obtained for different geometrical depths. In general, it is permissible to replace these depths in meters by pressures in decibars, since—as shown in Table 7.01—the numerical values agree within 1 to 2%. Vertical changes in temperature and salinity are mostly too small as to be of great importance, considering the small depth differences. Whether the calculations are based on temperature and salinity values obtained at a depth of 496 m (corresponding to the 500 dbar level) or on those obtained at a depth of 500 m, all this has hardly any noticeable influence on the result of the calculation of δ. Moreover, nearly the same systematic error will be included in the determination of δ for the neighboring stations. Hence, as far as horizontal differences in the δ values are concerned (which are the most interesting ones), such small errors cancel each other almost completely.

An example of the calculation of the anomalies of the specific volume is presented

Table 7.02. Example of the Calculation of Anomalies of Specific Volume and of Anomalies of Dynamic Depth[a]

1	2	3	4	5	6	7	8	9	10
Depth (m) P (dbar)	T (°C)	S (‰)	σ_T	$10^5 \Delta\sigma_T$	$10^5 \delta_{S,p}$	$10^5 \delta_{S,T}$	$10^5 \delta$	ΔD	$\sum_{p_0}^{p_1} \Delta D$
0	22.11	36.40	25.27	271.3	0	0	271.3		0
25	22.00	36.59	25.45	254.1	0.1	0.9	255.1	0.0658	0.0658
50	21.50	36.65	25.63	237.1	0.1	1.8	239.0	0.0618	0.1276
75	20.95	36.67	25.79	221.8	0.2	2.7	224.7	0.0579	0.1855
100	20.43	36.67	25.93	208.6	0.3	3.5	212.4	0.0547	0.2402
150	19.85	36.66	26.08	194.3	0.4	5.2	199.9	0.1031	0.3432
200	19.25	36.62	26.21	181.9	0.5	6.8	189.2	0.0973	0.4406
250	18.60	36.56	26.33	170.6	0.6	8.4	179.6	0.0922	0.5328
300	18.20	36.52	26.40	163.9	0.7	9.8	174.4	0.0885	0.6212
400	17.57	36.42	26.48	156.3	0.8	12.8	169.9	0.1722	0.7934
500	16.58	36.25	26.59	145.9	1.0	15.5	162.4	0.1662	0.9596
600	14.70	35.97	26.81	125.0	0.9	17.1	143.0	0.1527	1.1124
700	12.48	35.65	27.01	106.0	0.7	17.6	124.3	0.1336	1.2460
800	10.30	35.33	27.18	89.9	0.4	17.5	107.8	0.1160	1.3620
900	8.06	35.11	27.38	71.0	0.2	16.1	87.3	0.0976	1.4596
1000	5.90	35.02	27.60	50.1	0.0	13.8	63.9	0.0756	1.5352
1200	4.55	35.00	27.75	35.9	0.0	12.9	48.8	0.1128	1.6480
1400	4.09	34.98	27.78	33.1	0.0	13.8	46.9	0.0956	1.7436
1600	3.80	34.97	27.80	31.2	−0.1	14.7	45.8	0.0928	1.8364
1800	3.61	34.96	27.81	30.2	−0.1	15.7	45.8	0.0916	1.9280
2000	3.43	34.96	27.83	28.3	−0.1	16.6	44.8	0.0906	2.0186
2250	3.27	34.97	27.85	26.4	−0.1	17.7	44.0	0.1110	2.1296
2500	3.02	34.97	27.88	23.6	−0.1	18.1	41.6	0.1070	2.2366
3000	2.66	34.95	27.90	21.7	−0.2	18.9	40.4	0.2050	2.4416
3500	2.31	34.93	27.91	20.8	−0.3	18.8	39.3	0.1990	2.6406
4000	2.31	34.92	27.90	21.7	−0.4	21.1	42.4	0.2040	2.8446

[a] *Atlantis* Station 1227, $\varphi = 35°38'$N, $\lambda = 72°47'$W, April 21, 1932.

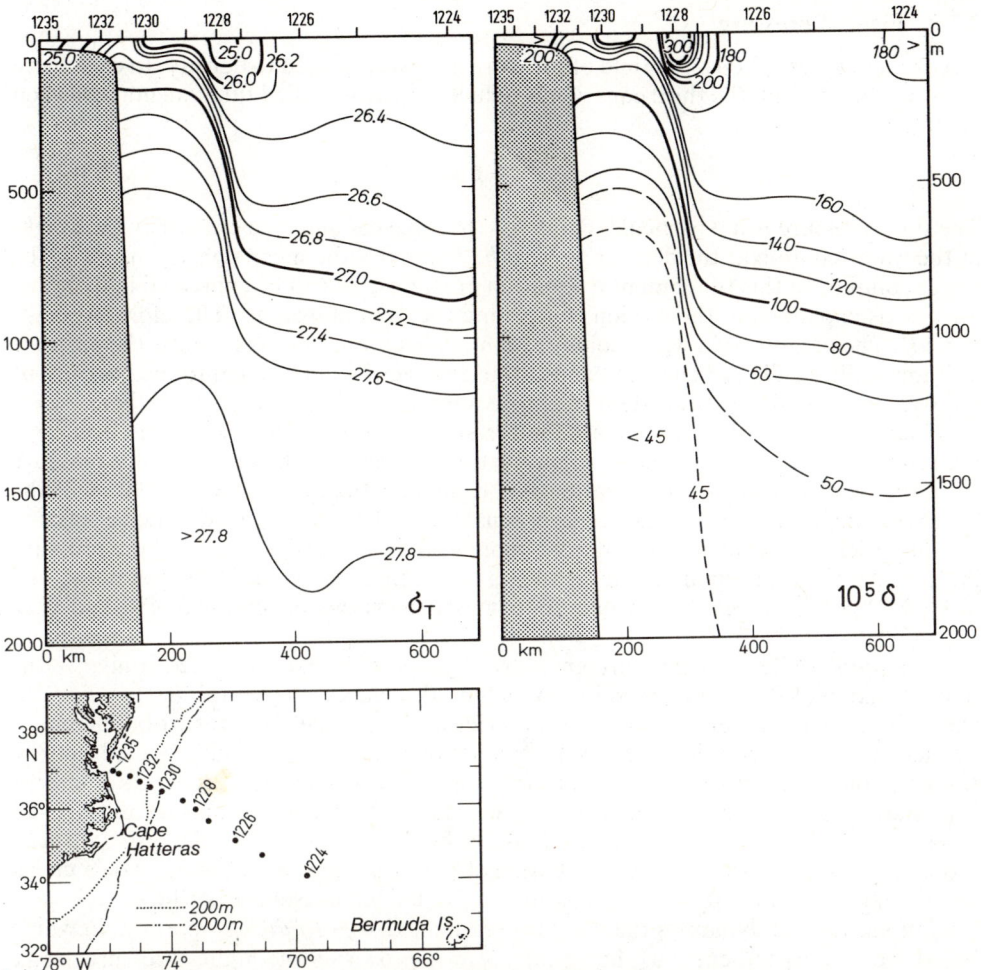

Fig. 7.06. Distribution of the density σ_T and the anomaly of specific volume in $10^5 \delta$ along a section perpendicular to the Gulf Stream (see sketch of location) based on observations by the research vessel *Atlantis*, April 19–23, 1932 (vertical exaggeration 500 times). Compare also Figs. 7.14, 7.16, and 10.42.

in Table 7.02. Column 1 contains the depths in meters as well as the pressures in decibars for the other columns, columns 2 and 3 give the corresponding temperature and salinity values, and column 4 contains the density values σ_T computed from them. Columns 5 to 7 show the different terms according to the tables by Sverdrup, and in column 8 the anomalies of the specific volume are given as the sum of those terms. The remaining columns will be dealt with in the following section of this book. As an example, Fig. 7.06 presents a cross section perpendicular to the Gulf Stream between Chesapeake Bay and the Bermuda Islands. In addition to σ_T and δ, the distributions of temperature, salinity, and density as observed by the research vessel *Atlantis* on this Gulf Stream cross section are given. The calculated current velocities (cf. Section 7.3.3) are shown too and—in order to make the picture complete—the distributions of oxygen and phosphate (cf. Figs. 7.14, 7.16, and 10.42).

7.2.2. Field of pressure

According to Eqs. (7.04), the field of pressure is closely related to density. In case there is no current present, the third equation is reduced to the so-called fundamental equation of statics:

$$\frac{\partial p}{\partial z} = g\rho \qquad (7.34)$$

The static pressure p at a particular depth h is the force per unit area exerted by the weight of the water column at the depth h: $p = \bar{\rho}gh$, where $\bar{\rho}$ is the mean density in situ of the water column. In this fundamental equation of statics, gh can be expressed in dynamic meters, as explained in the section concerning the field of gravity. Therefore, we have $p = \bar{\rho}D$. The pressure of a water column 1 d.m. in height has been chosen as the pressure unit and called 1 dbar. Ten dbar equal 1 bar and, hence, correspond to the pressure of 10^6 dyn cm^{-2}, which is equivalent to approximately 1 atm (1 atm = 1.013 bar).

Since it is not feasible to measure the pressure in the ocean with sufficient accuracy, the relationship $D = \bar{\alpha}p$ is used to calculate the dynamic depth of a particular pressure from the field of mass. In this relationship, $\bar{\alpha}$ denotes the mean specific volume of the water column in situ. Since $\bar{\rho}$ and $\bar{\alpha}$ nearly equal unity, there is no great difference between the numerical values of the gravitational potential in dynamic meters and the pressure in decibars. Such approximate agreement also includes the values of depth in meters. Table 7.01 contains the related values for a homogeneous water column of 0°C and 35‰ salinity.

The pressure field can be represented by a system of isobaric surfaces, similar to the representation of the gravity field by equipotential surfaces. In general, the two systems will intersect each other. The lines of intersection of isobaric surfaces with a potential surface give the distribution of pressure by lines of equal pressure, the so-called isobars. In meteorology this form of representation is used for plotting surface weather charts. The lines of intersection of equipotential surfaces with an isobaric surface are lines of equal gravity potential, called dynamic isobaths. In oceanography the latter form of representation is preferred. The distribution of the dynamic isobaths, plotted on a chart, results in the so-called dynamic topography of a particular isobaric surface.

For statics and dynamics the pressure gradient $G = -dp/dn$ is important, in which n is directed perpendicular to the isobaric surface towards the higher pressure. Since pressure represents the force per unit area (MLT^{-2} force, L^2 area, $ML^{-1}T^{-2}$ pressure), the dimension of the pressure gradient is that of a force per unit volume ($MLT^{-2}L^{-3}$). Its component perpendicular to the potential surface is nearly in equilibrium with the force of gravity, whereas its component parallel to the potential surface implies that the system cannot be at rest unless other external forces are present. Although the latter component of the pressure gradient is extremely small compared to the vertical component, it is nevertheless of decisive importance to ocean currents. It forms the internal field of force in the ocean, which can be described—if not by the pressure gradient—by the gradient of the dynamic topography of an isobaric surface.

However, we must keep in mind that the field of pressure obtained is referred to the ideal sea surface level of zero gravitational potential where isobaric and potential surfaces coincide. The topography of the physical sea surface level, from which the measurements of the mass structure are made, is unknown, however. In meteorology such a fundamental difficulty is not encountered because the height of the points of observation (where the atmospheric pressure is measured) above the main potential surface, that is, the sea surface, is very precisely known through height leveling. Furthermore, small differences in level are far from playing as important a part as they do in the ocean, since, in the atmosphere, pressure varies only slightly with height as compared with the ocean. Near the ground the changes amount to 0.13 mbar/1 m, whereas in the ocean we have 100 mbar/1 m. In the open ocean, however, a direct determination of the topography of the physical sea surface by means of height leveling is impossible. At best, such determination can be accomplished by fine leveling along the coast, if water levels recorded at tide gauges are geodetically connected. In general, only relative dynamic topographies, referred to the ideal sea surface level, can be determined from the mass field. Thus, only relative fields of pressure are obtained and likewise, only relative distributions of forces. It should be added that even an absolute internal field of pressure, which—with certain assumptions as to the field of motion (cf. Section 7.3.3)—can be derived from the relative field of pressure, does not yet denote the total field of pressure in the ocean, since the latter also includes the pressure field maintained by external forces such as wind and variations of atmospheric pressure.

7.2.3. Determination of the relative field of pressure

The relative field of pressure can be of value for dynamic considerations regarding the ocean, if one keeps in mind the restriction that it is only part of the total field of pressure. Provided the field of mass in the ocean is known, the relative pressure field can be determined with the aid of the fundamental equation of statics $D = \bar{\alpha}p$. In integral form, the equation reads as follows:

$$D_2 - D_1 = \int_{p_1}^{p_2} \alpha_{S,T,p} \, dp \qquad (7.35a)$$

or, if the anomaly of the specific volume is introduced,

$$(D_2 - D_1)_{35,0,p} + \Delta D = \int_{p_1}^{p_2} \alpha_{35,0,p} \, dp + \int_{p_1}^{p_2} \delta \, dp \qquad (7.35b)$$

Since the isobaric surfaces of a standard water column are parallel to each other, they do not contribute to the relative field of the internal horizontal forces. The relative field of forces, which is based on the field of mass, is completely described by the anomalies of the dynamic topography

$$\Delta D = \int_{p_1}^{p_2} \delta \, dp \qquad (7.36)$$

An example of this calculation is given in the last two columns of Table 7.02. Column 9 shows the anomalies of the dynamic depth in dynamic meters for the various layers. These values have been obtained by multiplying the mean anomalies of the specific volume of a particular layer—limited by two standard pressures—by the difference of pressure between these standard pressures. By cumulative addition of these values for the various layers, the anomaly of the dynamic depth—referred to the ideal sea surface level—is obtained for each standard pressure (cf. column 10).

With the anomalies given in columns 9 and 10 of Table 7.02, the specific volumes and the dynamic depths can also be determined by adding the standard values for a homogeneous water column $\alpha_{35,0,p}$ and $D_{35,0,p}$ computed by Bjerknes. For instance, the specific volume for 1000 dbar is $\alpha_{35,0,1000} = 0.96819$; in our example, therefore, $\alpha_{S,T,1000} = 0.96819 + 0.00064 = 0.96883$. Because $D_{35,0,1000} = 970.4032$ d.m. and $\Delta D = 1.5352$ d.m., we get $D_{S,T,1000} = 971.9384$ d.m.

7.2.4. Representation of the relative field of pressure

The relative internal field of pressure in the ocean can be determined if, in accordance with the example in Table 7.02, the mean vertical distribution of temperature and salinity is known at numerous points in an oceanic area. The relative dynamic topography of a particular isobaric surface can then be represented by lines of equal dynamic depth, as shown in Figs. 10.33(a) and 10.40.

The prerequisite for such a representation of the field of pressure, however, is that the mean field of mass must be known. This observational demand has so far very seldom been fulfilled in the ocean. If, in general, the dynamic topography of vast oceanic areas, based on single, widely spaced, synoptic data, does not show an irregular picture, this fact indicates that the stationary part of the relative internal pressure field must be the decisive factor in such cases.

7.2.5. Stability of water stratification

A water particle will remain at rest relative to its surroundings if it is lighter than the water mass below it and heavier than the water mass above it. In case such a particle is forced to shift vertically, it will tend to return to its initial position, thereby experiencing an acceleration which, according to the third component of the hydrodynamic equations of motion (Eqs. 7.04), with friction and the deflecting force of the earth's rotation neglected, is given by

$$\frac{dw}{dt} = +g - \frac{1}{\rho}\frac{\partial p}{\partial z} \tag{7.37}$$

For the surrounding water mass with the density $\bar{\rho}$, being in equilibrium, the fundamental equation of statics reads

$$g - \frac{1}{\bar{\rho}}\frac{\partial p}{\partial z} = 0 \tag{7.38}$$

If $\partial p/\partial z$ is eliminated, we get

$$\frac{dw}{dt} = -g\frac{\bar{\rho} - \rho}{\rho} \tag{7.39}$$

The acceleration with which the water particle tends to return to its initial position is proportional to the density difference between the particle and the surrounding water mass, and the expression $(\bar{\rho} - \rho)/\rho$ can be used as a measure of the stability conditions, if only the short vertical distance Δz is considered. Accordingly, Hesselberg (1918) defined the stability by the following expression

$$E = \lim_{\Delta z \to 0} \frac{1}{\rho}\frac{\bar{\rho} - \rho}{\Delta z} = \frac{1}{\rho}\frac{d\rho}{dz} \tag{7.40}$$

Depending on whether E is greater than, equal to, or less than zero, the distribution of mass is called stable, indifferent, or unstable, respectively. It should be noted that, in the case of vertical displacement, the water particle is subjected to a change of pressure and, due to its compressibility, also to changes of temperature. These changes are called adiabatic as long as they occur without any heat exchange with the surroundings. Since the compressibility of water is small, the adiabatic temperature changes are small, too, as can be seen from the numerical examples given in Chapter 2. At great depth, however,

Table 7.03. Examples of Stability

Layer (m)	Stability Meteor 248 3°30'S 22°36'W, Jan. 5, 1927 depth 5396 m	Meteor 122 55°3'N 44°46'W, March 9, 1935 depth 3381 m
0–25	-128×10^{-8}	-50×10^{-8}
25–50	59	− 3
50–75	1809	+36
75–100	**8109**	+29
100–150	504	−11
150–200	216	+11
200–300	165	+ 8
300–400	**200**	0
400–500	142	0
500–600	120	+ 2
600–700	85	+ 1
700–800	54	+ 7
800–900	53	+ 5
900–1000	51	− 5
1000–1200	**84**	+ 3
1200–1400	59	0
1400–1600	34	+ 3
1600–1800	22	0
1800–2000	19	+ 6
2000–2250	11	+ 1
2250–2500	10	+ 8
2500–3000	8	+17
3000–3500	6	
3500–4000	13	
4000–4500	**20**	
4500–5000	9	

the vertical temperature gradient in situ is also small, and therefore the adiabatic effect must not be neglected when stability is considered for deeper layers. Hesselberg and Sverdrup (1914–1915) have derived an expression for stability which shows this effect very clearly, and which is the basis for their tables:

$$E = \frac{1}{\rho}\left[\frac{\partial \rho}{\partial S}\frac{dS}{dz} + \frac{\partial \rho}{\partial T}\left(\frac{dT}{dz} - \frac{d\Theta}{dz}\right)\right] \qquad (7.41)$$

Here, $d\Theta/dz$ stands for the adiabatic temperature gradient; the factor $1/\rho$ may be neglected as it approximately equals 1.

Examples of stability conditions in the ocean are contained in Tables 7.03 and 7.04. The first example, based on a paper by von Schubert, published in the *Meteor* Report, is typical of the lower latitudes in the Atlantic Ocean. Apart from a thin near-surface layer, showing unstable conditions, the stratification of the entire water column is stable. It has four intermediate stability maxima coinciding with the transition zones between water masses of different origins. The uppermost maximum is found in the zone of transition between the surface layer, subjected to strong vertical turnover due to interaction with the atmosphere, and the lower oceanic warm-water sphere with very stable stratification. Further down, other maxima are recognizable: at 300 to 400 m, at the transition to the subantarctic intermediate water; at 1000 to 1200 m, at the transition

Table 7.04. Example of Stability and Potential Temperature[a]

1 Depth (m)	2 T (°C)	3 Θ (°C)	4 S (‰)	5 E
2,000	2.25	2.10	34.61	$+10 \times 10^{-8}$
3,000	1.64	1.41	34.66	$+1$
4,000	1.60	1.27	34.67	0
5,000	1.72	1.26	34.67	0
6,000	1.86	1.25	34.67	0
7,000	2.01	1.25	34.68	0
8,000	2.15	1.23	34.69	0
9,000	2.31	1.19	34.68	0
10,000	2.48	1.17	34.67	-1

[a] *Will. Snellius* 262, 9°41′N, 126°51′E, May 16, 1930.

to the North Atlantic deep water, and, finally, at 4000 to 4500 m, at the transition to the Antarctic bottom water.

The second example, based on data from an area south of Greenland in late winter, is characterized by low stability. Here we find the conditions for deep-reaching vertical turnover, the result of which is the formation of subarctic deep water in the North Atlantic Ocean.

Equation (7.41) indicates that, without any change of salinity with depth, stability depends only on the geometric and adiabatic temperature gradients. Depending on whether dT/dz is greater than, equal to, or less than $d\Theta/dz$, we find stable, indifferent, or unstable equilibrium, respectively. The example from the Philippine Trench, presented in Table 7.04, which is based on the deepest series of temperature and salinity observations so far obtained in the world ocean, shows that the increase of temperature measured beneath 4000 m does not indicate unstable stratification. The increase is caused by the adiabatic effect. The potential temperature in column 3, determined from the measured temperature by subtracting the adiabatic effect and referred to sea level, nearly remains constant. Below 3000 m, the stability E (column 5) is zero, at least within the limits of accuracy in the determination ($\pm 2 \times 10^{-8}$), which means that we have indifferent conditions. Neither by these nor by other observation series has it been proven conclusively that the lower layers may become unstable due to the effect of geothermal heat flux. This does not mean that the bottom water does not receive any heat from the interior of the earth. But even in the deepest layers we must expect random motion and, thus, some vertical heat exchange, which evidently is large enough to prevent the formation of a superadiabatic temperature gradient. The small decrease of Θ beneath 8000 m in the Philippine Trench indicates a lateral renewal at the greatest depths.

Negative stability values in the near-surface layer, occurring not only in the two examples in Table 7.03 but also representing a general characteristic of the world ocean, need further explanation because, according to the elementary stability criterion (Eq. 7.40), they simply cannot exist for any length of time. In fact, the stability criterion for thin layers calls for correction, since, under certain conditions, indifferent stability will occur not with $E = 0$, but with negative E values. At laboratory experiments, the physicist Weber (1855) already noticed that, in thin fluid layers, convection starts only when a certain instability of stratification is established. This convection is organized in regularly shaped convection cells, horizontal and vertical cross sections of which are shown in Fig. 7.07. They are called Benard Convection Cells, after Benard who thoroughly investigated them in 1901. They are hexagonal. The fluid rises in the center and sinks at the margins.

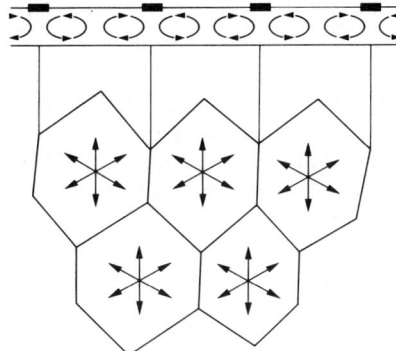

Fig. 7.07. Convection cells in a water layer, presented in vertical and horizontal cross sections. ■ Foam streaks with convective rolls.

The diameter of a cell is two or three times its height. In a current, these cells take the shape of longitudinal roll vortices.

Traces of convective rolls are frequently observed at the sea surface. They are manifest in elongated, parallel foam streaks accumulating in the region of convergence between two neighboring rolls (cf. Fig. 7.07). During the *Meteor* expedition of 1938, this phenomenon was very clearly observed in the northeast trade wind region of the North Atlantic Ocean. Here a vast area was covered by sargasso weed drifting in long parallel stripes aligned in the direction of the wind at a distance of about 100 m from each other.

The interpretation of those streaks as convective rolls has not yet been definitely assured. In regions with varying winds it has been observed that the streaks are adjusted to the new wind direction within a few minutes when a strong wind suddenly changes its direction. Welander (1963) connected this phenomenon with a surface film of organic material of varying concentration and with the friction conditions altered thereby. Krauss (1965) interpreted them as the result of a wave-induced friction coefficient depending on the crest length of the waves.

7.2.6. Representation of the field of motion

Currents in the ocean are determined directly by current measurements or indirectly by applying the hydrodynamic equations of motion. They can be described in two different ways. Either the paths of water particles during a given period of time are represented by so-called trajectories, or the distribution of direction and velocity at a given instant, that is, the current field, is determined. In the case of stationary currents both methods are identical.

Trajectories supply valuable data for the representation of currents if the positions of drifting floats are determined at short time intervals. This has repeatedly been achieved in the area of the Grand Banks where the International Ice Patrol Service follows the paths of marked icebergs (cf. Fig. 5.26) in order to be able to warn ships in the North Atlantic Ocean in good time. Drift observations of the ice of the Arctic Ocean furnish trajectories, at present the only source of our knowledge on surface currents in that particular ocean. Figure 10.51 contains ten such drifts, including those of two research vessels whose crews voluntarily agreed to let their vessels become beset by ice: Nansen's *Fram* from 1893 through 1896, and *Maud* from 1922 through 1924 with Sverdrup as the chief scientist. Altogether, the trajectories have given evidence that the surface currents in the inner part of the Arctic Ocean are anticyclonic, that is, rotate clockwise.

The state of motion can be represented in the form of current fields only if the direction and velocity of the current are known at a large number of points. Up to now, direct current measurements have not yet fully met this demand. The stream function is often used for the presentation of theoretical results. The stream lines indicate the direction of the current, while the strength of the current is determined by their distance from each other. Relevant examples are found in Section 7.3.7.

Another kind of representation is usually chosen in current atlases. Such charts are based on a great number of individual measurements mostly derived from ships' logs over long time periods.

Since, in general, the observational material does not suffice for synoptic current charts, the observations are summarized in monthly or seasonal charts over longer time periods. Such an averaging process reveals that the currents vary a great deal, and to a different extent in different regions. Any objective representation of current conditions cannot be obtained from direction and velocity alone, but must also take into account the degree of constancy. In the chart of ocean currents on Plate 6, mainly based on the more recent representations by Schott (1943), the direction of the current is indicated by arrows, the constancy by the length of their shafts, and the velocity by the thickness and the feathering of the arrows.

7.3. Stationary Currents

7.3.1. Geostrophic currents in a homogeneous ocean

The simplest model to be conceived of an ocean current is obtained by assuming that the water is homogeneous, friction is not influential, no other external force, apart from the ever-present gravity, is acting and, furthermore, that the motion is horizontal and of constant velocity. Then, the three equations of motion [Eqs. (7.04)] can be simplified to

$$2\Omega v \sin \varphi = -\frac{1}{\rho}\frac{\partial p}{\partial x}$$
$$2\Omega u \sin \varphi = \frac{1}{\rho}\frac{\partial p}{\partial y} \qquad (7.42)$$
$$g = -\frac{1}{\rho}\frac{\partial p}{\partial z}$$

The first two equations provide the information that, in this case, the deflecting force of the earth's rotation is balanced by the gradient force. If β denotes the slope angle between the isobaric surface p and a level surface in the x direction, as sketched in Fig. 7.08, we can also write instead of the fundamental equation of statics, which is the third equation of Eq. (7.42),

$$\frac{\partial p}{\partial x} = g\rho \tan \beta \qquad (7.43)$$

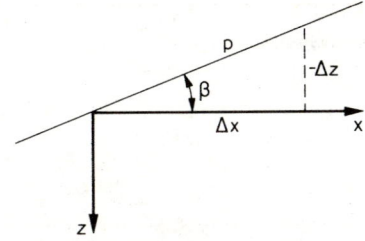

Fig. 7.08. Slope angle of the isobaric surface.

Elimination of $\partial p/\partial x$ from the first equation of motion [Eq. (7.42)] by means of Eq. (7.43) leads to

$$\tan \beta = -\frac{2\Omega \sin \varphi}{g} v. \tag{7.44}$$

A corresponding relation can be derived for the velocity component u. Accordingly, the theorem holds that the velocity is perpendicular *cum sole* to the inclination of the isobaric surfaces and that the inclination of these surfaces is a measure of the velocity.

The order of magnitude of the inclination of the isobaric surfaces (the sea surface also represents one of them) can be taken from a numerical example: let us assume $\varphi = 30°$ and $v = 10$ cm sec^{-1}, and furthermore that $\Omega = 7.29 \times 10^{-5}$ sec^{-1} and $g = 980$ cm sec^{-2}. Then, according to Eq. (7.44), $\tan \beta = -74 \times 10^{-8}$. This means that the sea surface rises by 74 cm, perpendicular to the current, over a distance of 10^8 cm (1000 km). Extremely small inclinations of the sea surface are therefore able to maintain ocean currents. It is quite obvious that direct measurements of such gradients, for example, by means of soundings, are impossible in the open ocean.

However, there is an example which, on the basis of observations, directly demonstrates the relation between current velocity and sea level inclination in the cross-stream direction. It is taken from the area of the Great Belt where the zero levels of the tide gauges at Slipshavn and Korsør are connected via the island of Sprogø by geodetic fine leveling. However, because of the long distance involved, the mean error of the leveling is relatively large (± 1.6 cm). At the lightship *Halskov-Rev*, located between the two gauges, the current is measured regularly. Figure 7.09 shows a sketch of the location and the relation between the daily average of the water level difference from Korsør to Slipshavn in centimeters and the current velocity in cm sec^{-1}, as observed at the lightship from June through September 1937. No particular significance should be attached to the apparent cross gradient present with zero current, which is reflected by the water level at Korsør being higher than at Slipshavn by 1.5 cm, as this value lies within the limits of error of the fine leveling. In this case the "quasi-leveling" is more accurate than geodetic fine leveling. Hence, we can assume the smooth line in Fig. 7.09 to be shifted parallel to itself so that it passes through the origin of the coordinates. The average relation shows that the water level on the right-hand side of the current is always higher than on the left-hand side (if we look in the direction of the current). For a velocity of 100 cm sec^{-1}, the water level difference between the two shores amounts to 12.7 cm. Of course, the current as measured at the lightship is not representative for the total distance from Korsør to Slipshavn. If we exclude the shallow reefs between the shore and the 6-m isobath, the

310 Theory of Ocean Currents

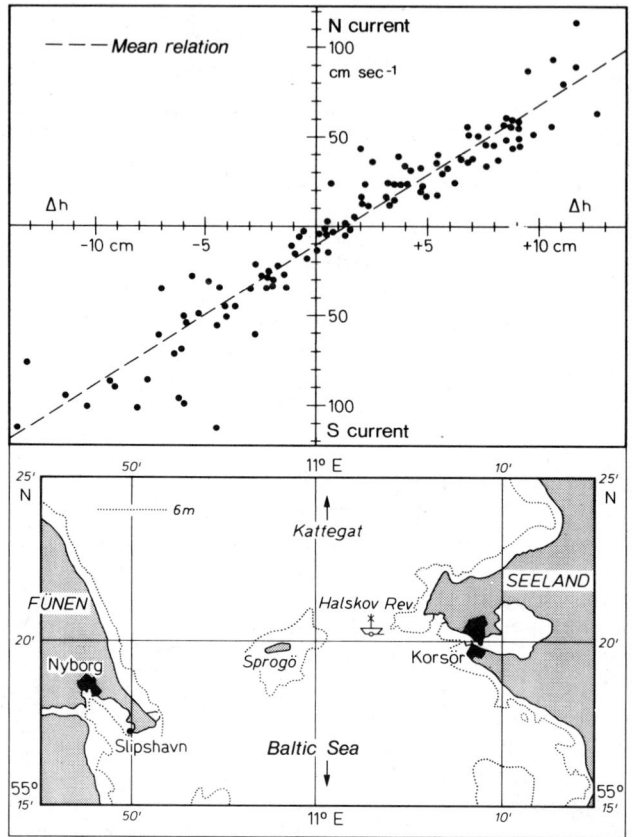

Fig. 7.09. Relation between the current, observed at lightship *Halskov-Rev*, and the sea level inclination measured between Korsør and Slipshavn in the Great Belt.

effective cross section of the current is 9.1 km wide. According to the observations, the ratio for this distance is $\tan \beta : v = 1.40 \times 10^{-7}$ cm^{-1} sec, which is in close agreement with the corresponding value obtained from Eq. (7.44), namely 1.22×10^{-7}.

7.3.2. Geostrophic currents in a two-layer ocean

As a modification of the case discussed above, dealing with a completely homogeneous ocean, let us consider two homogeneous water masses lying one on top of the other, the upper layer having the density ρ_1, the lower one ρ_2, with $\rho_1 < \rho_2$. As long as both water masses are at rest, the interface between them is horizontal. However, as soon as they move along at different velocities v_1 and v_2, the interface becomes inclined, depending on the state of motion. For a quantitative expression of this dependence it is sufficient to apply the simplified equations of motion as derived for a homogeneous ocean. They are valid separately for each individual water mass. In addition, the dynamic boundary condition stating that pressure and counterpressure at the interface are equal, must be taken into account. Assuming, for the sake of simplicity, that the motion is directed along the y axis only (that is, $u_1 = u_2 = 0$), we get the following relation for the slope angle γ

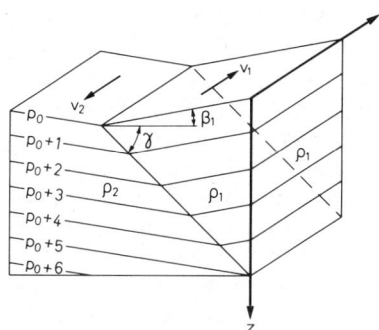

Fig. 7.10. Inclination of the isobaric surfaces and the interface of two homogeneous water masses in motion in the Northern Hemisphere.

of the interface:

$$\tan \gamma = - \frac{2\Omega \sin \varphi}{g} \frac{\rho_2 v_2 - \rho_1 v_1}{\rho_1 - \rho_2} \qquad (7.45)$$

known as the Margules' boundary equation. In 1906, Margules derived this equation for atmospheric conditions, and later on Defant (1929) applied it to the ocean. The following equations describe the inclination of the isobaric surfaces within the individual water masses:

$$\tan \beta_1 = - \frac{2\Omega \sin \varphi}{g} v_1, \quad \tan \beta_2 = - \frac{2\Omega \sin \varphi}{g} v_2 \qquad (7.46)$$

The interface remains inclined as long as $\rho_2 v_2 - \rho_1 v_1 \neq 0$, that is, as long as the momentum of the two water masses remains different and $\varphi \neq 0$. At the equator, the interface must always be horizontal. The block diagram in Fig. 7.10 illustrates the relationships valid for the Northern Hemisphere, although the slope angles β_1 and γ are greatly exaggerated. It can be adapted to the Southern Hemisphere by reversing the current arrows v_1 and v_2. Hence, if the equatorial region is excluded and if one looks in the direction of the current, it can generally be said that the isobaric surfaces and, thus, the physical sea surface level rise from left to right and that the interface beneath the lighter water mass drops from left to right. For the Southern Hemisphere, "left" and "right" must be exchanged in this rule.

Having introduced greatly simplifying assumptions, we accordingly obtain idealized relationships between the current and the inclination of the discontinuity surface and of the isobaric surfaces. Real discontinuities seldom exist in seawater. But observations in the world ocean show that water masses occur, more or less extensive ones, each of which is characterized by nearly uniform physicochemical properties and can therefore be considered an almost homogeneous water body. The transition between one water mass and the other is not given by an interface, but by a relatively thin boundary layer or pycnocline. Therefore, the schematic representation of stratification conditions by interfaces approximates reality better than the assumption of a homogeneous ocean. By applying the relations regarding the stable stratification of water bodies we easily come to understanding the structure and the movement of water masses in the ocean. This is very useful for a first orientation on current conditions. Two examples may demonstrate this.

To convey an impression of the order of magnitude of the slope angles β_1 and γ, an example of a two-layer configuration is presented which approximates the natural conditions in the area of the Gulf Stream in the cross section of Fig. 7.06: $\varphi = 36°$, $\rho_1 = 1.0266$, $\rho_2 = 1.0276$, $v_1 = 100$ cm sec^{-1}, and $v_2 = 0$ cm sec^{-1}. From Eq. (7.46) it follows that $\tan \beta_1 = -0.875 \times 10^{-5}$, and from Eq. (7.45) that $\tan \gamma = 898 \times 10^{-5}$. If these values of inclination are expressed in degrees, they correspond to the angles $\beta_1 = -0°0'1.8''$ and $\gamma = 0°31'$. Lighter water moves over heavier water with a very small angle γ, which, however, is still 1000 times greater than the slope angle of the isobaric surface and thus of the physical sea level. This fact is reflected in Fig. 7.06 by showing that—in spite of the 500-fold exaggeration of the vertical scale—the interface becomes almost vertical, but nevertheless the inclination of the sea surface remains so small that it is impossible to plot it in a diagram. If the inclination of the interface is characterized by the slope of the σ_T surface 27.2 (cf. Fig. 7.06) lying right in the middle of the pycnocline, there results a drop from *Atlantis* Station 1228 to Station 1227, that is, from 450 m depth to 810 m depth over a distance of 40.7 km. Such a drop corresponds to an inclination of $\tan \gamma = 884 \times 10^{-5}$, compared to 898×10^{-5} for two homogeneous water masses. In this example, it means that the assumption with respect to the velocity v_1 seems to approximate natural conditions quite well.

If the sloping interface reaches up to the surface, it is reflected there as a sharp boundary between the two water bodies. In the ocean, this is shown by sudden changes in the distribution of the physicochemical properties in the horizontal direction. The discontinuity of temperature and salinity (cf. Fig. 7.06), recognizable at the surface, is called "cold wall." As the so-called North Atlantic polar front, it forms the boundary between subtropical water on one side and polar and subpolar water on the other. It can be traced far into the Norwegian Sea. A similar boundary is found in the North Pacific Ocean, and in particular in the Southern Hemisphere where it is known as the Antarctic Convergence and can be followed all around Antarctica, since it is only slightly disturbed by topographical influences.

The tropical and subtropical parts of the oceans are most suitable for the application of the relationship between the inclination of the surfaces and the current system. In those areas, the supposition of a two-layer ocean is widely satisfied in the upper troposphere. In general, a nearly homogeneous warm surface layer is separated by a sharp discontinuity layer from a weakly stratified cold underlayer, which can be considered to be almost at rest. The position of this discontinuity layer is then determined by the currents in the surface layer. An example is given in Fig. 7.11 showing the depth distribution of the tropospheric discontinuity layer in the lower latitudes of the Atlantic Ocean. In accordance with the rule mentioned above, current arrows can be plotted along the isobaths. The crowding of the isobaths is a measure of the current velocity. This example is especially interesting because it shows that the depth distribution of the discontinuity layer is not symmetrical with respect to the equator but a current setting eastward is inserted between the North and South Equatorial Currents, both flowing westward. This eastward current is known as the Equatorial Countercurrent.

The conditions of stratification in an eddy are represented in Fig. 7.12. The cases where the upper layer rotates faster than the lower one are particularly interesting. The direction of rotation is of primary importance, that is, whether the motion is anticyclonic or cyclonic.

In the case of anticyclonic rotation ($v_2 - v_1 > 0$), schematically shown in Fig. 7.12(a), $\tan \gamma$ becomes negative, and the interface rises from the center of the eddy toward the margin; $\tan \beta_1$ becomes positive, and the isobaric surfaces descend from the center out-

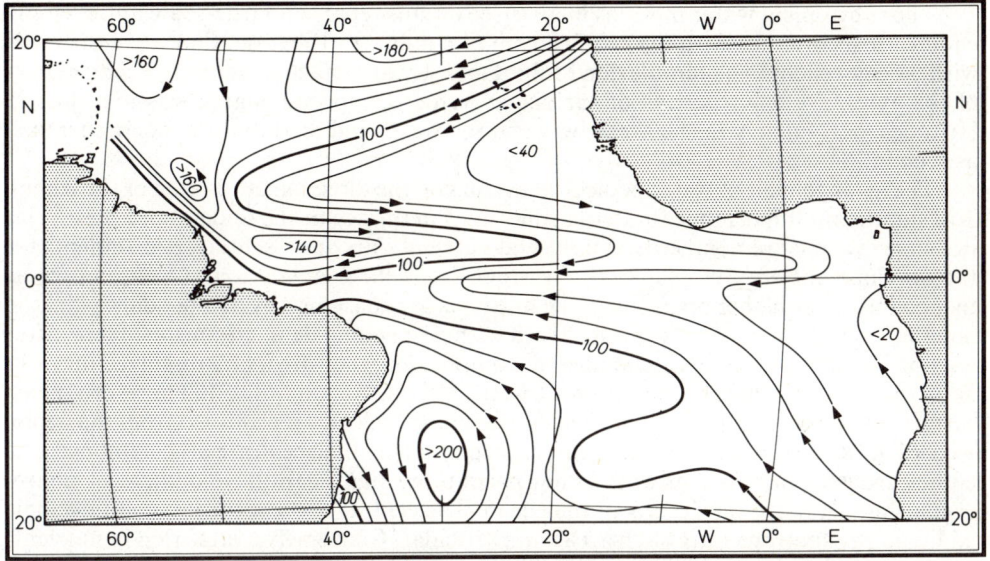

Fig. 7.11. Depth distribution (in m) of the tropospheric discontinuity layer in the tropical Atlantic Ocean. (According to Defant, 1938.) Current arrows for the near-surface layer are based on Margules' equation.

wards. Thus, the accumulation of lighter water in the center of the eddy is effected by anticyclonic motion. With cyclonic motion ($v_2 - v_1 < 0$), the slopes are just reverse. Then, $\tan \gamma$ becomes positive and $\tan \beta_1$ negative; the interface descends, and the isobaric surfaces rise from the center toward the margin, as indicated in Fig. 7.12(b). In the center of the eddy, the heavier water is accumulated in a convex form, whereas the lighter water is pressed toward the margin.

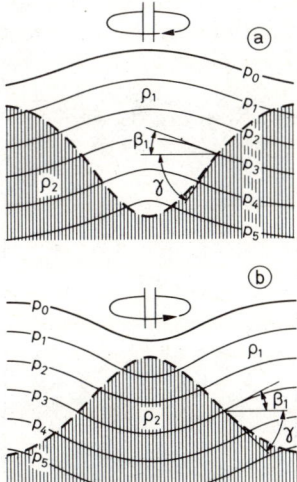

Fig. 7.12. Shape of the isobaric surfaces as well as of the interface between two water bodies if the upper one is rotating faster than the lower one: (a) with anticyclonic motion, (b) with cyclonic motion. (According to Defant, 1929.)

314   Theory of Ocean Currents

The subtropical water motions in all three oceans represent large-scale anticyclonic eddies. In accordance with the requirement of stable stratification, they are associated with a deep-reaching accumulation of light and warm surface water in the central part of the gyre. Cyclonic rotation of surface currents is a special characteristic of higher latitudes, particularly in the areas between the west wind drift and the polar currents, as seen in Plate 6.

The arrangement of water bodies, depending on the direction of rotation of the eddies, is of significant importance for living conditions in the ocean. To understand these relationships we must remember that light and nutrient salts are essential requirements for the development of phytoplankton, on which feeds the zooplankton, which, in turn, is the food basis for higher organisms. The nutrients are consumed by an abundant plankton population and transported into the depth with descending dead organisms. Here, after passing through processes of chemical decomposition, they are again dissolved in water and contribute to the accumulation of nutrients (cf. Section 6.2.5). Therefore, near-surface water becomes poor in nutrients if it is not sufficiently replenished from the depth. Accordingly, areas where surface water is accumulated in the central parts of anticyclonic eddies are poor in nutrients and do not permit any significant development of micro-organisms. They are the deserts of the ocean. The central parts of the large anticyclonic eddies in the subtropics are such areas, in particular. Conversely, water rich in nutrients is transported up to surface layers by the heavy water accumulated in cyclonic eddies. Thus, an important requirement is fulfilled for an abundant development of plankton in those regions, which serve as feeding grounds for higher organisms. These, in turn, are the basis of nutrition for fish eaten by man. Having followed these relationships, we understand why the fishing grounds of commercial fishery in the high seas are almost exclusively located in the area of subpolar cyclonic water motions.

7.3.3. Geostrophic currents in a continuously stratified ocean

Considerations valid for a two-layer ocean can also be transferred to a continuously stratified ocean. Figure 7.13 demonstrates that the inclination $\tan \beta_1$ of the isobaric surface p_1, relative to the inclination $\tan \beta_2$ of the isobaric surface p_2, is determined by

$$\tan \beta_1 - \tan \beta_2 = \frac{h_B - h_A}{l}$$

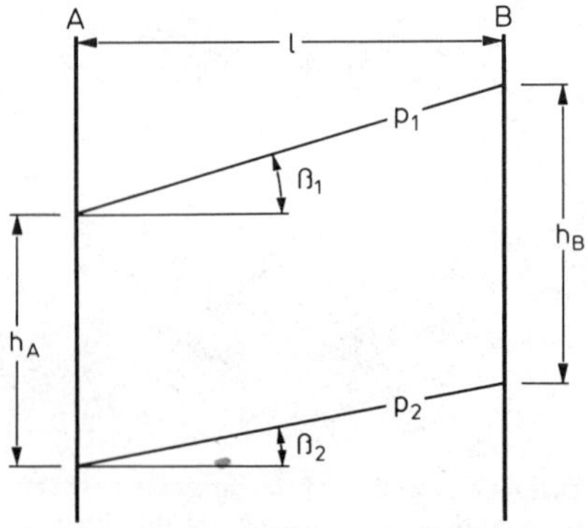

Fig. 7.13. Inclination of the isobars in a continuously stratified ocean.

where h_A stands for the geometric distance of the isobaric surfaces on the vertical A, and h_B for that on the vertical B, and l for the horizontal distance between A and B. According to Eq. (7.34), the geometric distance can be expressed by

$$h = \frac{10}{g} \int_{p_1}^{p_2} \alpha \, dp$$

Taking Eq. (7.46) into account, we get

$$-\frac{2\Omega \sin \varphi}{g}(v_1 - v_2) = -\frac{10}{gl}\left[\left(\int_{p_1}^{p_2} \alpha \, dp\right)_A - \left(\int_{p_1}^{p_2} \alpha \, dp\right)_B\right]$$

If the difference of the dynamic depths—given by the term in square brackets—is expressed in dynamic centimeters, it follows that

$$v_1 - v_2 = \frac{10}{2\Omega l \sin \varphi}(\Delta D_A - \Delta D_B). \tag{7.47}$$

This equation, derived by Helland-Hansen and Sandström in 1903 from Bjerknes' circulation theorem, has often been employed in investigations of oceanic water motion. If the vertical mass structure is known at two oceanographic stations A and B, the above equation gives the difference in the current velocities between two isobaric surfaces. One obtains the relative current components perpendicular *cum sole* to the line connecting A and B. According to the assumptions, this equation is valid only for frictionless and stationary currents if $\varphi \neq 0$.

Ever since this method of computing velocities (the so-called dynamic method) was introduced into oceanography, efforts have been made to transfer the relative values of velocity into absolute values. This was attempted in five different ways. One consisted of trying to determine the velocity at a given level directly from current measurements and to relate it to the calculated relative values. Wüst successfully applied this method in the Florida Current, utilizing earlier current measurements made by Pillsbury. In the open Atlantic Ocean in areas of weak gradient currents, this method fails, if one follows the guiding principle of the *Meteor* expedition and tries to relate the relative values of velocity to the current measurements taken at the anchor stations, because periodic tidal currents and other short-term disturbances impede any exact determination of stationary motion from short current records. Absolute current values from the sea surface are not usually suitable for this method, since they are widely subjected to the influence of the external wind stress. Moreover, current values obtained from ships' logs by relating dead reckoning positions to fixes are not accurate enough.

After the discouraging attempts at using direct current measurements taken in the open ocean for transforming the relative velocity values into absolute ones, some experience with regard to the vertical distribution of currents was used to the effect that the great depths were assumed to be free of currents. For near-surface layers, the errors introduced by this assumption do not carry much weight. Defant (1941) pursued a third way modifying the second procedure. In the Atlantic Ocean, he found a relatively thick intermediate layer with isobaric surfaces running parallel to each other, and he supposed that this layer was motionless. However, this hypothesis is not conclusive.

Furthermore, attempts have been made to deduce a "zero layer" of motion from the vertical distribution of the hydrographic parameters. This method, too, is applicable only with certain restrictions. Finally, the fifth method, suggested by Hidaka of Japan, makes use of the equation of continuity. In cases where the zero layer is located at great depth,

Table 7.05. Sample Calculation of Current Velocity and Volume Transport on the Basis of the Internal Pressure Field, between *Atlantis* Stations 1228 and 1227[a]

1	2	3	4	5
Pressure (dbar)	ΔD_{1228} (d.m.)	$\Delta D_{1228} - \Delta D_{1227}$ (d.m.)	$v_{0\,m} - v_{1300\,m}$ (cm sec^{-1})	M (10^6 m^3 sec^{-1})
0	0	0	114.2	0
25	0.0777	0.0119	117.6	1.18
50	0.1574	0.0298	122.7	2.40
75	0.2370	0.0515	129.0	3.69
100	0.3108	0.0706	134.5	5.03
150	0.4368	0.0936	141.1	7.84
200	0.5376	0.0970	142.1	10.72
250	0.6194	0.0866	139.1	13.58
300	0.6890	0.0678	133.7	16.36
400	0.8066	0.0132	118.0	21.49
500	0.9050	−0.0546	98.4	25.90
600	0.9820	−0.1304	76.6	29.46
700	1.0374	−0.2086	54.1	32.13
800	1.0832	−0.2788	33.9	33.92
900	1.1278	−0.3318	18.6	34.99
1,000	1.1712	−0.3640	9.4	35.56
1,200	1.2560	−0.3920	1.3	36.00
1,400	1.3424	−0.4012	−1.3	36.00
1,600	1.4308	−0.4056	−2.6	35.84
1,800	1.5212	−0.4068	−2.9	35.61
2,000	1.6132	−0.4054	−2.5	35.39

[a] For ΔD at Station 1227, see Table 7.02. $\varphi = 35°48'N$, $l = 40.7 \times 10^5$ cm, $2\Omega l \sin \varphi = 347.1$ cm sec^{-1}

the results of this method are rather uncertain because the distribution of density would have to be known much better than it is at present.

If the third method is applied to our example (Fig. 7.06) of the cross section from Chesapeake Bay to Bermuda, the zero level of motion would drop from 900 to 1600 m. Between *Atlantis* Stations 1228 and 1227 it would lie at 1300 m. An example showing the application of Eq. (7.47) and the calculation of the absolute velocity between those two stations is given in Table 7.05. Column 2 contains the anomalies of the dynamic depths for Station 1228, calculated in the same manner as for Station 1227 (cf. Table 7.02). Column 3 gives the differences in the dynamic depths between Stations 1228 and 1227. Column 4 shows the current velocities perpendicular to the base line of the two stations, referred to the zero level at 1300 m. The computed velocities of 114 cm sec^{-1} at the sea surface and 142 cm sec^{-1} at 200 m depth, representing mean values over a distance of 40.7 km, seem extremely high for currents in the open ocean. Nevertheless, such velocities are not unusual in the Gulf Stream (the core of which is concerned in this example), as has been shown by measurements of the surface current from a moving ship by means of a geomagnetic electrokinetograph. The velocity distribution, based on dynamic calculations along the entire cross section, is presented in Fig. 7.14. The high velocities of the Gulf Stream are concentrated in a narrow, relatively deep-reaching band.

Since the isobaric surfaces must be horizontal in layers of no motion, the relative values of the inclination of the isobaric surfaces can be transformed into absolute values, in a similar manner as the relative values of the velocity. In this way, the absolute dynamic topography of a particular isobaric surface is obtained. An example is given in Fig. 7.15

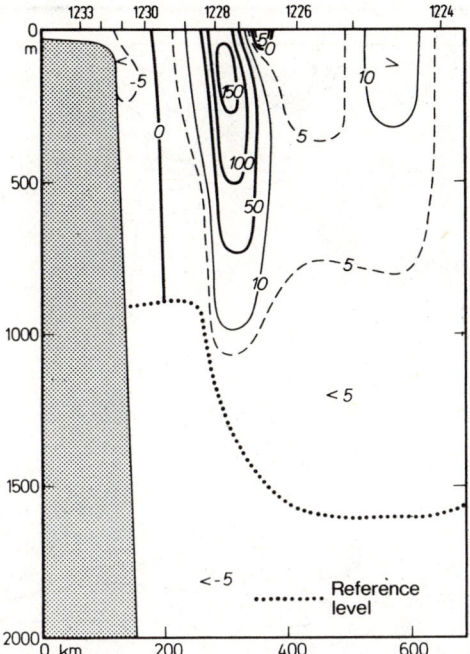

Fig. 7.14. Current velocity (in cm sec^{-1}) of the geostrophic current perpendicular to the cross section Chesapeake Bay—Bermuda Islands. Without sign: NE current, Gulf Stream direction; minus sign: SW current, countercurrents. .... Reference level.

showing the dynamic topography of the sea surface of the North Atlantic Ocean, based on the reference level determined by Defant according to the third method mentioned above. The sea level differences in the area of the North Atlantic Ocean reach approximately 150 cm.

The volume transport can easily be determined from calculated current velocities. Like the velocity, it is related only to the internal geostrophic current but not to the total flow. If the absolute velocity perpendicular to a cross section is known, then

$$M = l \int_0^d v \, dz$$

where d denotes the depth of the zero layer. Applying (7.47), we obtain the water transport M down to the depth h as amounting to

$$M = \frac{10}{2\Omega \sin \varphi} \int_0^h (\Delta D_A - \Delta D_B) \, dz - \frac{10h}{2\Omega d \sin \varphi} \int_0^d (\Delta D_A - \Delta D_B) \, dz$$

(7.48)

The volume transport is independent of the distance between the stations. Subsequent to the example of the calculation of velocities, column 5 in Table 7.05 contains the result of the computation of the volume transport. The individual values give the transport for the whole layer from the surface down to the respective pressure surface given in dbar and closely approximating the depth in meters. For instance, for the layer 0 to 1300 dbar

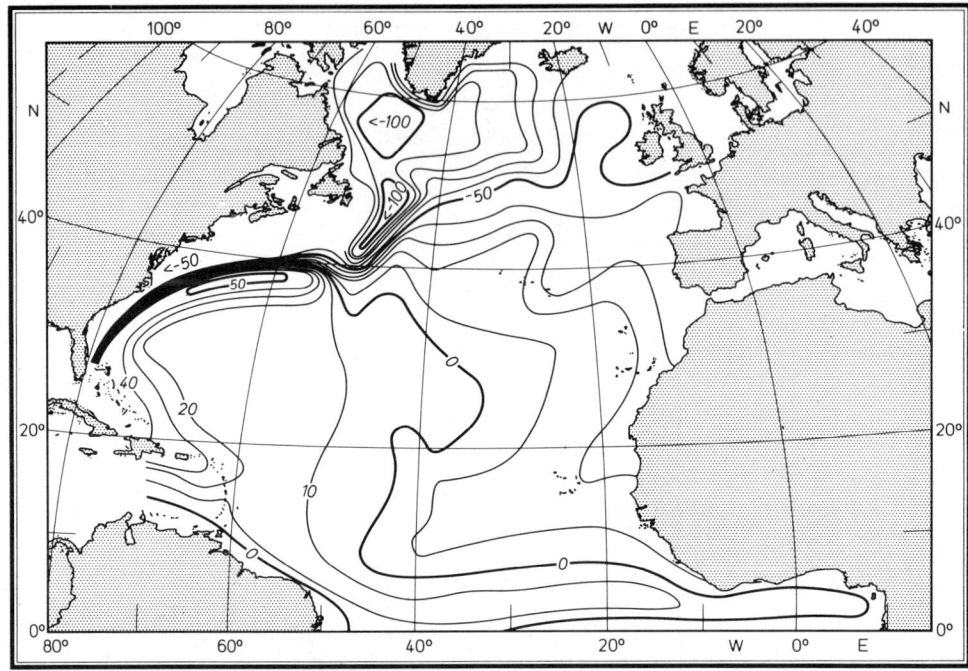

Fig. 7.15. Topography of the physical sea surface of the North Atlantic Ocean, given as a deviation from the ideal sea surface level, in d. cm. (According to Defant, 1941.)

or about 0 to 1300 m, a transport of 36×10^6 m^3 sec^{-1} is obtained. Figure 7.16 shows the distribution of the volume transport above the zero layer of motion. The water transport (roughly 57×10^6 m^3 sec^{-1}), concentrated in a narrow band within the Gulf Stream proper, is not equalled by any other oceanic current with the exception of the Kuroshio in the Pacific Ocean. It exceeds the runoff from all continental rivers into the world ocean, amounting to approximately 0.9×10^6 m^3 sec^{-1}, by a factor of around 65 (cf. Section 4.3.5). Consequently, the transport of heat and salt by the Gulf Stream exerts a long-range effect, compared to which the influence of the large continental rivers seems insignificant.

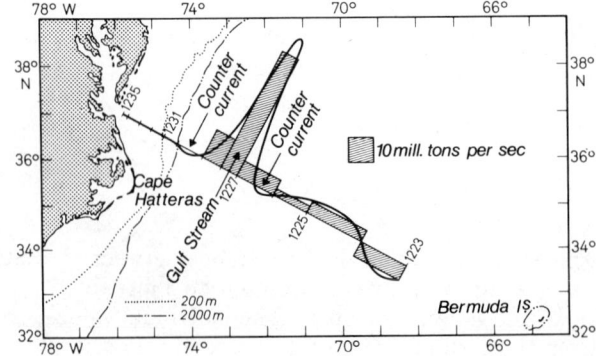

Fig. 7.16. Water transport perpendicular to the cross section Chesapeake Bay–Bermuda Islands.

7.3.4. Drift current in a homogeneous ocean

In order to investigate the effect on water masses exerted by wind alone, we neglect the internal pressure forces and assume a homogeneous, infinite ocean upon which a shearing stress, constant in time and space, is acting as external force. Taking into consideration the deflecting force of the earth's rotation and the frictional force, we can derive the following equations for horizontal motion from the equations of motion [Eqs. (7.04)]:

$$\frac{A_z}{\rho}\frac{\partial^2 u}{\partial z^2} + 2\Omega v \sin \varphi = 0$$

$$\frac{A_z}{\rho}\frac{\partial^2 v}{\partial z^2} - 2\Omega u \sin \varphi = 0 \tag{7.49}$$

According to Eq. (7.12), the boundary conditions are given by $T_x = -\mu \partial u/\partial z$ and $T_y = -\mu \partial v/\partial z$. The solution of these differential equations yields the vertical velocity distribution in a pure drift current. For an ocean of infinite depth we obtain

$$u = V_0 e^{-\pi z/D} \cos\left(45° - \frac{\pi}{D}z\right)$$

$$v = V_0 e^{-\pi z/D} \sin\left(45° - \frac{\pi}{D}z\right) \tag{7.50}$$

where

$$D = \pi \sqrt{\frac{A_z}{\rho \Omega \sin \varphi}} \tag{7.51}$$

$$V_0 = \frac{T}{\sqrt{\rho A_z\, 2\Omega \sin \varphi}} = \frac{\pi T}{D\rho\Omega\sqrt{2} \sin \varphi} \tag{7.52}$$

The detailed derivations were made by Ekman who published them in 1905. He had been stimulated by Nansen who, during his polar drift on the *Fram* (1893–1896), had regularly observed deviations of the direction of the ship's drift from the wind direction. In the equations for the pure drift current (Eqs. 7.50), V_0 denotes the velocity of the forced current at the sea surface. The corresponding current direction is turned 45° *cum sole* from the direction into which the wind is blowing. If the resultants of the components u and v are represented in a three-dimensional model as functions of the depth z (cf. Fig. 7.17), the vectors of the drift current form a spiral staircase tapering off downwards. In the Northern Hemisphere, the current vectors turn to the right-hand side when depth increases. Projected on a xy plane, they represent a logarithmic spiral. At the depth D, which—according to Ekman—is called the depth of frictional resistance of the drift current, the current vector is turned by 180° against the vector at the sea surface. Down there, the velocity is reduced to only $e^{-\pi} = 1/23$ of the surface value V_0. D increases with the square root of the vertical austausch coefficient A_z and decreases with the square root of the sinus of the latitude.

The water transport by the pure drift current down to the depth of frictional resistance is obtained by integrating Eqs. (7.50):

$$M_x = \frac{T_y}{\rho 2\Omega \sin \varphi} \qquad M_y = -\frac{T_x}{\rho 2\Omega \sin \varphi} \tag{7.53}$$

The transport depends linearly on the wind stress but is independent of the depth of frictional resistance and is directed perpendicular *cum sole* to the wind direction.

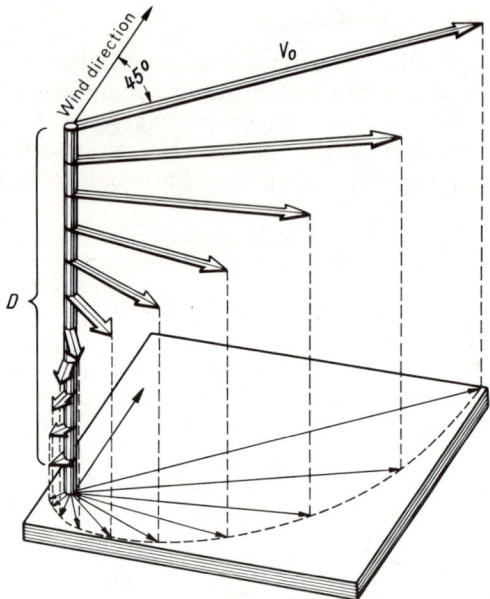

Fig. 7.17. Vertical current distribution in a pure drift current in the Northern Hemisphere. (According to Ekman, 1905.)

With finite water depth, noticeable changes in the relations of the pure drift current do not occur as long as the water depth considerably exceeds the depth of frictional resistance D. With decreasing water depth, the angle of deflection between the wind and the surface current becomes smaller than 45°. Ekman gives the angle of deflection as a function of the ratio of water depth to frictional depth, assuming the austausch coefficient A_z as constant. In reality, the vertical austausch becomes zero at the bottom, and the angle of deflection is influenced by the dependence of the austausch on the distance from the bottom, as shown by Fjeldstad (1929). The variation of A_z with depth, however, is not known well enough.

As in shallow water, the pure drift currents are modified when density is stratified. The mixing length and thus the vertical austausch decrease with increasing stratification and may cease completely if the density changes abruptly in the vertical (in a so-called pycnocline), in accordance with the criterion for the maintenance of vertical turbulence by Richardson (Eq. 7.25). Such a pycnocline impedes the full development of a pure drift current as much as the sea floor does.

It is difficult to check the theory of drift currents against observations in the ocean, because the simplifications introduced in the theory by certain assumptions cannot be strictly fulfilled in the ocean. Shorelines obstruct the water transport of the drift currents and cause a pile-up of water, which, in turn, induces currents that will be included in the observations. Furthermore, even in the climatological average, the wind is not uniform over large distances of the oceans. Therefore, some piling-up of water must always be expected in the open ocean. If special regions are chosen for observation, such difficulties will possibly be avoided, and the results of the drift current theory, as advanced by Ekman, can be confirmed to a certain extent.

7.3.5. Ekman's elementary current system

Since a characteristic current structure is established under the influence of the frictional force of the wind, a similar effect must be expected for the layer near the ocean floor where bottom friction is effective.

The vertical current distribution within the bottom current can be derived from the structure of the pure drift current by a simple deduction. First, let us consider a frictionless, homogeneous current, briefly called deep current, which extends to the bottom and has the velocity V. If the velocity $-V$ is ascribed to the deep current and to the bottom, the water will be at rest, and the bottom will move with the velocity $-V$. When friction comes into play, a bottom current must develop that corresponds to a pure drift current, except for the fact that top and bottom are reversed and the surface velocity of the drift current is present at the bottom, pointing in the negative x direction. If, then, $+V$ is again superimposed geometrically on the entire system, the actual motion is obtained in which no current exists at the bottom, and the bottom current passes into the homogeneous deep current at the lower depth of frictional resistance D'. The deep current may, for instance, be caused by a pressure gradient.

In a homogeneous ocean, the entire current system usually consists of three regimes, as indicated in Fig. 7.18(a): (1) a deep current, determined by the pressure gradient, (2) an upper current, superimposed on the deep current down to the upper depth of frictional resistance D and representing the geometrical vector sum of the deep current and the pure drift current, and (3) a bottom current, beneath the lower depth of frictional resistance D', which includes the influence of bottom friction on the deep current. Figure 7.18(b) shows the corresponding distribution of current direction projected on a horizontal plane. This fundamental form of current stratification is called "elementary current" according to Ekman. At the same time, Fig. 7.18(b) is valid for the vertical structure of currents with winds blowing parallel to the coast. The water carried away from the coast by the surface current must be replaced by the bottom current, because the deep current does not contribute to the compensation of mass perpendicular to the coast. However, since the upper current carries light surface water and the bottom current transports the heavy water of deeper layers, heavy, mostly cold water, rising for reasons of continuity, accumulates near the shore. Part of this so-called upwelling water reaches the surface. In particular, this phenomenon is a characteristic feature of the areas of origin of the trade winds along the western sides of the continents where winds prevail that blow parallel to the coast or offshore. Pertinent examples are the coasts of Southwest Africa and Northwest Africa as well as the coasts of California and Peru. As the water of the deeper layers is especially rich in nutrients for reasons mentioned previously, rich marine life develops in those areas of upwelling.

7.3.6. Sverdrup regime

The theory of the elementary current (cf. Section 7.3.5) is not applicable, if the assumption of a homogeneous ocean, unlimited in the horizontal, does not hold true any more, and if the water transport of the drift current contributes to changes in the field of mass and, thus, to the development of an internal pressure field. Then, internal gradient currents will be generated. The application of Bjerknes' circulation theorem to such currents yields useful results if the supposition of stationarity is fulfilled, as confirmed, for example, by the agreement of the computed water transport of the Florida Current with the value

322 Theory of Ocean Currents

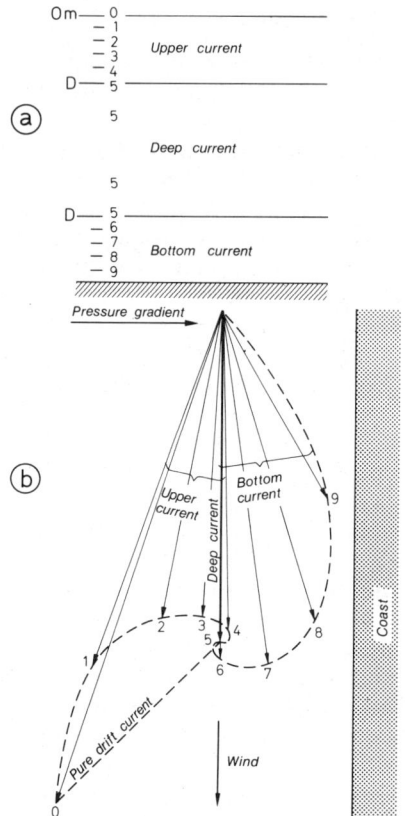

Fig. 7.18. Ekman's elementary current system.

measured directly. For the former Wüst (1924) computed 25.6×10^6 tonnes sec^{-1}, and for the latter he calculated 25.0×10^6 tonnes sec^{-1} from observational data. While the confidence in the applicability of Bjerknes' circulation theorem on quantitative computations of velocity and transport has been supported, especially by this example, although external forces such as the action of wind and friction are neglected in the theorem, this does not imply that no external forces are acting. It indicates a close mutual relationship between the distribution of mass in a stratified ocean and, thus, the internal field of force and the actual current generated by external forces. If the current is changed by the influence of external forces, a redistribution of the mass field will take place, which is reflected by circulations transverse to the main current direction. This process will continue until a new equilibrium has been reached. Such cross circulations must be associated with vertical components of motion which, though small compared to the horizontal current components, cannot be left out of consideration if the mass distribution is to be understood.

Particular progress in the general theory of oceanic circulation in the surface layers has been made since 1947 after the hydrodynamic equations of motion had been simplified in some way other than in Section 7.3.4. For the large-scale circulation, the dependence on latitude of the deflecting force of the earth's rotation must be taken into account. The

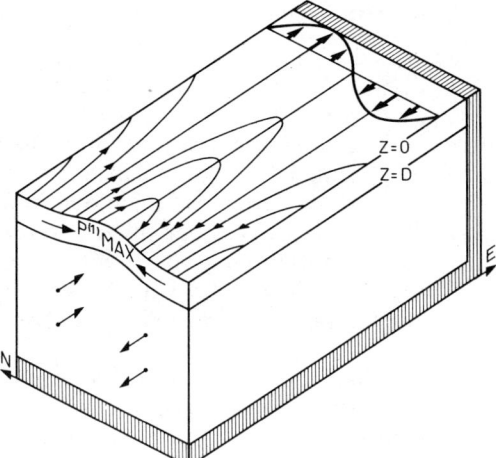

Fig. 7.19. Wind field (thick arrows) and current system in the Ekman friction layer ($0 < Z < D$) and in the depth. (According to Stommel, 1965.)

equations can be considerably simplified if we renounce any information on the vertical velocity distribution of the currents, which means integration over the entire water column down to a layer assumed not to be in motion. In this way, an equation for the vorticity, the vertical component of the curl of the velocity vector, can be derived from the equations of motion. It is the sum of three vorticity quantities representing the effects of wind, internal friction, and of the deflecting force of the earth's rotation. Since the latter quantity is well known, since internal friction can be estimated roughly, and since the wind field is given by observation, the horizontal mass transport by ocean currents can be determined approximately.

Neglecting internal friction, Sverdrup (1947) reduced the vorticity equation for zonal wind conditions to

$$\frac{\partial M_x}{\partial x} = -\frac{1}{\beta}\frac{\partial^2 T_x}{\partial y^2} \qquad (7.54)$$

Here, M_x is the zonal mass transport, T_x the zonal shear stress, and β the change of the Coriolis parameter with latitude; x points toward the west and y toward the north. The solution of this equation for a wind field corresponding to that of the lower latitudes is given in Fig. 7.19. The thick arrows and the curve enveloping them represent the wind field of the lower latitudes (trades) and the middle latitudes (west wind drift). The mass transport in the Ekman layer between the depths 0 and D is directed perpendicular to the wind direction. This creates a pressure field that drives the circulation in the depth. The mass transport, integrated vertically, has one component directed toward the equator and another one directed westward, which grows with increasing distance from the coast. This corresponds to the conditions in the world ocean. On the eastern sides of the oceans, the motions are directed toward the equator, but, with increasing distance from the coast, they become zonal currents parallel to the wind. The system, expressed by Eq. (7.54), is called the Sverdrup regime. Because of the unlimited growth of the solution toward the west, its validity is restricted to the eastern and central parts of the oceans. In the western parts, friction is dominant.

7.3.7. Linear theory of the western boundary currents

The frictionless Sverdrup model does not offer any solution for enclosed oceanic areas. On the western sides of the oceans the zonal current, increasing toward the west, must comply with the kinematic boundary condition, that is, decrease to zero. This is impossible unless the frictional force is taken into consideration. Then, the balance equation will represent the equilibrium among frictional vorticity, planetary vorticity, and the vorticity of the tangential shearing stress. In the lower and middle latitudes, the ocean receives anticyclonic vorticity from the wind field. Without any friction, vorticity would continuously increase in the ocean and an anticyclonic vortex with steadily increasing velocity would develop. There are three possibilities of vorticity dissipation: (1) dissipation at the sea floor, (2) dissipation by internal friction, (3) dissipation by lateral friction along coasts and shelves.

The wind-driven circulation does not reach down to the sea floor. Therefore, the first possibility is inapplicable. With a view to the general circulation, internal and lateral friction basically have the same result. The latter model, however, permits the representation of the current system in greater detail. First of all, we are going to discuss the results obtained under consideration of internal friction (Stommel, 1948). They are represented in Figs. 7.20(a) and 7.20(b).

Figure 7.20(a) shows the circulation in a rectangular basin under the assumptions of a constant Coriolis parameter, of easterly winds in the lower latitudes, and of westerly winds in the middle latitudes. In this case, the frictional vorticity and the vorticity of the tangential shearing stress are in balance, and the anticyclonic cell is symmetrical.

Figure 7.20(b) represents the analogous result in case the Coriolis parameter increases linearly with latitude (β plane), as is approximately true on the earth. This increase of the Coriolis parameter corresponds to a field of cyclonic planetary vorticity in the Northern Hemisphere. At the equator, this field is zero, and it increases toward the poles. There is now a vorticity balance among frictional vorticity, planetary vorticity, and the vorticity of the tangential shearing stress. On the eastern sides of the oceans, the water moves toward the equator, where the cyclonic planetary vorticity is smaller than in the middle latitudes. This decrease of cyclonic planetary vorticity corresponds to the added anticyclonic vorticity of the wind field. Thus, these two forms of vorticity are practically balanced. Since the currents are weak, the frictional vorticity is small enough to be neglected. This part corresponds to the Sverdrup regime described in the previous section.

In an enclosed oceanic area, the mass transport on the eastern side toward the equator must be compensated on the western side by a mass transport toward the poles. But there,

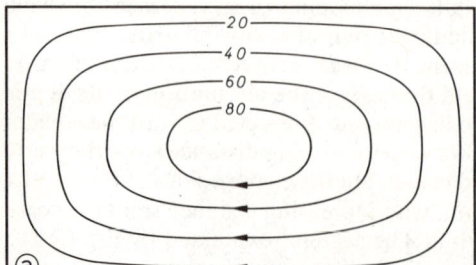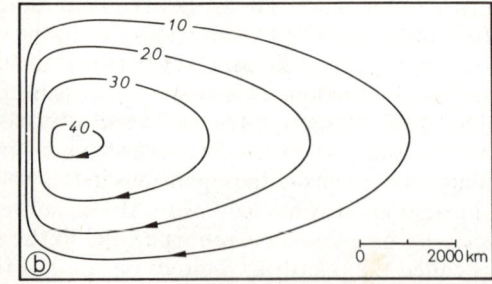

Fig. 7.20. Stream lines in a rectangular basin. Left-hand side: for a constant Coriolis parameter. Right-hand side: with consideration of the dependence on latitude of the Coriolis parameter (β plane).

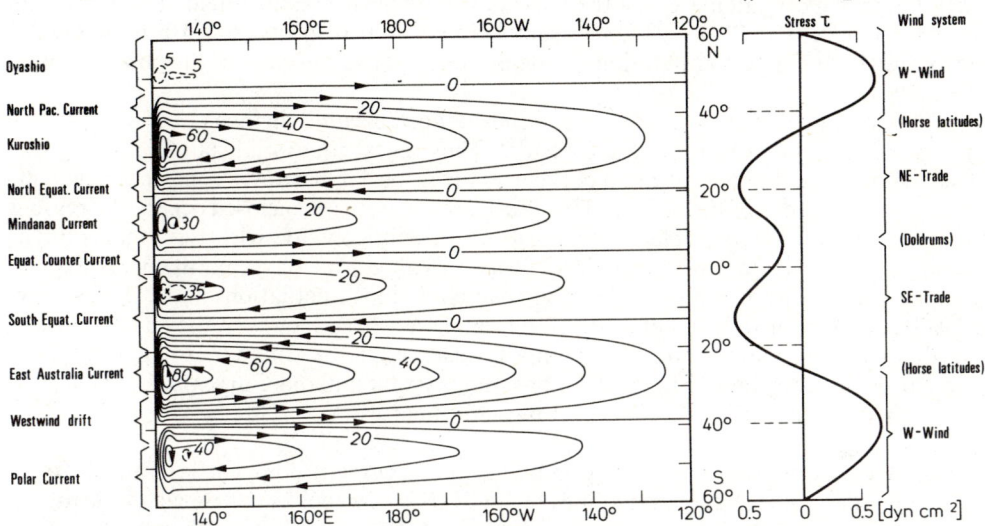

Fig. 7.21. Mean mass transport in a rectangular ocean (position and dimensions adapted to the Pacific Ocean) for a given zonal wind system (water transport between two lines amounts to approximately 10×10^6 m^3 sec^{-1}). (According to Hidaka, 1951.)

the planetary vorticity cannot balance the vorticity of the wind field. A water particle moving from lower to higher latitudes passes from regions of little planetary vorticity into others with greater planetary vorticity. In higher latitudes, therefore, it possesses anticyclonic vorticity, relative to its surroundings, by which the effect of the anticyclonic vorticity of the wind field is intensified even more. This can be balanced only by frictional vorticity. But as friction is proportional to velocity, greater velocities must occur on the western sides of the oceans than on the eastern sides in order to compensate those two forms of vorticity. The result is the asymmetrical cell shown in Fig. 7.20(b), with a weak and wide current, directed toward the equator, on the eastern side and a narrow but strong current on the western side, which, in the North Atlantic Ocean, corresponds to the Gulf Stream and in the North Pacific Ocean to the Kuroshio.

In contrast to Stommel's model, Munk (1950) was the first to consider lateral friction. His model also yields the countercurrent formed at the right-hand margin of the Gulf Stream.

The equations have been applied to a great number of ocean models. If analytical solutions are dispensed with, and the equations are integrated numerically, the actual distribution of the shear stress can also be taken into account.

Relevant results are illustrated by the example of the Pacific Ocean in Fig. 7.21. To simplify matters, a rectangular basin of constant depth is considered on the rotating earth. The dimensions of the basin are adapted to those of the Pacific Ocean. Furthermore, a purely zonal wind system is assumed, rather similar to the actual conditions in the Pacific Ocean. The shear stress at the sea surface is plotted in the right-hand part of the figure. A horizontal austausch coefficient of $A_h = 6.5 \times 10^7$ g cm^{-1} sec^{-1} is used. The resulting system of surface currents is sharply concentrated on the western side of the ocean. The geographic names of the various parts of this system are given on the left-hand side of the figure. Comparison with the current chart on Plate 6 shows that all significant parts of the circulation are represented; even the mass transport seems reasonable if compared

326 Theory of Ocean Currents

with the transport calculated on the basis of the internal pressure field. The Kuroshio, for which more than 70×10^6 tonnes sec^{-1} are given, transports 65×10^6 tonnes sec^{-1} according to dynamic computations made by Sverdrup on the basis of the observed mass distribution.

The computed large-scale forms of wind-induced oceanic surface circulation, shown in Fig. 7.21, contain two remarkable results. They reveal equatorial countercurrents and also demonstrate the concentration of the water transport in narrow, strong ocean currents on the western sides of the oceans. This illustrates that, in an enclosed ocean, the regions with the highest current velocities do not necessarily coincide with the regions of the strongest wind action. Furthermore, the theoretical results may quantitatively explain the pulsations of water transport in strong currents, since fluctuations in the wind system must be accompanied by fluctuations of the water transport. However, they do not explain the details in the course of ocean currents as observed in the Gulf Stream, partly also in the Kuroshio, and slightly in the Agulhas and East Greenland Currents. In particular, three phenomena cannot be explained by such theories.

1. **Meandering currents.** As soon as the main current does not laterally lean against a continent, streamlines begin to show a wavelike course which seems to resemble the meandering of a continental river. An idea of such meandering currents is conveyed by Fig. 10.44, based on synoptic temperature observations in the Gulf Stream, if it is noted that the axis of the main current coincides with the strong congestion of isotherms.

2. **Propagation of current meanders.** Meanders of currents are not fixed locally like meanders of rivers on the continents, but move slowly with the current. This phenomenon is recognized only after repeated observations in a major oceanic area and is illustrated in Fig. 7.22 by the example of the Gulf Stream, from which two such sets of observations of the temperature distribution, obtained at a 2 weeks' interval, are available.

3. **Separation of cyclonic eddies.** If the amplitude of a current meander exceeds

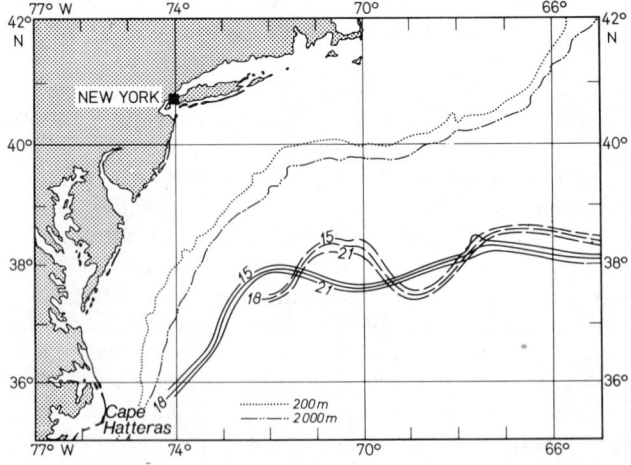

Fig. 7.22. Displacement of the Gulf Stream within two weeks, deduced from the average temperature of the uppermost 200 m on June 8 (solid isotherms for 15, 18, and 21°C) and on June 21–22, 1950 (dashed isotherms). (After Fuglister and Worthington, 1951.)

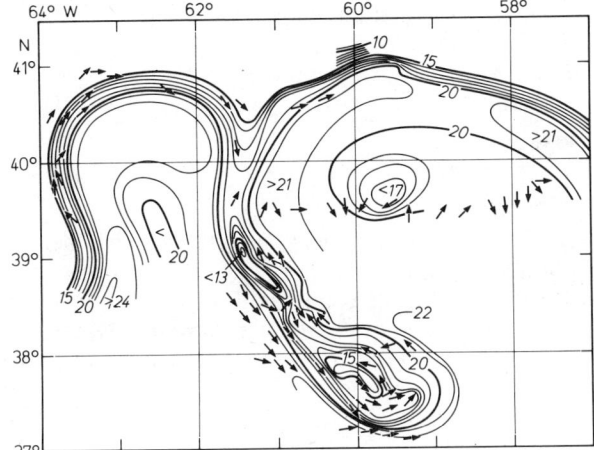

Fig. 7.23. Example of the separation of a cyclonic eddy from the Gulf Stream, deduced from the average temperature of the uppermost 200 m on June 17, 1950, and the current direction based on observations made with the GEK from the moving vessel. (According to Fuglister and Worthington, 1951.)

a critical value, large water masses are cut off in cyclonic eddies. Such a process has been observed in the Gulf Stream and is reproduced in Fig. 7.23. The separation of a cyclonic eddy south of the Gulf Stream margin is almost completed. Eventually, the eddy will gradually vanish as a consequence of mixing. The accumulation of cold water, observed east of the developing eddy (cf. Fig. 7.23), was probably caused by a previous eddy of the same kind.

Each of these phenomena is found in the so-called jet stream in the higher troposphere of the atmosphere in moderate latitudes. From the similarity of the phenomena in ocean and atmosphere it can be concluded that we deal with processes characteristic of the internal dynamics of jet streams in a stratified medium on the rotating earth.

7.3.8. Nonlinear theory of the western boundary currents

There is no doubt that the asymmetry of large-scale circulation cells is a consequence of the earth's rotation. What they gain in anticyclonic vorticity by the wind field and the earth's rotation, the water masses, when flowing poleward, can balance only at high velocity. But it is open to question whether the nonlinear terms of the equations of motion (Eq. 7.04) can be left out of consideration.

Charney (1955) and Morgan (1956) have shown that the vorticity balance can also be maintained by means of the nonlinear terms (instead of the friction terms); they have developed theories in which the friction terms are not taken into account. In a completely analogous way, western boundary currents, resembling the great ocean currents, result from the nonlinear terms. It is an essential contrast to linear models that the core of the anticyclonic cell in the Northern Hemisphere is shifted toward the northwest so that the current is concentrated in the northwestern part of the ocean.

For the time being, the complete hydrodynamic equations of motion can only be integrated numerically. By applying austausch coefficients of different sizes, we can investigate which details of currents will result from the combined effect of friction and nonlinearity. As regards the results, attention may be drawn to Bryan (1963) and Veronis (1966a).

7.4. Nonstationary Currents

7.4.1. Currents and waves

In Section 7.3 only stationary currents have been dealt with. Those currents, however, are subject to temporal variations because they depend on the fields of wind and air pressure varying with time. For geostrophic equilibrium an exact balance is required between the horizontal component of the pressure gradient and the Coriolis force as well as between the vertical component of the pressure gradient and gravity. This is not the case with variable currents, for which deviations from such equilibrium will cause accelerations, and the water particles will oscillate around the position of equilibrium. Here, two different types of waves can be distinguished:

1. Inertial waves and planetary waves (Rossby waves) result from the Coriolis force and from the dependence of the Coriolis parameter on latitude.
2. Gravity waves are generated by the gravity force.

The former usually represent strictly horizontal motions while vertical motions occur with the latter so that also the sea surface is subjected to variations.

For a survey, the currents varying with time can be demonstrated best in the form of a current spectrum. In such a spectrum the current is represented by the sum of currents varying periodically and having different frequencies (or periods). Each period can be associated with a particular amplitude. If those amplitudes are plotted as a function of the frequency $\omega^* = 1/\tau$ (where τ is the period), the current spectrum is obtained. Figure 7.24 gives a schematic example largely corresponding to the current spectra observed in the ocean (cf. Fig. 10.14) (thin line). The thick lines represent the amplitudes of the mean current, of the tidal currents, and of the inertial currents. The current field would be composed of these components, if nothing but tidal forces and highly variable meteorological power fields were the causes of currents. But, in fact, the meteorological fields, variable in time and space, have a broad spectrum of motion of their own. All those me-

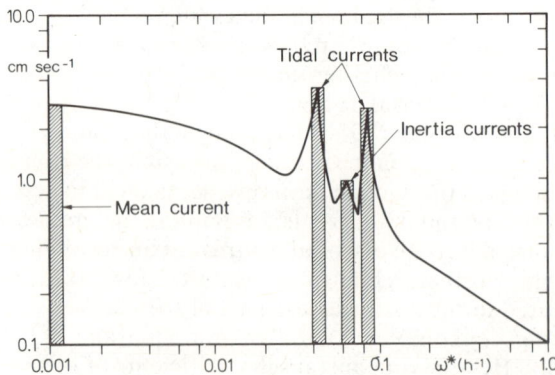

Fig. 7.24. Schematic representation of a current spectrum.

teorological fluctuations generate motions in the ocean so that the oceanic spectrum of motion also is continuous (thin line). The eddies contribute to the spectral distribution as well. The fact that the spectrum of motion in the ocean is a continuous one prevents any clear separation into waves and currents. Pure, undamped waves do not occur in nature, waves rather grow to different magnitudes and fade away later on. Thus, their amplitudes vary with time. Often, there are only a few wave trains. If several such waves of different frequencies simultaneously superimpose each other, very often the wavelike nature of the process is no longer recognizable in this superposition which is the only object of observation. Therefore, waves and nonstationary currents often are nothing but formal classifications of the same phenomenon.

7.4.2. Inertial waves

A water particle which is governed only by the Coriolis force obeys the equations

$$\frac{\partial u}{\partial t} + 2\Omega v \sin \varphi = 0$$

$$\frac{\partial v}{\partial t} - 2\Omega u \sin \varphi = 0 \quad (7.55)$$

if the nonlinear terms in Eq. (7.04) are neglected. The solution of the above equations reads as follows:

$$u = u_0 \sin (ft) \qquad v = -u_0 \cos (ft), \quad (7.56)$$

where $f = 2\Omega \sin \varphi$ is the Coriolis parameter. Integration of (7.56) leads to

$$(x - x_0)^2 + (y - y_0)^2 = u_0^2/f^2, \quad (7.57)$$

that is, the particle describes a circle with the radius u_0/f. This circle is called inertial circle. It is passed by the particle within the period $\tau = 2\pi/f = \pi/(\Omega \sin \varphi)$, which corresponds to half the time required for the complete turning of the plane in which a Foucault pendulum oscillates; therefore, the period is also called half a pendulum day.

On the rotating earth, a point of mass can move along a straight line with the velocity V_0 only if a constant force is acting on it which is antipodal to the deflecting force of the earth's rotation. Therefore, this force must also be directed perpendicular to the velocity, and it must have the magnitude of $2\Omega V_0 \sin \varphi$ per unit mass. For any other velocity V, the motion will consist of two parts—a motion along a straight line with uniform velocity V_0 and another one along the inertial circle, the radius of which is determined by $V - V_0$. The result is a cycloidal trajectory of the water particle.

These predictions are confirmed by the example in Fig. 7.25 showing a trajectory based on current observations in the Baltic Sea west of Gotland over a period of 7 days. Inertial motion is superimposed upon the general motion directed toward the northwest and the north. This inertial motion turns *cum sole* after being initiated by some spontaneous change of the current-generating force which, in the above case, is unknown. The oscillation period of 14 hr corresponds to half a pendulum day for the latitude of the point of observation, namely $\tau = 14$ hr 8 min. Inertial oscillations of this kind occur very frequently in the ocean. A review of the results of observations has been given by Webster (1968). Within the near-surface layer, inertial oscillations are directly generated by wind, as already shown by Fredholm (see Ekman, 1905). For the problem of the Ekman drift current, discussed in Section 7.3.4, stationary wind is assumed. Even with variable wind

330 Theory of Ocean Currents

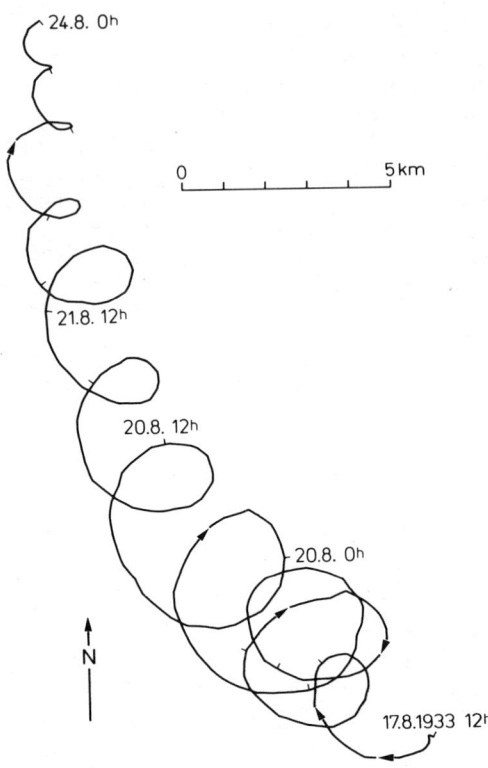

Fig. 7.25. Trajectory based on current measurements from August 17–24, 1933, at 57°49′N and 17°49′E, west of Gotland. (After Gustafson and Kullenberg, 1936.)

fields, a drift current can be formed in an ocean of infinite extension. It is deflected by 45° *cum sole* against the wind direction, but this state will be reached only after an infinitely long period of time.

Figure 7.26 shows how the vector of the surface current eventually approaches the stationary direction if, after a period of calm, wind of constant direction and strength

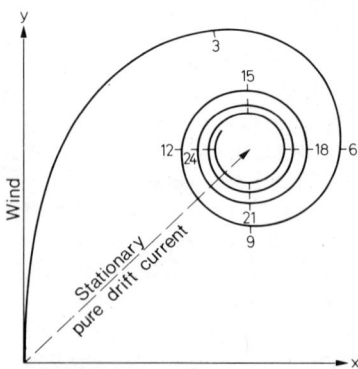

Fig. 7.26. Adjustment of the current vector at the sea surface to the stationary position after the onset of the wind, in pendulum hours. (According to Ekman, 1923.)

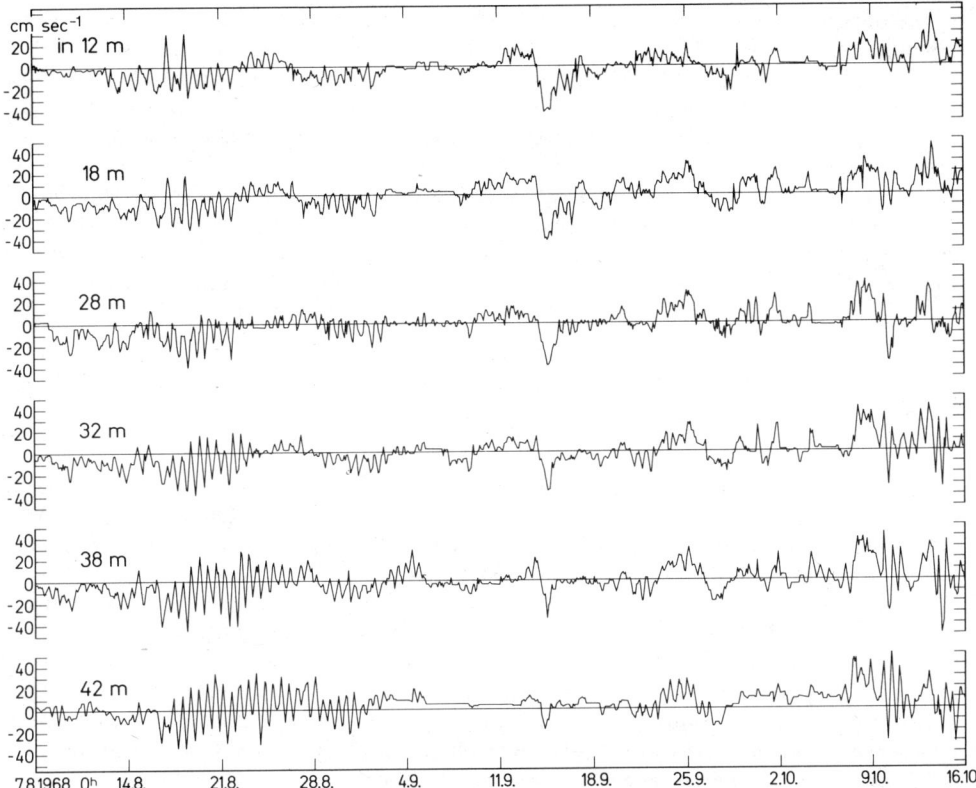

Fig. 7.27. East component u of the current at depths of 12, 18, 28, 32, and 42 m at a permanent station in the Arkona Basin (Baltic Sea) ($\varphi = 55°02.4'$N, $\lambda = 13°50.6'$E) from August 7 to October 16, 1968. The short oscillations are inertial waves.

starts blowing. The curve represents the path of the tip of the current vector originating at the center of the coordinate system. The numbers give the position of the tip of the vector in pendulum hours after the onset of the wind. The vector describes damped oscillations with the period of half a pendulum day, that is, with the period of inertial oscillations.

Inertial waves mostly occur when the wind suddenly starts blowing, particularly at the passage of meteorological fronts. Such waves are damped only slightly and therefore they often last longer than one week. An example is given in Fig. 7.27 representing the east component of a velocity record in the Baltic Sea over a period of several weeks. An explanation of the little spatial coherence of the oscillations has not yet been found. In general, observations at different locations yield very different records. This may indicate that inertial waves represent local oscillations of water particles around their state of equilibrium.

7.4.3. Planetary waves

A role similar to that of the inertial waves, caused by the Coriolis force as repelling force, is probably played in the large-scale circulation by planetary waves, which are caused by the dependence on latitude of the Coriolis parameter. How great an influence this dependence on latitude has on the stationary general circulation has been described in Section 7.3.7. Processes of beginning oscillation, corresponding to those of the previous section, occur in nonstationary wind fields. But here, the waves are phenomena of very large scale with wavelengths on the order of magnitude of 1000 km and periods of several days or weeks.

Planetary waves, also called Rossby waves, were discovered through the solution of the wave equation for a globe completely covered with water. They are designated as solutions of second order of the tidal equation. Rossby (1939) has investigated them in greater detail on the so-called β plane. The β plane is the horizontal plane of a coordinate system in which the dependence on latitude of the Coriolis parameter $f = 2\Omega \sin \varphi$ is approximated by a linear relation $f = f_0 + \beta y$.

The phase velocity of planetary waves is given by

$$c = \frac{\beta}{k^2} \tag{7.58}$$

where k is the absolute value of the wavenumber vector ($k = \sqrt{\kappa^2 + \eta^2}$). The waves generally move westward. Their energy, however, can be transported in different directions. The decisive factor for this is the group velocity. Short waves transport their energy toward the east, long waves toward the west. The vector of the group velocity is also decisive for the reflection of those waves. In contrast to normal waves, whose reflection is governed by Snell's law stating that the angle of incidence equals the angle of reflection, the reflection of planetary waves at the coasts occurs in such a way that the vector of the group velocity complies with Snell's law. Consequently, short planetary waves with the group velocity vector pointing eastward are reflected at the east coasts of the oceans as long waves, whereas long waves, the energy of which is transported toward the west, are reflected as short waves at the west coasts of the oceans.

In enclosed oceanic areas, the possible periods of such waves are given by

$$\tau_{mn} = \frac{4\pi}{\beta} \sqrt{\left(\frac{m\pi}{L}\right)^2 + \left(\frac{n\pi}{B}\right)^2} \tag{7.59}$$

where L and B denote the length and the width of the oceanic area, and m and n are integers.

As mentioned before, planetary waves are generated by fluctuations of the large-scale wind fields. Although up to now no definite observational data are available, these waves can be expected especially in the monsoon areas where the direction of the entire wind system changes with the season.

7.5. The Influence of Bottom Topography on Ocean Currents

7.5.1. Potential vorticity

In the theories mentioned so far, the water depth is always assumed to be constant. Actually, however, the bottom topography has a great influence on the general circulation. This holds true not only in the case of a deep-sea ridge interfering with a current, but quite in general, as pointed out already by Ekman (1923). Considering a water column, that is, a water mass between sea floor and sea surface, we can approximately state that the

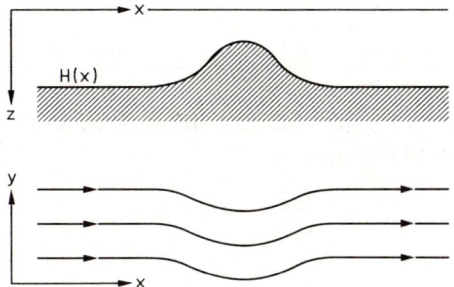

Fig. 7.28. Overflow of a deep-sea ridge. Above: vertical cross section in the xz plane. Below: horizontal current in the xy plane.

potential vorticity $(2\Omega \sin \varphi + \zeta)/H$ of this column remains constant. Here,

$$\zeta = \frac{\partial v}{\partial x} - \frac{\partial u}{\partial y}$$

is the so-called relative vorticity, that is, the z component of the curl of the velocity vector in a relative coordinate system rotating with the earth. The equation of the potential vorticity reads as follows:

$$\frac{d}{dt} \frac{2\Omega \sin \varphi + \zeta}{H} = 0 \qquad (7.60)$$

This equation was first formulated by Rossby (1940). It represents a special case of Ertel's vorticity theorem. According to this equation, the potential vorticity is a constant quantity for any water column. If the water column moves into another latitude of the same water depth, the relative vorticity must change. Zonal currents, when crossing a deep-sea ridge, also change their relative vorticity. The consequence is that the current is deflected from the zonal direction, as shown in Fig. 7.28.

For the large-scale current in any direction, the large-scale distribution of depth is less essential than the distribution of $(2\Omega \sin \varphi)/H$, because the current tries to follow the isolines of this quantity. Welander (1969) has shown that charts of this quantity differ from bathymetric charts, in some parts rather considerably.

The influence of bottom topography is particularly obvious in the course of the Gulf Stream. South of Cape Hatteras, the Gulf Stream runs close to the coast of North America. Then, it follows the shelf edge and is strongly deviated especially in the region of the Grand Banks. When flowing over this bank, the Gulf Stream turns southward and, after reaching deep water, it again flows in a more northerly direction.

7.5.2. Topographic Rossby waves

What has been said in Section 7.5.1 shows that the same effects as those caused by the dependence on latitude of the Coriolis parameter can also be produced by a change of water depth. This also applies to wave solutions. In Section 7.4.3 Rossby waves have been described that result from nothing else but the dependence on latitude of the Coriolis parameter. The same waves can also be generated by simple change of depth. In this case they are called topographic Rossby waves. The analogy is evident in the following example.

In the simplest case, Rossby waves in an ocean of constant depth obey these equations:

$$\frac{\partial u}{\partial t} + fv = -\frac{1}{\rho}\frac{\partial p}{\partial x}, \quad \frac{\partial v}{\partial x} - fu = -\frac{1}{\rho}\frac{\partial p}{\partial y}, \quad \frac{\partial u}{\partial x} + \frac{\partial v}{\partial y} = 0 \qquad (7.61)$$

If pressure is eliminated and, for reasons of simplification, the current is assumed to depend on y only, we obtain the equation

$$\frac{\partial^2 v}{\partial x \partial t} - \beta v = 0 \qquad (7.62)$$

The solution is

$$v = v_0 \cos(\kappa x - \omega t)$$

where, in accordance with $\omega = \beta/\kappa$, the frequency ω depends on the wavenumber κ. These are the planetary waves of Section 7.4.3. The prerequisite for their formation is that, according to $f = f_0 + \beta y$, the Coriolis parameter depends on latitude, that is, on the y direction. If, on the other hand, f is assumed to be constant, and a variable depth distribution is conceded, the two equations of motion, Eqs. (7.61), are valid again as well as the continuity equation in the form:

$$\frac{\partial u H}{\partial x} + \frac{\partial v H}{\partial y} = 0 \qquad (7.63)$$

With pressure eliminated, we get the following equation for the same type of motion as mentioned above, in case the sea floor changes in accordance with $H = H_0 \exp(-\alpha y)$:

$$\frac{\partial^2 v}{\partial x \partial t} - f\alpha v = 0. \qquad (7.64)$$

Equations (7.62) and (7.64) show that β and $f\alpha$ correspond to each other. Thus, the same kinds of waves are generated by a change of depth as are by the change in latitude of the Coriolis parameter, provided $\alpha = \beta/f$ is valid.

7.6. Thermohaline Circulation

7.6.1. Large-scale thermohaline processes

In the ocean, water motion is not only generated by wind. The horizontal temperature and salinity differences, maintained by climatic influence at the sea surface, cause density differences which will initiate circulation. As shown in experiments by Sandström (1908) in particular, and in thermodynamic considerations by Bjerknes (1936) in general, such circulation requires that expansion should take place at higher pressure than contraction does. Referred to the ocean and to water temperature, this means that the heat source must lie at a lower level than the cold source, as illustrated by case (b) in Fig. 7.29. Conversely, if the heat source is situated at a higher level [case (a)], thermal circulation will not develop. In the ocean, the heat and cold sources are located at the same level, namely at the sea surface [cf. Fig. 7.29(c)]. Thus, conditions do not favor strong thermal circulation, in sharp contrast to the atmosphere. However, such circulation is not completely lacking in the ocean. Water close to a heat source attains a lesser density than water close to a cold source. It becomes lighter and spreads at the surface in the direction of the cold source. For reasons of continuity, water below the heat source will ascend,

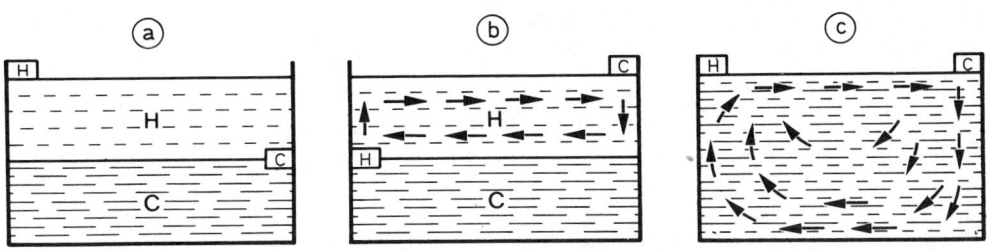

Fig. 7.29. Thermal circulation in basins filled with water with different positions of heat and cold sources.

and water below the cold source will descend and, while subsequently spreading in deeper layers, it will become warmer through heat conduction and mixing. In this way, some circulation is induced, but it is not very effective because of the slow energy transfer via mixing and thermal conduction.

For a long time, such circulation, caused by temperature differences alone, had been considered the main factor in oceanic deep-sea circulation, which was supposed to form two vortices arranged symmetrically to the equator. Accordingly, in the lower layers, cold water was believed to move from the higher latitudes to the lower ones and, in the upper layers, warm water was supposed to move in the opposite direction. This hypothesis of oceanic vertical circulation dates back to von Humboldt who, in this way, interpreted the low water temperature he had observed in the deeper layers of the equatorial zone. Von Humboldt's hypothesis was widely propagated by Maury (1856) in his book *The Physical Geography of the Sea* published in numerous editions. It was accepted until 1922, although there were contradictions between this hypothesis and the results of a thorough analysis of the observations made on *Challenger* as well as on *Planet* and *Deutschland*. Merz and Wüst were the first to draw a conclusion from those earlier observations when clearly showing by the example of the Atlantic Ocean that hardly a trace is found of a simple symmetrical vertical circulation. The new hypothesis was formulated in an exceptionally plausible way by Merz (1925) and formed the working basis for the "Deutsche Atlantische Expedition *Meteor* 1925-1927." The basic concept of the hypothesis was largely confirmed by the results of this expedition. Analogous considerations can be made with regard to the haline circulation.

The water budget (cf. Section 4.3.6) indicates that, besides areas rich in salinity, there are others with deficiencies of salinity at the sea surface. The former with their excess of evaporation and ice formation represent a cold source, whereas the latter with excess precipitation, melting of ice, and continental runoff take the place of a heat source. Hence, there is also a haline circulation in the ocean, in addition to thermal circulation.

In some areas, both circulations move in the same direction, thus reinforcing their effect; in others, they act against each other, and their effect is weakened. In Fig. 7.30, the near-surface branches of circulation are schematically indicated in a meridional section approximating the conditions in the South Atlantic Ocean at 30°W. Thermal and haline circulations reinforce each other in the area between the warm doldrums near the equator with their heavy precipitation and, therefore, low salinity, and the subtropical high with higher salinities and lower temperatures. They act against each other in the area between the subtropical high and the belt of heavy precipitation of the west wind zone south of the polar front. Toward the pole, both circulations again reinforce each other, but the haline circulation is insignificant there because of the small differences in salinity.

In addition, the wind-induced vertical circulation should be mentioned as it might

336 Theory of Ocean Currents

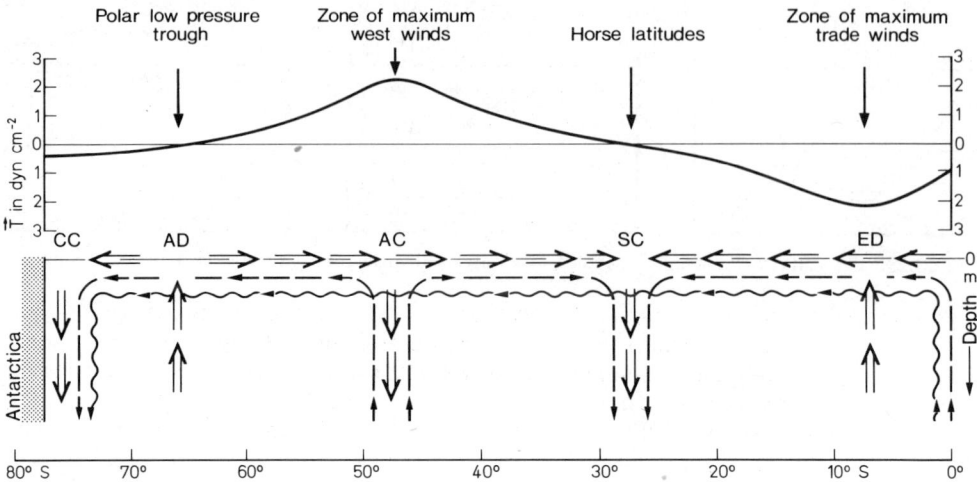

Fig. 7.30. Scheme of the thermal (∿), haline (---→), and wind-induced (⇒) meridional components of circulation near the sea surface. Above: Tangential shear stress T of the zonal wind component in the South Atlantic Ocean at 30°W in August.

be expected in an ocean without any density stratification. In the upper part of Fig. 7.30, the tangential shearing stress of the zonal wind components is indicated schematically as is, in the lower part, the corresponding meridional component of the mass transport by wind-induced currents. In the region of the polar low pressure trough, divergent mass transport prevails (AD: Antarctic Divergence); at the continent the motion is forced downward (CC: Continental Convergence); in the zone of the strongest west winds, convergence of water motion occurs (AC: Antarctic Convergence); in the region of the subtropical high we again find convergence (SC: Subtropical Convergence); and in the area of the strongest east winds, the center of the trades, divergence is dominant (ED: Equatorial Divergence).

In principle, this permits us to devise three models regarding the development of thermohaline circulation.

1. Differences in the horizontal distribution of temperature and salinity are the driving mechanisms of thermohaline circulation. Mixing processes are of minor importance.

    Under such assumptions, Goldsbrough (1933) computed the stationary system of haline circulation by taking into account precipitation and evaporation as influx or loss of water, respectively, at the sea surface. Since, on the rotating earth, horizontal velocities can be expected to prevail—due to the deflecting force of the earth's rotation—he applies the vertically integrated equations of motion, that is, he dispenses with the representation of the vertical structure of the solution.

    Figure 7.31 shows the resulting horizontal circulation (streamlines) in a spherical sector between equator and pole, limited by two meridians. As the consequence of the distribution of precipitation and evaporation, depending only on latitude and being in rough agreement with the distribution in nature, a western boundary current flows northward while a broad return current

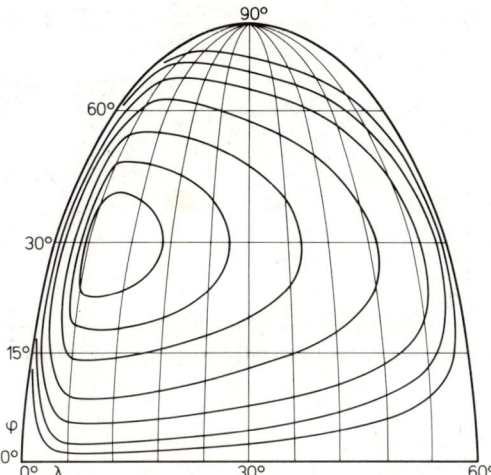

Fig. 7.31. Streamlines of the horizontal current in a spherical sector between equator and pole as a consequence of precipitation and evaporation. (After Goldsbrough, 1933.)

appears on the eastern side of the ocean. The distribution is very similar to that obtained as the result of the large-scale wind distribution [cf. Fig. 7.20(b)], except that it is weaker.

Stommel (1957), starting from the same assumptions as Goldsbrough, has developed a hypothetical model, but mainly with a view to vertical circulation. On the understanding that a vast downwelling area exists in the Arctic Ocean, he found a deep current on the western sides of the oceans, which—in the Atlantic Ocean, for instance—is opposed to the direction of the Gulf Stream.

2. It is true that the differences in the horizontal distribution of temperature and salinity are the driving mechanisms, but mixing processes are also of deep-reaching influence and quite essential for the distribution of temperature and salinity. Mixing processes alone (Krauss, 1958) would, however, cause the transition layer between warm-water and cold-water spheres to lie rather deep in the lower latitudes, whereas in the middle latitudes, it would lie somewhat higher, which is in contrast to reality. Therefore, mechanisms of circulation must be involved as well.

3. Eventually, deep-sea circulation is driven by the wind-induced surface circulation. As shown in Fig. 7.30, the convergences and divergences of the large-scale, wind-induced current field of the near-surface layer generate vertical motions which disappear at the sea surface but must not be neglected at the lower boundary of Ekman's friction layer. Circulation in the depth is driven by the vertical velocity field and by the density distribution at the lower boundary of the Ekman layer. The models developed by Welander (1959, 1969), Robinson and Welander (1963), Blandford (1965) and others are based on the above assumption. In some cases, mixing is also taken into account. The equations are nonlinear, and the way to solve them is very complicated. Figure 7.32 shows the density distribution on a meridional section. The wide transition layer between the warm-water and cold-water spheres is recog-

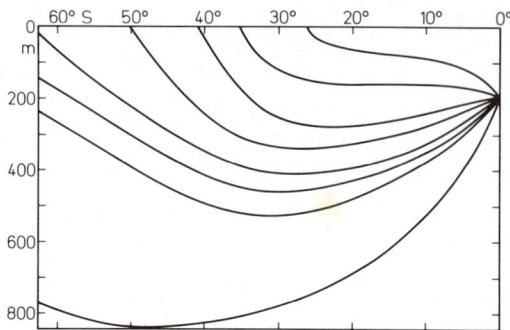

Fig. 7.32. Theoretical distribution of density at a meridional section as the results of thermal circulation. (After Welander, 1969.)

nizable. At the equator, it lies at about 200 m depth, but it reaches considerably deeper in the middle latitudes. Advection plays an important role in such models.

Dealing theoretically with thermohaline circulation is much more difficult than with wind-induced circulation because, in addition to the hydrodynamic equations of motion and the continuity equation, the equations of thermal conduction and of diffusion as well as the equation of state must also be taken into account.

7.6.2. Coastal currents in higher latitudes

In higher latitudes, generally belonging to the regions of humid climate, rivers transport large amounts of fresh water into the ocean. Consequently, salinity is greatly diminished along the coasts as opposed to the salinity in the open ocean. The result is a haline circulation (Krauss, 1955). Under the influence of the rotation of the earth, currents are formed that generally flow in such a way that (in the Northern Hemisphere) the low-salinity water lies on the right-hand side if one looks in the current direction. Since the less saline water is found just offshore, the current flows around the islands with the land always on its right-hand side.

The same is true for shallow banks on the shelf within the region of coastal water. Since, due to tidal currents, turbulence, and variable wind-induced currents, mixing is more intensive in shallow water than in deep water, the salinity in a water column over such banks, down to greater depth, is lower than that in deeper water. Thus, such banks act rather like islands; in the Northern Hemisphere water flows around them in the same anticyclonic fashion (cf. Fig. 7.33). If the coast acts like a fresh-water line source, salinity is considerably reduced in that offshore region, as illustrated in Fig. 7.33(a) with horizontal and vertical austausch coefficients of 10^8 and 10^2, respectively, assumed on the shelf. Figure 7.33(b) represents the actual salinity distribution in the Norwegian Current.

7.6.3. Compensation currents in ocean straits

The thermohaline currents easiest to deal with are those in ocean straits, because here nothing but continuity must be taken into account. Let us consider a channel with two-layered water under stationary conditions, as indicated in the longitudinal section of Fig.

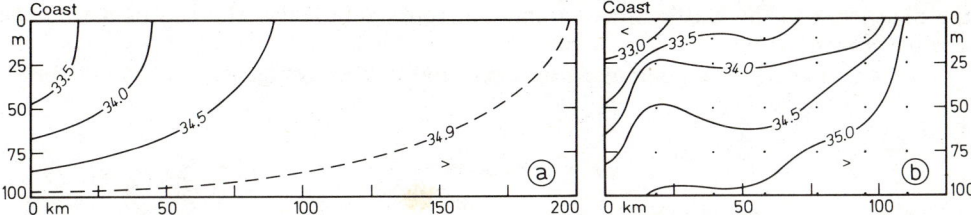

Fig. 7.33. Distribution of salinity, theoretical (a) and observed (b), off the Norwegian coast. (According to Krauss, 1955.)

7.34. Let A_1 and A_2 be the areas of the upper parts of cross sections "1" and "2," A'_1 and A'_2 the areas of the lower parts, and F the surface area. The mean current velocities perpendicular to the cross sections are given by u_1, u'_1, u_2, and u'_2, and r stands for the inflow of water through F. The S values represent salinities. The continuity of the water volume requires that the inflow must equal the outflow:

$$A_1 u_1 + A'_2 u'_2 + rF = A'_1 u'_1 + A_2 u_2 \tag{7.65}$$

Because of the continuity of salinity under stationary conditions the following equations are valid:

$$S_1 A_1 u_1 = S'_1 A'_1 u'_1 \text{ and } S_2 A_2 u_2 = S'_2 A'_2 u'_2 \tag{7.66}$$

If, for example, the terms for the volume transport in the lower layer are eliminated in Eq. (7.65) with the aid of Eq. (7.66), the following relation is obtained:

$$\left(1 - \frac{S_2}{S'_2}\right) A_2 u_2 = \left(1 - \frac{S_1}{S'_1}\right) A_1 u_1 + rF \tag{7.67}$$

which is known as the "Hydrographic Theorem" of Knudsen.

Knudsen, for example, limits the Kattegat and the straits between the Danish Islands by two cross sections, the first in the south across the Sound and the Narrows of Darss (corresponding to section "1" in Fig. 7.34) with the values $S_1 = 8.7‰$ and $S'_1 = 17.4‰$. The other cross section lies at the northern exit of the Kattegat with $S_2 = 20‰$ and $S'_2 = 33‰$. Here, rF approximately equals zero. Then, $A_1 u_1 = 0.8 A_2 u_2$. The average outflow from the Kattegat into the Skagerrak is therefore larger by 12.5% than the inflow from the Baltic Sea, a fact that can be explained only if it is assumed that water carried from

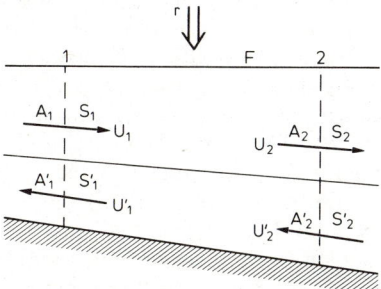

Fig. 7.34. Condition of continuity in two-layered water.

the Skagerrak to the Kattegat as an undercurrent again leaves the Kattegat with the surface current.

If a channel with an opening at only one end is considered, Eq. (7.67) is reduced to

$$rF = -A_1 u_1 \left(1 - \frac{S_1}{S'_1}\right) \tag{7.68}$$

This relation can be applied to the European Mediterranean Sea with the Strait of Gibraltar as the opening. According to observations, two layers of water exist there: $S_1 = 36.25‰$ and $S'_1 = 37.75‰$. After Schott, $A_1 u_1 = 1.75 \times 10^6$ m³ sec⁻¹. Therefore, $rF = -7 \times 10^4$ m³ sec⁻¹. This loss of water at the surface of the Mediterranean, resulting from an excess of evaporation over the gain by precipitation and continental runoff, would be equivalent to the annual loss of a water layer of 881 mm, if related to the total area of the Mediterranean Sea (2507×10^9 cm²). A detailed calculation by Wüst (1952) of the different components of the water budget, namely precipitation over the sea plus continental runoff minus evaporation from the sea, has led to the value of 965 mm yr⁻¹. If the Strait of Gibraltar could be closed by a dam, the water level of the Mediterranean Sea would be lowered, after 100 yr by slightly less than 100 m.

As useful as such considerations of continuity may be, they always permit only very limited conclusions. Better understanding of the internal processes of motion involved is possible only if, in addition to the equation of continuity, the hydrodynamic equations of motion are also taken into account.

7.6.4. Thermoclines

One of the most conspicuous phenomena with regard to the vertical variation of temperature in the ocean and the marginal seas is the summer thermocline at approximately 30 to 50 m depth. It develops in spring, gradually descends deeper until late summer, and dissipates in the fall. Furthermore, a warm-water sphere has been found in the world ocean as a top layer of about 300 to 600 m thickness. This layer is the consequence of the general thermal circulation and has already been discussed in Section 7.6.1. The summer thermocline is a wide-spread phenomenon. Because of the processes of mixing one might expect any differences in the ocean to have a general tendency toward smoothing. If, nevertheless, discontinuity layers, that is, pronounced boundaries are formed, the conclusion must be drawn that dynamic processes play an important role.

There is no doubt that the summer thermocline at about 30 to 50 m depth is mainly the result of intensive mixing processes in the uppermost water layers which are exposed to wave motion. Rouse and Dodu (1955) have proven by experiment that a sharp interface between two water layers does not dissipate when strong turbulence is produced in the top layer by artificial whirling. On the contrary, the interface is strengthened by water from the sublayer being continuously entrained into the highly mixed top layer, where it is also subjected to whirling. Thus, the thickness of the top layer is increased. The phenomenon that a sharp boundary is formed between turbulent motion and a fluid in the state of rest and that some fluid from the resting environment is entrained into the turbulent area, is also known in other fields of hydrodynamics. With regard to the thermocline, Turner and Kraus (1966) have carried out experiments yielding some relevant information. By adding lighter water to the surface (thus simulating the summer heating) and by inducing permanent mixing near the surface, they were able to develop thermoclines which, similar to the summer thermoclines in the ocean, were gradually shifted into the depth.

7.6.5. Numerical solutions regarding the general circulation

Direct numerical solutions of the system of hydrothermodynamic equations have been given, in particular, by Bryan and Cox (1967). Analogous computations have been carried out by Sarkisyan (1969) and Friedrich (1967).

The direct integration of the system of equations can begin at any initial state. Under the influence of the actual (or hypothetical) mean external forces, the integration, after a sufficient length of time, results in a quasistationary terminal state which is independent of the initial values.

Bryan and Cox investigated the oceanic circulation as the result of a given field of wind and temperature at the sea surface. Some findings with a view to the stream function at about 200 m depth are represented in Figs. 7.35(a) and (b). Figure 7.35(a) shows the stream function as a result of the temperature field alone. A circulation pattern, similar to that obtained in Section 7.3.7 as the result of a wind field, can be recognized. It extends much farther northward, however. Thus, the temperature distribution at the sea surface, as imposed by meteorological conditions, can also enforce a large-scale circulation corresponding to observations, at least in principle. However the velocities are too small. The same current system also occurs as a consequence of precipitation and evaporation, according to analytical studies carried out by Goldsbrough (1933) (cf. Fig. 7.31). Figure 7.35(b) shows the stream function as result of the temperature field and the wind. The flow in the large-scale anticyclonic gyre is considerably intensified. Furthermore, a cyclonic vortex is formed in higher latitudes as the result of the given wind field.

The anticyclonic gyre induced by temperature in the upper layer [cf. Fig. 7.35(b)] corresponds to a cyclonic vortex in the lower layers. Hence, we find a bottom current directed southward on the western side of the ocean as already supposed by Stommel (cf. Section 7.6.1) and partly confirmed by observations beneath the Gulf Stream.

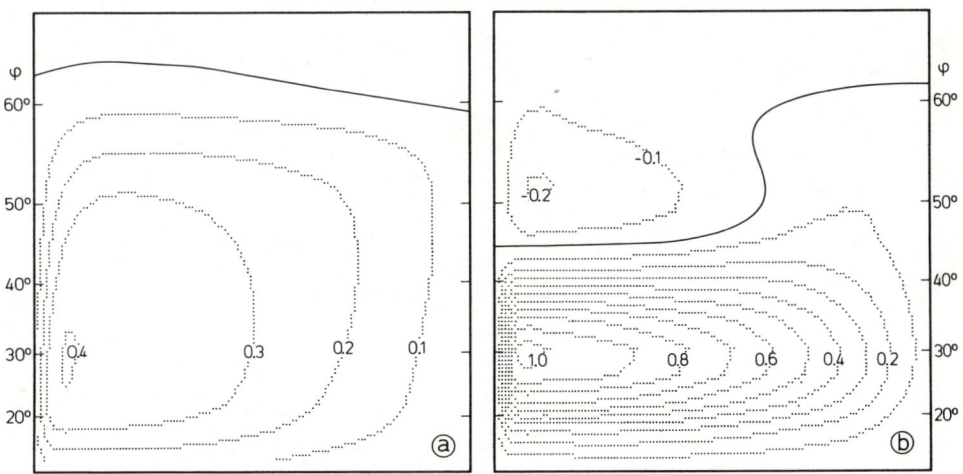

Fig. 7.35. Streamlines of the horizontal motion as the consequence of the temperature distribution at the sea surface (a) as well as of the combined effect of the distributions of temperature and wind (b). (According to Bryan and Cox, 1967.)

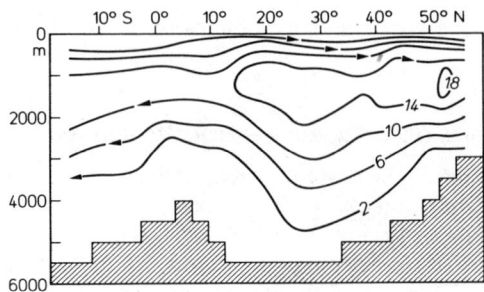

Fig. 7.36. Streamlines (numerical values in 10^6 m^3 sec^{-1}) of the meridional motion based on numerical integration of the hydro-thermodynamic equations. (After Friedrich, 1970.)

The extension of such computations is possible with the aid of a multilayer model permitting us to calculate three-dimensional processes of motion. The basic methods, applied in analogous models of the atmosphere, have been adapted to the ocean by Bryan (1969).

In a multilayer model the ocean is divided into numerous layers described by vertically integrated quantities. The system of equations obtained thereby is integrated numerically and, thus, mean values are obtained for each layer.

Figure 7.36 shows the meridional circulation in the Atlantic Ocean as yielded by a 14-layer model (Friedrich, 1970). The mass flow in the top layer from the South Atlantic Ocean into the North Atlantic Ocean is compensated by a countercurrent in the depth.

The analytical solutions of the previous section as well as the numerical solutions have provided significant insight into thermohaline circulation. But, at present, it is hardly possible to give more than a qualitative representation. Processes that are of fundamental importance for the description of the general circulation are still unknown to a large extent. There is particular need of studies on turbulence and mixing in the interior of the ocean as well as on interaction at the ocean–atmosphere interface.

8 Surface Waves and Internal Waves

8.1. Classification of Waves

8.1.1. Progressive and standing waves, surface waves and internal waves

An inexperienced observer on shipboard or at the shore, watching the waves perpetually passing him, will be easily misled into believing that the water mass of each wave crest actually rolls past him. However, if the trajectory of a body floating in the water is followed, it becomes obvious that this body is not advancing with the waves but is only moving back and forth and, at the same time, also up and down, describing a so-called orbital path. It is the shape of the sea surface and not the water that progresses.

In addition to describing the configuration of the sea surface, one of the main objects of wave theory is to explain the orbital paths of the water particles and the propagation speed of such progressive waves. Since Newton's time, the difficulties encountered in the exact description of the relevant processes have attracted many eminent mathematicians like Laplace, Lagrange, Bernoulli, Euler, Airy, Cauchy, and others.

Questions regarding the generation, growth, transformation, and decay of waves, especially of the short waves of the sea state, belong to the second major scope within the theory of progressive waves. But it was not before the past century that these problems were taken up, in particular by Scott Russell, Lord Kelvin, von Helmholtz, and Jeffreys. Quite recently, increased efforts to interpret the observations quantitatively have been initiated by scientists in Germany, in the United Kingdom, and the United States. They have succeeded in achieving significant progress in this field, especially with respect to the understanding of a phenomenon that is called the state of the sea.

Besides progressive waves, standing waves, too, are of great importance. In general, these are long waves, that is, oscillations of an enclosed or nearly enclosed oceanic area which are induced by variations of the fields of wind and atmospheric pressure.

Water waves are observed not only at the sea surface but also where adjacent water masses of different densities border on each other. These so-called boundary surface waves of the ocean will be dealt with separately in Section 8.5. If density changes gradually in the ocean, internal waves will occur instead of boundary surface waves. Surface waves have their largest amplitudes at the sea surface, internal waves at any depth, depending mainly on the density and on the generating forces.

The totality of progressive and standing waves can be classified from different points of view. The most important ones will be described in the following sections.

8.1.2. Classification with respect to restoring forces

Waves are generally described by oscillation equations obtained by specialization from the total system of hydrodynamic equations. In these equations, so-called restoring forces occur: if a water particle is pushed out of its position of rest, such forces try to drive it back to its initial position. During this process the particle usually overshoots its initial position; it oscillates around its position of rest. If forces inducing the oscillation are not taken into account, such waves are considered free waves.

The restoring forces acting on ocean waves result from the following.

1. Surface tension. It is effective with the curvature of the sea surface and important only for very short waves. The resulting waves are called capillary waves.
2. Gravity. This force acts on all water particles and tries to keep the sea surface and the surfaces of equal density level when currents are not present. Vertical displacement of the sea surface results in surface waves; that of isopycnic surfaces results in boundary surface waves or internal waves. If merely under the influence of gravity, both these types of waves represent gravity waves.
3. Coriolis parameter. Since gravity is involved in all vertical water motions, the Coriolis force frequently leads only to a modification of gravity waves, for instance, to Kelvin waves. Besides, waves also occur in which the Coriolis parameter alone leads to a restoring force. This is true for the inertial waves already discussed in Section 7.4.2.
4. Coriolis parameter depending on latitude. A force results that permits planetary or Rossby waves (cf. Section 7.4.3).
5. Changes of depth. A special type of waves (planetary or Rossby waves) is not only due to the Coriolis parameter depending on latitude but generally to the factor f/H, where f stands for the Coriolis parameter and H for the water depth. Waves generated by changes of H alone are called topographic Rossby waves. They have been described in Section 7.5.2.
6. Compressibility. Although this quantity is very small in water, it is decisive for another class of waves, the sound waves.

From the equations on which all the types of waves mentioned above are based, relationships can be established between wavelength and wave period (or wave frequency and wavenumber). Such relations are called equations of dispersion. If represented in a frequency–wavenumber diagram they provide a good survey of the whole of the possible free waves. Figure 8.01 shows the dispersion curves of the aforementioned types of waves after Magaard (1973). From this diagram it can be taken which wavelengths and periods characterize the individual types of waves, and how the period, as a function of wavelength, varies within each class of waves.

In this chapter, only capillary and gravity waves will be considered, as well as the influence exerted on them by Coriolis force and friction. A detailed derivation of these results has been given by Kinsman (1965) with regard to the sea state, and by Krauss, volume 1 (1973) and volume 2 (1966) with regard to the other waves.

8.1.3. Classification of gravity and capillary waves with respect to generating forces

The dispersion curves in Fig. 8.01 show periods and wavelengths of possible free waves. In nature, first of all, generating forces are needed to deflect the water particles from their rest position. Such generating forces are the following.

1. Tidal forces. They are described in Chapter 9, also containing a description

Classification of Waves 345

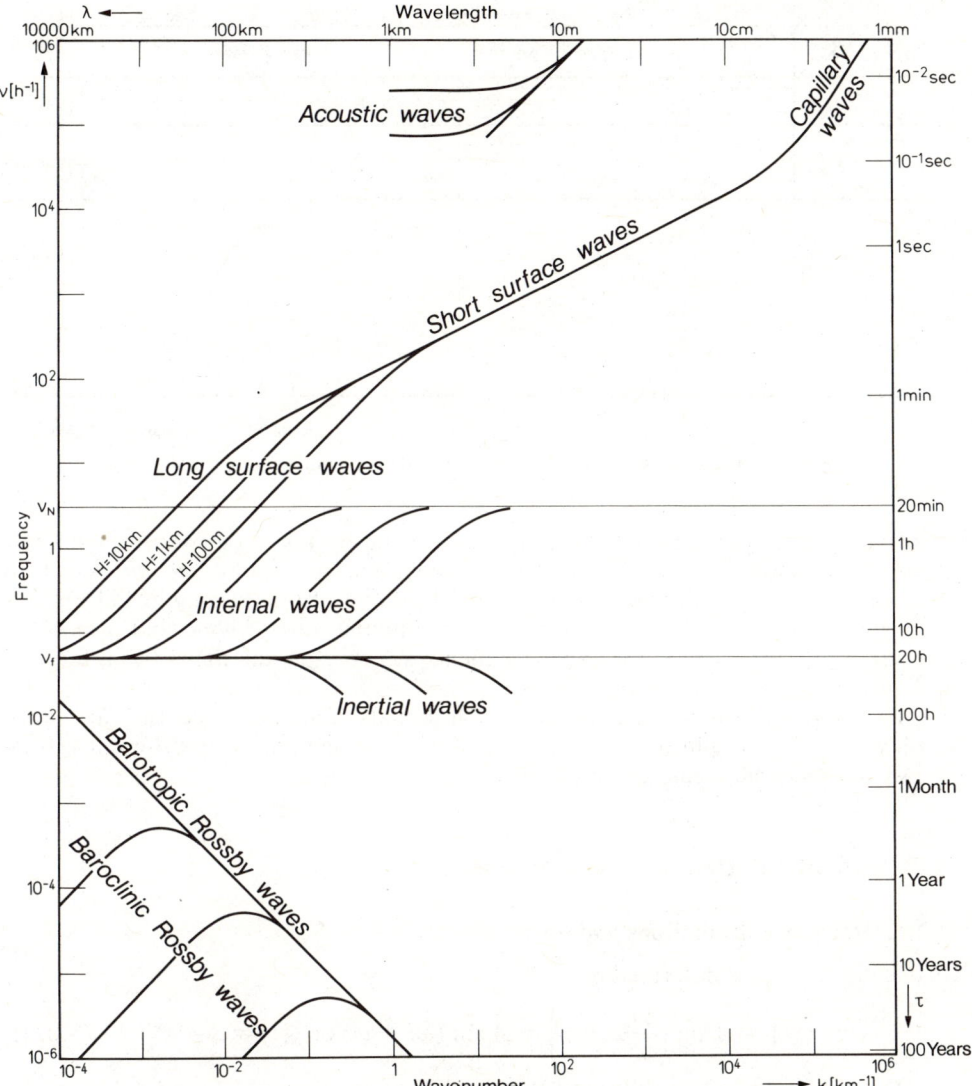

Fig. 8.01. Dispersion curves of waves in the ocean. (After Maagard, 1973.)

of the tidal waves at the sea surface that are caused by such tidal forces. Internal tides are dealt with in Section 8.5.3.

2. Meteorological power fields. They are understood as fields of wind and atmospheric pressure varying in space and time. Their nonperiodic character brings about the great variability in the ocean.

3. Earthquakes. Their foci, when lying beneath the ocean floor, will cause sudden deformations of the ocean bottom which can lead to disastrous flood catastrophes. The waves generated are called tsunamis.

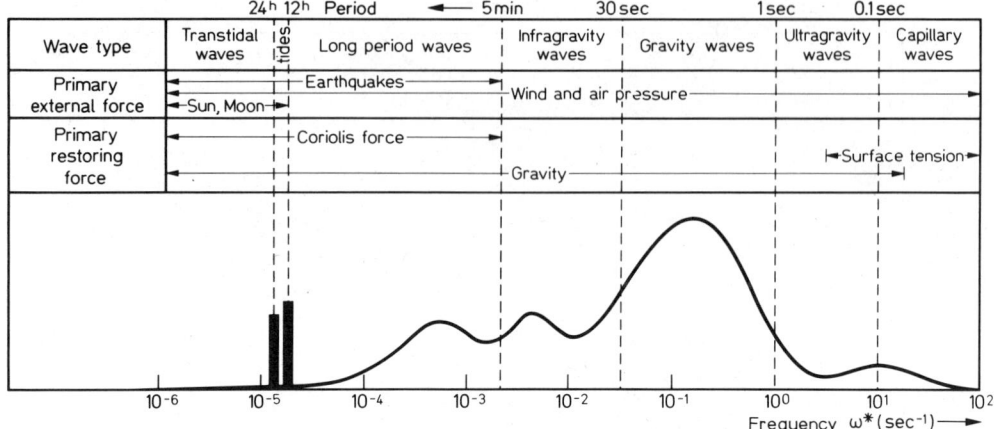

Fig. 8.02. Amplitude spectrum of surface waves (hypothetic) for wave classification.

A survey of the spectrum of surface waves is given by Fig. 8.02, where (hypothetic) mean amplitudes have been plotted as a function of the frequency $\omega^* = 1/\tau$. Apart from pure capillary waves, all other waves are gravity waves. In the second line of Fig. 8.02, the primary generating forces are indicated. Fluctuations of wind and atmospheric pressure are effective over the entire frequency range. Sun and moon (tidal forces) as well as earthquakes induce long waves.

The largest mean amplitudes occur with periods of about 10 sec; they are sea and swell waves. The amplitudes indicated for the tides, assuming very different values in the various oceanic areas, are hypothetical.

8.2. Kinematic Properties of Waves

8.2.1. Harmonic oscillations and wave fields

A harmonic wave is described by

$$\zeta(t) = A \sin\left(\frac{2\pi}{\tau} t + \varphi\right) = A \sin(\omega t + \varphi) \text{ or } \zeta(t) = A e^{i(\omega t + \varphi)} \quad (8.01)$$

(cf. Fig. 8.03). The amplitude A of the wave is the maximum value of ζ. The range of positive deflection is called *wave crest*, that of negative deflection is the *wave trough*.

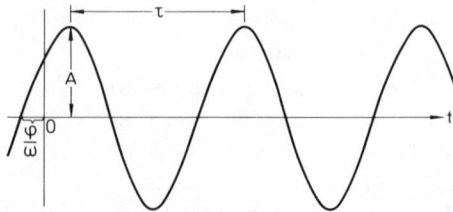

Fig. 8.03. Harmonic wave.

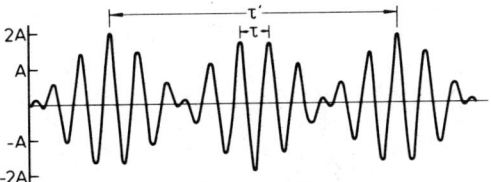

Fig. 8.04. Beat as a consequence of the superposition of two waves with only slightly differing periods.

The vertical distance between wave crest and wave trough is called *wave height*. In a harmonic wave this corresponds to twice the amplitude.

The time interval after which the wave process is repeated is called the *wave period* τ. It is related to the angular frequency ω according to the equation $\omega = 2\pi/\tau$. The inverse of the period, $\omega^* = 1/\tau$, is called *frequency*. φ stands for the phase angle, giving the displacement of the wave with respect to $t = 0$.

If two waves of the type of Eq. (8.01), with the frequencies ω_1 and ω_2 differing only slightly, and different phase angles φ_1 and φ_2 are superimposed, the result obtained is a beat:

$$\zeta(t) = 2A \cos\left(\frac{\omega_1 - \omega_2}{2}t + \frac{\varphi_1 - \varphi_2}{2}\right) \sin\left(\frac{\omega_1 + \omega_2}{2}t + \frac{\varphi_1 + \varphi_2}{2}\right) \quad (8.02)$$

represented in Fig. 8.04. This oscillation has two periods: the long beat period $\tau' = 4\pi/(\omega_1 - \omega_2)$ and the short carrier period $\tau = 4\pi/(\omega_1 + \omega_2)$. By such superposition, individual groups of waves are formed which are periodically repeated after each beat period τ'.

With progressive waves, the wave train proceeds at the phase velocity $c = \lambda/\tau$, where λ denotes the wavelength, that is, the distance between two wave crests. Such a wave is described by

$$\zeta(x,t) = A \sin(\kappa x - \omega t) \text{ or } \zeta(x,t) = A e^{i(\kappa x - \omega t)} \quad (8.03)$$

$\kappa = 2\pi/\lambda$ is the wavenumber. For $\kappa x - \omega t = \text{const}$, the wave always has the same value, that is, for the locations $x = (\omega/\kappa)t + \text{const}/\kappa$. These locations are dependent on time, and the phase velocity is obtained by differentiation:

$$c = \frac{dx}{dt} = \frac{\omega}{\kappa} = \frac{\lambda}{\tau} \quad (8.04)$$

This is the velocity with which the wave proceeds. The ratio wave height to wavelength, $\delta = h/\lambda$, is called the steepness of the wave.

If, in analogy to Eq. (8.02), two progressive waves of adjoining wavelengths and periods are superimposed, one also obtains beats, as represented in Fig. 8.04. They are described by

$$\zeta(x,t) = 2A \cos\frac{(\kappa_1 - \kappa_2)x - (\omega_1 - \omega_2)t}{2} \sin\frac{(\kappa_1 + \kappa_2)x - (\omega_1 + \omega_2)t}{2} \quad (8.05)$$

Individual waves as well as wave groups have distinct propagation velocities, generally of different values. The propagation velocity of the wave group results from $(\kappa_1 - \kappa_2)x - (\omega_1 - \omega_2)t = 2 \times \text{const}$ and amounts to

$$c_{gr} = \frac{dx}{dt} = \frac{\omega_1 - \omega_2}{\kappa_1 - \kappa_2} \quad \text{or} \quad c_{gr} = \frac{d\omega}{d\kappa} \quad (8.06)$$

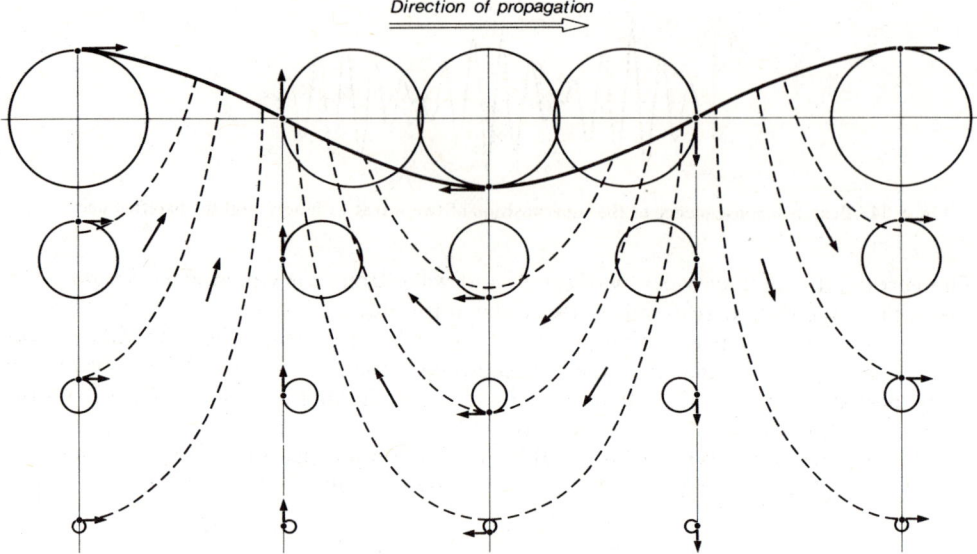

Fig. 8.05. Orbital paths, streamlines, and direction of motion of particles in a deep-water wave.

Instead of this we can also write

$$c_{gr} = c - \lambda \frac{dc}{d\lambda} \qquad (8.07)$$

This group velocity is of great importance for the dynamics of the sea state because it gives the velocity with which the energy propagates. Waves whose group velocity is smaller than the phase velocity c, and whose phase velocity therefore increases with increasing wavelength ($dc/d\lambda > 0$), show normal dispersion. But if $c_{gr} > c$ holds, the dispersion is anomalous. With $c_{gr} = c$, the waves are free of dispersion. In deep water, the individual water particles follow circular orbits when a progressive wave is passing, while in shallow water they follow ellipses, as shown in Fig. 8.06. A more detailed representation of waves, the wavelength of which is small against the water depth, is contained in Fig. 8.05.

According to the results of classical hydrodynamics, the orbital paths of water particles in surface waves are ellipses covered within the time interval $\tau = 2\pi/\omega = \lambda/c$, that is, within the same time interval during which the wave travels over the distance of a wavelength. If the water depth exceeds half the wavelength, the elliptic orbit changes into a circular orbit, provided the amplitude can be considered very small compared to the wavelength. The radius of the circular orbit, given by $r = Ae^{-(2\pi/\lambda)z}$, decreases from $r = A$ at the surface, that is, from the size of the amplitude of the wave, to $r = Ae^{-\pi}$, that is to $1/23$ of the amplitude, at the depth of half the wavelength. Similarly, the velocities of the water particles, amounting to $v = 2\pi r/\tau = (2\pi/\tau)Ae^{-(2\pi/\lambda)z}$, very rapidly decrease with depth. Consequently, the motion of the surface waves is restricted to a thin surface layer, as illustrated in Fig. 8.05. The thick arrow indicates the propagation direction of the wave. At the wave crest the particles move in the same direction as the wave, but at the wave trough in the opposite direction. They ascend at the front side and descend at the back.

Harmonic waves, moving in any direction in the horizontal plane, can be described by

$$\zeta(x,y,t) = Ae^{i(\kappa x+\eta y-\omega t)} \tag{8.08}$$

κ and η are the components of the wavenumber vector $\mathbf{k}$ in the x and y directions. The direction of propagation depends on the ratio η/κ. The wavelength is given by $k = \sqrt{\kappa^2 + \eta^2}$, where $\lambda_x = 2\pi/\kappa$ and $\lambda_y = 2\pi/\eta$ are the wavelengths with respect to the x and y directions.

Figure 8.04 shows that even the mere superposition of two waves of the same amplitude results in wave forms differing considerably from the pure sine wave. But the topography of the sea surface is much more complicated and can be described only by an infinite sum of partial waves of the form given in Eq. (8.08):

$$\zeta(x,y,t) = \sum_{n=-\infty}^{+\infty} A_n e^{i(\kappa_n x+\eta_n y-\omega_n t+\varphi_n)} \tag{8.09}$$

Each partial wave has its individual amplitude A_n, its frequency ω_n, and its phase angle φ_n. Its wavelength is given by the wavenumber $k_n = \sqrt{\kappa_n^2 + \eta_n^2}$, while its direction of propagation is determined by the ratio η_n/κ_n. For the deterministic description of the configuration of the sea surface within an ocean wave field, we would have to know the wavenumbers, frequencies, phase angles, and amplitudes of all partial waves. But that is impossible. Hence, we must restrict ourselves to giving a statistical description of the sea state and assume that the same dispersion relation between the frequency ω_n and the wavenumber k_n applies to each partial wave, as found for infinitesimal sine waves. For the phases φ_n it is presupposed that they will take all the values of the interval from 0 to 2π with equal probability. A more detailed discussion will be given in Section 8.3.6.

8.2.2. Standing waves

In a progressive wave, all water particles at the same depth describe the same orbital path within one wave period [cf. Fig. 8.06(a)], but the phases are different for each of the particles. In standing waves, all the water particles move along straight lines [cf. Fig. 8.06(d)]. The phases of their motions equal each other. The crest and the trough of a standing wave are stationary. Here, the vertical motion of the particles within one period reaches its maximum, forming the antinodes. Halfway between crest and trough, the particles do not move in vertical direction at all but only horizontally; here, we have the nodes.

Standing waves are described by

$$\zeta(x,t) = A \sin(\kappa x) \sin(\omega t) \tag{8.10}$$

They can be understood as the superposition of two progressive waves moving in opposite directions. Figure 8.07 shows a standing wave in an enclosed basin. The particles move along short distances of the streamlines, plotted as thin lines, which then can be considered as straight lines.

8.2.3. Damped waves and forced waves

Due to friction, all waves in the ocean are damped. The simplest form of a damped wave is given by

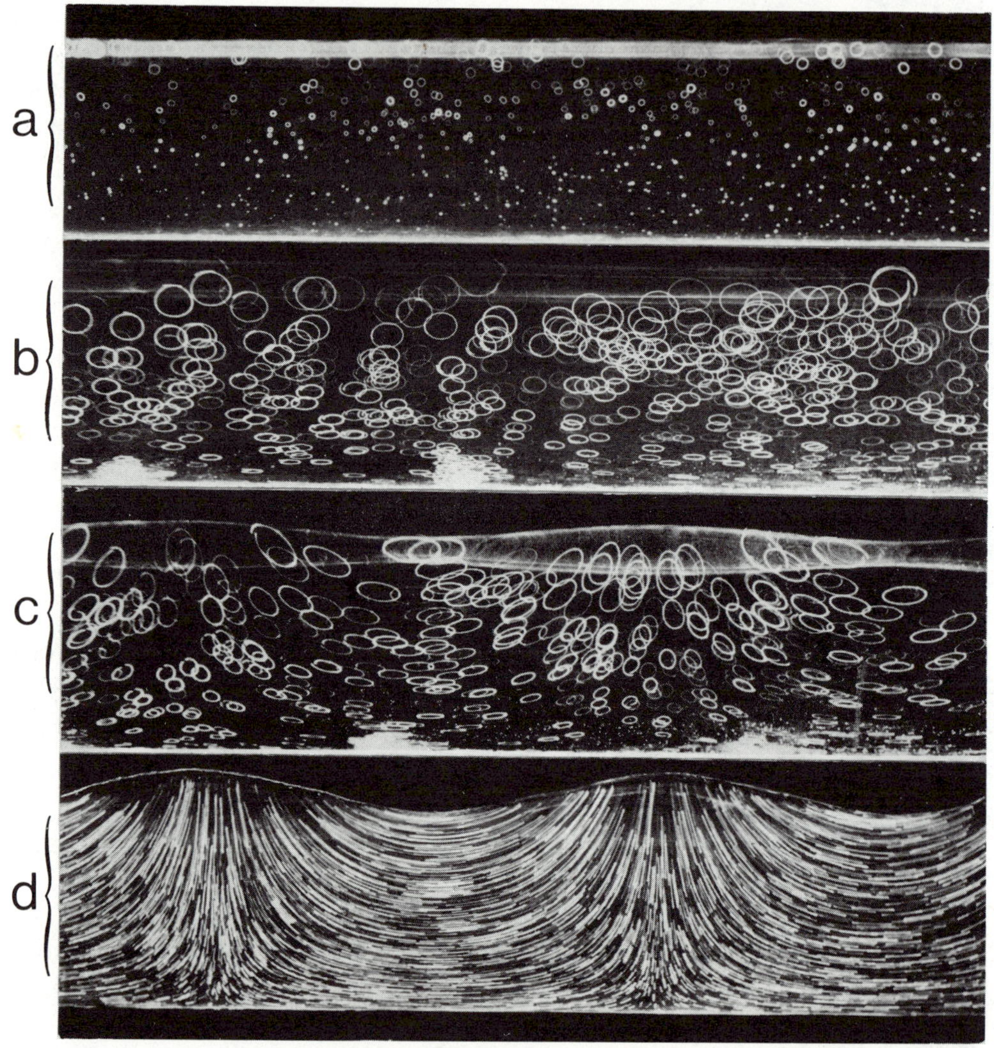

Fig. 8.06. Trajectories of water particles in a wave during one wave period. Photograph taken in a laboratory wave tank. (*a*) Progressive wave in deep water. (*b*) Progressive wave in shallow water. (*c*) Superposition of two propressive waves in shallow water with opposite directions of propagation (amplitude of one wave is 38% of the amplitude of the other progressive wave). (*d*) Standing wave. (Photos: Laborat. Dauphinois d'Hydraulique, Grenoble.)

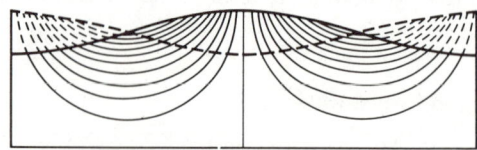

Fig. 8.07. Standing wave.

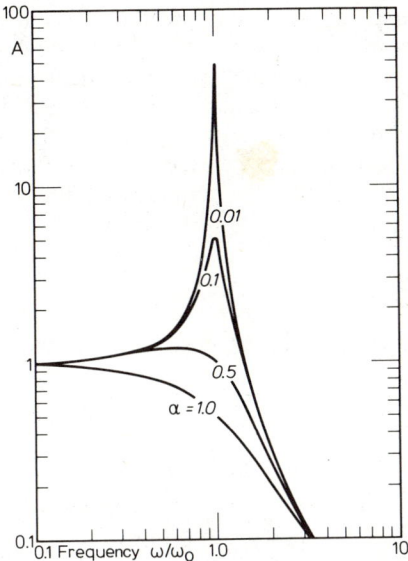

Fig. 8.08. Amplitude amplification at the resonance frequency as a function of the friction coefficient α.

$$\zeta(x,t) = Ae^{-\beta t} \cos(\kappa x - \omega t) \qquad (8.11)$$

The amplitude values ζ_1 and ζ_2, when succeeding one another, decrease exponentially. β stands for the damping factor. In practice, the logarithmic decrement is usually applied as a measure of damping:

$$d = \ln \left|\frac{\zeta_1}{\zeta_2}\right| = \frac{\beta \tau}{2} \qquad (8.12)$$

where $\tau = 2\pi/\omega$ is the period.

Damping is a very essential measure for the description of forced waves.

Free waves and forced waves differ in so far as the periods of forced waves are in agreement with the periods of the wave-generating force, while the periods of free waves depend on the dimensions of the oscillating water mass and on friction.

In forced waves, friction influences only the amplitude of the waves, as illustrated

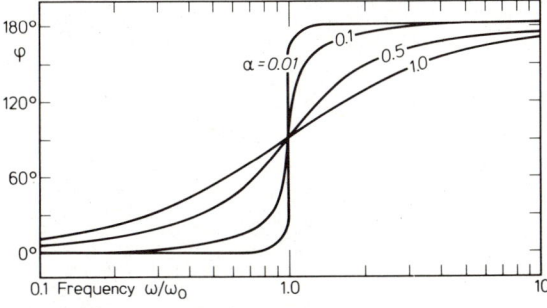

Fig. 8.09. Change of phase as a consequence of friction.

in Fig. 8.08, where the amplitude $A(\omega)$ of a forced oscillation is represented as a function of ω/ω_0. ω_0 is the frequency of free waves. If this frequency agrees with that of the generating force, we have $\omega/\omega_0 = 1$, and the amplitude $A(\omega)$ would become infinitely large in a frictionless system. The actual resonance amplification depends on the damping $\alpha = \beta/\omega_0$. The greater the damping factor is, the smaller the resonance amplification.

The phase of the oscillation is represented in Fig. 8.09. In a medium without any friction, oscillations are in phase with the generating force if $\omega < \omega_0$. If $\omega > \omega_0$, a phase difference of 180° occurs (inverse oscillation). The modifications indicated in the figure are caused by friction.

8.3. Short Surface Waves or Deep-Water Waves

8.3.1. Gravity waves

The classical theory of gravity waves with infinitely small amplitudes leads to the following dispersion relation:

$$\tanh \kappa H = \frac{\omega^2}{g\kappa} \quad \text{or} \quad c = \sqrt{\frac{g}{\kappa} \tanh \kappa H} = \sqrt{\frac{g\lambda}{2\pi} \tanh \frac{2\pi H}{\lambda}} \quad (8.13)$$

where H stands for the water depth, ω for the angular frequency, κ for the wavenumber, g for the acceleration of gravity, and c for the phase velocity. The first equation indicates how the wavenumber (or wavelength) depends on frequency (or period), as represented in Fig. 8.10. This relation is valid for short waves as well as for long waves. Short waves are defined by a very small ratio of wavelength to water depth ($\ll 1$). Then, $\kappa H = 2\pi H/\lambda$ is very large so that $\tanh \kappa H \approx 1$ holds true. Hence, with regard to short waves Eq. (8.13) is simplified to

$$\omega^2 = g\kappa \quad \text{or} \quad c = \sqrt{\frac{g}{\kappa}} = \sqrt{\frac{g\lambda}{2\pi}} \quad (8.14)$$

So we see that their phase velocity is independent of the water depth. Very short waves travel more slowly than longer waves. With λ expressed in meters, $c = 1.25\sqrt{\lambda}$ (in m sec^{-1}) applies approximately. In a field of ocean waves, the longer waves outrace the shorter ones.

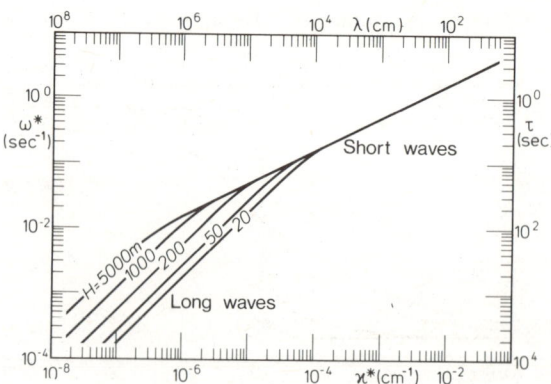

Fig. 8.10. Dependence of the period τ on the wavelength λ with respect to gravity waves.

According to Eq. (8.06), the group velocity of short surface waves results from Eq. (8.14) and reads as follows:

$$c_{gr} = \frac{c}{2} \qquad (8.15)$$

thus being half as large as the phase velocity. In a wave group, energy is propagating at this velocity. Half of the energy consists of potential energy and the other half of kinetic energy. The total energy of a wave, as determined per unit sea surface, amounts to

$$E = \frac{\rho g A^2}{2} \qquad (8.16)$$

that is, the energy of a system of progressive waves with the amplitude A equals the work needed to lift a water layer of the thickness A by the height $A/2$.

8.3.2. Capillary waves

The shortest waves at the sea surface are the so-called capillary waves. Here the curvature of the sea surface is so great that surface tension becomes the dominating force. The surface tension σ at the water–air interface amounts to 72 dyn cm^{-1}.

If gravity as well as surface tension are taken into account, the dispersion relation, analogous to Eq. (8.13), is given by

$$\tanh \kappa H = \frac{\omega^2}{g\kappa + (\sigma/\rho)\kappa^3} \quad \text{or} \quad c = \sqrt{\left(\frac{g}{\kappa} + \frac{\sigma \kappa}{\rho}\right) \tanh \kappa H} \qquad (8.17)$$

from which there follows for short waves

$$\omega^2 = g\kappa + \frac{\sigma}{\rho}\kappa^3 \quad \text{or} \quad c = \sqrt{\frac{g}{\kappa} + \frac{\sigma \kappa}{\rho}} = \sqrt{\frac{g\lambda}{2\pi} + \frac{2\pi\sigma}{\rho\lambda}} \qquad (8.18)$$

ρ stands for the density of water. In Fig. 8.11 the phase velocity c is represented as a

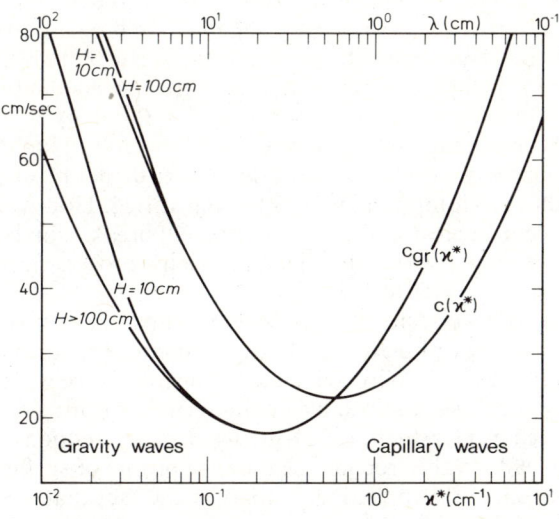

Fig. 8.11. Phase velocity $c(k^*)$ and group velocity $c_{gr}(k^*)$ of capillary waves and short gravity waves as a function of the wavelength.

function of the wavenumber $\kappa^* = 1/\lambda$. Accordingly, the smallest phase velocity with which surface waves can travel is given by $c = 23.05$ cm sec^{-1}. This phase velocity is found for waves with a wavelength of 1.67 cm and a period of 0.072 sec. The first term in Eq. (8.18) is predominant with longer waves, and the phase velocity increases with increasing wavelength λ, as already mentioned in the previous section. For shorter waves the second term in Eq. (8.18) is predominant, and the phase velocity also increases, but with decreasing wavelength. With regard to the right-hand part in Fig. 8.11, the first term can finally be neglected, and pure capillary waves are obtained:

$$\omega^2 = \frac{\sigma}{\rho}\kappa^3 \quad \text{or} \quad c = \sqrt{\frac{\sigma\kappa}{\rho}} = \sqrt{\frac{2\pi\sigma}{\rho\lambda}} \tag{8.19}$$

In Fig. 8.11 the group velocity is also plotted. Gravity waves have a normal dispersion, because their phase velocity increases with wavelength. Capillary waves, however, show anomalous dispersion. Their phase velocity decreases with increasing wavelength, and one obtains

$$c_{gr} = \frac{3}{2}c \tag{8.20}$$

Hence, the energy of a capillary wave travels faster than the wave itself. In this case, too, half of the total energy consists of potential energy and half of kinetic energy. The total energy per unit area amounts to

$$E = \frac{\sigma\kappa^2 A^2}{2} \tag{8.21}$$

8.3.3. Waves of finite amplitude

For the surface waves described in the two previous sections it is presumed that the steepness of the waves $\delta = h/\lambda$ is very small. Such waves are called *infinitesimal waves*. With regard to these waves, the nonlinear terms of the hydrodynamic equations of motion can be neglected so that such waves satisfy the linearized equations of motion. Waves of finite amplitude can be described by means of the perturbation technique. There, the functions occurring in the hydrodynamic equations are expanded into perturbation series of the steepness δ. One begins, for instance, with $\zeta = \zeta^{(0)} + \delta\zeta^{(1)} + \delta^2\zeta^{(2)} + \cdots$, where the steepness is still considered a small quantity, and the terms of higher order in δ become very small. In fact, observations have shown that waves do not reach a steepness of more than 1:8. So the above assumption appears to be justified. However, when waves with greater steepness occur in nature, the wave crests will break, which possibly represents an important process of energy transfer. To date, such a process cannot yet be described theoretically.

In 1847, Stokes, in his classical theoretical investigation of waves, showed that the profile of gravity waves of finite amplitude approximates the form of a trochoid, if the water is deep enough and the amplitude is small compared to the wavelength. As a matter of fact, such a wavy curve is described by a point on a disk rolling on a level surface. An older theory, advanced by Gerstner (1802), giving an exact solution of the hydrodynamic equations, directly leads to the trochoid as the true profile of surface waves. However, this result, though frequently applied, is limited by the fact that the motion is not irrotational. The trochoidal form results in narrower wave crests and flatter wave troughs than would correspond to a sinusoidal wave. With increasing amplitude the crests tend

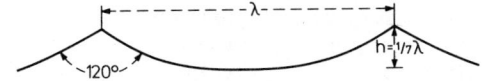

Fig. 8.12. Theoretical limiting values of surface waves.

to approach a limiting angle of 120°, as shown in Fig. 8.12. According to Michell (1893), the corresponding limiting ratio of wave height to wavelength is 1:7. Furthermore, it is remarkable that the orbital trajectories of water particles in a Stokes' wave are no longer closed circular orbits. Near the sea surface, water is transported horizontally in the direction of wave propagation:

$$u_0 = \pi^2 \delta^2 c^{(0)} e^{-2\kappa z}, \tag{8.22}$$

where $c^{(0)}$ is the phase velocity of the linear theory [cf. Eq. (8.14)]. With greater amplitudes the phase velocity increases according to

$$c = \sqrt{\frac{g}{\kappa}\left(1 + \frac{1}{2}A^2\kappa^2 + \cdots\right)} \tag{8.23}$$

Analogous calculations can also be carried out for capillary waves, whose behavior is just the opposite of that of the gravity waves: with increasing amplitude, the wave crests become wider, the troughs more pointed, and the phase velocity decreases.

Standing waves of finite amplitude have also been investigated. Longuet-Higgins (1950) has shown that here a pressure fluctuation occurs with twice the frequency than would correspond to the wave considered. After heavy gales, such fluctuations are occasionally observed with the aid of seismographs at nearshore stations (microseism).

8.3.4. Nonlinear interactions

The Stokes' wave is a special type of nonlinear waves in so far as the form of the wave is maintained, that is, the wave when travelling keeps up its form. This is possible because all partial waves contributing to the wave profile travel at the same phase velocity. However, if it is assumed that a specific dispersion relation, as that corresponding to the theory of infinitesimal waves, applies to each individual partial wave, interactions will take place among the individual partial waves because of the nonlinear terms. The result of such interactions is a continuous transformation of the wave profile.

The theory of these weak interactions is a perturbation calculation of higher order. It has been introduced into oceanography by Hasselmann (1960) and Phillips (1960), independently. Interaction among partial waves is defined as weak because the wave field changes only very slowly. This can also be taken from the fact that, in many respects, ocean waves have already been well described by the classical linear wave theory. So, the dispersion relation, for instance, is valid to a good approximation. If second-order perturbation terms are taken into account, only minor changes will result compared with the linear theory. It is only under consideration of third-order perturbation terms that the energy transfer among the individual partial waves will start.

The theory of the weak interactions shows that waves whose wavenumber vectors are $\mathbf{k}_1$, $\mathbf{k}_2$, and $\mathbf{k}_3$ and whose frequencies ω_1, ω_2, and ω_3 are connected with the wavenumbers according to Eq. (8.14), $\omega_i^2 = g|\mathbf{k}_i|$, are in resonance with a fourth wave if Eqs. (8.24) are satisfied:

$$\mathbf{k}_1 \pm \mathbf{k}_2 \pm \mathbf{k}_3 = \mathbf{k}_4, \quad \omega_1 \pm \omega_2 \pm \omega_3 = \omega_4, \quad \omega_i^2 = g|\mathbf{k}_i| \tag{8.24}$$

Then, energy flows from the first three waves to the fourth, causing the amplitude of the latter to increase with time. In the general Eqs. (8.24), any combination can be chosen with regard to signs. Also, two waves of the first three waves may be identical so that the new wave results from the interaction of only two waves.

In a continuous field of ocean waves, an infinite number of partial waves is present, as described by Eq. (8.09). From these an infinite number of combinations of the form (8.24), fulfilling the resonance condition, can be found. Thus, a continuous flow of energy to other waves can take place. We have learned from the classical theory of the sea state that friction causes some damping of the wave amplitude. In deep water, the damping factor is $\beta = 2\nu\kappa^2 = 8\nu\pi^2/\lambda^2$, that is, it is inversely proportional to the square of the wavelength λ. Therefore, only the very short waves among the deep-water waves are subject to noticeable damping. In shallow water, the effect is greatly intensified, but here, too, it is mostly the short waves that are damped. Hence, the mechanism of the decay of a wave field is to be understood like this: first, due to nonlinear interaction, energy must be transferred from long waves to very short waves, where the energy is then dissipated. The breaking of the waves, as well as the turbulence caused thereby can be taken into account by introducing apparent friction. This process is strongly nonlinear and certainly of great influence on the waves.

The fact that turbulence acts to damp waves can easily be recognized by observing water where more turbulence is induced artificially than is present in neighboring areas, as, for example, is the case in the wake of a ship. There, the water remains smooth and free of small waves for quite some time. Also, when a ship is drifting athwart with respect to wind direction, some damping of waves will occur on the windward side, owing to the increased turbulence in the upwelling water. Some captains use this phenomenon by not heading the ship into the sea when riding out a storm but drifting with the ship's axis parallel to the wave crests.

Some nonlinear effects can be directly described without applying the theory of weak interactions. This includes the influence of short waves on considerably longer ones or on the mean current.

In Section 7.1.8 Reynolds' friction term has been introduced. It concerns a transport of momentum caused by small-scale turbulent processes of motion. This transport of momentum appears in the equations of smoothed motion and indicates the effect of small-scale processes of motion on the mean current.

In an analogous way, a flux of momentum can be determined that is based on the field of motion of short waves and that has an effect on longer waves similar to that of the Reynolds' transport of momentum on the smoothed current. This momentum flux of a wave field is called *radiation stress,* according to Longuet-Higgins and Stewart (1960). It causes, for example, the water level within the region of the nodes of a standing wave to lower, but to rise within the region of the antinodes. When a group of high waves passes by, the water level will drop as well. This lowering of mean water level is coupled with a current opposed to the propagation direction of the waves.

Figure 8.13 shows such a wave group with the envelope a, the mean water level $\bar{\zeta}$, the radiation stress S_x, as well as the current v belonging to the mean water level $\bar{\zeta}$. This process offers an interpretation of the phenomenon of the so-called *surf beats,* concerning waves with periods from about 20 sec to 8 min. Such waves usually appear on the shelf together with groups of extremely high swell waves and are frequently observed at the coasts of the oceans. When surf beats are approaching the coast together with the wave group, the swell waves are dissipated by breaking, whereas the long surf beats are reflected and travel back into the sea. Their amplitude amounts to about one tenth of that of the

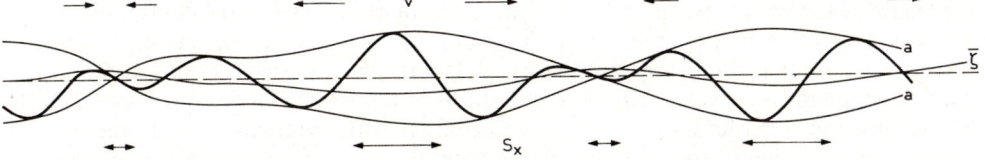

Fig. 8.13. Surf beats. Lowering of mean water level and mass transport beneath a wave group. (After Longuet-Higgins and Stewart, 1964.)

swell, so they cannot be perceived with the naked eye. They are clearly visible, however, on wave records.

8.3.5. Properties of sea and swell

Everyone is familiar with the light ripples that will cover a completely smooth water surface when a wind gust moves over it. Such small ripple waves—called "cat's paws" by sailors—represent the initial stage of surface waves. Although this phenomenon is quite a common one and can be observed not only at the sea surface but also on any rain puddle, the quantitative explanation of the generation of waves on the basis of the acting forces involves considerable difficulties, some of which have not yet been solved.

Neumann (1949) formulated an equation giving the conditions for the generation of surface waves as well as the relation between the wavelength λ and the wave height h as a function of the wind velocity W. The numerical results are summarized in Fig. 8.14. The lower limit of the wind speed that must be exceeded if surface waves are to be generated lies at $W = 69.5$ cm sec^{-1}. This is in agreement with careful observations made by Roll (1951) on the tidal flat of Neuwerk on the German North Sea coast. According to computation, the height of the first initial waves, also representing the slowest waves possible ($c = 23.1$ cm sec^{-1}) with a wavelength of $\lambda = 1.72$ cm, amounts to $h = 0.022$ cm. At higher wind speed, two different kinds of waves are formed by the combined action of gravity and surface tension; namely capillary waves, whose wavelengths and amplitudes decrease with increasing wind speed, and gravity waves, whose wavelengths, amplitudes and, also, steepness increase with increasing wind speed. For a wind speed of 1 m sec^{-1},

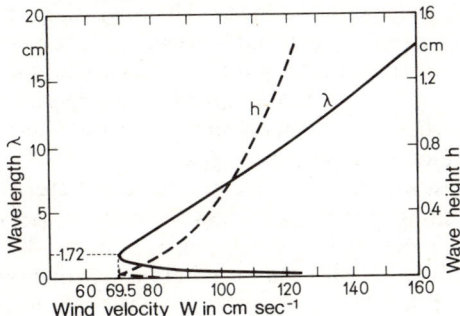

Fig. 8.14. Wavelength λ and wave height h of most strongly induced initial waves at different wind speeds. (After Neumann, 1949.)

the length of the gravity waves is $\lambda = 6.7$ cm, of capillary waves $\lambda = 0.4$ cm (cf. Fig. 8.14). The corresponding heights are $h = 0.49$ cm and $h = 0.002$ cm, respectively. Capillary waves have practically disappeared. Also, calculations and observations agree in this respect that capillary waves adjust to the speed of the onsetting wind within fractions of a second; the water level reacts almost immediately with slight vibration. Larger gravity waves, however, begin to grow out of the rippled sea surface only after several seconds.

If the wind becomes stronger, the waves quickly reach a critical steepness which, according to observations, is 1:8 and thus approximates the value of 1:7 expected according to theory. The crests become more and more pointed and of glassy appearance and foam begins to form. With further supply of energy, the wavelength increases faster than the height. The waves have passed through the initial stage proper; they become what is called sea state.

If the wind, with direction and speed remaining constant, acts long enough on a large oceanic area, a certain typical sea state develops. The sea surface assumes a characteristic appearance which has been made the basis for a scale of the sea state. Today, this scale is no longer of great importance because, as will be shown later, quantitative comprehension of the characteristic elements of the waves has been attempted in more recent wave research rather than only a pure description of the sea. For this reason, reporting the sea state according to that scale has been omitted in the international weather telecommunication service since January 1, 1949. Nevertheless, the scale is presented in Table 8.01 because it is still important for two reasons. First, it conveys a short, pertinent description of the appearance of the sea surface at different wind velocities to nonseafaring people. Second, the appearance of the sea surface serves as a help in estimating the wind force at sea. Introduced by Admiral Beaufort during the age of sailing ships, the twelve-part wind scale was originally based on the sails set in order to fit the respective wind force. After the disappearance of sailing ships, it would have lost its meaning if it had not been connected with other characteristic phenomena. The state of the sea surface at fully developed sea has become an approved measure for estimating the wind force at sea (cf. Figs. 8.15–8.19).

At a fully developed sea the configuration of the sea surface is a very complicated function.

The two wave diagrams in Fig. 8.20 and the corresponding photograph in Fig. 8.18 give an idea of the topography of the sea surface during a storm of force 9 to 10 Beaufort. They represent not only the first evaluation of stereophotogrammetric wave pictures of such high seas, but this is also the first picture series with a 1 sec time lapse, permitting the determination of the propagation velocity. The wave diagrams and the corresponding wave profiles (Fig. 8.20) demonstrate that waves of different lengths are present simultaneously. The longest wave has a length of approximately 420 m. Two directions of propagation can also be recognized. In this case, the superposition of different wave trains leads to wave heights of 16.5 m at the maximum. Therefore, descriptions of "mountainous waves" should not be rejected as sailors' yarns.

More than mere description, measurement of characteristic wave quantities contributes to the understanding of the process how waves grow as a result of increasing wind. What is particularly important is the relationship, derived from observation, of the steepness of the wave $\delta = h/\lambda$ with the ratio between propagation velocity c and wind velocity $W (\beta = c/W)$. For a "young" sea β is smaller than for a fully developed sea. Hence, β is called the "age" of the sea. In addition to the values that have been observed, Fig. 8.21 shows the average relationships between steepness and age as used by Sverdrup

Table 8.01. Sea State and Wind[a]

Wind			Sea State		State of the Sea Surface for Fully Developed Sea
Descriptive Term	Beaufort	Velocity (m sec^{-1})	Force	Descriptive Term	
Calm	0	0.0–0.2	0	Smooth	Mirror-like
Light air	1	0.3–1.5	1	Rippled	Small, scaly-looking ripples without foam crests (Fig. 8.15)
Light breeze	2	1.6–3.3			Small wavelets, still short but more pronounced; crests have a glassy appearance and do not break
Gentle breeze	3	3.4–5.4	2	Gentle	Crests begin to break; foam mostly of glassy appearance; occasional small white foam crests (white horses) (Fig. 8.16)
Moderate breeze	4	5.5–7.9	3	Light	Waves still small but becoming longer; fairly frequent white horses
Fresh breeze	5	8.0–10.7	4	Moderate	Moderate waves taking a pronounced long form; white horses everywhere; chance of some spray (Fig. 8.17)
Strong breeze	6	10.8–13.8	5	Heavy	Large waves begin to form; crests break and form large areas of white foam; some spray
Neal gale	7	13.9–17.1	6	Very heavy	Sea heaps up; white foam from breaking waves begins to be blown in streaks in direction of wind
Gale	8	17.2–20.7	7		Moderately high waves with crests of considerable length; edges of crests begin to break into spindrift; foam is blown in well-marked streaks in wind direction
Strong gale	9	20.8–24.4		High	High waves, dense foam streaks in wind direction; wave crests begin to roll over; spray may affect visibility (Fig. 8.18)
Storm	10	24.5–28.4	8	Very high	Very high waves with long overhanging crests; sea surface white from foam; heavy, shocklike tumbling of the sea; visibility affected by spray
Violent storm	11	28.5–32.6	9		Exceptionally high waves; edges of wave crests are blown into froth; visibility reduced by spray (Fig. 8.19)
Hurricane	12	32.7–36.9		Exceptionally high sea	Air filled with foam and spray; sea completely white; visibility very seriously reduced; no distant view at all

[a] According to the Petersen Scale of the Deutsche Seewarte.

360 Surface Waves and Internal Waves

Fig. 8.15. Sea state at wind force 1 to 2 Beaufort. Photograph taken in the Northeast Atlantic Ocean on board the fishery protection vessel *Meerkatze* from a height of 7.5 m. (Photo: Roll.)

and Munk (1947) and Neumann (1952) in their wave theories. After having passed through the initial stage, the waves go through various phases of development. Three types of waves seem to occupy preferred positions.

1. Short but steep waves with a propagation velocity of $c = \frac{1}{3}W$. They appear as a certain roughness of the wave, but get lost in the large-scale wave profile, as indicated in Fig. 8.20(c). Although they are of no consequence to seafaring people, they certainly are very important for the energy transfer from wind to water.

2. Longer and flatter waves than those described above with the steepness δ decreasing with increasing age β. With a constant wind blowing, they grow out of the mixture of waves of the first type. They constitute the "sea" which is of prime interest to the sailor because it influences the speed of the ship. In fully developed state, they reach a certain maximum steepness. Beyond that, they become unstable and form breakers that may be dangerous to small ships. During this stage of breaking "seas," a third type of waves develops; these are described below.

3. Long, flat waves of constant steepness, for which $c = 1.37W$. An inexperienced observer will not always distinguish them because they are superimposed by the "seas." Since their propagation velocity is larger than the wind speed, they precede the wind field that has generated them. As so-called swell they can travel over vast oceanic areas and produce surf at distant coasts.

Fig. 8.16. Sea state at wind force 3 Beaufort. Photograph taken in the Northeast Atlantic Ocean on board the fishery protection vessel *Meerkatze* from a height of 7.5 m. (Photo: Roll.)

Fig. 8.17. Sea state at wind force 5 to 6 Beaufort. Photograph taken in the Northeast Atlantic Ocean on board the fishery protection vessel *Meerkatze* from a height of 7.5 m. (Photo: Roll.)

362 Surface Waves and Internal Waves

Fig. 8.18. Sea state at wind force 9 Beaufort. Photograph taken on board the liner *Europa* from a height of 26.5 m (compare corresponding wave diagram in Fig. 8.20). (Photo: Schumacher.)

However, height and period of waves do not depend on wind speed alone. The length of the time during which the wind acts (called *duration, T*) and the distance along which it acts (called *fetch, F*) are also significant.

A few important results will be mentioned here. Some of them are listed in Table 8.02

Fig. 8.19. Sea state at wind force 11 Beaufort. Photograph taken off Cape Horn. (Photo: v. Larisch-Moennich.)

Short Surface Waves or Deep-Water Waves 363

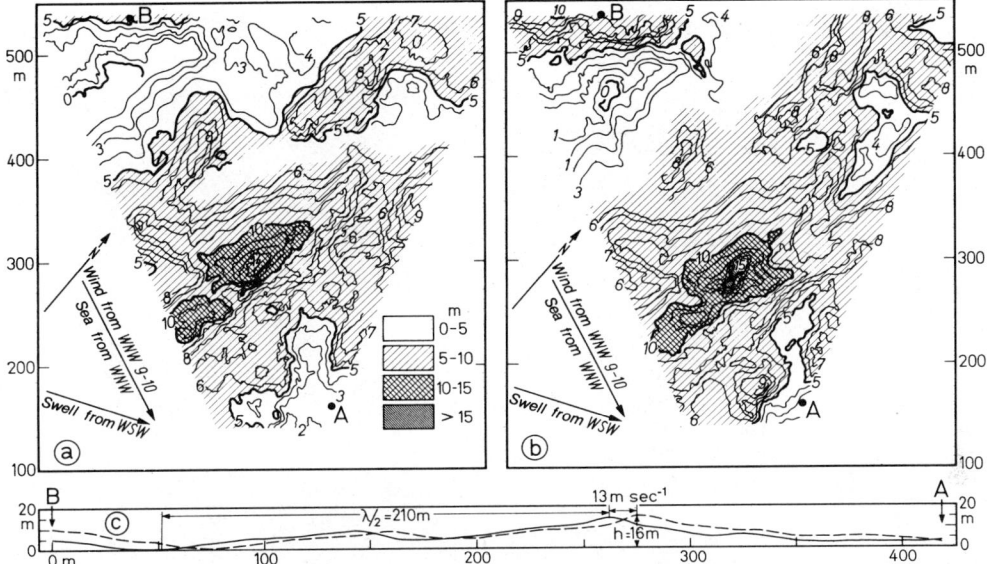

Fig. 8.20. Topography of the sea surface (in m) during a storm in the North Atlantic Ocean, based on cine-stereophotogrammetric wave pictures taken on board the liner *Europa*. (After Schumacher, 1952.) Position: 47.5°N, 26.8°W. Time: April 3, 1939, 1700 hr. Time lapse between (a) and (b): 1 second. (c) Longitudinal profile of the wave train in (a) and (b) along the line A–B (no vertical exaggeration).

and show the relationship between wind and waves for fully developed sea. Others are summarized in Fig. 8.22 permitting us to recognize the growth of the sea up to its fully developed state. Table 8.02 indicates, for instance, that for wind force 7 (corresponding to a wind velocity of 28 to 33 knots), the sea is fully developed as soon as this wind has blown with constant direction along a fetch of $F_u = 290$ nautical mi for a duration of $T_u = 24$ hr. This table further reveals that, in this case, the main part of the wave spectrum, containing 92% of the total energy of the waves, is found in the period range from $\tau_l = 4.8$ to $\tau_u = 17.0$ sec; that the mean wave period amounts to $\tau_m = 8.7$ sec, and that the waves containing most of the energy have a period of $\tau_{max} = 12.4$ sec. Incidentally, the period $\tau_m = 8.7$ sec corresponds to the quantity that an observer with a stop watch would

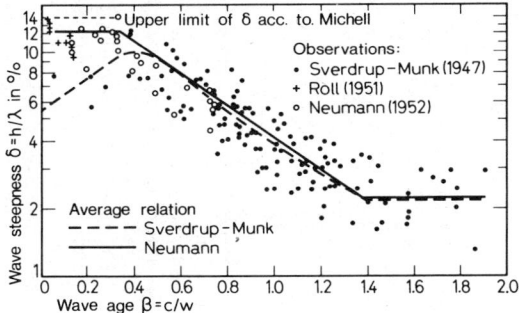

Fig. 8.21. Wave steepness δ as a function of wave age β. (After Sverdrup and Munk, 1947 and Neumann, 1952.)

Table 8.02. Characteristic Quantities of Fully Developed Sea for Different Wind Velocities[a,b]

Beaufort	Wind W (knots)	F_u (nm)	T_u (hr)	Wave Periods (sec) τ_l	τ_u	τ_m	τ_{max}	Wavelengths (m) λ_l	λ_u	λ_m	λ_{max}	Wave Heights (m) h_m	h_t
1	1–3	0.06	0.07	—	1.0	0.33	0.47	—	0.9	0.12	0.34	0.011	0.023
2	4–6	0.56	0.7	0.4	2.8	1.4	2.0	0.16	8.1	2	6	0.055	0.112
3	7–10	5.9	2.3	0.8	4.9	2.4	3.4	0.7	25	6	18	0.182	0.365
4	11–16	24	4.8	1.6	7.6	3.9	5.4	2	60	16	47	0.55	1.12
5	17–21	65	9.2	2.8	10.6	5.4	7.7	8	116	31	93	1.3	2.7
6	22–27	140	15	3.8	13.6	7.0	9.9	15	193	51	153	2.5	5.2
7	28–33	290	24	4.8	17.0	8.7	12.4	24	300	80	240	4.5	8.8
8	34–40	520	37	6.0	20.5	10.5	14.9	37	440	115	345	7.0	14.2
9	41–47	960	52	7.0	24.2	12.5	17.7	51	610	163	490	11.0	22.2
10	48–55	1570	73	8.0	28.2	14.7	20.8	66	830	225	675	15.8	32
11	56–63	2500	101	10	32	17	24	104	1060	301	900	22.2	45
12	>63	—	—	10	(35)	>18	>26	105	(1280)	>337	1050	(23)	(45)

[a] After Neumann, 1954.
[b] Definitions for column headings are as follows:

W: Wind velocity in knots (nautical mi hr^{-1}).
F_u: Minimum value of fetch in nautical miles, that is, the distance along which a wind of the velocity W must act on the sea surface to produce a fully developed sea at the end of this distance (provided T_u is satisfied).
T_u: Minimum value of duration in hours needed by the wind of the velocity W to generate a fully developed sea at the end of the minimum fetch F_u.
τ_l, τ_u: Lower and upper limits of the wave periods (in seconds) in the wave spectrum containing 92% of the total wave energy.
τ_m: Mean value of wave periods.
τ_{max}: Period of the waves containing most of the energy in the wave spectrum.
λ_l, λ_u: Lower and upper limits of the wavelengths (in meters) in the wave spectrum containing 92% of the total wave energy.
λ_m: Mean value of wavelengths.
λ_{max}: Wavelength of the waves containing most of the energy in the wave spectrum.
h_m: Mean height of all waves present (in meters).
h_t: Mean height of the 10 highest out of 100 consecutive waves (an approximate value for the maximum wave height).

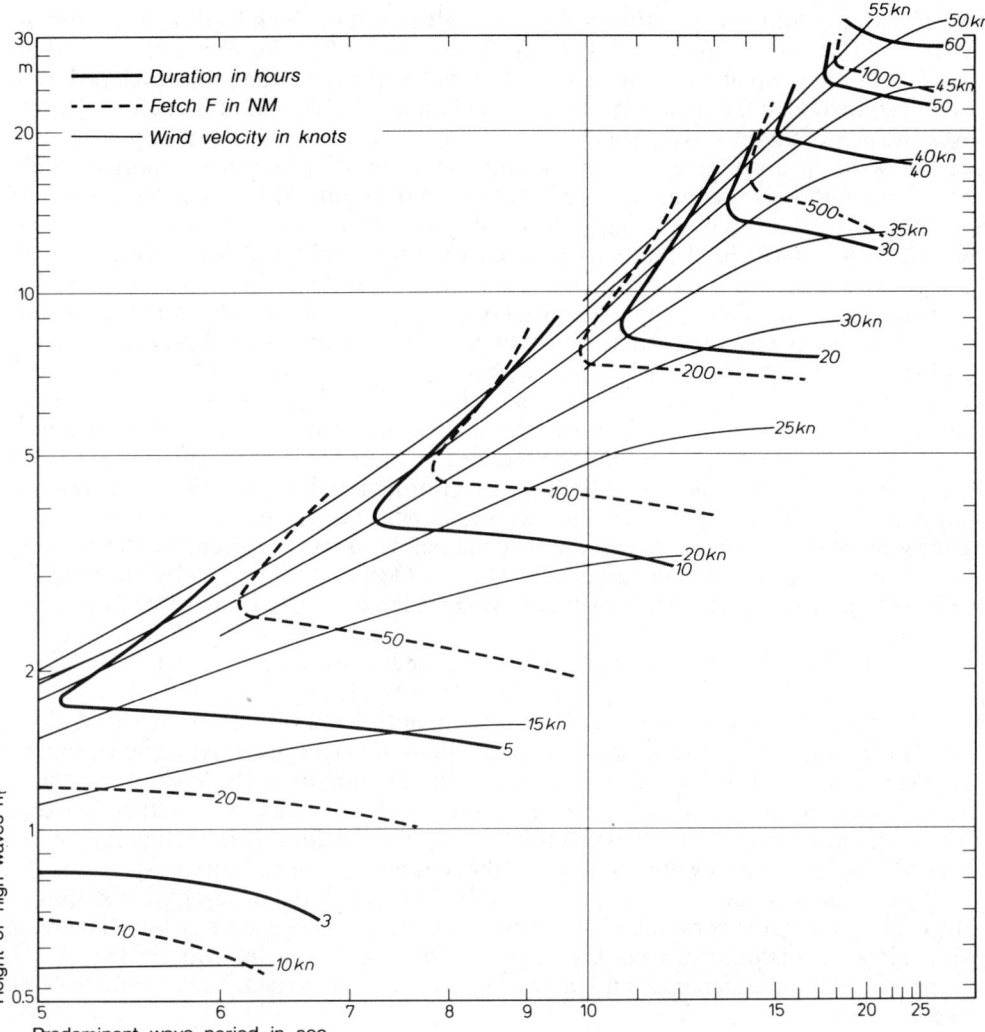

Fig. 8.22. Growth of height and predominant period of high waves as a function of duration and fetch of the wind for different wind velocities. (Based on Neumann, 1953.)

determine from a longer series of observations following the ups and downs of drifting objects. The example also suggests that, at wind force 7, the wavelength lies between 24 and 300 m, that the average wavelength is 80 m, and that the length of those waves where most of the energy is concentrated amounts to 240 m. Moreover, we learn that the average height of all waves is $h_m = 4.5$ m, and that the average height of the ten highest of a hundred consecutive waves is $h_t = 8.8$ m. All figures in this example refer to the fully developed final state. Regarding the development up to that state, Fig. 8.22 gives the following information: a wind of 30 knots acting along a fetch of $F = 50$ nautical mi, creates waves of only $h_t = 2.5$ m, provided it has blown for at least 7 hr. If the duration is shorter, for example, $T = 5$ hr, h_t would amount to only 2.0 m. The decisive one of the two effective quantities, duration and fetch, is always that which leads to a smaller h_t. After $F = 100$ nautical mi and $T = 12$ hr, h_t would be 4.5 m; for $F = 200$ nautical mi and $T = 18$ hr, h_t would be 7.0 m, and for $F = 290$ nautical mi, the fully developed state is reached with $h_t = 8.8$ m, a value that is also found in Table 8.02. Figure 8.22 further gives the predominant wave period for a developing sea. For a not yet fully developed sea, the wave spectrum is cut off at a certain upper limit of the period approximately corresponding to the predominant wave period. In the figure, this value can be read off the scale of the abscissa by moving vertically downward from the point where the critical quantity (duration or fetch) intersects the curve of the wind speed. Accordingly, for $W = 30$ knots and $F = 100$ nautical mi, the predominant wave period would be 8 sec.

Figure 8.22 offers much information. If only high waves h_t are represented here, this has been done because these are the most important ones for practical problems of navigation. According to Fig. 8.22, wave heights of more than 20 m will probably be encountered only rarely; a storm with a wind velocity of 42 knots would have to blow for two full days with constant direction over the same oceanic distance of at least 800 nautical mi before such wave heights would be produced in the sea state. In some winter storms of the west wind zones, such conditions may be fulfilled. In the example of the wave diagram in Fig. 8.20, a height of 16.5 m was measured. The greatest wave heights known so far were observed by officers on the British liner *Majestic* in December 1922 during a storm of hurricane force in the North Atlantic Ocean. After critically allowing for typical observational errors, Cornish (1934) considered heights of up to 27 m possible in this case.

Table 8.02 and Fig. 8.22 can be used for wave predictions under given wind conditions. For practical application the method has been described in detail in a manual by Pierson et al. (1953). This method seems to have proved quite successful, as shown by Walden (1954) who compared observations and predictions of waves in the Atlantic Ocean.

Apart from wind velocity, duration, and fetch, the stability of the lower atmospheric layer also has considerable influence on the growth of waves, as suggested by Seilkopf (1940) and proven by Roll (1952) on the basis of observations. If the temperature difference between air and water is taken as the measure of the stability of stratification of an air mass over the ocean, the data collected on North Atlantic weather ships (cf. Fig. 8.23) show that at constant wind velocity, wave heights and wavelengths will increase with increasing instability. Since the height of a wave grows faster than the length, the steepness of the wave increases with increasing instability of the near-surface atmospheric layers.

If the wind subsides, or if the waves travel out of the area on which the wind is acting, the energy supply is discontinued, and the waves are subject only to nonlinear interaction and to energy dissipation strongly depending on the steepness δ.

The sea state, due to the influence of effective friction, undergoes a selective process as soon as it is no longer under the influence of the wind. The short and steep forms of the sea gradually disappear with increasing distance from the generating area, and the long, rounded forms of the swell remain, which, with their great wavelengths, have large propagation velocity and travel through vast oceanic areas within relatively short time periods. Such cases are illustrated in Fig. 8.24. Accordingly, swell emanating from a storm area dominated by a wind velocity of 20 m sec^{-1} would arrive after 80 hr at a distance of 2800 km with a wave height of 2 m and a wave period of 16 sec. By using this diagram, one can determine the geographic position of the generating area and the wind force that has been acting there from the time of arrival of the swell, from its amplitude and period,

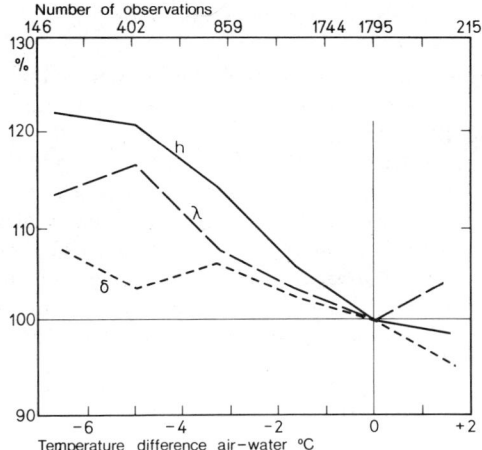

Fig. 8.23. Change of height h, length λ, and steepness δ of surface waves as a function of the stability of atmospheric stratification. (After Roll, 1952.)

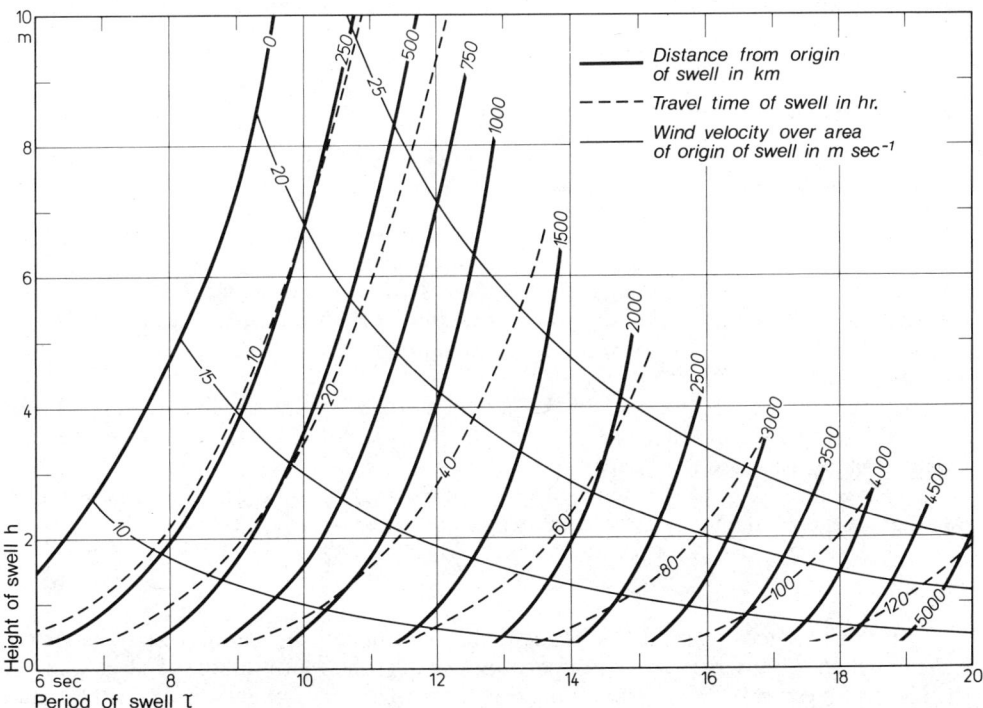

Fig. 8.24. Relationship between height and period of swell and the distance from the area of origin as well as the travel time of swell, valid for fully developed sea at different wind velocities in the generating area. (According to Sverdrup and Munk, 1947.)

and, additionally, from the direction of its crests. This method has been successfully applied for storm prediction at sea.

For practical purposes, only those waves that exceed a certain value and may cause damage are of interest. Smaller waves can be left out of consideration here. In order to obtain well-defined data about the wave height in a record of the sea state, comprising N waves, the wave heights of the individual waves are arranged according to their size, and the average value of pN waves is calculated, where p is a number between 0 and 1. In practice, $p = 1/3$ and $p = 1/10$ are the values commonly used. $h^{(1/3)}$ is the wave height of the so-called significant wave, that is, the average of the heights of the one-third highest waves. Analogously, $h^{(1/10)}$ represents the average of the one-tenth largest wave heights. Comprehensive measurements in various oceanic areas have shown that the ratio of the maximum wave height h_{max} as well as of the average wave height to the significant wave height is a fairly constant one. Some average values are given here

$$h_{max}/h^{(1/10)} \approx 1.45, \quad h_{max}/h^{(1/3)} \approx 1.6, \quad h^{(1/10)}/h^{(1/3)} \approx 1.26, \quad h^{(1/3)}/h_m \approx 1.56$$

Since significant wave heights are averages obtained from a large number of measured data, they are characteristic of an oceanic area under certain wind conditions. The above numerical ratios permit us to calculate, for example, the maximum wave height, which is 1.45 times larger than the average of the heights of the one-tenth highest waves, etc. Furthermore, these ratios also show that the undulations of the sea surface obviously are subject to certain statistical laws. This has initiated a purely statistical description of the sea state.

8.3.6. Statistical description of the sea state

The fluctuations of the sea surface with time can, at a point of observation, be represented by the sum of harmonic oscillations

$$z(t) = \sum_{n=1}^{N} A_n \cos(\omega_n t + \varphi_n) \qquad (8.25)$$

with definite amplitudes A_n, angular frequencies ω_n, and phases φ_n. But such a single record does not provide sufficient information on the fluctuations of the sea surface in general. This is obtained only when a large number of such "realizations" has been measured.

According to Eq. (8.16), each of the partial waves with the amplitude A_n can be associated with an average energy $E_n = g\rho A_n^2/2$. The distribution of $A_n^2/2$ on the discrete frequencies ω_n is called the energy spectrum.

$\overline{z(t)} = 0$ is valid for the temporal average of the realization $z(t)$, whereas the square mean gives the total energy E^* of the measured series:

$$\overline{z^2(t)} = \sum_{n=1}^{N} \frac{A_n^2}{2} = E^* \qquad (8.26)$$

Since their place of generation is accidental within certain limits, the partial waves, passing a point of observation, can superimpose each other with random phases φ_n from the interval $(-\pi, \pi)$. Thus, for any possible combination of phases, a different course of the function $z(t)$ will be recorded, as also shown by repeated measurements carried out under otherwise equal conditions. According to Eq. (8.26), however, all those possible combinations have the same total energy E^*, since the latter depends only on the amplitudes of the partial waves, not on their phases. Hence, it suggests itself to consider the total of all possible records with the total energy E^* instead of considering the individual records $z(t)$. Such totality is called random function $\zeta(t)$. As the phases φ_n can be distributed at random, some pertinent assumption must be made. The most reasonable assumption is that the phases are uniformly distributed in the interval $(-\pi, \pi)$. Now we can represent the sea state at the point of observation as a random function with equally distributed phases:

$$\zeta(t) = \sum_{n=1}^{N} A_n \cos(\omega_n t + \varphi_n) \tag{8.27}$$

Consequently, the relations $[\zeta(t)] = 0$ and $[\zeta^2(t)] = \sum_n A_n^2/2$ hold for the statistical averages [ ] as well as for the temporal averages $z(t)$. The second term gives the total energy per unit area. If $E(\omega)\,d\omega$ represents the energy in the infinitesimal frequency interval $d\omega$, we can also say

$$[\zeta^2(t)] = \sum_n \frac{A_n^2}{2} = \int E(\omega)\,d\omega \tag{8.28}$$

$E(\omega)$ is also called energy spectrum. It gives the energy as a function of frequency. Some insight into the statistics of the sea state can already be obtained from Eq. (8.27). If the state of the sea is considered at an arbitrary time $t = 0$, the following equation is valid:

$$\zeta = \sum_{n=1}^{N} A_n \cos \varphi_n \tag{8.29}$$

This is the sum of N independent random quantities (because of φ_n), all of them having the average value zero and the variance $A_n^2/2$. In accordance with the central limit theorem of probability theory, the sum [Eq. (8.29)] for $N \to \infty$ then has the normal distribution

$$f(z) = \frac{1}{\sqrt{2\pi E^*}} e^{-z^2/2E^*} \tag{8.30}$$

with the average value zero and the variance E^*. In fact, frequency distributions of vertical sea surface deflections show such a normal, that is, Gaussian distribution. In Fig. 8.25, the relative frequency of occurrence of vertical deflections of the sea surface in the North Atlantic Ocean is represented after Carlson et al. (1967). It is obvious that the Gaussian distribution approximates the measured data rather well. This is generally the case if the wave spectrum comprises a large number of different waves as it occurs in the sea state. On the other hand, if swell waves prevail, the frequency distribution of the maximum vertical deflection becomes narrower and passes into the Rayleigh distribution

$$f(z_{max}) = \frac{2z_{max}}{E^*} e^{-z_{max}^2/2E^*} \tag{8.31}$$

as shown by Longuet-Higgins (1952). For each of such distributions the significant wave heights $h^{(1/3)}$ etc., mentioned in the previous section, can be determined. The result is a surprisingly good agreement between the theoretical values and the measured values.

As an extension of the model Eq. (8.27) describing the sea state at a given point, the sea state in an oceanic area is described by

$$\zeta(x,y,t) = \sum_n A_n \cos(\kappa_n x + \eta_n y + \omega_n t + \varphi_n) \tag{8.32}$$

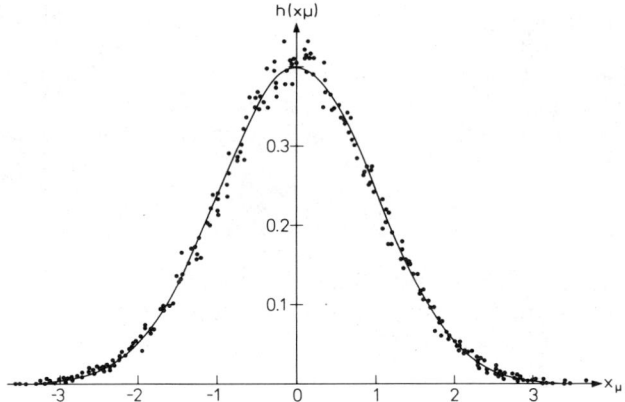

Fig. 8.25. Relative frequency of occurrence of the vertical deflection of the sea surface normalized to the variance 1. The symbols correspond to observations. The curve represents a normal distribution. (After Carlson et al., 1967.)

Here, it is assumed that the following relations are valid:

1. The wavenumbers κ_n, η_n are densely distributed over the whole range of wavenumbers, that is, all wavenumbers will occur.
2. ω_n is connected with κ_n and η_n by the dispersion relation for gravity waves Eq. (8.14),

$$\omega_n^2 = gk_n, \quad k_n^2 = \kappa_n^2 + \eta_n^2$$

3. In analogy to Eq. (8.26), the following equation shall hold in each range $d\kappa\, d\eta$:

$$\sum_\eta^{\eta+d\eta} \sum_\kappa^{\kappa+d\kappa} \frac{A_n^2}{2} = E(\kappa, \eta)\, d\kappa\, d\eta$$

4. The phases ω_n are uniformly distributed in the interval $(-\pi, \pi)$.

The additional assumption 2 has proven true. Measurements (Longuet-Higgins et al., 1961) have shown that the relation $\omega_n^2 = g\kappa_n \tanh \kappa_n H$ holds with an inaccuracy of 5%. This can also be taken from the fact that the velocity at which wave fields travel out of the generation area equals the group velocity, as Munk et al. (1963) could show by the example of swell waves that had travelled over a distance of nearly 12,000 nautical mi.

Furthermore, measurements of the sea surface slope have shown that $\partial \zeta/\partial x$ and $\partial \zeta/\partial y$ are distributed according to Gauss. Thus, the assumption that the sea state is a random function with Gaussian normal distribution can be considered as confirmed to a great extent. Only very short waves show a certain deviation from that distribution. The surface slope in the direction of the wind exhibits a skewness unequal to zero.

The spectrum $E(\kappa, \eta)$ as a function of the wavenumbers κ, η can also be represented as the spectrum $E(k, \theta)$, if polar coordinates according to $\kappa = k \cos \theta$, $\eta = k \sin \theta$ are introduced. Due to the relationship between ω and k, $E(\omega, \theta)$ can also be given, in particular. Such a spectrum is called directional spectrum. Here,

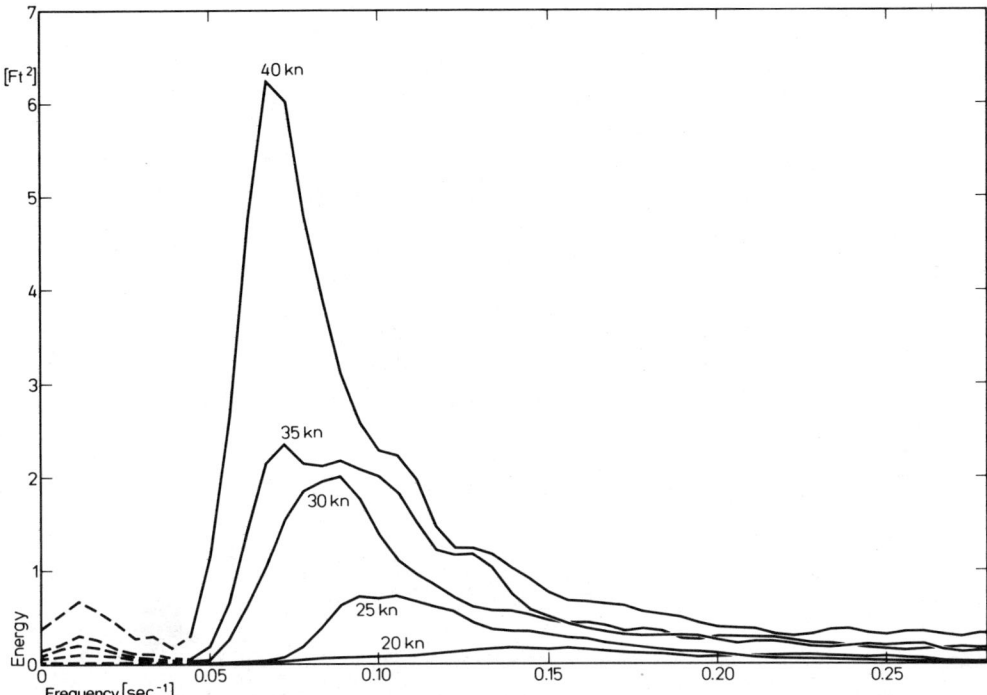

Fig. 8.26. Energy spectrum of the fully developed sea at different wind velocities. (After Moskowitz, 1964.)

$$E(\omega, \theta) = \frac{2k^2}{\omega} E(\kappa, \eta)$$

holds true. In an extension of $E(\omega)$, $E(\omega, \theta)$ does not give the energy only as a function of frequency, but, in addition, also as a function of the directions in which the waves of such energy travel. Furthermore,

$$E(\omega) = \int_0^{2\pi} E(\omega, \theta) \, d\theta$$

is valid.

Measuring the spectrum $E(\omega)$ is comparatively simple. Only measurements taken at time series of sufficient length at one location are required. In the past decade, numerous general laws for such spectra were derived, partly with very significant differences. From groups of about 10 spectra, obtained under equal conditions, Moskowitz (1964) has derived mean spectra. The result is represented in Fig. 8.26. With increasing wind velocity the energy maximum is shifted to lower frequencies. The areas under the curves, each of them representing the respective total energy of the sea state, is proportional to the fourth power of the wind velocity. The dimensionless form of those spectra can be expressed by a general law

$$E(\omega) = \frac{\alpha g^2}{\omega^5} e^{-\beta(\omega_0/\omega)^4} \tag{8.33}$$

as shown by Pierson and Moskowitz (1964). Here, g stands for the acceleration of gravity,

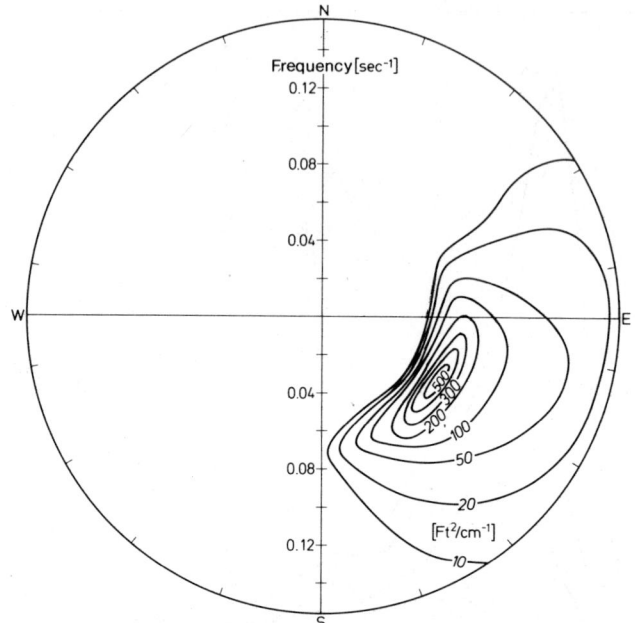

Fig. 8.27. Observed directional spectrum of the sea state. (According to Cartwright, 1963.)

$\omega_0 = g/W$; W is the wind velocity in cm sec^{-1}, and α and β are dimensionless constants ($\alpha = 8.10 \times 10^{-3}$, $\beta = 0.74$).

For high frequencies, that is, in the right-hand half of the figure, the spectra, therefore, decrease proportionally to ω^{-5}. In theory, this was already derived by Phillips (1958) for a fully developed sea. The reason herefor is that waves of a certain frequency can only have enough energy to keep their steepness ($\delta = h/\lambda$) smaller than $1/7$, as mentioned in Section 8.3.3. If this value is exceeded, the waves will break, thus releasing their excessive energy to other frequency ranges.

Measuring the directional spectrum is much more difficult. For this purpose, buoys are usually employed to measure the vertical acceleration and the slope of the sea surface in two directions perpendicular to each other. A result is given in Fig. 8.27 (Cartwright, 1963). It shows the energy concentrated at about 0.07 Hz or 14 sec. The waves mostly travel in a southeasterly direction. In a first approximation, the distribution is symmetrically arranged around this preferred direction. For theoretical investigations it is therefore mostly assumed that the directional spectrum is proportional to $\cos^n\theta$ with a suitable power figure n.

8.3.7. Generation of the sea state

The generation of the sea state and the modification of its spectrum during growth and decay are not yet completely understood. Numerous processes are involved. Of the older theories on the generation of sea surface waves there are worth mentioning, in particular, the instability theory by Lord Kelvin (1871) and von Helmholtz (1868) as well as the so-called sheltering theory by Jeffreys (1924).

The Kelvin–Helmholtz instability concerns the occurrence of waves at an interface.

These waves grow exponentially with time. If friction in air and water is neglected, and a constant wind is assumed to blow over a level water surface at the velocity W, the interesting question is whether this surface remains stable against small disturbances, with acceleration of gravity g and surface tension σ being taken into consideration. If the small disturbance is assumed to be a wave traveling, according to $\cos \kappa(x - ct)$, with a given wavenumber κ in the x direction, the following phase velocity will result:

$$c = \frac{\rho_L W}{\rho + \rho_L} \pm \sqrt{\frac{g}{\kappa} \frac{\rho - \rho_L}{\rho + \rho_L} + \frac{\sigma \kappa}{\rho + \rho_L} - \frac{\rho \rho_L}{(\rho + \rho_L)^2} W^2} \qquad (8.34)$$

where ρ_L stands for the density of the air. As long as c is real, the wave is stable, but if c gets complex, the amplitude of such a wave increases exponentially until nonlinear effects become influential. Instability will occur when the expression below the square root sign becomes negative. Kelvin has shown that this is the case for a certain range of wavenumbers as soon as the wind velocity exceeds 650 cm sec^{-1}. With increasing wind velocity, this applies to a larger and larger range of wavenumbers. The most unstable wave is that with a wavelength of $\lambda = 1.7$ cm; at the critical wind speed, its phase velocity amounts to 0.8 cm sec^{-1}.

Hence, the Kelvin theory states that a wind velocity of at least 6.5 m sec^{-1} is needed to induce growing waves. This corresponds to Beaufort force 3 to 4, which is much too high as compared to observations. Helmholtz had already before pointed out that pure gravity waves can be generated at any wind speed provided the wavelength is sufficiently small. It is not justified, however, to neglect surface tension when dealing with such short waves.

The sheltering theory by Jeffreys is based on the observation that a current will not flow smoothly around a body of irregular form but break off behind this body and form eddies, thus producing a pressure difference between the windward side and the leeward side. In analogy, Jeffreys presumed that the wind cannot follow the irregular sea surface, but will break away at each crest and smoothly lean against the wave only before the next crest. Thus, a pressure difference between front and back of the crest is caused, and if the wind velocity is greater than the phase velocity, the wind will release energy to the wave. Should this energy exceed the energy dissipated in the wave by molecular friction, the wave will grow. It can be shown that the prerequisite for waves to grow in such way is the condition

$$c(W - c)^2 \geq \frac{4\nu \rho g}{s \rho_L} \qquad (8.35)$$

where s is a resistance coefficient which is principally unknown; ν is the kinematic viscosity. The minimum wind speed at which waves are generated is given by

$$W_{min} = 3 \left(\frac{\nu \rho g}{s \rho_L}\right)^{1/3} \qquad (8.36)$$

According to observations by Jeffreys, wave formation begins at $W_{min} = 110$ cm sec^{-1}, which yields $s = 0.3$. Later measurements at wave profiles, however, have shown that s is smaller by approximately one power of ten, whereby the whole mechanism of energy transfer also becomes weaker by one order of magnitude.

Phillips (1957) has introduced statistical viewpoints into the theory on the generation of sea surface waves. His theory is based on the fact that the wind always is turbulent to some extent. This turbulence is associated with pressure fluctuations acting on the sea surface. In contrast to the sheltering theory, those turbulent air pressure fluctuations

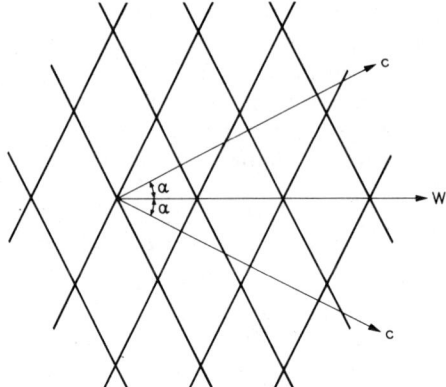

Fig. 8.28. Generation of two wave systems in direction c caused by a turbulent air pressure fluctuation traveling at wind velocity W.

are completely independent of the waves. If W denotes the mean wind velocity at a given height above the sea surface with which the turbulent pressure fluctuations travel (frozen turbulence), the directional spectrum of the sea state is obtained as follows:

$$E(\omega, \theta) \propto \frac{k^2 \omega t}{2(g\rho)^2} \int_0^\infty P(\mathbf{k}, \zeta) \cos\left[\left(\frac{W \cos \theta}{c} - 1\right)\omega\zeta\right] d\zeta \qquad (8.37)$$

Here $\mathbf{k}$ stands for the wavenumber vector and $P(\mathbf{k}, \zeta)$ for the spectrum of the air pressure fluctuations at the sea surface at the time ζ within the wind field traveling at wind velocity.

If it is assumed that there is no special direction preferred by the turbulent pressure fluctuations, the integral in Eq. (8.37) will become a maximum when the cosine function equals unity, that is, when $W \cos \theta = c = g/\omega$ is valid. This can be interpreted as follows.

Let us consider a pressure disturbance with the wavenumber k. It will generate waves of the same wavenumber, which travel in different directions θ. However, if the wind vector is projected upon the propagation direction of these waves, all waves whose phase velocity equals $W \cos \theta$ are in resonance with the pressure disturbance. According to the above equation, these are the waves which form the angle $\theta_{res} = \alpha = \arccos(c/W)$ with the wind direction. Thus, for $W > c$ we can expect two wave systems to develop, forming the angle $\pm \alpha$ with the wind direction, and, according to Eq. (8.37), growing linearly with time since their phase velocity agrees with the travel velocity of the turbulent pressure field. The two wave systems are plotted in Fig. 8.28. The initial stage of wave generation is described by this theory.

At the same time as Phillips, Miles (1957) published a wave theory that was based on a completely different mechanism. Miles assumed that small waves are already present, and, like Jeffreys, he dealt with pressure disturbances produced by the wavy sea surface, which, in turn, do work on the waves. Miles determined the resistance coefficient directly from theory, whereas Jeffreys assumed it to be a known quantity.

It is supposed that the wind velocity increases with height. The energy flow from wind to the wave with the phase velocity c depends on the curvature of the wind profile at the critical height where the wind speed equals the phase velocity of the wave. This theory resembles that of Kelvin–Helmholtz, but it results in a much more effective energy flow.

This flow is all the larger, the smaller the critical height is. It is particularly effective for waves whose phase velocity is smaller than the wind speed. Energy is supplied to the waves only from the critical height. The combined theory by Phillips and Miles was first tested by Snyder and Cox (1966). There it was shown that Phillips' theory gives a good description of the initial stage of wave generation. When the energy of the sea state increases, Miles' mechanism becomes dominant. However, the process of growing up to the fully developed sea state is not described even by this theory.

Nonlinear interaction, described in Section 8.3.4, plays an important part with regard to the form of the spectrum. If partial waves are generated by the mechanisms mentioned above, they are, with growing amplitude, subject to the continuous process of redistribution of energy among the various partial waves of the spectrum.

The change of the energy spectrum is described by the following equation

$$\frac{\partial E}{\partial t} + \mathbf{c}_{gr} \cdot \nabla E = S \qquad (8.38)$$

where $\mathbf{c}_{gr}$ stands for the vector of the group velocity, ∇E for the horizontal energy gradient and S for a source function. The latter includes all processes supplying or removing energy. This mainly concerns three components: $S = S_1 + S_2 + S_3$. The component S_1 is the energy supply from the atmosphere according to the mechanisms mentioned above. S_2 includes the nonlinear interactions, and S_3 represents the energy dissipation by breaking of waves, bottom friction, turbulence, etc. Equation (8.38) can be employed as predictive equation. At present, however, too many questions regarding generation of sea surface waves and dissipation are still unsolved.

8.3.8. Wave transformation in shallow water; surf

When swell or wind waves reach an area shallower than half their wavelength, the waves begin to "feel" the bottom and start to undergo a transformation. For wave periods of 5, 10, 15, and 20 sec, this process sets in at water depths of 19.5, 78, 175.5 and 312 m, respectively. It is reflected in the orbits of the water particles changing from circular to elliptical shapes, and finally, when the water depth has decreased to approximately $\frac{1}{20}$ of the wavelength, the particles begin to move in predominantly linear paths. Thus, the surface wave assumes the character of a "long wave" with the propagation velocity of $c = \sqrt{gH}$ diminishing with decreasing depth. As a result, the wave crests move closer together, that is, the wavelength λ decreases. At the same time, the wave height h increases, provided the energy content of the wave is not changed. Consequently, the steepness δ increases. When the particle velocity v at the wave crest gets larger than the propagation velocity c, the crests break. Over a shallow, evenly sloping bottom, this occurs in the beach surf when the wave height becomes $h = 1.3H$, with H indicating the water depth as measured in the wave trough. Over steep, irregular rocky bottom, breaking takes place in the sea cliff surf. Hereby, part of the energy can be reflected.

An observer will note that, before the waves reach the surf zone near a beach, even in an irregular sea coming from different directions ("cross sea"), the short-crested waves disappear and wave trains with long crests develop. Jeffreys (1924) explained this phenomenon as follows: in shallow water short seas become unstable before longer ones do so.

Since waves advance more slowly over shallow water than over deeper water, the near-shore part of a long crest is more retarded than the seaward part. Thus, even waves with crests originally progressing parallel to the shore in the open sea, eventually turn

376 Surface Waves and Internal Waves

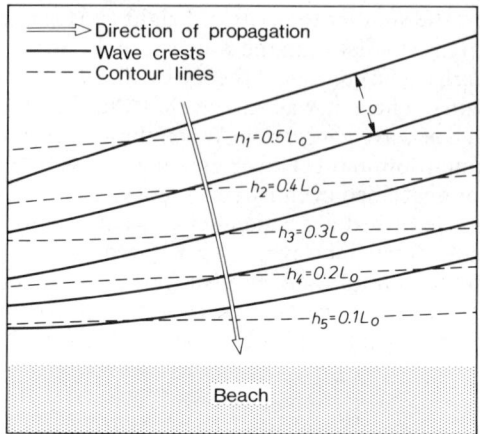

Fig. 8.29. Schematic representation of the refraction of surface waves over an evenly sloping bottom.

toward the coast. This means that the waves are subjected to refraction, as schematically illustrated in Fig. 8.29 and as can be recognized in the aerial photograph of Fig. 8.30.

Surf waves are no longer of a periodic nature. They transport a considerable amount of water masses towards the shore. These water masses do not swing back seaward with

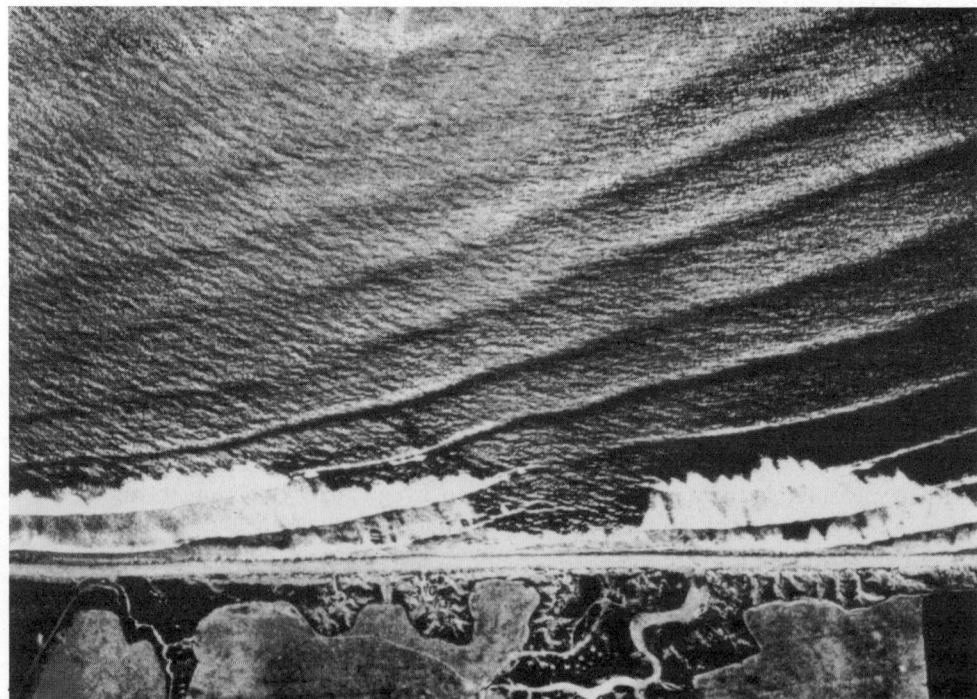

Fig. 8.30. Refraction and beach surf of long swell waves. Aerial photograph taken at the California coast. (Photo: U.S. Navy (from Munk and Taylor, 1947).)

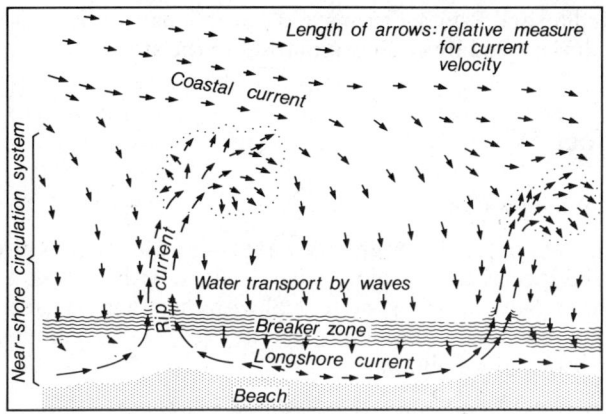

Fig. 8.31. Schematic representation of surface currents in the vicinity of the beach under surf conditions. (According to Shepard and Inman, 1951.)

the wave but flow back along the bottom as a current. Any person who has ever stood in the surf zone has experienced this current well enough. This so-called undertow can cause a person to lose balance and may easily be fatal for him in heavy surf. Furthermore, if the surf waves run up the beach at an angle, a current along the shore will develop which intermittently forces its way back into the sea in so-called rip currents, another danger to swimmers. Figure 8.31 gives a schematic picture of the system of near-shore surface currents under surf conditions.

Longshore currents are of special importance for the sand transport along the beach and, thus, also for the morphology of sandy shores. Their velocity v, derived by Putnam et al. (1949) from energy considerations, amounts to

$$v = K \sqrt[3]{\frac{mH^2 \sin 2\epsilon}{\tau}} \qquad (8.39)$$

where K is a friction coefficient which, for a sandy beach, is $K = 8.0$; m stands for the bottom slope and ϵ for the angle formed by the wave crests in the surf zone with a straight coast. According to Dietrich and Weidemann (1952), the application to conditions in the Lübeck Bay (in the Baltic Sea) for $m = 0.03$ and $\epsilon = 10°$ results in $v = 1.3$ m sec^{-1} at wind force 7. This is a multiple of the current velocity of other coastal currents and is decisive for the material transport near the beach.

Particularly effective and impressive surf develops where swell originating from the large areas with frequent storms hits the coast. The surf at the African west coast in the Gulf of Guinea is notorious, and the same is true for the so-called Kalema farther south which, at that coast with only a few harbors, is a very troublesome nuisance for ship-to-land traffic of ships that anchor offshore. The Kalema is a breaking swell generated in the stormy west wind zone in the South Atlantic Ocean, the "roaring forties," at about 40° latitude. Like those storms, the Kalema reaches its maximum during the southern winter from June to September. The "rollers" off St. Helena and Ascension Island, as well as the "Raz de Marée" along the Atlantic coast of Morocco can be equally unpleasant. These are caused by breaking swell generated by northwest storms in the North Atlantic Ocean during the stormy period from December to April. The name of "Raz de Marée" is misleading since this phenomenon has nothing to do with tides (in French:

la marée). In the Indian and Pacific Oceans, too, exceptional surf is found at certain parts of the coasts, which is caused by swell originating in the stormy west wind zones.

8.4. Long Surface Waves

8.4.1. Properties of long waves

The dispersion relation Eq. (8.13) for gravity waves of any length has already been given in Section 8.3.1. The special case of short waves, that is, with $\lambda/H \ll 1$ or $\kappa H \gg 1$, has been considered there. Now we are going to deal with the other extreme: the wavelength λ will be large compared to the water depth, that is, $\kappa H \ll 1$. Then $\tanh \kappa H$ in Eq. (8.13) can be replaced by κH and we obtain

$$\omega^2 = \kappa^2 g H \quad \text{or} \quad c = \sqrt{gH} \tag{8.40}$$

Apart from the acceleration of gravity, the phase velocity of long waves thus depends only on the depth of the water. The condition $H \ll \lambda$ is already well satisfied for $H < \lambda/10$; then, the phase velocity differs from the exact value by not more than about 6%.

Therefore, the term "long waves" always refers to the local water depth. Waves with wavelengths of more than 10 km generally are long waves, thus including, for example, all tides, which will be treated comprehensively in Chapter 9. Swell and sea can also become long waves and behave like these when they enter shallow water.

Progressive long waves of the form $\zeta = A \cos(\kappa x - \omega t)$ have a velocity field given by

$$u = \sqrt{\frac{g}{H}} \zeta \qquad w \propto (H - z) \tag{8.41}$$

that is, the horizontal velocity is the same at all depths, and the vertical velocity decreases linearly from the sea surface down to the bottom. When a long wave is passing, the water particles move along elongated ellipses.

Consequently, in a progressive wave, the flow reaches a maximum at the wave crest in the direction of propagation, as well as at the wave trough in the opposite direction. At the mean water level the motion is zero; this is where the current turns.

The group velocity of long waves equals the phase velocity. In such waves, the vertical acceleration is very small and can therefore be neglected so that the pressure is given by the fundamental equation of statics. At the depth z, the pressure amounts to $p = g\rho(z + \zeta)$, if ζ stands for the deflection of the sea surface from its position at rest. Thus, long waves can be recorded in a simple way by means of pressure measuring devices (gauges), for example, at the sea floor. With long waves, too, special consideration is required for the case that the wave height can no longer be assumed as infinitesimally small. Such waves travel with the phase velocity

$$c = \sqrt{g(H + \zeta_0)} \tag{8.42}$$

where ζ_0 is the amplitude of the wave. So far, two types of waves have been investigated: the cnoidal waves and the solitary waves. They are represented in Fig. 8.32. The cnoidal wave has been named after the elliptic function cn describing the wave profile. If the wavelength of such a cnoidal wave gets infinitely long, a solitary wave is obtained which consists of just one single wave crest. Solitary waves, in particular, play a role as individual

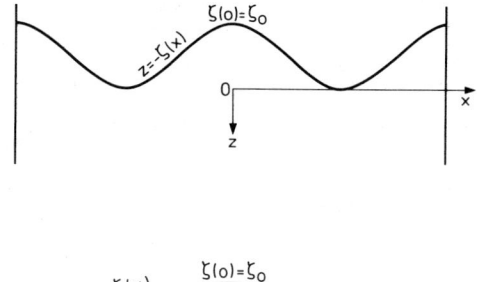
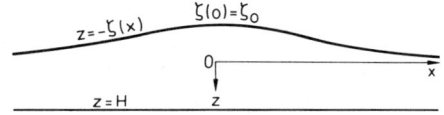

Fig. 8.32. Cnoidal wave (above) and solitary wave (below).

phenomena of piling-up. Korteweg and De Vries (1895) were the first to investigate them, after they had been described in detail by Lord Russel (1838, 1845) as waves in channels.

8.4.2. Tsunamis and storm surges

Long progressive waves are frequently observed in the ocean. They develop in areas of intense cyclones as the result of rapid changes of atmospheric pressure. In typhoons and hurricanes such changes may amount to more than 100 mbar within a few hours. They can also be generated by winds of hurricane force acting on the sea surface, and precede the actual sea and swell. These long waves are believed to produce clapotis, that is, standing waves, at steep coasts, as well as to be responsible for the generation of seismic unrest at the sea floor which can be recorded far inland as tiny bottom oscillations. Since these waves are caused by atmospheric influences, it is difficult to assess details in their behavior because the area of disturbance is extended in space besides being in motion. Both factors are insufficiently known for any particular case in the ocean.

As an introduction to studying the behavior of undisturbed progressive long waves, tsunamis are more suitable. This Japanese term is used for progressive long waves generated by submarine earthquakes and volcanic eruptions and spreading annularly from the epicenter. In the immediate vicinity of the epicenter, large amplitudes of tsunamis have been observed which, when transformed by breaking processes in shallow water, may reach enormous dimensions. They result in disastrous floods, and the damage is often heavier than that caused by the actual earthquake or volcanic eruption. The greatest wave heights so far observed were on the Sunda Islands after the Krakatao eruption in the Sunda Strait on August 26 and 27, 1883. They reached heights of up to 35 m. 36,830 lives were lost. Tsunamis occur most often in the Pacific Ocean, which can be explained by the great frequency of earthquakes at the marginal zones of that ocean (cf. Plate 2). In a summary, Heck (1947) has listed tsunamis caused by several hundred seaquakes during the past centuries. Among all Pacific coasts, the Japanese coasts are most frequently hit. On the average, tsunamis with heights of more than 7.5 m will occur there once every 15 yr. Since 684 A.D., four tsunamis of more than 30 m height have been recorded.

In 1856, when not much was known about the depth distribution of the Pacific Ocean,

380 Surface Waves and Internal Waves

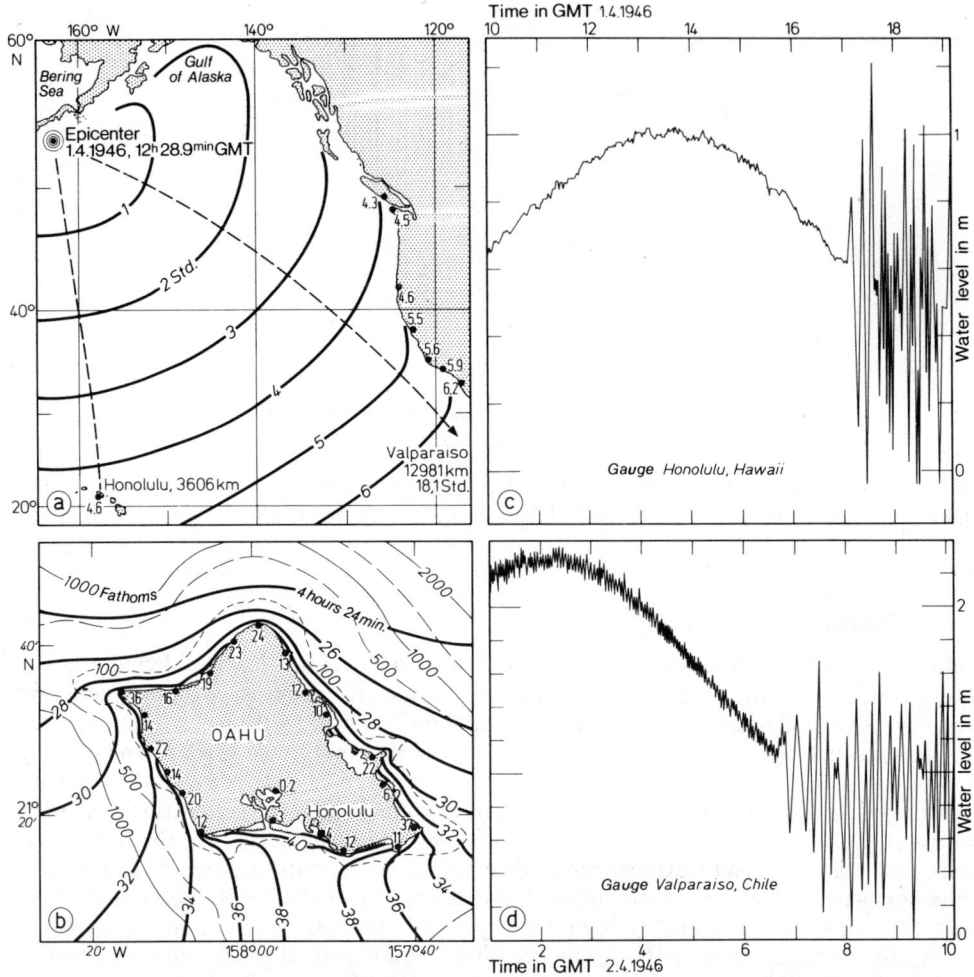

Fig. 8.33. (a) Travel time of the tsunami in the North Pacific Ocean, expressed in hr after the quake in the Aleutian Trench on April 1, 1946, 12h 58.9min GMT. (b) Maximum range of tsunami (in feet) on the Hawaiian island of Oahu on April 1, 1946 and the calculated travel time from the epicenter in hr and min. Depth in fathoms. (c) (d) Gauge records of the tsunami in Honolulu and in Valparaiso. (a, c, d after Green, 1946; b after Shepard et al., 1950.)

Bache already computed the mean depth of the North Pacific Ocean from the equation for the propagation velocity of long waves [Eq. (8.40)]. He used the travel time of Pacific tsunamis caused by a Japanese quake and observed in California. He arrived at mean depths from 4200 to 4500 m. Compared with earlier estimates by Laplace yielding 18,000 m depth, Bache's result represented rather a good value at that time.

The propagation of more recent tsunamis has been studied intensively, especially that of the tsunamis of April 1 and 2, 1946 (Shepard et al., 1950). They were caused by a heavy quake in the Aleutian Trench on April 1 at 12 hr 58.9 min GMT, and led to the greatest devastations on the Hawaiian Islands that ever occurred in the history of that group of islands. In Fig. 8.33(a) the epicenter of the quake is given, as well as the computed travel time in hours, and the travel time as observed at several coastal tide gauges in the northeastern Pacific Ocean. The distance along the great circle from the epicenter to Honolulu (3606 km) was covered in 4.6 hr, and to Valparaiso (12,981 km) in 18.1 hr. According to Green (1949), the travel times computed by Eq. (8.40) differ from the observed ones only by 1.2% on the average. On the greatly reduced sections of tide gauge records from Honolulu and Valparaiso, given in Figs. 8.33(c) and 8.33(d), the sudden arrival of the tsunamis can well be recognized in both cases. The period of the first waves amounted to 15 min in Honolulu, and to 18 min in Valparaiso. As shown in Fig. 8.33(b), Honolulu is protected against tsunamis coming from the north by the configuration of the island; therefore, the observed wave heights of 4 ft (1.2 m) remained relatively small. At other coastal sections of the same island, wave heights of 35 ft (10.7 m) were observed. These, however, were improper waves at shallow coasts, which run up the beach like a bore.

It may be mentioned that in addition to gravity waves in the form of tsunamis, compression waves also originate at the epicenter. They spread in water at sound velocity, that is, at 1500 m sec^{-1}, and are considerably faster than tsunamis; for example, they would need 40 min to cover the distance between the epicenter and Honolulu [cf. Fig. 8.33(a)], whereas the tsunamis travelled for 4.6 hr. If such sound waves were observed in the ocean by means of the SOFAR method, a warning system for tsunamis could possibly be established. The present warning service, built up by the U.S. Coast and Geodetic Survey after the catastrophe on the Hawaiian Islands on April 1, 1946, is based on registrations of seismic waves by means of seismographs. It proved successful when the Hawaiian Islands were hit by large tsunamis on November 4, 1952. Such warning systems can be of value only in places where the tsunamis are generated far away from the coast. In Japan, which is most heavily endangered by tsunamis, such a system must fail, because the largest tsunamis usually originate in the immediate vicinity of the Japanese islands.

In areas of seismic activity, the compression waves caused by seaquakes are sometimes perceived on ships as sound or shock, which, in former times, occasionally produced the impression that the ship had hit the bottom. This explains why reports of shoals were entered in sea charts which, later on, when more soundings were available, turned out to be nonexistent.

Long waves produced by meteorological factors can also cause disastrous devastations when they enter shallow areas of the ocean. Their height usually amounts to several meters. It has been estimated that, in the shallow Bay of Bengal, 250,000 people were killed during the storm surges of the years 1864 and 1876. In 1780, 15,000 people died in a storm surge that flooded the islands of Santa Lucia and Martinique. In the Holland storm surge during the night of January 31 to February 1, 1953, the water level at the Dutch coast rose by 3.0 to 3.5 m. Together with the waves of the sea state, with heights from 6 to 7 m, this storm surge caused 400 breaches of dykes. It came to be the most disastrous storm surge catastrophe of the Netherlands in historic times. Land extending over 25,000 km^2 was flooded, and 2000 people were killed. 600,000 people had to leave their homes.

In Fig. 8.34 weather charts covering the British Isles and the North Sea are shown for the period of a weaker storm surge on January 7 to 8, 1949. As a consequence of a depression over the northern North Sea, traveling northeastward, the wind suddenly turned from southwest to northwest and continued to blow from this direction for quite a long time. In Fig. 8.35, the variations of the water level during the period from January 6 to 10 have been plotted for numerous stations bordering the North Sea. Tidal effects

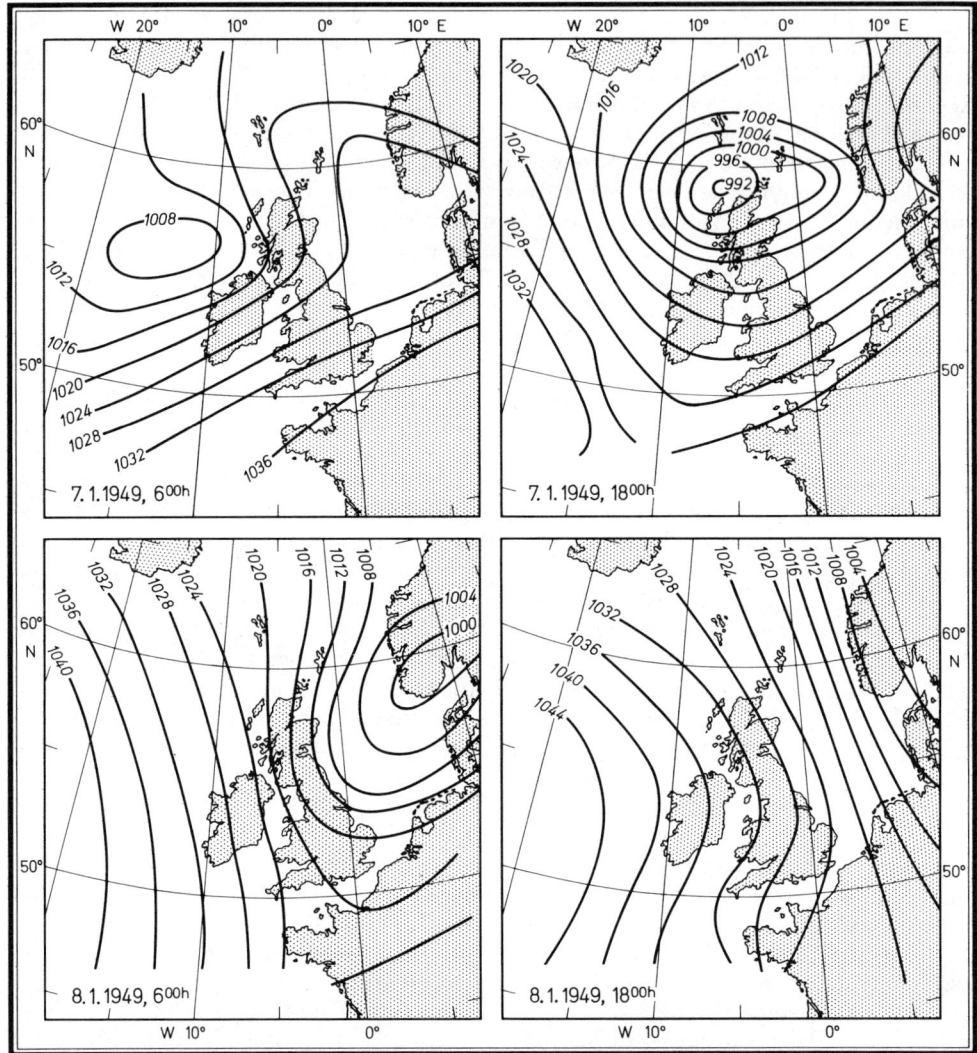

Fig. 8.34. Weather charts for the British Isles and the North Sea during the storm surge of January 7–8, 1949. (According to Corkan, 1950.)

have been eliminated. It can be seen that, in the northwestern North Sea, a wave developed as the result of the storm, which travelled counterclockwise along the coastline of the North Sea from Aberdeen to Bergen within 24 hr (after Corkan, 1950). Even small-scale water level fluctuations of this kind can occur in the North Sea and show large amplitudes. In German they are called "Seebären," wherein "bär" is a mispronounciation of the word "boeren" in low German, meaning "to lift." But these are comparatively rare phenomena. It was only after a detailed investigation of the wave processes in the German Bight on August 19, 1932 that Gaye and Walther (1934) could find a definite explanation. In this special case, it was a heavy squall during a thunderstorm northwest of Heligoland that

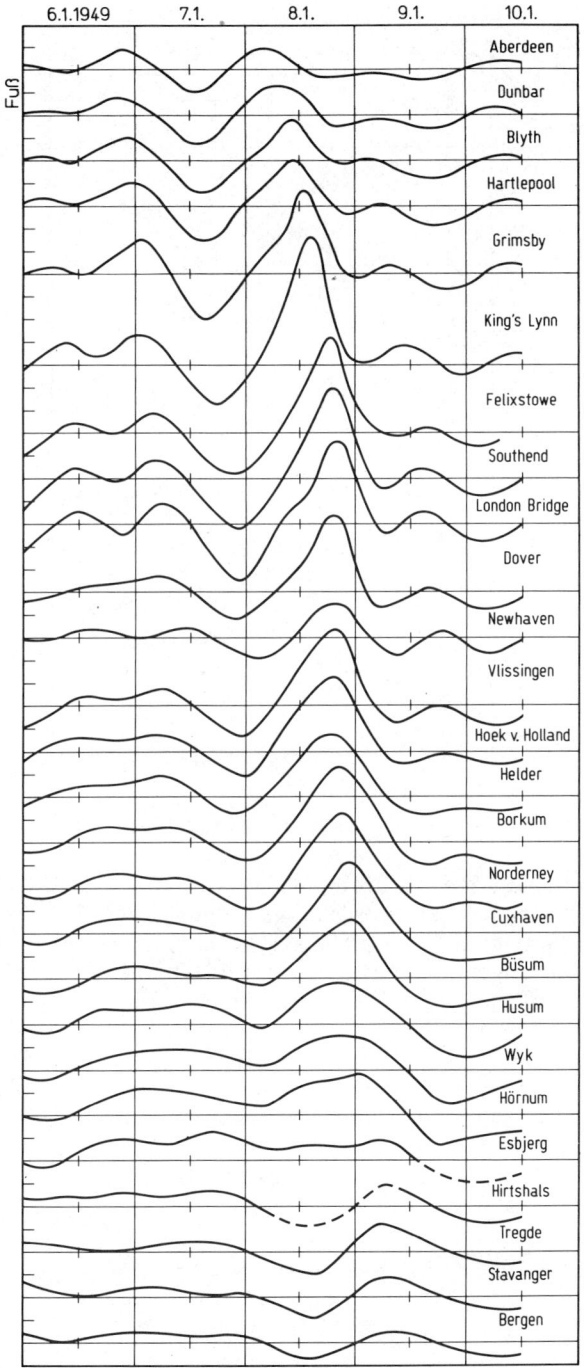

Fig. 8.35. Water level variations at a number of stations bordering the North Sea for the period from January 6–10, 1949. (After Corkan, 1950.)

384 Surface Waves and Internal Waves

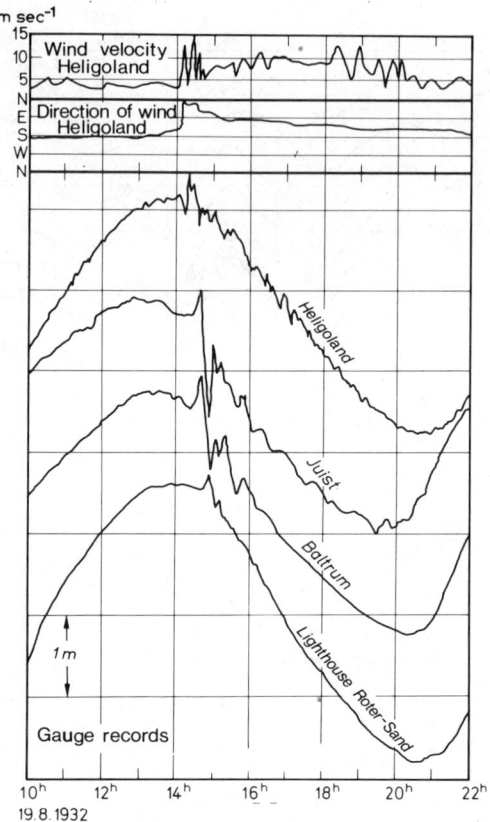

Fig. 8.36. So-called "Seebär" in the German Bight on August 19, 1932. (After Gaye and Walther, 1934.)

caused the progressive wave. Figure 8.36 shows the wind conditions on the island of Heligoland and the water level records taken at four gauges in the German Bight. The wave period was approximately 15 min, and a sea level rise of almost 1.5 m was recorded at Juist. The 31 min travel time of the wave from Heligoland to the lighthouse of *Roter Sand* at the mouth of the river Weser corresponds to the propagation velocity of long waves for the given depth distribution.

Numerous methods have been developed for the computation of such fluctuations of the water level. In recent times, numerical integration is usually applied. By integration over the entire water column, the system of hydrodynamic equations can be arranged for long waves in the following form:

$$\frac{\partial U}{\partial t} + fV = -g\frac{\partial \zeta}{\partial x} + \frac{T_0^{(x)}}{\rho H} - \frac{T_H^{(x)}}{\rho H}$$

$$\frac{\partial V}{\partial t} - fU = -g\frac{\partial \zeta}{\partial y} + \frac{T_0^{(y)}}{\rho H} - \frac{T_H^{(y)}}{\rho H} \qquad (8.43)$$

$$\frac{\partial \zeta}{\partial t} + \frac{\partial (UH)}{\partial x} + \frac{\partial (VH)}{\partial y} = 0$$

Here, U and V are the velocities averaged over the entire water column, ζ denotes the

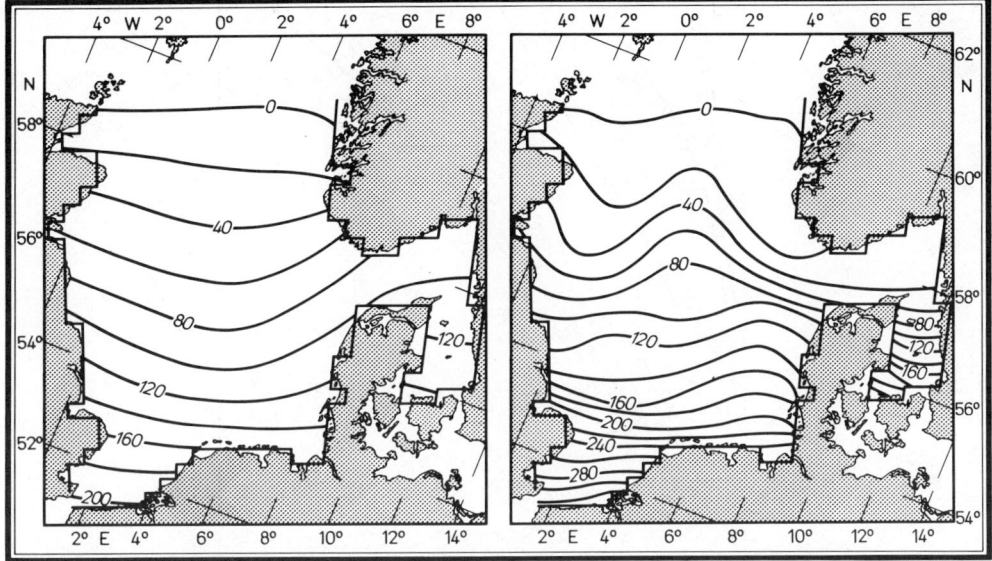

Fig. 8.37. Water level of the North Sea with northerly winds, under the assumption of constant depth (left-hand side), and with actual depth distribution. (According to Sündermann, 1966.)

fluctuation of the water level, H the water depth, f the Coriolis parameter, $T_0^{(x)}$ and $T_0^{(y)}$ stand for the tangential shear stress at the sea surface, and $T_H^{(x)}$ and $T_H^{(y)}$ for the shear stress at the sea floor. The numerical treatment of this system of equations has been particularly advanced by Hansen (1966). Among other results it has been shown by these computations that the natural bottom topography is of great influence on the water level. For comparison, in Fig. 8.37 are given the water levels caused by a northerly wind of 23.3 m sec^{-1} in a hypothetical North Sea with a constant water depth of 80 m, and for a North Sea with a depth distribution corresponding to nature (after Sündermann, 1966). As can be expected, large deviations of the water level occur in the shallow regions. Therefore, for the prediction of the water level, the exact depth distribution and the configuration of the coast in rather a large surrounding area must be taken into account. Water level prediction must aim—together with numerical weather forecasting—at directly computing in advance water level charts for those sea areas endangered by storm surges.

8.4.3. Edge waves, Kelvin waves, and double Kelvin waves

In Section 8.1.2, where the classification of waves is described, it was already mentioned that waves can also be generated by a change of water depth. The most illustrative example is given by the edge waves and shelf waves which have been investigated in great detail by Munk et al. (1959).

Fluctuations of the water level caused by storms on the open ocean can travel over long distances as free waves (similar to the example in Fig. 8.35) and eventually reach the shelf and the coast. Here, they either are reflected and travel back into the open ocean or they are "trapped" by the shelf and travel along the shelf as shelf waves or edge waves. They have characteristic wavelengths from 5 to 100 km, which corresponds in many areas to the width of the shelf. The shelf acts as a wave guide. With a view to their propagation

386 Surface Waves and Internal Waves

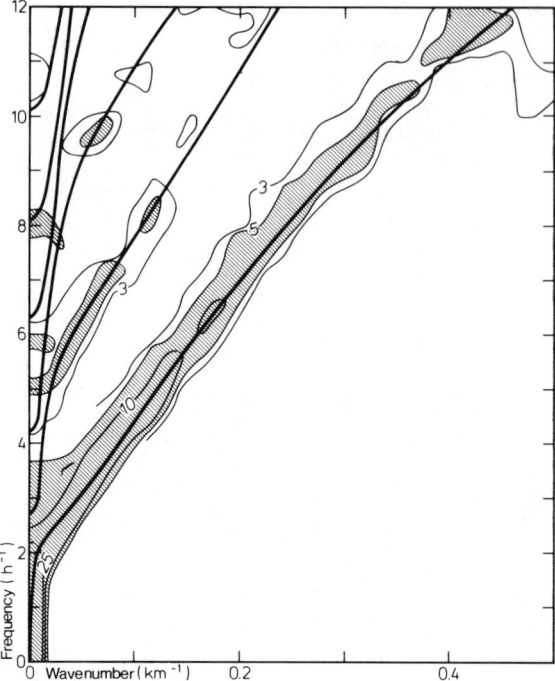

Fig. 8.38. Dispersion relations for edge waves. Theory: thick lines. Observations on the shelf off California: hatched areas. (After Munk et al., 1964.)

direction (parallel to the coast), edge waves are harmonic waves. Their crest heights, however, usually decrease with growing distance from the coast. As compared with tides, their amplitudes are small and often amount to no more than a few centimeters, but they may also be a great deal larger. These waves have been described theoretically first by Stokes (1847) and later, in more detail, by Reid (1958). At the coast of California, they have been intensively investigated by Munk et al. (1964). Figure 8.38 shows the theoretical dispersion relation of these waves for the shelf off California (thick lines) as well as the frequency–wavenumber ranges within which they were observed at the Californian coast (hatched areas). Observation and theory agree rather well.

The shape of the Kelvin waves resembles that of the edge waves. However, their form is not a result of the bottom topography but rather of the Coriolis force.

Let us consider the effect of Coriolis force by studying the simplest case possible, namely a channel of infinite length and constant depth. If friction is neglected, the equations for long waves (8.43) are simplified to

$$\frac{\partial U}{\partial t} = -g\frac{\partial \zeta}{\partial x}, \quad -fU = -g\frac{\partial \zeta}{\partial y}, \quad \frac{\partial \zeta}{\partial t} + H\frac{\partial U}{\partial x} = 0 \quad (8.44)$$

If marginal conditions are taken into account, the following solution is obtained:

$$\zeta = \zeta_0 e^{(f/c)y} \cos(\kappa x - \omega t), \quad U = \sqrt{\frac{g}{H}} \zeta, \quad c = \sqrt{gH} \quad (8.45)$$

This solution, describing a progressive long wave under the influence of Coriolis force,

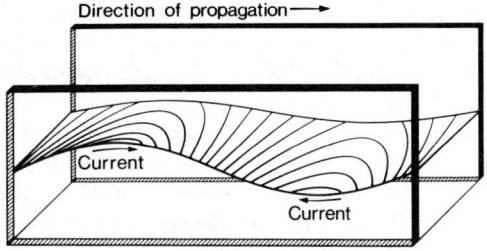

Fig. 8.39. Topography of the sea surface for a Kelvin wave in a wide channel in the Northern Hemisphere.

has been derived by Thomson (Lord Kelvin) (1879). Therefore, such a wave is also called Kelvin wave.

From the solution we recognize that the rotation of the earth does not influence the propagation velocity c but the amplitude and the current velocity U. The crest of such a wave changes like an exponential function. Looking in the direction of propagation we notice that in the Northern Hemisphere the water in the wave crest is forced to the right-hand side and that in the wave trough to the left-hand side of the channel, as indicated perspectively in Fig. 8.39. Therefore, the amplitude on the right-hand side is larger than that on the left-hand side. The currents that occur are only alternating ones, reaching their maximum at high tide and low tide. On the right-hand side they are larger than on the left-hand side. Like the amplitudes, they depend on the exponent f/c. The changes of amplitude remain small as long as y is small (i.e., if the channel is narrow) and f/c is small (i.e., if the wave period is small compared to half a pendulum day). That is why the Coriolis force does not exert any significant influence on tsunamis, which, in general, have periods between 10 and 60 min.

Longuet-Higgins (1968) has shown that similar waves may also travel along the shelf edge or in corresponding regions where the water depth varies considerably. There, they are like double Kelvin waves, one developing over deep water and another one over the shelf. The largest amplitudes will occur over the shelf edge from which they decrease toward the deep sea as well as toward the coast. Such waves can also be trapped by the shelf base of an island, in which case they will move around the island. Their period is always greater than the inertial period.

8.4.4. Seiches

A progressive long wave hitting a perpendicular wall at $x = 0$ will be totally reflected. This reflected wave, now advancing in the opposite direction, superimposes the incoming wave, and both form a standing wave. Let

$$\varphi_1 = A_0 \cos(\kappa x - \omega t)$$

be the incoming wave, then

$$\varphi_2 = A_0 \cos(\kappa x + \omega t)$$

describes the reflected wave and

$$\varphi = \varphi_1 + \varphi_2 = 2A_0 \cos(\kappa x) \cos(\omega t)$$

is the wave resulting from them. If n is an integer and λ is the wavelength, the resulting wave has antinodes at $x = 2n\lambda/4$ and nodes at $x = (2n + 1)\lambda/4$. In contrast to progressive waves, the horizontal motions of the water particles in standing waves are largest at the nodes; at the wave crests they are zero, as clearly demonstrated by the photo of a standing wave in a laboratory tank shown in Fig. 8.06. This picture gives only part of the tank, at the ends of which (not shown) there must be antinodes.

The period of a standing wave in a rectangular tank of the length L and the depth H can be easily calculated. Let us consider a long progressive wave generated at one end of the tank; with its velocity $c = \sqrt{gH}$ the wave will reach the other end after the time $L/\sqrt{gH}$ and, after reflection, it will return to the starting point after an equal period of time. Hence, the total period of the oscillation is

$$\tau_0 = \frac{2L}{\sqrt{gH}} \qquad (8.46)$$

as already given by Merian (1828). This oscillation, the period of which is determined by the dimensions of the oscillating systems, is also called *eigen oscillation* or *free oscillation* of the system. It has one node and is the longest of all possible free oscillations. For n nodes, $\tau = \tau_0/n$ holds. If the vertical displacement ζ is no longer very small compared with the water depth, the propagation velocity will also depend on ζ. The period of such free oscillations is reduced:

$$\tau_S = \tau_0 \left(1 - \frac{3}{2} \frac{\zeta}{H}\right) \qquad (8.47)$$

These considerations hold true for enclosed water basins. For bays open at one end, it should be noted that there is always a node at the open end to the ocean, because there is always sufficient water available to be included in the horizontal oscillation. The period of the fundamental mode of a bay is therefore twice the period of the fundamental mode of an enclosed basin of the same length and depth, namely

$$\tau_0 = \frac{4L}{\sqrt{gH}} \qquad (8.48)$$

If the opening of the bay is large in comparison to the length of the basin, a mouth correction must be applied to the period, by which the period is prolonged. In analogy to acoustic processes in pipes open at one end, the prolongation will reach 32% of τ_0, according to Lord Rayleigh, if the width of the opening equals the length of the bay.

Standing waves as free oscillations occur in all natural waters. They were first recognized as such in the Lake of Geneva by the Swiss physician Forel. The name of "seiches" which, for generations, has been in use there for the fluctuations of the water level of the lake, has generally been adopted as a term for free oscillations of more or less enclosed water basins. Ever since the classical monograph on the seiches in the Lake of Geneva by Forel (1895), numerous lakes and ocean bays have been investigated and uninodal and multinodal oscillations have been revealed.

Forel already recognized that the period observed in the Lake of Geneva, as amounting to 73 min for the fundamental mode, is only in poor agreement with the Merian formula [Eq. (8.46)]. According to this equation, a period of $\tau_0 = 59$ min has been calculated for the length of the lake of $L = 70$ km and a mean depth of $H = 160$ m. This is where the theory of seiches, dealing with oscillation problems for narrow channels with arbitrary bottom profile, becomes effective. The basic idea was suggested by Chrystal (1905). The considerations start from the hydrodynamic equations of motion which are transformed by integration over the depth and the width of the respective sea area. Concerning the derivations, the reader is referred to Krauss (1973). This method has been used most

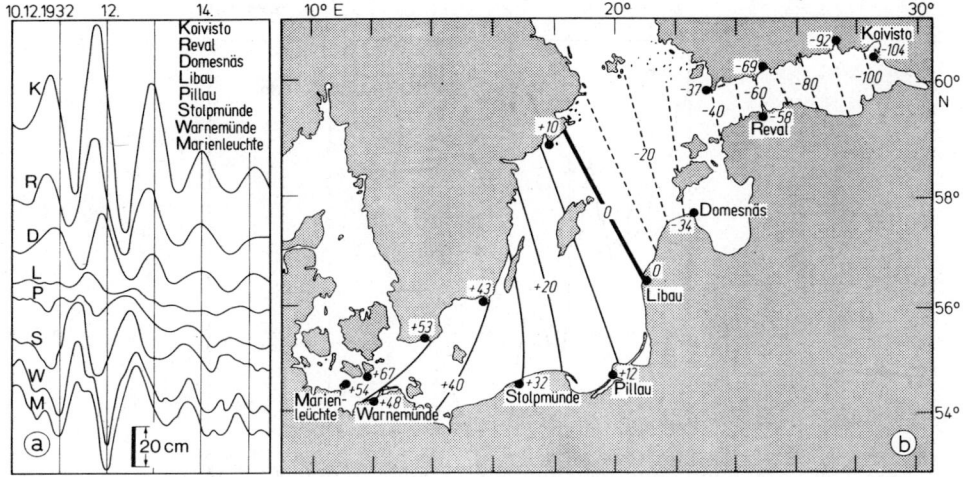

Fig. 8.40. Seiches in the Baltic Sea. (According to Neumann, 1941.) (a) Water level fluctuations on eight Baltic Sea gauges between December 10, 1932, 0^h GMT and December 15, 1932, 10^h GMT. (Locations of the gauges are given in (b).) (b) Ranges (in cm) between consecutive extremes of the water level on December 11 and 12, 1932. Broken lines and negative values indicate a lowering of the sea level; solid lines and positive values show a rise of the sea level.

frequently by Defant. It is based on numerical integration and permits the exact computation not only of the periods but also of the distribution of the water surface deviations from the equilibrium level, and of the horizontal displacement of the water particles even for basins with complicated bottom topography.

The theory of seiches has helped to interpret short-period oscillations, appearing as vibrations superimposed on normal tide records, as free oscillations of small basins, as well as to recognize periods of many hours as free oscillations of large oceanic basins. It is difficult, however, to apply this theory to areas without definite limits, which, to a certain degree, also holds true for the Baltic Sea.

Sometimes high uninodal seiches are observed in the Baltic Sea, which cause heavy inundations at the antinodes at the ends, especially in Leningrad. The example of Fig. 8.40, illustrating the case of oscillation from December 10 to 15, 1932, represents a uninodal free oscillation of the Baltic Sea–Gulf of Finland system. When the water level rises in one half of the system, it drops in the other one; only in the central part, near Libau, does the water level remain nearly unchanged. Figure 8.40 gives the water level difference between the consecutive extremes of the water level on December 11 and 12, 1932, as recorded by numerous gauges. The lowering of the sea surface by more than 100 cm in the Bay of Kronstadt is opposed by a rise of more than 50 cm in the western Baltic Sea. The nodal line extends between Libau and Stockholm. Any possible deviations from a simple longitudinal oscillation will be discussed later on.

Theoretical methods of computation yield a period of 27.5 hr when the oscillating system extends from the Fehmarn Belt to the Bay of Kronstadt. For the average depth of the Baltic Sea amounting to 55 m and the length of the whole system of 1450 km, the Merian formula [Eq. (8.46)] would give $\tau_0 = 34.8$ hr. This is in contrast to the particular case in Fig. 8.40 with an observed period of 27.3 hr. These values show that the complicated natural bottom topography is of considerable influence on the period, and that the theoretical methods provide satisfactory results. If, after the analysis of numerous seiches

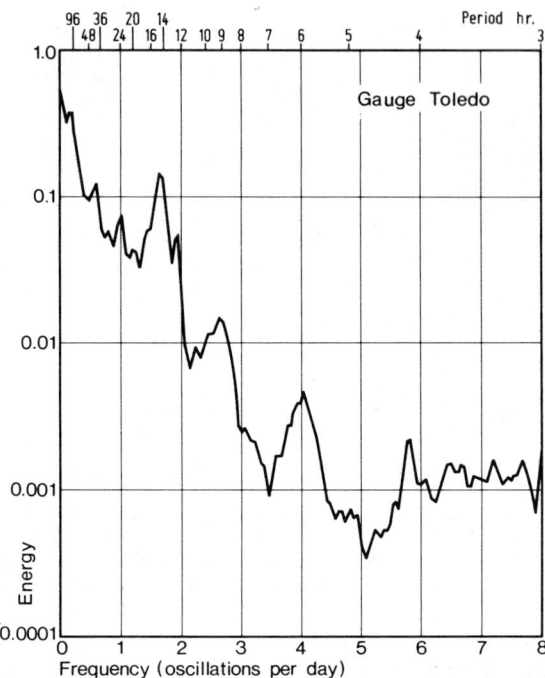

Fig. 8.41. Spectrum of the water level at the gauge of Toledo on Lake Erie. (After Platzman and Rao, 1964.)

of the Baltic Sea by Neumann (1941), the observed periods of the fundamental mode have proven to be not completely constant but to vary between 26 and 28 hr, it is conceivable that the oscillating system has not always been the same. Since the Baltic Sea has passages in the north and in the west, the extension of the system may depend on the particular excitation.

Oscillations of the Baltic Sea–Gulf of Bothnia system, with a calculated period of $\tau_0 = 39.1$ hr, have been observed only rarely.

Krauss and Magaard (1962) have also calculated the periods of multinodal oscillations. Analogous studies are available for the Great Lakes. Figure 8.41 shows the spectrum of the water level at the gauge of Toledo on Lake Erie. The energy maxima at 14.4, 9.1, 5.9, and 4.2 hr are in good agreement with the theoretical seiches periods of 14.1, 8.9, 5.7, and 4.1 hr. Thus it is shown what a great contribution seiches provide for the water level spectrum.

The causes of seiches have often been discussed. In particular, the wind-induced piling-up of water at the end of a bay or basin leads to swinging motions of the water masses when the wind decreases. Seiches can also be generated by air pressure fluctuations, especially if the period of the pressure fluctuation agrees with the free period of the oscillating system, because, in this case, the seiches are reinforced by resonance. In our example regarding the Baltic Sea in December 1932 (cf. Fig. 8.40), such coincidence has contributed to the high seiches.

The effect of the Coriolis force on seiches can be estimated if we dispense with an exact derivation. If u denotes the horizontal displacement of water in the fundamental mode

$$\zeta = \zeta_0 \cos \frac{2\pi}{\tau} t$$

the following equation is obtained:

$$u = \frac{d\zeta}{dt} = -\frac{2\pi}{\tau} \zeta_0 \sin \frac{2\pi}{\tau} t = \frac{2\pi}{\tau} \zeta_0 \cos \frac{2\pi}{\tau} \left(t + \frac{\tau}{4}\right) \qquad (8.49)$$

The Coriolis force with the absolute value of uf acts transverse to the fundamental oscillation. A water particle is subjected not only to the gradient force in the direction of the fundamental oscillation but also to the Coriolis force, under the influence of which the sea surface assumes the inclination β transverse to the fundamental oscillation: $\tan \beta = -uf/g$ [cf. Eq. (7.46)]. Since u changes periodically, this inclination will also oscillate periodically. Thus, due to the Coriolis force, a transverse oscillation is generated, the period of which is equal to the period τ of the longitudinal oscillation, but with a phase difference of $\tau/4$ with respect to the phase of the longitudinal oscillation.

We find again that a noticeable transverse oscillation occurs only for longer periods of the longitudinal oscillation. In the case of the example from the Baltic Sea (cf. Fig. 8.40), such a transverse oscillation is evident. Its amplitude, reaching its maximum where the maximum of the horizontal displacement of water occurs, that is, in the nodal line of the longitudinal oscillation, amounts to about 20% of the amplitude of the seiches at the ends of the Baltic Sea. Hence, the transverse oscillation is of minor significance in the elongated Baltic Sea. The resultant of transverse and longitudinal oscillations is a rotating or amphidromic wave (see Section 8.4.5), although a very asymmetrical one where the cotidal lines are crowded together in the area of the nodal line of the fundamental oscillation between Libau and Stockholm. In Fig. 8.40 this can be recognized by the fact that high tide and low tide do not occur simultaneously at all points in one half of the Baltic Sea. Due to the turning of the amphidromic wave, which, like all normal amphidromic systems in the Northern Hemisphere, occurs counterclockwise, high tide and low tide are observed at Reval 2 hr earlier than at Koivisto.

8.4.5. Amphidromic systems

In wide basins standing waves may occur along the longitudinal axis as well as along the transverse axis. The interference of the two oscillations no longer represents a simple oscillating motion but results in a rotating wave which, according to Harris (1904), who was the first to study such waves in detail, is also called amphidromic wave or amphidromic system. Conditions are particularly simple if transverse and longitudinal oscillations have the same periods and if the phases of the two waves differ by one quarter of the period, that is, if during one oscillation the sea surface reaches its maximum deviation from the mean water level when in the other one it coincides with the mean water level.

The course of the resulting oscillation is illustrated in Fig. 8.42 by showing the configuration of the sea surface and the corresponding currents. A square basin of constant depth is assumed, the sides of which are oriented north–south and east–west. Each partial oscillation has the same period of 12 hr and the same amplitude of 50 cm. The meridional oscillation has a phase difference of 3 hr as compared with the zonal oscillation. When, for example, high water occurs on the eastern side at 3^h [cf. Fig. 8.42(a)], it will take place on the northern side at 6^h (Fig. 8.42(d)). In the meantime, the high water travels coun-

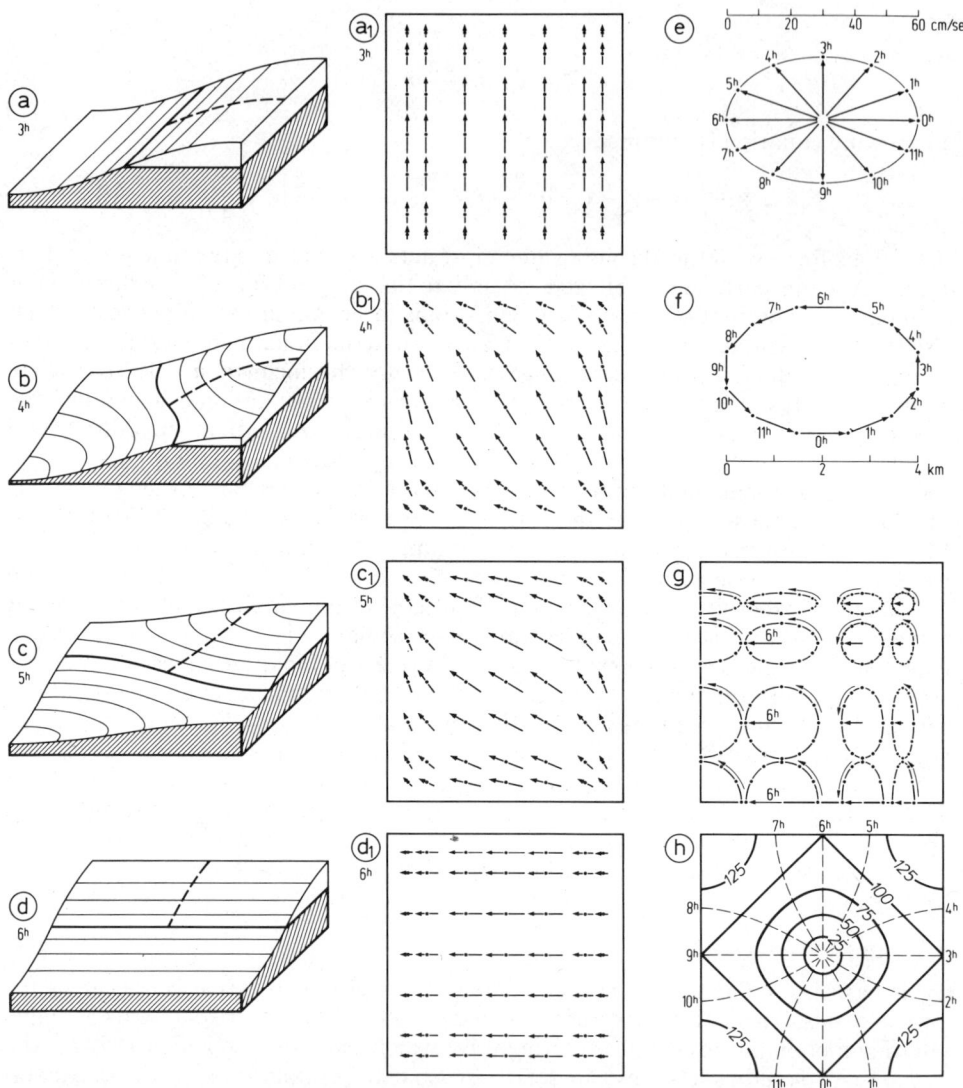

Fig. 8.42. Rotating wave (amphidromic wave) in a square basin at the interference of two standing waves with a period of 12 hr, crossing each other perpendicularly. (Following Thorade, 1931.) (a)–(d): Topography of the sea surface from hour to hour over $1/4$ of a period, presented in the form of block diagrams. (a_1)–(d_1): Currents corresponding to (a)–(d). (e): Current ellipse at one particular point. (f): Trajectory of a water particle for case (e). (g): Distribution of current ellipses in the northeast quadrant of the basin. (h): Cotidal lines and tidal ranges (in cm) for one period.

terclockwise, as indicated in Figs. 8.42(b) and (c) for 4^h and 5^h. Within 12 hr it has completed the circle around the basin. A nodal line no longer exists, but there is just one single point without any vertical displacement, the *nodal point* or *amphidromic point*, which, in our simple case, is located in the center of the basin. High water no longer occurs simultaneously in one half of the basin, but only along a line starting from the nodal point as represented in Figs. 8.42(a)–(d) by a dashed line. This is the so-called *cotidal line*. The distribution of the cotidal lines for the entire period and the distribution of the resulting tidal ranges reached within this time are contained in Fig. 8.42(h). This figure shows the typical picture of an amphidromic wave rotating towards the left: with the tidal range vanishing at the nodal point, with the largest displacement at the corners, and with cotidal lines arranged radially.

A current field [Figs. 8.42(a_1)–(d_1)] can be related to each representation of the configuration of the sea surface in Figs. 8.42(a)–(d). But it must be taken into account that, in a single standing wave, the maximum current is found when the water level passes the equilibrium level and that there is no current at the time of the largest deviation from the equilibrium level. At 3^h the current is dominated by the meridional oscillation, at 6^h by the zonal oscillation. In between, the transition takes place: the current turns from the northerly into the westerly direction, that is, it turns continuously. Thus, the rotating wave is associated with a rotating current. If the current at one point of the basin is considered, the tip of the current vector generally describes an ellipse within one entire period of oscillation. This is the so-called current ellipse. In Fig. 8.42(e) it is illustrated for a selected point. The vector in this ellipse turns counterclockwise (*contra solem*). If, for an entire period, a water particle is followed which, at a certain moment, was located at the same selected point, an elliptic trajectory results that has the same direction of rotation. This is represented in Fig. 8.42(f). When the current approaches the boundaries of the basin, the current ellipses become narrower and narrower and, finally, they pass into a straight line. Thus, the current becomes alternating. Conversely, the current ellipses take a circular form along the diagonals of the basin. In Fig. 8.42(g), covering only the northeast quarter of the basin, such locally different current forms can be recognized.

At first sight, our example in Fig. 8.42 may seem to be a very special one, particularly due to the assumption that the two interfering partial oscillations are supposed to differ in phase by a quarter of the period in such a way that the oscillation, which is delayed by one-quarter period, is oriented *contra solem* to the other if one goes around the basin. But it is exactly these particular assumptions that are satisfied to a large degree under natural conditions on earth: owing to the Coriolis force, a certain transverse oscillation is induced in addition to the longitudinal oscillation; both oscillations have the same period but differ in phase by a quarter of a period. Hence, in nature, eventually the Coriolis force is mainly responsible for the transformation of long waves into amphidromic waves. On the earth, progressive and standing waves are subject to the deflecting force of the earth's rotation. This influence, however, is significant only for waves, whose periods range in the order of magnitude of the period τ of the inertial oscillation, that is, of half a pendulum day. This is true for seiches of larger oceanic areas as well as for tides, but not for tsunamis and seiches in smaller bays.

The exact computation of long waves in enclosed oceanic areas on the rotating earth involves great mathematical difficulties. The general course of the water level distribution in an amphidromic wave indicates that Kelvin waves are largely responsible for the generation of such a system. However Kelvin waves cannot be fully reflected at the end of a basin. A solution presented by Taylor (1922) seems to be particularly informative. It concerns a channel open at one end, with a rectangular cross section. Taylor discussed the complicated results numerically by giving this example: he assumes a bay at 53°N, the length of which is twice its width, which is supposed to be 465 km. The depth is taken as 74 m everywhere. The period of the wave reflected at the end of the channel is assumed to equal the period of the semidiurnal tide. Thus, the numerical example approximates the North Sea, with a tidal wave entering from the north. The cotidal lines and the tidal

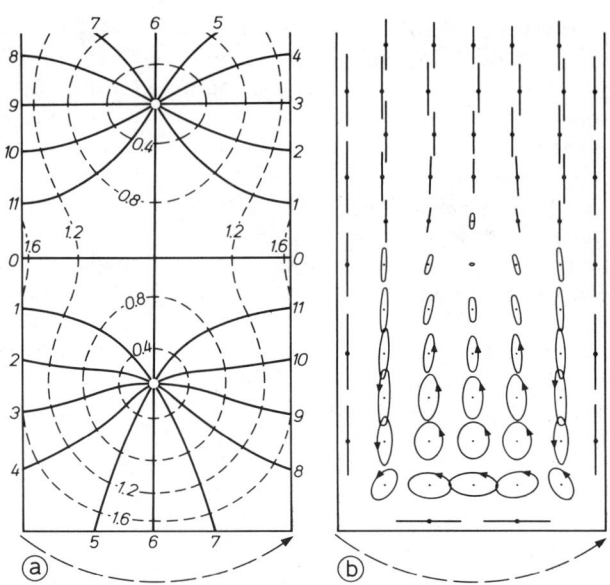

Fig. 8.43. Reflection of Kelvin waves (period of 12 hours) in a channel open at one end. (After Taylor, 1922.) Arrow at the lower edge indicates the direction of the earth's rotation. Left: cotidal lines and lines of equal tidal range. Right: trajectories of water particles.

ranges are given in Fig. 8.43(a); Fig. 8.43(b) shows the trajectories of the individual water particles. Two rotating waves develop, which turn *contra solem;* the outer (upper) one corresponds to the superposition of two Kelvin waves advancing in opposite directions. Transverse currents do not occur in this case. The inner (lower) amphidromic wave, however, shows transverse currents becoming the larger the closer they approach the end of the channel. This causes the trajectories of the water particles in the inner part of the bay to take an elliptical shape. At the margins of the bay, where the strongest currents occur, alternating currents remain in existence. This example demonstrates that, under the influence of the Coriolis force, amphidromic waves can develop, the course of which superficially resembles that of amphidromic systems resulting from interference of standing waves (cf. Fig. 8.42). The currents, however, are distinctly different. Therefore, a conclusion as to whether we deal with an interference of standing waves or of Kelvin waves cannot be based only on the existence of an amphidromic wave in the ocean; the current distribution must also be taken into consideration. In enclosed basins numerous amphidromic systems may occur. Their number essentially depends on the relation of the oscillation period to the inertial period. Rao (1966) has carried out extensive calculations in this regard, to which attention may be drawn here.

8.4.6. The influence of friction on long waves

All long waves are associated with horizontal displacement of water particles reaching down to the bottom. In addition to the Coriolis force, such motion is subject to friction, not so much to internal friction but to the friction resulting from the roughness of the ocean floor. The dissipation of kinetic energy manifests itself in the modification of undisturbed long waves and, eventually, leads to their extinction, unless, as with the tides, a periodically acting force continues to generate waves over and over again.

With standing waves, the damping of the oscillations recognizable in the tide gauge records of seiches in the Baltic Sea (cf. Fig. 8.40), is clearly an effect of friction. Such a damped oscillation is represented by Eq. (8.11).

Table 8.03 contains some data regarding damping and friction, which have been derived from observations of seiches. Here H denotes the mean depth and L the length of the basin. τ_{obs} is the observed oscillation period and τ_{comp} the oscillation period computed by different scientists. d stands for the logarithmic decrement and β for the damping factor. Although the basins differ fundamentally with respect to their position, location, and dimension, the orders of magnitude of the friction coefficients are in remarkable agreement. This may be an indication that it is the bottom friction and not internal friction which causes the damping of seiches. By observation of d, the mean value of the coefficient of turbulent friction A/ρ (cm^2 sec^{-1}) is also determined, as shown by Lettau (1934). These data for the various basins have also been listed in Table 8.03. Their dependence on the mean depth is evident. In addition to the damping of standing waves, turbulent friction also causes (1) the amplitude in the nodal lines or at nodal points—in case amphidromic waves are concerned—not to decrease to zero but a zone of minimum amplitude to be established; (2) the phase in this zone to change gradually rather than suddenly; (3) the period of the fundamental mode to be somewhat increased.

The modification of progressive long waves under the influence of friction can be studied by following tidal waves in the lower courses of rivers. It is illustrated in Fig. 8.44 by the example of the tidal wave in the River Elbe. On its way up the river, the progressive wave that has entered the River Elbe at Cuxhaven becomes more and more unsymmetrical. On the average, the high water needs 4 hr 14 min for the distance from Cuxhaven to Hamburg, whereas the low water needs 4 hr 58 min. Thereby the duration of rise, amounting to 5 hr 39 min at Cuxhaven, is shortened to 4 hr 55 min at Hamburg; the duration of fall, on the other hand, is increased, namely from 6 hr 46 min to 7 hr 30 min. At the same time, the tidal range decreases upstream and finally vanishes at Geesthacht, 141 km upstream of the mouth of the River Elbe. There, the so-called tide limit is found. Another fact, confirmed by observation, is indicative for the modification of progressive waves under the influence of friction: at Cuxhaven, the current turns from incoming to outgoing tide 1 hr 10 min after high tide on the average; further upstream this time difference decreases more and more until at Hamburg it amounts to only 15 min. The wave, which even at Cuxhaven is no longer an undisturbed progressive wave (if it were undisturbed, the turning would occur 3 hr 6 min after high tide), approaches the characteristics of a standing wave, as far as the current is concerned; but with respect to the shape of its surface it behaves like a progressive wave. Furthermore, Fig. 8.44 demonstrates that only the rise of the water level still has wave characteristics, though already somewhat disturbed. The fall of the water level, however, shortly after high tide, shows an almost linear decrease at all gauges, very much like a water-filled basin that is being emptied.

The explanation of the modification of such a progressive wave in shallow water is mainly found in the effect of bottom friction F_b assumed to be proportional to the square of the current velocity:

$$F_b = \kappa \rho u_b^2 \qquad (8.50)$$

Here, κ depends on the roughness of the bottom; according to empirical determinations

Table 8.03. Uninodal Free Oscillations and their Damping in Various Basins[a]

Water Mass	H (m)	L (km)	τ_{obs}	τ_{comp}	d obs.	A/ρ (cm² sec⁻¹)	β (10⁻⁵ sec⁻¹)	Source
Lake of Garda	136.1	52.2	42.92 min	42.83 min	0.023	41.6	0.89	1
Baltic Sea (Fehmarn Belt to Kronstadt Bay)	55	1450	27.3 hr	27.5 hr	0.50	(17.6)	0.51	2
Lake of Starnberg	54	19.5	25.0 min	—	0.03	21.7	2.0	3
Kurisches Haff (Lagoon of Courland)	4.0	85	4.12 hr	4 hr	0.33	1.48	2.22	4
Lake of Balaton	3	77.2	9 hr	8 hr	(0.9)	1.9	2.78	5
Frisches Haff (Fresh Lagoon)	2.6	89	8 hr	—	0.46	0.52	1.60	4

[a] Sources: 1, Defant (1916); 2, Neumann (1941); 3, Emden (1905); 4, Lettau (1934); 5, Endros (1930).

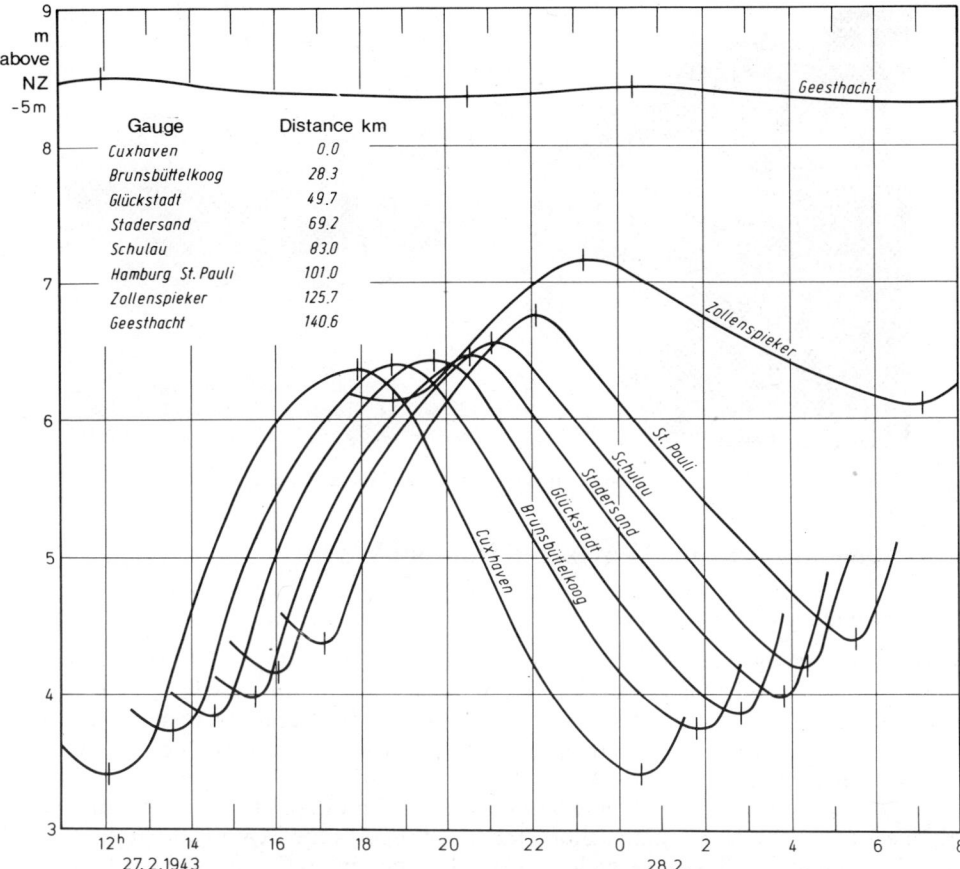

Fig. 8.44. Course of the tidal wave entering the River Elbe at Cuxhaven up to the tide limit at Geesthacht. (From Mügge, 1952.) The tidal curves of February 27–28, 1943 almost correspond to the average tidal curves for the period 1941–1945. Curves are based on normal zero (NZ) − 5.00 m. Vertical marks on the curves indicate times of arrival of high or low water, respectively.

with the aid of models, κ ranges between $\kappa = 0.0025$ for smooth bottoms and $\kappa = 0.3$ for very rough bottoms with coarse debris.

In some rivers, resembling funnels, that is, becoming narrower upstream, and being shallow at low tide, the duration of rise of the tidal wave decreases so much that the rise occurs suddenly in the form of a breaking wave, which travels upstream, covering the total width of the river with its steep, wall-like front. The English name of "bore" (in French: Mascaret) has been adopted for this type of wave. Smaller bores are known to occur in the English rivers Severn and Trent and in the French rivers Seine and Gironde. Krümmel (1911) has described in detail a bore, 5 m high, in a tributary of the Amazon, called Pororoca there, and another bore, occasionally 8 m high, in the Tsien-tang kiang, south of Shanghai, as particularly impressive phenomena of nature. Figure 8.45 gives an impression of the bore in the Tsien-tang kiang. Such a bore is a degenerated solitary wave rather resembling an eddy with horizontal axis than a wave and behaving like a breaker in the surf zone of a beach.

Fig. 8.45. Tidal bore in the Tsien-tang kiang at Hangtschou Bay. (Photo: from Thorade, 1931.)

8.5. Boundary Surface Waves and Internal Waves

8.5.1. Boundary surface waves

At the boundary of two water layers of different densities, short and long waves may appear as they do at the interface between water and air. They are called boundary surface waves. In general, they cannot be perceived directly in the ocean. Systematic measurements at short time intervals are required to detect those waves in the fluctuations of the vertical structure of water masses.

When "dead water" was observed, this was a first hint that internal waves might exist in the ocean. In the Norwegian sailors' language there is the expression "to lie in dead water" which means "to make no headway." In Norwegian fjords and in other coastal waters, where a thin layer of light water (melt water, river water) lies on top of heavier seawater, boats and small craft sometimes are kept there as if by magic. After Nansen, with the *Fram,* had observed the same phenomenon north of the Taimyr Peninsula during his North Pole expedition, he encouraged Ekman to investigate the problem experimentally and theoretically. In 1904, Ekman was able to explain the "dead water" by processes associated with short boundary surface waves. Since that time, numerous observations have proven that waves do not only exist at sharply defined discontinuity surfaces but also in continuously stratified water; they are then called internal waves. Besides short progressive waves, short standing internal waves, associated with stability oscillations, have also been discovered, as described by Neumann (1946). Helland-Hansen and Nansen (1909) were the first to recognize long progressive internal waves during extensive serial observations on board the Norwegian research vessel *Michael Sars* in the North Atlantic Ocean. In several papers, Defant has contributed to the explanation of such long internal waves, especially of those with tidal character which were observed on the research vessels *Meteor* and *Altair*. Standing long internal waves were also observed in the ocean, for the first time by Pettersson (1909) in the Gullmar Fjord at the Swedish Skagerrak coast, after they had been found in lakes.

If, according to observations, internal waves and boundary surface waves of various lengths, sometimes even with very large amplitudes, are among the common phenomena in the ocean, they must be generated easily. Indeed, even small forces are sufficient to generate them, owing to the relatively small density difference between two water layers. Compared with the energy of surface waves, their total energy, at the same wave height, usually is much smaller.

From the theory of progressive boundary surface waves in a two-layer ocean, as advanced by Stokes (1847), the important general relation for the propagation velocity c may be cited here:

$$c^4 \left(\coth \kappa h + \coth \kappa h' + \frac{\rho'}{\rho} \right) - c^2 (\coth \kappa h' + \coth \kappa h) \frac{g}{\kappa} + \frac{\rho - \rho'}{\rho} \frac{g^2}{\kappa^2} = 0 \quad (8.51)$$

The ratio of the amplitude at the sea surface to that at the interface is

$$\frac{A'}{A} = \frac{\kappa c^2}{\kappa c^2 \cosh \kappa h' - g \sinh \kappa h'} \quad (8.52)$$

where h' and h are the thicknesses of the upper and the lower water layers, respectively, and ρ' and ρ the corresponding density values. Since Eq. (8.51) is quadratic in c^2, two wave systems result for each wave period. Their significance is most simply grasped by examination of two special cases, which will be mentioned here because they approximate conditions existing in the ocean: the short and the long boundary surface waves.

The first case is concerned with the propagation velocity of short internal waves, which occurs when h is large and h' is small as compared to the wavelength, that is, when a thin, light top layer lies on deep, heavy water. For this case, the solutions of Eq. (8.51) read as follows:

$$c_1^2 = \frac{g\lambda}{2\pi} \quad \text{and} \quad c_2^2 = gh\frac{\rho - \rho'}{\rho} \quad (8.53)$$

The propagation velocity c_1 is identical with that of surface waves; c_2 belongs to the boundary surface waves which have their maximum amplitude at the boundary surface. At the sea surface, their amplitudes are negligibly small. A numerical example adapted to conditions in Norwegian fjords yields $c_2 = 100$ cm sec^{-1}, if a thin fresh water layer ($h' = 4$ m) is assumed over deep saline water, with a density difference of $\rho - \rho' = 0.025$. A vessel proceeding with a speed of approximately c_2 generates a wave at the boundary surface, as indicated in Fig. 8.46. To generate such a wave, a certain amount of energy is required which must be provided by the vessel. From this context, the "dead water" can be explained. Only if the vessel succeeds in attaining a speed higher than the greatest

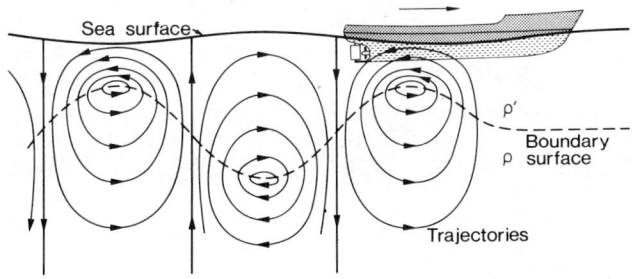

Fig. 8.46. Progressive internal wave and the phenomenon of "dead water."

possible propagation velocity of the boundary surface waves, will the generation of those waves be stopped and the abnormal resistance cease.

Solutions for the second special case of Eq. (8.51), which are particularly important for application to the ocean, are obtained if the total water depth $h + h'$ is assumed to be small compared to the wavelength. The relevant solutions are

$$c_1^2 = g(h + h') \quad \text{and} \quad c_2^2 = \frac{ghh'}{h + h'}\left(1 - \frac{\rho'}{\rho}\right) \tag{8.54}$$

The propagation velocity c_1 refers to long waves at the sea surface that have already been dealt with. c_2 is the propagation velocity of long boundary surface waves, where the ratio of the amplitude at the sea surface to that at the boundary surface is given by

$$\frac{A'}{A} = -\left(\frac{\rho - \rho'}{\rho'}\right)$$

This ratio is negative because the phase of the wave at the surface is opposite to the phase at the boundary surface. If, for instance, the density difference at the boundary surface amounts to $\rho - \rho' = 2 \times 10^{-3}$ (a value already very large for oceanic conditions) and if the height of the internal wave is $H = 10$ m, the corresponding wave height at the sea surface amounts to only 2 cm. In general, long internal waves as well as short ones remain hidden to an observer at the sea surface unless he is following them in deeper layers with a measuring instrument. Furthermore, if it is assumed in the numerical example that the density discontinuity is found at 100 m depth and the total depth is 200 m, we get $c_2 = 99$ cm sec^{-1} compared with $c_1 = 4430$ cm sec^{-1}. Provided the wave period is $\tau = 1$ hr, the corresponding wavelengths are $\lambda_2 = 3.6$ km and $\lambda_1 = 159.5$ km. Therefore, long internal waves advance much more slowly than long waves at the sea surface and, for the same period, the former are much shorter than the latter.

Standing long boundary surface waves have a very long period of oscillation and are, therefore, observed only rarely. The free period of a channel open at one end is, for instance, given by

$$\tau = \frac{4L}{n}\sqrt{\frac{\rho}{\rho - \rho'}\frac{h + h'}{ghh'}}, \tag{8.55}$$

where L is the length of the channel and n the number of nodes of the standing wave. Pettersson's observation of a two weeks' period in the Gullmar Fjord would correspond, as shown by Wedderburn (1909), to the free period of the uninodal oscillation $\tau = 13.9$ days, where, according to observation, $L = 200$ km, $\rho - \rho' = 4 \times 10^{-3}$, $h = 100$ m, and $h' = 20$ m.

8.5.2. Internal waves

In Section 7.2.5 the stability of a water particle in stratified water has been described. There, it is shown that a particle of the density ρ, when shifted from its position at rest, is subject to a restoring force of the magnitude of $g(\bar{\rho} - \rho)/\rho$, and that its motion can be described by Eq. (7.39)

$$\frac{dw}{dt} = -g\frac{\bar{\rho} - \rho}{\rho}$$

where $\bar{\rho}$ is the density of the surroundings. For small motions, it can be assumed that $\bar{\rho}$ changes linearly, $\bar{\rho} = \rho(z_0) - \zeta(d\bar{\rho}/dz)$. Since, in its resting position, the density of the particle must be equal to that of the surroundings, we have $\rho = \bar{\rho}(z_0)$ and $w = -d\zeta/dt$ and arrive at the following equation:

$$\frac{d^2\zeta}{dt^2} + g\frac{1}{\rho}\frac{d\bar{\rho}}{dz}\zeta = 0 \tag{8.56}$$

The solution of this equation is

$$\zeta = \zeta_0 \cos\left(\sqrt{\frac{g}{\rho}\frac{d\bar{\rho}}{dz}}\, t\right) \tag{8.57}$$

that is, the particle oscillates with the frequency

$$N = \sqrt{\frac{g}{\rho}\frac{d\bar{\rho}}{dz}} \tag{8.58}$$

N is called Väisälä frequency after Väisälä (1925) who was the first to mention such oscillations in connection with meteorological radiosonde balloons, also carrying out oscillations of this kind when ascending into the atmosphere. Those are the shortest oscillation periods τ_N that a water particle can have in stratified water; depending on stratification, they may amount to several minutes. Larger water bodies may be subjected to analogous oscillations. But their frequency is always smaller than N, that is, the period is always larger than the Väisälä period. For free waves, however, it cannot become larger than the inertial period τ_E. Hence, the period of free internal waves lies within the range of $\tau_N < \tau < \tau_E$.

As shown in the sections dealing with surface waves, the vertical velocity (and, thus, the amplitude) of short surface waves decreases exponentially with depth. With long waves, the decrease is linear. Boundary surface waves have their largest amplitude at the boundary surface, from which it decreases exponentially towards either side. With regard to internal waves, the change of the amplitude with depth depends on the density stratification. The amplitudes disappear at the sea surface and at the bottom. In between, they may have one maximum or several maxima. According to the number of such maxima, the waves are called internal waves of first order or second order, etc. An example is given in Fig. 8.47, where the amplitudes of internal waves of first to fourth order, W_1 to W_4, are plotted. It can be taken from this figure that internal harmonic waves of first order have a maximum at medium depth, and that the amplitude decreases toward the sea floor as well as toward the sea surface. Second-order internal waves have a maximum, nearly equally large, at about $3/4$ of the water depth, where they are in phase with first-order waves. Their amplitudes decrease to zero towards the sea floor and towards the depth where first-order waves have a maximum. A counteroscillation appears in the entire upper half of the water column.

In the ocean, numerous orders are usually excited simultaneously. Their superposition results in the total field of internal waves. In the case of equal periods, first-order waves have the largest wavelengths. With growing order, the wavelength decreases. Further properties have been described by Krauss (1966).

The generation of internal waves is as little understood as that of surface waves. Variations of wind and atmospheric pressure might be the chief causes. However, the small number of observational data available so far does not permit us to establish a definite relationship.

402 Surface Waves and Internal Waves

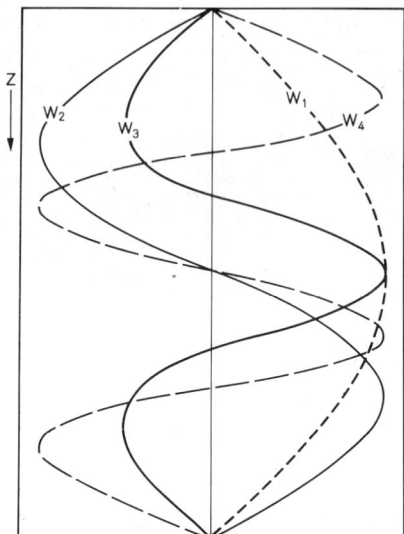

Fig. 8.47. Internal waves of first to fourth order (W_1-W_4).

8.5.3. Internal tidal waves and inertial waves

Among the long internal waves those with tidal periods play a special role. Such *internal tidal waves* can be recognized in each of the numerous continuous observations carried out so far in the ocean from moored research vessels, like *Michael Sars, Armauer Hansen, Meteor, Dana, Willebrord Snellius,* and *Atlantis.* They are common phenomena, like tides at the sea surface. A particularly impressive example is represented in Fig. 8.48. It is based on observations in the central North Atlantic Ocean gained at a three-day anchor station of *Meteor* during the German North Atlantic Expedition in 1938, and

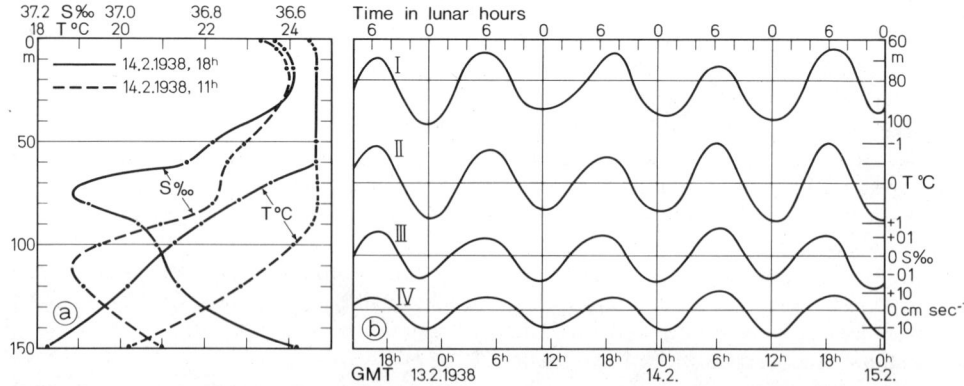

Fig. 8.48. Internal tide based on observations at *Meteor* Station 385 in the North Atlantic Ocean at 16°48′N and 46°17′W, February 12–14, 1938. (a) Vertical distribution of temperature and salinity at two time instances; (b) time series of: I, depth of the 24°C isotherm; II, deviation of the average temperature in the layer 70–120 m from the average; III, deviation of the salinity at 80 m depth from the average; IV, deviation of the north component of the current velocity at 50 m depth from the average.

analyzed by von Schubert (1944). The water depth was 2950 m. Figure 8.48(a) shows how, within half a tidal period, the vertical distribution of temperature and salinity changed from the sea surface down to a depth of 150 m. Figure 8.48(b) reproduces the change in time, in a somewhat smoothed fashion, with curve I representing the depth of the 24°C isotherm. This shows that, in this case, the height of the internal wave reaches 40 m. The changes of the average temperature of the layer from 70 to 120 m (curve II), as well as the changes of salinity at 80 m depth (curve III) and those of the current velocity at 50 m depth (curve IV) occur all of them in strict agreement with the lunar period of 12 hr 25 min.

The question as to how those internal tidal waves originate has repeatedly been investigated. The explanation is difficult since the tide-producing force has a very large wavelength, while the wavelengths of internal tides are only of the order of magnitude of 100 km. It is possible to interpret them in the following way: they may be generated by tidal waves of the sea surface which release part of their energy to internal tides when crossing rough ground, especially at the continental margin. Hence, internal tides are supposed to originate in particular at the shelf edges (Cox and Sandström, 1962). Observations, however, have shown that they are general phenomena in the ocean. The problem has not yet been fully solved.

Long internal waves also include *internal inertial oscillations* occurring when the equilibrium state of stratification is disturbed by some external influence, for example, by rapid changes of wind or of atmospheric pressure. Under the influence of the deflecting force of the earth's rotation, the displacement of water is transformed into an inertial oscillation with the period of half a pendulum day, as already demonstrated in Fig. 7.25 by means of current measurements. From the theory of internal waves it follows that free internal waves of large wavelengths always have the period of inertial oscillations.

In the ocean, internal inertial oscillations and internal tidal waves superimpose each other and, because of their amplitudes (which are sometimes large), they contribute to the fact that changes with time of the vertical structure of water masses may appear rather confused. This is especially disturbing if the internal pressure field in the ocean is to be calculated from the mass field. In general, only the vertical mass distribution at a particular time is known, but not the phase and amplitude of internal waves. As shown by Dietrich (1937) in an example from the Atlantic Ocean, about 10 dynamic centimeters (d.cm.) of vertical displacement of the sea surface were due to the effect of internal waves. In some cases, this would correspond to the horizontal slope of the physical sea surface over several hundred kilometers, as indicated in Fig. 7.15. In smaller oceanic areas, internal waves may make the determination of the pressure field (and thus of the current field) questionable unless special observations enable them to be eliminated numerically, as Defant (1950) succeeded in demonstrating for Californian waters.

In general, internal waves have symmetric wave profiles like those shown in Fig. 8.48. But there are also exceptions. When internal waves proceed from deep water into shallow areas of the sea, the amplitude becomes much steeper, and the wavelength decreases, as found by Magaard (1962). Repeatedly, asymmetry could be observed; in rare cases even a breaking of the wave profile indicating the existence of *internal surf*. A relevant example is presented in Fig. 8.49 based on hourly observations from the Danish research vessel *Dana* in the Strait of Gibraltar. The slow rise of the isopycnal lines is followed by a sudden fall of the pycnocline by about 100 m, 4 hr after high tide in Cadiz. This internal surf, however, does not show the same vehemence as the surf of surface waves. Since internal waves have a small propagation velocity, the breaking process occurs in "slow motion," so to speak. However, it was a danger to submarines when crossing the Strait

404 Surface Waves and Internal Waves

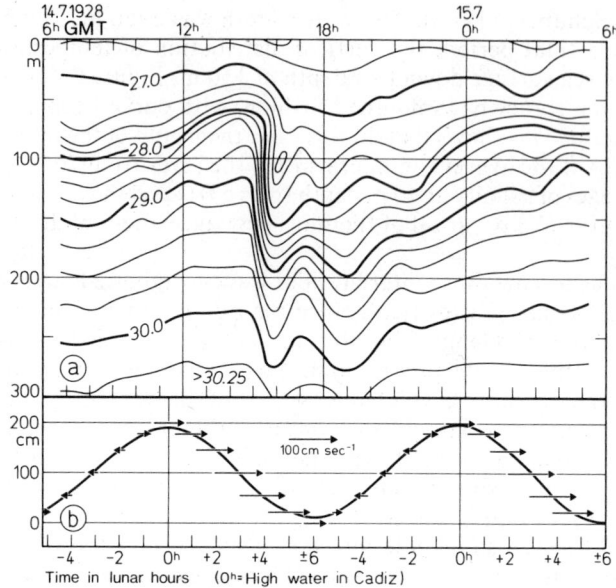

Fig. 8.49. Example of internal surf: (a) Change of the vertical density distribution ($\sigma_{S,T,p}$) with time in the Strait of Gibraltar, based on hourly observations from the research vessel *Dana* on July 14–15, 1928. (After Jacobsen and Thomsen, 1934.) (b) Tidal curve at Cádiz and surface currents in the center of the Strait of Gibraltar for the same time period. (Arrow toward the right: current toward the east; arrow toward the left: current toward the west.)

of Gibraltar under water because it forced them either to the surface or to deeper layers.

In individual cases, the internal surf may become evident at the sea surface in narrow straits, as, for instance, in the Strait of Messina. Here, especially at spring tide, a large internal breaking wave of about 60 m height develops between the heavier water masses of the Ionian Sea and the lighter ones of the Tyrrhenian Sea, lying on top of the former.

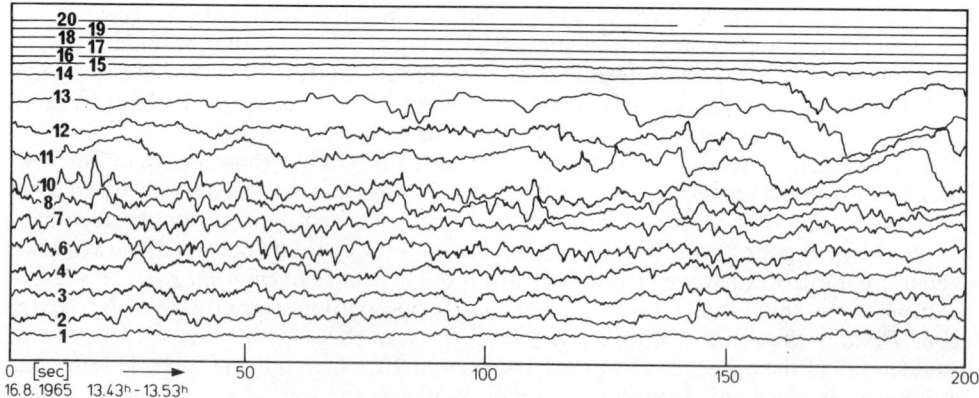

Fig. 8.50. Temperature fluctuations in a discontinuity layer. The thermistors 1–20 have a vertical distance of 20 cm each. (According to Krauss et al., 1973.)

Since in the northern part of the strait the cross section becomes very narrow, these water masses come to lie one beside the other, and the eddy motion of the internal surf wave reaches the sea surface. These phenomena are known as "Scylla and Charybdis." The danger they represented to seafarers is poetically reflected in Homer's Odyssey; today, however, it hardly exists any more, not even for small craft. After Defant (1940), who has interpreted the oceanographic conditions of these phenomena, it is conceivable that the cross section of the Strait of Messina, lying in a zone of frequent volcanic activity, has been enlarged in historic time. According to theory, such widening could have brought about a weakening of Scylla and Charybdis. The story has been handed down that, during the earthquake of 1883, the rocks near the village of Scylla disappeared into the sea along a wide stretch of coastline, and that ever since, the Scylla whirlpool is said to have weakened considerably.

Evidence has not yet been found that internal surf with its strong effect on the mixing of water masses also occurs in the open ocean. As far as its causes are known today, this seems improbable. At the shelf edge, however, conditions are different.

8.5.4. Stability oscillations

From Section 8.5.2 we know that short internal waves have a period chiefly depending on the vertical density gradient present in the stratification. This Väisälä period is given

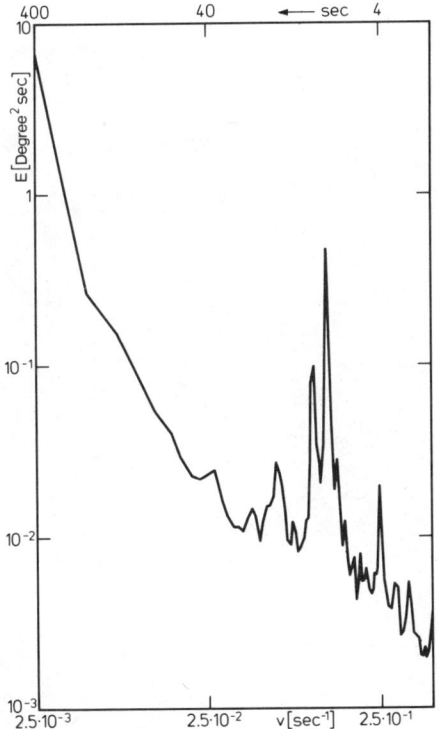

Fig. 8.51. Spectrum of the temperature fluctuations in a discontinuity layer. The relative energy maximum at about 10 sec corresponds to the Brunt-Väisälä frequency.

by

$$\tau_{\min} = 2\pi \sqrt{\frac{\bar{\rho}}{g\, d\bar{\rho}/dz}} \qquad (8.59)$$

Since these fluctuations are oscillations around the stable resting position, they are also called stability oscillations. They often occur in discontinuity layers of temperature and salinity, especially if currents of different strength prevail at either side. Then they are to be understood as Kelvin–Helmholtz instabilities which possibly also play a role in the generation of sea surface waves (cf. Section 8.3.7.).

Figure 8.50 shows an example of such oscillations taken from thermistor records within the near-bottom discontinuity layer of temperature and salinity in the Baltic Sea. In this layer 20 thermistors were fixed to a rigid observation mast at an average distance of 20 cm each. Figure 8.51 gives the spectrum of such a registration. With increasing frequency the spectrum slopes approximately in accordance with a $5/3$ power law, as is known from turbulence theory. At the Väisälä frequency, however, a peak of energy has been observed about 50 times larger than the energy of the surroundings. Krauss et al. (1973) have carried out a great number of such measurements, according to which time intervals of complete rest in the thermocline alternate with others of intensive fluctuations.

9 Tidal Phenomena

9.1. Definitions Concerning Tidal Phenomena

9.1.1. Tides and tidal currents

Tides occur in the world ocean as progressive or standing long waves modified by reflection, the Coriolis force, and friction. Of all oceanic long waves, tides are of the greatest practical importance for the coastal population as well as for seafaring people. Among other reasons, this is why it seems necessary for us to deal with these waves in more detail. Moreover, tides hold an exceptional position among all waves occurring in the ocean; this also requires a special treatment. They are the only waves generated by periodic disturbances of the terrestrial gravity field, caused by sun and moon. Thus, a definite period of oscillation is forced upon them. These periodic forces must be known if the tidal phenomena are to be understood. To begin with, a general survey of the most important tidal phenomena in the ocean will be given.

The water level rises and falls regularly in an approximately semidiurnal or diurnal rhythm along all coasts of the world ocean. The water off the coasts moves back and forth with the same regularity. The rise and fall of the water level is called tide; the back and forth motion of the water is called tidal current. Tide gauges serve to measure the tides; there are coastal gauges (cf. Section 3.2.6) and high-sea gauges (cf. Section 3.2.6). Tidal currents are determined by current meters operated from anchored ships (cf. Section 3.2.12.1) or by moored current meters (cf. Section 3.2.12.1).

Water level records over longer periods show that, on the average, 12 hr 25 min pass from the beginning of one water level fall to the next along most coasts of the world ocean. This means that the tide is delayed from day to day by 50 min. The relation to the apparent movement of the moon in the sky is evident, because a full revolution of the moon around the earth requires 24 hr 50 min on the average, a so-called lunar day. However, longer records of water level and current also show that tides and tidal currents do not repeat themselves exactly at a given location and, furthermore, that there are great differences from one place to the other. These inequalities are responsible for the complicated nature of tides and tidal currents. Because of the great importance of tidal phenomena for shipping industry and for the inhabitants of coastal areas, it is necessary to provide means which may enable us to predict tides and tidal currents for all locations of interest. For this purpose, tide tables are issued by the hydrographic offices of the maritime nations. They contain the tides predicted for a small selection of seaports and also the differences of the tides at numerous other locations in comparison to these ports. Moreover, information on tidal currents is given, for the same purpose, in nautical charts and sailing directions as well as in atlases of tidal currents, for example, in the *Atlas der Gezeitenströme für die Nordsee, den Kanal und die britischen Gewässer* (*Atlas of Tidal Currents for the North Sea, the English Channel and British Waters*), published by the German Hydrographic Institute in Hamburg.

Some main characteristics and terms concerning tides are represented in Fig. 9.01

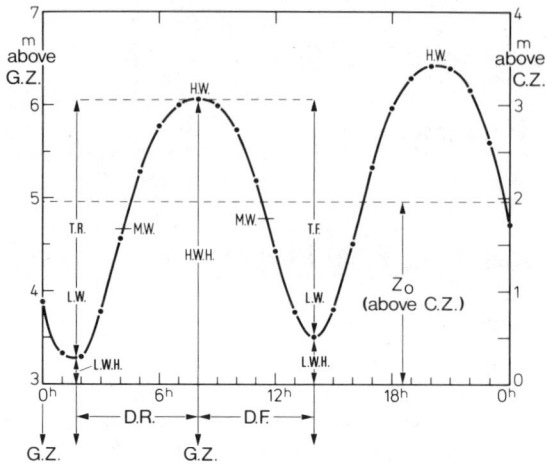

Fig. 9.01. Some important terms used in connection with tides: HW = high water, LW = low water, HWH = height of high water, LWH = height of low water, TR = tide rise, TF = tide fall, DR = duration of rise, DF = duration of fall, MW = mean water, Z_0 = height of mean sea level, CZ = chart zero or chart datum, GZ = gauge zero = NZ − 5.00 m, NZ = normal zero.

by the example of a water level curve covering 24 hr. Low water (LW) and high water (HW) can be recognized twice each. The rise and fall of the water level from one LW to the following is called a tidal cycle. The rise is called *flood* and the fall, *ebb*. Therefore, the duration of a tidal cycle is given by the sum of the duration of flood or rise (DR) plus the duration of ebb or fall (DF). DR and DF vary, within certain limits, from one tidal cycle to the other. So does their sum. Hence, we speak of "inequalities in time." In Fig. 9.01, the "inequalities in height" are more readily recognizable than inequalities in time. Neither the individual heights of high water (HWH) nor those of low water (LWH) are identical. Thus, there is a different tide rise (TR) and tide fall (TF). The average of TR and TF within one tidal cycle is called tidal range. It varies from one tidal cycle to the next. Therefore, the mean water (MW), computed from the mean HWH and the preceding or succeeding LWH also varies. This mean water (MW) must be carefully distinguished from the mean sea level (Z_0), determined by averaging hourly readings, if available, for 1 month and, in general, referring them to CZ (chart zero or chart datum).

Gauge records over longer time periods suggest that the inequalities are, in a certain way, related to the revolution of the moon around the earth. A section of a tidal record, obtained at the gauge of Cuxhaven over a period of 14 days, is reproduced in Fig. 9.20 on a greatly reduced scale. The most important inequalities are the following.

1. *Semimonthly or fortnightly inequality.* One or two days after full moon and new moon, at the so-called spring tide, the tide range reaches a maximum: the spring range. The delay with respect to full moon and new moon is called spring delay or also "age of tide." With approximately the same delay, the so-called neap delay, a relatively small tidal range, the neap range, occurs after the first and last quarter of the moon at the so-called neap tide. The time passing between a new moon and the following full moon, which is also the period of the semimonthly inequality, is, on the average, equal to 14.7653 days,

that is, half a synodic month. ("Synodic" means "meeting"; the term refers to the meeting of sun and moon at new moon.)

2. *Parallax or monthly inequality:* The tidal range varies with the distance of the moon from the earth. With a certain delay after the moon has passed through perigee (point of the moon's orbit closest to earth), the tidal range becomes especially large. After the moon's passage through apogee (point of the moon's orbit most distant from earth) the tidal range becomes especially small. The term "parallax inequality" results from the horizontal parallax of the moon being a measure for the moon's distance from the earth. The duration of one revolution of the moon on its elliptical orbit around the earth from one perigee to the next, and, hence, the period of the parallax inequality is, on the average, equal to one anomalistic month of 27.5546 days. ("Anomalistic" means "irregular"; the term refers to the deviation of the moon's orbit from a circle.) Since the anomalistic month is about two days shorter than the synodic month, the moon's passage through perigee lags somewhat behind the appearance of the new moon. If the moon's passage through perigee coincides with the full moon or the new moon, the spring range is especially large, as shown in Fig. 9.06 by the example of Immingham in the second half of March, where new moon and perigee coincided on March 23, 1936.

3. *Declination inequality.* This is exhibited by the fact that, after the greatest northern and southern declinations of the moon, a relatively small tidal range follows and, after the moon's passage of the equator, a relatively large tidal range. The inequality, therefore, is independent of the sign of declination. The time passing between the greatest northern and the greatest southern declination of the moon gives the period of the declination inequality; it amounts to half a tropical month or 13.6608 days. ("Tropicus" means "turn"; the term refers to the return of the celestial body after reaching the greatest declination.)

4. *Diurnal inequality.* Its effect can be recognized in Figs. 9.20 and 9.06 in the changing height of high water and low water. These height differences of successive high and low waters reach their maximum with a certain delay after occurrence of the greatest northern or southern declination of the moon. They disappear with the same delay after the lunar transit of the equator. Closer studies reveal that the inequality does not only depend on the degree, but also on the sign of the declination of the moon. Thus, its period is given by the full tropical month of 27.3216 days.

There are some further inequalities of tides, among them also some with longer periods; but the four mentioned above are of the greatest influence. The most important of the longer inequalities has the period of a saros cycle of about 18 yr 11 days. (Saros is the Chaldean name for the return of solar and lunar eclipses in this cycle.) This period corresponds to a full revolution of the ascending moon's node along the ecliptic.

The aforementioned inequalities of tides with regard to range and time depend on certain astronomical quantities, like the true solar time of the lunar transit of the meridian, the horizontal parallax, and the declination of moon and sun. They can be determined and tabulated for any location if a sufficiently long series of observations is available, preferably covering more than 19 yr. Conversely, the tides for a given location can be predicted with the aid of such tables and the predicted astronomical quantities published

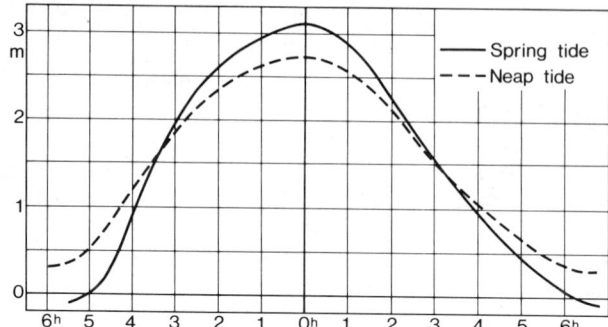

Fig. 9.02. Water level curves at Cuxhaven for mean spring tide and mean neap tide from 6 hr before high water to 6 hr after. Heights of water level referred to chart datum. (After German Hydrographic Institute, 1971.)

regularly in astronomical yearbooks. Laplace (1775) and Lubbock (1831) used such ideas in their prediction scheme. In contrast to the more recent harmonic method, this empirical, so-called nonharmonic, method is quite elementary. But for shallow water areas, like the North Sea, it provides better predictions than the harmonic method (cf. Section 9.3.1), especially if, following Horn (1948), suitable astronomical variables are chosen. Therefore, this method is also applied to the calculation of tide tables for the German North Sea coast. Predictions, however, can be made only with a view to high and low water levels and time of arrival but not for the entire tidal curve. Furthermore, this method is only applicable for locations with semidiurnal tides (cf. Section 9.3.4), where the diurnal inequality, mentioned above, is hardly noticeable. This can be corrected by attaching mean tidal curves for spring and neap tides to the tide tables, as illustrated, in simplified form, by Fig. 9.02 for Cuxhaven. Taking into account the phase of the moon and the spring delay, the most probable tidal curve for each day can be predicted in this way.

9.1.2. Reference levels of tides

Water level data always require a reference level. In Germany, the gauge zero (GZ) of coastal tide gauges is set 5 m below the normal zero (NZ) of the geodetic survey in order to avoid negative water level values. Some time ago, NZ was determined by the mean sea level of the Baltic Sea at the gauge of Swinemünde. This NZ is not identical with the chart zero (CZ) or chart datum in nautical charts. CZ indicates a height beneath which the water level drops only in rare cases. This reference level is defined differently by the various nations. Detailed information is found in the summaries of CZ contained in German tide tables. At the German North Sea coast CZ is equal to the mean low water springs (MLWS) as it is at the coasts of Argentina and Chile. At the French, Spanish, and Portugese Atlantic coasts, CZ equals the lowest possible low water.

The mean water level Z_0 as the statistical average of local water level records, superimposed by tides, is subject to seasonal, long-term, and secular variations. Monthly Z_0 values of the most important gauges along the coasts of the world ocean are collected and published from time to time by the Association Internationale des Sciences Physiques d'Océan. Besides, this organization has issued a complete bibliography in two volumes (1719–1958 and 1959–1969).

Variations of Z_0 are primarily caused by seasonal changes in the prevailing winds, in atmospheric pressure, the density stratification of the sea and the continental runoff.

Wemelsfelder (1971) has described as many as 22 different effects combined in action. It is difficult to clearly relate all effects to one instant as Dietrich (1954) tried to do for the Esbjerg gauge.

In addition, coastal lifting or subsidence as well as changes in the total water volume of the world ocean are also hidden in the long-term variations of Z_0. The former, the so-called isostatic changes of the water level, can reach -1.0 m per 100 yr in exceptional cases as at the Swedish coast of the Gulf of Bothnia. At present, the land there is rising at this rate as an aftereffect of the inland ice which receded in the recent geological past. These rates quickly decrease with greater distance from the Gulf of Bothnia. At the German Baltic Sea coast the rate is already zero. Independent of this phenomenon, the entire water volume of the world ocean has increased in the past decades; so the mean water level Z_0 of the world ocean rose by 1.1 mm yr^{-1} during the period from 1900 through 1936 (Gutenberg, 1941). This is the so-called eustatic change of the sea level, which is part of the post-glacial rise of the sea surface in the world ocean, as represented in Fig. 1.22 for the past 20,000 yr. The total rise amounts to 100 m; it was caused by the recession of the masses of inland ice, that is, by a change of climate. In addition to these variations and changes of Z_0, precise leveling has also shown that the Z_0 values of gauges along a coast do not always lie exactly on the same potential surface. In such cases, ocean currents must be present in order to maintain the inclination of the Z_0 surface. Along the Baltic Sea coast, from Memel to Travemünde, the Z_0 surface descends by 15 cm; along the United States coast from Portland, Maine to St. Augustine, Florida, by 39 cm. Precise leveling between zero points of gauges can be carried out only on land. Therefore, the topography of the sea surface can be determined only along the coasts but not perpendicular to them.

The position of the mean physical sea level of the entire world ocean, corresponding to a Z_0 surface determined only by the density stratification of the ocean, with all other effects disregarded, has been cartographically represented by Lisitzin (1965). The dynamic topography of the sea surface, as given in Fig. 10.33(a) for the North Pacific Ocean, corresponds to such a representation. Accordingly, level differences up to approximately 1.4 m occur, especially in the region of the Kuroshio.

9.2. Tide-Producing Forces

9.2.1. Description of the system of tide-producing forces

The tidal phenomena described in Section 9.1.1 are the effects of the tide-producing forces of moon and sun. Newton (1687) was the first to explain these forces as resulting from the general earth–moon and earth–sun mass attraction.

In order to arrive at an understanding of the system of tide-producing forces, let us first consider the influence of the moon alone. The distance between the earth and the moon is just as large that the total of the attracting forces among all mass particles of earth and moon is in complete equilibrium with the total of all centrifugal forces resulting from the motion of both celestial bodies around their common center of gravity. This equilibrium applies only for the total of both forces, if earth and moon are considered as an entity, but not for individual points on the earth's surface. This fact provides the basis for the tide-producing forces. It is schematically indicated in Fig. 9.03. Since all points on the earth describe the same path during a revolution around the common earth–moon center of gravity (in terms of mechanics, earth and moon perform a "revo-

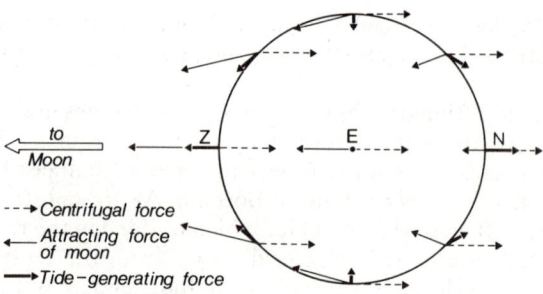

Fig. 9.03. Tide-producing force as resultant of attracting force and centrifugal force along a meridional section through the earth. Z, N: moon in zenith or nadir, respectively; E: center of the earth.

lution without rotation" around their common center of gravity), the centrifugal force has the same magnitude and direction everywhere on earth. The attracting forces, however, have slightly different directions and, above all, differ in their magnitude, depending on the distance of the points on earth from the moon. The attraction of the moon is largest at the point Z, which is closest to the moon, and weakest at the point N on the earth's surface, which is farthest from the moon, as indicated in Fig. 9.03. The attracting force is dominant on the half of the earth facing the moon, whereas the centrifugal force is dominant on the half facing away from the moon. Only at the center of the earth E do both forces exactly cancel each other. At all other points, small residual forces are left as resultants of these two forces. These are the tide-producing forces.

The tide-producing forces are very small compared to gravitational forces. It is appropriate to separate the tide-producing forces into vertical and horizontal components. Then, the former will act in the direction of the gravitational force, which it can change, at most, by 1/9,000,000. For tidal phenomena in the ocean, this vertical component is of no importance, quite in contrast to the horizontal component which has the same order of magnitude as other forces acting in the ocean in the same direction. Figure 9.04 shows, in perspective, the distribution of these horizontal components for the case where the moon is at the equator and at the zenith of point Z. On the side facing the moon, all force vectors converge at point Z, and likewise, at point N on the side facing away from the moon. Figures 9.03 and 9.04 illustrate a special case inasmuch as, here, the moon is assumed to be at the equator. If the moon has a certain declination, the system of forces is also displaced, but always in such a way that the moon is at the zenith over the point of convergence Z.

The sun, like the moon, is also the source of a tide-producing force, which, however, is not quite half that of the moon. Since the tide-producing force is given by the difference between the force of mass attraction and the centrifugal force, and since these forces depend on the distance of the earth from sun and moon, the system of tide-producing forces will vary in accordance with the changing distances of sun and moon from the earth. Such variations are partly reflected in the inequalities of tides mentioned before.

In the above discussion of the tide-producing forces, the daily rotation of the earth around its axis has been completely neglected. In fact, it does not have any influence at all on the formation of these forces. Its only consequence is that the system of forces, as indicated in Fig. 9.04, is continually displaced and that, to an observer on the surface of the earth, the same tide-producing force of the moon or sun appears to act periodically after half a lunar day or half a solar day has passed.

This short description is restricted to the most significant points necessary for an

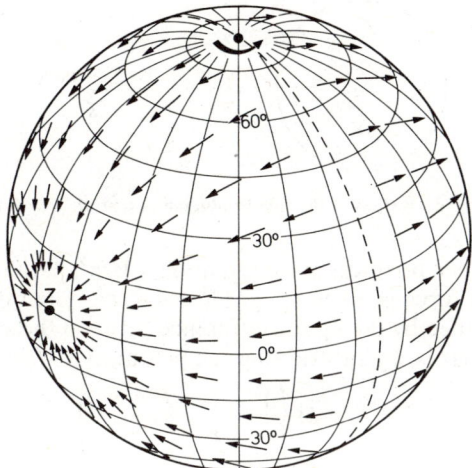

Fig. 9.04. Distribution of the horizontal component of the tide-producing force over the globe. (Position of the tide-generating celestial body at the equator in the zenith of Z; arrow at the pole indicates sense of earth's rotation.)

understanding of the tidal forces. More detailed descriptions have been published by Defant (1961). This short introduction precedes the quantitative description presented in the two following sections, because misleading conceptions frequently exist concerning the influence of the earth's rotation and the revolution of earth and celestial bodies around their common center of gravity. Such misconceptions may be avoided even without applying mathematical tools.

9.2.2. Derivation of the tidal potential

The quantitative description of the tide-producing forces and the fundamentals of their prediction require considerable mathematical effort. They can be outlined here only briefly. A detailed treatment has been given by Bartels (1957).

The tide-producing forces at any given point P on the earth's surface are defined for any given instant by the difference in the attracting forces exerted by moon and sun on the earth's center of gravity and on the considered point of mass on the earth's surface. Instead of specifying the tide-producing forces superimposed on the gravitational force, in terms of direction and magnitude, it is advantageous to derive an expression for their potential V. The forces can, then, be expressed as gradients of V.

To begin with, some quantities used in the derivation of the tide-producing potential V will be defined. As indicated in Fig. 9.05, these are the following: OA = r = true distance between the center of the earth and that of the celestial body, P is an arbitrary point on the earth's surface. Then, OP = R = radius of the earth, ϑ = geocentric zenith distance of the celestial body, r_1 = true distance PA. Furthermore, g_0 = acceleration of gravity at the equator, γ = gravitational constant; E, M, and S are the masses of earth, moon, and sun, respectively; c_M and c_S are the mean distances between the centers of the earth and the moon (c_M = 384,000 km) and of the earth and the sun (c_S = 149.5 × 10^6 km).

The gravitational potential of the moon at point P is equal to $\gamma M/r_1$. For the total

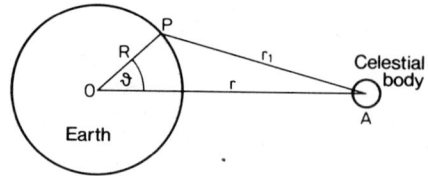

Fig. 9.05. For the derivation of the tide-producing potential. (For symbols, see text.)

mass of the earth, the moon produces an acceleration $\gamma M/r^2$, parallel to OA. The potential of this uniform field of force is given by $\gamma MR \cos \vartheta / r^2$. The tide-producing potential of the moon V_M is obtained by taking the difference between those two potentials relative to the earth's center, that is, by subtracting the potential $\gamma M/r$

$$V_M = \gamma M \left(\frac{1}{r_1} - \frac{1}{r} - \frac{R \cos \vartheta}{r^2} \right) \qquad (9.01)$$

From the triangle OPA in Fig. 9.05 it is evident that

$$r_1^2 = r^2 + R^2 - 2rR \cos \vartheta$$

By means of this relation, r_1 can be eliminated in Eq. (9.01) and the resulting equation can be expanded in powers of R/r, the horizontal parallax of the moon. Since R/r is very small (approximately $1/60$), the higher terms can be neglected, and Eq. (9.02) results

$$V_M = \frac{\gamma M R^2}{r^3} \frac{1}{2} (3 \cos^2 \vartheta - 1) + \cdots \qquad (9.02)$$

The expression for the tidal potential of the sun V_S has the same form, except that M must be replaced by S and r is to be understood as the true distance between the centers of earth and sun. The tidal potential, produced by moon and sun together, is $V = V_M + V_S$.

The radial component Z of the tide-producing force, also considered as the tidal perturbation of the gravity force g, is then given by

$$Z = -\frac{\partial V}{\partial R} = -\frac{2G}{R} \left(\cos 2\vartheta + \frac{1}{3} \right) \qquad (9.03)$$

The tangential component is

$$H = -\frac{\partial V}{R \partial \vartheta} = \frac{2G}{R} \sin 2\vartheta \qquad (9.04)$$

Here, G stands for the constant in the expression of the tide-producing potential at the distance R from the earth's center:

$$G_{moon} = \left(\frac{3}{4}\right) \gamma M \frac{R^2}{c_M^3} \qquad (9.05)$$

$$G_{sun} = \left(\frac{3}{4}\right) \gamma S \frac{R^2}{c_S^3} \qquad (9.06)$$

Since the acceleration of gravity deviates only very slightly from the radial direction, Z and H can also be considered vertical and horizontal components of the tide-producing force.

The ratio of the tide-producing forces of sun and moon is given by $G_S:G_M = 0.46051$. Accordingly, the effect of the sun on the tide is not quite half as large as that of the moon.

If the tide-producing forces are to be predicted, the zenith distance ϑ and the distances of sun and moon in Eq. (9.02) must be expressed as functions of the commonly used mean solar time. We use the following relation

$$\cos\vartheta = \sin\varphi \sin\delta + \cos\varphi \cos\delta \cos(\tau_1 - 180°) \qquad (9.07)$$

where φ stands for the geocentric latitude, δ for the declination, and τ_1 for the hour angle of the celestial body, counted from the lower culmination. Substituting Eq. (9.07) in Eq. (9.02) we obtain

$$V_M = \frac{\gamma M R^2}{r^3} \frac{1}{4} [(3\sin^2\varphi - 1)(3\sin^2\delta - 1) \qquad (9.08)$$
$$+ 3\sin 2\varphi \sin 2\delta \cos\tau_1$$
$$+ 3\cos^2\varphi \cos^2\delta \cos 2\tau_1]$$

The period of the declination δ of the moon equals one tropical month of 27.32 days; the period of the hour angle τ_1 is one lunar day = 24 hr 50.47 min. Hence, the expression in the square brackets of Eq. (9.08) separates the effect of the moon on the tide into three terms of approximately semimonthly, diurnal, and semidiurnal periods. Since the declination δ of the sun has a period of 1 yr, the corresponding tidal potential of the sun contains three terms of semiannual, diurnal, and semidiurnal periods.

9.2.3. Harmonic expansion of the tidal potential

Our aim is to describe the tide-producing potential V as the sum of sine and cosine terms of constant amplitude and frequency. For this purpose, Eq. (9.08) must be further transformed in order to avoid the time occurring twice, namely in δ as well as in τ_1. The position of moon and sun with respect to the earth is described by the distance r from the earth's center and by the latitude and longitude as referred to the ecliptic. At a particular time t, all three quantities are periodic functions of five angles—of the mean longitudes of the moon (s), of the sun (h), of the perigee of the moon's orbit (p), of the ascending node of the moon's orbit (N), and of the perigee of the sun's orbit (p_s). They are measured along the ecliptic from the simultaneous mean vernal equinox and change almost uniformly with the time t.

Numerical values of the changes σ of these five angles within one mean solar hour and of the period in units of mean solar days or solar years, respectively, are given in Table 9.01.

According to evidence given by celestial mechanics, the motions of the true sun and the true moon, relative to the earth, can be expressed by infinite series for s, h, p, N', and p_s, where $N' = -N$. Such series have been derived for the sun as well as for the moon. For the sun, the series are comparatively simple. The motion of the moon, however, is very complicated because of the simultaneous influence of the earth and the sun. The relevant series contain several hundred terms.

Doodson (1921) has used these series for the most extensive expansion so far, of the tide-producing potential. Since the hour angle τ is added to the five other variables, Doodson's expression assumes the form of a six-dimensional Fourier series with terms

416 Tidal Phenomena

Table 9.01. Numerical Values of the Five Variables in the Coordinates of Moon and Sun[a]

Variable σ (° hr^{-1})		Period = 360°/σ	
s	0.549017	27.321582 days	(tropical month)
h	0.041069	365.242199 days	(tropical year)
p	0.004642	8.847 yr	(revolution of the mean perigee)
N	−0.002206	18.613 yr	(revolution of the mean node)
p_s	0.000002	20,940 yr	(revolution of the perihelion)

[a] In degrees per mean solar hour and as the full time of revolution.

called "tidal constituents" or "partial tides." They have nothing in common with the tide as defined in Section 9.3.1. The number of partial tides is infinitely large. Doodson's expansion contains 396 partial tides, arranged in the order of importance. Each partial tide of the tide-generating potential has this form:

$$K \cdot Q \cdot F(\varphi) \cdot G_{\sin}^{\cos} (A\tau + Bs + Ch + Dp + EN' + Fp_s) \quad (9.09)$$

where A is a positive number; B to F are positive or negative numbers which exceed the value 5 only in exceptional cases. If the numbers from B to F are increased by 5 each, in general only positive values will be obtained. The six digit characteristic number in the form

$$A(B + 5)(C + 5).(D + 5)(E + 5)(F + 5)$$

the so-called argument number, was used by Doodson to characterize the individual partial tides. In Table 9.02, 29 of the most important of the 396 partial tides are presented, arranged according to their argument numbers (column 1). A lucid summary of all 396 partial tides was published by Bartels and Horn (1952). The arrangement in the order of argument numbers represents at the same time an arrangement in the order of the angular velocities of partial tides (column 5), if the angular velocity of a partial tide expresses the change in the argument of the partial tide in degrees per hour mean solar time. This change is given by the sum in the parentheses of Eq. (9.09) (cf. column 4 in Table 9.02). Hence, we get the most important division of the partial tides into four classes: constant and long-periodic, diurnal, semidiurnal, and ter-diurnal tides, depending on whether the argument number begins with 0, 1, 2, or 3, respectively.

The classes are divided into groups according to the second digit in the argument number, the groups into principal tides according to the third digit. The partial tides, which are "related" to the principal tides but do not appear in Table 9.02, are distinguished only by their last three digits. The products $Q \cdot F(\varphi) \cdot G$ are called "geodetic functions," according to Doodson. $F(\varphi)$ denotes a function varying with geocentric latitude; G is the constant for the expression of the tidal potential as given for moon and sun in Eqs. (9.05) and (9.06). The factors Q are chosen in such a way that all geodetic functions assume the same largest value of G.

The factor K in Eq. (9.09), the so-called coefficient of the partial tide, listed in column 6 of Table 9.02, gives a measure of the importance of the individual partial tides, if they contain the same geodetic function. Since the different geodetic functions assume their maximum values at different geographic latitudes, the coefficient is not suitable to indicate the importance of a partial tide at a particular place, but it only is a measure of the importance that a tide may reach under the most favorable conditions on earth. Of greatest significance is the semidiurnal principal lunar tide, the so-called M_2 tide, which has the characteristic number 255.555, with $K = 0.90812$. The expansion by Doodson contains partial tides with $K > 0.0001$. Table 9.02 is restricted to partial tides with $K > 0.01$. From this table, the sequence of partial tides according to their importance can be obtained: $M_2, S_2, K_1, O_1, P_1, N_2, K_2, \mu_2, \cdots$, of which the first four are by far the most predominant.

The intuitive interpretation of the individual partial tides in Table 9.02 is restricted to the short remarks in column 7, where hints are given with respect to the effects of the different elements of orbital motion, namely, obliquity of the ecliptic, ellipticity, nodal motion, evection, and variation. Detailed descriptions have been given by Bartels (1957).

To present a clearer survey, a distinction is made between the lunar and solar terms in column 2 of Table 9.02. It is evident that there are partial tides which are of different origin but have the same angular velocity and which, therefore, cannot be separated in an analysis of tidal forces. In Table 9.02 such components are the K_1 and K_2 partial tides.

If the numerical values of the tidal arguments in Eq. (9.09) are to be determined for an arbitrary time, it is of advantage to replace τ by $\tau = t - s + h$ and to calculate it for the Greenwich meridian for $t = 0^h$ GMT. This results in the so-called astronomical argument V_0 as tabulated by Schureman (1941). The argument of the partial tide of the angular velocity σ at a time that is t hours later, is then given by $V_0 + \sigma t$. A principal tide (9.09) assumes the form

$$K \cdot Q \cdot F(\varphi) \cos (V_0 + \sigma t) \qquad (9.10)$$

The superposition of a principal tide and its related tides, with very little difference in angular velocity and, in general, considerably smaller coefficients K, can be considered as a beat effect of the principal tide with a long period, for example, of 19 yr. Therefore, it is appropriate if, for the prediction of the tide-generating potential for a limited time period (e.g., 1 yr), the expression for the principal tide (9.10) is replaced by

$$f \cdot K \cdot Q \cdot F(\varphi) \cos (V_0 + u + \sigma t) \qquad (9.11)$$

where, in general, f differs very little from 1 and u is a small value. Extensive auxiliary tables for the determination of f and u have also been given by Schureman (1941).

Equations (9.10) and (9.11) are valid for the Greenwich meridian. For a location situated L degrees west of Greenwich, where the mean solar time of the meridian with the length S is introduced, the argument V', for 0 hr of this time, is obtained by

$$V' = V_0 - pL + (\sigma S/15) \qquad (9.12)$$

where p is the so-called index of the tide and equal to A in Eq. (9.09).

9.3. Representation of Ocean Tides

9.3.1. Harmonic representation of tides and tidal currents

In principle, it should be possible to arrive at a representation of oceanic tides if the derivation of the tidal potential is introduced into the general hydrodynamic equations [Eqs. (7.04)]. In these equations, the horizontal and vertical components of the tide-producing forces then assume the position of the components X, Y, and Z of the external force. Integration of these equations should yield the periods, amplitudes, and phases

Table 9.02. The Most Important Principal Tides of the Tide-Generating Potential[a]

1	2	3	4	5	6	7
Argument Number	Origin	Symbol	Composition of the Argument	Frequency	Coefficient	Generation

I. Constant and Long-Periodic Tides

055.555	M	M_0	0	0.0000000	0.50458	Constant lunar tide
055.555	S	S_0	0	0.0000000	0.23411	Constant solar tide
056.554	S	Sa	$+h\ -p_s$	0.0410667	0.01176	Elliptical tide of first order to S_0
057.555	S	Ssa	$+2h$	0.0821373	0.07287	Declination tide to S_0
063.655	M	MSm	$+s\ -2h\ +p$	0.4715211	0.01578	Evection tide to M_0
065.455	M	Mm	$+s\ -p$	0.5443747	0.08254	Elliptical tide of first order to M_0
073.555	M	MSf	$+2s\ -2h$	1.0158958	0.01370	Variation tide to M_0
075.555	M	Mf	$+2s$	1.0980331	0.15642	Declination tide to M_0

II. Diurnal Tides

127.555	M	σ_1	$-4s\ +3h$ $-90°$	12.9271398	0.01153	Variation tide to O_1
135.655	M	Q_1	$-3s\ +h\ +p$ $-90°$	13.3986609	0.07216	Elliptical tide of first order to O_1
137.455	M	ρ_1	$-3s\ +3h\ -p$ $-90°$	13.4715145	0.01371	Evection tide to O_1
145.555	M	O_1	$-2s\ +h$ $-90°$	13.9430356	0.37689	Diurnal principal lunar tide
155.655	M	NO_1	$-s\ +h\ +p\ +90°$	14.4966939	0.02964	Elliptical tide of first order to K_1
163.555	S	P_1	$-h$ $-90°$	14.9589314	0.17554	Diurnal principal solar tide
165.555	M	K_1	$+h$ $+90°$	15.0410686	0.36233	Diurnal principal declination tide
165.555	S	K_1	$+h$ $+90°$	15.0410686	0.16817	Diurnal principal declination tide
175.455	M	J_1	$+s\ +h\ -p\ +90°$	15.5854433	0.02964	Elliptical tide of first order to K_1
185.555	M	OO_1	$+2s\ +h$ $+90°$	16.1391017	0.01623	Diurnal declination tide of second order

418

III. Semidiurnal Tides

235.755	M	$2N_2$	$-4s+2h+2p$	27.8953548	0.02301	Elliptical tide of second order to M_2
237.555	M	μ_2	$-4s+4h$	27.9682084	0.02777	Large variation tide to M_2
245.655	M	N_2	$-3s+2h+p$	28.4397295	0.17387	Large elliptical tide of first order to M_2
247.455	M	ν_2	$-3s+4h-p$	28.5125831	0.03303	Large evection tide to M_2
255.555	M	M_2	$-2s+2h$	28.9841042	0.90812	Semidiurnal principal lunar tide
265.455	M	L_2	$-s+2h-p+180°$	29.5284789	0.02567	Small elliptical tide of first order to M_2
272.556	S	T_2	$-h+p_s$	29.9589333	0.02479	Large elliptical tide of first order to S_2
273.555	S	S_2	0	30.0000000	0.42286	Semidiurnal principal solar tide
275.555	M	K_2	$+2h$	30.0821373	0.07858	Semidiurnal declination tide to M_2
275.555	S	K_2	$+2h$	30.0821373	0.03648	Semidiurnal declination tide to S_2

IV. Ter-diurnal Tide

355.555	M	M_3	$-3s+3h$	43.4761563	0.01188	Ter-diurnal principal lunar tide

[a] Explanations to Table 9.02:
Column 1: argument number after Doodson;
Column 2: origin from lunar potential (M) or solar potential (S);
Column 3: common abreviation of individual tides;
Column 4: composition of the astronomical argument V_0;
Column 5: numerical value of the angular velocity σ in deg hr^{-1};
Column 6: coefficient K of the "geodetic function" by Doodson;
Column 7: interpretation of the individual tides.

420 Tidal Phenomena

of the oscillations generated at each location, particularly for the tide and tidal currents. In this way, complete prediction of tidal phenomena would have been achieved deductively.

In practice, however, great mathematical difficulties arise: the extremely complicated configuration of the ocean bottom has to be included in the boundary conditions. Further difficulties turn up if the convective terms in the hydrodynamic equations and friction, important in shallow waters, are also to be considered. Intuitively speaking, this means that the prediction of tidal phenomena in the ocean, if based only on the effect of tide-producing forces, is not fully satisfactory because the modifying influences originating from the complicated bottom topography cannot be described completely. In Section 9.4 it will be shown that even the deductive representation of tidal phenomena in a basin with a geometrically very regular bottom presents considerable mathematical difficulties.

The dynamic tide theory of Laplace has offered proof that every partial tide of the tide-generating potential produces an oscillation in the ocean with the period of the partial tide. These are the so-called astronomic tides. However, this result is valid only on condition that the hydrodynamic equations of the tidal motions remain linear and that, therefore, the tidal waves generated by the different partial tides of the tide-producing potential can be considered to be independent of each other. This is no longer the case in shallow shelf areas, particularly not in estuaries where the tidal amplitude is not very small compared to the water depth, and bottom friction becomes effective. Hence, in addition to astronomic tides, supplementary tides must occur in the tides of shallow seas—the so-called shallow-water tides, similar to the higher harmonics in acoustics. Their angular velocity either is an integer multiple of the arguments of astronomic tides (in which case they are called *higher harmonics*), or they are combinations of different astronomic tides (in which case they are called *combination tides* or *mixed tides*). A detailed investigation of such shallow-water tides was carried out by Rauschelbach (1924). Some of the most important out of the great number of such tides are contained in Table 9.03. The short name indicates the chief cause: $M_4 = 2M_2$ and $M_6 = 3M_2$ represent higher harmonics of the semidiurnal main lunar tide M_2, whereas $MS_4 = M_2 + S_2$, $2SM_2 = 2S_2 - M_2$, $2MS_6 = 2M_2 + S_2$ are combination tides of M_2 and S_2 tides. From the corresponding composition, the angular velocity σ is obtained in deg hr^{-1} (given in column 2), and the astronomical argument is listed in column 3 of Table 9.03. The angular velocities of some individual shallow-water tides are identical with those of astronomic tides, for example, $2MS_2 = \mu_2$.

If the tidal phenomena in the ocean are considered a process depending only on the positions of moon and sun with respect to the point of observation, the harmonic development also contains effects on the water level which have nothing to do with the tide-

Table 9.03. Some Important Shallow-Water Tides[a]

Name	σ (deg hr^{-1})	V_0
$2SM_2$	31.0158958	$+2s - 2h$
M_4	57.9682084	$-4s + 4h$
MS_4	58.9841042	$-2s + 2h$
S_4	60.0000000	$0°$
M_6	86.9523127	$-6s + 6h$
$2MS_6$	87.9682084	$-4s + 4h$

[a] After Rauschelbach, 1924.

producing forces. These are, in particular, seasonal fluctuations in the water level caused by seasonal variations of wind and atmospheric pressure, by variations in the density stratification of the ocean and in the runoff of continental rivers. They are called "meteorological tides." The most important have the arguments h and $2h$ and are therefore completely or nearly identical with the astronomical tides Sa and Ssa in Table 9.04.

According to the results of the theory of tides, oceanic tides may be considered as the sum of harmonic terms, completely analogous to the harmonic expansion of the tidal potential in the form of Eq. (9.10). Hence, the shallow-water tides and the meteorological tides can be treated as astronomical tides. Each single term has the form

$$H \cos (V_0 + \sigma t - \kappa) \tag{9.13}$$

and is called a partial tide as in the series expansion of the tidal potential. H and κ stand for the still unknown amplitude and phase for each location. They are called harmonic constants.

9.3.2. Analysis of the tide observations

The harmonic constants of the individual partial tides can be determined from water level observations in two ways: (1) according to the harmonic analysis by Thomson (Lord Kelvin) (1868); (2) by means of the spectral analysis as applied by Munk and Cartwright (1966).

1. The *harmonic analysis* consists in determining H and κ from water level registrations at a given point with σ values provided astronomically. Hourly values are desirable. Extensive literature regarding the practical execution of the harmonic analysis is available; the most detailed publications have come from Rauschelbach (1924), Doodson (1928, 1957), and Schureman (1941), to which the reader is referred.

    A strictly formal analysis of the most important principal tides and related tides in the ocean should be extended over a time period given by the revolution of a node of the lunar orbit, that is, over approximately 19 yr. Because of the extraordinary tedious computations and the high requirements placed on tide observations, such detailed analyses are exceptions. Usually, we are satisfied with observations extending over the period of 1 yr, from which only principal tides can be analyzed in a form analogous to Eq. (9.11):

$$f \cdot H \cdot \cos [V_0 + \sigma t - (\kappa - u)] \tag{9.14}$$

    The quantities f and u, necessary for the reduction of H and κ, are the same as in Eq. (9.11). In order to include at least the harmonic constants of the most important tides related to the principal tides, the assumption is made that the astronomical principal tide and its related tide have the same ratio of amplitudes and the same phase difference as the corresponding tides in the development of the tidal potential. This is permissible because of the small differences between the periods of principal tides and related tides. If observations extending over a period of one month are the only ones available, only groups of tides can be separated, that is, those tides in Table 9.02 which differ in the second digit of the argument number (column 1). From observations of a few days, only classes of tides can be separated.

Table 9.04. Harmonic Constants of the Four Seaports in Fig. 9.06. Angular Velocities of the Tides, and Importance of the Tides According to the Equilibrium Theory, Referred to the M_2 Tide ($M_2 = 100.0$)[a]

Location	Immingham (England) 53°38′N 0°11′W		San Francisco (California) 37°48′N 122°27′W		Manila (Philippines) 14°36′N 120°57′E		Do-Son (Indochina) 20°43′N 106°48′E		Angular Velocity, (deg hr^{-1})	Importance of Tides ($M_2 = 100.0$)
Duration of Observation	1 yr		10 yr		2 yr		2 months			
Tide	Phase, κ	Ampl. (cm)	Phase, κ	Ampl. (cm)	Phase, κ	Ampl. (cm)	Phase, κ	Ampl. (cm)		
Sa	185	7.0	214	1.7	152	11.8			0.0411	1.3
Ssa	95	3.9	292	3.6	42	2.0			0.0821	8.0
Mm	137	1.6							0.5444	9.1
Mf	163	1.8							1.0980	17.2
MSf	35	1.8							1.0159	1.5
K_1	279	14.6	106	37.0	320	29.7	91	72.0	15.0411	58.4
O_1	120	16.4	89	23.0	279	28.3	35	70.0	13.9430	41.5
P_1	257	6.4	104	11.5	317	9.3	91	24.0	14.9589	19.3
Q_1	64	4.8	82	4.0	254	5.5	(7)	(13.6)	13.3987	7.9
J_1	29	1.5	114	2.1	341	1.7			15.5854	3.3
M_1	299	0.2	104	1.2	306	0.8			14.4921	0.7
S_1	164	2.3	146	0.3	352	0.9			15.0000	0.6

									Frequency	%
M_2	161	223.2	330	54.2	305	20.3	113	4.4	28.9841	100.0
S_2	210	72.8	334	12.3	338	6.8	140	3.0	30.0000	46.6
N_2	141	44.9	303	11.5	291	3.8	(99)	(0.8)	28.4397	19.1
K_2	212	18.3	328	3.7	325	2.1	140	1.0	30.0821	12.7
ν_2	131	9.4	310	2.3	292	0.8			28.5126	3.6
μ_2	215	6.8	226	0.9	272	0.5			27.9682	3.1
L_2	173	12.1	350	1.8	320	0.6			29.5285	2.8
T_2	203	4.3	278	0.5	338	0.4			29.9589	2.7
$2N_2$	142	6.5	278	1.2	276	0.5			27.8954	2.5
λ_2	156	6.1	336	0.7	321	0.2			29.4556	0.7
$2SM_2$	47	4.4							31.0159	
M_4	161	2.6	25	2.3	352	0.5			57.9682	
MS_4	222	3.3	29	0.9					58.9841	
M_6	79	1.1	14	0.2	285	0.2			86.9523	
$2MS_6$	150	1.9							87.9682	

$2(M_2 + S_2)$	592.2 cm	133.0 cm	54.2 cm	14.8 cm
$2(K_1 + O_1)$	62.0 cm	120.0 cm	116.0 cm	284.0 cm
$\dfrac{K_1 + O_1}{M_2 + S_2}$	0.11	0.90	2.15	18.9

[a] From Dietrich, 1944b.

424   Tidal Phenomena

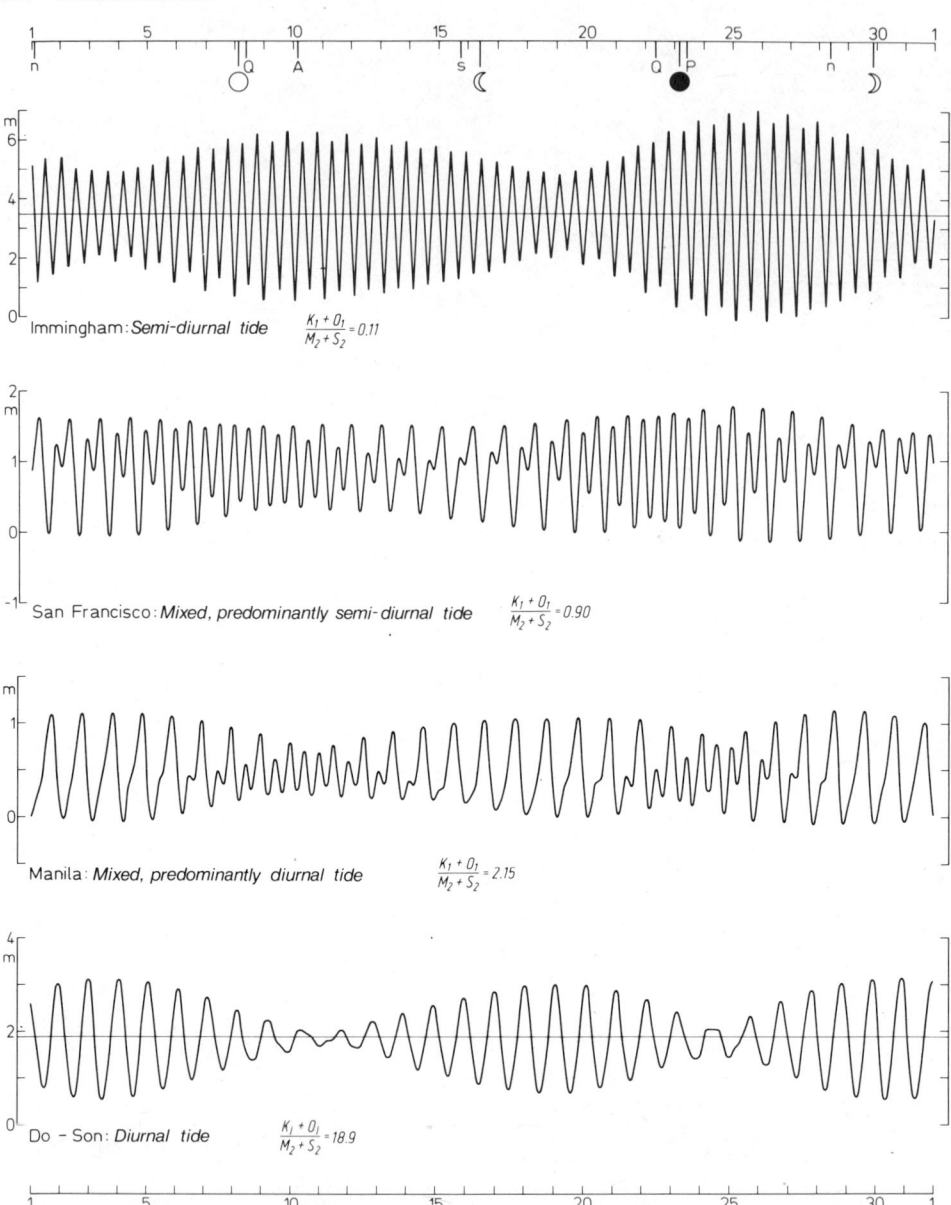

Fig. 9.06. Tidal curves for the month of March, 1936, referred to the corresponding chart datum. (From Dietrich, 1944b.) ○ ☽ ● ☾ = phases of the moon; n, largest northern; s, largest southern declination of the moon; Q, equator passage of the moon; A, apogee; P, perigee of the moon.

2. Munk and Cartwright (1966), assuming that the tidal energy in the ocean is not concentrated only in the frequencies given astronomically, have analyzed the whole spectrum of angular velocities. In this way it is possible to identify the shallow-water tides and compound tides in shallow seas. On the basis of

the most detailed spectral analysis of water levels so far, namely at Anchorage, Alaska, Zetler and Cummings (1967) have given the harmonic constants for 114 tides, where the largest (M_2) has an amplitude of 3.366 m and the smallest (S_6) an amplitude of 0.00183 m.

There are limits to this method, since, strictly speaking, observational series of infinite length would be required for a continuous analysis of the period spectrum. In the practical calculation, two tides of different periods can be separated from each other if, within the time of observation, the number of periods differs by one. Here, a significant role is played by the energy of the "background noise," as shown by Munk and Hasselmann (1964). When approaching the period of a principal tide, this "background noise" energy is increased. Hence, an ideal line spectrum of tides does not exist, but "cusps" occur, covering a certain interval of angular velocities (Munk et al., 1965). A physical interpretation of those cusps has not yet been given. Apparently, interaction with the mean water level, still assumed by Munk and Hasselmann (1964), does not exist as shown by Cartwright (1968) on the basis of a spectral analysis of the water levels at various British seaports. Since there is not even a significant coherence between the cusps of various locations, Cartwright has supposed that local effects are influential during the formation of the cusps.

All the harmonic constants of tides and tidal currents known so far are continually augmented and corrected. They are published by the International Hydrographic Bureau in Monaco (1930–1968); at present, they include data for approximately 4000 locations. This important compilation is the source for the harmonic constants listed in Table 9.04 for four seaports. The harmonic constants are of special value for two reasons. They allow us to elucidate the extraordinarily complicated pattern of oceanic tides at one location and to follow their changes from place to place. This will be discussed in Section 9.3.4. Furthermore, they enable us to predict the tides at a particular location for any time interval. The method usually employed for this will be dealt with in Section 9.3.3.

9.3.3. Harmonic method of tide prediction

In principle, this method consists in the summation of the individual partial tides of the form of Eq. (9.14). The water level z at a particular time t is given by

$$z(t) = Z_0 + \sum \{f \cdot H \cdot \cos [V_0 + \sigma t - (\kappa - u)]\}_\nu \qquad (9.15)$$

where Z_0 denotes the mean water level (cf. Section 9.1.2). The index ν indicates the different partial tides. Although the numerical evaluation is simple, it is very tedious. For this reason, Thomson (Lord Kelvin), shortly after the introduction of the harmonic analysis of tide observations, proposed the construction of a tide-predicting machine that should perform the summation automatically by mechanical means. Such a machine was built in 1876. It could sum up 10 astronomical tides. The hydrographic offices of several nations installed improved mechanical tide-predicting machines (the first German machine, introduced in 1916, included 20 tides, the last, of 1938, 62 tides), but at present they are no longer in use. Today, the prediction of tides, though still based on the established harmonic method, is done by electronic computers. The mechanical prediction yields satisfactory results for oceanic coasts, but not for the coasts of large shelf seas, like the North Sea, nor for estuaries. In the open ocean and at its coasts, the relations between the tidal potential and the oceanic tide are approximately linear: the "response" of these tides is nearly linear. In the large shallow seas, however, it is mostly nonlinear. This reflects the great influence of numerous shallow-water tides (cf. Section 9.3.1), the harmonic constants of which are not sufficiently known, and which, in addition, have high angular velocities. At present, to predict tides for these areas, one has to rely on applying the nonharmonic method (cf. Section 9.1.1) (Horn, 1960).

9.3.4. Tide characteristics

The harmonic constants permit us to recognize easily the characteristic features in the temporal course of the tides at a particular location. For locations with semidiurnal tides, the phase of the M_2 tide gives the mean time of arrival of high water after the moon's transit of the local meridian, a value sometimes referred to as "lunitidal interval" or "establishment of the port." According to Table 9.02, the angular velocity of the M_2 tide is smaller by $30.00° - 28.98° = 1.02°$ hr^{-1} than that of the S_2 tide. Hence, the M_2 tide, in its temporal course, falls behind the S_2 tide by $1.02°$ hr^{-1}. If, at any time, both tides are at the same phase, the amplitude H of M_2 is added to the amplitude H of S_2, which results in the range $2(M_2 + S_2)$. This is the time of the spring tide, and the spring range is reached. The process is repeated 14.77 days later. In the meantime the moment occurs when the amplitudes of both tides result in $M_2 - S_2$; this is the neap tide. Consequently, the amplitude of the S_2 tide determines the "semimonthly inequality" mentioned in Section 9.1.1. The phase difference $S_2 - M_2$, divided by the difference between the angular velocities of both tides, represents the spring delay, the *phase age,* in hours. The arrival time of high water of the M_2 tide in the world ocean as well as the spring tidal range $2(M_2 + S_2)$ are given in Plate 7 for a selected number of locations (cf. also Section 9.5.1).

The arrival time of high water for the diurnal tide is determined mainly by the phase of the K_1 tide in conjunction with the phase of the O_1 tide. The K_1 tide, with the O_1 tide not taken into account, gives the approximate time of arrival of high water, local time, on June 21. Each following day, high water occurs 4 min earlier. The O_1 tide is related to the K_1 tide in a similar way as the S_2 tide is to the M_2 tide. It periodically weakens or reinforces the tidal range and thus produces the spring and neap times of the diurnal tide, which follow each other every 13.66 days. The arrival time of high water of the K_1 tide in the world ocean as well as the spring tidal range $2(K_1 + O_1)$ are represented in Plate 8 for the same selected locations as in Plate 7 (cf. also Section 9.5.1). Analogous to the *phase age* of the semidiurnal tide, the phase difference $K_1 - O_1$, divided by the difference between the angular velocities of both tides, gives the spring delay of the diurnal tide after the time of the greatest declination of the moon. The superposition of semidiurnal and diurnal tides results in the *diurnal inequality* already mentioned in Section 9.1.1. Of the two other important inequalities mentioned there, namely the *parallax inequality* and the *declination inequality*, the former is caused mainly by the superposition of the M_2 and N_2 tides, and the latter by the superposition of the M_2 and K_2 tides.

The diurnal inequality becomes the greater, the greater the ratio between the sum of the amplitudes of the diurnal tides $K_1 + O_1$ and the sum of the amplitudes of the semidiurnal tides $M_2 + S_2$. This ratio

$$F = \frac{K_1 + O_1}{M_2 + S_2} \qquad (9.16)$$

referred to as *form number,* indicates the form of the tidal curve during one day. However, a strict classification based on these form numbers cannot be made, for example, in the sense that exclusively semidiurnal tides occur in one area and only diurnal tides in another. This limitation is explained by the fact that the form number is valid only for one instant, that is, for the spring times of semidiurnal and diurnal tides and thus only after syzygies (conjunction and opposition of sun and moon) and after the moon has passed its maximum declination. In the following, the classification of tidal forms will be used, for which Courtier (1938) has given the reasons.

> F is 0 to 0.25: Semidiurnal tide. Two high and two low waters of nearly the same height occur each day. The time of high water follows in nearly constant time interval after the moon's transit of the local meridian. The mean range at spring tide is given by $2(M_2 + S_2)$.
>
> F is 0.25 to 1.5: Mixed, predominantly semidiurnal tide. Two high and two low waters occur each day, but with large inequalities in range and time. They assume their maxima when the declination of the moon has passed its maximum value. The mean spring range is $2(M_2 + S_2)$.
>
> F is 1.5 to 3.0: Mixed, predominantly diurnal tide. At times only one high water occurs each day, namely after the extremes of the moon's declination. Otherwise, there are two high waters each day which, however, show large inequalities in range and time, especially when the moon has passed over the equator. The mean spring range is $2(K_1 + O_1)$.
>
> F is greater than 3.0: Diurnal tide. Only one high water occurs each day. At neap tide, when the moon has crossed the equatorial plane, two high waters may occur. The mean spring range is $2(K_1 + O_1)$.

To illustrate the four different forms of tides, the tidal curves for four seaports, predicted for March 1936, are represented in Fig. 9.06. They are based on the German tide tables for the year 1940, volume II. Immingham, on the English North Sea coast, shows a semidiurnal tidal form with $F = 0.11$; San Francisco has mixed, predominantly semidiurnal tides with $F = 0.90$. Manila exhibits a mixed, predominantly diurnal tidal form with $F = 2.15$, and Do-Son, on the Gulf of Tonking in Indochina, shows a typical form of diurnal tide with $F = 18.9$.

The harmonic constants, from which the tidal curves in Fig. 9.06 were derived, are listed in Table 9.04. κ denotes the phases referred to the local meridians; the amplitudes are given in centimeters. The harmonic constants of the most important partial tides are presented, as they were published by the International Hydrographic Bureau in Monaco (1930–1968). It is obvious to what large extent the four principal tides M_2, S_2, K_1, and O_1 dominate the tidal phenomena. In each of the four seaports, the sum of the amplitudes of these four tides amounts to about 70% of the sum of all amplitudes.

The geographic distribution of the four tidal forms is emphasized in Plate 7 by different colors. This representation is restricted to the areas for which harmonic constants are available. It is conspicuous how much the semidiurnal tidal form prevails in the entire Atlantic Ocean. There, the form number remains considerably below the ratio of the tide-generating forces of the partial tides which amounts to $(K_1 + O_1)/(M_2 + S_2) = 0.68$. The special position of the Atlantic Ocean with respect to tidal forms is based not so much on exceptionally high semidiurnal tides as on unusually small diurnal tides. In contrast, the mixed, predominantly semidiurnal tidal forms prevail at the coasts of the

Pacific Ocean. The tides along the coasts of the Indian Ocean occupy an intermediate position. The mixed, predominantly diurnal tidal form and the pure diurnal tidal form are found only rarely. They are important only in some mediterranean and marginal seas, in particular in the Gulf of Mexico, the South China Sea, the Gulf of Siam, the Gulf of Tonking, in the Borneo and Java Seas as well as in the Sea of Okhotsk and in the Bering Sea.

9.4. Theory of Ocean Tides

9.4.1. The scope of the theory of ocean tides

The determination of the system of tide-producing forces is an astronomical problem that has been solved with greatest accuracy, including all details (cf. Section 9.2.3). The experimental verification on earth has been achieved by measurements with horizontal pendulums and gravimeters, although with considerably less precision than by computation on the basis of astronomical data. The way that the water masses of the world ocean react on the tide-producing forces is also well known from observations along the coasts and partly also on the shelf. The results are manifest primarily in the harmonic constants obtained by analyzing local observations. Matters are quite different with regard to our knowledge on how the tidal amplitudes and phases, with great local differences, are formed along the coasts and how they are distributed in the open ocean. The reaction of an irregularly shaped oceanic area on the system of forces is the main problem of the theory of ocean tides. This is no longer an astronomical problem but a hydrodynamic one.

Since Newton, the theory of ocean tides has attracted great mathematicians like Bernoulli, Laplace, Airy, Poincaré, and others. Fundamental problems have been elucidated. However, in order to overcome the mathematical difficulties, idealized ocean areas of simple geometrical shapes were considered, and the solutions do not apply to the irregularly shaped natural seas. Some results of those classic theories will be dealt with in Section 9.4.3.

In recent decades, significant new attempts have been made to explain the tides of natural oceanic areas. These theories are based on the numerical integration of the hydrodynamic equations and take into account the observations of tides along the coasts as well as of tidal currents at the boundary between a marginal sea and the open ocean. Relevant contributions have chiefly been made by von Sterneck, Defant, Thorade, Proudman, and Hansen. They will be discussed in Section 9.4.4.

9.4.2. Equilibrium theory of tides

The first fundamental investigation of the tides after Newton goes back to Bernoulli (1740). His theory is based on the assumption that the entire earth is covered by water and that, under the effect of the tide-producing forces, the sea surface instantaneously assumes the configuration of a level surface, that is, is always in equilibrium. The height of the sea surface $\bar{\zeta}$ above the undisturbed potential surface of an earth assumed to be rigid, is then given by

$$\bar{\zeta} = V/g \qquad (9.17)$$

where $\bar{\zeta}$ is the height of the so-called equilibrium tide. With the aid of Eq. (9.02) this height can be calculated for moon and sun. For the lunar equilibrium tide, we obtain

$$\bar{\zeta}_M = \frac{1}{2}\frac{M}{E}\left(\frac{R}{r}\right)^3 (3\cos^2\vartheta - 1) \qquad (9.18)$$

The maximum range of the lunar tide, therefore, amounts to 55 cm; for the solar tide it is 24 cm.

Many peculiarities of ocean tides can be explained by the equilibrium theory, above all the periods of the observed water level oscillations and all inequalities contained there (cf. Section 9.1.1). But this theory completely fails in explaining the time of occurrence of high water and the tidal ranges. According to the theory, the high water of the lunar tide at a particular place should occur when the moon passes through the meridian of that location. Actually, observations have shown that the high water arrives at a place with a certain delay after the moon's transit of the meridian, and that this delay may vary considerably from place to place, in some areas by several hours for locations only a very short distance apart. The tidal range, too, is in no way related to the equilibrium tide [Eq. (9.18)]. Observations along the coasts of the world ocean have established that the lunar tide disappears at some places and may reach heights of several meters at others.

The assumption of an earth completely covered by water is less responsible for the disagreement between observation and theory than the fact that the equilibrium position at any instant, required by the theory, cannot be established. To fulfill this requirement, an enormous displacement of water would be necessary, owing to the fast propagation of the equilibrium tide on earth. The establishment of an equilibrium position is never really achieved, at least not for the diurnal tides of the tide potential or tides of even shorter periods. The equilibrium theory may find application only for long-periodic terms (cf. Table 9.02). The main problem with regard to ocean tides is not a static one but a hydrodynamic one, amounting to the problem of water motion under the action of periodic tide-generating forces.

9.4.3. Classic hydrodynamic theories

Newton already recognized that ocean tides must, in essence, be considered a hydrodynamic problem. Merit is due to Laplace (1775) for having derived the equation of tides and presented solutions for the first time. The comprehension of his mathematical treatment is rendered difficult because he connected this problem with other problems of celestial mechanics. Therefore, the reader is referred to the modern representation of Laplace's theory of tides given by Poincaré (1910), who has also extensively described the further development of the theory suggested by mathematicians of the 19th century and by himself. A shorter treatment of the classic theory can be found in the textbook by Lamb (1931).

If the nonlinear terms of the equations of motion as well as the frictional terms are neglected, and if quasistatic conditions are assumed, the equations of tides in spherical coordinates (r = distance from the earth's center, θ = polar distance, λ = geographic longitude, u = east component of the velocity and v = north component of the velocity, ζ = water level) read as follows

$$\frac{\partial u}{\partial t} - 2\Omega v \cos\theta = -\frac{g}{R\sin\theta}\frac{\partial}{\partial\lambda}(\zeta - \bar{\zeta}) \qquad (9.19)$$

$$\frac{\partial v}{\partial t} + 2\Omega u \cos\theta = \frac{g}{R}\frac{\partial}{\partial\theta}(\zeta - \bar{\zeta}) \qquad (9.20)$$

430   Tidal Phenomena

$$\frac{\partial \zeta}{\partial t} + \frac{1}{R \sin \theta} \left[\frac{\partial}{\partial \theta} (-Hv \sin \theta) + \frac{\partial}{\partial \lambda} (Hu) \right] = 0 \qquad (9.21)$$

From these equations an equation for ζ can be derived, which is called the *equation of tides*. For ζ the equilibrium tide has to be inserted. In spite of the factors neglected in these equations, so far nobody has succeeded in giving general analytical solutions. The classic theories of Laplace (1799) and Hough (1898) are based on series expansions of ζ. Laplace expanded the water level ζ into power series; Hough applied spherical functions.

Because the equation of tides represents an equation of oscillation, the height of the water level is mainly determined by the frequency ratio between tide-generating forces and free oscillations. At certain critical depths, the free oscillations are in resonance with the generating forces. Because friction is neglected, the amplitudes then become infinitely large. At the same time, the direct tides turn into reversed tides. For a globe completely covered by water, these critical depths lie at 1965 and 7937 m for the M_2 tide, and at 2248 and 8894 m for the S_2 tide. A reversed M_2 tide and a direct S_2 tide would occur in an ocean covering the whole globe and having a constant depth of, say, 2000 m. For some special cases of depth distribution, analytical solutions can also be found. With diurnal tides, this is possible, that is, for a sea where, in accordance with $H = H_0 \cos^2 \varphi$, the water depth depends on the geographic latitude. The solution contains a critical water depth. Namely, if

$$h_0 < \frac{R^2 \Omega^2}{2g} = 11$$

(h_0 in kilometers, $R = 6383$ km = radius of the earth, $\Omega = 7.29 \times 10^{-5}$ sec^{-1} = angular velocity of the earth, $g = 981$ cm sec^{-2} = gravity), reversed tides result. This implies that, in contrast to the equilibrium theory, a wave trough—and not a wave crest—occurs at that point where the tide-generating celestial body is at the zenith or nadir. This would be the normal case for the oceanic depths commonly found on earth.

For constant depth, diurnal tides disappear as far as the tidal range is concerned; horizontal motions remain. Laplace himself greatly emphasized this result, because it showed that his theory, in contrast to the equilibrium theory, could explain the conspicuously small diurnal tides observed at Brest on the coast of the Atlantic Ocean even at that time. Observations (cf. Section 9.3.4) have confirmed this for the Atlantic Ocean but not for the Indian and Pacific Oceans. However, considering the natural bottom topographies, this fact is more likely to be explained by the unfavorable resonance conditions in the entire Atlantic Ocean. For the depth distribution $H = H_0 \cos^2 \varphi$, an analytical solution can also be found for semidiurnal tides. It equally shows that the tides would be reversed.

It was early recognized that the shape of the natural seas exerts an extraordinary influence on the tides. Hence, a solution of the tidal problems has been sought not only for an earth completely covered with water, but also for basins of geometrically simple shapes. Fundamental in this respect was the *canal theory* by Airy (1842) resulting in forced oscillations for various canals of different position and length, particularly for those open at both ends or at one end as well as those closed at both ends. The restriction to canals implies that traverse oscillations and the deflecting force of the earth's rotation are negligible.

Here only the simplest case may be mentioned, that is, a circular canal going around the earth at the equator. Let the moon's orbit be in the equatorial plane, and let the angular velocity of the moon relative to a point on the earth be n; then solving the equations of motions yields forced waves of the following form

$$\zeta = \frac{1}{2} \frac{c^2 \bar{\zeta}}{c^2 - n^2 R^2} \cos 2 \left[nt + \frac{x}{R} + \epsilon \right] \quad (9.22)$$

where $c = \sqrt{gH}$ and $\bar{\zeta}$ is the range of the equilibrium tide [cf. Eq. (9.17)], x/R is the arc corresponding to the longitude x, counted positively toward the east, R is the radius of the earth, and ϵ represents the zenith distance of the moon for $x = t = 0$. The solution shows that the tide is semidiurnal. If the moon is at the zenith, $nt + x/R + \epsilon = 0$, and ζ is equal to the factor of the angular function in Eq. (9.22). If the denominator $c^2 - n^2 R^2$ is larger than zero, the tide is direct; if it is smaller than zero, the tide is reversed. Accordingly, the calculated tide would always be reversed for the natural depth distribution in the ocean, because the velocity nR, with which a point at the sea surface rotates away from the moon, is always greater than the velocity c of the free wave. This reflects a general physical law for forced oscillations, stating that reversed oscillations will occur whenever the propagation velocity of the forced oscillations is greater than that of the free oscillations.

The further development of the classic theories by Poincaré (1910), Goldsbrough (1915, 1928), and Taylor (1920) has been temporarily concluded in the detailed treatment of Proudman (1936) and Doodson (1936, 1938). Doodson considered an ocean of constant depth, bounded by two meridians separated by a longitudinal distance of 90°. The mathematical effort for solving the equations of motion has grown tremendously. The numerical results can only be understood if presented graphically. Three instructive examples are reproduced in Fig. 9.07, two of which refer to semidiurnal tides by the example of the K_2 tide for constant water depth of 4420 m and 4920 m; a third example demonstrates the behavior of diurnal tides, in this case the K_1 tide for 4420 m depth. The first two examples show, in a remarkable way, how strongly the semidiurnal tides react to a change in depth of only 500 m. In addition to the two amphidromic waves turning to the left, a third develops that turns toward the right, whereby the distribution of amplitudes is modified considerably. The diurnal tide [shown in Fig. 9.07(c)] disappears at the equator for all depths. In the special case presented here for a depth of 4420 m, the maximum range is found at the pole. The values for the range are given as fractions of the maximum range, which is set equal to 1 and which is related in a certain way to the equilibrium tide [cf. Eq. (9.17)].

9.4.4. Co-oscillating tides

The classical theories have provided fundamental insight into the behavior of water masses in response to tide-producing forces in oceans of simple shapes. The application of the results obtained to the explanation of the tides observed in natural oceans is difficult and has remained unsatisfactory in many respects. In contrast, a geophysical concept, including tide observations to a greater extent, but also using the proven hydrodynamic fundamentals, has led to remarkable success.

The decisive impetus came from the research on seiches in lakes (cf. Section 8.4.4) by von Sterneck (1919) and especially by Defant (1919). They showed that oceanic tides in bays and channels have much in common with seiches. The difference between the two is that tides are forced oscillations with periods τ_k determined by the tide-producing forces, whereas seiches are free oscillations with periods τ_f determined by the length and

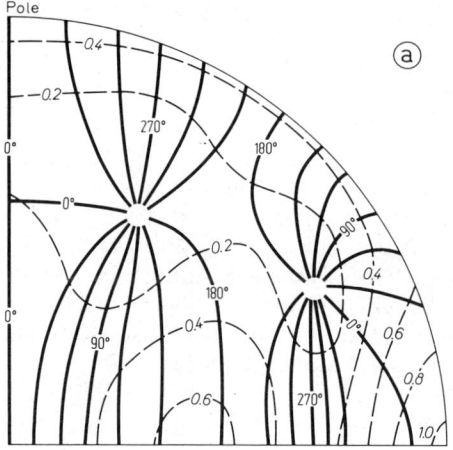
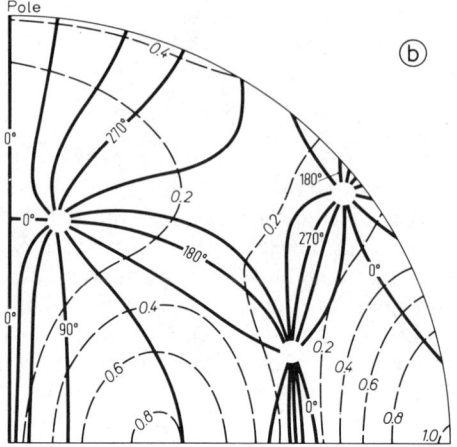
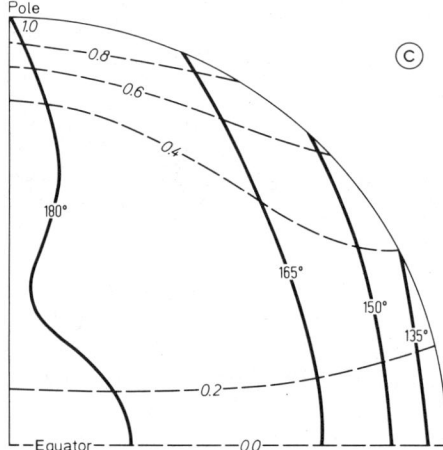

Fig. 9.07. Distribution of phases and ranges of semidiurnal and diurnal tides in an ocean of constant depth, bounded by two meridians (meridional difference 90°). (a) K_2 tide for a depth of 4420 m, (b) K_2 tide for a depth of 4920 m, (c) K_1 tide for a depth of 4420 m. Tidal range is expressed in fractions of the largest occurring range. (After Doodson, 1936, 1938.)

the depth of the basin. This τ_f, however, is decisive for the distribution of amplitudes of the tides in marginal and adjacent seas.

This can be demonstrated quite simply by considering a rectangular ocean bay with the length L and constant depth H. At first, friction and Coriolis force will be neglected. In such a bay, two kinds of forced tidal oscillations can be distinguished, both of which are standing waves resulting from the reflection at the end of the bay. As long as tides are present in the open ocean at the entrance of the bay, the water masses of the bay will be induced to co-oscillate. The period of such co-oscillating tides is necessarily equal to the tidal period $\tau_k = 2\pi/\omega$. The uninodal free oscillation of the bay, however, is $\tau_f = 4L/\sqrt{gH}$, according to Eq. (8.46). The ratio $\tau_f/\tau_k = \nu$ largely determines the horizontal

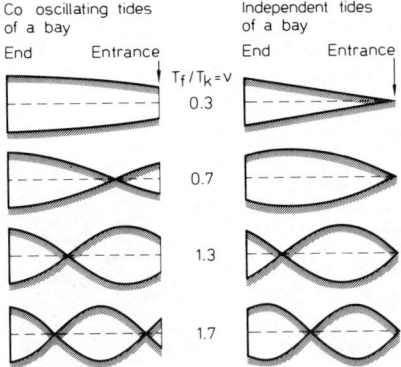

Fig. 9.08. Extreme positions of the sea surface for co-oscillating tides and independent tides in a rectangular bay for different values of ν. (After Defant, 1925.)

and vertical displacement of the water particles ξ and ζ within the bay. According to the derivation by Defant (1925), the equations for the co-oscillating tides can be written as follows

$$\xi = -Z \frac{L}{\nu\pi H} \frac{\sin \nu\pi x'}{\cos \nu\pi} \cos(\omega t + \epsilon) \tag{9.23}$$

$$\zeta = Z \frac{\cos \nu\pi x'}{\cos \nu\pi} \cos(\omega t + \epsilon)$$

where $x' = x/L$ and Z = amplitude of the oceanic tides outside the bay. The equations show that whenever $\nu = 1/2, 3/2, 5/2$, etc., the amplitudes of the co-oscillating tide become infinitely large, that is, resonance occurs. Figure 9.08 presents the extremes of the sea surface for a longitudinal section through the bay, as they are valid for selected values of ν.

Under the action of the periodic force $X = X_0 \cos(\omega t + \epsilon)$, this same bay also has independent tides. The corresponding equations, according to Defant, are

$$\xi = \frac{2X_0}{\sigma^2 \cos \nu\pi} \sin\left(\frac{\nu\pi}{2} x'\right) \sin \nu\pi \left(1 - \frac{x'}{2}\right) \cos(\omega t + \epsilon)$$

$$\zeta = \frac{X_0 L}{g\nu\pi} \frac{\sin \nu\pi(x' - 1)}{\cos \nu\pi} \cos(\omega t + \epsilon) \tag{9.24}$$

For independent tides, too, resonance occurs when $\nu = 1/2, 3/2, 5/2$, etc. In the longitudinal section of the bay in Fig. 9.08, the extreme positions of the sea surface for these independent tides are confronted with the extremes of the co-oscillating tides. The co-oscillating tide always has an antinode at the end of the bay, while the independent tide always has a node at the entrance of the bay. In each case, the position and the number of further nodes depend on ν. In Fig. 9.08 they are indicated for selected values of ν. For a bay with $L = 800$ km, $H = 80$ m, approximately corresponding to the North Sea, we have $\tau_f = 15.9$ and, for semidiurnal tides, $\nu = 1.3$. In such a bay, the independent tides have two nodes, one at the entrance, the other one at a distance of 200 km from the end of the bay. The amplitude is given by the factor

$$\frac{X_0 L}{g\nu\pi} = \frac{X_0}{2g\pi} \tau_k \sqrt{gH} = 0.00273\sqrt{H}$$

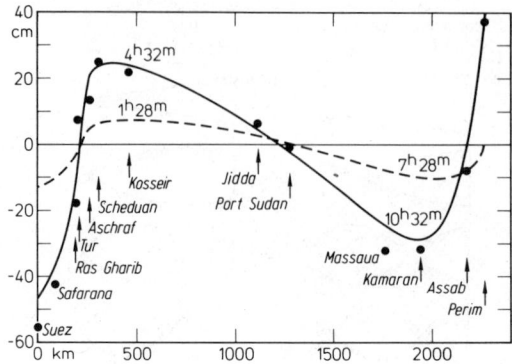

Fig. 9.09. Semidiurnal tides (M_2 tide) of the Red Sea. Dots: observed amplitudes; broken line: independent tide of the Red Sea; solid line: co-oscillating tide with the Indian Ocean; time: phases of independent tide and co-oscillating tide in lunar hr. (After Grace, 1930.)

if $X_0/g = 1.235 \times 10^{-7}$ is used as the maximum value for tides and H is expressed in meters. It follows that, in bays and marginal seas, in addition to independent tides, co-oscillating tides exist that are far greater than the independent tides. In the North Sea, independent tides only reach about 2 cm.

When investigating the tides of bays and marginal seas, one is not restricted to the simple shapes of basins from which the discussion has started. Numerical integration of the equations of motion, as performed in the studies on seiches, also permits the consideration of a complicated bottom topography if independent tides and co-oscillating tides are to be determined. In this way, Defant (1925) succeeded in providing a satisfactory explanation of the tides for a great number of bays and marginal seas.

An example is presented by reproducing the results obtained by Grace (1930) for the Red Sea, which is especially suitable for the application of these methods. The calculated tide of the semidiurnal M_2 tide, co-oscillating with the Indian Ocean, shows three nodes, namely at Assab, Port Sudan, and Tur, as indicated in Fig. 9.09. The amplitude of the co-oscillating tide is about three times as large as that of the independent tide. However, the two tidal components are not at the same phase but have a phase difference of about 3 hr and therefore they cannot be simply added. The dots in Fig. 9.09 indicate the observed amplitudes and demonstrate the good agreement between observations and theory. Results of similar quality have been obtained for a great number of oceanic areas (cf. e.g., Defant, 1961, volume 2).

If the period of the free oscillation τ_f in a marginal sea is known, the ratio $\tau_f/\tau_k = \nu$ serves to give us a rough idea of the distribution of the amplitudes. Some relevant examples are presented in Table 9.05. Relatively high tidal ranges develop for near resonance. The Bay of Fundy, the Aegean Sea, and the English Channel with the Hoofden (eastern approaches to the Channel) are examples of such areas. The highest semidiurnal tides on earth occur in the Bay of Fundy, owing to near resonance and, additionally, to the narrowing of the cross section. The spring tidal range $2(M_2 + S_2)$ amounts to 14.4 m. Although only 42 cm are reached in the Aegean Sea at Saloniki, this represents a remarkable increase in the tidal range compared to that of 6 cm at the southern entrance near the Sporade Islands. In the English Channel, a spring range $2(M_2 + S_2)$ of 8.04 m has been recorded at Dieppe and of 10.58 m at St. Malo as compared to 4.60 m at the Channel entrance.

Table 9.05. Semidiurnal Co-oscillating Tides of Some Bays and Marginal Seas in Longitudinal Direction

Sea Area	τ_f (hr)	$\tau_f/\tau_k = \nu$	Number of Nodes
Bay of Fundy	6.5	0.5	1 ($\approx$ resonance)
Gulf of St. Lawrence	7	0.6	1
Gulf of California	7.7	0.6	1
White Sea	10	0.8	1
Aegean Sea	12.0	1.0	1 ($\approx$ resonance)
North Sea	15.9	1.3	1
Persian Gulf	22.6	1.8	2
English Channel and Hoofden	29.5	2.4	2 ($\approx$ resonance)

The explanation of the tides in bays and marginal seas remains incomplete unless friction and Coriolis force are taken into account. Their influence on long waves has already been discussed in Sections 8.4.3 to 8.4.6. Tidal waves, of course, are subject to the same influence. Friction prevents the occurrence of complete resonance because the effect of friction increases with growing amplitudes and, hence, the period of the free oscillation becomes greater (cf. Section 8.4.6). Thus, resonance conditions are no longer satisfied.

Under the influence of the Coriolis force, progressive tidal waves assume the character of Kelvin waves. Amphidromic systems develop, as illustrated in Fig. 8.42. Therefore, the nodes listed in Table 9.05 for some bays and marginal seas are almost exclusively transformed into amphidromic points.

9.4.5. Numerical integration of the tide equations

In wide oceans, considerable difficulties arise if attempts are made to determine the course of tides in the open ocean merely on the basis of water level observations at the coast, as in Defant's method mentioned above. In shallow offshore areas, namely, tides may be very much disturbed by friction. Since, in the equations of motion, the vertical displacement of water is intimately related to the horizontal displacement, it seemed reasonable to use tidal current observations for the determination of tides in the open sea. Defant (1923) and Thorade (1924) as well as Proudman and Doodson (1924) have proceeded this way.

The modern theory of tides is based on the direct integration of the system of equations, Eqs. (9.19) to (9.21) with additional consideration of the frictional force. This is necessary for the stability of the solutions as well as for the exact determination of amplitudes, because, in many oceanic areas, the periods of the free oscillations are close to those of the tide-producing force. In shallow waters, where the horizontal velocities of tidal currents become large, the nonlinear terms must sometimes also be taken into account. The frictional force is usually assumed to depend linearly on the velocity, in shallow waters a quadratic function is also used, which seems justified by experiments in flumes.

The numerical integration of the equations of tides is mostly carried out as a so-called initial boundary value problem. It is a boundary value problem because at the coasts, solutions must satisfy the kinematic boundary condition, according to which the normal component of the velocity must disappear there. At open boundaries, for example, at the boundary between a marginal sea and the open ocean, either the water level or the tidal currents are assumed to be known (e.g., by observation).

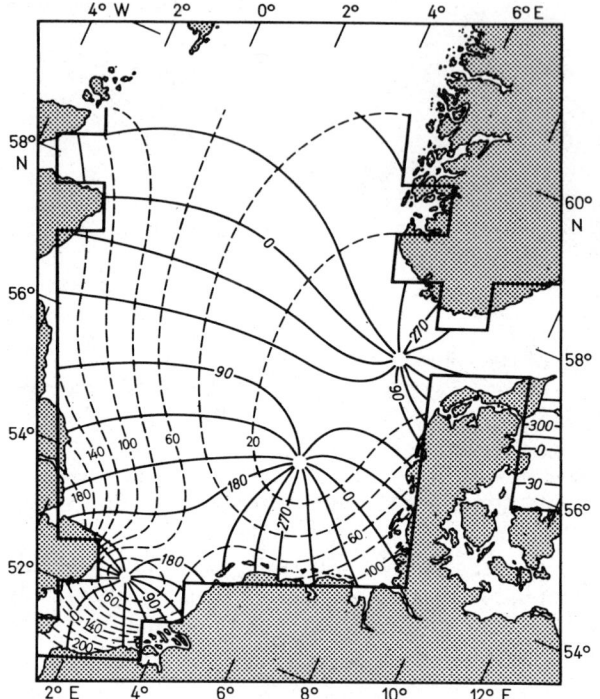

Fig. 9.10. M_2 tide in the North Sea, computed theoretically. Broken lines: amplitude in cm. Solid lines: phase in degrees, referred to the moon's transit of the Greenwich meridian. (After Brettschneider, 1967.)

The system of differential equations is transformed into a system of difference equations, and the oceanic area is covered with a net of grid points at which the unknown values of u, v, and ζ are calculated. If we start from arbitrary initial values of these functions, the values of u, v, and ζ at a later time can be computed from the difference equations. Owing to friction, the influence of the arbitrary initial values fades away after some time, and the solutions depend only on the tidal forces. This method has been systematically applied in particular by Hansen (1948, 1966). Figure 9.10 shows the lines of equal tidal range in centimeters (broken lines) and the cotidal lines in degrees (solid lines), referred to the meridian of Greenwich, for the M_2 co-oscillating tide of the North Sea after Brettschneider (1967). Under the influence of the earth's rotation three amphidromic systems develop: in the Hoofden (southwestern part of the North Sea), in the southern North Sea, and at the entrance of the Skagerrak. The latter is responsible for the co-oscillating tides of the Baltic Sea being so very small.

Another example of a numerical integration of the tide equations is given in Fig. 9.11 showing the distribution of the M_2 tide for the entire world ocean after Pekeris and Accad (1969). The model is based on a one-degree grid, which requires an enormous amount of computer time. But such a grid, when applied to shelf areas, is still too wide-meshed for the calculation of the tides of the world ocean. The dissipation of tidal energy chiefly takes place in shallow waters, that is, on the shelves. The period of the M_2 tide is almost in resonance with the free waves of the world ocean. Hence, the amplitudes of the tides mainly depend on friction. Satisfactory agreement between the theoretical solution and

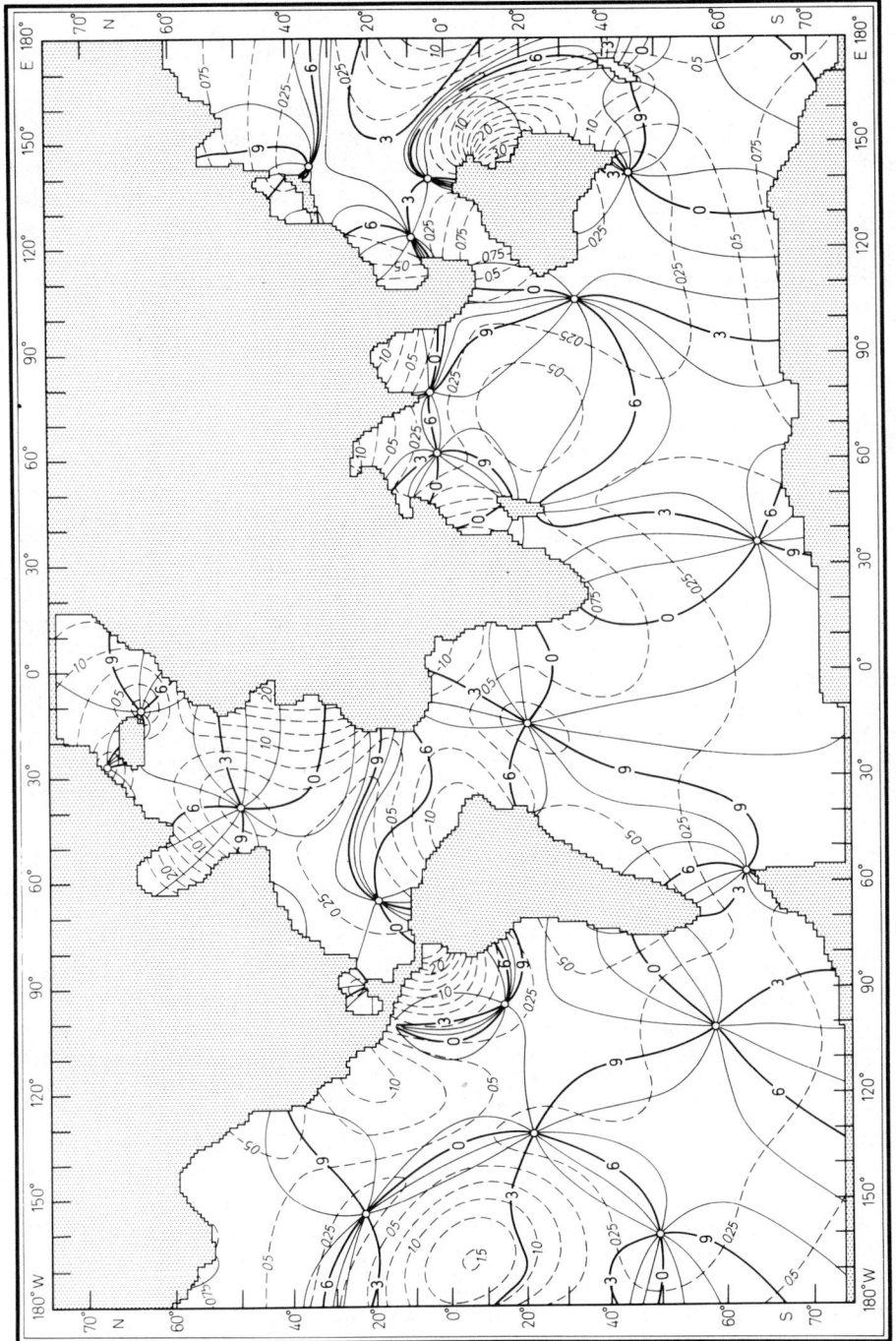

Fig. 9.11. M_2 tide in the oceans, computed theoretically. Broken lines: range in m. Solid lines: phase in hr, referred to the moon's transit of the Greenwich meridian. (After Pekeris and Accad, 1969.)

the distribution of tides observed at the coasts has not yet been achieved, not even with the largest computers available. For such a purpose, we would have to know much more about friction conditions as well as about the transformation of the energy of surface tides into internal tides. Furthermore, the grid must be refined adequately.

Figure 9.11 contains corange lines in meters (broken lines) and cotidal lines (solid lines). The distribution reveals numerous amphidromic systems which had already been gathered from observations along coasts and on islands. But a comparison with the tidal chart (Plate 7) will also show considerable deviations. In general, it is difficult to decide in which particulars the distribution derived from observation is better than that computed theoretically because, due to the transformation of the tides on the shelf, observational data obtained on islands and at coastal stations can differ a great deal from those won in the offshore open ocean. Munk et al. (1970) have shown for the northeastern Pacific Ocean which different results can be attained in detailed representations.

9.5. Tidal Phenomena of the World Ocean

9.5.1. Oceanic tides

If we want to get a survey of the actual tides in the world ocean we must at least know the distribution of phases and amplitudes of a semidiurnal tide and a diurnal tide. Since, in general, the amplitudes and phase differences of semidiurnal as well as of diurnal tides along the coasts of the world ocean have proven to be essentially constant, the fundamental features of the tidal phenomena would thus already be determined. Plates 7 and 8 represent such surveys although merely in part, since they contain only the cotidal lines of the M_2 and the K_1 tides, that is, the time of arrival of high water for the respective tide. The distribution of amplitudes in the entire open world ocean can not yet be represented, although some first steps have been taken (cf. Table 9.06). The information given in Plates 7 and 8 is restricted to the values of the spring tidal range, that is, the residual ranges $2(M_2 + S_2)$ and $2(K_1 + O_1)$ for a small selection of locations.

Plates 7 and 8 are based on the complete harmonic constants of about 3500 locations. They are supplemented by related harmonic constants derived for 3000 other locations, as published in English tide tables. Apart from the Antarctic coasts, these data suffice to consider our knowledge on the course of tides along the coasts of the continents as correct. The harmonic constants of oceanic islands are of particular value for the determination of oceanic tides. Unfortunately, the opportunities for observation on such islands have as yet been used only partly. Wherever this was done, the harmonic constants do not always meet the necessary requirements. Tidal observations in the deep ocean are still more valuable. Here, we are faced with great difficulties regarding the measuring technique. Recently, however, such difficulties have been partly overcome by four different developments, namely by Sutton et al. (1965), Eyries (1968), Snodgrass (1968), and Filloux (1969, 1971). A first series of measurements was provided by Nowroozi et al. (1966) from the Pacific Ocean, 100 nautical mi off the California coast at a water depth of 3900 m. It has been evaluated and published by Fliegel and Nowroozi (1970). Other series of measurements include a profile of four measuring stations perpendicular to the Californian coast. They were obtained by Munk et al. (1970) in the year 1968. The global investigation of oceanic tides can be tackled now and is pursued by an international team under the initiative of Munk (1967). There is hope that it will be possible to represent the course of oceanic tides within the next decades.

The distribution of cotidal lines in the open ocean as given in Plates 7 and 8, is mainly based on a study by Dietrich (1944a). In order to include the areas of the open oceans, from which observations were not available, he has drawn on the improved canal theory by Airy and the theory by Doodson (cf. Section 9.4.3). In the meantime, the harmonic constants of numerous other tide gauges at oceanic coasts have been published by the International Hydrographic Bureau in Monaco (up to 1970). They have been taken into account, without any essential alterations being needed, in Plates 7 and 8. Thus, the concept on the course of tides along the coasts has been confirmed. Since 1944, five other ingenious investigations of oceanic tides have become known: Villain (1952), Hendershott (1966), Bogdanov (1967), Perkeris and Accad (1969), and Zahel (1970). But these studies have not been taken into consideration for the representation in Plates 7 and 8.

From the historical point of view, Dietrich's (1944a) attempt must be considered the fourth, so far, with the aim of representing the tides of the entire world ocean. After the older classic theories could only partly solve the tidal problem, and phases and amplitudes of the actual oceanic tides could not be computed, Whewell (1833, 1836) made the first attempt to explain the progress with time of oceanic tides from observations. In June 1835, following his recommendations, tide observations over a period of three weeks were taken at 666 coastal points along the North Atlantic Ocean. His cartographic representation of the cotidal lines for semidiurnal tides found wide dissemination, although, as early as 1848, Whewell, influenced by Airy's canal theory (1842), gave up his original concept of considering tides as pure progressive waves, and subsequently revoked his chart of oceanic tides.

The next attempt was made by Harris (1904). He could already base his studies on an unobjectionable foundation available in the form of harmonic constants. However, he had data from no more than 183 locations along all oceanic coasts. Harris, in contrast to Whewell, considered the tides as pure standing waves and tried to find oscillating areas in the world ocean, the free period of which was in resonance with the period of the tide-producing force. Subjective interpretation in defining the boundaries of those areas could not be avoided, and Krümmel's criticism (1911) of the oscillating areas as being "a completely arbitrary division of the water masses" of the world ocean seems to be justified.

Von Sterneck (1920, 1921) was the third to attempt a representation of tides for the entire ocean, although harmonic constants of only 204 locations along the oceanic coasts were available to him. In addition to a semidiurnal tide, a diurnal tide was derived for the oceans for the first time. Von Sterneck used a purely formal mathematical approach by regarding the tidal wave as the sum of two standing waves displaced against each other by one quarter of a period. Later on (1926), he tried to provide a physical interpretation for his concept, using the Atlantic Ocean as an example. In one wave he considered the longitudinal oscillation as predominant, which he believed to be induced by co-oscillation with the tide of the Antarctic water belt as well as by the effect of tide-producing forces. The other wave was attributed by him mainly to transverse oscillations.

In the fifth attempt (Villain, 1952), based on the same observational data of 1944 as Plates 7 and 8, these charts were confirmed. A numerical integration of the tidal equations is common to four more recent attempts (from Hendershott to Zahel). In order to achieve this, idealized boundaries must be taken as a basis instead of the natural configuration of the coasts and the actual bottom topography. Nevertheless, a representation of the tides has been derived only by the treatment of the mathematical–hydrodynamic problem, taking into account the tide-producing forces but without using any observational data. This was done most extensively by Pekeris and Accad (1969) (cf. Fig. 9.11) and by Zahel (1970) for the M_2 tide which, in some oceanic areas, shows a surprising agreement with Plate 7. To date, the capacity of even the largest electronic computers will suffice only for a grid with 1° mesh size. If the grid can be made denser and if the consideration of boundary and frictional conditions is improved, in particular in the shelf seas, it will be possible to evaluate a representation permitting to ascertain Plates 7 and 8 better also on the part of theory.

After this short review of the development of the concepts concerning the actual motion of tides let us discuss the tides of the world ocean as they appear in Plates 7 and 8. With few exceptions, semidiurnal and diurnal tides, caused by tide-producing forces, develop as rotational waves or amphidromic systems, resulting from the influences of reflection, Coriolis force, and friction. In the Northern Hemisphere, they generally rotate counterclockwise and in the Southern Hemisphere clockwise. In most cases, the shape of the amphidromic systems is not completely symmetrical. Crowding of cotidal lines occurs and, in extreme cases, the cotidal lines change into nodes. However, exact nodes do not exist in the ocean but rather, owing to the influence of friction, narrow regions where the phase changes rapidly within short distances. The rotating waves of semidiurnal and diurnal tides in the North Atlantic Ocean are shaped symmetrically. For the semidiurnal tide, phase jumps in the form of nodes are found off the Gulf of Bengal, between Japan and New Guinea, and between Tasmania and New Zealand. The amphidromic systems off the Gulf of Panama and near the Solomon Islands almost represent nodes.

The distribution of tidal ranges is closely related to the crowding of cotidal lines. In the centers of the amphidromic regions and at the nodes, the range decreases to nearly zero. With increasing distance from these singular points and lines, the range increases, as confirmed by direct observations at the Micronesian Islands. In the immediate vicinity of the node of the semidiurnal tide between Japan and New Guinea, the smallest spring tidal range known so far amounts to only 16 cm; toward the east of this node, it increases with growing distance from the node to 44, 100, 120, and 158 cm. Corresponding ranges are found to the west of the node: 60, 89, 116, 136, 152, and up to 176 cm. The peculiarity of the distribution of the tidal range within the center of an amphidromic region becomes evident if it happens to lie near an island. At the Solomon Islands, the spring range amounts only to 12 cm, at Puerto Rico only to 3 cm, at New Zealand (for the diurnal tide) only to 10 cm. But, since the amphidromic points are generally located in the open ocean, the highest values of the tidal range are always to be expected along the coasts of the continents. Nodes, however, may reach the coast and cause the tidal range to practically disappear. Thus the range of the spring tide of the semidiurnal tide at the coast of southern Brazil decreases to 16 cm, and of northwestern Sumatra to 18 cm.

The results of the series of measurements at the deep-sea floor, extending along a profile perpendicular to the coast for more than 1000 km into the ocean, have been compiled in Table 9.06. This is the very first profile of this kind which shows how the tides change when approaching the continent from the deep sea. The respective current measurements are also available. According to these measurements, the spring tidal range of the semidiurnal tides $2(M_2 + S_2)$, amounting to 114.7 cm off La Jolla, decreases along the four observation points toward the open ocean to 120.4 to 117.4 to 84.8 to 61.0 cm. For the diurnal tides $2(K_1 + O_1)$ the corresponding spring ranges are 112.4 cm at La Jolla and 105.5 to 109.6 to 94.8 to 86.8 cm on the deep sea. These results reveal that at coasts without any pronounced shelf, like those off southern California, oceanic tides are hardly modified—in contrast to shelf coasts (cf. Section 9.5.2). Towards the open

Table 9.06. Results of the First Recorded Profile of Oceanic Tides[a,b]

Location	Position φ (N)	Position λ (W)	Depth (m)	Tides O_1 H (cm)	O_1 G (°)	K_1 H (cm)	K_1 G (°)	M_2 H (cm)	M_2 G (°)	S_2 H (cm)	S_2 G (°)
La Jolla, off Scripps Pier											
"Josie"	31°02′	119°48′	3640	21.8	192.0	34.4	206.8	51.0	142.7	21.3	137.6
"Flicki"	32°14′	120°51′	3700	20.1	195.2	32.7	208.7	42.6	142.1	17.6	135.3
"Kathy"	27°45′	124°26′	4200	22.0	197.0	32.8	217.0	42.5	149.6	16.4	145.0
"Filloux"	24°47′	129°01′	4400	17.5	199.0	29.9	212.9	28.6	128.0	13.8	115.8
				15.6	201.3	27.8	222.3	18.8	107.1	11.7	94.9

[a] By Munk et al. (1970).
[b] Amplitudes H are represented in cm, phases G in degrees, referred to Greenwich, as evaluated from the records of deep-sea gauges in 1968, distributed on a profile as far as 750 nautical mi seaward off Southern California.

ocean, the amplitudes decrease to e^{-1} of the amplitude at the coast, as shown by Munk et al. (1970). With regard to phases, the data in Table 9.06 can be compared directly with those in Plates 7 and 8 if expressed in hours. According to the table, the phase value at La Jolla is 4.9 hr and for those four deep-sea sites they are 4.9, 5.2, 4.4, and 3.3 hr. In Dietrich's charts (1944) the phases for the four deep-sea points are 4.9, 5.1, 3.8, and 2.8 hr. There is some little difference only in the latter two values. This has been taken into account by shifting the cotidal lines for 3 and 4 hr in Plate 7 slightly toward the west. This correction does not cause any change in the system of oscillations. Even better agreement is found for the phases of the K_1 tide: for La Jolla the phase is 13.7 hr and for the four deep-sea points the corresponding values are 13.9, 14.4, 14.2, and 14.8 hr. According to Dietrich (1944b) the values are, in good agreement, 13.8, 14.2, 14.2, and 14.6 hr.

9.5.2. Tides of adjacent seas

The relations between the distribution of ranges and phases are valid also for mediterranean, marginal, and adjacent seas. But owing to the small scale of Plates 7 and 8 they cannot be represented there. With regard to the North Sea, the Baltic Sea, and the East China Sea the reader is referred to Figs. 9.12 to 9.14. A summarizing treatment of the tides in adjacent seas has been given, in particular, by Defant (1961) who investigated a great number of adjacent seas already earlier (1925). Further detailed studies on the North Sea have been supplied by Sager (1964) and, on the Gulf of St. Lawrence, by Farquarson (1970). On broad shelf areas, the connection between the distribution of tidal ranges and the oceanic oscillation system is partially obscured and, in mediterranean and marginal seas, it is even completely unrecognizable. This is effected by two processes.

1. The range of a tidal wave greatly increases when the wave approaches shallow water, especially if, in addition, the coastal outline is funnel shaped. From the principle of conservation of energy it follows that the tidal range grows approximately inversely to the fourth root of the water depth and inversely to the square root of the width. Reflection and friction counteract any increase. Examples of this phenomenon can be found at many coasts. It is particularly conspicuous for the semidiurnal tide in the Bristol Channel (spring tidal range at the entrance near the Scilly Islands 452 cm, in the interior near Bristol 1147 cm), in the Bay of St. Malo (off the entrance at Alderney 530 cm, within the Bay at Granville 1158 cm), and in the Gulf of Cambay in the Arabian Sea (at the entrance at Diu Head 140 cm, in the interior near Bhaunagar 887 cm).
2. As soon as the free period of the water masses over the shelf approaches the tidal period, resonance occurs. If the free period is larger than the tidal period, some nodes and amphidromic systems are formed on the shelf. Such co-oscillating tides, with one or several nodes or amphidromic systems, respectively, are listed in Table 9.05. They are found in all mediterranean, marginal, and adjacent seas, and are particularly numerous in the East China Sea, as shown in Fig. 9.14. They also occur on broad oceanic shelves: on the Patagonian shelf with two amphidromic points for the semidiurnal tides and one amphidromic point for the diurnal tide, off the Gulf of Maine with a nodal line for the semidiurnal tide, and on the shelf of Nova Scotia with one amphidromic point for the diurnal tide.

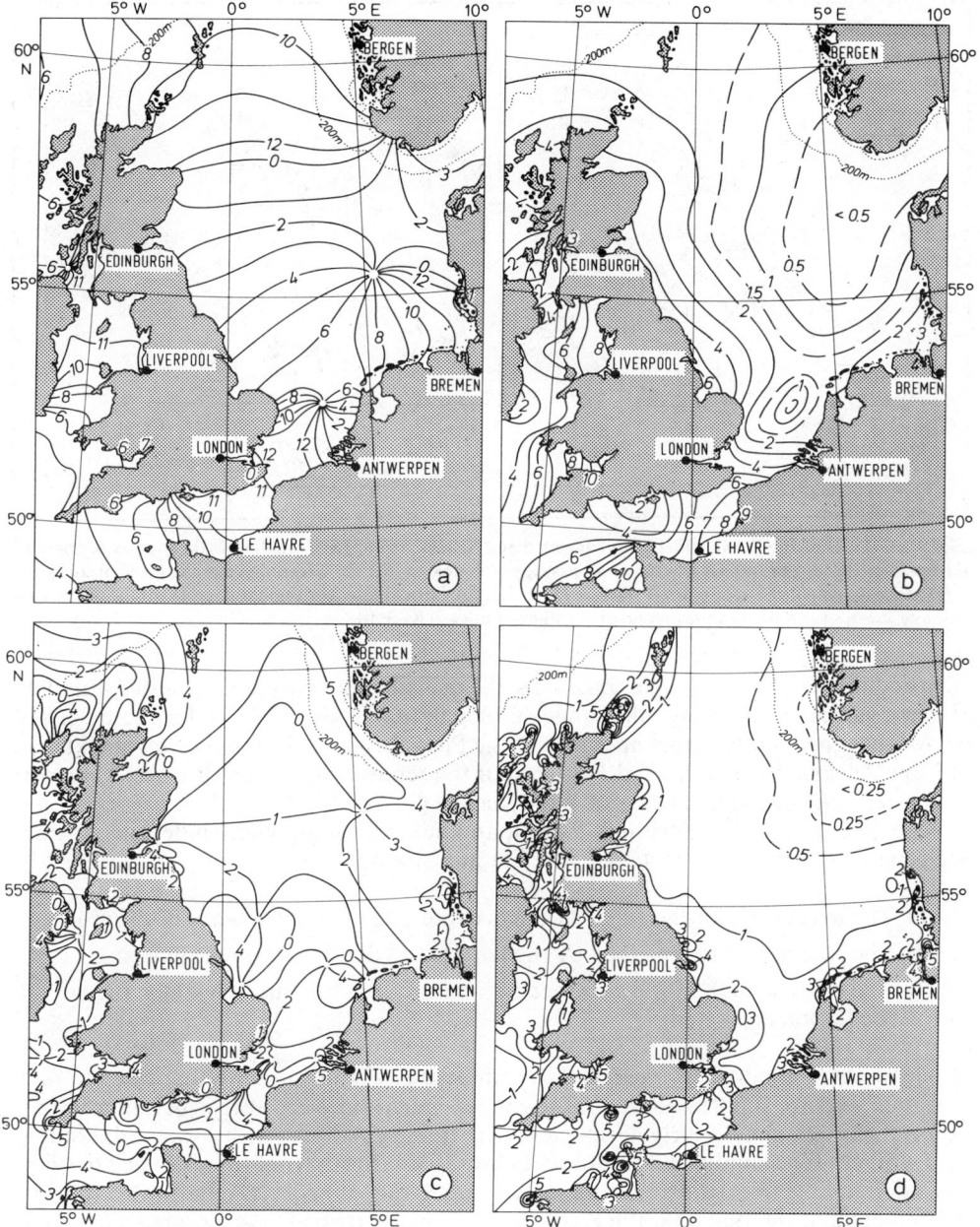

Fig. 9.12. Tides and tidal currents of the North Sea. (a) Cotidal lines of the semidiurnal M_2 tide, referred to the mean passage of the moon over the Greenwich meridian (after German Hydrogr. Institute, 1970). (b) Mean spring tidal range $2(M_2 + S_2)$ in m (after German Hydrogr. Institute, 1970). (c) Lines of equal time of occurrence of the maximum tidal current (in hr) during mean spring tide, referred to the moon's mean passage over the Greenwich meridian (after Sager, 1967). (d) Mean maximum velocity (in knots) of the tidal current at spring tide (after Sager, 1967).

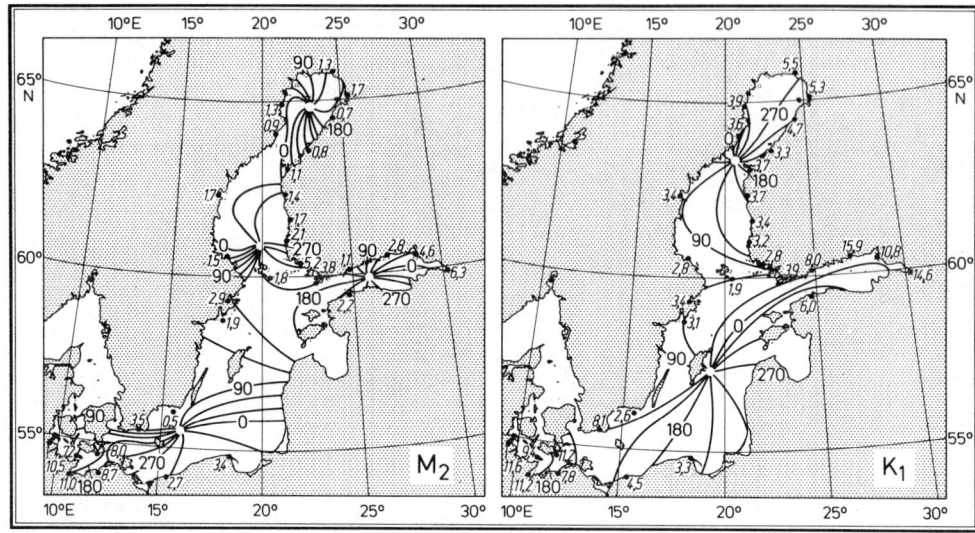

Fig. 9.13. Tides of the Baltic Sea. (After Magaard and Krauss, 1966.) (a) Cotidal lines (in angular degrees) of the semidiurnal M_2 tide, referred to the mean passage of the moon over the Greenwich meridian, and mean spring tidal range (in cm) of the semidiurnal tides $2(M_2 + S_2)$. (b) Cotidal lines (in angular degrees) of the diurnal K_1 tide and mean spring tidal range (in cm) of the diurnal tides $2(K_1 + O_1)$.

If, in addition to good resonance conditions, the rising bottom creates a favorable situation, extreme values of the spring tidal range develop in funnel-shaped bays; for example, 814 cm for the diurnal tide in the inner part of the Penzhinskaya Gulf in the Sea of Okhotsk at Cape Astronomitscheski, and 1414 cm for the semidiurnal tide at Burncoat Head at the eastern end of the inner Bay of Fundy in the Gulf of Maine.

For tides in mediterranean, marginal, and adjacent seas, where independent tides are generally of minor importance, three factors are decisive.

1. The magnitude of the range of the oceanic tide off the entrance of the sea.
2. The free period determined by the length and depth of the marginal sea.
3. The cross section of the entrance to the marginal sea through which the co-oscillation is excited.

If one of these conditions is particularly unfavorable, the co-oscillating tides of such marginal seas are suppressed. Since their independent tides are always of minor importance, these seas show poorly developed tides. If marginal seas with spring tidal ranges below 50 cm for semidiurnal tides and below 20 cm for diurnal tides are classified as having negligible tides, the following seas belong to this category: the Baltic Sea (cf. Fig. 9.13) including the Belt Sea, the Kattegat, and the Skagerrak, the European Mediterranean Sea (with the exception of the northern Adriatic Sea and the northern Aegean Sea), the Sea of Japan, and the Arctic Ocean (including the north Siberian marginal seas). In the American Mediterranean Sea it is only the semidiurnal tide that is poorly developed, especially in the Gulf of Mexico (Zetler and Hansen, 1970).

The reason for the small tides in the European Mediterranean Sea can be found in the fact that the entrance through the Strait of Gibraltar is rather narrow. In the

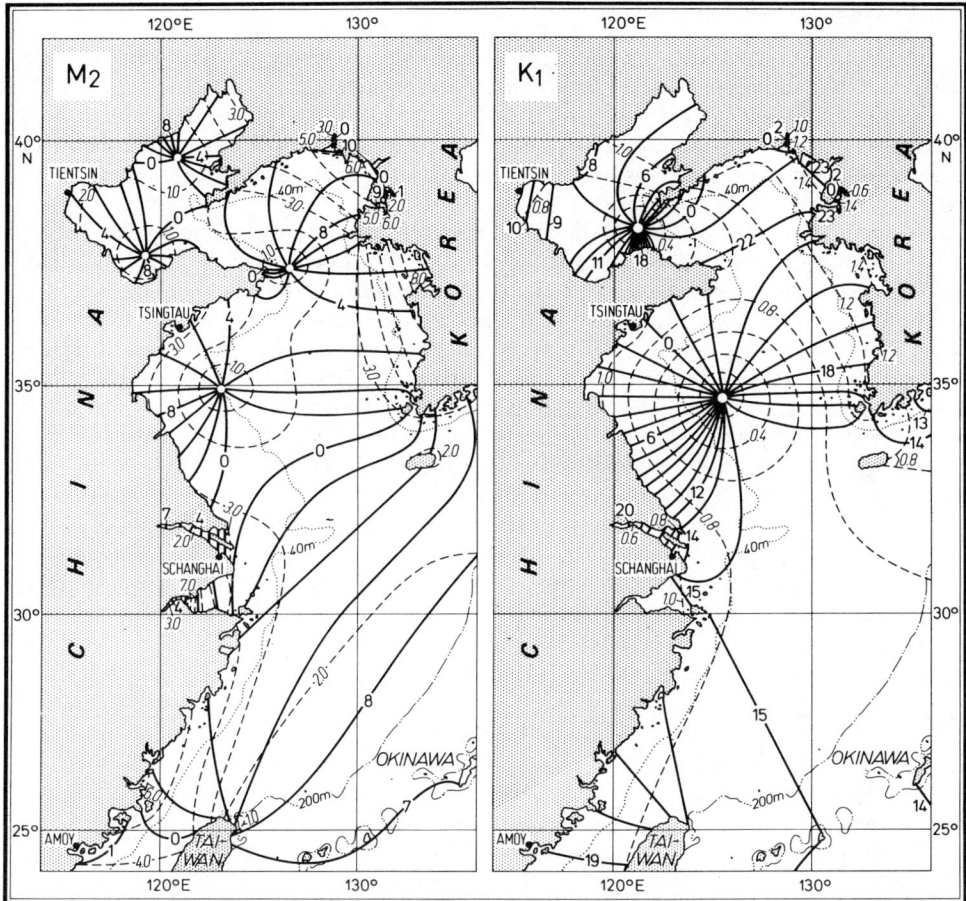

Fig. 9.14. Tides of the East China Sea. (After Ogura, 1933.) Left-hand side: cotidal lines of the M_2 tide (referred to 135°E) and spring tidal range (in m): $2(M_2 + S_2)$. Right-hand side: cotidal lines of the K_1 tide (referred to 135°E) and spring tidal range (in m): $2(K_1 + O_1)$.

American Mediterranean Sea (only with regard to semidiurnal tides), and in the Skagerrak (cf. Fig. 9.12), and the Kattegat, strong co-oscillating tides cannot be excited because of the small tidal range off the wide entrances. In the Baltic Sea (cf. Fig. 9.13), both factors are effective, namely, the slight excitation from the Kattegat and the narrow cross sections of the Belts and the Sund. The small tides of the Japan Sea can also be attributed to the narrow passages preventing the penetration of sufficient tidal energy to excite strong co-oscillation. In the Arctic Ocean, wide passages and the moderate semidiurnal tides of the Norwegian Sea and the Barents Sea should favor the excitation of co-oscillation. It still remains an unsolved question to what extent unfavorable resonance conditions and decay of tidal energy resulting from friction under the pack ice sheet contribute to the small tides in this ocean. In the shallow marginal seas off northern Siberia, the small tides can be explained by the weak excitation from the Arctic Ocean and by the strong friction under the ice. This was shown especially by the result of the *Maud* expedition (Sverdrup, 1926).

9.5.3. Tidal currents, in general

Tides and tidal currents are two manifestations of the same process. They are most closely coupled. Observations of tidal currents collected with current meters (cf. Section 3.2.12.1) require much greater effort and costs than those of tides. Therefore, the number of observations in the world ocean is relatively small. Exceptions are the North Sea, the British waters, and numerous estuaries, because there, tidal currents are of great practical interest to shipping and coastal engineering.

Such current observations, especially the few extending over longer time periods, suggest that tidal currents no more repeat themselves at one place than tides do. In addition, they differ from place to place. The same inequalities (cf. Section 9.1.1) encountered with tides are also found in tidal currents. However, the phenomena are yet more complicated, because tidal currents generally represent a three-dimensional process, whereas tides are two-dimensional. Only in exceptional cases do tidal currents become two-dimensional, namely in rivers and in the immediate vicinity of the coast where the bottom topography permits only alternating motion. In the open sea, the current direction, in general, changes simultaneously with the current velocity. Hence, the current figures, described by the tip of the current vector, deviate from a straight line.

Tidal currents, like tides, can be subjected to harmonic analysis. This is done most conveniently by separating the current vector into two perpendicular components, for example, a north and an east component, and by subjecting each component separately to harmonic analysis. Table 9.07 gives an example of the harmonic constants of a tidal current as found in the water of North Australia. Then, the harmonic constants of tidal currents, like those of tides, may be used for prediction purposes in accordance with the harmonic method. The tide form $F = (K_1 + O_1):(M_2 + S_2)$, in most cases, corresponds approximately to the same tidal current form. Hence, the distribution of tidal forms in Plate 7 also holds for tidal currents.

Predictions of tidal currents indicate that, in general, the current figure is an ellipse only in areas with pure semidiurnal or diurnal tides. The magnitude and ratio of the axes of this ellipse vary from tide to tide in accordance with the inequalities. In areas with mixed tides, the tidal currents become very complicated. An example of such a case is given in Fig. 9.15 showing the current figures predicted for the time period from March 1 to 15, 1936, on the basis of the harmonic constants in Table 9.07. The tip of the current vector describes the hodograph in the direction of the little arrows. The position of the current vector for each full hour is given by the points, and the origin of the vector is always located at the center of the coordinate system. The configuration of the rather complicated current figures changes from day to day. Only after half a month does the figure become similar to that of the first day.

Even for simple elliptic tidal currents, one cannot immediately derive such currents from common tidal charts giving range and phase distributions. Maximum velocity and time of occurrence of a tidal current do not depend on range and phase of the tide alone, but, among other things, also on the water depth and the Coriolis force. If progressive waves—often resembling tidal waves in the world ocean—are assumed, the important quantity, the maximum velocity and, thus, the maximum displacement of a water particle within half a tidal period, can be calculated from the amplitude ζ_0 of the tide. If ζ denotes the vertical displacement and u the tidal current velocity in the direction of propagation of the wave, the following equation holds for a progressive wave (as in Section 8.2.1):

$$\zeta = \zeta_0 \cos(\omega t - \kappa x) \tag{9.25}$$

Table 9.07. Harmonic Constants of Tidal Currents at Proudfoot Shoal ($\varphi = 10°31'S$, $\lambda = 141°29'E$)[a]

		Harmonic Constants								
		M_2	S_2	N_2	K_2	K_1	O_1	P_1	M_4	MS_4
North component	H (nautical mi hr^{-1})	0.24	0.04	0.05	0.01	0.24	0.17	0.08	0.01	0.02
	g (°)	45	90	346	90	67	335	67	158	110
East component	H (nautical mi hr^{-1})	0.48	0.25	0.10	0.06	0.40	0.25	0.13	0.01	0.01
	g (°)	61	189	353	189	127	61	127	304	247

[a] H: Amplitude in nautical mi hr^{-1}, g: phases in degrees referred to the time meridian 150°E.

Fig. 9.15. Daily current figures of the mixed, predominantly semidiurnal tidal currents of Proudfoot Shoal, Strait of Torres (10°31'S, 141°29'E) between March 1 and March 15, 1936. nm h^{-1}: velocity in nautical mi hr^{-1}. (According to Horn, 1952.)

From the equations of motion and continuity it follows that

$$u = \sqrt{\frac{g}{H}}\, \zeta_0 \cos(\omega t - \kappa x) \tag{9.26}$$

Hence, the maximum tidal current U is determined by

$$U = \sqrt{\frac{g}{H}}\, \zeta_0 \tag{9.27}$$

Then the horizontal displacement ξ_m of a water particle during half a tidal period is

$$\xi_m = U \int_0^\pi \cos \omega t\, dt = 2U\frac{1}{\omega} = U\frac{T}{\pi} \tag{9.28}$$

To obtain a quantitative idea of the magnitudes of U and ξ_m in the world ocean at different water depths, let us assume $\zeta_0 = 100$ cm. Then, for a water depth $H = 10$ m, the maximum tidal current $U = 99.0$ cm sec^{-1}; for $H = 100$ m, we obtain $U = 31.3$ cm sec^{-1}; for $H = 1000$ m, $U = 9.9$ cm sec^{-1}; and for $H = 4000$ m, $U = 4.9$ cm sec^{-1}. The corresponding displacements amount to $\xi_m = 7.1, 4.4, 1.4,$ and 0.7 km, respectively. This estimate shows that, in oceans with appreciable tides, the velocity of tidal currents generally exceeds that of all other currents. This also holds true for the open ocean.

9.5.4. Tidal currents of the oceans

It should be noted that uniform conditions with respect to tidal current ellipses do not exist in the vertical at all, as it might be expected for a simple long wave. Significant modifications are caused by bottom topography and stratification. Examples are given in Figs. 9.16 and 9.17.

Oceanic tidal currents result from superposition of barotropic and baroclinic tides. While barotropic tides are modified only by friction, several factors seem to play a role with baroclinic tides. To date, a final answer to this question has not yet been found.

9.5.5. Tidal currents of adjacent seas

The small velocities of tidal currents over oceanic depths may increase wherever the cross section narrows, either because the depth decreases from the open ocean toward the shelf area (Fig. 9.17) as well as over submarine ridges and seamounts, as over the Great Meteor Bank (Fig. 9.16), or because of lateral constraint in oceanic straits, as in the example of the Strait of Gibraltar (Fig. 10.21) or of Bab al Mandab (Fig. 10.24). Conditions of tidal currents in straits without any stratification are relatively simple to recognize. The

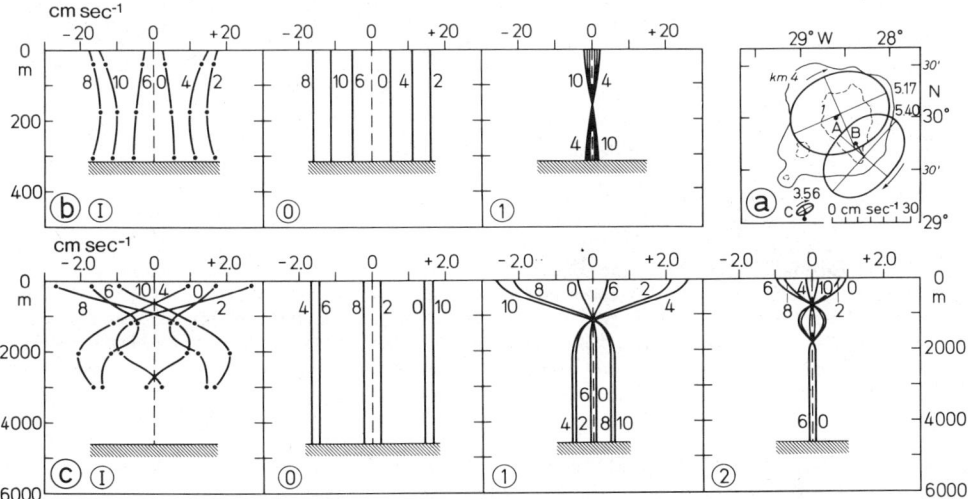

Fig. 9.16. Examples of semidiurnal tidal currents in the eastern North Atlantic Ocean in the area of the Great Meteor Seamount. (After Meincke, 1971.) (Cf. also Figs. 1.06 and 10.14) (a) A, B, C: positions of simultaneous current measurements by research vessel *Meteor* at 3 to 4 different depth levels from April 13 to 27, 1967. Tidal current ellipses of the M_2 surface tide with sense of rotation and time of occurrence of the maximum tidal current in lunar hr after the moon's passage over Greenwich. Residual current: thick arrow between station point and intersection of the axes of the ellipse. (b) Vertical profiles of the north component of the M_2 tidal current at position A. I. Observed current profiles according to harmonic analysis. 0–1: separation of I into internal tidal waves of zero order (surface tide) and first order. (c) Vertical profiles of the north component of the M_2 tidal current at position C. I. Observed current profiles according to harmonic analysis. 0–2: separation of I into internal tidal waves of zero order (surface tide) and first and second order.

450 Tidal Phenomena

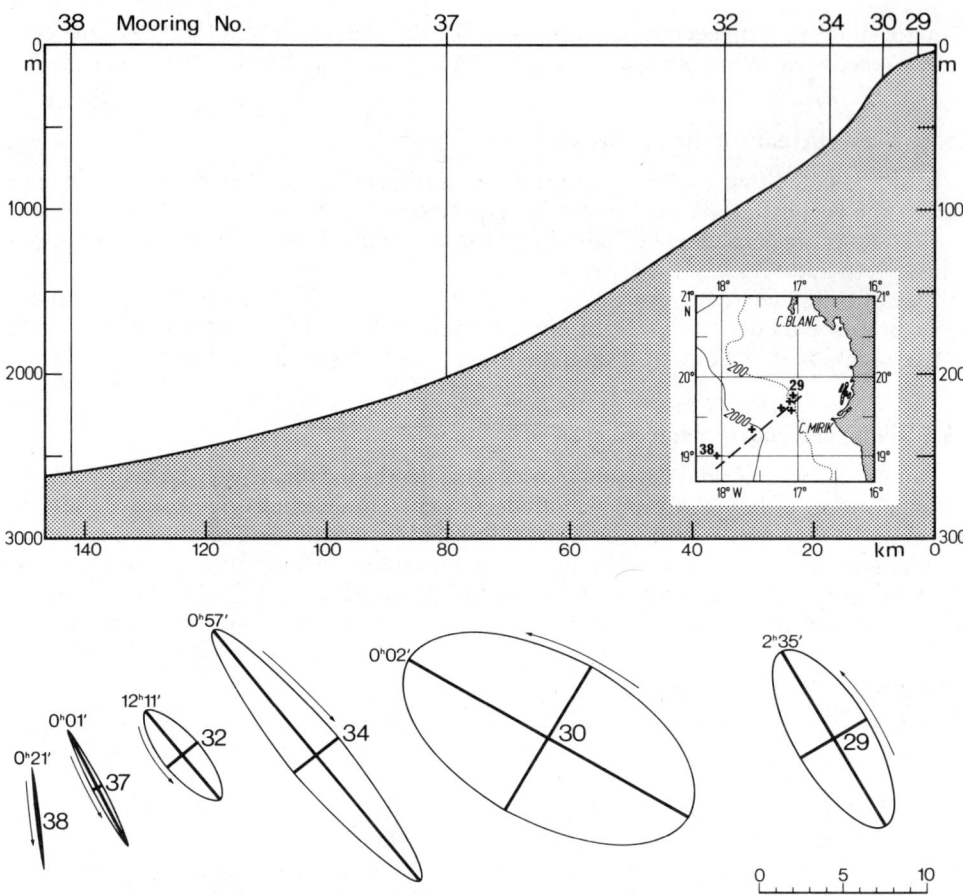

Fig. 9.17. Current ellipses of the M_2 surface tide along a section perpendicular to the continental slope off NW Africa. Result of simultaneous current measurements within the period from 26 February to 13 March, 1972 at three depth levels (Station 29), four depth levels (Stat. 30), seven depth levels (Stat. 34), and six depth levels (Stats. 32, 37, 38). The analysis was carried out according to Fig. 9.16. Sense of rotation of the ellipse and time of occurrence of the maximum tidal current in lunar hr after the moon's passage over Greenwich are given. (After Meincke et al., 1975.)

maximum values of tidal currents are obtained in the following way: let S be the cross section of the strait, A the area of the bay separated by the strait, and $2\zeta_0$ the mean tidal range in the bay. Then the water volume flowing through the strait into the bay during half a tidal period, that is, from low water to high water, is given by $2\zeta_0 A = S\bar{u}T/2$, if $\bar{u}$ stands for the mean velocity of the tidal current in the strait during half a tidal period. Since the maximum velocity of the tidal current is $U = \bar{u}\pi/2$, it follows that

$$U = \frac{A\pi}{ST} 2\zeta_0 \qquad (9.29)$$

For a strait of 100 m width and 30 m depth, with $A = 250$ km² and $2\zeta_0 = 2.0$ m, U is equal to 7.5 m sec⁻¹ for the semidiurnal tide. These numerical values reflect the conditions in the Skjerstad Fjord at Bodø in northern Norway. According to observations, the velocities in the narrows between Skjerstad Fjord and Salt Fjord, in the so-called Saltstraumen, reach $U = 8$ m sec^{-1}. They probably are among the highest current velocities in the world ocean. The water shoots jetlike through the narrow of the fjord, and intense eddies with deep whirlpools accompany the current along its sides. This tidal current represents an impressive natural phenomenon frequently visited by tourists traveling in northern Norway.

The stronger tidal currents are of special interest to the shipping industry. They influence the speed of the ship relative to the ground. Therefore, atlases of tidal currents have been compiled, for example, for the North Sea, the English Channel, and the waters west of Great Britain, which are kept up to date by the German Hydrographic Institute. The representation of the mean maximum tidal currents at spring tide in Fig. 9.12(d) is based on such material. After this, the tidal currents in the North Sea reach up to 5 knots near the coast. In the major part of the open North Sea, they are below 1 knot, and in the northeastern part below 0.5 knot.

Strong tidal currents are not without any danger to small craft, namely when high, wind-generated waves run in opposite direction to the tidal current. Tidal current and orbital motion superimpose each other; thus, the wavelengths of the surface waves become shorter, the waves steeper, and the wave crests sharper. Small vessels may literally be rolled over by the sea before they can rise from a wave trough to the next wave crest. These combined effects of tidal current and sea may be the explanation of the frequently cited maelstrom, which is identical with the "Mosken Straumen" running along Moskenesøy Island off the outer Lofoten. When the water of the West Fjord ebbs seaward in westerly direction during half a tidal period (at spring tide, speeds of up to 4.5 m sec^{-1} have been recorded), and when, at the same time, a strong westerly gale is blowing, short, breaking waves develop, greatly endangering small fishing vessels. Similar conditions are encountered in the Pentland Firth, north of Scotland, and between the Channel island Alderney and Cape de la Hague in Normandy. In both cases, the spring current also amounts to 4.5 m sec^{-1}.

Enormous energies are associated with tidal currents. Between high water and low water they perform work amounting to

$$E = mgh = 8g\rho A \zeta_0^2 / T \qquad (9.30)$$

Applied to our schematic example reflecting the dimensions of the Skjerstad Fjord, a power of 600,000 hp is obtained. So it is not surprising that for a long time, attempts have been made to utilize the energy of tidal currents for technical purposes. As early as in the Middle Ages, so-called flood mills operated the wheels of grain mills, gypsum mills, and sawmills in some English rivers, where tides are effective. Such flood mills functioned as water mills and were anchored in the tidal current. The first large-scale tidal power plant was built in the Bay of St. Malo, France, in the estuary of the river Rance and set in operation in 1966. With 24 turbines utilizing ebb and flood currents, it annually supplies 544×10^6 kWh, which corresponds to three times the capacity of the Walchensee power plant in Bavaria. Similar power stations are being planned in several countries (Sager, 1968a).

In the stratified ocean, tidal currents are rather complicated since the barotropic tide, which would also occur in a homogeneous ocean, can be superimposed additionally by internal tidal waves of various orders, that is, by baroclinic tides. Hence, due to the influence of stratification, tidal currents can greatly vary within small vertical distances as well as in time, with the depth of observation remaining constant. This may be illus-

452 Tidal Phenomena

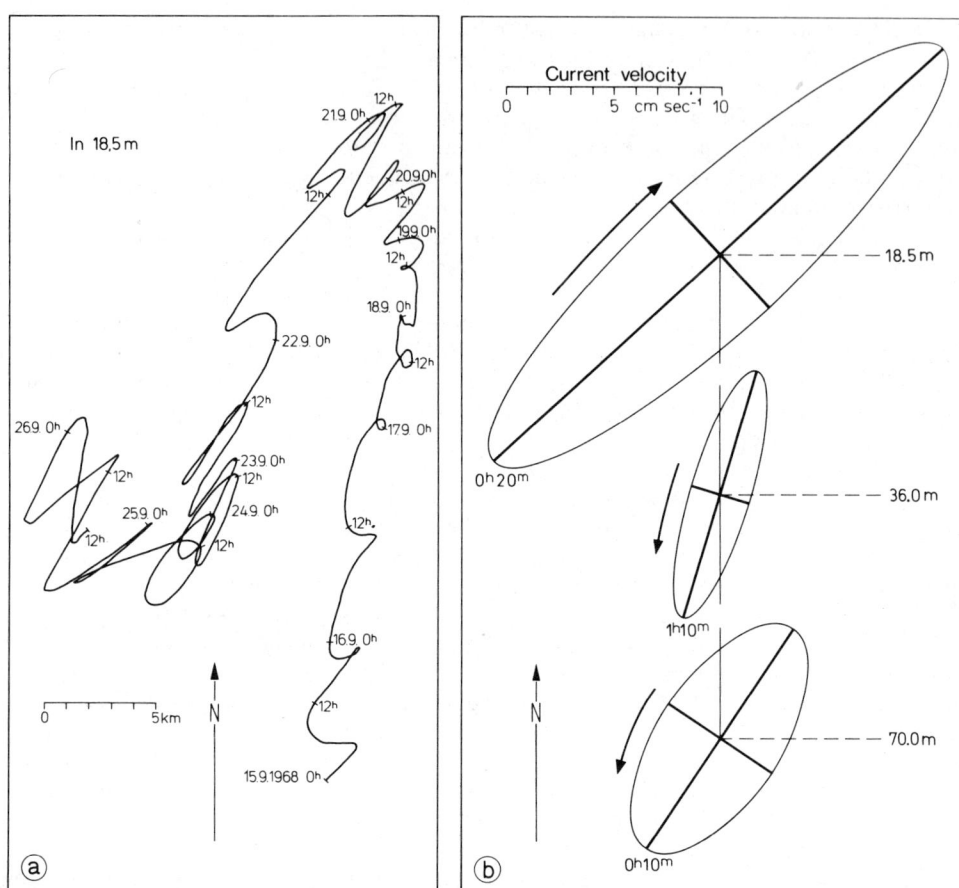

Fig. 9.18. Tidal currents in a stratified sea in the example of the North Sea. (After Schott, 1972.) Current meter mooring at $\varphi = 56°20'N$, $\lambda = 1°0'E$ from September 14 to 26, 1968, water depth 82 m. Pronounced thermocline at 35 m (cf. Fig. 10.57). (a) Progressive vector diagram from September 15, 0^h to September 26, 12^h according to hourly averaged current measurements in the top layer (recording depth 18.5 m). (b) Tidal current ellipses from records obtained from September 14, 15^h to September 16, 16^h40^{min}, 1968, at 18.5 m, 36 m, and 70 m depth; neap tide September 16, 1968, 9^h32^{min} GMT.

trated by an example from the northwestern North Sea, obtained by the research cutter *Alkor* with moored current meters. A very strongly pronounced thermocline was found at 35 m depth, as shown in Fig. 10.57 for a section. The result of the current measurements at 18.5 m depth from September 15 to 26, 1968 is represented in Fig. 9.18(a); Figure 9.18(b) shows the tidal current ellipses for 18.5, 36, and 70 m depth, evaluated from the first section of current measurements which included four tidal cycles. Besides, the semidiurnal tidal currents are superimposed by a residual current, which is evident from the spiral path of the water particles shown in Fig. 9.18(a). What distinguishes these results, obtained in a stratified sea with baroclinic tides, from those in an unstratified sea with barotropic tides, are the following observational facts.

1. The sense of rotation of the current curve at a fixed observational depth varies

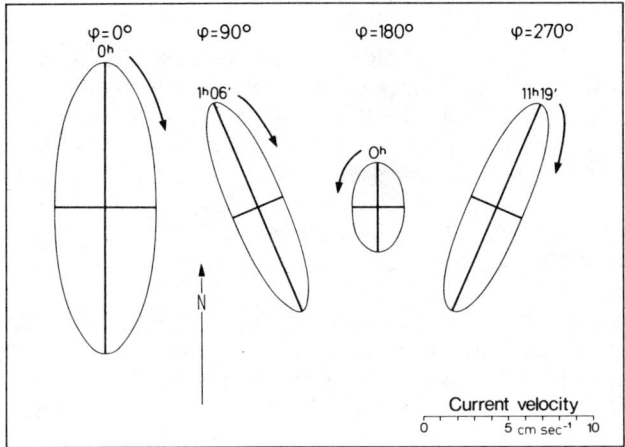

Fig. 9.19. Current ellipses, as calculated for the superposition of surface tide (maximum current $v_0 = 10$ cm sec^{-1}, $u_0 = 0$ cm sec^{-1}) and internal tidal wave ($v_1 = 7$ cm sec^{-1}, $u_1 = 6$ cm sec^{-1}) for phase differences of 0°, 90°, 180°, and 270° of the v components. (After Schott, 1972.)

in Fig. 9.18(a): in spite of the superposition of variable residual currents, a rotation to the right-hand side is clearly recognized for September 15 to 17, whereas for September 18, 20, and 23 to 24, it is a rotation to the left.

2. The maximum tidal current strongly changes at the thermocline [Fig. 9.18(b)].
3. The sense of rotation of the tidal current ellipse changes at the thermocline [Fig. 9.18(b)].

Because of the pronounced two-layer stratification in the observational area of Fig. 9.18, the interpretation of the observed facts is less difficult here than in the case of a complicated density stratification; namely, here only one internal tidal wave occurs, that of the first order. By an example corresponding to conditions in the northwestern North Sea, Schott (1972) has shown that the observations represented in Fig. 9.18 can be explained by the superposition of a barotropic tide and an internal tidal wave. Figure 9.19 exhibits the current ellipses resulting for various phase differences of the meridional currents if a barotropic tide with merely a north–south component and a free tidal wave, propagating meridionally, superimpose each other. Let the amplitude of the barotropic wave be 10 cm sec^{-1} and the baroclinic wave have a meridional component of 7 cm sec^{-1}. As a free internal tidal wave, the latter also has a transverse component of 6 cm sec^{-1}. From the result (cf. Fig. 9.19) it is obvious that the position of the current axes varies with the phase difference, although both waves propagate meridionally, and that there may be a phase range where the current ellipse is passed in opposite sense. Furthermore, the time of occurrence of the current maxima varies and so does the length of the axes of the ellipse. Since the internal tidal wave has a phase difference of 180° between upper and lower layer, the example of Fig. 9.18 would correspond to an ellipse with $\varphi = 0°$ in the upper layer and $\varphi = 180°$ in the lower layer.

In addition, this variability of tidal currents is increased by the alternation between neap tide and spring tide and by changes in stratification.

9.5.6. Superposition of astronomic tides

With a view to their causes, tides hold a special position in the ocean; they are due to tide-producing cosmic forces. Their periods are given; only their phases and amplitudes depend on local conditions. In the ocean, they appear as long waves (cf. Section 8.4) and are superimposed by other long waves, the generating forces of which are found on the earth. Among the latter are tsunamis (cf. Section 8.4.2), edge waves (cf. Section 8.4.3), seiches (cf. Section 8.4.4), and "meteorological tides" (cf. Section 9.3.1). Particularly frequent and sometimes very strong disturbances of the water level are caused by the wind-induced setup of water which may result in storm surges if stormy winds are blowing off the coasts of shallow seas.

Since the general public is greatly interested in reliable storm surge forecasts, a large number of relevant papers have been published. An earlier, very complete bibliography, covering the years 1726–1955, was edited by the International Association for the Physical Sciences of the Ocean (IAPSO) in 1957. A more recent survey on investigations of storm surges has been published by Heaps (1967). Most recent publications have been taken into account by Duun-Christensen (1971) when dealing with the numerical solution of the hydrodynamic equations for the computation of water level variations. Among others he used various studies made in the Institut für Meereskunde (Institute for Marine Research) in Hamburg in connection with the paper by Hansen (1956).

Water level registrations at the gauge of Cuxhaven are reproduced in Figs. 9.20 and 9.21. In the first case, the tides are only slightly superimposed by wind-induced setup, in the second case, they are very much superimposed. With persistent storm from the west and northwest, the wind pile-up lasted from December 20 to 24, 1954 and reached a maximum value of 2.4 m. Tomczak (1955), using the example of the wind setup at Cuxhaven, has shown for this storm surge that four-fifths of the water level rise, as compared to the tides predicted astronomically, must be attributed to the effect of the

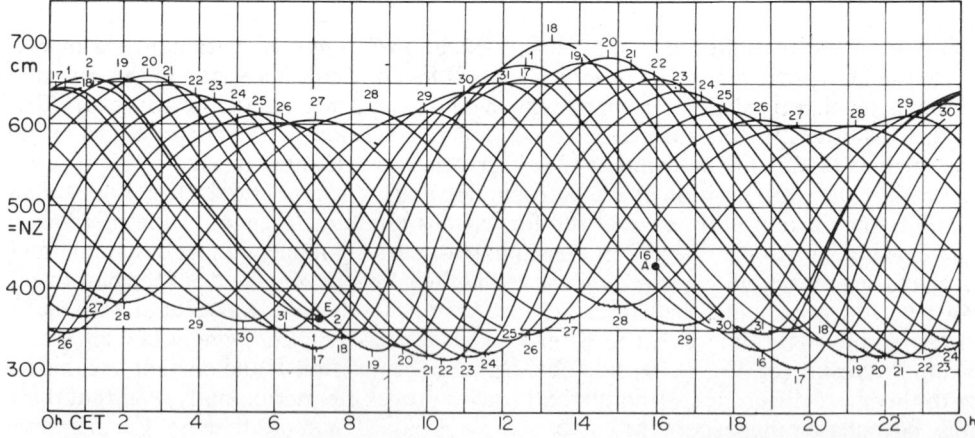

Fig. 9.20. Example of tides slightly disturbed by wind setup. Water level record at the gauge of Cuxhaven, August 16 to September 2, 1955. (Date is marked along the curves at occurrence of high water and low water.) A: beginning, E: end of record. – Height in cm above gauge zero (NZ – 500 cm). New moon, August 17, 20^h58^{min}. First quarter, August 25, 9^h51^{min}. Full moon, Sept. 2, 8^h59^{min}. Perigee of moon, August 14, 19^h. Apogee of moon, August 26, 16^h. Largest northern declination of moon, August 13, 7^h. Largest southern declination of moon, August 26, 20^h. Passage of the moon across the equator, August 19, 15^h, Sept. 3, 2^h (Central European Time). Mean spring delay, 72 hr. (Observed by Wasser- und Schiffahrtsamt, Cuxhaven.)

Tidal Phenomena of the World Ocean   455

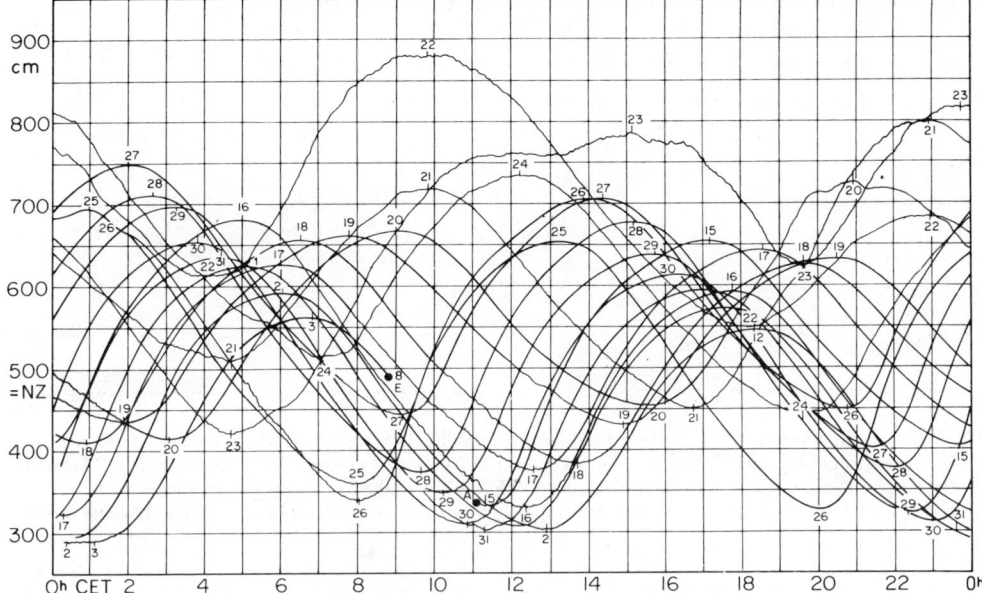

Fig. 9.21. Example of tides very much disturbed by wind setup. Water level record at the gauge of Cuxhaven, December 15, 1954 to January 3, 1955. (Date is marked along the curves at occurrence of high water and low water.) A: beginning, E: end of record. — Height in cm above gauge zero (NZ — 500 cm). Last quarter, Dec. 17, $3^h 21^{min}$. New moon, Dec. 25, $8^h 33^{min}$. First quarter, Jan. 1, 1955, $21^h 29^{min}$. Perigee of moon, Dec. 9, 3^h. Apogee of moon, Dec. 21, 10^h. Largest northern declination of moon, Dec. 10, 3^h. Largest southern declination of moon, Dec. 24, 4^h. Passage of the moon across the equator, Dec. 16, 15^h, Dec. 31, 6^h (Central European Time). Mean spring delay, 72 hr. (Observed by Wasser- und Schiffahrtsamt, Cuxhaven.)

wind over the German Bight. The remaining fraction was the result of the influence of wind action over the northern North Sea. This example in Fig. 9.21 indicates that tide prediction alone is not sufficient to forecast the actual water levels along coasts of shallow seas. For this reason, a public warning system for wind setup and storm surges was established in 1924 for the German Bight. Since 1945, this service has been maintained by the German Hydrographic Institute in Hamburg. Its activity has been described by Tomczak (1954).

9.5.7. Friction of tidal currents

So far, the discussion of tidal currents has not included friction. Friction acts on the current at the bottom; it influences the tidal currents in the entire water column, owing to the turbulence in the water, which at the same time depends on the density stratification. The general considerations regarding the effect of friction on long waves, presented in Section 8.4.6, apply also to tidal waves. A few remarks may be added here.

The influence of friction on tidal currents was discussed mainly by Sverdrup (1926) and Fjeldstad (1929, 1936) on the basis of the *Maud* observations in the north Siberian shelf seas. Thorade (1928) investigated it in the German Bight, and Grace (1936) in British waters. Frictional influences are most conspicuous in the north Siberian marginal seas because there, friction at the pack ice sheet is effective in addition to bottom friction.

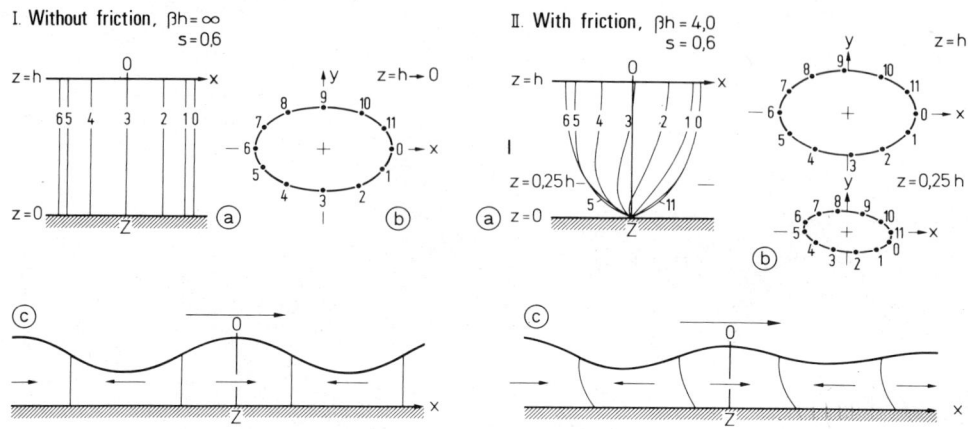

Fig 9.22. Semidiurnal tidal currents in shallow seas. (According to Sverdrup, 1926.) I, without friction; II, with friction. (a) Current velocity along the vertical OZ in the direction of propagation, represented for 7 lunar hr. (b) Tidal current ellipses ($z = h$ is the surface, $z = 0.25\,h$ is 1/4 of the water depth above the bottom). (c) Longitudinal section through the wave, showing surface configuration and current distribution at the time $t = 0$.

Thus, tidal waves propagating from the Arctic Basin toward the Asiatic continent become practically obliterated before they reach the coast of northern Siberia.

The various effects exerted by friction on tidal currents are illustrated in Fig. 9.22, where a tidal wave without friction is compared with a tidal wave with friction. The complete mathematical treatment of both cases, which deals with a progressive tidal wave in an ocean laterally unbounded, has been given by Sverdrup (1926). From the solution of the hydrodynamic equations it follows that rotational currents occur with a ratio of axes

$$s = \frac{2\Omega \sin \varphi}{2\pi} \tau = \frac{\tau}{\tau_E}$$

where τ_E denotes the duration of half a pendulum day and τ stands for the tidal period. These Sverdrup waves can be considered counterparts to Kelvin waves, although there is no pile-up at lateral walls; their crests are horizontal. Their propagation velocity amounts to

$$c = \sqrt{\frac{gH}{1 - s^2}} \qquad (9.31)$$

Such waves can only develop if $s < 1$, that is, if the tidal period τ is shorter than half a pendulum day. Other results are summarized in Fig. 9.22 for the case of $s = 0.6$. If s is referred to the semidiurnal tide with $\tau = 12.4$ hr, Fig. 9.22 represents conditions for the latitude $\varphi = 35°$. From Fig. 9.22 the following can be deduced for a frictionless tidal current: at any instant, the current is equal in magnitude and has the same direction throughout the entire water column. It is an amphidromic current which, in the Northern Hemisphere, turns to the right. Maximum current velocities occur beneath the wave crest and beneath the wave trough. In the latter case, the current direction is opposite to the propagation direction of the wave. The current velocity also depends on s. The orientation of the major axis of the current ellipse coincides with the direction of propagation. The ratio of the axes of the current ellipse is equal to s.

In Fig. 9.22, the case without friction (I) is confronted with case II in which bottom friction is taken into account. The vertical austausch coefficient A_v is contained in $\beta = \sqrt{\omega/2A_v}$. The case $\beta H = 4.0$, represented here, means that $A_v = 110$ g cm^{-1} sec^{-1} for $H = 50$ m and $\tau = 12.4$ hr. With respect to the tidal current, the results indicate the following: for large water depths, that is, if the water depth is greater than the lower depth of frictional influence, the upper layers behave as in the first case without friction, whereas the current in the lower layers decreases towards the bottom. It still remains an amphidromic current. The maximum current velocities occur in front of the wave crests or wave troughs, respectively, that is, noticeably earlier than in the upper layer. The orientation of the major axis of the current ellipse no longer coincides with the direction of propagation of the wave, but is deflected to the right. The ratio of axes is smaller than s; hence, the current ellipses become the more elongated the closer they are to the bottom.

These theoretical results have been confirmed by the few observations of the vertical current distribution obtained on broad shelf seas at great distances from the coast. According to Sverdrup, the vertical austausch (*virtual viscosity*) of the lower layer assumes values between 500 and 1000 g cm^{-1} sec^{-1} and, as far as the order of magnitude is concerned, is in agreement with the vertical austausch in wind-generated currents. For Sverdrup-type waves, one more fact may be mentioned: the propagation velocity is more strongly reduced by friction in higher latitudes than in lower latitudes.

Tidal friction can also help to explain why the earth's rotation is slowing down so that the length of the day is prolonged by 0.002 sec per century. According to the mechanics of rotating bodies, the loss of rotational energy amounts to 3×10^{19} erg sec^{-1}. Since Taylor (1919) and Jeffreys (1921) it has been assumed that this energy, which is continuously supplied to the world ocean by the tide-producing forces, is dissipated by the bottom friction of tidal currents, in particular in the shelf seas with their strong tidal currents. The decisive contribution was supposed to come from the Bering Sea. Munk and MacDonald (1960) have disputed this explanation. Brosche and Sündermann (1971) also have some doubt about this concept, which would be correct if tidal currents were directed from east to west everywhere. Only in this case, would their friction have a delaying effect. Since the direction of tidal currents varies regionally, the rotation of the earth in some areas is even accelerated by tidal friction. What is important for the effect of tidal friction is the direction of the torque. The final answer depends on the detailed knowledge—which we do not yet have—on tidal currents in the world ocean and its shelf seas.

9.5.8. Turbulence of tidal currents and its consequences

The vertical velocity distribution in tidal currents, which, owing to the effect of bottom friction, is already a complicated phenomenon, becomes even more difficult to understand if the density stratification has to be taken into account (cf. Figs. 9.16 and 9.18). This is especially true for shallow waters. Some observational facts and their qualitative interpretation may be mentioned here, because they elucidate an interesting phenomenon of far-reaching consequences.

In the vicinity of the coast, because of the fresh-water influx from land, saline oceanic water often borders on low-salinity coastal water which, being lighter, tends to move over the heavier seawater. As long as the tidal currents remain parallel to the coast, they do not affect these conditions very much, unless turbulence of tidal currents with high velocity destroys the stratification. However, if rotating tidal currents are present, a remarkable process sets in as ascertained by observation (cf. Fig. 9.23). Northwest of the Dutch island

458   Tidal Phenomena

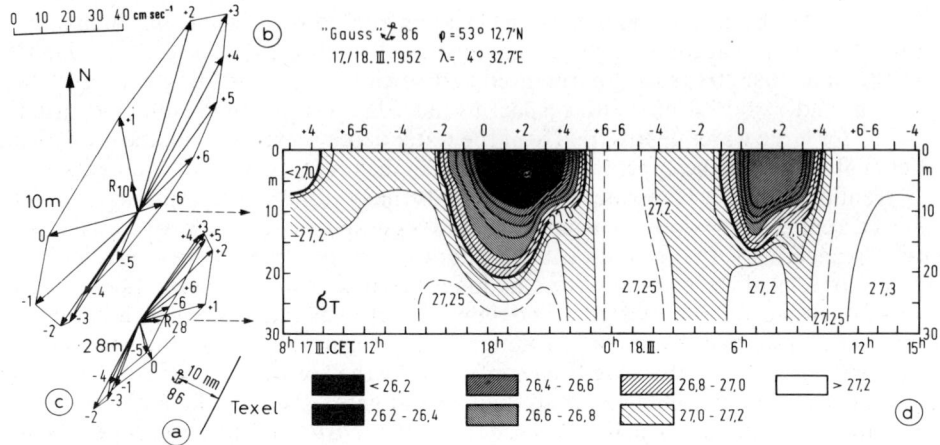

Fig. 9.23. Tidal currents in stratified water near the shore. (After Dietrich, 1953.) (a) Location of point of observation. (b) and (c) Currents observed with paddle-wheel current meter at 10 m and 28 m depth at full hr before and after the moon's passage through the Greenwich meridian. Water depth 30 m. Vectors are averages of two tides on March 17–18, 1952. R_{10} and R_{28}: residual currents. (d) Density stratification for the period March 17, 1952, 8^h to March 18, 15^h.

of Texel, the current was measured at depths of 10 and 28 m with paddle-wheel current meters. Simultaneously, the density stratification was recorded from the anchored ship. As soon as the tidal current gets an offshore component in the upper layer (2 hr before the moon's transit through the Greenwich meridian), lighter coastal water moves seaward. A pycnocline is formed, separating the coastal water from the lower layer. Turbulence is considerably reduced in this pycnocline; the discontinuity layer becomes a layer in which mixing is inhibited and where water masses may glide along each other. Due to reduced friction, more water flows out than can be replenished from behind, because of the coast. The necessary replacement is made from the lower layer. Observations have shown that the current in the lower layer is directed toward the coast and is strongest half an hour after the moon's passage over the Greenwich meridian, just when the offshore component of the current in the upper layer is most pronounced. As soon as the tidal current again flows parallel to the coast, turbulence—because of the high velocity—destroys the stratification. This process begins about 3 hr after the lunar transit of the Greenwich meridian and is completed within 2 hr. Hence, tidal currents in stratified coastal water have two remarkable effects: (1) they reinforce the offshore component in the upper layer and the onshore component in the lower layer. Accordingly, they favor a transverse circulation perpendicular to the coast. The residual currents, resulting if the observed currents are averaged over a tidal period, are illustrated in Fig. 9.23 and denoted by R_{10} and R_{28}. The residual current in the lower layer explains why all suspended matter near the bottom, like fine-grained sand and fry, is transported toward the shore. (2) They favor rapid mixing, especially in areas with rotating currents as, for example, in the North Sea off Texel. Synoptic investigations of the vertical velocity distribution in tidal currents, carried out in the German Bight by three research vessels in the spring of 1955 (Dietrich, 1957), have confirmed that the conditions in Fig. 9.23 are not exceptional; rotating tidal currents in stratified water in the vicinity of coasts always represent highly complicated phenomena. They all show a tendency toward a transverse circulation perpendicular to the coast.

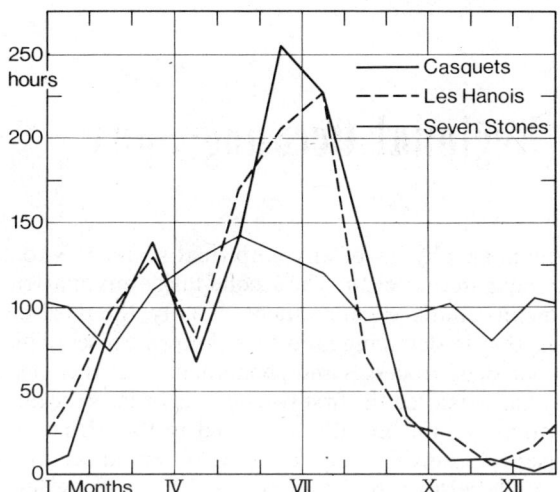

Fig. 9.24. Influence of tidal current mixing on the frequency of fog. (After Dietrich, 1951.) Average number of hours of fog per month in the central part of the English Channel (lighthouses Casquets and Les Hanois) and at the western entrance of the English Channel (lightship *Seven Stones*).

With stronger tidal velocities, vertical turbulence and therefore mixing increase. At velocities of about 1 m sec^{-1}, no stratification in the water can be maintained down to a depth of around 100 m. The summer thermocline, which is otherwise typical for oceans in higher and moderate latitudes, does not develop, as can be recognized from Fig. 5.08, showing the example of the North Channel of the Irish Sea. In these cases, the amount of heat received by the sea surface during the warming in spring is distributed over a relatively deep water layer. In this way, the annual range of surface temperature is reduced, and the time of occurrence of the temperature maximum is delayed, as has been shown for the North Sea and the British waters (Dietrich, 1953a). Thus, in these areas, the temperature difference between water and air becomes strongly negative in early summer, but strongly positive in late fall. In the first case, formation of sea fog is favored; in the second case, it is impeded. Hence, relatively deep shelf seas with pronounced tidal mixing are characterized by a marked annual march in the occurrence of fog. The fog maximum occurs in early summer, the minimum in late fall. Figure 9.24 represents some relevant examples. Owing to the strong tidal mixing in the central part of the English Channel, fog appears very frequently in June to July, with all its dangers to shipping. This contrasts with the extreme scarcity of fog in October through December. At the western entrance to the English Channel, the frequency of fog is distributed much more uniformly over the year since tidal mixing is very much weaker there. This phenomenon is part of the widely ramified chain of causal connections, which are ultimately related to tidal currents. It is especially noteworthy that a periodic wave process such as a tidal current exerts a long-lasting influence on hydrodynamic and climatic conditions.

10. Regional Oceanography

The first task of oceanography, as of any empirical science, is to observe exactly the processes and phenomena in the ocean. The second task aims at attributing them to their primary causes by means of analytical methods. Finally, the third task—the goal of any science—is synthesis, that is, deducing facts from known causes. This third task includes the prediction and control of processes and phenomena. But the question is open to what extent oceanography has mastered the first two tasks in order to be able to tackle the third. Certainly, this question cannot be fully answered in the affirmative. The principles, however, of the causal relationships regarding the observed oceanic processes and phenomena are known. Therefore, the attempt can be made to develop the science of oceanography systematically from its base.

As in the atmosphere, solar radiation must ultimately be considered the primary cause of numerous processes and phenomena in the ocean. Its influence, however, is not restricted to conditions of light and heat in the ocean, and the water exchange between ocean and atmosphere alone. Under the influence of the deflecting force of the earth's rotation, this energy source maintains atmospheric circulation and, thus the near-surface wind system. This, in turn, significantly determines the development of the oceanic surface current system, which together with mixing and currents caused by the density distribution, also exerts its influence upon the circulation in the deep sea. Further physical, chemical, and biological phenomena in the ocean depend on these processes of motion. In oceanography, therefore, processes of motion and the spreading of water masses must be given a central position. The other physicochemical processes and phenomena in the ocean can be divided into four groups, among which some close relations exist. These are (1) the radiation budget, (2) the water budget, (3) the budget of substances dissolved or suspended in water, and (4) waves and tides. Since the following is intended as a survey in which the entire world ocean will be considered, we will restrict the description to an outline of large-scale relations.

In Section 10.1, the processes of motion and the spreading of characteristic water masses in the deep layers of the three oceans will be discussed briefly. In Section 10.2, the large and deep mediterranean seas will be dealt with. In addition, the near-surface layers will be treated separately in Section 10.3, not only because processes and phenomena are much more manifold here than in the depth but, above all, because the interest of man is mainly centered on those near-surface layers.

10.1. Stratification and Circulation in the Deep Layers of the Three Oceans

10.1.1 Water masses of the cold-water sphere

Observations of the vertical temperature distribution (cf. Section 5.1.4 and Plates 3, 4, and 5) have shown that to a first approximation the ocean can be considered to consist of two parts, namely a thin warm-water sphere between the oceanic polar fronts in the Northern and Southern Hemispheres, and a cold-water sphere occupying the entire remaining world ocean. Owing to the strong density stratification, mixing and turbulence are reduced in the boundary layer between the two spheres, such that, in general, we deal with separate circulations.

Our first subject concerns stratification and circulation of the cold-water sphere. Poleward of the polar fronts, this sphere extends up to the surface where it is directly exposed to the influence of heat and water exchanges between ocean and atmosphere. These exchanges cause the surface temperature in particular areas to decrease so far that, in spite of the relatively low salinity, water masses are formed that are heavier than the water masses surrounding them. Furthermore, in zones of frequent storms, especially in the areas of the west wind belts containing the paths of atmospheric cyclones, descending motion is induced and develops because of the adjustment of ocean currents to the changing atmospheric wind field. Thus, thermohaline circulation is reinforced. The downwelling water masses move towards a depth level whose density corresponds to their own density, and, at this level, they begin to spread horizontally. In this way, the cold-water sphere gets a laminated structure. Based on the distribution of a hydrographic factor especially characteristic for the water mass under consideration in its area of formation, like temperature, salinity, content of oxygen or silicate, the spreading of such a water mass can be followed from its area of origin throughout the ocean. In some cases, the internal field of force may become so large that deductions as to the velocities at particular levels can be made from the pressure gradient.

Since 1957 (Revelle and Suess), numerous other authors, among them Lal (1969) as well as Broecker and Li (1970), have employed natural radioactive substances, especially ^{14}C, as tracers for the determination of origin and renewal ("age") of the deep water.

Four main water masses are formed in the cold-water sphere as a result of the contact with the atmosphere: the *polar bottom water,* including (1) the Antarctic bottom water (AABW) and (2) the Arctic bottom water (ABW); and the *subpolar intermediate water* where we can distinguish (3) the sub-Antarctic intermediate water (SIW) and (4) the sub-Arctic intermediate water (NIW).

10.1.2. Antarctic bottom water

Antarctic bottom water is predominantly generated in the Weddell Sea along the slope of the Antarctic continent, where the water temperature under the shelf ice amounts to −1.9°C in winter and the salinity to 34.62‰. It represents the heaviest water in the Antarctic Ocean and descends to the bottom of the Atlantic–Indian South Polar Basin. Owing to mixing with neighboring warmer water masses, it arrives at the sea floor with a temperature of −0.9°C (if the adiabatic effect is neglected) and a salinity of 34.65‰. This was already recognized by Brennecke (1921) during the German Antarctic Expedition with the research vessel *Deutschland* in 1911–1912. More advanced concepts have been contributed by several authors (Fofonoff, 1956; Deacon, 1963; Mosby, 1966; Foldvik and Kvinge, 1969; Seabroke et al., 1971; and for the Pacific–Antarctic sector in particular, Gordon, 1966). The cross section through the Weddell Sea shown in Plate 4 is based on more recent data obtained by the American icebreaker *Glacier* and the Argentine research vessel and ice breaker *General San Martín.* Thus, it has been confirmed that the Antarctic continental slope of the Weddell Sea is to be considered the area of origin of the Antarctic bottom water. The volume of the water transported away from there must be considerable, about 24×10^6 m^3 sec^{-1} (Mosby, 1971); it should approximately equal the volume transport of the high-salinity deep water of the North Atlantic Ocean (cf. Section 10.1.8). For further elucidation, measurements beneath the ice of the Weddell Sea would be required in the southern winter. Experiments by an American–Norwegian expedition deploying moored instruments for long-term continuous recording have yielded relevant data.

The abyssal basins of the oceans are open to the inflow of the Antarctic bottom water. Its spreading can be roughly followed if the potential bottom temperature θ is used as tracer and if its geographic distribution is plotted on the basis of observations (Fig. 10.01). Under the influence of the deep-reaching, wind-generated Circumpolar Current, the near-bottom water in the Antarctic water belt is forced to spread zonally. On its way around the Antarctic continent, it is warmed by mixing with the overlying warmer water, namely from $-1.0°$C in the Weddell Sea to about $+0.4°$C in the Drake Passage between Tierra del Fuego and Graham Land. This has been verified by numerous careful measurements of the American vessel *Eltanin* (cf. Fig. 10.02). The arrows in this figure qualitatively indicate the spreading of the bottom water. According to the analysis by Gordon (1971), it can be recognized in this figure that the water at the bottom consists of, at least, two water masses of different origin, to be distinguished by their salinities:

1. one of lower salinity of about 34.66‰ which, however, is dominant and originates from the Weddell Sea;
2. another one whose higher salinity of about 34.75‰ is influenced by the Ross Sea.

The spreading of the Antarctic bottom water into the individual oceans (cf. Fig. 10.01) starts from the circumpolar motion and closely depends on the location of the ridges and sills on the abyssal floor. In the Atlantic and Indian Oceans, the elongated Mid-Ocean Ridge causes mainly meridional spreading, separately in the eastern and western basins. Within these basins, more or less pronounced transverse sills impede the spreading (cf. Plates 3 to 5). Some of them can be overflown by the bottom water; in this case, increased mixing with overlying water will take place above the sills and contribute to a marked temperature increase, as, for example, is the case for the flow from the Brazil Basin into the Guiana Basin, from the Agulhas Basin into the Cape Basin, and from the Madagascar Basin into the Mascarenes Basin. Other ridges act as bars, such as the Walvis Ridge running between Southwest Africa and the Mid-Atlantic Ridge. This transverse ridge prevents direct inflow of Antarctic bottom water into the East Atlantic Basin. Only on a detour through a narrow passage in the Mid-Atlantic Ridge at the equator (the Romanche Deep) can some admixture of Antarctic bottom water penetrate into the eastern basin from which it spreads northwest and southward. The ultimate extensions of Antarctic bottom water, though already greatly mixed with the overlying warmer water, still do not completely conceal their origin and can be traced in the Atlantic Ocean up to 45°N, that is, up to the continental rise of the Grand Banks. The first current measurements in this water (Zimmermann, 1971), taken at 38°N 2 m above the sea floor at more than 5000 m depth, yielded current velocities of up to 16 cm sec^{-1} within recording periods of 7 days. In the Pacific Ocean, Antarctic bottom water can be traced as far as 50°N, that is, up to the vicinity of the Aleutians. In spite of some minor contribution of water from the Sea of Okhotsk, the circulation of the North Pacific bottom water remains small; according to Burkow (1969) it amounts to about 1 cm sec^{-1}.

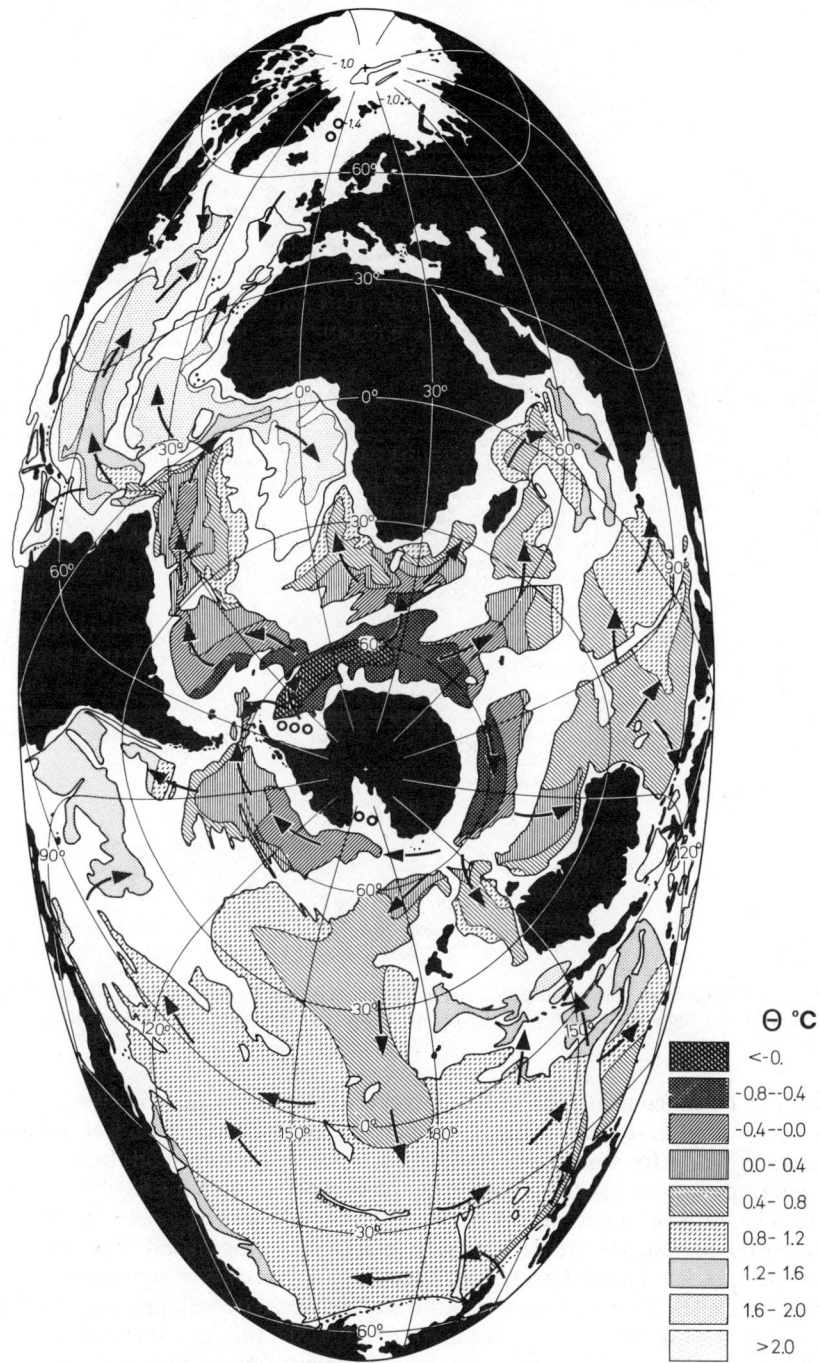

Fig. 10.01. Potential bottom temperature (θ °C) as well as areas of origin (○) and spreading (→) of Antarctic and Arctic bottom waters in the world ocean at depths of more than 4000 m. (After Wüst, 1938; with amendments by Gordon, 1966, 1972b; Olson, 1968; Lynn and Reid, 1968; and Dietrich, 1969a.)

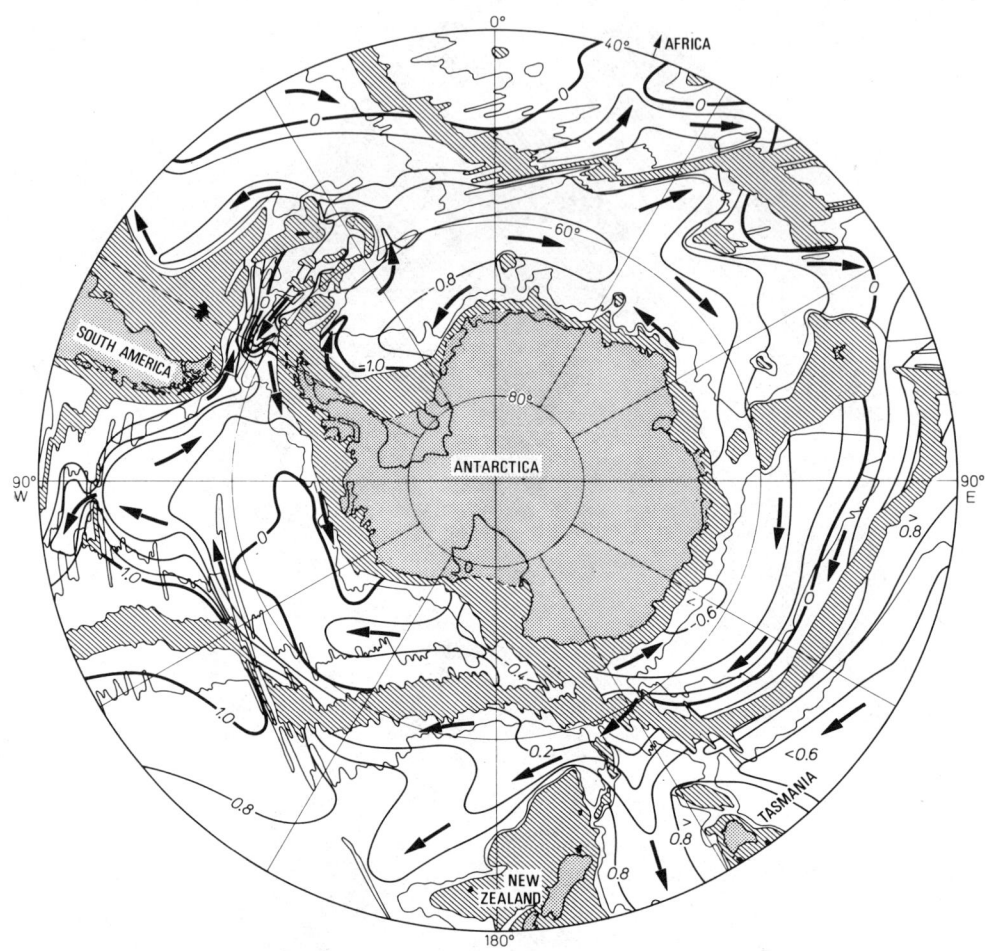

Fig. 10.02. Potential bottom temperature (°C) and spreading of Antarctic bottom water south of 40°S. (After Gordon, 1966, 1976; Gordon and Goldberg, 1970.) Hatched area: 0–3000 m depth; other line in the ocean: 4000 m depth.

What can be called a current with measurable velocities within the Antarctic bottom water is only found in the Antarctic water belt and on the western side of the South Atlantic Ocean. Since, for stationary conditions, advective inflow must be in equilibrium with mixing, the ratio of the austausch A to the velocity u can be determined. According to Defant (1936b), this ratio ranges between 2 and 3 in the abyssal basins and between 5 and 6 over the bottom sills on the western side of the Atlantic Ocean. Wattenberg (1935) found an austausch value of $A = 4 \text{ g cm}^{-1} \text{ sec}^{-1}$, based on chemical processes of calcium dissolution at the sea floor by near-bottom water masses. Accordingly, the velocities have an order of magnitude of $u = 0.5$ to 2 cm sec^{-1}, that is, the Antarctic bottom water on the western side of the Atlantic Ocean, where the bottom current is developed rather strongly, requires from 10 to 30 yr to travel from 50°S to the equator. Along this path, the water masses lose 40% of their characteristic properties because of mixing. In the axis of the Antarctic bottom current, flowing in a narrow band at the base of the South

American continental slope, Wüst (1955) computed a mean velocity of 7.2 cm sec^{-1} and a maximum velocity of 12.5 cm sec^{-1}. Owing to the turbulence of the stronger bottom currents, increased turbidity is maintained by suspended matter in the near-bottom layer. In the western North Atlantic Ocean, this layer has an average thickness of 1 km (Eittreim and Ewing, 1972); the average size of the particles amounts to about 12 μm. The scatter of light is about 10 times stronger in the middle of the layer than in the ocean water on top of it. The cause of this may be found in the turbidity currents (cf. Section 1.3.3) developing at the continental slope. To date, it has not yet been examined to what extent erosion by tidal currents acting together with turbulent diffusion might explain the increased turbidity in the lowest 1000 m.

10.1.3. Arctic bottom water

For the formation of Arctic bottom water only small areas are available at the sea surface. Their importance is furthermore diminished by the fact that high-rising submarine ridges (the Greenland–Scotland Ridge in the Atlantic Ocean and the shallow Bering Strait in the Pacific Ocean) separate the Arctic water masses from the Atlantic and Pacific deep seas. In the Norwegian Sea and the Arctic Ocean, conditions favor the formation of heavy bottom water in winter only outside the polar ice sheet, that is, in the area north of Jan Mayen (cf. Fig. 10.30). Stratification and circulation in this sea area will be described in Section 10.2.6. This water—with $T = -1.2°C$, $S = 34.92‰$, and $\sigma_T = 28.12$—is the heaviest water anywhere in the deep sea. This cold water flows intermittently through the Faeroe Channel, over the central part of the Greenland–Scotland Ridge, and through the transverse channel in the Greenland–Iceland Rise. Investigations with the research vessel *Anton Dohrn* (Dietrich, 1956b) and during the subsequent international Overflow Expedition of 1960 [Tait (Ed.), 18 authors, 1967] have provided evidence of the overflow of the Iceland–Faeroe Ridge. Because of its high density, the Arctic bottom water very rapidly flows down the southern slope of the ridge [cf. Fig. 10.03(b)], thus intermittently renewing the bottom water in the eastern basin of the North Atlantic Ocean. As can be taken from the temperatures below 0°C in Fig. 10.03(a), the bottom water simultaneously flows out through the narrow but deep Faeroe Channel, southwest of the Faeroe Islands, with a sill depth of 800 m. In this passage, Crease (1965) determined velocities of as high as 100 cm sec^{-1} at a depth of 760 m in the cold deep water flowing out of the Norwegian Sea. The residual currents recorded near the bottom show that the cold water when crossing the ridge generally flows parallel to the depth contour lines. The overflow is not only a cascade-like overspilling of the ridge, but—because of the Coriolis acceleration—it follows the bottom topography.

Besides, the overflow is not a stationary phenomenon. This was revealed by the first continuous recordings of the current in the open ocean carried out during the Overflow Expedition of 1960 (Joseph, 1967). The tidal currents are relatively strong (cf. Fig. 10.04 and Table 10.01), which can be explained by the fact that there is an amphidromic system of the semidiurnal tide on the Iceland–Faeroe Ridge, which is degenerated to nearly a nodal line (Plate 7). The tidal current is greatly dependent on the depth, which is evident from the velocity in the direction of the main axis of the current ellipse given in Table 10.01. In addition, the overflow is subject to further variations with time. Stratification and overflow are highly variable; hence, a large-scale, intensive measuring program with devices for continuous registration is required in order to grasp the physical relationships.

With a view to variability, the renewal of the bottom water in the northern part of

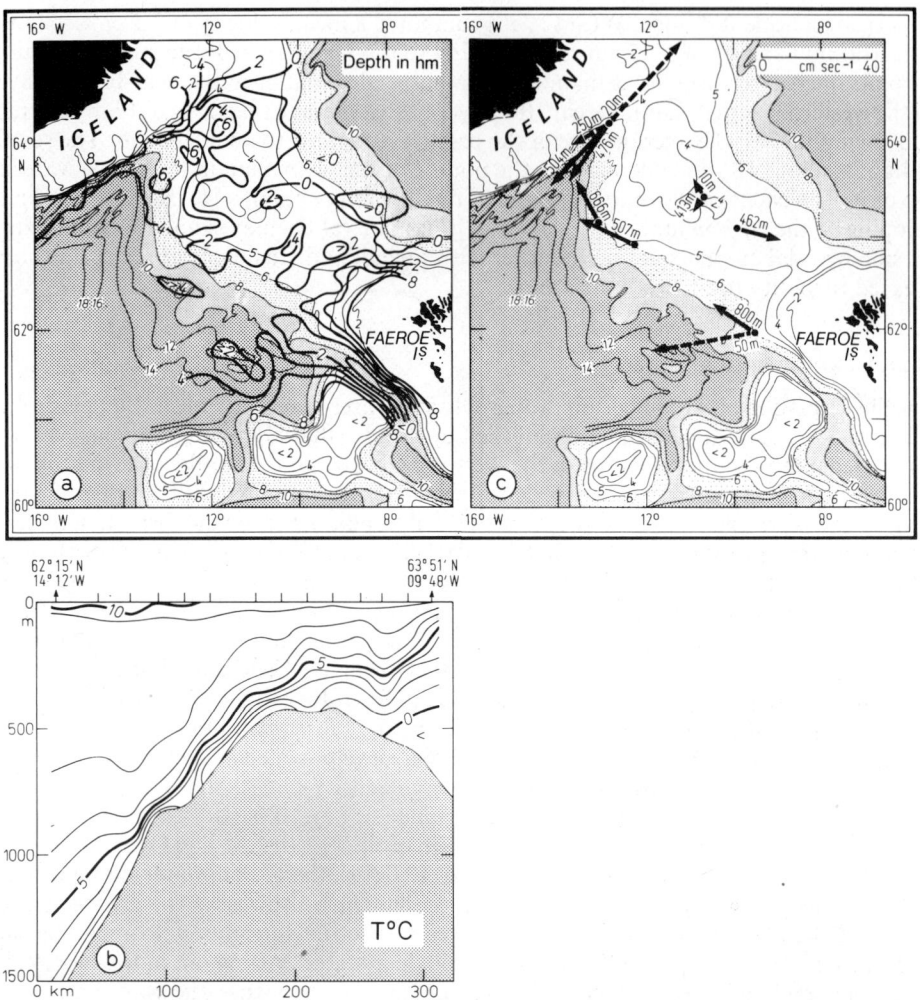

Fig. 10.03. Overflow of the Iceland-Faeroe Ridge during the international Overflow Expedition in June 1960. (a) Distribution of bottom temperature from May 30 to June 2, 1960. (b) Temperature distribution on a cross section perpendicular to the ridge, June 6 to 8, 1960. (c) Residual currents near the surface and near the bottom, based on continuous records from June 9 to 12, 1960. Depth of measurement (in m) marked at the current arrows. ((a) and (b) after Tait et al, 1967; (c) after Joseph, 1967.)

the West Atlantic Basin resembles that occurring across the Iceland–Faeroe Ridge in the East Atlantic Basin, as shown by the results of the *Anton Dohrn* Expedition to the Irminger Sea (Dietrich, 1956b). Here, the inflow comes through the Denmark Strait across the Greenland–Iceland Rise. The cross section of the Denmark Strait in Fig. 10.52 exhibits the cold bottom water in the transverse channel. Since such inflow also occurs irregularly there, the circulation in the depth is excited irregularly, too. Such processes impede tracing the formation of bottom water and deep water in the Northwest Atlantic Ocean (Lee and Ellett, 1967). Furthermore, water from the overflow of the Iceland–Scotland Ridge follows the island basis of Iceland and the Reykjanes Ridge as far as 53°N

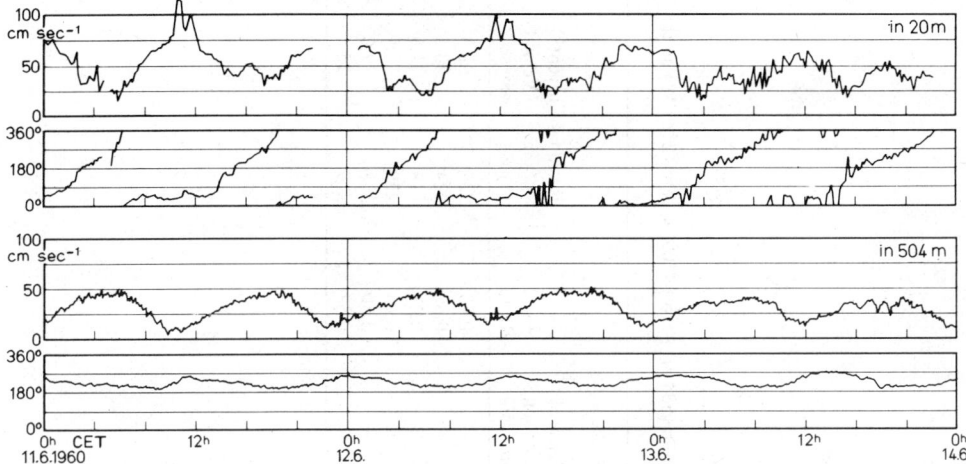

Fig. 10.04. Part of the continuous registration of the current at 20 and 504 m depth on the Iceland-Faeroe Ridge by the research vessel *Gauss* ($\varphi = 64°09'N$, $\lambda = 12°47'W$, depth = 504 m) from June 11–14, 1960 during the international Overflow Expedition. Position: see Fig. 10.03c. (After Joseph, 1967.)

and goes through a passage in the ridge westward into the Labrador Basin, where it mixes with the Arctic bottom water that has overflown the Greenland–Iceland Rise. According to Lee and Ellett (1965), it can be traced as far as about 43°N.

As already shown in Fig. 5.20, the Sea of Okhotsk also holds a special position by forming bottom water of its own.

In summary, we can state that, according to Fig. 10.01, the Arctic proportion of the polar bottom water outside the Arctic Mediterranean Sea is only rudimentary and restricted to the northern parts of the North Atlantic and North Pacific Oceans. Thus, water of Antarctic origin determines the physicochemical factors of the near-bottom water in the deep-sea basins.

10.1.4. Subpolar intermediate water

Poleward of the oceanic polar fronts, the sea surface is the place where the third and fourth main types of water of the cold-water sphere originate, that is, the sub-Antarctic and sub-Arctic intermediate waters (SIW and NIW in Plates 3, 4, and 5). Since precipitation greatly exceeds evaporation in these areas (cf. Fig. 4.19), the intermediate water has relatively low salinity. In the Southern Hemisphere, where the polar front is well developed all around Antarctica, as shown in Plate 6, the intermediate water is formed in a narrow band around the earth. In the Northern Hemisphere, however, its formation is restricted to the western sections of the polar front; in the North Atlantic Ocean to the Labrador and Irminger Seas, and in the North Pacific Ocean to a zone south of the Kuril Islands at 45°N (Roden, 1970). In the regions where the intermediate water originates, salinities amount to about 33.8‰, with the exception of the North Atlantic Ocean where they amount to 34.8‰.

As can be taken from the longitudinal sections in Plates 3, 4, and 5, the low-salinity intermediate water of all three oceans (with the exception of the North Atlantic Ocean) descends within a small distance from the surface to about 900 m depth and then slightly

Table 10.01. Examples of Tidal Currents and Residual Currents Near the Surface (20 m) and Near the Bottom (504 m) ($\varphi = 64°09'N$, $\lambda = 12°47'W$) on the Iceland–Faeroe Ridge during the International Overflow Expedition of 1960[a]

Periods Analyzed June 12 to 13 1960 (CET) Day	Time Period	Measuring Depth (m)	Phase hr	Phase min	Current Ellipses Main Axis Direction (°)	Main Axis Velocity (cm sec⁻¹)	Sense of Rotation *cum sole*	Axial Ratio	Residual Current Direction (°)	Residual Current Velocity (cm sec⁻¹)
12th	0700 h to 1900 h	20	+2	36	229	46.9	−	0.35	37	38.3
12th		504	+2	23	190	15.8	+	0.60	224	33.5
12th	2000 h to 0800 h	20	+1	50	197	54.0	−	0.41	19	21.6
13th		504	+2	59	171	16.0	+	0.44	229	27.9
13th	0900 h to 2100 h	20	+1	49	211	49.5	−	0.50	338	7.0
13th		504	+3	10	154	18.3	+	0.37	229	25.1

[a] According to records of research vessel *Gauss* (after Joseph, 1967).

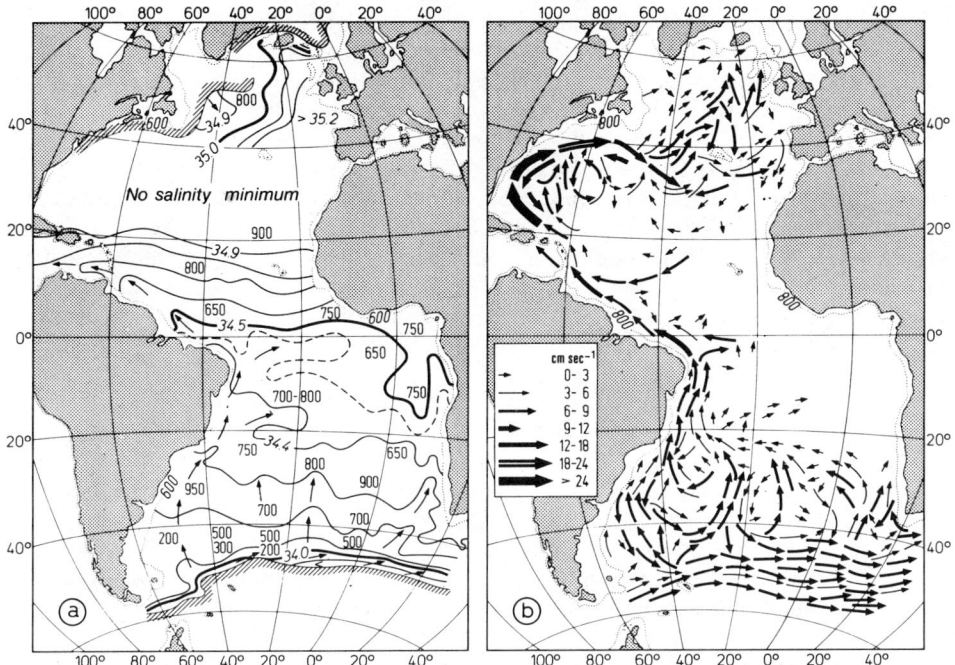

Fig. 10.05. Spreading of the sub-Antarctic and sub-Arctic intermediate waters in the Atlantic Ocean. (a) Represented by the salinity distribution in the layer of the salinity minimum (at about 500–900 m depth). Vertical figures: approximate depth of the layer (after Wüst, 1936); line with hatched screen: polar front; dotted line with oblique figure: 600 m depth contour line. (b) Represented by the current field at 800 m depth (after Defant, 1941). Dotted line: 800 m depth contour line.

ascends toward the equator. While, in the Atlantic Ocean, it can be traced from the region of its origin at 45 to 55°S across the equator up to 25°N, the intermediate water of the Pacific Ocean is distributed symmetrically with respect to the equator. According to the results obtained up to now, it is believed that neither the sub-Arctic nor the sub-Antarctic intermediate water of the Pacific Ocean crosses the equator in significant volumes. The same is true for the sub-Antarctic intermediate water in the Indian Ocean. In the North Atlantic Ocean, the intermediate water is not very conspicuous, owing to the high salinity prevailing in the region of its formation. As indicated in Plate 4, this water descends to greater depth, due to its relatively high density, and spreads southward together with the North Atlantic deep water (DWM and DWL), which will be described in the following section.

For the Atlantic Ocean, a three-dimensional concept of the spreading of the subpolar intermediate water has been deduced on the basis of the "core layer method" by Wüst (1936) and also by computations of the geostrophic currents by Defant (1941). The salinities of the core of the intermediate water are given in Fig. 10.05(a), together with the preferred direction of spreading which is recognized from it. The current field at 800 m depth is shown in Fig. 10.05(b). Only on the western side does the spreading, as "western boundary current," assume the character of a current with velocities of 6 to 9 cm sec^{-1}. (The same is true for the spreading of the Antarctic bottom water.) According to Stommel and Arons (1971), the width of the current is approximately ten times larger than that of the near-surface "western boundary currents" (Gulf Stream, Kuroshio). The high velocities of more than 24 cm sec^{-1} in the western North Atlantic Ocean must be attributed to the deep-reaching Gulf Stream and not to the intermediate water. They have been confirmed there by direct current measurements (cf. Fig. 10.09). Exceptionally high values were obtained with the direct current measurements carried out by the USSR research vessel *Michael Lomonossov* during the international Equalant III Expedition in 1964 (Boguslawkii and Belyakov, 1966). On a section east of Cape San Roque, six current meter arrays had been moored for 5 days. Maximum values of 40 cm sec^{-1} were measured at a distance of 100 nautical mi from the continental slope. The mass transport of the intermediate current was reported to be 30×10^6 m^3 sec^{-1}, which is four to five times higher than Defant computed from mean geostrophic currents. It is still an open question whether the total transport could be reduced by variability in time and by countercurrents.

10.1.5. Deep water

Depending on the magnitude and direction prevailing in the spreading of polar bottom water and subpolar intermediate water, as the two primary branches of deep-sea circulation, the necessity of compensating motion arises (cf. Plate 3, 4, 5). Where present, it occurs within the layer between the two water masses at depths from about 1000 to 4000 m. In general, this so-called deep water (DW) represents a mixed-water type, formed of bottom water and intermediate water. It is therefore very uniform with regard to the distribution of hydrographic factors. The deep waters of the North Atlantic and the Northwest Indian Oceans (DWU) are exceptions. Here, lateral injections of characteristic water masses from adjoining mediterranean and adjacent seas interrupt the uniformity in these deep layers.

In the North Pacific Ocean, deep water is not clearly distinguishable (Plate 3). The sub-Arctic intermediate water, weakly developed anyhow, which spreads toward the equator, evidently has sufficient compensation in the Antarctic bottom water, slowly moving northward. The result is a very sluggish deep-sea circulation. This also is reflected in the fact that the oxygen consumed during the oxidation of dead organisms, sinking from near-surface layers, is replenished only very slowly. In the cold-water sphere, the North Pacific Ocean has, by far, the lowest oxygen content of all oceans. In the South Pacific Ocean, the intermediate and bottom waters, though only weakly developed but both directed toward the equator, enforce a southward-directed compensation within the deep water.

In the Indian Ocean, the deep water is more recognizable than in the South Pacific Ocean (cf. Plate 5), which is to be expected because of the noticeable spreading of the bottom water and the intermediate water toward the equator. This deep water contains admixtures of high-salinity water originating in the Red Sea and entering the Indian Ocean through the Strait of Bab al Mandab.

A cross section through this deep water is presented in Fig. 10.06(c). It is one out of a number of cross sections taken by the research vessel *Meteor* off Somalia during the Indian Ocean Expedition in 1965 [cf. sketch of positions of stations in Fig. 10.06(a)]. The vertical records of temperature and salinity at Stations 132 to 138 are given as *TS* diagrams [cf. Fig. 10.06(b)], while the smoothed salinity registrations are represented as cross section in Fig. 10.06(c). Figure 10.06(b) shows remarkable differences in the fine structure of the stratification from one station to the next. The intermediate salinity maximum has already vanished at the continental slope; the core of the high-salinity water

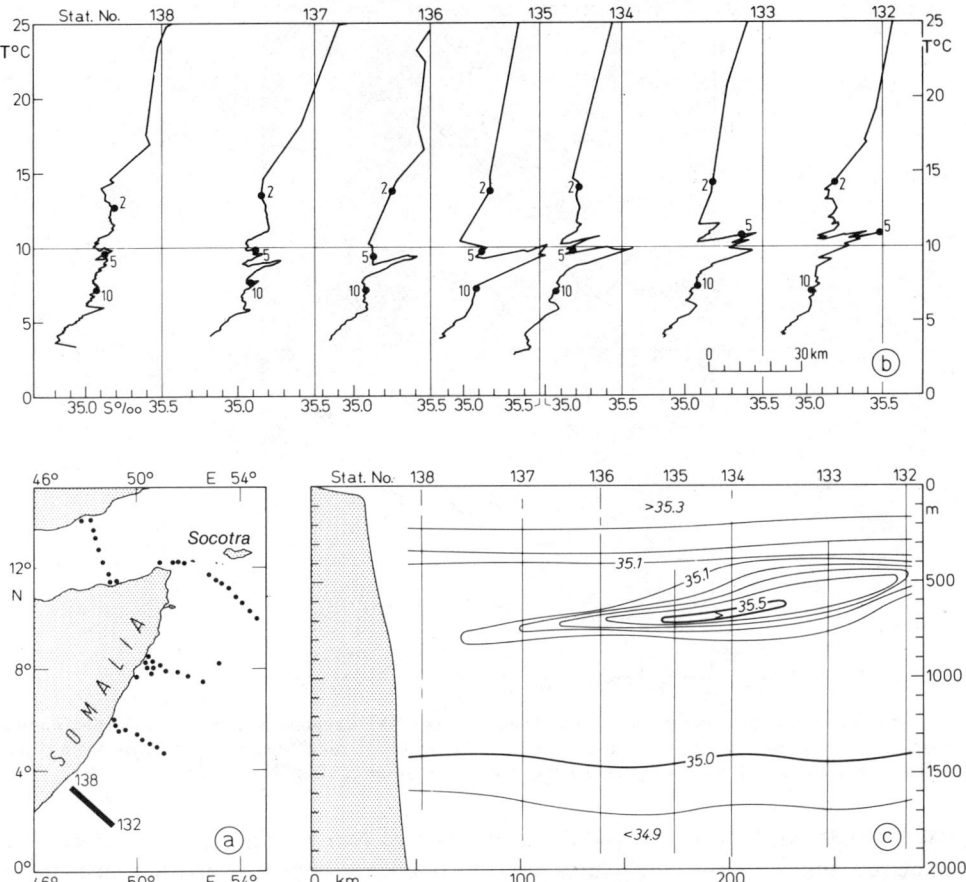

Fig. 10.06. Spreading of Red Sea water in the Arabian Sea, based on bathysonde records by the research vessel *Meteor* at Stations 132–138 in January 1965. (After Krause, 1968.) (a) Positions of stations with location of cross section (132–138) in (b) and (c). (b) TS diagrams. 2, 5 and 10 give depths in hectometers. Each station is referred to 35.5‰ salinity. The distance between the 35.5‰ lines is proportional to the distance of the stations from each other (scale in km). (c) Salinity cross section from averaged bathysonde data.

from the Red Sea lies 150 km farther seaward. This inflow is also very poor in oxygen (about 0.5 ml liter^{-1}), and thus contributes to intensifying the oxygen minimum formed anyhow in the poorly ventilated northern Indian Ocean. During the Indian Ocean Expedition of *Meteor* in 1964–1965, Grasshoff (1969) was able to show that beneath 200 m depth the water is practically free of oxygen and that nitrite has already been oxidized to nitrate.

Antarctic bottom and sub-Antarctic intermediate water, both of which—as already mentioned—show a specially pronounced, northward directed spreading in the Atlantic Ocean, require a correspondingly strong southward compensation in the deep water. Consequently, in contrast to the Indian Ocean and especially the Pacific Ocean, the Atlantic Ocean is exceptionally well ventilated, and its cold-water sphere has a very high oxygen content, as illustrated in Plate 4 compared with Plates 3 and 5. Wüst (1936) has proven in detail that three levels can be distinguished in the Atlantic deep water: the upper

472 Regional Oceanography

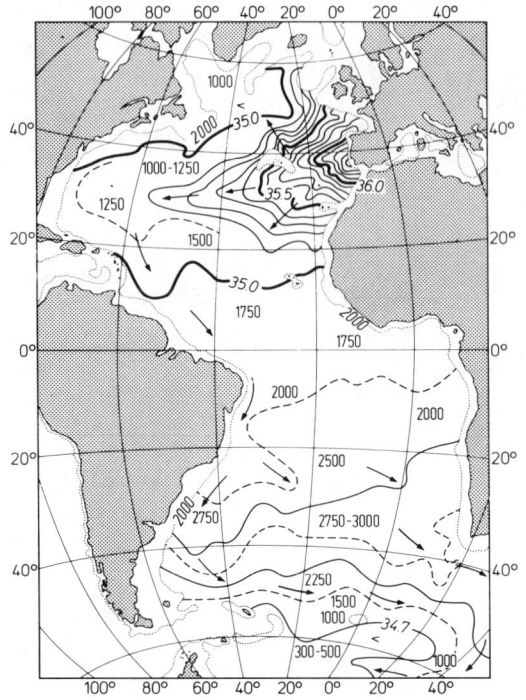

Fig. 10.07. Spreading of the upper North Atlantic deep water (DWU), represented by the salinity distribution in the layer of the salinity maximum (at about 1000–3000 m depth). Vertical figures: approximate depth of this layer in m; dotted line with oblique figure: 2000 m depth contour line. (After Wüst, 1936.)

(DWU), the middle (DWM), and the lower (DWL) North Atlantic deep water. They can be recognized by the admixtures of water masses with characteristic properties in accordance with their regions of origin. The upper North Atlantic deep water receives an injection of warm water of particularly high salinity but of low oxygen content, propagating from the European Mediterranean Sea through the Strait of Gibraltar, which is 350 m deep (cf. Figs. 10.20 to 10.22). At first, it spreads radially at a depth of 1000 to 1250 m, as can be seen in Fig. 10.07. However, it quickly loses its marked high salinity by mixing with water masses of lower salinity between which it is advancing. In spite of this, it remains clearly distinguishable because of its relatively high salinity and can be followed into the Antarctic regions (cf. Plate 4). The middle and lower North Atlantic deep waters are characterized by high oxygen contents. As far as the middle deep water is concerned, this can be attributed to admixtures of subpolar intermediate water formed north of the polar front in the Labrador and Irminger Seas in winter. The lower deep water is enriched with oxygen by admixtures of Arctic bottom water penetrating from the Norwegian Sea across the Greenland–Scotland Ridge. On the basis of comprehensive observational data, Reid and Lynn (1971) have traced the spreading of the lower North Atlantic deep water right into the North Indian and North Pacific Oceans.

Due to the tongue-like spreading of the highly oxygenated water, as shown in Fig. 10.08(a), the propagation of the middle North Atlantic deep water can be followed southward. Moreover, the current field at 2000 m depth, as presented in Fig. 10.08(b), permits a quantitative concept of the current velocities in the North Atlantic deep water.

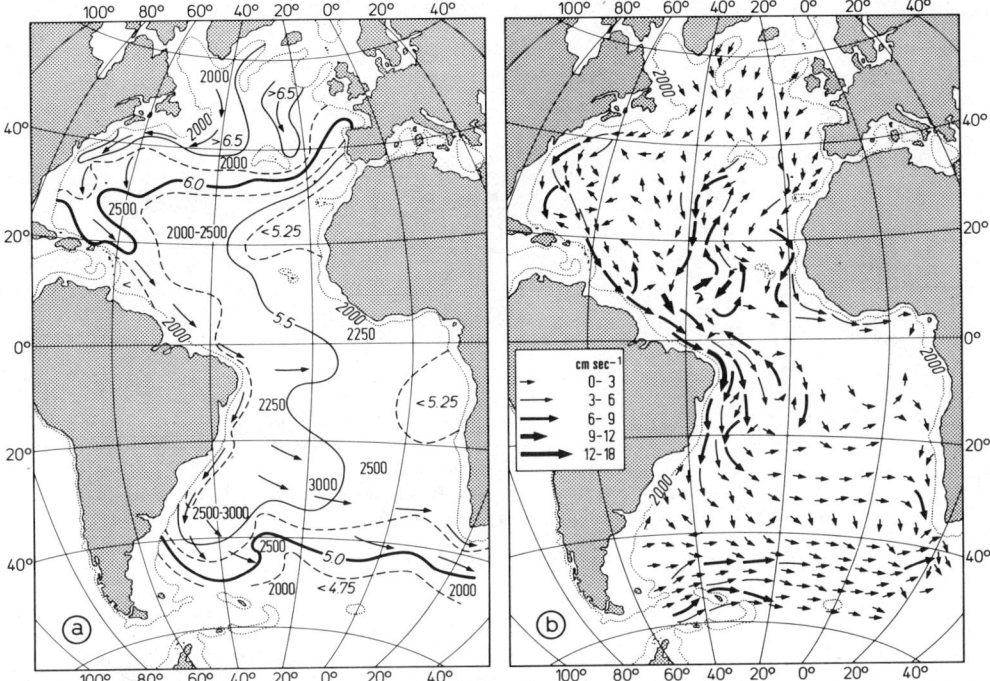

Fig. 10.08. Spreading of the middle North Atlantic deep water (DWM) in the Atlantic Ocean. (a) Represented by the oxygen distribution in the layer of the oxygen maximum (at about 2000–3000 m depth). Vertical figures: approximate depth of this layer in m. (After Wüst, 1936.) (b) Represented by the current field at 2000 m depth. (After Defant, 1941.) Dotted line: 2000 m depth contour line.

In general, velocities are less than 3 cm sec^{-1}, sometimes they amount to 7 cm sec^{-1}, with a maximum of approximately 15 cm sec^{-1} south of the equator, where the cross section of the current is narrowed. Similar to conditions in the bottom and intermediate waters, it is again the western side of the ocean that is characterized by the strongest current. The geostrophic current field shows a countercurrent beneath the Gulf Stream (cf. Fig. 10.08), which has been confirmed by direct measurements as demonstrated in Fig. 10.09. Such single values of current measurements also illustrate how confusing the results are. Even in the deep sea we must expect a great temporal variability of motion (cf. Section 10.1.9), particularly in the vicinity of strong, variable currents such as the Gulf Stream in the overlying layers.

Along its way south, the North Atlantic deep water mixes with the Antarctic components of the intermediate and bottom waters, and south of 40°S it ascends to higher levels. When reaching the Antarctic water belt, it is included in its entire vertical extent in the Antarctic Circumpolar Current. In this way, the relatively high salinities of the North Atlantic deep water penetrate into the South Indian Ocean and even into the South Pacific Ocean (Plate 5). Ultimately, they represent a large-scale remote effect originating at great depth in the North Atlantic Ocean.

A simple mathematical model of the North Atlantic Ocean (Coriolis parameter, depth distribution of the actual ocean, homogeneous ocean, one single source of deep water, and constant upwelling), as computed by Welander (1969), is in remarkable agreement with the chart of geostrophic currents given in Fig. 10.08.

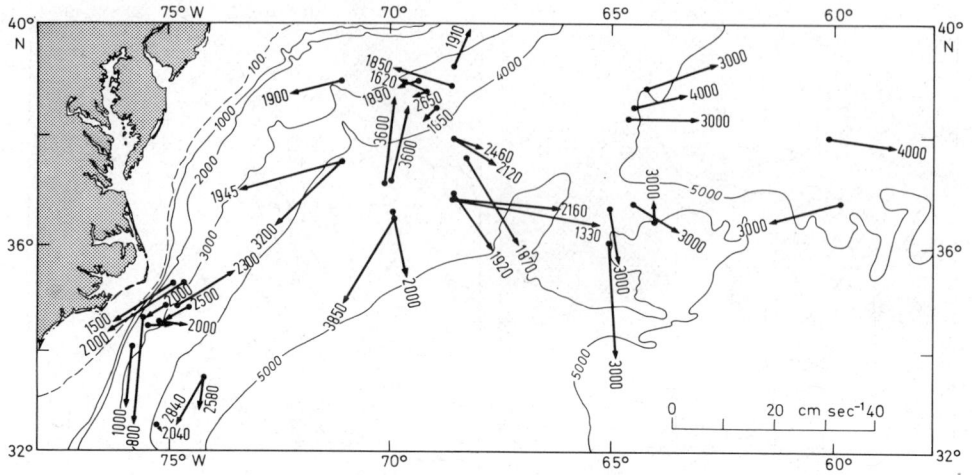

Fig. 10.09. Mean drift of deep-sea drifting buoys (Swallow Floats) at different depths (recording depth in m at the respective arrow head). (After various authors in Stommel, 1965.)

10.1.6. Antarctic water masses

The circulation of water masses in the Antarctic water belt must be called unique if compared to the other parts of the world ocean. (1) Its flow is ring-shaped and not interrupted by continental barriers. (2) Conditions are nearly barotropic due to the weak stratification of temperature and salinity (cf. Plates 3 to 5). Pressure disturbances, occurring at the sea surface, can reach the abyssal floor, which is not the case, due to baroclinic conditions, in the highly stratified warm-water sphere of the lower and moderate latitudes. (3) The winds are strong (cf. Fig. 4.16) and in general directed zonally towards the east. Under such conditions, a mighty and deep-reaching circumpolar current is enforced, comprising the entire water column. A comprehensive description has been given by Gordon (1971). The mass transport is represented in Fig. 10.10 in accordance with calculations by Kort (1964). The 3000 dbar surface has been chosen as reference level, that is, it has been assumed that there is no current at about 3000 m depth. Between adjacent isolines, a water volume of 20×10^6 m^3 sec^{-1} each is transported. Hence, the total transport amounts to 150×10^6 m^3 sec^{-1}. According to results obtained recently, this mass transport seems to increase considerably toward the deep-sea floor. Reid and Nowlin (1971) have computed the water transport through the Drake Passage as amounting to 237×10^6 m^3 sec^{-1}, which means that not 20 but 32×10^6 m^3 sec^{-1} flow between two adjacent isolines (cf. Fig. 10.10). This result is based on direct measurements at six stations, carried out 300 m above the abyssal floor by the American research vessel *Thomas Washington* in 1969. The current measurements served as reference values for computations of the geostrophic currents. From the shear stress of the wind, Ichiye (1970) calculated an amount of 260×10^6 m^3 sec^{-1}. Even without any direct current measurements, Gordon (1966) arrived at similar transport values: 218×10^6 m^3 sec^{-1}. So we may say that the Antarctic Circumpolar Current is the ocean current that achieves the largest water transport on earth.

What would happen to oceanic circulation if this current were interrupted, by closing the Drake Passage for example? Gill and Bryan (1971) have given some relevant hints based on several numerical models with different geometric features of

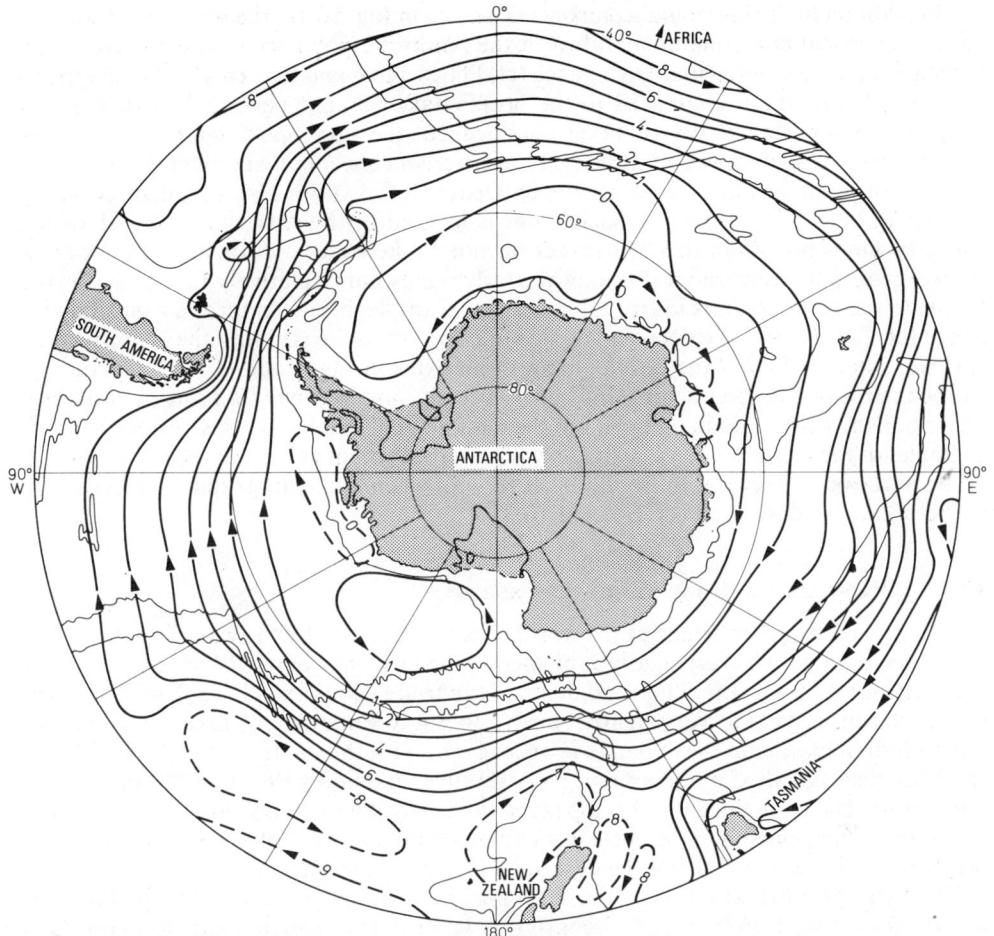

Fig. 10.10. Water transport in the Antarctic water belt. (After Kort, 1964.) Between two lines the water transport amounts to about 20×10^6 m^3 sec^{-1}, if it is assumed that there is no current at 3000 m depth. The only depth contour line: 3000 m.

the Antarctic region. Since the Drake Passage is rather narrow, not only concentration of the water transport is enforced, but also meridional circulation is increased, which is reflected by downwelling motion north of the passage at 50°S, that is, in the sub-Antarctic intermediate water (cf. Fig. 10.05), and by upwelling motion south of the passage at 60°S in the upper North Atlantic deep water (cf. Fig. 10.07 and Plate 4). The influence of bottom topography is clearly evident from the course of the deep-reaching water transport. Before overflowing bottom obstacles, the current deviates toward the left, and after passing them, it turns toward the right, as is to be expected for deep-reaching currents on the rotating earth because the potential vorticity must be conserved. Five large obstacles, namely the South Antillean Arc, the South Atlantic Ridge, the Kerguelen Ridge, the Macquarie Rise, and the South Pacific Ridge must be crossed by the circumpolar current. The result is a wavelike course of the large-scale water transport with a mean direction around the west–east axis.

In addition to the horizontal components, shown in Fig. 10.10, the total mass transport contains vertical components manifest in the Antarctic polar front, the Antarctic Divergence, and the Continental Divergence (cf. Plates 3 to 6 and Fig. 10.46). These vertical components are small compared to the horizontal ones, but nevertheless they are of far-reaching importance. Wyrtki (1960) has deduced the circumpolar course of the fronts from the wind distribution and explained the different concepts concerning the Antarctic polar front described as a convergence by Deacon (1937) but also as a divergence by Wexler (1959). The Antarctic polar front is located in the west wind belt. If strong westerly winds prevail in this belt rather far north, the water motion at the polar front is divergent. It is convergent if strong westerly winds blow relatively far south. Owing to the vertical components in frontal zones, for example, in the Antarctic zone of divergence (cf. Fig. 10.46), the deep water, rich in nutrients, is carried into the vicinity of the surface, where it favors the development of an extraordinary abundance of plankton. In the course of the food chain, this benefits fairly small crabs (*Euphausia superba*), the so-called "krill," which, in turn, are the main food of whales. This is one of the many examples of the close relations existing between physicochemical processes and conditions of life in the ocean; several more will be dealt with when the hydrographic regions of the world ocean are described (cf. Section 10.3).

10.1.7. Water masses of the warm-water sphere

The largest part of the surface of the world ocean is occupied by the warm-water sphere, the top layer of which is especially subjected to the influence of winds—the trade winds being the most important. The wind drifts of the trade winds exist in each of the oceans during the entire year as very steady North and South Equatorial Currents; only in the North Indian Ocean do they change in accordance with the monsoon winds. These currents are distinguished by being relatively shallow, extending down to 200 m depth at the utmost. The phenomena in this top layer will be dealt with in Section 10.3. The above fact is only mentioned here because it is explained by the thermohaline structure of the warm-water sphere, which will be described in the following.

The hydrographic structure of the three oceans in the tropics and subtropics is characterized by a well-pronounced discontinuity layer in the vertical distribution of temperature and thus of density, which is associated with a conspicuous distribution of salinity and oxygen content. The vertical curves of the four factors, shown in Fig. 10.11(a), give typical examples of tropical waters, one for each of the three oceans, and illustrate the considerable agreement. The discontinuity layer separates the homogeneous top layer, extending to 80 m in the first example (a_{Atl}), to 100 m in the second (a_{Pac}) and to 25 m in the third (a_{Ind}), from a nearly homogeneous lower layer which passes into the cold-water sphere further down. The resemblance to summer conditions of stratification in moderate latitudes is striking (cf. Section 5.1.4). Although this tropical–subtropical discontinuity layer is present in the world ocean all year round, its formation is similar, in principle, to that of the summer thermocline. The downward directed heat transport, induced by the forced turbulence of wind-driven currents, decreases with increasing depth. This happens because vertical turbulence is weakened by the density difference between the upper water mass, maintained at higher temperature owing to the heat exchange at the sea surface, and the lower water mass, which is relatively cold due to lateral advection from higher latitudes. The vertical heat flux ceases when the density gradient and the turbulent austausch motions in the vertical direction reach a critical limiting value (Richardson number, cf. Section 7.1.8). Thus, the discontinuity layer may become a layer

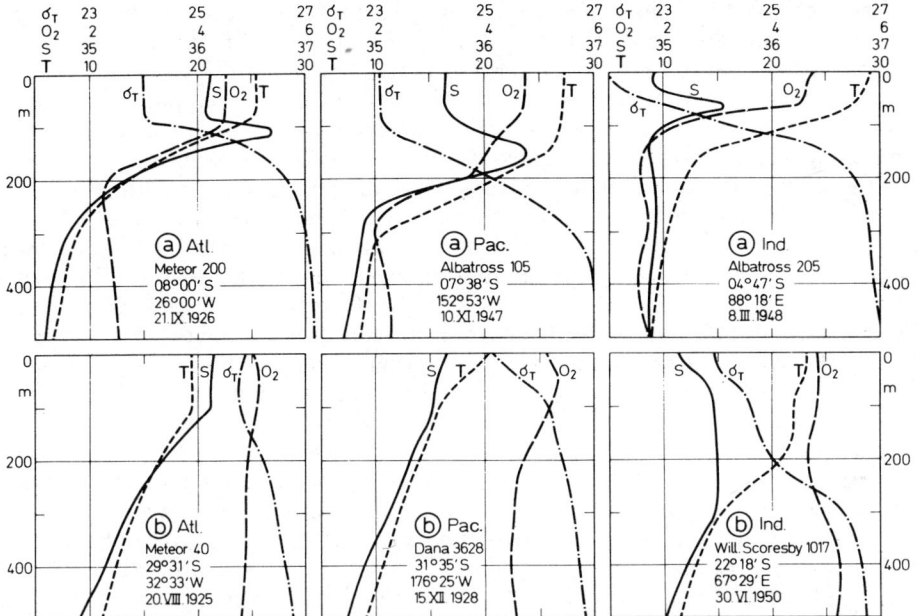

Fig. 10.11. Characteristic distribution of temperature (T in °C), salinity (S in ‰), density (σ_T), and oxygen (O_2 in ml l^{-1}) in the warm-water sphere of the tropics (a) and subtropics (b) of the Atlantic (Atl.), Pacific (Pac.), and Indian (Ind.) Oceans.

barring any exchange and almost completely separating the cold, dense sublayer from the homogeneous top layer, as is the case in the examples cited in Fig. 10.11(a).

Closer investigation of the depth distribution as well as of the intensity of the tropical–subtropical discontinuity layer reveals that distinct regional differences occur within this layer, which are not based on corresponding differences of the wind-induced vertical heat flux. More likely they are caused dynamically by water motions. According to Margules' boundary equation (cf. Section 7.3.2) this is required for a two-layer ocean. The reader is referred to the depth distribution of the tropical–subtropical discontinuity layer in the eastern equatorial Pacific Ocean, presented in Fig. 10.35. A corresponding position of the discontinuity layer is found in the other two oceans. The discontinuity layer rises from about 200 m depth in the subtropics to 20 to 40 m depth toward the eastern side of the equatorial zone. Saddles are noticeable north of the equator in the Atlantic and Pacific Oceans and south of the equator in the Indian Ocean, although there only during the southern summer. These saddles extend over the entire width of the oceans and are closely related to the eastward directed Equatorial Countercurrent, which is embedded in the North and South Equatorial Currents directed westward (cf. Section 10.3.3).

Further remarkable features are recognized if the three examples of the hydrographic structure in the tropics are confronted with three characteristic examples from the subtropics of the oceans, as done in Fig. 10.11(b). If examined more closely, these phenomena will shed light on the motions in the lower warm-water sphere. In all three examples from the equatorial zone, the vertical salinity distribution in the upper part of the density discontinuity layer or pycnocline contains a narrow but pronounced salinity maximum. Thanks to the investigations by Defant (1963a), the formation and distribution of this maximum can well be followed in the Atlantic Ocean. It can be shown that this type of water must be understood as a subtropical water mass spreading under the less saline water of the top layer. Similar conditions are found in the Pacific Ocean (cf. Fig. 10.34). It is not only the direction preferred for spreading that can be recognized, but also the equatorial zone where an intermediate salinity maximum is lacking. This zone coincides with the highest position of the tropical–subtropical discontinuity layer, which can be seen if Fig. 10.35 is viewed for comparison. Here the horizontal spreading of the high-salinity water in the upper part of the discontinuity layer comes to an end. The water masses evade upwards, that is, motions converge at the level of the high-salinity water and diverge in the top layer. The ascending water motions in this equatorial zone of divergence carry water rich in nutrients to the sea surface and are of great biological significance because they thus provide favorable living conditions within this zone (cf. Section 10.3.3). The cause of stratification and circulation in the upper warm-water sphere of the tropics is to be found in the wind effect. Relevant relations are elucidated in Fig. 10.12.

It is a peculiarity of the Atlantic Ocean, as well as of the Pacific Ocean, that the wind system is asymmetrical with respect to the equator. The zone of equatorial calms, the convergence zone of the trade wind, is not situated at the equator but at 5 to 10°N, as indicated schematically in Fig. 10.12(a). The wind-induced water transport in the top layer, the so-called *Ekman transport,* running at 90° to the wind direction, on the right in the Northern Hemisphere and on the left in the Southern Hemisphere, results in a divergence of water motion at the equator, in a convergence on the equatorward side of the calms, and in a divergence on their poleward side. Conditions at the sea surface are illustrated in Fig. 10.12(a); those at a meridional cross section in Fig. 10.12(b). This cross section shows the vertical circulation associated with the system of divergences and convergences. It follows from the meridional sea surface gradient that an eastward current must exist in the zone of the calms. This is the Equatorial Countercurrent (cf. Section 10.3.3). The Equatorial Undercurrent at the equator, also directed eastward, is only understandable from the schematic picture of Fig. 10.12 if the piling-up of the trade wind drift along the eastern sides of the continents is taken into account. The Equatorial Undercurrent is a slope current, without any action of Coriolis acceleration (cf. Section 10.3.2). The Indian Ocean holds a special position. Only in the northern winter does a zone of calms exist here, which is, however, located in the Southern Hemisphere; this is where the Equatorial Countercurrent flows (cf. Plate 6).

If a layer of high-salinity water with a vertical thickness of merely 20 to 50 m [cf. Fig. 10.11(a)] is maintained in large areas of tropical waters, special circumstances must be influential. They have the effect that vertical turbulence almost ceases in this layer and that mixing with the less saline water masses above and below hardly takes place. Nearly laminar motion, without appreciable mixing over long distances, is possible only within a well-defined pycnocline. In fact, observations show that the sandwiched tropical salinity maximum is restricted exclusively to the discontinuity layer.

Beneath the tropical–subtropical discontinuity layer, there is great uniformity of hydrographic factors, indicating only slight water motion [cf. Fig. 10.11(a)]. Owing to the blocking effect of the discontinuity layer, vertical convection and turbulence are prevented almost completely. Thus, the renewal of the water masses of this sublayer can be achieved only in the horizontal direction. At the same time, oxygen is subjected to depletion due to chemical decomposition processes of the sinking detritus. The weak lateral rejuvenation and the great distances over which it takes place explain the extraordinarily low oxygen content beneath the discontinuity layer in the tropics and sub-

Stratification and Circulation in the Deep Layers of the Three Oceans 479

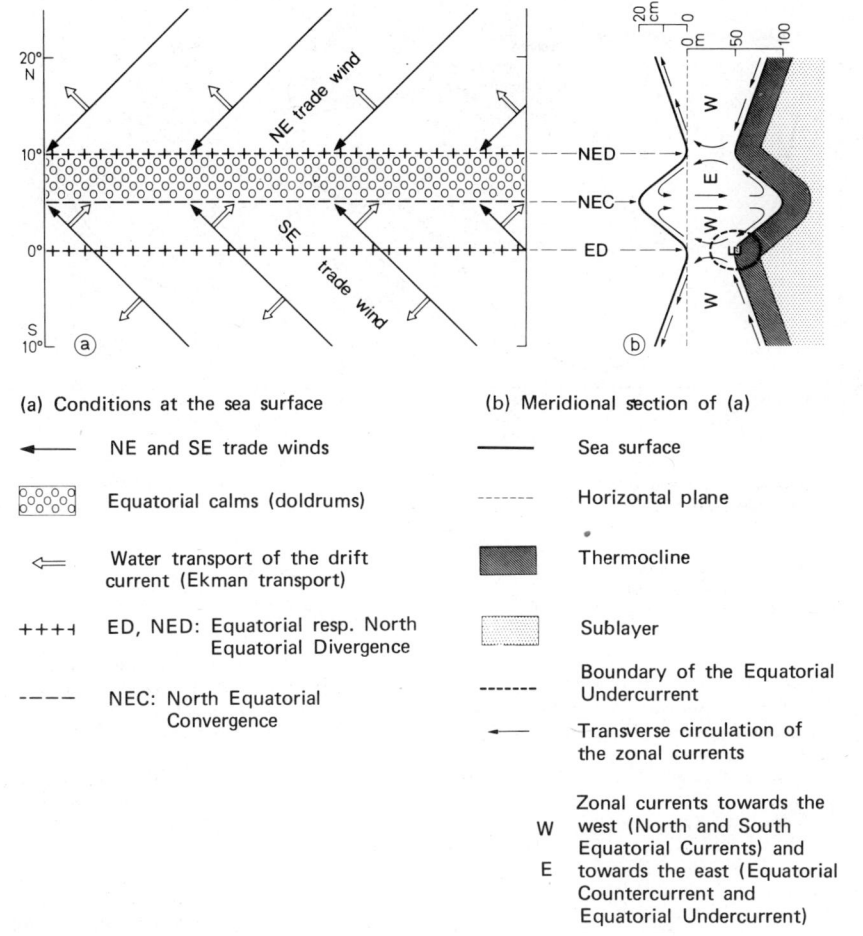

(a) Conditions at the sea surface (b) Meridional section of (a)

⟵	NE and SE trade winds	——	Sea surface
∘∘∘∘	Equatorial calms (doldrums)	------	Horizontal plane
⇐	Water transport of the drift current (Ekman transport)	▇	Thermocline
++++	ED, NED: Equatorial resp. North Equatorial Divergence	░	Sublayer
----	NEC: North Equatorial Convergence	------	Boundary of the Equatorial Undercurrent
		⟵	Transverse circulation of the zonal currents
		W E	Zonal currents towards the west (North and South Equatorial Currents) and towards the east (Equatorial Countercurrent and Equatorial Undercurrent)

Fig. 10.12. Scheme of the wind-induced circulation in the upper tropical warm-water sphere. (In accordance with Neumann, 1968.)

tropics of the three oceans. In the examples given in Fig. 10.11(a), the minimum values for oxygen amount to about 2 ml liter^{-1}. In some regions, especially on the eastern sides of the oceans, the water beneath the discontinuity layer is practically free of oxygen (cf. Fig. 6.12).

In all three oceans, the strong density stratification diminishes poleward of 20° latitude. In the area of the subtropical convergence, from which the three stations of Fig. 10.11(b) have been taken, the discontinuity layer is hardly recognizable. In these regions, a tendency for deep-reaching downward water motion prevails, which explains not only the high oxygen content but also the low nutrient content of the warm-water sphere. This nutrient deficiency contributes to the very unfavorable ecological conditions in these oceanic areas, where, therefore, very little plankton can develop. This is reflected by the fact that, in these regions, the clearest water of the world ocean is found with the greatest range of visibility and cobalt-blue color. Near the surface, these regions are part of the horse latitudes (cf. Section 10.3.5).

480 Regional Oceanography

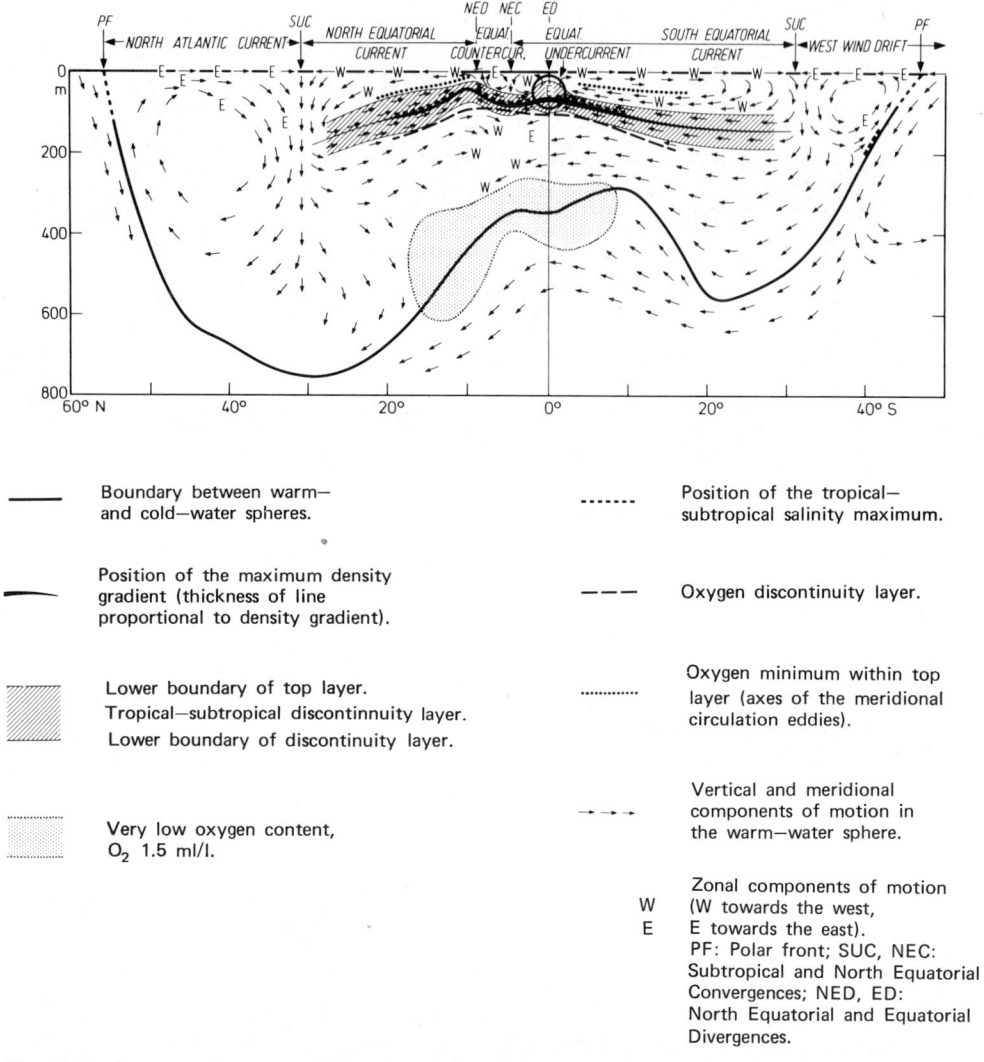

Fig. 10.13. Structure and circulation in the warm-water sphere along the central axis of the Atlantic Ocean. (After Defant, 1936, with amendments by Dietrich.)

In the somewhat schematic meridional cross section along the central axis of the Atlantic Ocean, represented in Fig. 10.13, the various phenomena of structure and motion described above form a coherent picture. The cross section extends from the southern to the northern polar front. An oxygen minimum denotes the lower boundary of the warm-water sphere against the cold-water sphere. The hatched area within the warm-water sphere indicates the thickness of the tropical–subtropical discontinuity layer in true scale. Above, we find a homogeneous top layer, below, a weakly stratified sublayer. The line within the discontinuity layer represents the depth of the maximum vertical density gradient; its thickness is a relative measure of the intensity of the gradient. The dotted line within the top layer denotes a weak oxygen minimum, which—although not present in our examples [Fig. 10.11(a)]—has been observed in some areas. It can be considered the axis of the meridional circulation eddy indicated by arrows. Poleward of 20 to 30° latitude, descending motion begins and, in addition, a motion sets in within and beneath the discontinuity layer, which is directed toward the equator. This results in the intermediate salinity maximum and the oxygen minimum at the lower boundary of the discontinuity layer. Both are especially marked in Fig. 10.13. With this picture in mind, one should remember that a zonal motion is superimposed on the entire meridional circulation, as indicated by arrows. In general, this zonal motion is considerably stronger than the meridional component. Its direction is indicated by W for the current toward the west and by E for the current toward the east.

10.1.8. Circulation gyres of water masses

In combination with mixing processes within the oceanic circulation, the spreading of water masses in some favored layers, as described in Sections 10.1.2 to 10.1.7, leads to the formation of gyres. Neumann (1968) has given a summary of recent results. Various methods have been applied to investigate the circulation. (1) Observation of the distribution of selected parameters by which water masses of different origin can be characterized (temperature, salinity, and content of oxygen, nutrients, and radioactive isotopes, in particular ^{14}C and ^{3}H). (2) Geostrophic computation of currents on the basis of the observed density structure. (3) Direct current measurements. (4) Observation (by means of bottom photography) of ripple marks and erosion forms as indicators of bottom currents. (5) Computation of numerical models under different boundary conditions, such as published, for example, by Bryan and Cox (1967) and by Gill and Bryan (1971). So far, the methods mentioned under (1) and (2) have contributed best to the improvement of our knowledge, although restricted to average circulation conditions. For a long time, direct long-term and continuous current measurements at oceanic depths were problematic as far as the measuring technique is concerned; they have become possible only since the early 1960s (cf. Section 10.1.3). The large increase of current measurements that has been going on in many parts of the world ocean ever since indicates that the circulation is greatly variable in time (cf. Section 10.1.9).

Concepts regarding the average oceanic circulation in the three oceans are based on the observations of numerous expeditions. Some of these concepts have been summarized in the representation of gyres, especially those for the Atlantic Ocean, where the gyres are more pronounced than in the Indian and Pacific Oceans. A schematic block diagram of the Atlantic circulation, presented by Wüst (1950), has often been used in the relevant literature (cf. Fig. 204 in the first edition of this book). This block diagram has proven useful, as shown by the most detailed numerical computation so far of a 14-layer model carried out by Friedrich (1970) for a grid of 3° mesh size in the North Atlantic Ocean. The refinement into a grid of 1° mesh size, as intended by Friedrich, promises that stationary conditions will be reproduced still better, although the demands on computing time and storage capacity of large electronic computers will be high. In future, such modeling studies will become more important if we succeed not only in confirming the observed facts but also in quantitatively determining the reaction of the circulation to various factors by changing the boundary values.

The following will deal with the meridional circulation, as represented for each of the oceans in Plates 3 to 5. With regard to those meridional cross sections, it should be remembered that the depth scale is considerably larger than the horizontal scale, and that oceanic circulation actually takes place in water layers whose vertical thickness is

Table 10.02. Mean Meridional Water Transport through the Western and Eastern Atlantic Basins according to *Meteor* Cross Sections (5–35°S)[a,b]

	Western Atlantic Basin				Eastern Atlantic Basin			
Water Mass Depth	$\varphi°$ (S)	T (10^6 m³ sec⁻¹)	ΣT (10^6 m³ sec⁻¹)	Water Mass Depth	$\varphi°$ (S)	T (10^6 m³ sec⁻¹)	ΣT (10^6 m³ sec⁻¹)
W + SIW				W + SIW			
0–1400 m	9.5	17.5	7.0	0–1200 m	9.5	3.0	15.8
	15.5	13.2			15.5	8.3	
	22	13.2			22	12.9	
	28	3.5			28	22.1	
	33	−12.3			33	32.5	
DW	9.5	−57.2		DW + AABW	9.5	46.9	4.8
	15.5	−38.6			15.5	3.9	
1400–3750 m	22	−26.0	−27.5	1200–5000 m	22	3.5	
	28	−12.0			28	−9.9	
	33	−3.9			33	−20.4	
AABW	9.5	0.0					
	15.5	8.6					
	22	0.9	2.0				
3750–5000 m	28	−2.0					
	33	2.6					
ΣT (0–5000 m)			−18.5	ΣT (0–5000 m)			20.6

[a] After Wüst, 1957.
[b] W: Warm-water sphere. SIW: Sub-Antarctic intermediate water. DW: North Atlantic deep water. AABW: Antarctic bottom water. +: northward; −: southward.

extremely small compared with their horizontal extent. For the motion, this means that the vertical components of motion are negligibly small compared with the horizontal components. In spite of this, their importance should not be underestimated because, in general, the vertical gradients of hydrographic factors are very much larger than the horizontal gradients. Hence, small vertical components of motion may have far-reaching consequences for the horizontal distribution of the hydrographic factors. This is especially evident at the sea surface. Conversely, processes and phenomena in the ocean cannot be sufficiently understood unless the weak components of the vertical circulation are taken carefully into account. From this point of view it is justified, to a certain degree, that the weak vertical components of motion in Plates 3 to 5 have been strongly emphasized by great exaggeration of depth.

The importance of the individual branches of circulation becomes still more obvious when the three plates are supplemented by a list of mass transport values, as given in Table 10.02 for the South Atlantic Ocean. The data are based on five *Meteor* cross sections from South America to Africa, according to calculations by Wüst (1957).

Regarding the average total values, the southward transport of North Atlantic deep water (DW) in the western basin, amounting to 27.5×10^6 m^3 sec^{-1}, has proven to be the dominating part of the circulation. It is contrasted by a northward transport of 7.0×10^6 m^3 sec^{-1} in the upper levels from 0 to 1400 m depth and of 2.0×10^6 m^3 sec^{-1} in the bottom water. No pronounced water transport has been found in the deep and bottom waters of the eastern basin; in the upper level of this basin, however, a northward water transport takes place as a compensation to the southward transport in the western basin.

The transport calculations by Wüst (1957) have been confirmed surprisingly well by investigations during the International Geophysical Year 1957 (Fuglister, 1960; Wright, 1970) and by the work carried out by the research vessels *Akademik Kurchatov* and *Dimitriy Mendeleev* from February to May 1969 (Kort, 1972). This can be taken as an evidence that the stratification is largely stationary. Direct current measurements by *Akademik Kurchatov*, however, show transport values that are considerably higher than those computed geostrophically. This indicates that strong fluctuations in time, summarized under the term variability (cf. Section 10.1.9), are more effective than the average oceanic circulation.

10.1.9. Variability in the ocean

Variability in the ocean is a very comprehensive term for variations of stratification and motion in time and space. Since the middle of the 1960s it has become a central topic of marine research (Dietrich, 1966). Intensive investigations have been carried out, and international symposia have taken place [Sears (Ed.), 1969; Deacon (Ed.), 1971; Lee (Ed.), 1972]. This does not mean that, formerly, stratification and currents were considered to be stationary; but the measuring technique was not so well developed that variability could be studied systematically. Repeated attempts were made to provide the necessary fixed point in the ocean by anchoring a research vessel from which individual measurements could be taken. As an example of this method, ten 3 day anchor stations may be mentioned that were carried out by *Meteor* during the German Atlantic Expedition of 1925–1927. Defant (1932) made use of them when investigating the variability in time of currents and stratification. The weak points of this measuring technique were the strong motions of the vessel itself when anchored in the deep sea, and the great time intervals between the individual measured values. Both these disadvantages have been eliminated by modern measuring technique employing moored instrument arrays. Each of the instruments supplied continuous records of current direction and velocity as well as of temperature at selected levels (cf. Section 3.1.2.1). If several measuring arrays are moored at selected distances, the variability in space can also be determined. During the "Norwegian Sea 1969" investigation, relevant experiments were made in a limited area with 48 current meters at 6 moorings (Dietrich and Horn, 1973). The same was done during the "Upwelling Expedition 1972" with 10 moorings carrying 59 recording instruments at short distances.

Nonstationary currents manifest themselves in the energy spectrum (cf. Section 7.4.1). Periodic and nonperiodic processes can be recognized, their properties can be described, their mechanism can be explained, and their interrelations with other outward phenomena, such as wind and sea state, can be determined. But with regard to the distribution of measuring points and the duration of registration, the new series of data do not yet suffice for the complete understanding of the variability. As long as this is the case, it remains difficult for us to recognize to what degree a series of measurements is representative. The variability and its consequences have been discussed by Dietrich (1969b). In the following, some significant phenomena of variability will be illustrated by three examples in Figs. 10.14 to 10.16.

The energy spectra (cf. Section 7.4.1) for selected horizons throw novel light on the dynamics of oceanic motion processes. In Fig. 10.14, spectra from the western North Atlantic Ocean are confronted with others from the eastern North Atlantic Ocean. In Fig. 10.14(a), they are based on continuous current records at position "D," obtained at several horizons in water 2600 m deep by the Woods Hole Oceanographic Institution since 1965. Figure 10.14(b) represents continuous current records taken by a current meter mooring at 4600 m water depth, anchored by *Meteor* at the Great Meteor Seamount over a period of two weeks in 1967.

Some observational facts may be emphasized here.

1. Variability is governed by four periods: (a) stability oscillations (Brunt–Väisälä periods), the shortest of which amounts to about 1 hr or less, according to the vertical density gradient (not represented in Fig. 10.14); (b) semidiurnal tidal periods; (c) diurnal tidal periods; (d) inertial period, which is particularly pronounced in Fig. 10.14(a)—in the case of Fig. 10.14(b) concerning 30° geographical latitude, it is identical with the period of the diurnal tide.
2. The kinetic energy of the fluctuations is considerably greater than that of the mean current.
3. The ratio of the energy of fluctuations to the energy of mean currents is conspicuously similar in both oceanic areas.

Table 10.03 shows that, according to registrations made over several years, the kinetic energy of mean currents and of variable currents greatly decreases with depth but increases again near the bottom. A highly important fact is elucidated here, namely that the kinetic energy of mean currents at all levels is surpassed many times by that of currents variable in time. The conclusion is obvious that changes in the kinetic energy of fluctuations might affect the long-term variations of mean currents.

To complete the results of the spectral analysis of continuous current registrations, given in Fig. 10.14, progressive vector diagrams of selected sections of the records, based on hourly mean values, are represented in Fig. 10.15 for the same positions and horizons.

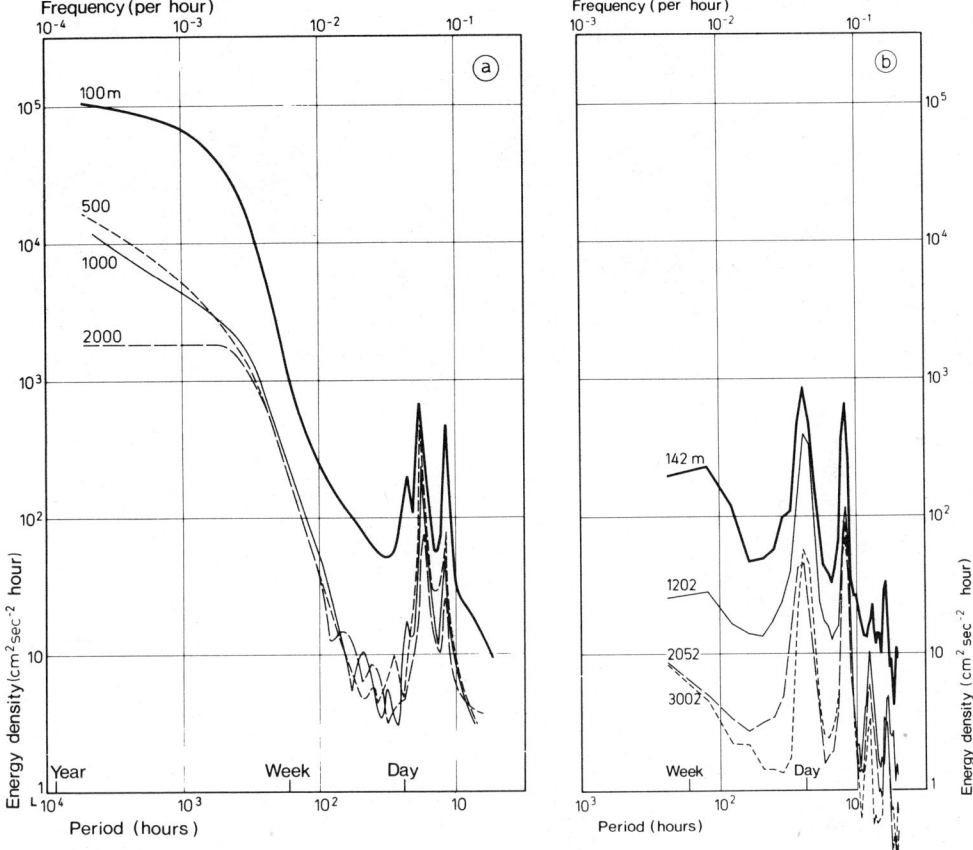

Fig. 10.14. Horizontal kinetic energy density in cm² sec⁻² multiplied by hr as a function of the period (in hr) and frequency (per hr), respectively. (a) At position "D" ($\varphi = 39°20'N$, $\lambda = 70°0'W$) at 100, 500, 1000, and 2000 m depths with a water depth of 2600 m. (After Thompson, 1971.) The current records were obtained during the years from 1965 to 1969 and include several hundred recording days at each depth level. (b) At position "C" (cf. Fig. 9.16; $\varphi = 29°5'N$, $\lambda = 29°2'W$) during the period April 13 to 29, 1967 at 142, 1202, 2052, and 3002 m depths with a water depth of 4600 m. (After Meincke, 1971.) Compare also Figs. 10.15 and 9.16.

It is characteristic that, apart from the periods of tides and inertial waves, abrupt changes of direction do occur, most of them at all levels.

Rossby and Webb (1971) adopted a new way of determining long-term variations of oceanic motions. By employing the acoustic fixing and ranging method (SOFAR), they traced the trajectory of a floating body—a special kind of Swallow Float—from the coast over a period of 4 months. The result is represented in Fig. 10.16. The float was drifting at a depth of about 1100 m at the level of the Mediterranean water (cf. Fig. 10.07). The transport, generally directed westward, is in agreement with the direction of the geostrophic motion, but not so the velocities: those that were observed directly range between 0 and 15 cm sec⁻¹ and amount to 2.8 cm sec⁻¹ on the average, whereas the values computed geostrophically only amount to about 0.3 cm sec⁻¹. One might suspect that planetary waves (cf. Section 7.4.3) are involved to a great extent in the oceanic basins.

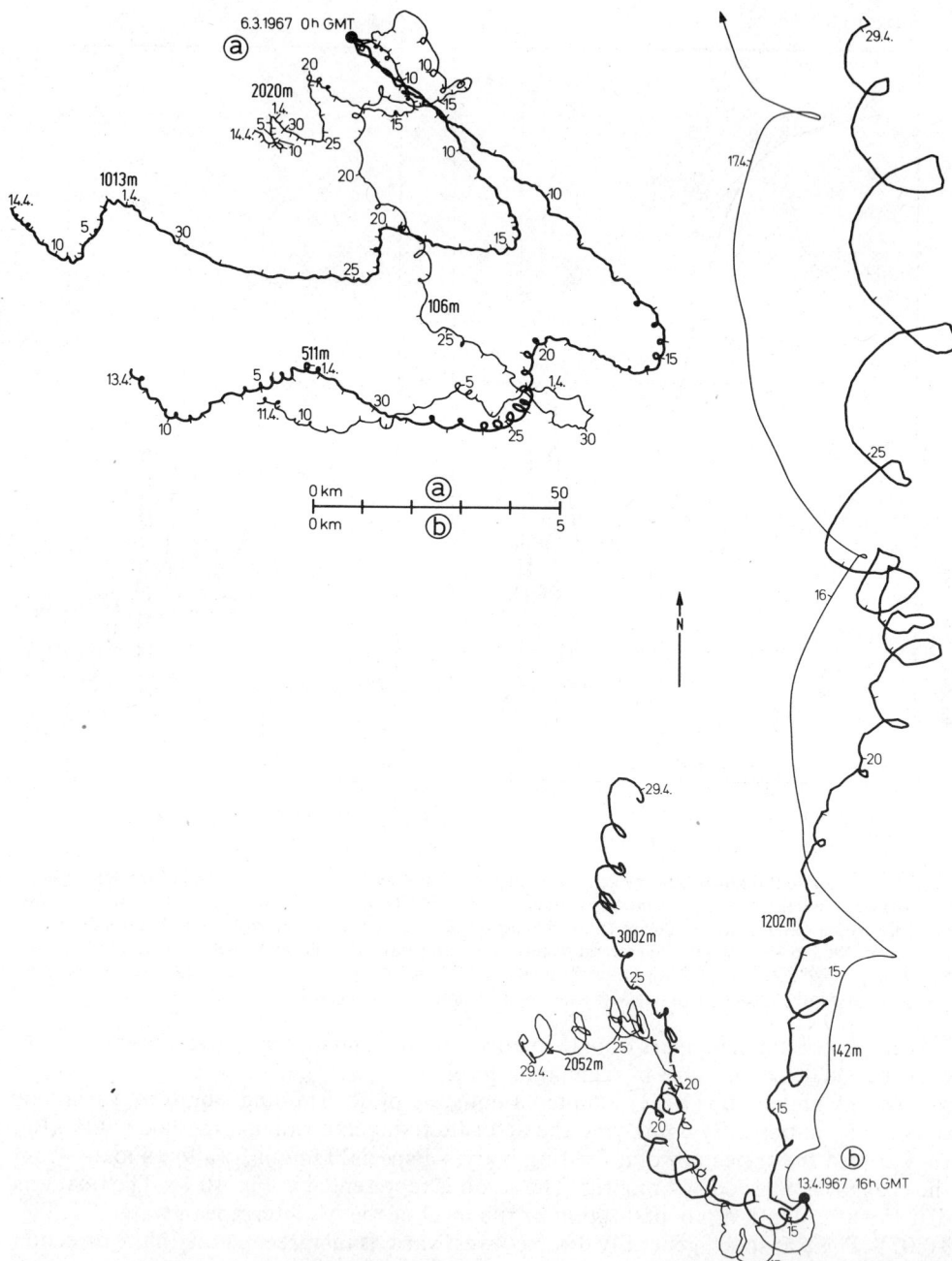

Fig. 10.15. (a) Water displacement at position "D" (cf. Fig. 10.14) at 106, 511, 1013, and 2020 m depths from March 6 to April 14, 1967, represented as a progressive vector diagram. (After Fofonoff, 1969.) (b) Water displacement according to current meter moorings of the research vessel *Meteor* at 142, 1202, 2052, and 3002 m depths from April 13 to 29, 1967, at the Great Meteor Seamount, represented as a progressive vector diagram. (After Horn et al., 1971.)

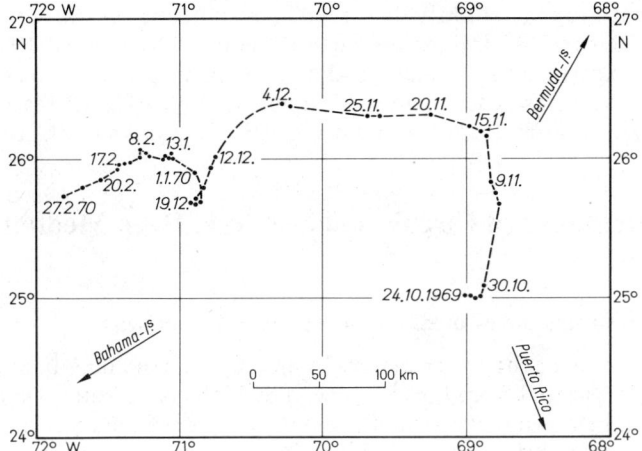

Fig. 10.16. Trajectory of a Swallow Float at about 1100 m depth in the region between Puerto Rico, the Bermuda and Bahama Islands from October 24, 1969 to February 27, 1970. (After Rossby and Webb, 1971.) Dots: well-established positions.

The cause of such planetary waves could be found in the influence of wind and atmospheric pressure. Quantitative relations are not yet known, since an observational network that is dense enough is still lacking, but it is expected that such a network will be provided by the great international program of IGOSS (Integrated Global Ocean Station System). As long as it has not yet been established, investigations are concentrated on the variability in selected regions. The key problems refer to the dynamics of the following phenomena:

1. Jet stream in the open ocean. Test objects: the Gulf Stream between Cape Hatteras and the Grand Banks as well as the Equatorial Undercurrent.
2. Overflow of ocean margins. Test object: the overflow of the Greenland–Scotland Ridge.
3. Upwelling at ocean margins. Test objects: upwelling off the coasts of Northwest Africa, Peru, and California; in the first case, in particular with the following research vessels taking part: *Meteor, Planet, Discovery,* and *Jean Charcot*.

Table 10.03. Kinetic Energy of Mean and of Variable Currents at Position "D" ($\varphi = 39°20'N$, $\lambda = 70°0'W$)[a]

Depth (m)	Kinetic Energy (cm² sec⁻²)		Ratio Variable/Mean
	Mean Current	Variable Current	
10	85	404	5
100	25	197	8
500	7.0	34	5
1000	6.5	20	3
2000	1.2	16	13
2500	7.9	71	9

[a] After Fofonoff and Webster, 1971.

4. Areas that are only slightly disturbed. Test objects: central western North Atlantic Ocean (MODE: Mid-Ocean Dynamics Experiment 1973) and the Polygon Experiment at around 16°30'N, 33°30'W over a period of 6 months in 1970. The latter experiment was carried out by six USSR research vessels with 170 current meters on 17 moorings (Brekhovskikh et al., 1971).

10.2. Stratification and Circulation of Large, Deep Mediterranean Seas

10.2.1. Circulation scheme of mediterranean and adjacent seas

Structure and motion of the water masses in mediterranean and adjacent seas, that is, in oceanic areas separated from the open ocean by high-rising bottom sills, show special characteristics of their own that are fundamentally different from those of the neighboring ocean. Such seas will be dealt with in this section; but the discussion will be restricted to the four large mediterranean seas which also are intercontinental (cf. Table 1.01) and to the Red Sea. Here, details cannot be considered, and only the significant fundamental features will be described.

Two phenomena are of special importance with regard to hydrographic conditions at great depths of all mediterranean and adjacent seas: (1) the horizontal density difference between the marginal sea and the open ocean at the level of the sill depth; (2) the inclination of the sea surface in the area connecting marginal sea and open ocean, which is maintained by the water exchange W_Σ within the marginal sea.

If the heat exchange Q_Σ (cf. Section 4.2.6) and the water exchange W_Σ (cf. Section 4.3.6) permit a deep-reaching thermohaline convection in a marginal sea for longer periods of the year, the density of this sea will assume higher values at the level of the sill depth than in the open ocean. Water will flow over the separating bottom sill into the ocean. Conversely, water from the ocean will flow over the sill into the marginal sea if the density is not increased there by thermohaline convection but decreased by dynamic convection. Two fundamentally different cases result: (A) outflow from the marginal sea over the saddle; (B) inflow into the marginal sea over the saddle of the separating sill.

Depending on the inclination of the sea surface, maintained between the marginal sea and the ocean, two cases (I) and (II) can be distinguished within (A) and (B): (I), the inclination is of no importance; (II), the inclination determines the current velocities in the connecting strait. The inclination of the sea surface will be larger if, within the marginal sea, the ratio of the water exchange W_Σ per year to the cross section F_S of the connecting strait (corresponding to the velocity V_S) results in measurable currents in this strait. In the Caribbean Sea ($W_\Sigma = -2.16 \times 10^{12}$ m^3 yr^{-1}, $F_S = 480 \times 10^6$ m^2), $V_S = -4.5$ km yr^{-1} = -0.014 cm sec^{-1}, whereas in the European Mediterranean Sea ($W_\Sigma = -2.42 \times 10^{12}$ m^3 yr^{-1}, $F_S = 4.2 \times 10^6$ m^2), $V_S = -576$ km yr^{-1} = -1.9 cm sec^{-1}. In accordance with the classification mentioned above, four categories of mediterranean and adjacent seas (A_I, A_{II}, B_I, B_{II}) can be distinguished. These four types are sketched schematically in Fig. 10.17, and three examples are given for each of them.

A_I: At times, especially in winter, climatic conditions cause deep-reaching convection and the generation of heavy bottom water. The formation and spreading of polar and subpolar bottom water, illustrated in Fig. 10.01, repre-

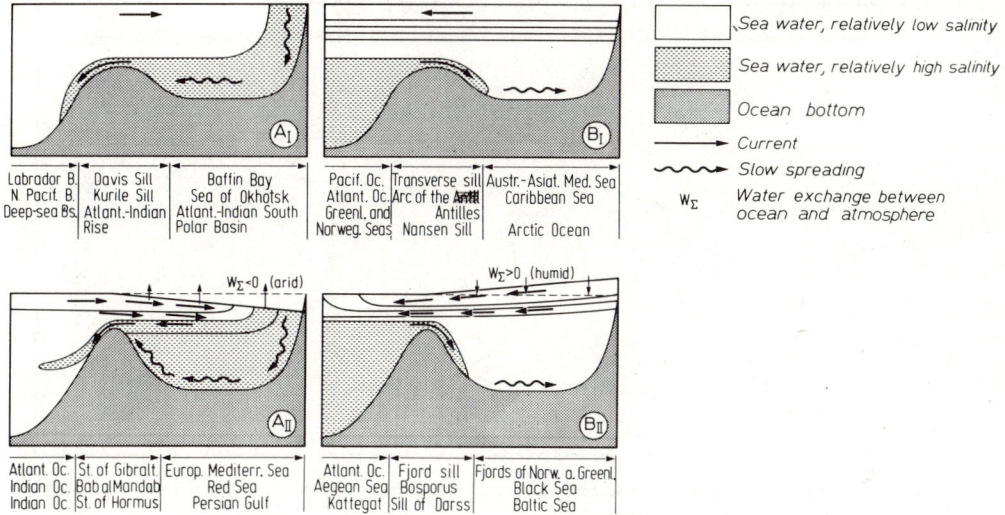

Fig. 10.17. Circulation schemes for mediterranean and adjacent seas with three examples each. (A) with, (B) without deep-reaching thermohaline convection. I, for small V_S; II, for large V_S.

sent the most common example of this case. The Greenland and Norwegian Seas (cf. Section 10.2.6) belong to the same category.

A_{II}: A pronounced excess of evaporation, as it occurs in arid climates, tends to increase the salinity at the sea surface and facilitates deep-reaching thermohaline convection in winter. Consequently, the water in the marginal sea has a higher density at the level of the sill depth than the water in the neighboring ocean. The result is an outflow over the bottom sill. Owing to the negative water balance of the marginal sea, an inflow exists in the upper layer, its velocity depending on this water balance and the cross section of the connecting strait. At the sea surface, the inflowing lighter water can be followed far into the marginal sea. The European Mediterranean Sea (cf. Section 10.2.2), the Red Sea (cf. 10.2.3), and the Persian Gulf (cf. 10.3.9) represent the main examples.

B_I: If, in contrast to A_I, a pycnocline is maintained throughout the entire year, the vertical exchange causes a density decrease in the deep layers. Hence, the water of the marginal sea is less dense at the level of the sill depth than the water in the neighboring ocean. The heavier ocean water flows across the bordering bottom sill into the marginal sea and spreads there in the layer between the level of the sill depth and the bottom of the deep-sea basin, while being subjected to permanent mixing. This category includes the Austral–Asiatic (cf. Section 10.2.4) and the American Mediterranean Seas (cf. Section 10.2.5), as well as the Arctic Ocean (cf. Section 10.2.6).

B_{II}: B_{II} comprises the mediterranean and adjacent seas in humid climates favoring the development of a pycnocline, which, to a large extent, prevents thermohaline convection. At the exits of such seas, the outflow in the upper layer is fairly large, due to the positive water balance W_Σ. It transports the lighter water of marginal seas far beyond the exits into adjacent oceanic areas.

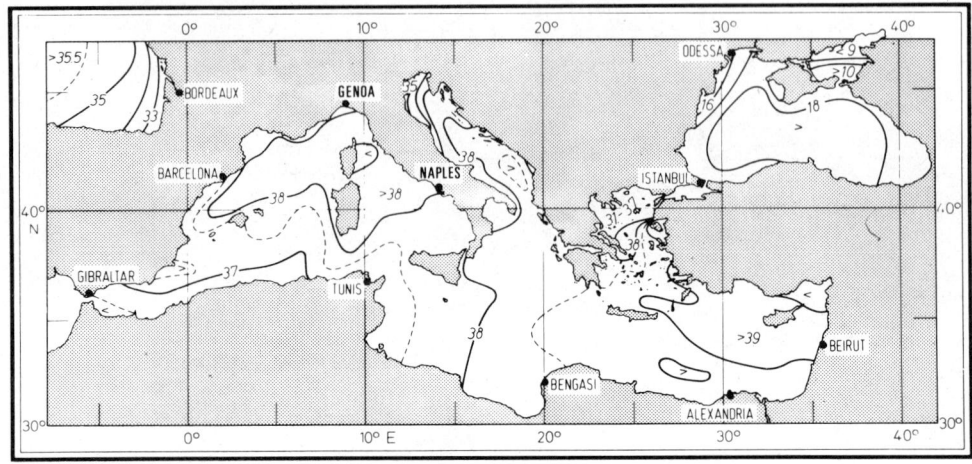

Fig. 10.18. European Mediterranean Sea and Black Sea. Mean surface salinity (‰). (After German Hydrogr. Inst., 1959, and U.S. Navy Oceanogr. Office, 1967.)

The fjords of Norway and Greenland (cf. Section 10.3.10), the Baltic Sea (cf. Section 10.3.9), and the Black Sea (cf. Section 10.2.2) are representatives of this category.

10.2.2. The European Mediterranean Sea

With a view to circulation and stratification, the European Mediterranean Sea belongs to category A_{II}. A summarizing survey regarding stratification is found in the *Atlas of the Mediterranean Sea* by Miller et al. (1970). Deep-reaching thermohaline convection can occur in winter at four different locations where, due to high salinity, strong cooling is favored by cold northerly gales. The chart of surface salinity (Fig. 10.18) indicates where those locations may be found. This has been confirmed by Ovchinnikov and Plakhin (1965) on the basis of measurements taken by the Soviet research vessel *Akademik S. Vavilov* in 1963:

1. The Ligurian Sea and the area off the Lion's Gulf. Here, the surprisingly large vertical component was directly measured during the Medoc investigation in January 1969 (Voorhis and Webb, 1970). The formation of vertical convection, investigated during this study, has been dealt with by Stommel (1972).
2. The southern Adriatic Sea.
3. The southern Aegean Sea.
4. The waters between Rhodes and Cyprus.

Furthermore, the distribution of the potential bottom temperature, as determined by Wüst (1961) for depths of more than 1000 m, also shows the main spreading of bottom water (cf. Fig. 10.19). The deep water of the eastern Mediterranean Sea is separated from the western Mediterranean by the Sicilian Sill with a saddle depth of only 350 m in the Strait of Tunis. Within each of these basins, the water is of extreme homogeneity

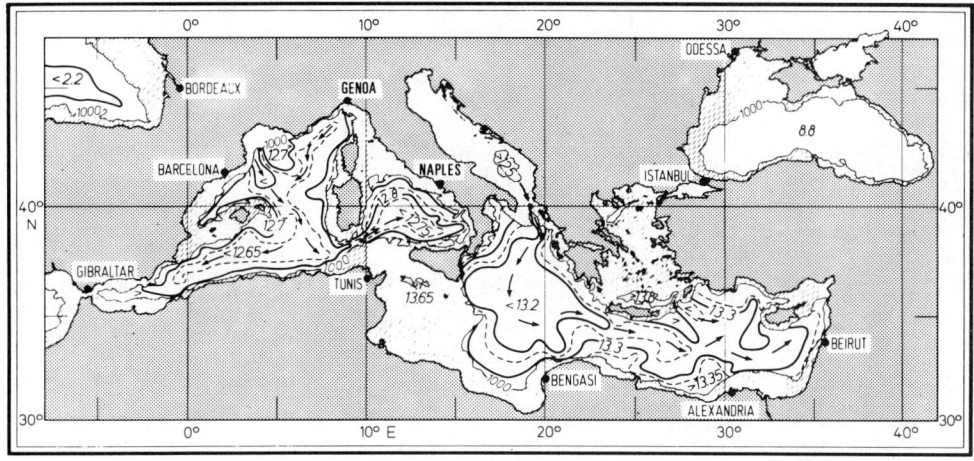

Fig. 10.19. European Mediterranean Sea and Black Sea. Temperature (°C, potential) of the bottom water and its spreading. (After Wüst, 1961.) Hatched screen: 0–1000 m.

and shows a constant composition of isotopes (Menaché, 1970). Therefore, this water is already being used at several institutes (Cox, McCartney and Culkin, 1970; Kremling 1971) as reference water for precision measurements of the relative density of seawater. This is rather surprising, because deep-reaching vertical convection, as observed during Medoc in 1969, suggests that these water masses are not of great age.

Another water body, formed in the Mediterranean Sea, is the Levantine intermediate water, which, in winter, assumes its peculiar features at the sea surface in the area of Cyprus by interacting with the atmosphere. Such properties include particularly high salinity (about 39.1‰), as shown by Morcos (1972). The core of this water body can be traced as intermediate salinity maximum (cf. Fig. 10.20), crossing the Sicilian Sill with a salinity of 38.7‰ and leaving the Mediterranean Sea over the saddle of the Gibraltar Sill with a salinity of 38.4‰. Since the currents, flowing in a pronounced tidal rhythm, are extremely strong in the Strait of Gibraltar (Fig. 10.21), extensive mixing takes place with the less saline, colder Atlantic water. Thus, salinity and temperature of the outflowing Mediterranean water greatly decrease with increasing distance from the Strait. Nevertheless, the mixed water west of the Strait is still heavier than the surrounding water. It descends to a depth of approximately 1000 m, where it encounters water corresponding

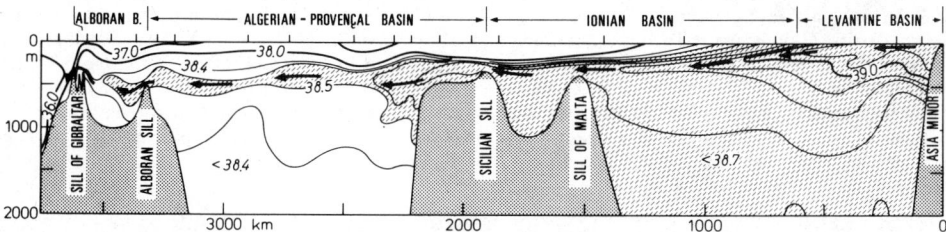

Fig. 10.20. Longitudinal section of salinity through the Mediterranean Sea along the axis of the Levantine intermediate water in winter. (After Wüst, 1961.) Arrows: main spreading of the Levantine intermediate water. Light screen: salinity > 38.5‰. Exaggeration of vertical scale: 2700-fold.

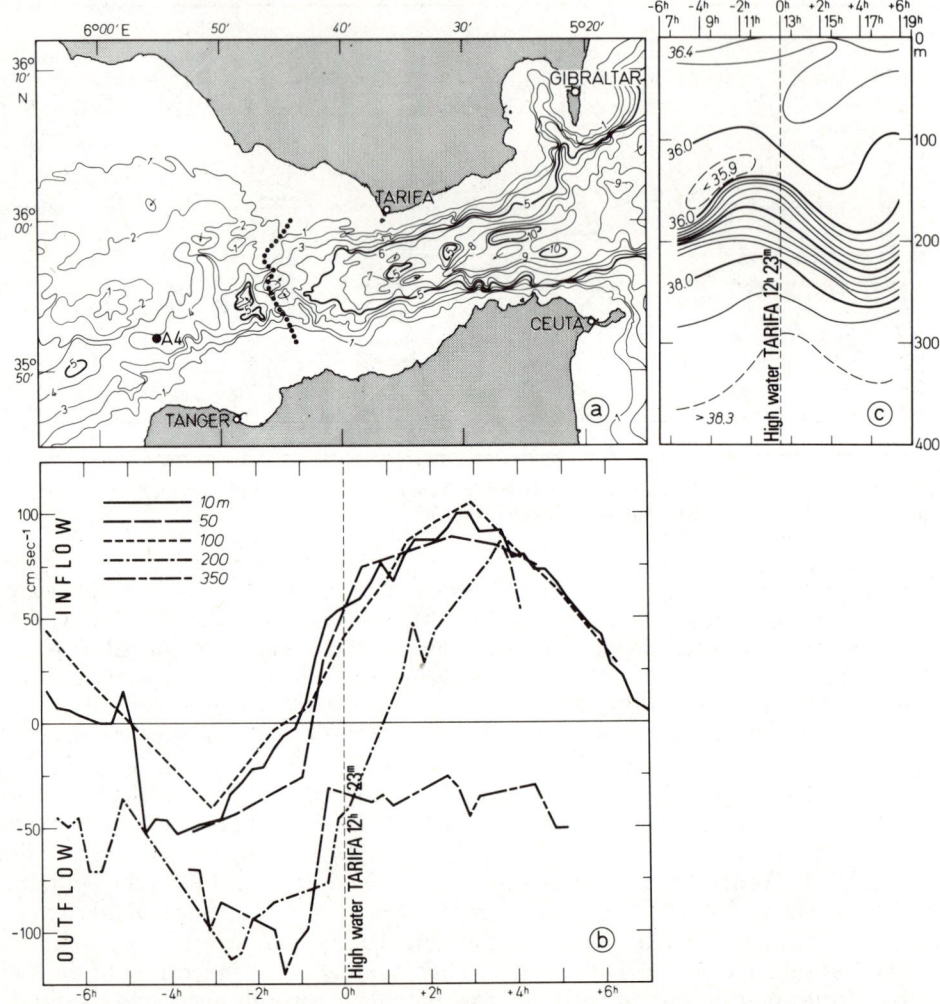

Fig. 10.21. Temporal course of current and stratification in the Strait of Gibraltar on September 18, 1960 at anchor station A 4 ($\varphi = 35°52.4'$N, $\lambda = 5°55.2'$W), referred to the time of high water at Tarifa (12 hr 23 min CET). (After Lacombe et al., 1964.) (a) Bathymetric chart of the Strait (depth in hm) after Giermann and Pfannenstiel, 1961, with station A 4. Dotted line: crest of the sill; sill depth approximately 286 m. (b) Currents at five different depths in the direction of the longitudinal axis of the Strait. Time scale: hr before and after high water at Tarifa. High water at Tarifa on September 18, 1960: 12 hr 23 min CET. (c) Vertical distribution of salinity during the course of the tide. Time scale: hr in CET. Above: hr before and after high water at Tarifa. Tidal coefficient C = 67. The "tidal coefficient" is a very useful relative measure of the tidal range. It is mainly applied by French oceanographers. It indicates the tidal range in percent of the spring tidal range at the time of the equinoxes. C = 67 roughly corresponds to the mean tidal range of C = 70; the mean spring tidal range is C = 96, the mean neap tidal range is C = 45.

to its own density. At this level, the water spreads fan-like. This "Mediterranean water" can clearly be recognized in Fig. 10.22 showing cross sections of temperature, salinity, and density by Zenk (1971), based on continuous bathysonde records of the research vessel *Meteor* between Portugal and Northwest Africa. Under the influence of the Coriolis force, the outflowing water with its intermediate maxima of salinity (greater than 36.5‰)

and of temperature (12.5°C) leans against the continental slope. Its velocity is so great that investigations of sediment, made during the same *Meteor* cruise, suggest erosion at the sea floor (Giesel and Seibold, 1968). This intermediate salinity maximum is the characteristic feature of the core of the upper North Atlantic deep water, which, as shown in Fig. 10.07, can be traced throughout the Atlantic Ocean and also in the South Indian and South Pacific Oceans (cf. Plates 5 and 3). This phenomenon may be considered as a long-distance effect of the European Mediterranean Sea.

Furthermore, attention is drawn to the unusual stratification shown in Fig. 10.22. Warm, high-salinity water from the Mediterranean Sea lies on top of colder, low-salinity deep water of the Atlantic Ocean. Since molecular thermal conduction occurs about 100 times faster than molecular diffusion of salinity, the stratification at depths from 1200 to 1500 m becomes unstable. "Salt fingers" (i.e., chimney-like vertical convection) develop and are manifest in a staircase-like structure of stratification with steps about 30 m high. Zenk (1971) has described and studied this phenomenon on the basis of *Meteor* observations obtained in January 1967.

Since the water balance of the Mediterranean Sea is negative, there must be a strong inflow of Atlantic surface water through the Strait of Gibraltar. It can be recognized by its relatively low salinity when flowing along the North African side of the Mediterranean Sea (cf. Fig. 10.18), where it is subjected to mixing, as investigated by Lacombe and Tchernia (1960).

With the completion of the Aswân Dam in 1965, a change in the regime of the water masses has taken place in the eastern Mediterranean Sea. Although it is a small-scale event with regard to the entire Mediterranean Sea, its effect is of general interest as an example of the interference of man with the water budget of nature. Relevant contributions have been given by Morcos (1967) and Oren (1969); besides, Oren (1970) has treated this problem in a monograph on the waters of the eastern Levantine Sea. Since 1965, the water of the Nile, rich in nutrients, has failed to appear in the Mediterranean Sea, and the Egyptian sardine fishery off the delta of the Nile has broken down. This is but one effect of the damming of the Nile; another can be seen in the fact that water from the Nile no longer enters the northern section of the Suez Canal (El Sabh, 1969). Thus, water and organisms of the Red Sea can get through the Canal into the Mediterranean Sea without any difficulty. As a result, increased immigration of Indo-Pacific organisms takes place in the Mediterranean Sea. Steinitz (1967) has determined 115 different new species of animals, among them immigrants of interest to commercial fishery.

The *Black Sea* is generally regarded as part of the Mediterranean Sea. From the hydrographic point of view, it belongs under category B_{II}, according to Section 10.2.1. Its water in the deep layers is only weakly renewed, because the highly saline undercurrent, flowing through the Bosporus into the Black Sea, supplies an annual average of merely 193 km^3 yr^{-1}, as shown by Merz and Möller (1928). At this rate, it would take 2500 yr before all the water of the Black Sea were replaced. Consequently, the salinity in the deep layers remains low and represents the equilibrium between influx and vertical convection. The vertical distribution of salinity is illustrated in Fig. 10.23. The renewal is not sufficient to compensate for the oxygen depletion. Therefore, it is a characteristic feature of the Black Sea, as shown in Fig. 10.23, that, from a depth of about 130 m down to the greatest depth of 2244 m, the water contains no oxygen, but large amounts of hydrogen sulfide, making any life in the deeper layers impossible. These relations between the physicochemical structure and the conditions of motion in the Black Sea are the subject of a detailed oceanographic monograph by Neumann (1944). Marine chemical conditions have been dealt with by Skopintsev (1962). A summarizing description has been given by Zenkovitch (1966). Detailed oceanographic investigations of the Black

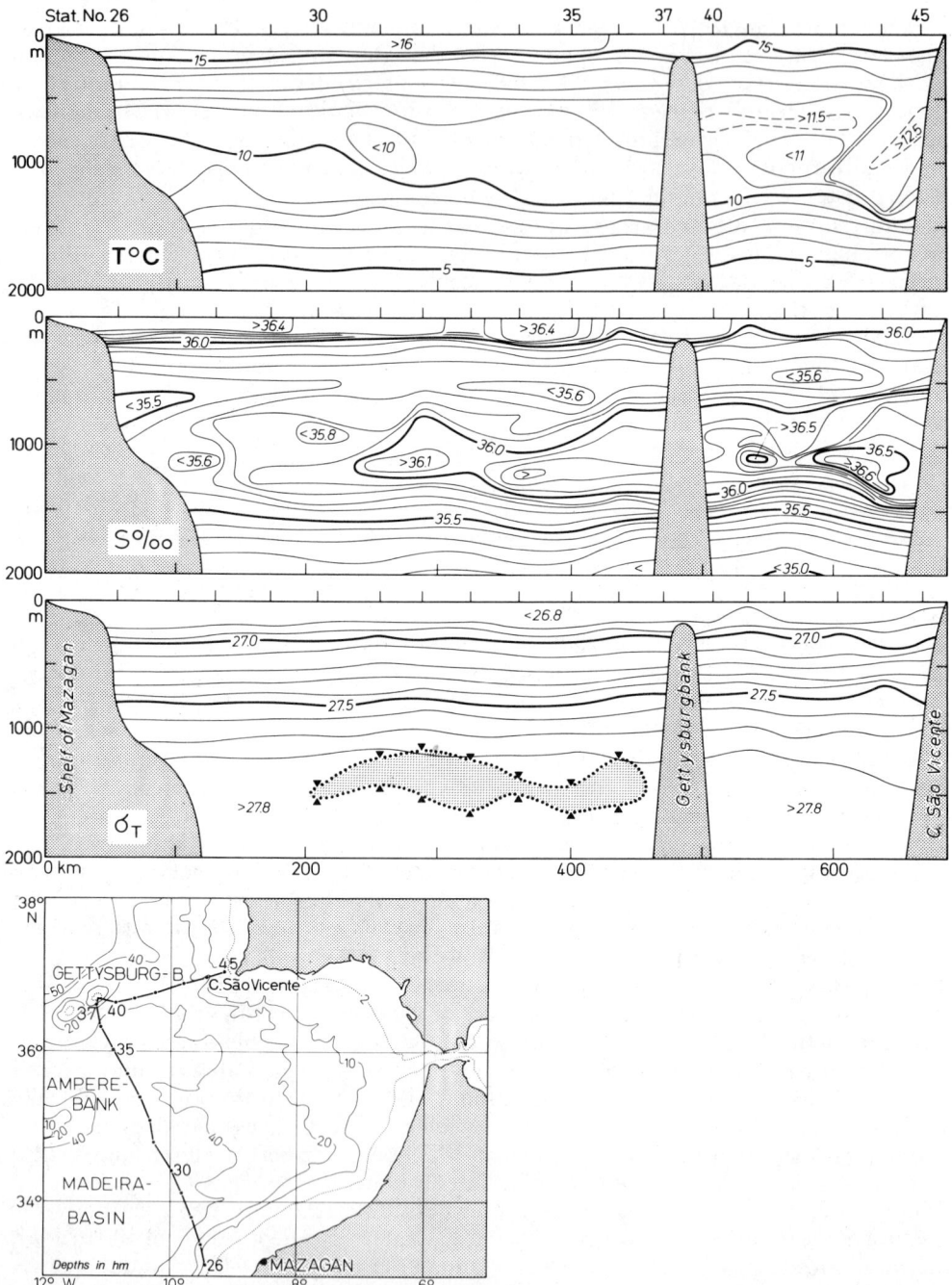

Fig. 10.22. Outflow from the Mediterranean Sea. Stratification from the surface to 2000 m depth between NW Africa near Mazagan and Portugal near Cape São Vicente (see sketch of location), based on bathysonde records of the research vessel *Meteor* in January 1967. (After Zenk, 1971.) Screen in σ_T section: Step-like stratification of salinity.

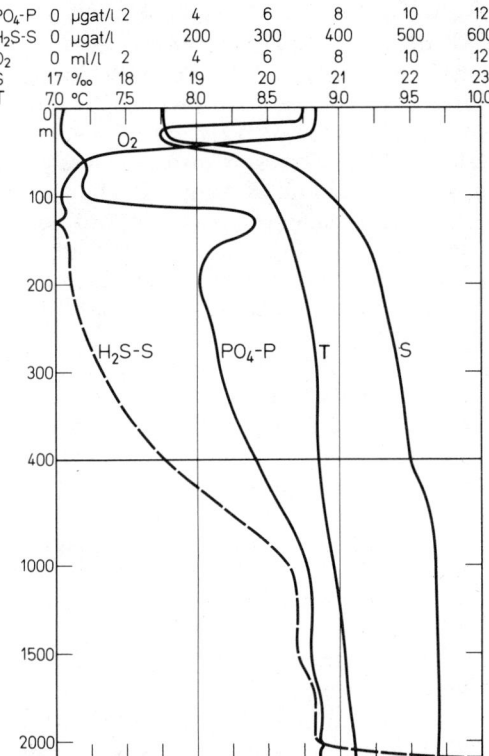

Fig. 10.23. Vertical distribution of temperature, salinity, oxygen, hydrogen sulfide, and phosphate in the Black Sea at *Atlantis II* Station 1466 ($\varphi = 43°01'$N, $\lambda = 38°30'$E, water depth 2106 m) on April 16, 1969. (After Grasshoff, 1971.) Note change of depth scale at 400 m.

Sea were carried out by the research vessel *Pillsbury* of the Institute of Marine Sciences, Miami in July to August 1965 (Sen Gupta, 1971) and by the research vessel *Atlantis II* of the Woods Hole Oceanographic Institution from March through May 1969 (Degens and Ross, 1970). During the latter period, the data shown in Fig. 10.23 were obtained by Grasshoff (1971).

Unlike the European Mediterranean Sea, the Black Sea is not situated in the region of arid climate. The large influx of fresh water from the rivers Danube, Dniester, Dnieper, and Don cause a water surplus, thus contributing to the freshening of the near-surface top layer to less than 19‰ salinity (cf. Fig. 10.18). The bottom salinity at the great depths beneath 1000 m amounts to 22.0 to 22.5‰ (cf. Fig. 10.23). The water surplus of the Black Sea runs off through the Bosporus as a low-salinity surface current (348 km³ yr⁻¹); a highly saline undercurrent enters the Black Sea in the opposite direction (193 km³ yr⁻¹). This peculiarity of the water exchange was already recognized by the ingenious Italian Marsili (1681) through his measurements in the Bosporus. Therefore, he is regarded as the founder of modern quantitative physical oceanography.

10.2.3. The Red Sea

Like the European Mediterranean Sea, the Red Sea belongs to category A_{II}; the conditions, however, are much more extreme: almost no precipitation, no continental runoff, but very high evaporation (about 2 m yr^{-1}, Patzert, 1972). The Red Sea has a mean depth of 490 m and a maximum depth of 2604 m; at the connection to the open ocean at Bab al Mandab, the sill depth amounts to 175 m. The wind is characterized by a regular seasonal change according to the monsoons. In winter, due to the northeast monsoon, the water in the Gulf of Aden is piled up off the exit of the Red Sea, while, in summer, negative setup is the result of the southwest monsoon. Consequently, the inflow into the Red Sea is especially large in winter and extremely small in summer. A detailed survey regarding the physicochemical oceanography of the Red Sea has been presented by Morcos (1970). A special study on conditions in winter, supplied by Maillard (1971), is based on the observations of the French research vessel *Commandant Robert Giraud* in 1963; a comprehensive monograph on the circulation in the Red Sea has been compiled by Patzert (1971).

In the Red Sea, deep-reaching thermohaline circulation is especially favored in the Gulf of Suez in winter. At this time of high evaporation and considerable cooling of the shallow Gulf, water of high density is formed and descends to deeper layers. Together with the longitudinal section through the western Indian Ocean, a profile through the Red Sea is shown in Plate 5. It is based on observations by *Atlantis II* in the summer of 1963. The high-salinity deep water of the Red Sea with a salinity of more than 40‰ flows through the narrows of Perim and descends in the Gulf of Aden. Siedler (1968) investigated the mechanism on the basis of observations by *Meteor* during the Indian Ocean Expedition 1964–1965, whereas Grasshoff (1969) studied the chemical problems involved therein.

Information on the results of continuous records obtained by *Meteor* during a period of strong winds in December 1964, is presented in Fig. 10.24 (a), (b), and (c). An intermediate layer of complicated and variable structure is inserted between a homogeneous top layer from 0 to about 60 m and a nearly homogeneous lower layer extending from 200 m depth down to the bottom at 267 m. On the average, the current boundary between the inflow into and the outflow from the Red Sea lies in the upper part of the intermediate layer at approximately 80 m. This depth level varies greatly in detail; it partly follows the rhythm of semidiurnal and diurnal tides, and it partly varies irregularly with the considerable scattering range of 50 cm sec^{-1}. A reversal of the current in tidal rhythm does not take place in this case, in contrast to conditions in the Strait of Gibraltar shown in Fig. 10.21.

The Red Sea has a further peculiarity, unique so far in the world ocean. In a very limited area of this sea, between Jidda and Port Sudan, hot brines emerge from the sea floor at about 2150 m depth. The first records of the depth distribution of temperature, turbidity, and sound reflection were obtained by the research vessel *Meteor* in December 1964 during the Indian Ocean Expedition (Krause and Ziegenbein, 1966). This peculiarity is illustrated in Fig. 10.25. Below 1960 m depth, two temperature jumps were observed before a bottom temperature of 58.4°C was reached. At the bottom of the Red Sea, several other depressions have been recorded by *Meteor* with the aid of a special echo sounder (Dietrich and Krause, 1969). Some of them are even deeper than that with the hot brine, but no special reflections were recognizable in the echo soundings. Meanwhile, this area of hot brines in the Red Sea has been intensively investigated. Relevant reports have been compiled in a volume edited by Degens and Ross (1969). The value of important minerals, present in the uppermost 10 m of the bottom deposits in the region of the hot water, is estimated at around 5×10^9 dollars. Questions concerning the commercial utilization were investigated by the German research vessel *Valdivia* in 1971.

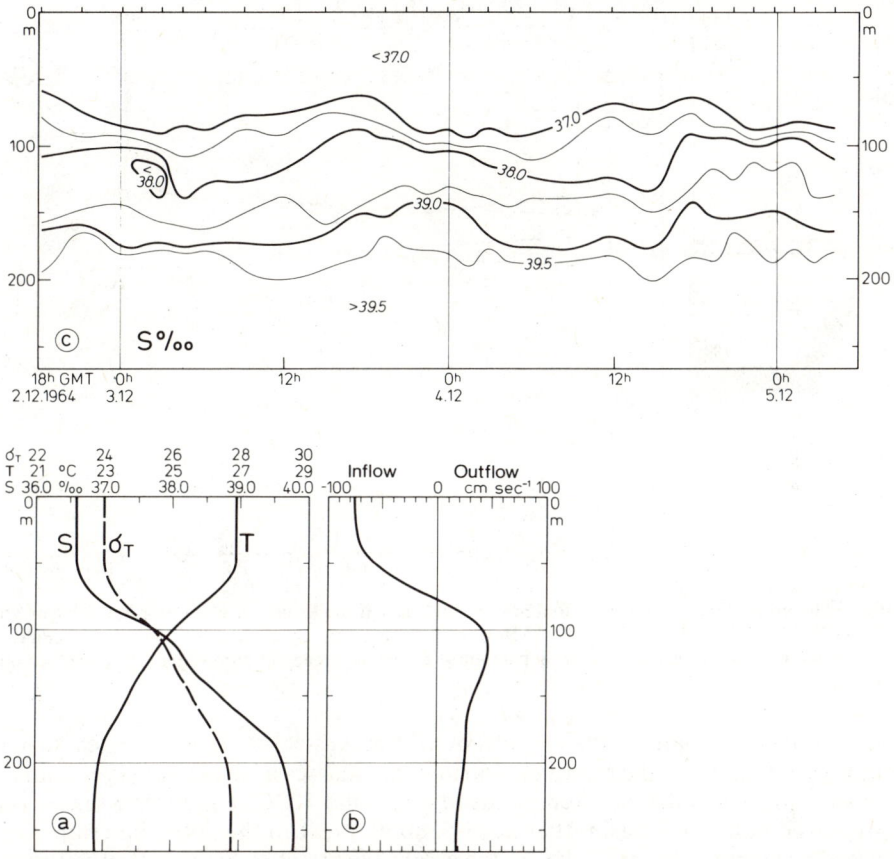

Fig. 10.24. Stratification at the narrowest passage of the southern exit of the Red Sea, based on *Meteor* anchor station No. 62 from December 2 to 5, 1964 ($\varphi = 12°36.5'N$, $\lambda = 43°16.3'E$, water depth 267 m). (After Siedler, 1968a.) (a) Mean vertical distribution of temperature, salinity, and density; (b) mean vertical distribution of current velocity; (c) temporal variation of salinity distribution. Marks at the upper edge: dates of registration from sea surface to bottom with the bathysonde.

10.2.4. The Austral–Asiatic Mediterranean Sea

Stratification of the water and processes of spreading in the deep layers of the Austral–Asiatic Mediterranean Sea can be classified under category B_I. The deep water originates from the neighboring Pacific Ocean. The paths along which the water moves because of the amply structured bottom relief, have proven to be rather complicated. Only the results of the Dutch *Snellius* Expedition in 1929–1930 have elucidated the connections between bottom topography and spreading. As shown in Fig. 10.26, the entire Mediterranean Sea, with the exception of the southern part including the Timor Trench and the Aru Basin, is influenced by water masses of Pacific origin. The water of one basin renews the water of the next at the level of the sill depth. Due to mixing with the overlying warmer water masses, the water becomes warmer from basin to basin and, at the same time, poorer in oxygen, as can be seen in Table 10.04 supplementing Fig. 10.26. According to Table 10.04, the Sulu Basin, the deep water of which is renewed at the north from the South

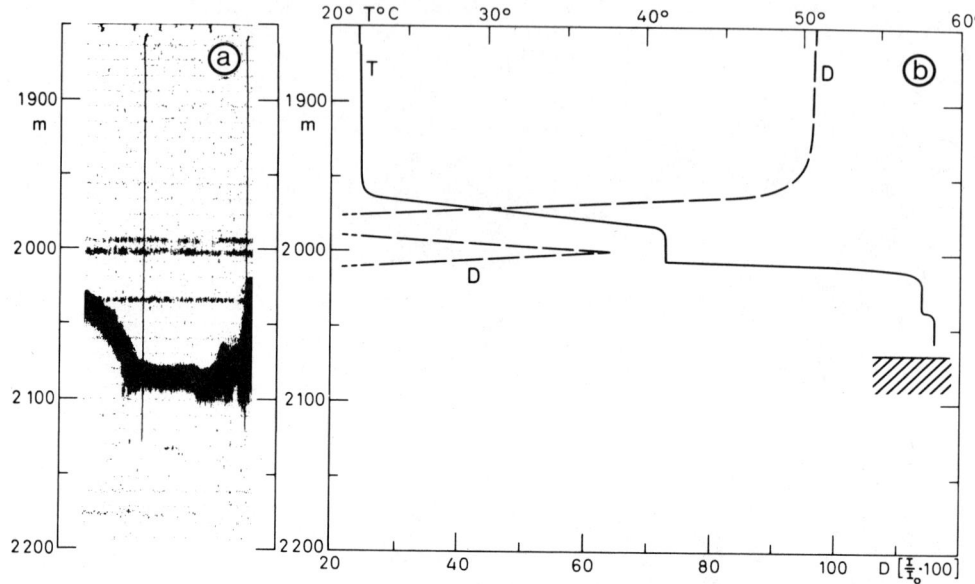

Fig. 10.25. Hot brine at the bottom of the Red Sea. (From Dietrich and Krause, 1969.) Records at *Meteor* Station 384 (21°33′N, 38°3.8′E on April 27, 1965). (a) Bottom and scattering layers in the echogram. (b) Temperature (T), T_{max} = 58.4°C, and transparency (D) in percent of the transparency of pure water. "Salinity" amounted to 325.5‰.

China Sea, has the shallowest sill depth of all the Austral–Asiatic deep-sea basins. It amounts to 400 m. Only the relatively warm water masses of the upper layers can enter this basin unobstructed. Their temperature of around 10°C remains the same even at the greatest depth of the basin, that is, at 5580 m. Next to the potential temperatures in the Red Sea, amounting to 21.4°C at the greatest depth of 2600 m, and in the European Mediterranean Sea with θ = 13.3°C at the greatest depth of 4755 m, they are the highest anywhere in the deep-sea basins of the world ocean, if the hot brines of the Red Sea, as local phenomena, are left out of consideration (cf. Fig. 10.25).

The circulation of the near-surface layers is governed by the influence of the monsoons (cf. Fig. 10.39). It is represented according to the monograph on southeast Asiatic waters by Wyrtki (1961). At the time of the winter monsoon (NE monsoon), a strong current runs counterclockwise along the Asiatic continent and the northern side of Java to the area west of New Guinea. At the time of the summer monsoon (SW to S monsoon), the direction of the current is the opposite.

10.2.5. The American Mediterranean Sea

The American Mediterranean Sea, a notation used as a general name embracing the Caribbean Sea, the Yucatán Sea, and the Gulf of Mexico, belongs to category B_1 like the Austral–Asiatic Mediterranean Sea. The deep water is renewed from the neighboring ocean, that is, from the Atlantic Ocean. According to Dietrich (1937a), the most important passages for the renewal of the deep water of the Caribbean Sea are the Virgin Islands—Anegada Passage east of Puerto Rico. The Windward Passage between Cuba and Haiti serves the same purpose for the Yucatán Sea and the Gulf of Mexico. The

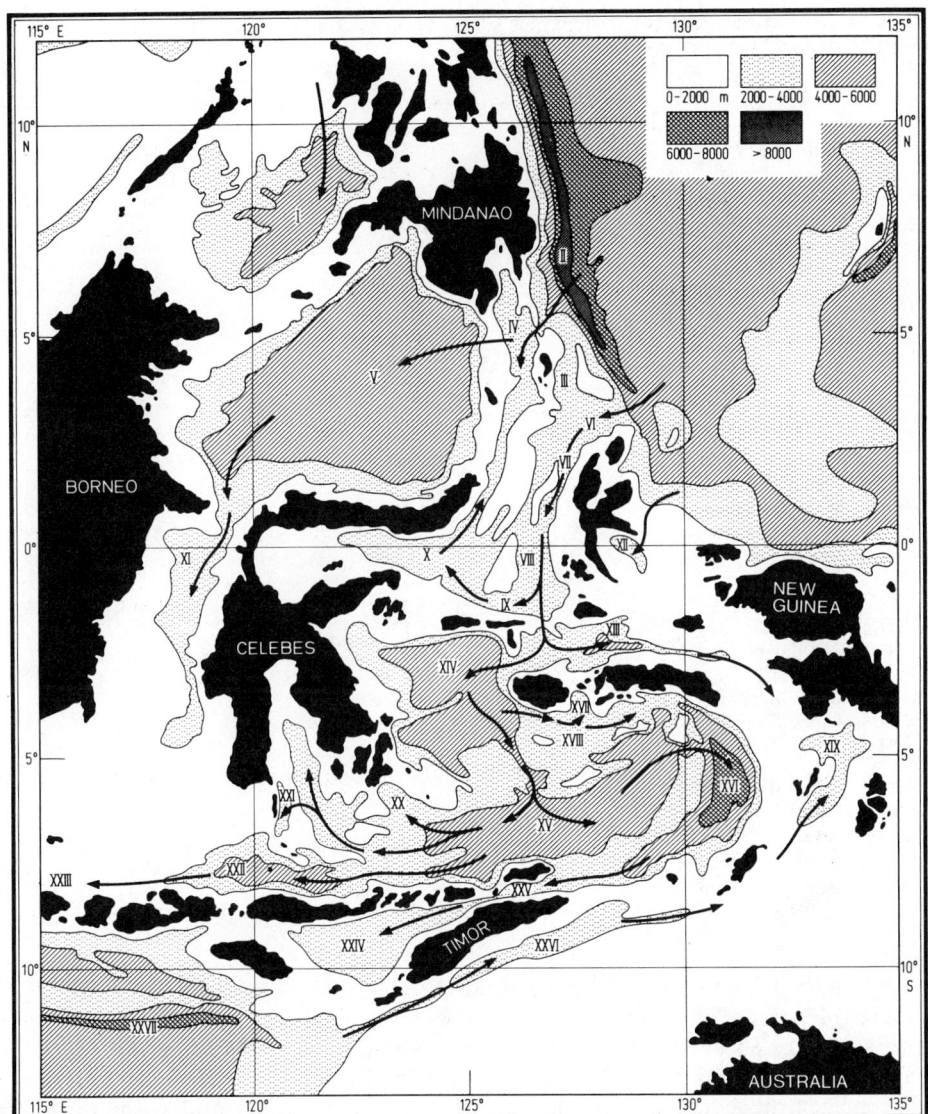

Fig. 10.26. Spreading of deep water in the basins and trenches of the Austral-Asiatic Mediterranean Sea. (After van Riel, 1934.) Roman numerals I–XXVII: Basins, see Table 10.04.

circulation of the deep water is slow and cyclonic as derived by Dietrich (1939) in a monograph on the American Mediterranean Sea. More recent studies on the circulation by Wüst (1964), Gordon (1967), Ichiye and Sudo (1971) have confirmed these results, as can be seen in Fig. 10.27 showing the distribution of the potential bottom temperature. The passages, through which the deep water is renewed, have repeatedly been investigated, in particular with a view to the decisive sill depth. For the Virgin Islands–Anegada Passage, Dietrich (1937c) gave a sill depth of 1630 m; Wüst (1963) found 1750 m, and

Table 10.04. Temperature, Salinity, and Oxygen Content near the Bottom in the Basins and Trenches of the Austral–Asiatic Mediterranean Sea [a,b]

1	2	3	4	5	6	7
I	Sulu Basin	400	5,580	9.84	34.47	8.9
II	Philippine Trench	—	10,500	1.16	34.67	44.5
III	Talaud Trough	3,130	3,450			
IV	Sangihe Trough	2,050	3,820	2.15	34.65	35.7
V	Celebes Basin	1,400	6,220	3.26	34.59	28.7
VI	Morotai Basin	2,340	3,890	1.55	34.68	39.5
VII	Ternate Trough	2,710	3,450	1.65	34.66	39.3
VIII	Batjan Basin	2,550	4,810	1.78	34.66	34.1
IX	Mangole Basin	2,710	3,510			
X	Gorontalo Basin	2,700	4,180	1.95	34.65	
XI	Makassar Trough	2,300	2,540	3.39	34.51	27.4
XII	Halmahera Basin	700	2,039	7.54	34.60	43.2
XIII	Buru Basin	1,880	5,319	2.66	34.63	33.8
XIV	Northern Banda Basin	3,130	5,800	2.73	34.60	32.6
XV	Southern Banda Basin	3,130	5,400	2.75	34.62	32.6
XVI	Weber Deep	3,130	7,440	2.75	34.63	31.6
XVII	Manipa Basin	3,100	4,360	2.85	34.60	32.1
XVIII	Ambalau Basin	3,130	5,330	2.75	34.61	33.4
XIX	Aru Basin	1,480	3,680	3.62	34.62	22.6
XX	Butung Trough	3,130	4,180			
XXI	Salajar Trough	1,350	3,370	3.66	34.60	28.9
XXII	Flores Basin	2,450	5,130	2.96	34.60	30.4
XXIII	Bali Basin	—	1,590	3.46	34.61	26.9
XXIV	Savu Basin	2,100	3,470	3.14	34.56	27.1
XXV	Wetar Basin	2,400	3,460	2.92	34.61	31.1
XXVI	Timor Trench	1,940	3,310	2.57	34.68	34.2
XXVII	Sunda Trench	—	7,140	0.77	34.67	59.5

[a] According to van Riel, 1934, 1943, 1950.
[b] Column 1: Number (see Fig. 10.26).
Column 2: Name of basins and trenches.
Column 3: Sill depth.
Column 4: Greatest depth in meters.
Columns 5, 6, 7: Potential temperature, salinity in ‰, and oxygen content in % of saturation near the bottom.

Sturges (1970) 1860 m. The sill depth in the Windward Passage lies at about the same level. On the basis of investigations carried out by the Soviet research vessel *Michael Lomonossov* in February 1965, Sukhovey and Metalnikov (1968) determined it to be 1688 m. The renewal of the cold-water sphere in the Caribbean Sea is illustrated in Fig. 10.28 representing a cross section through the arc of the Antilles and giving the distribution of the potential temperature. In September 1966, Worthington (1966) investigated the passages with the research vessel *Crawford,* taking new cross sections. According to his results, the overflow does not always occur in the same manner as described by Wüst (1963), who assumed the current to flow cascade-like into the basins (cf. Fig. 10.28). From the cross section in Fig. 10.28, it should neither be concluded that the overflow is stationary nor that the renewal has come to an end, as Worthington believed. By taking direct current measurements over a period of several days in February 1969, Sturges (1970) found evidence of large temporal variations in the tidal rhythm of the inflow.

In the warm-water sphere (cf. Section 10.1.7), the American Mediterranean is fed by the North Equatorial Current passing between the Lesser Antilles and flowing through the Caribbean Sea as the Caribbean Current. Then, it enters part of the southeastern

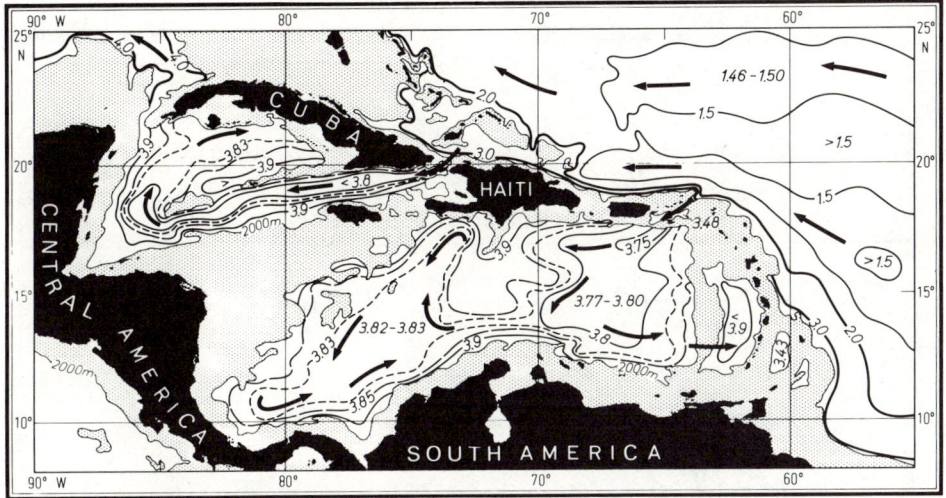

Fig. 10.27. Distribution of the potential bottom temperature (θ °C) in the deep-sea basins of the Caribbean and Yucatán Seas below 2000 m depth, and main spreading of the bottom water. (After Wüst, 1963.)

Gulf of Mexico as the Yucatán Current and leaves it as the Florida Current, eventually forming the main source of the Gulf Stream (cf. 10.3.6).

10.2.6. The Arctic Mediterranean Sea

The Arctic Mediterranean Sea is one of the large and deep mediterranean seas of type A_1 in Fig. 10.17. It comprises eight large deep-sea basins (cf. Dietrich and Ulrich, 1968), four in the Arctic Ocean (the Siberian, Canadian, Eurasian, and Fram Basins) and four in the Greenland and Norwegian Seas (the Greenland, Iceland, and Norwegian Basins as well as the Lofot Basin). At the sea surface, it can be divided into the seas Nos. 5 to 17 in Plate 1, among which only Nos. 5, 6, and 17 contain deep-sea basins, the others being marginal shelf seas. The hydrographic conditions near the surface will be included in the descriptions of Sections 10.3.7 and 10.3.8.

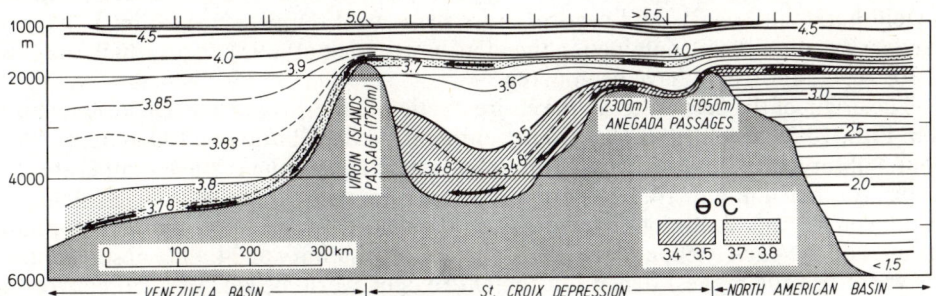

Fig. 10.28. Vertical distribution of potential temperature (θ °C) along the axis of the deep and bottom currents below 1000 m from the North American Basin through the Anegada–Virgin Islands Passages into the Caribbean Sea. (After Wüst, 1963.) Sill depths (in m) of the passages in brackets. Marks at the upper edge: oceanographic stations. Vertical exaggeration 71-fold.

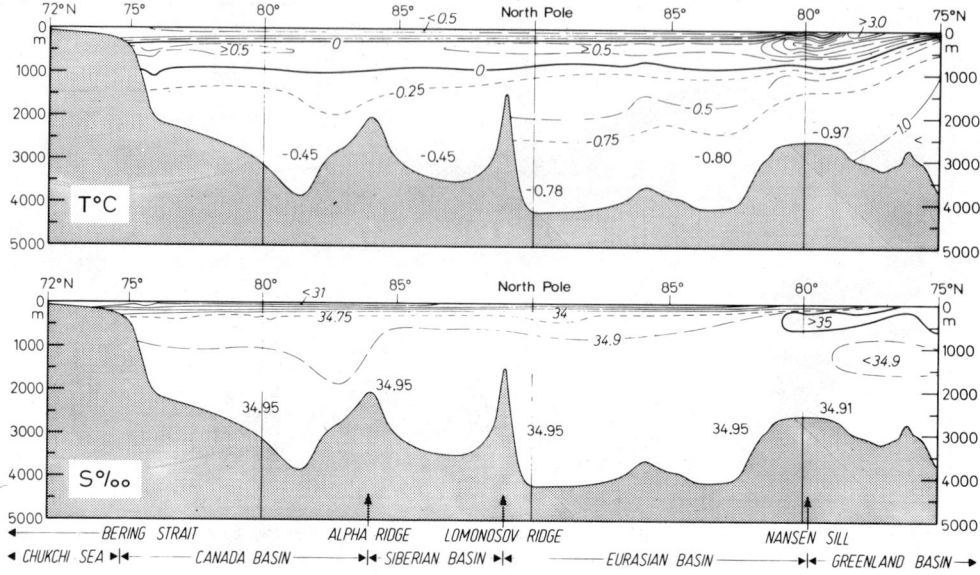

Fig. 10.29. Vertical distribution of temperature and salinity in the Arctic Mediterranean Sea from the Greenland Sea across the North Pole in the direction of the Bering Strait. (After Tripp, 1971.) Vertical exaggeration 185-fold.

In spite of the difficult natural conditions of the polar and subpolar environment, intensive investigations have been carried out in the deep sea of the Arctic Mediterranean Sea. Research was started by Nansen in 1893–1896 with the famous *Fram* drift. It was on this expedition that the polar deep sea was first discovered. Owing to the increased use of manned drifting stations (cf. Fig. 10.51), three deep-sea ridges were found which divide the Arctic Ocean into four basins. The Greenland and Norwegian Seas have been explored predominantly by Norwegians. One of the relevant publications is the classic monograph on the Norwegian Sea by Helland-Hansen and Nansen (1909). Information about the more comprehensive recent literature can be found in the publications by Lee (1963), Aagaard (1968), Worthington (1970), and Mosby (1962, 1970). The data collected systematically during the International Geophysical Year 1958 have been evaluated to yield horizontal charts and vertical cross sections of temperature, salinity, oxygen, and density, all of them contained in the atlas by Dietrich (1969a), from which the Figs. 10.30, 10.47, and 10.52 have been taken.

The origin of the deep water in the entire Arctic Mediterranean Sea can be recognized from Figs. 10.29, 10.30, and 10.47. The temperature distribution at 200 m depth, even though observed in summer, approximately corresponds to the surface temperature in winter. As shown in Fig. 10.47, the temperature minima of about −0.8 to −1.4°C (depending on weather conditions in winter) are found north of Jan Mayen in the Greenland Sea outside the pack ice boundary. With the uniform salinity of 34.92‰ observed here, this water has the highest density of the entire open world ocean. Vertical convection sends it down to the abyssal floor. Figure 10.30 shows a *Johan Hjort* cross section of September 1958, running from East Greenland across Jan Mayen to middle Norway. Due to its constant salinity of 34.90 to 34.92‰, the area of origin of this deep water can be approximately derived from the temperature distribution. The actual center of the

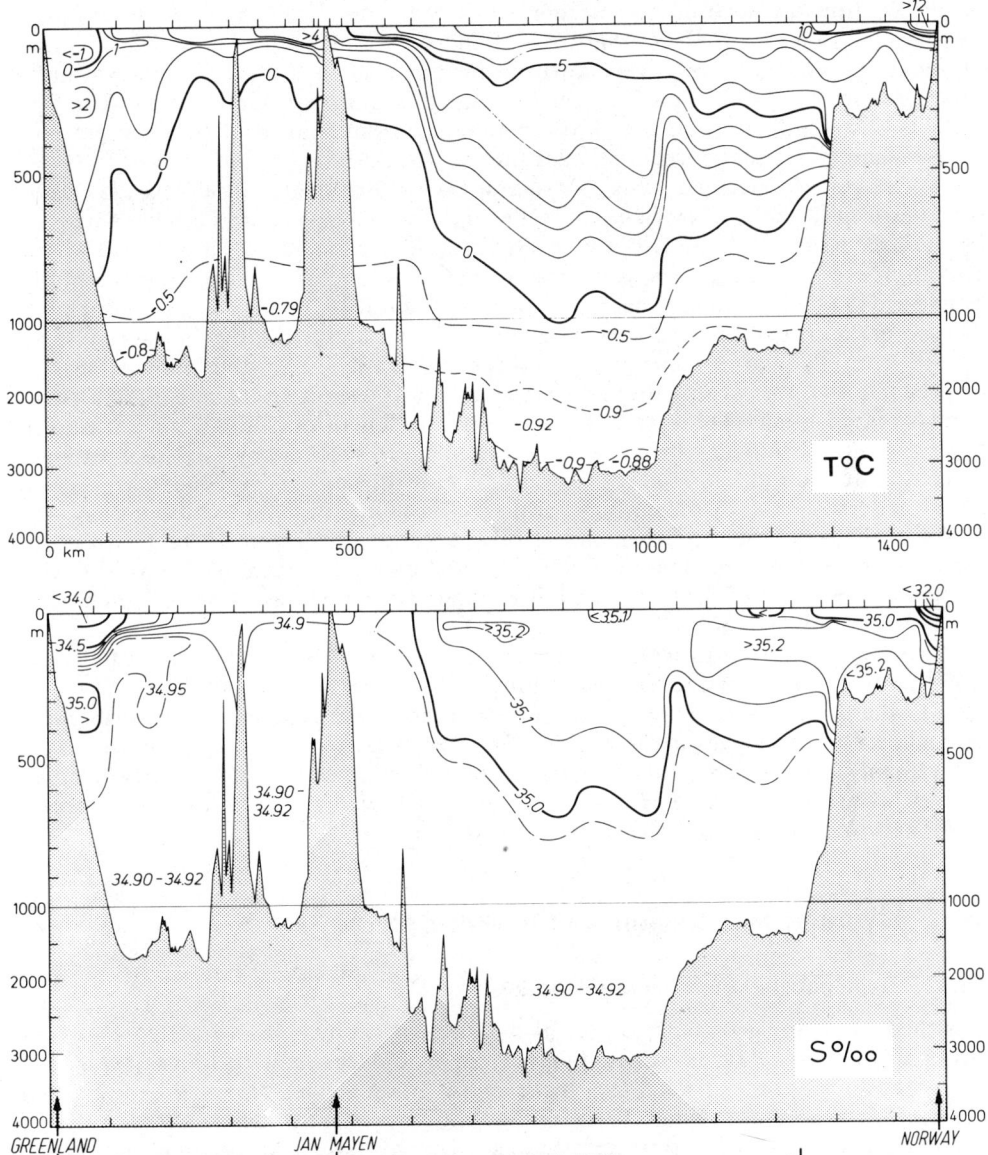

Fig. 10.30. Vertical distribution of temperature and salinity in the Greenland and Norwegian Seas in the winter of 1958, according to observations by the Norwegian research vessel *Johan Hjort* from East Greenland ($\varphi = 70°56'N$, $\lambda = 21°35'W$) via Jan Mayen to West Norway ($\varphi = 66°06'N$, $\lambda = 11°41'E$). (From Dietrich, 1969a.) Exaggeration 0–1000 m: 500-fold; 1000–4000 m: 2000-fold.

sinking processes is located 200 nautical mi north of Jan Mayen in the central Greenland Basin. Since the sill depths of the eight Arctic basins lie relatively deep (all of them beneath 1000 m), the heavy deep water of the Greenland Basin can penetrate into all the other basins. This is demonstrated in Fig. 10.29 for the polar basins. Even the deep water on the continental slope off the Bering Strait originates from the Greenland Basin. Furthermore, Fig. 10.29 shows that beneath the low-salinity top layer with a thickness of about 100 to 200 m, relatively warm water (warmer than 0.5°C), rich in salinity (greater than 35.0‰), spreads from the Greenland Sea into the Arctic Ocean. The high stability of stratification in the top layer, maintained by melting ice in summer, prevents deep-reaching vertical convection under the ice of the Arctic Ocean.

The Arctic Mediterranean Sea is—so to speak—a large bight of the Atlantic Ocean, open between Greenland and Scotland. Of course, this opening is considerably narrowed by the Greenland–Scotland Ridge with sill depths of around 800 m (cf. Figs. 10.03 and 10.52). The Bering Strait, the passages of the Canadian Archipelago, and the Smith Sound are too shallow and too narrow as to play a significant role in the water exchange. The water balance of the Arctic Mediterranean Sea is highly positive in the vicinity of the surface. The factors contributing to this positive balance are the following: excess of precipitation over evaporation, continental runoff, especially from Siberian rivers carrying large amounts of water (cf. Table 4.05), inflow through the Bering Strait (cf. Plate 6), by the Irminger Current west of Iceland, and especially by the North Atlantic Current northwest of Scotland (cf. Figs. 10.49 and 10.50). This excess of water is balanced, to some extent, by the outflow within the East Greenland Current, but especially by the overflow across the Greenland–Scotland Ridge (cf. Section 10.1.3) with water temperatures below zero. This very cold deep water mainly originates from the downwelling winter water of the Greenland Sea (about 8×10^6 m^3 sec^{-1}), as schematically represented in Fig. 10.50.

The variability of stratification and currents is very pronounced even in the deep sea. Long-term changes are due to downwelling caused by weather conditions (Aagaard, 1968), whereas short-term changes are due to internal waves. During the Norwegian–German cooperative investigation called "Norwegian Sea 1969," such processes in the Norwegian Current were studied by means of numerous continuous records taken by the research vessels *Helland-Hansen* and *Planet* (Dietrich and Horn, 1973; Keunecke, 1973; and Leinebö, 1973), (cf. also Section 10.1.9).

10.3. Hydrographic Regions of the World Ocean

10.3.1. Regional classification of the oceans

It is the task of any geoscience, hence also of oceanography—apart from carefully investigating individual phenomena—to give a survey of large-scale relationships. If such a survey is not to be restricted to a mere enumeration of facts, it should consist in a comparison of oceanic regions. Therefore, we must apply the principle of the theory of cognition stating that everything can be recognized by looking at its opposite. The difficulty lies in the fact that diverse boundaries of oceanic regions are conceivable; for instance, they may be determined by bottom topography, or one might proceed from considerations of marine physics, marine chemistry, or marine biology. A further difficulty is that within each system of classification any number of oceanic regions can be distinguished, depending on the gradation chosen. Consequently, there is no general classification principle of absolute validity, but there are only systems of classification from which we can choose that which is appropriate to our specific problem.

Regional boundaries mentioned so far have been determined either by the conditions observed or by the causes from which certain conditions result. The first group includes the "natural regions," as defined by Schott (1935, 1942), and the boundaries given by Sverdrup et al. (1942). Schott, as he stated himself, arrived at a "highly subjective" division, since he attempted to satisfy oceanographic, climatic, and biological viewpoints all by one single classification. Sverdrup based his textbook "The Oceans" on the limits of similar water bodies. This classification has been continued in more detail by Stepanov (1965, 1969) and Mamayev (1969). The second way starts from causes leading to a natural classification, like the bottom topography (Wüst, 1939) and the three-dimensional near-surface current field (Dietrich, 1956a). The latter approach has been chosen in the following (Fig. 10.31).

In spite of the somewhat one-sided principle of classification applied here, the division is not a purely formal one, but it emphasizes those natural conditions in the ocean which are important to man, because the relationships of current conditions with physical, chemical, and biological phenomena near the sea surface are so very many. Only in higher geographical latitudes, where ice becomes the dominating factor, would such a classification, based on current conditions, frequently be meaningless. Therefore, ice conditions have been chosen as a means for drawing the boundaries in these latitudes.

It should be noted that any natural division of the world ocean that is based on water bodies will always contain a number of uncertainties. In the ocean, distinct border lines do not exist but at best border zones, which, however, are surprisingly well defined in some cases. Nor do such border zones remain stationary. They are subject to temporary dislocation, especially in an annual rhythm, and nonperiodically in connection with large-scale eddy motions in the water. Finally, we should not forget that our knowledge on surface currents is almost exclusively based on ships' logs, that is, on the differences between positions obtained by dead reckoning and true positions, thus being rather incomplete in less frequently navigated oceanic areas.

If local details, especially in the vicinity of coasts and in marginal and adjacent seas (cf. Section 10.3.9), are neglected, we can distinguish seven major regions of the world ocean, four of which should be subdivided for obvious reasons. Their characterization in the following compilation also provides a clue to the principles applied in drawing the boundaries in Fig. 10.31.

- T: *Region of trade wind currents:* Fairly to very persistent currents directed westward during the entire year.
  - T_E: With a strong velocity component directed toward the equator (deviation of the current direction from due west more than 30°).
  - T_W: With current moving strictly westward.
  - T_P: With a strong velocity component directed toward the poles (deviation of the current direction from due west more than 30°).
- E: *Region of equatorial countercurrents:* Currents in the vicinity of the equator directed eastward at times or during the entire year.
- M: *Region of monsoon currents:* Regular reversal of the current system in spring and fall.
  - M_L: In lower latitudes (connected with small annual variation of surface temperature).
  - M_M: In moderate and higher latitudes (connected with large annual variation of surface temperature).
- H: *Region of horse latitudes:* Weak currents of variable directions at times or during the entire year.

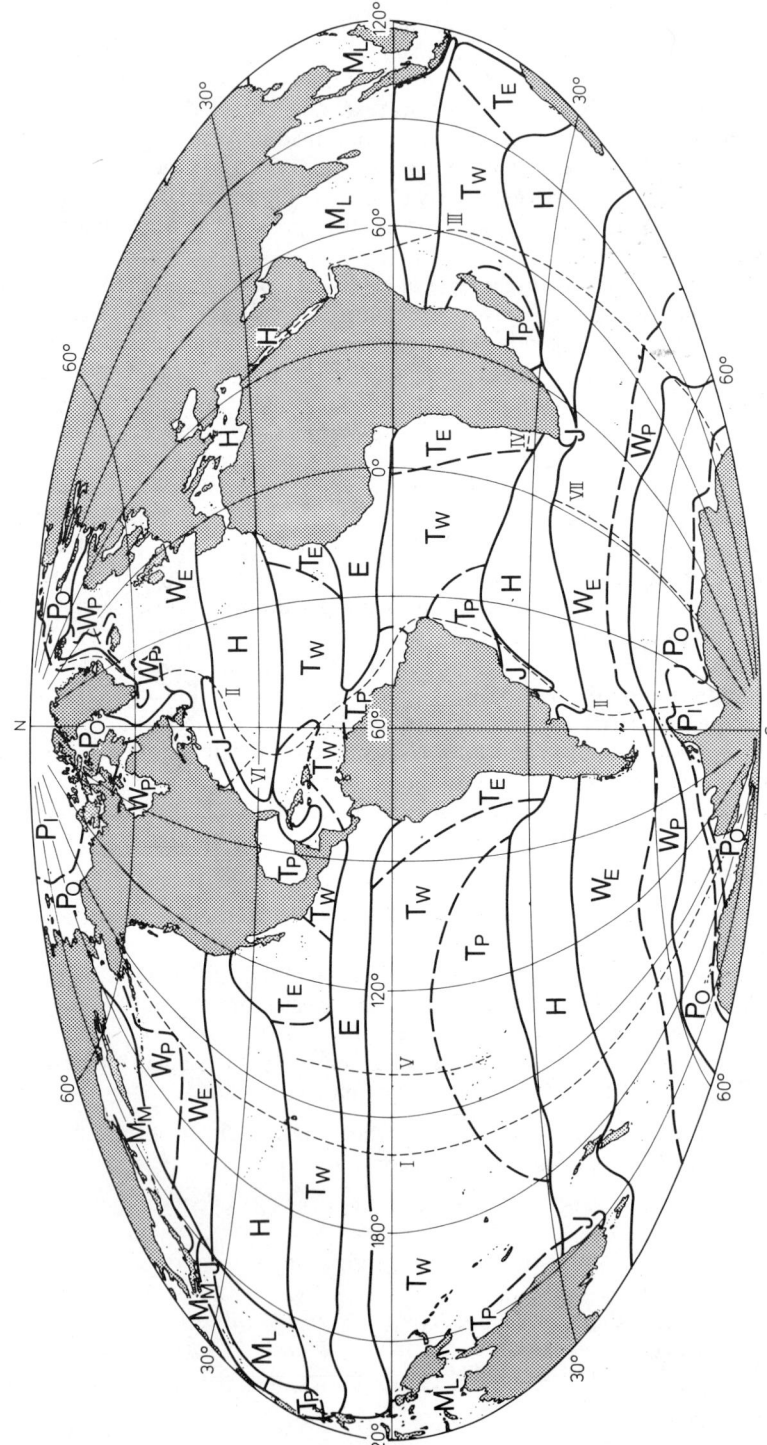

Fig. 10.31. Regional classification of the world ocean. (After Dietrich, 1956a.) *T*: Region of trade wind currents. *E*: Region of equatorial countercurrents. *M*: Region of monsoon currents. *H*: Region of horse latitudes. *J*: Jet stream region. *W*: Region of west wind drift. *P*: Polar region. For explanation of indices see Section 10.3.1. I–VII: vertical cross sections; for I–III see Plates 3–5, for IV see Fig. 10.32, for V see Fig. 10.34, for VI see Fig. 10.42, for VII see Fig. 10.46.

J: *Jet stream region:* Strong, narrow currents during the entire year as a result of discharge from the regions of trade wind currents.

W: *Region of west wind drift:* Variable, predominantly easterly currents during the entire year.
W_E: Equatorward of the oceanic polar front.
W_P: Poleward of the oceanic polar front.

P: *Polar region:* Covered with ice at times or during the entire year.
P_O: Outer polar region: always or frequently covered with ice in winter and spring.
P_I: Inner polar region: covered with ice during the entire year.

It is no coincidence that the distribution of the oceanic regions and of the areas with major wind systems over the world ocean show a great similarity. Certainly, the wind is the decisive factor in the development and maintenance of surface currents. Such relationship has been emphasized here by the nomenclature of the various regions. It becomes evident in detail if the wind systems over the world ocean, as presented in every major school atlas, are inspected for comparison. The agreement in certain regions reaches such proportions that the oceanic boundaries in Fig. 10.31 nearly coincide with the climatological boundaries, as drawn, for example, by Köppen (1936) in his chart of the wind areas of the world ocean contained in the *Handbuch der Klimatologie* (*Encyclopedia of Climatology*), volume I: *E* corresponds to the doldrums in the atmosphere, that is, the equatorial regions with weak winds, *T* to the trade wind zones with steady, usually fresh winds, *M* to the monsoon zones with regular change of wind direction in spring and fall, *H* to the weak-wind zones of the subtropical high-pressure areas, *W* to the west wind zones with variable, predominantly westerly winds, *P* to the zones of polar east winds, the extent of which, however, is considerably smaller than that of our oceanic polar regions defined by the ice limits. Only the jet stream region *J* does not have a counterpart in the wind areas over the world ocean. The currents in these zones are primarily under the influence of obstacles, namely the continents standing in the way of the westward trade wind currents, whereas the deflecting force of the earth's rotation, which increases polewards, contributes to the jetlike concentration of the currents. Currents that are caused orographically occur at every coast but cannot be dealt with in more detail here. The currents of the *J* regions are the largest and most conspicuous disturbances of such kind.

It is true that many differences exist among the corresponding regions of the individual oceans, but, nevertheless, the similarities are so dominating that even a single example can illustrate many typical properties of the relevant region. This is the purpose of a small number of vertical cross sections which—mostly for the layer from 0 to 300 m—show the distribution of some important hydrographic factors. Their geographical location can be taken from Fig. 10.31. To prevent any impression of rigidly stationary conditions, typical variations with time and displacements in space within the regions themselves are indicated in the following sections and in further figures.

10.3.2. The regions of trade wind currents *T*

The regions of trade wind currents, marked by *T*, are characterized by persistent currents flowing westward all year long. They largely coincide with the areas of steady trade winds, which embrace the largest part of the world ocean (31%) and thus become the active zones of surface circulation.

Two regions of trade wind currents exist in each of the oceans, with the exception of the Indian Ocean, where only one such region is found in the Southern Hemisphere. The corresponding region in the Northern Hemisphere cannot develop because of the seasonal change of the monsoon winds. If current direction is taken as criterion of useful classification, each region of trade wind currents can be subdivided into an eastern, a central, and a western region, abbreviated by T_E, T_W, T_P, respectively. The central region T_W is least subject to the influence of continental boundaries and shows a purely zonal westward motion. In the regions of the eastern (T_E) and western (T_P) trade wind currents, however, the currents deviate from the zonal direction by more than 30°. In the eastern region (T_E), they have a component of motion directed toward the equator, and in the western region (T_P) a component of motion toward the poles. This anticyclonic turning of current direction, as experienced when the ocean is crossed from east to west in the region of the trade wind currents, is not only determined by the distribution of the continents surrounding the ocean but is repeated in the sense of turning of the trade winds, such that the wind contributes to the deviations from the zonal water motion in the eastern and western regions of the trade wind currents. If, in addition to these characteristic changes, near-surface stratification is included in the consideration, a great number of phenomena occurring in this region become comprehensible. Below the homogeneous top layer, a strong pycnocline is present during the entire year, which—under the influence of the persistent current and the deflecting force of the earth's rotation—is subject to a dynamically enforced inclination (cf. Fig. 10.12), such that this pycnocline rises towards the left-hand side in the Northern Hemisphere and towards the right-hand side in the Southern Hemisphere, if one looks in the direction of the current.

10.3.2.1. Regions of trade wind currents T_E with components of motion strongly directed towards the equator

The currents of these regions are known as the Canary and Benguela Currents in the North and South Atlantic Ocean, as the California and Peru (Humboldt) Currents in the North and South Pacific Ocean, and as the West Australia Current in the South Indian Ocean (cf. Plate 6). Since they flow from higher to lower latitudes, they carry water that is colder than the average for the corresponding latitudes. These temperature anomalies are increased at the sea surface because, in the areas where the trade winds originate, the winds blowing offshore or parallel to the coast contribute to the formation of a cross circulation in the currents mentioned above. This causes cold water from deeper layers to ascend to the surface at the shelf edge. These so-called zones of upwelling play a special role in the T_E regions, with the exception of the West Australia Current. Probably, the area of influence of the offshore trade winds, which is restricted to northwestern Australia, is too small.

Some typical phenomena of the upwelling zones are given in Fig. 10.32, showing an example from South Africa. The main current moving toward the equator, in this case the Benguela Current, is superimposed by a transverse circulation transporting cold and, therefore, specifically heavy water, which also is poor in salt and oxygen but rich in nutrients, to near-surface layers. It would be a mistake to assume that this water must come from great depths. It probably rises from a depth of 100 to 300 m. The velocities of the vertical motion are very small: 0.001 to 0.005 cm sec^{-1}. On the basis of theoretical and empirical investigations, Hidaka (1954), using the upwelling water of the California

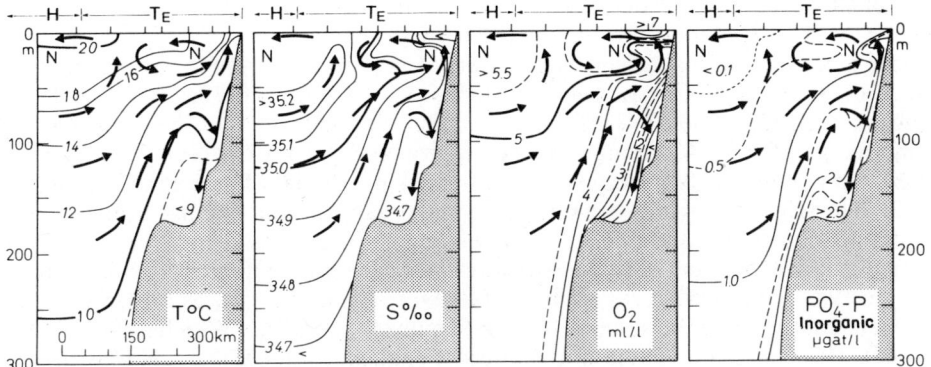

Fig. 10.32. Example of stratification and motion within the region of trade wind currents T_E. (After Dietrich, 1956a.) Distribution of temperature, salinity, oxygen, and phosphate along a cross section through the Southwest African upwelling region at 28°40'S, based on observations of the research vessel *William Scoresby* on March 12–14, 1950. N: main current directed towards the north (perpendicular to cross section): Benguela Current. Arrows: transverse circulation of the main current. Marks at the upper edge: oceanographic stations. Vertical exaggeration of the cross section: 2500-fold. For geographic position of the cross section: see Fig. 10.31 (line IV).

Current as an example, estimated 80 m month^{-1}. This corresponds to mere 0.0031 cm sec^{-1}. Similar vertical velocities can be expected for the example in Fig. 10.32. In the long run, however, this upwelling has lasting effects reflected, for example, in the great horizontal temperature differences maintained at the sea surface. According to Fig. 10.32, temperatures of 12°C were observed at the coast, whereas those of 20°C were measured only 250 km farther seaward. Another effect is manifest in the abundance of nutrients in the upwelling water, as shown in Fig. 10.32 by the example of the phosphate content contributing to the extraordinarily strong development of plankton. Like the abundance of nutrients, this is not restricted to some places but extends over the entire coastal region off Southwest Africa and Angola. During the Danish *Galathea* Expedition, Steemann Nielsen and Jensen (1957) measured averages of the total organic production in the upwelling area off Southwest Africa as amounting to 0.5 g C m^{-2} day^{-1}. Maximum values were determined to be 3.8 g C m^{-2} day^{-1}. This value corresponds to about 10 g of organic substance m^{-2} day^{-1}, which is equivalent to the annual yield of intensive agriculture in Europe. In recent time, during the EASTROPAC cruises in the eastern tropical Pacific Ocean, thorough investigations have revealed that seasonal variations of primary production, secured statistically, also occur in tropical waters (Owen and Zeitzschel, 1970).

As can be seen in Fig. 10.32, oxidation of sinking dead organisms results in strong oxygen consumption already at shallow depths. Occasionally, even hydrogen sulfide has been observed, which is quite unusual in the world ocean, apart from enclosed adjacent seas. The consequence is a great mortality of fish. The abundance of plankton is reflected by a strongly diminished transparency of the water in the T_E region and by the green color of seawater in contrast to the cobalt-blue color of the neighboring horse latitudes.

Conditions similar to those indicated by the cross section taken in the upwelling region off Southwest Africa (cf. Fig. 10.32), have been observed in the upwelling areas off Northwest Africa and California, as well as in an especially pronounced form off the coasts of Chile and Peru. Evidence with regard to the latter two areas is given in Fig. 10.33 showing that high content of nutrients coincides with an abundance of zooplankton. Many relevant investigations have been carried out at sea (Gunther, 1936; Schweigger, 1949; Schott, 1951; Hart and Currie, 1960; Yoshida, 1967; Wooster, 1970a; Wooster, 1970b; Tomczak, jr., 1973; Jones, 1971; Bang, 1971) as well as by means of theoretical models (Tomczak, jr., 1970a). A summary with extensive bibliography has been given by Smith (1968). Upwelling and its consequences are treated in a special volume edited by Dietrich (1972). The relationship between upwelling and organic production has been discussed in great detail by Cushing (1971). The upwelling zone that has been investigated most thoroughly lies off California. In the Northwest African upwelling area, systematic studies began in 1937 with the first research vessel *Meteor* (Tomczak, jr., 1970b). Since 1968 they were continued with the new *Meteor* (Tomczak, jr., 1973; Weichart, 1970; Mittelstaedt, 1972) and culminated in the international expedition CINECA (Cooperative Investigations of the northern part of the Eastern Central Atlantic). The results of CINECA were reviewed at the Symposium on the Canary Current Upwelling Resources in Las Palmas, Canary Islands, in 1978.

Investigations in the 1960s have confirmed that, apart from general concepts of currents as demonstrated in Fig. 10.32, there are three nonstationary processes which will have to be the subjects of further studies in the near future. They concern the following phenomena:

1. Countercurrent (polewards) under the main current at the continental slope below 200 m depth.
2. "El Niño" off the coast of Peru.
3. Small-scale variability of upwelling.

So far, evidence of countercurrents, as under 1, has been found beneath the Humboldt Current (Wooster and Gilmartin, 1961) and under the California Current (Reid, 1962a; Wooster and Jones, 1970). According to a theoretical model of the circulation of the eastern tropical Pacific Ocean (Yoshida, 1967), such countercurrents could be expected. The "El Niño" Current (under 2) indicates that warm tropical surface water, poor in nutrients, advances poleward. This occasionally occurs in summer, as in the years of 1891, 1925, 1941, 1953, 1957, and 1965. Those were years connected with climatic and ecologic catastrophes (see below). Bjerknes (1966) understood the "El Niño" phenomenon as a result of anomalously weak trade winds in the Pacific Ocean, whereby heat is accumulated in the near-surface layer of the T_W region. In addition, the wind-driven, cold Humboldt Current, directed toward the equator, and its transverse circulation are weakened; instead, warm tropical surface water advances polewards and toward the coast. Small-scale variability, as under 3, could be recognized in its full extent only after new measuring methods were introduced which permitted continuous records of chemical factors at the sea surface from aboard a ship underway (Grasshoff, 1965; Armstrong et al. 1967; Weichart, 1970). According to such observations, upwelling takes place in limited eddies drifting along in the main current, while water is ascending in them from the depth as in a chimney.

Oceanographic conditions in the T_E regions are not without any effect on the atmosphere and life in the ocean. In general, surface temperatures are lower than air temperatures, especially in the vicinity of coasts. Hence, formation of persistent coastal fog, known under the name of *Garua* at the Peruvian coast, is favored. The air, which is dry

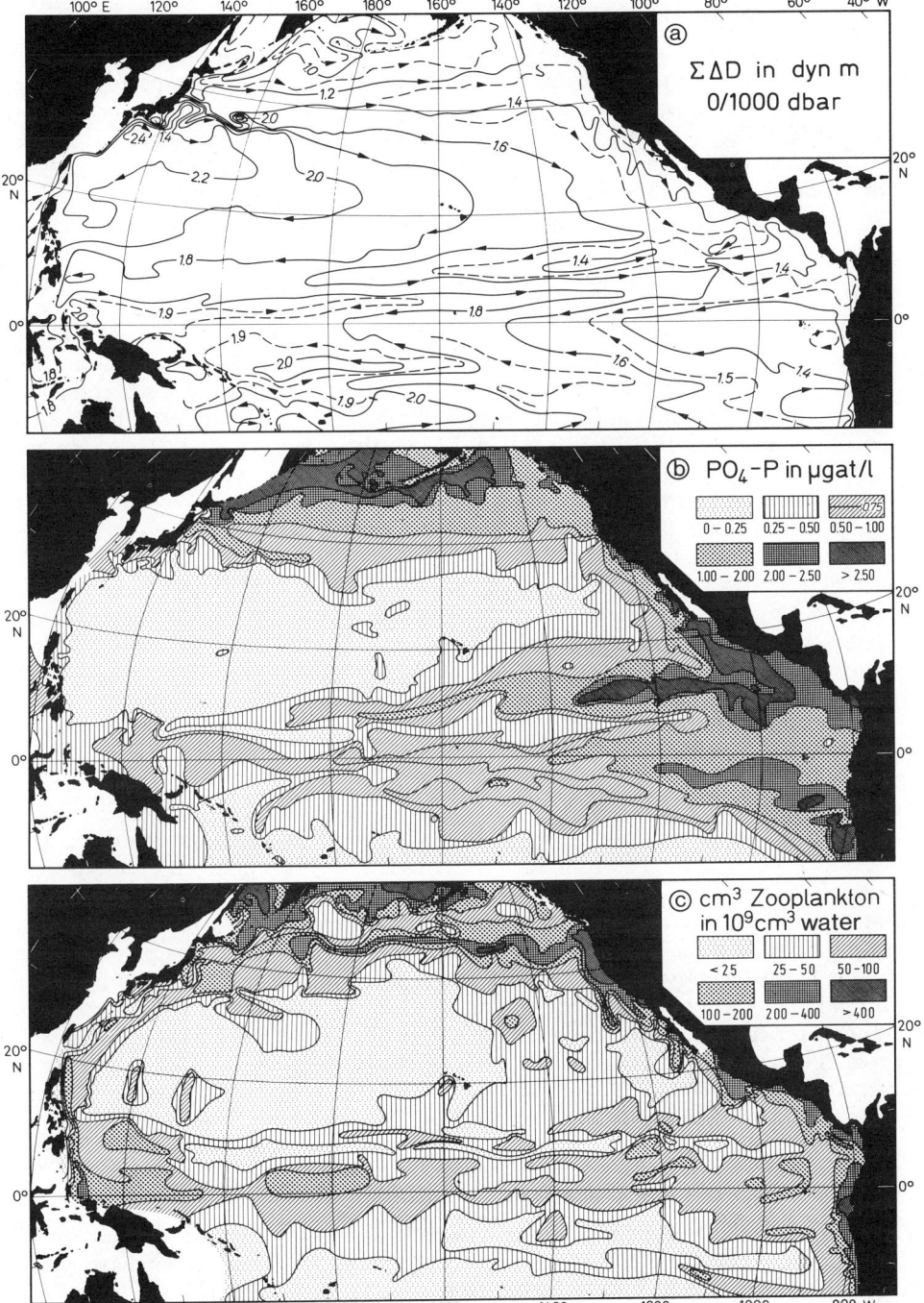

Fig. 10.33. Surface currents (a), nutrient content (b), and biomass (c) in the North Pacific Ocean. (After Reid, 1961, 1962b.) (a) Surface currents, geostrophic (represented as dynamic topography of the sea surface in dyn m, referred to the 1000 dbar surface); (b) distribution of PO_4-P (in μg at/l) at 100 m depth; (c) distribution of the zooplankton volume (in cm^3 per 10^9 cm^3 of water) in the uppermost 150 m.

anyhow in those areas of origin of the trade winds, owing to their offshore component, remains unsaturated in water vapor, because it is warmed along its path toward the equator, and because evaporation is only slight, which is a common characteristic of the T_E regions (cf. Fig. 4.11). Consequently, there is very little precipitation everywhere in the T_E regions; in fact, these are the areas with the smallest amount of precipitation on earth. Oceanic islands in these regions, like the Cape Verde Islands in the North Atlantic Ocean or the Galapagos Islands in the South Pacific Ocean, have a climate typical of deserts.

Although, in general, there are no storms in the T_E regions, these areas are exposed to the influence of high swell arriving from the stormy west wind zones, the "roaring forties" and the "howling fifties," and greatly impeding commercial traffic at the coasts of South America and Africa, where few harbors exist, and boats shuttling back and forth from the shore to the ships, riding at anchor far out, are the only means of transport.

The unusual profusion of fish in the upwelling areas can be explained by the abundance of plankton [cf. Fig. 10.33(c)]. Until 1958, these riches were little used by man, except in California, since the adjacent coastal deserts are but scarcely populated. The seabirds, rather than man, took advantage of the riches; they are responsible for the formation of the large guano deposits along those coasts. In the upwelling area off Peru, a fundamental change took place in the years from 1958 through 1964: The fishing yield increased from 0.9×10^6 tonnes in 1958 to 9.1×10^6 tonnes in 1964 and remained at this level for several years (9.5×10^6 tonnes in 1969). Peru suddenly became the country with the highest yield of fish on the globe. It is true, however, that the catch consisted almost exclusively of small anchovies, generally not eaten by man but used only for the production of fish meal, which is exported by Peru as livestock feed.

Far-reaching consequences of such a vast increase of fishery off Peru did not fail to appear. In 1957, the number of seabirds, above all of cormorants, which live on anchovy, was estimated at roughly 20×10^6. But in 1958, an unusual "El Niño" Current caused their number to decrease to 6×10^6. The reason was not that the anchovies had died but that they had escaped from the warm, plankton-depleted "El Niño" water into the depth. Hence, they were no longer within the reach of seabirds, thus damned to starvation. In the period of 1958 through 1969, this decimated bird population increased only by 20%, owing to very intensive anchovy fishing since 1964. It seems that after the strong interference by man catching 10×10^6 tonnes of fish per yr, a sufficient food supply is no longer available for 20×10^6 seabirds, which would need about 5×10^6 tonnes of fish per yr. Optimum exploitation of fish is determined by constant control of the environmental conditions for fish and birds. Even the phantastically fertile Peruvian waters, covering an area 800 nautical mi long and 30 nautical mi wide, that is, 0.02% of the entire world ocean, and yielding 15% of the total catch of commercial fisheries in the world, are now threatened by overfishing.

10.3.2.2. Regions of trade wind currents T_W with strictly westward motion

In these regions, the North and South Equatorial Currents flow westward rather steadily. This uniformity of current conditions corresponds to the uniformity of wind and weather that made this region ideal for navigation during the time of sailing ships. This also was the reason why, in the age of the great discoveries, the North Atlantic trade wind region was given the surname "El golfo de las damas" (Ladies' Sea). By giving it this name the gallant Spaniards wanted to indicate that navigation in this region was so easy that the heavy rudder of a ship could be handled by the weak hands of a lady. Rare exceptions of such uniformity of wind and sea are the tropical revolving storms. Improved meteorological observations, like cloud photographs from weather satellites, point to irregularities in the wind field of the trades (Bjerknes, 1969), which must affect the equatorial currents including their drain currents, like the Gulf Stream. Little is known as yet about such relationships, but systematic investigations have been suggested (Tsuchiya, 1970).

Slight variations in surface circulation result in changes of the heat storage with far-reaching consequences, as Bathen (1971) has shown for the North Pacific Ocean.

There is little precipitation in the T_W regions, although they are not so extremely dry as the T_E regions. Since evaporation is strong, the salinity of the surface water is relatively high. An impression of the typical stratification in these regions is conveyed by the left- and right-hand sides of Fig. 10.34. North of 10°N and south of the equator, the thermocline descends to greater depths. On a large scale, this can be recognized in the example of the eastern Pacific Ocean in Fig. 10.35. Hence, the thickness of the near-surface, warm top layer with high salinity increases polewards. Since in the T_W regions, except in those near the equator, the vertical components of the currents are descending ones, and since, furthermore, the annual variation of surface temperature nearly always amounts to less than 4°C (cf. Fig. 5.05) and therefore, deep-reaching thermal vertical convection cannot be induced, the renewal of nutrients from deeper layers is insufficient [cf. Fig. 10.33(b)]. Consequently, organic production of phytoplankton is greatly restricted. It lies below 0.1 g C m^{-2} day^{-1}, in contrast to more than 0.5 g in the T_E region. The T_W regions, in particular their poleward sides, are among the areas of the world ocean that are poorest in plankton [cf. Fig. 10.33(c)]. This is reflected by clear, deeply cobalt-blue seawater and a great deficiency of commercial fish.

The uniformity of the T_W regions is interrupted occasionally and in limited areas by tropical revolving storms (cf. Section 4.3.2). They are the most violent of all the storms on earth. In East Asia, they are called typhoons, in the South Indian Ocean, Mauritius cyclones, and in the North Atlantic Ocean, hurricanes. Their devastating force is feared alike by sailors and by people living on the coast. As shown in Fig. 4.15, they mostly originate in the T_W regions, outside an equatorial zone from 5°N to 5°S, and at a distance from the shore of at least 250 nautical mi. Furthermore, the air temperature in the area of their formation must exceed 27°C in the fall. Hence, the warm western sides of the oceans are particularly afflicted, whereas the eastern South Pacific Ocean and the entire South Atlantic Ocean are not affected by tropical cyclones.

The Equatorial Undercurrent represents a peculiarity of the T_W regions and partly of the M_L region in the Indian Ocean. This current was discovered in the Pacific Ocean in 1952 by Cromwell et al. (1954). There, it is also called Cromwell Current after the scientist who first described it and later died in an accident. Detailed observations have been contributed in particular by Knauss (1960, 1963, 1966). Four meridional sections through equatorial currents were carried out by the research vessel *Vitiaz* in 1961 (Koshlyakov and Neyman, 1965). Figures 10.34 to 10.37 show some significant facts characterizing the Cromwell Current.

1. The Cromwell Current is an eastward undercurrent at the equator, which is embedded in the westward equatorial currents (Fig. 10.34).
2. It is a very narrow current between 2°N and 2°S, arranged in a remarkably symmetrical fashion with respect to the equator (cf. Fig. 10.34).
3. It extends from a depth of about 20 m down to 300 m (cf. Fig. 10.36).

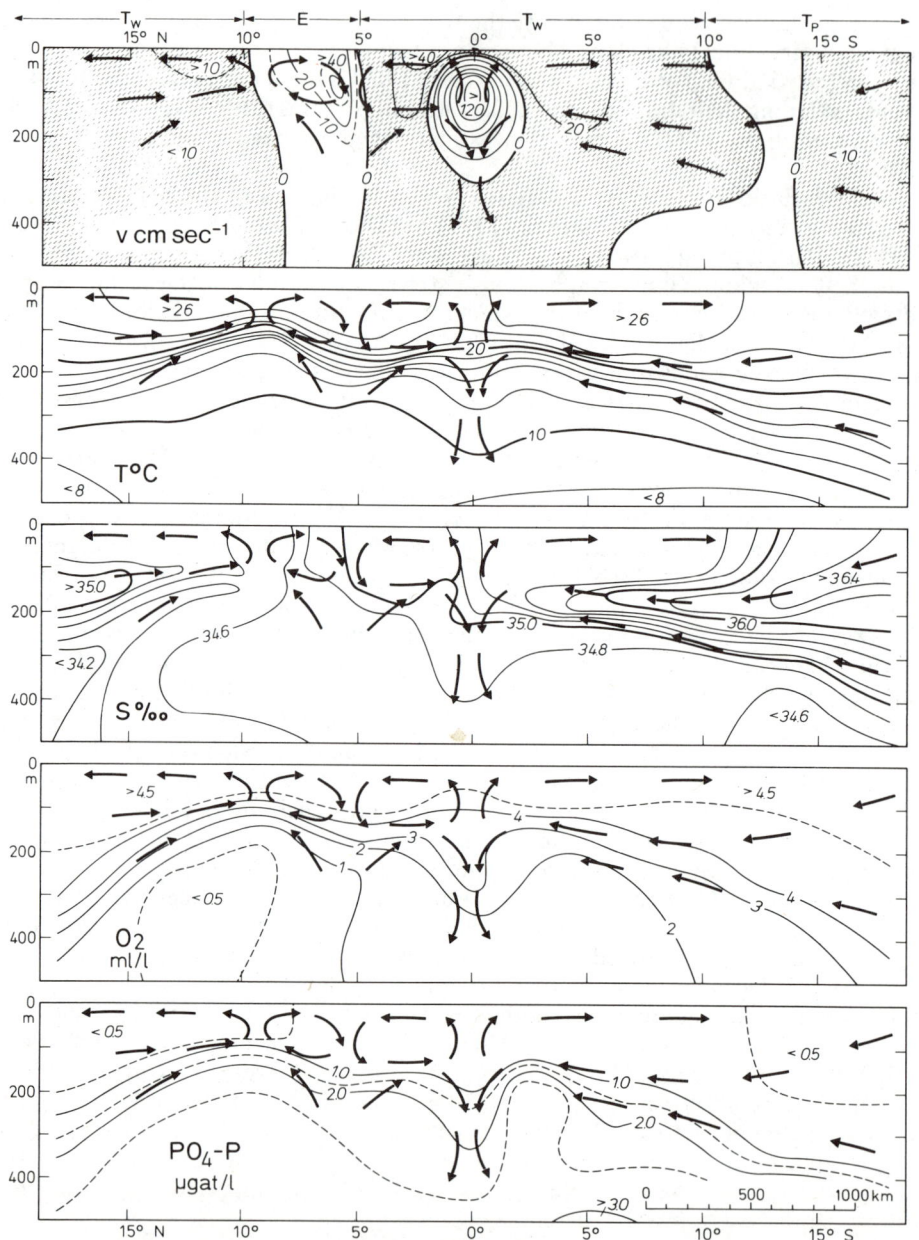

Fig. 10.34. Example of stratification and motion in the regions of the trade wind currents T_W and of the equatorial countercurrent E. (After Dietrich, 1970.) Distributions of current velocity, temperature, salinity, oxygen, and phosphate on a meridional section in the Pacific Ocean at 140°W, based on observations of the research vessel *Hugh M. Smith* (following Knauss, 1963, with cross circulation supplemented). With screen: main westward current, perpendicular to section: north and south equatorial currents. Without screen: main eastward current, perpendicular to section: Cromwell Current and equatorial countercurrents. Long arrows: transverse circulation of the main currents. Vertical exaggeration of cross section: 2000-fold. For location of cross section see Fig. 10.31 (line V).

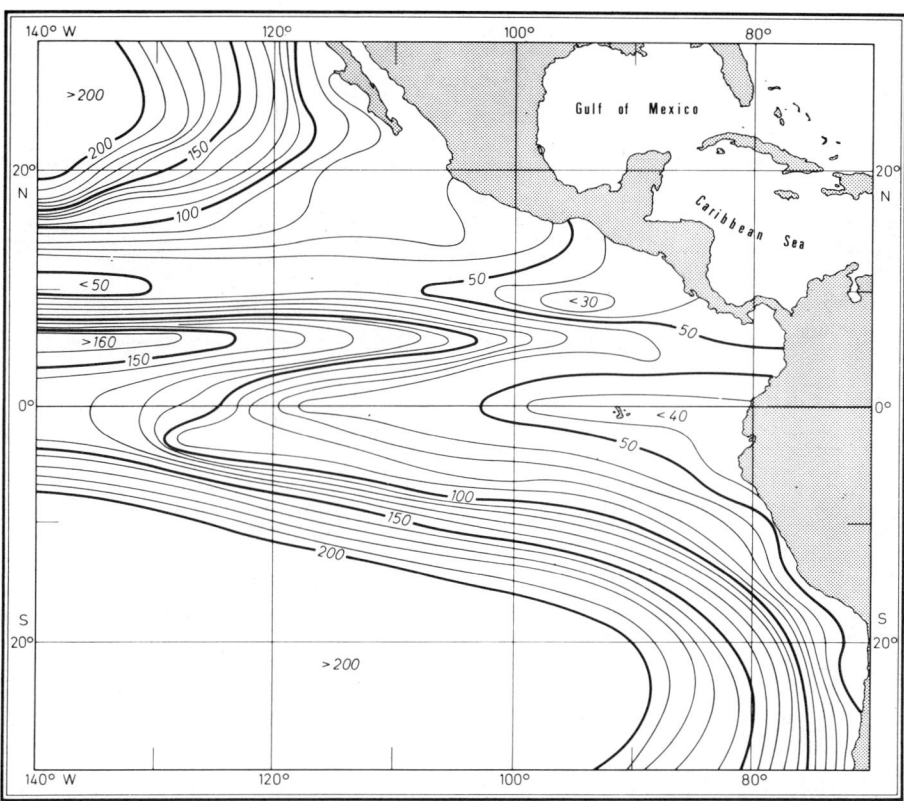

Fig. 10.35. Depth (in m) of the seasonally constant thermocline (middle of the layer) in the eastern equatorial Pacific Ocean in October. (After Wyrtki, 1965b.)

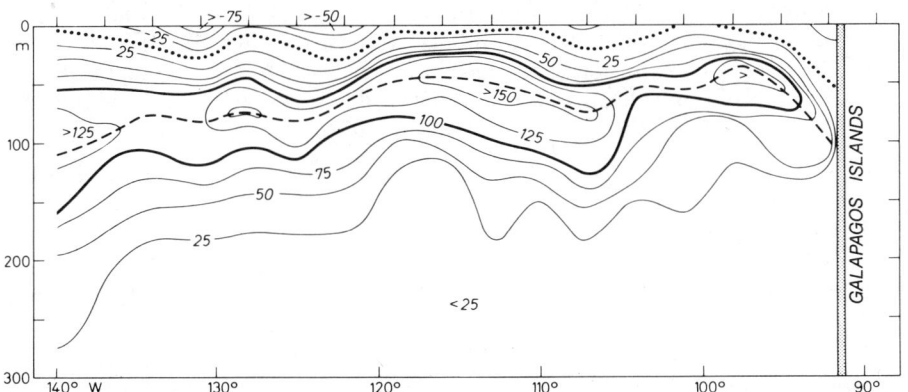

Fig. 10.36. Equatorial Undercurrent (Cromwell Current). Current velocity (east-west component in cm sec^{-1}) in the eastern Pacific Ocean on a cross section along the equator (140°–90°W, from north of the Marquesas to the Galápagos Islands). (After Knauss, 1960.) Marks at the upper edge: stations with direct current measurements at short distances of the research vessels *Horizon* and *Hugh M. Smith* in 1958. .... Boundary between North Equatorial Current and Cromwell Current. - - - - - - - - Location of the axis of maximum current velocity.

515

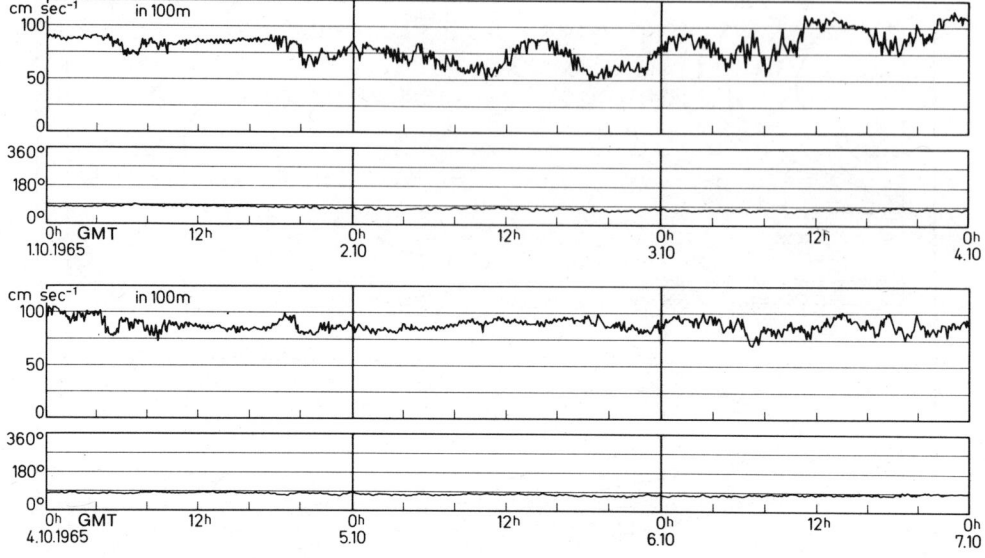

Fig. 10.37. Part of a continuous current registration in the Equatorial Undercurrent in the Atlantic Ocean by the research vessel *Meteor* ($\varphi = 0°36'S$, $\lambda = 29°28'W$, recording depth 100 m, water depth 4200 m) from October 1, 0^h to October 7, 0^h, 1965. (After Neumann, 1972.)

4. It has a conspicuous core ascending at the equator from about 100 m (140°W) up to a depth of 40 m (100°W) (cf. Fig. 10.36).

5. In its core it has maximum velocities of more than 100 cm sec^{-1}, partly of more than 150 cm sec^{-1} (cf. Figs. 10.34, 10.36), and its total water transport amounts to approximately 40×10^6 m^3 sec^{-1}, almost corresponding to the capacity of the Gulf Stream.

6. Its core lies in the upper part of the thermocline (Fig. 10.34) and, thus, of the pycnocline, which rises from west to east (cf. Fig 10.35) East of the Galápagos Islands, where the discontinuity layer lies at the depth of only 20 m, the Cromwell Current is but weakly developed, with a water transport of about 3×10^6 m^3 sec^{-1} (Christensen, 1971).

7. The thermocline, which is very pronounced poleward of 2° latitude, shows a weaker gradient near the equator (Fig. 10.34). Hence, surface temperatures at the equator are relatively low (cf. Fig. 5.05).

8. What the figures presented here do not show but what was proven by observations of the French research vessel *Coriolis* in 1967 is that, at 170°E, the undercurrent is split into two cells (Hizard et al. 1970).

In the 1960s, the Equatorial Undercurrent, which oceanographers had almost failed to notice until 1952, has become a phenomenon thoroughly investigated in all three oceans. Some further facts have resulted from this.

9. Studies of pertinent literature, made by Neumann (1960), have shown that equatorial undercurrents flowing eastward had already been described in 1886 (Buchanan) and 1895 (Puls) for the Atlantic Ocean as well as in 1895

(Puls) for the Pacific Ocean. However, no attention was paid to them. Other earlier hints regarding this undercurrent were mentioned by Matthäus (1969). Thus, the old experience becomes even more obvious which says that discoveries not attracting the attention of contemporaries simply do not exist.

10. In the Atlantic Ocean, Voigt (1961) was able to rediscover the undercurrent on the basis of his measurements on the research vessel *Michael Lomonossov* in 1959. Numerous measurements followed from this: for example, by Metcalf et al. (1962) in 1961, by Neumann and Williams (1965), and Stalcup and Metcalf (1966), by *Meteor* in 1965 and 1969 (Neumann, 1972), by Brosin and Nehring (1968), by LeFloch (1970) on the basis of investigations with *Jean Charcot* in 1968, which, for the first time, gave some information on where the undercurrent disappears on the eastern side of the Atlantic Ocean. A westward undercurrent has been found off the Ivory Coast. The first continuous records of the Equatorial Undercurrent were carried out on the research vessel *Meteor* in 1965. Figure 10.37 exhibits a section from these registrations, covering 6 days near the equator where the water is 4200 m deep. Here, records were taken at 100 m depth at 5 min intervals over a period of 11 days. The current is distinguished by a great constancy with regard to direction (steadily eastward) as well as to velocity, which amounts to around 75 cm sec^{-1} on the average (maximum 130 cm sec^{-1}, minimum 50 cm sec^{-1}). Occasionally semidiurnal tidal currents occur, particularly obvious on October 2 and 3, 1965. For further clarification of the mechanism of this specific current in the world ocean, records over a period of several months would be required at several measuring depths with a number of moorings arranged along a cross section perpendicular to the equator.

In the Indian Ocean, Ovchinnikov (1961) has found some indication of an equatorial undercurrent that must exist at least at times. During the International Indian Ocean Expedition, detailed investigations, especially by Taft and Knauss (1967), by Düing et al. (1967), as well as by Swallow (1967) have shown that the equatorial undercurrent is not always present. The seasonal change of the monsoon winds creates initial conditions different from those provided by the steady trade winds in the other two oceans. After Schmitz (1964) determined that the time interval within which the sea level becomes adjusted to the large-scale wind field amounts to several months, it cannot be expected that the undercurrent is coupled with the wind in a simple way. It seems to reach its greatest constancy towards the end of the period of the northeast monsoon. This is in agreement with theoretical investigations by Robinson (1966), showing that the undercurrent is supported by zonal components of the shear stress. In the Indian Ocean, these are most pronounced at the end of the period of the northeast monsoon.

An interpretation of the Equatorial Undercurrent that would include all the facts observed has not yet been fully achieved (Fofonoff and Montgomery, 1955; Yoshida, 1959). Several models have been dealt with in a special issue of Deep-Sea Research 6 (1960) (Charney et al.), by Ichiye (1964), and by Gill (1971). The undercurrent is explained by the inclination of the sea surface and the fact that the Coriolis force is missing at the equator. The westward blowing trade winds build up the inclination of the sea surface from west to east. In the Pacific Ocean, it amounts to approximately 70 d.cm. at the equator between 130°E and 85°W [cf. Fig. 10.33(a)]. In the central Pacific Ocean, it is most pronounced between 180 and 120°W. Beneath the drift current in the top layer, the water follows the direction of the inclination from west to east as an undercurrent with a lower boundary in the highly stratified water. To understand the sharp lateral limitation is a more difficult problem. The explanation might be found in the possibility that—because of the different direction of the Coriolis force on either side of the equator—the divergent water transport by drift currents near the equator (Ekman transport, cf. Fig. 10.12) may cause convergence below the drift current and, thus, the concentration of the undercurrent. Any meandering of the undercurrent—if it should develop at all—cannot be maintained because the Coriolis force becomes effective with growing distance from the equator. Hence, eastward currents are led back to the equator and forced to channel into a jet stream.

10.3.2.3. Regions of trade wind currents T_P with components of motion strongly directed poleward

These regions differ from the T_W regions by their anomalously high surface temperature (cf. Fig. 5.05) which, together with strong evaporation, contributes to making the lower atmospheric layers unstable. Thus, the tendency for precipitation is increased. If, furthermore, orographic rainfall is added along the coasts, this will suffice for an exuberant tropical vegetation to develop on oceanic islands as, for example, in Polynesia, in northern Australia, on the eastern side of Madagascar, or on the Brazilian coast south of Bahia. Such feedback effects on the atmosphere are especially obvious if the natural conditions of the T_P regions are compared with those of the T_E regions, where coastal deserts lie at nearly the same geographical latitude, as in western Australia, in Southwest Africa (the Namib Desert), and in northern Chile (the Atacama Desert). The stratification resembles that in the T_W region (cf. the example in Fig. 10.34), the only difference being that the discontinuity layer lies at a deeper level (cf. Fig. 10.41) and that the salinity is lower (Masuzawa, 1964).

10.3.3. The regions of equatorial countercurrents E

In these regions, equatorial countercurrents, directed eastward, occur permanently or at certain times of the year. They are among the most conspicuous components of oceanic surface circulation [cf. Plate 6, Figs. 10.33, 10.34, 10.35, 10.38(a)]. They move in narrow bands, that is, narrow if compared to their length, and reach very high velocities with an average of 40 cm sec^{-1} and a maximum of more than 150 cm sec^{-1}. Their location largely coincides with the zone of equatorial calms or doldrums.

In the Atlantic Ocean, the Equatorial Countercurrent flows north of the equator all year along. It is best developed in northern summer when it extends from the vicinity of the coast of South America at 50°W to the inner part of the Gulf of Guinea, where it is called Guinea Current. In northern winter, the countercurrent begins only at 25°W (cf. Plate 6). The Equatorial Countercurrent is most impressive in the North Pacific Ocean, where, in northern summer, it can be traced between 5 and 9°N with a width of only 300 to 700 km over a distance of 15,000 km from the Philippines in the west to the Gulf of Panama in the east. In northern winter it is shifted toward the south by up to two degrees of latitude. Kendall (1970) was the first to describe this current in a monograph based on 262 meridional sections. According to this description, the current is most strongly developed in the west, with an average water transport of approximately 45×10^6 m^3

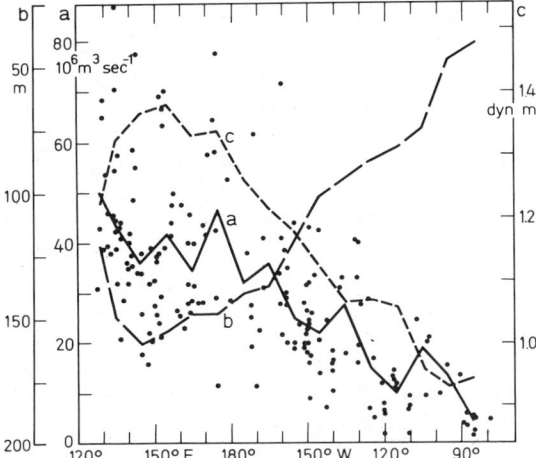

Fig. 10.38. The Pacific Equatorial Countercurrent. (After Kendall, 1970.) (a) Zonal water transport directed eastward in 10^6 m^3 sec^{-1}. Solid line: mean transport, averaged over 10 degrees of longitude, from the western to the eastern side of the Pacific Ocean. Dots: transports based on individual cross sections. (b) Mean thickness (in m) of the upper warm-water layer ($T > 20°C$), referred to the axis of the countercurrent. (c) Mean sea surface level in dyn m over 500 dbar, referred to the axis of the countercurrent.

sec^{-1}. How it decreases on its path from west to east across the Pacific Ocean can be seen in Fig. 10.38 (line a). In the Gulf of Panama, the weak offshoot of the current is turned southward. With the water transport diminishing from west to east, the warm top layer with temperatures of more than 20°C becomes thinner, that is, goes from 160 to 40 m (cf. Fig. 10.38, line b). Simultaneously, the sea surface is lowered by approximately 50 cm (cf. Fig. 10.38, line c). The decrease of the Equatorial Countercurrent from west to east is superimposed by a great variability, as can be seen in Fig. 10.38. Such variation may occur within a few days (Knauss, 1961; Montgomery and Stroup, 1962), or it may even lead to complete interruption (Wyrtki, 1965a). In the first representation of the dynamic topography of the sea surface in the entire Pacific Ocean (Reid, 1959, 1961), another, much weaker equatorial countercurrent can be recognized; it is indicated at about 10°S as a geostrophic current flowing from the Solomon Islands toward Peru [cf. Fig. 10.33(a)]. The meridional section in Fig. 10.34 also suggests the existence of such a second countercurrent, and so does the cross section at 170°E, taken by Merle et al. (1969) for 9°S. At the surface, this countercurrent cannot be distinctly recognized from ships' set and drift. The observational data are too incomplete for the purpose, and the velocities are too small. In the Indian Ocean, the Equatorial Countercurrent is present only in the northern winter, and merely south of the equator. In northern summer, it moves to the Northern Hemisphere and amalgamates with the southwest monsoon current.

The first rational explanation of equatorial countercurrents was given by Neumann in 1947. Up to that time, the piling-up of water masses at the continental margins on the western sides of the oceans, caused by trade winds, was believed to be the decisive factor. The countercurrents were supposed to follow the inclination of the sea surface from west to east. Hence, the Coriolis force was neglected, which is permissible only in the immediate vicinity of the equator, where, in fact, a countercurrent corresponding to this concept is formed. This is the Equatorial Undercurrent, also called Cromwell Current (cf. Section 10.3.2). Neumann (1947) has explained the Equatorial Countercurrent as the result of water masses piled up in the open ocean. Such setup is locally maintained by the peculiarity of the tropical wind system, as schematically represented in Fig. 10.12(a). The asymmetry with respect to the equator of the trade winds that exists over the oceans causes a convergence of surface currents on the equatorial side of the doldrums and a divergence on their poleward flank. The result is a poleward inclination of the sea surface in this zone. Under the influence of the deflecting force of the earth's rotation, which, according to this concept, can no longer be neglected at 5 to 10° latitude, this inclination brings about an eastward current, that is, the Equatorial Countercurrent. Following this explanation, an equatorial countercurrent should exist in any ocean, even without the continental boundary in the west and, hence, also on a globe covered with water, provided the planetary wind system is arranged asymmetrically to the equator.

According to Margules' boundary equation (cf. Section 7.3.2), the stationary current field of equatorial currents in the stratified ocean must be associated with a certain field of mass, the typical distribution of which was first recognized by Sverdrup (1932) on the basis of a section obtained by the research vessel *Carnegie* in the Pacific Ocean. Defant (1936a) found similar conditions in the Atlantic Ocean. By also considering friction, he arrived at the conclusion that a transverse circulation, consisting of eddies with horizontal axes, must be associated with the whole current system, which, however, did not yet include the Cromwell Current, only discovered as late as 1952. The complete system of equatorial zonal circulation, with the superimposed meridional cross circulation and the inclination of the sea surface as well as of the associated thermocline, are represented schematically in Fig. 10.12(b). Details of the typical relations between the distribution of hydrographic factors and the current system can be seen in a cross section through the Equatorial Countercurrent in Fig. 10.34. The transverse circulation, indicated by arrows, consists of several eddies, into which the divergence and convergence zones at the sea surface fit quite easily. The zones of divergence are characterized by a relatively low surface temperature (cf. also Fig. 5.05). In contrast to Fig. 10.34, the surface temperature may also distinctly decrease on the poleward flank of the countercurrent, as is the case at the Gold Coast, that is, at the northern flank of the Atlantic Equatorial Countercurrent, here called Guinea Current. With regard to the example of Fig. 10.34, it must be noted that the high precipitation rate in the doldrums at 10°N contributes to a strong decrease of salinity in the surface water. The result is a pronounced pycnocline at small depth, acting as a blocking layer against cold upwelling water. Strong upwelling on the northern side of the Pacific Equatorial Countercurrent is demonstrated in Fig. 10.35. From 5 to 10°N, the permanent thermocline rises, perpendicular to the current direction, by more than 100 m, namely from about 160 to 50 m depth. This figure, which is based on a great number of observational data from various years, confirms that the details of the cross section in Fig. 10.34 are fairly representative and typical.

The water masses ascending at the equator and at the northern flank of the equatorial countercurrents transport water that is comparatively poor in oxygen, but rich in nutrients, to the vicinity of the sea surface. Thus, a rich development of plankton can take place in the euphotic zone. In 1963–1964, thorough investigations of the Atlantic E region were carried out during the ICITA cruises (International Co-operative Investigations of the Tropical Atlantic). This exercise was divided into three parts: Equalant I, II, and III (Zeitzschel, 1969). Like the T_E regions, these equatorial areas of upwelling are also characterized by the green color of seawater and an abundance of fish quite in contrast to the neighboring cobalt-blue, desert-like regions of the trade wind currents T_W.

In recent years, the equatorial divergence zones of all three oceans have proved to be rich catching grounds for the valuable tunafish. Japanese fishermen are particularly active in this field. Scientific investigations are mostly carried out by the United States of America and by Japan. Results have been compiled by FAO (Food and Agriculture Organization of the United Nations) (1963) and by Blackburn (1965). The ecological conditions of tuna are not easy to understand because they vary from species to species. The species most important for fishing are the following: bluefin (*Thunnus thynnus*), albacore (*T. alalunga*), yellowfin (*T. albacares*), bigeye (*T. obesus*), and skipjack (*Euthynnus pelamis*). The temperature range plays a decisive role for the distribution of the various species. There are differences: the yellowfin prefers 20 to 28°C, the albacore 15 to 21°C, the skipjack 19 to 23°C. Furthermore, it must be taken into account that the upwelling in the divergence zones certainly causes great production of phytoplankton, but that time is needed for its transformation into tuna food in the form of micronekton. Hence, it is not the equatorial divergence zone itself but its immediate surroundings that must be considered the preferred habitat of tunafish.

10.3.4. The regions of monsoon currents M

The regular change of the direction of monsoon winds (cf. Section 4.3.2 and Fig. 4.14) in spring and fall causes a complete reversal, or at least a considerable change of direction, of the surface currents in these seasons, as clearly shown for southeast Asiatic waters in Fig. 10.39. A change in the distribution of temperature and salinity as well as in the hydrographic stratification in the uppermost 100 to 200 m of the M regions is associated with this. In lower latitudes, however, the effects of this change differ from those in moderate latitudes. Therefore, we distinguish between tropical, low-latitude monsoon current regions M_L and extratropical monsoon current regions M_M in moderate latitudes.

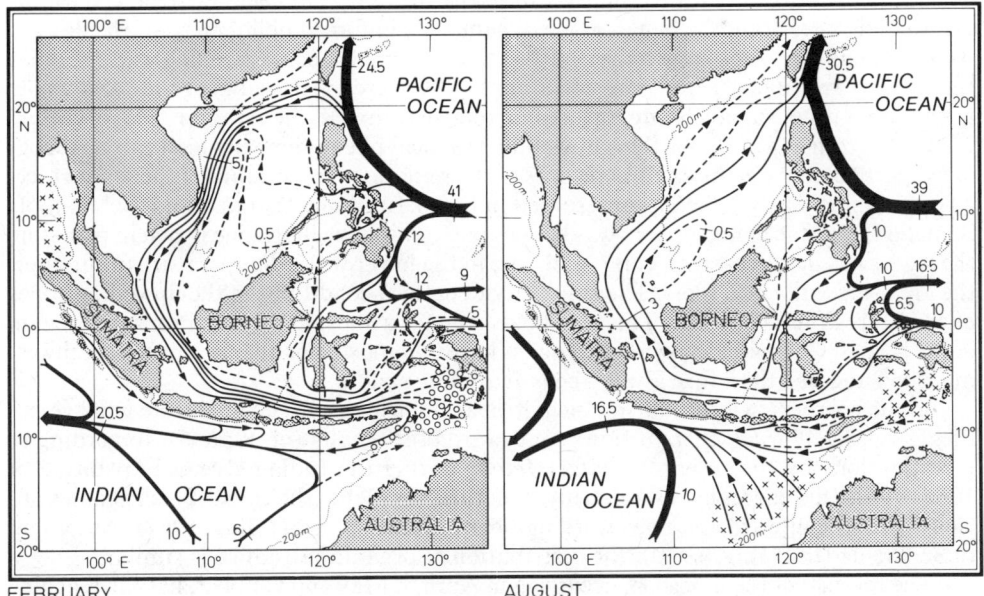

Fig. 10.39. Surface circulation in southeast Asiatic waters in February and August. (After Wyrtki, 1961.) Water transport in 10^6 m^3 sec^{-1}. x x x x x Upwelling areas. O O O O O O O Downwelling areas. .... 200 m depth contour line.

From observations and theoretical considerations (Stommel and Veronis, 1956; Lighthill, 1969) it can be deduced that the oceanic circulation in lower latitudes reacts to large-scale reversals of the wind system very much faster than the circulation in higher latitudes, especially in baroclinic processes of motion. As shown in Fig. 10.31, the M_L currents are mainly present in south Asiatic waters and the M_M currents in east Asiatic waters.

The season of the winter monsoon lasts from November to March or April. Under the influence of the Coriolis force, the winds blow from the northeast over the North Indian Ocean, from the north over the waters around Indochina, and from the north or the northwest over the east Asiatic marginal seas as well as over the adjacent open Pacific Ocean. In every case, the winds always blow offshore from the Asiatic continent. During this season, hydrographic conditions in the North Indian Ocean resemble those of a region of trade wind currents: the surface water flows westward in the northeast monsoon current (cf. Plate 6), while its salinity increases along its way toward the west, because the dry continental air of the northeast monsoon does not show any tendency toward precipitation, and evaporation alone is effective in the water exchange between ocean and atmosphere.

The season of the summer monsoon lasts from May to September. During this time, the wind blows as a strong southwesterly over the North Indian Ocean, and from south to southeast over east Asiatic waters; that is, the wind always has an onshore component toward the Asiatic continent.

A special position within the M_L region must be assigned to the Somali Current present only in northern summer. It develops as a gradient current because, at the east African coast, the sea level, due to the pile-up by the southeast trade winds, is higher south of the equator than north of the equator, where it is kept low by the southwest monsoon. In the vicinity of the equator, the water follows the inclination, and only north of about 8°N does the current turn to the right into the southwest monsoon current. With increasing distance from the equator, the dynamically enforced obliqueness of the density stratification increases so much that, at the coast of Somalia, cold upwelling water reaches the surface. In August 1964, temperatures of 14°C were measured near the coast while, at the same time, surface temperatures farther seaward, in the Arabian Sea, exceeded 25°C (Warren et al., 1966). This upwelling is a local phenomenon, which can be recognized particularly well in satellite records of the radiation temperature of the sea surface. The example of July 3, 1966, represented in Fig. 4.13, shows that at 7 and 10°30′N off Somalia the surface temperature was below 19°C. This upwelling zone is rich in plankton, the water has an olive-green color, and, due to the lower water temperature in summer, the area is free of coral reefs. The similarity to natural conditions in the upwelling zones of the T_E regions is obvious. This also holds true for atmospheric conditions: the region of the Somali Current is known for frequent haze and fog during the time of the southwest monsoon and has therefore always been feared by sailors.

In the northern winter, during the northeast monsoon with its moderate wind velocities, a relatively weak current flows southward off Somalia (cf. Plate 6). According to investigations by the research vessel *Meteor* during the Indian Ocean Expedition of 1964–1965, this current shows a stripy structure parallel to the coast. This becomes obvious when the upward and downward motions are determined (Düing, 1967). According to Szekielda (in Düing, 1970), the distribution of organic substance is similar.

The reversal of the monsoon, occurring in April to May and October, has far-reaching consequences in the entire M_L region, particularly in the Gulf of Bengal, the Andaman Sea, and the waters around Indochina. At the time of the winter (northeast) monsoon current, the low-salinity surface water of the Gulf of Bengal and the Andaman Sea, formed during the rainy season of the summer monsoon, regionally spreads westward into the Arabian Sea. In summer, on the other hand, the southwest monsoon current pushes the low-salinity surface water back into its area of origin, where large amounts of precipitation and continental runoff contribute to a strong salinity decrease. The final result is that the monsoon current region M_L, especially in the Gulf of Bengal, in the Andaman Sea, in the waters around Indochina, and in the South China Sea, are characterized by unusually high annual amplitudes in salinity. While such amplitudes amount to 0.1 to 0.2‰ in the central North Atlantic Ocean, they reach 1 to 3‰ in the M_L region and are even higher in the Gulf of Martaban. Since marine fauna and flora are very sensitive to variations in salinity, these regions must occupy a special position from the biological point of view.

Studies on processes of reversal and adaptation of oceanic circulation to the seasonal changes of the monsoons were among the main objectives of the International Indian Ocean Expedition of 1959 to 1965 (Dietrich, 1965). Fundamental problems concerning the dynamics of oceanic circulation were touched upon there. In addition to numerous individual papers, among which contributions based on the *Meteor* expedition of 1964–1965 and published in the *Meteor* Research Reports, the data obtained during this international expedition, the largest to date, with 44 research vessels from 25 nations participating, have been summarized in two atlases, an oceanographic one by Wyrtki (1971) and a climatological one by Ramage (1970).* Furthermore, special topics have been treated in a number of monographs, one of which, by Düing (1970), deals with the reaction of currents on the monsoons in the Indian Ocean. Some of the findings are rather surprising; the most important result is demonstrated in Fig. 10.40 showing that the near-surface circulation in the Indian Ocean differs from the other parts of the oceans in the Northern as well as in the Southern Hemisphere. It is not characterized by a large anticyclonic gyre in the tropics and subtropics, as is the case, for example, in the North Pacific Ocean (cf. Fig. 10.33). The circulation in the North Indian Ocean shows a cellular structure of cyclonic and anticyclonic eddies. Figure 10.40 is based on highly valuable data collected by the research vessels *Atlantis II, Anton Bruun,* and *Discovery* during a short period in July, August, and September 1963. A similar cellular structure is also shown in representations for other seasons with less homogeneous observations.

Another important result, obtained by Düing, is the temporal course of the adaptation process of the circulation. At present, such studies can only be based on theoretical models. The decisive assumption in this regard is this: dissipative processes play a less important role than they do in stationary circulation models. The model computations mentioned above (Düing, 1970) were carried out for a vertically homogeneous ocean (barotropic case). A baroclinic circulation model (Lighthill, 1969), however, with dissipative processes being neglected, has yielded the result that baroclinic processes of motion are more strongly induced in the Somali Current than barotropic processes. At the boundaries, the adaptation of the sea to the wind occurs within a period of about 10 days. Nobody yet knows how much time this needs in the open ocean. According to model computations (Schmitz, 1964), even a whole monsoon season may not be sufficient. Oceanographic measuring methods, as used up to now, with stations of individual vessels, are not suitable for studying such large-scale, time-dependent processes. Satellite records of the radiation temperature of the sea surface represent an alternative (cf. Fig. 4.13). Düing and Szekielda (1971) have investigated this possibility by evaluating a series of radiation records.

Translators' remark: The Phytoplankton Production Atlas (Krey and Babenerd, Eds.) was published in 1976 while the Geological-Geophysical Atlas (Udintsev, Ed.) appeared in 1975.

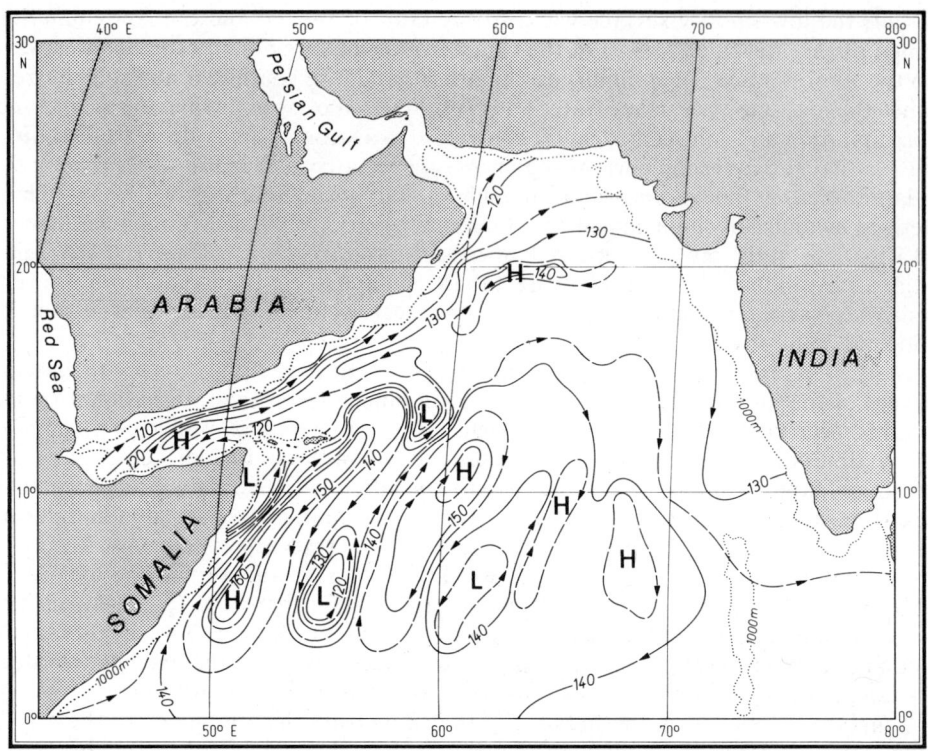

Fig. 10.40. Surface currents (geostrophic) in the Arabian Sea in the summer of 1963. (After Düing, 1970.) The dynamic topography of the sea surface is represented in dyn cm, referred to the 800 dbar surface. H: elevations of the sea surface, centers of anticyclonic eddies. L: depressions of the sea surface, centers of cyclonic eddies. ..... 1000 m depth contour line.

At the onset of the southwest monsoon, the horizontal temperature gradient is at first proportional to the wind velocity with a phase lag of 12 days; later on, the reaction time increases to 40 days, and large-scale geostrophic effects become dominant in the monsoon current.

The boundary between the tropical monsoon region M_L and the extratropical monsoon region M_M has been assumed to lie in the Strait of Formosa because, north of it, strong thermal redistribution occurs in the water, connected with the seasonal changes of wind and current conditions. This is common to all east Asiatic marginal seas, that is, the East China Sea and the Yellow Sea, as well as the Sea of Japan, the Sea of Okhotsk, and the Bering Sea. The offshore winds of the winter monsoon carry cold air from the Asiatic continent out to the sea, whereas the onshore winds of the summer monsoon bring warm oceanic air masses into those regions. Owing to such atmospheric influences, the sea surface temperatures have a mean annual variation of more than 10°C, in most areas of more than 15°C, in North Korean and Manchurian waters of more than 20°, sometimes even of more than 25°C. Annual variations of as much as 20°C are not found in any other oceanic area of the globe.

Beside these common features, temperature differences of course exist due to the different latitudes at which the marginal seas of the M_M region are located. In the Sea of Japan, the Sea of Okhotsk, and the Bering Sea, the water temperatures drop to the freezing point in winter. The typical ice of adjacent seas is formed. Even Vladivostok, situated at the same latitude as Florence, Italy, is not completely free of ice in winter. Farther north, ice conditions become more severe. In spring and summer, when maritime air masses saturated with water vapor arrive monsoon-like over those cold oceanic areas, persistent sea fog develops. Ice in winter and fog in summer, both equally feared by sailors, make great parts of the Sea of Okhotsk and of the Bering Sea most unhospitable areas.

10.3.5. The regions of horse latitudes H

The definitions regarding the region of the trade wind currents T and the region of the west wind drift W give the boundaries of the transitional area H lying between them (cf. Fig. 10.31). During part of the year, currents occur in this H region that either assume the characteristics of trade wind currents (this happens in summer in those parts of the H regions facing the equator) or of currents of the west wind drift. This is true for the poleward parts of the H regions in winter. The H regions also include the water masses with weak and variable motion in the subtropical high pressure area with its weak winds—the actual horse latitudes.

Trade wind currents and west wind drift flow anticyclonically around the interior regions of the horse latitudes. In this way, deep-reaching accumulation of light surface water is dynamically enforced in the center of the current eddy, as shown in Fig. 10.41 by the example of the North Pacific Ocean. The density surface for $\sigma_T = 26.2$, which roughly corresponds to a surface with a temperature of 11.5°C, reaches a depth of 500 to 600 m in the west, near the Japanese islands, as opposed to 100 to 200 m north and south of that area. The surface currents, converging in the center of the subtropical current gyre, maintain the accumulation of surface water. Thus, a deep-reaching, homogeneous, warm top layer is established, which is also very saline because evaporation exceeds precipitation and, consequently, salinity is increased at the sea surface (cf. Fig. 4.19).

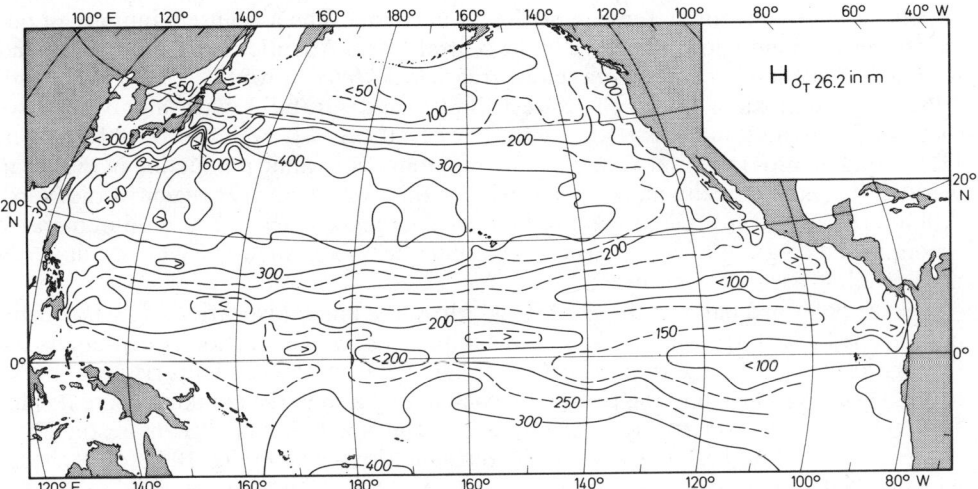

Fig. 10.41. Mean depth (in m) of the density surface $\sigma_T = 26.2$ in the North Pacific Ocean in October–December. (After Barkley, 1968.)

After recent Japanese investigations (Uda, 1955) have drawn attention to deviations from a simple, stationary anticyclonic gyre in the H regions, a subtropical countercurrent is supposed to be present at 20 to 25°N in the North Pacific Ocean, at least in the months of spring. On the basis of measurements of the International CSK 1965-1966 (Co-operative Study of the Kuroshio), Uda and Hasunuma (1969) have proved the existence of the countercurrent from 122 to 160°E. Independently, Seckel (1968) found this current north of the islands of Hawaii. Robinson (1969), basing his studies on comprehensive material of bathythermograph records, has confirmed it for the western Pacific Ocean as far as 160°W, especially for the time from January through May. If this current does not clearly appear in Fig. 10.41 by Barkley (1968), namely by a ridge at 20°N, it should be noted that conditions are represented as they are in the fall. The explanation of the subtropical countercurrent has been given by Yoshida and Kidokoro (1967). A weak trough in the subtropical atmospheric high at 20 to 25°N, developing particularly in spring, causes westerly wind components. The large anticyclonic subtropical gyre is split into two anticyclonic eddies, separated from each other by a weak cyclonic motion. The whole pattern is variable, and no definite evidence of it has been found in the other oceans so far. In the North Atlantic Ocean, there are indications in the Sargasso Sea that such a countercurrent may exist, at least for limited periods (Voorhis and Hersey, 1964; Voorhis, 1969), and that it may be associated with a thermal front in the uppermost 200 m (Katz, 1969). In the South Pacific Ocean, investigations by the French research station at Nouméa on New Caledonia have shown a subtropical countercurrent on a meridional section at 170°E, and that at 17°S (Merle et al., 1969).

There is no other region in the open world ocean where temperatures and salinities at depths between 200 and 400 m are higher than in the five regions of the horse latitudes. This can be seen in Plates 3, 4, and 5 for the H regions of the individual oceans. The water of the top layer, which is in relatively fresh contact with the sea surface, is characterized by a high oxygen content. This fact is particularly striking when compared with the poor ventilation conditions in the regions of trade wind currents (cf. Fig. 10.11). The H regions represent an accumulation of surface water, in which nutrients have been consumed by plankton and carried into the depth by dead organisms sinking down. Hence, these regions are characterized by an extreme depletion of nutrients. The absolute minimum of nutrients, reported until now, is located in the central North Atlantic H region. It was found during an 8 day anchor station of the research vessel *Meteor* (30°18′N, 29°25′W, April 1967). Concentrations lay at the lowest limit of measurability with highly sensitive methods (Grasshoff, in Closs et al., 1969): $PO_4 < 0.03$, $Si < 0.2$, $NO_3 < 0.1$, $NO_2 < 0.1$, $HN_3 < 0.3$ μg atom per liter. The absolute minimum of organic production in the ocean was determined by Steemann Nielsen (1957) in this same central H region of the North Atlantic Ocean. Referred to the production of organic carbon, the minimum value amounts to 0.05 g C m^{-2} day^{-1}, as contrasted to the maximum of 3.8 g C m^{-2} day^{-1} in the upwelling area off Southwest Africa.

In the poleward parts of the H regions, where the annual variation of surface temperature is fairly large, and thus deep-reaching, thermal convection is made possible during the winter cooling, the surface water benefits from the nutrients rising from the depth, which is reflected in the phosphate content and also in the plankton content. This fact is clearly demonstrated by an example from the North Pacific Ocean, for the phosphate content in Fig. 10.33(b), and for the zooplankton volume in Fig. 10.33(c). It should be noted that 31 mg m^{-3} P = 1 μg atom l^{-1} P. An idea of the hydrographic structure of the H regions, as expressed in temperature, salinity, density, and horizontal currents, is given in Fig. 10.42. In addition, the distributions of oxygen and phosphate as well as

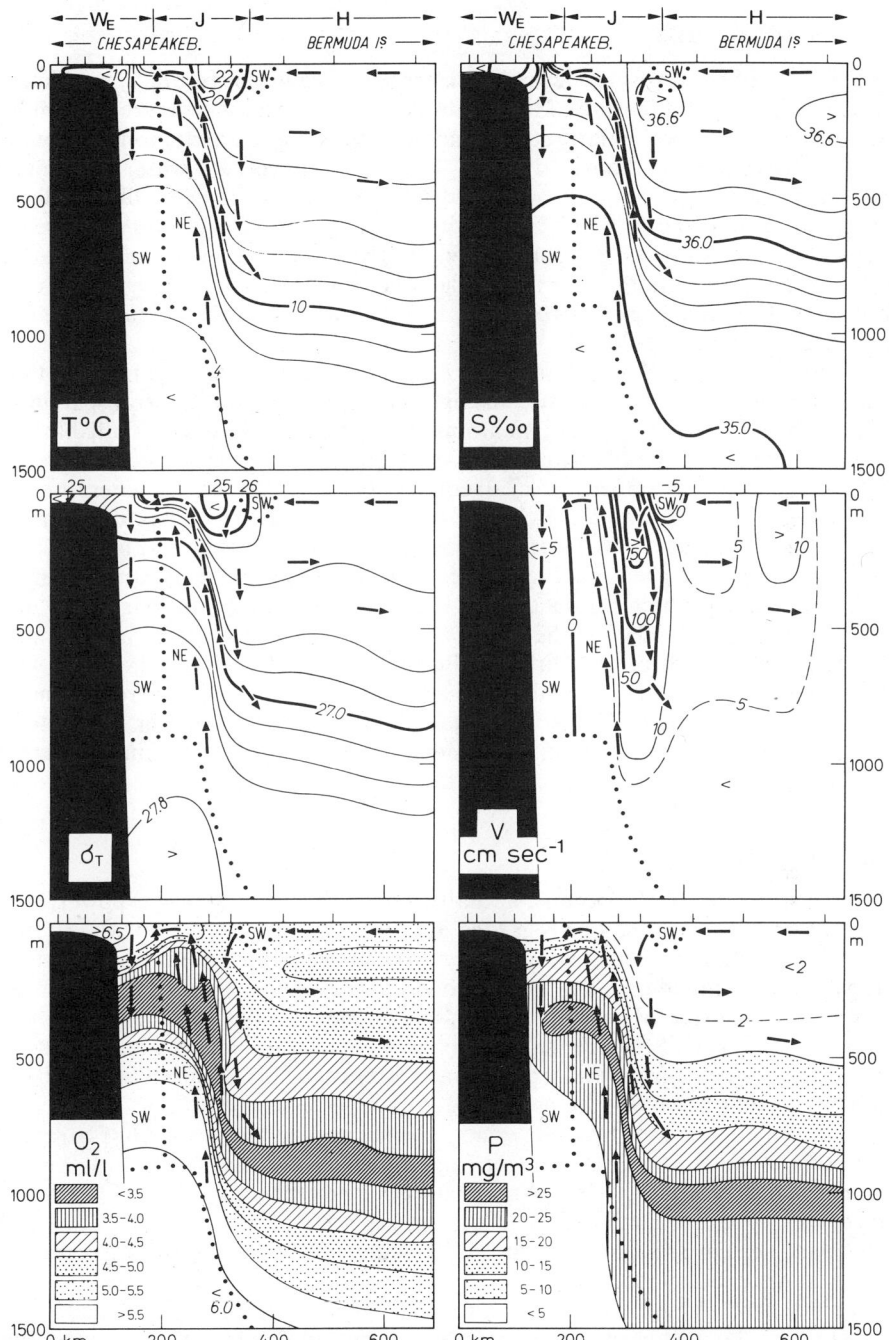

Fig. 10.42. Example of stratification and motion in the jet stream region *J* and in the region of horse latitudes *H*. (After Dietrich, 1956a.) Distribution of temperature, salinity, density, current velocity, oxygen, and phosphate on a cross section through the Gulf Stream and its surroundings near Cape Hatteras, based on observations of the research vessel *Atlantis* in April 1932. NE, SW: main current towards northeast or southwest, respectively, perpendicular to the cross section. NE: corresponds to the Gulf Stream. Arrows: transverse circulation of the main currents. Dotted lines: current boundaries. Marks at the upper edge: oceanographic stations. Vertical exaggeration of cross section: 500-fold. Location of cross section: see Fig. 10.31 (line VI).

some estimates of the transverse circulation are provided in this figure. A restrictive remark concerns a special feature of this example: in the Sargasso Sea, the accumulation of subtropical water is particularly strong and deep-reaching to an extent not encountered in the other H regions of the world ocean. The slight, convergent water movements in the Sargasso Sea make it possible that a golden-yellow plant, with berry-like air bladders, the sargasso weed, can exist there. These brown algae comprise several pelagic species of the genus *Sargassum,* which reproduce vegetatively. They appear in clusters, usually arranged in long streaks in the direction of the wind. It is mere sailor's talk when the speed of ships is said to have been obstructed by sargasso weed. On the average, one cluster can be expected in every 100 m^2.

The distinct scarcity of plankton in the H regions, especially in the parts lying towards the equator, is the reason why they are distinguished by extremely clear, transparent water untroubled by any turbidity worth mentioning and, hence, of a deep cobalt-blue color. Calm seas, rare occurrence of storms, good visibility, little cloudiness, moderate precipitation, and mild air temperatures, all contribute to the conviction, generally held by mariners, that the equatorial flanks of the H regions are favored by nature to a degree unknown in any other part of the world ocean. The oceanic islands that lie in these regions, though few in number, are veritable cradles of health and well-being, quite apart from their natural beauty which makes them also a place favored by tourists. The "insulae fortunatae" (happy islands), as the Canary Islands were already known in the time of the Romans; the islands of Madeira, the Bermuda Islands, and the Hawaiian Islands all fall within this category. The poleward areas of the H regions, where, for instance, the Azores lie, enjoy the full benefits of their situation only in summer. In winter they fall under the influence of variable, but sometimes powerful westerly winds which bring them great cloudiness and heavy precipitation.

10.3.6. The jet stream regions J

On the western sides of the oceans, actually within the T_W regions, a powerful slope current is formed, which may be regarded as a drain for those water masses in the T_W regions that are driven westward by the drag of the trade winds. These currents show a notable feature: outside the trade wind zones, in the subtropics with their light winds (that is, where the external energy supply is small), they do not spread fan-like, as might be expected as a result of turbulence and mixing. Instead, they concentrate into a narrow current band whose edges are sharply defined laterally. This is equally true where the bottom topography in no way helps to produce a jet effect.

Such drain currents of the trade wind currents are known as the Gulf Stream and the Kuroshio in the Northern Hemisphere and as the Brazil Current, the East Australian Current, and the Agulhas Current in the Southern Hemisphere (cf. Plate 6). All of them can be combined under the heading: jet stream regions J. This term is used to describe the fact that, here, we are dealing with jetlike currents flowing at high speed through relatively undisturbed water.

The five currents mentioned above are not all equally strongly developed. The two in the Northern Hemisphere are the most important and have always aroused special interest because of their manifold effects on natural conditions. In the J regions of the Southern Hemisphere, the basic characteristics of dynamics and volume transport are approximately known. This is least true for the Brazil Current, somewhat more for the East Australian Current (Hamon, 1965 and 1970; Rotschi and Lemasson, 1967). The latter current leaves the Australian continental slope at the latitude of Sydney, at 33 to

34°S, and turns from a southward current into a current directed northward. According to calculations of the geostrophic current for the years 1960–1964, the volume transport above 1300 m ranges between 12 and 43 × 10^6 m^3 sec^{-1}. Altogether, the East Australian Current, as the western boundary current of the entire South Pacific Ocean, is relatively weak, compared with the Gulf Stream on the western side of the very much smaller North Atlantic Ocean. Much of the energy is probably consumed by the circumpolar west wind drift. Lord Howe Island, situated at 32°S, 300 nautical mi east of Australia, sometimes lies to the east and sometimes to the west of the East Australian Current flowing off northward. So this island offers an observational basis singular in the *J* regions, which is used for water level registrations. The object of the most thorough investigations in the *J* regions of the Southern Hemisphere is the Agulhas Current (Dietrich, 1935; Darbyshire, 1964; Anderson, 1967; Duncan, 1970). In his dissertation, Duncan (1970) has shown that the current flows along the continental slope of South Africa and that its mean volume transport ranges between 80 × 10^6 m^3 sec^{-1} in summer and 100 × 10^6 m^3 sec^{-1} in winter, if the layer from 0 to 2500 m is included. What Dietrich (1935) observed has been confirmed; namely that the current, south of the Agulhas Bank, almost completely turns back into the South Indian Ocean.

In recent decades, numerous Japanese scientists have contributed to the exploration of the Kuroshio. A bibliography containing 800 relevant publications has been supplied by Yoshida and Shimizu in Stommel and Yoshida (1972). A comprehensive description by Uda (1964), with a large number of references cited, clearly demonstrates that the large-scale features of oceanic surface circulation in the north Pacific Ocean, in general (cf. Fig. 10.33), as well as the fine structure of the Kuroshio Current, in particular, are largely identical with the phenomena observed in the Gulf Stream. Hence, the following short description is mainly restricted to the Gulf Stream, which shows the typical features of a jet stream region in a particularly conspicuous way.

Ever since the Gulf Stream was discovered by the Spaniard Ponce de Leon in 1513, it has frequently been the object of investigation. A detailed history of early investigation has been given by Kohl (1868) and another one of more recent studies by Stommel (1950). These studies are of special interest because they reflect the history of marine research in all its stages of development. This is true for the methods of analysis as well as for the concepts about the origin and structure of oceanic currents. The Gulf Stream research was intensified when a large oceanographic institute was founded at Woods Hole, Massachusetts in 1930 (Woods Hole Oceanographic Institution). Since its research vessel *Atlantis* was commissioned in 1931, the Gulf Stream has been the main object of the investigations by this institute. The monograph *Gulf Stream* by Stommel (1965, 2nd ed.) not only represents a summary of our knowledge on the dynamics of the Gulf Stream, but it also refers to unsolved problems and, thus, is a guide for present and future marine research. More recent findings have been summarized by Robinson (1971).

The name "Gulf Stream," originating from the first speculative interpretations made in the 16th century, expresses the hypothesis that the Gulf of Mexico should be considered the basin where the water masses accumulate and from which the Gulf Stream is discharged. Even though this concept is out of date now, and although it has been recognized (Dietrich, 1937a; Leipper, 1970) that only the southeast corner of the Gulf is included in the discharge of the trade wind current entering through the Yucatán Channel, the conventional name of Gulf Stream has remained in common usage. Moreover, it has become customary in the past decades to speak of the Gulf Stream as part of the greater "Gulf Stream system," as suggested by Iselin (1936), the long-time director of the institute at Woods Hole.

Figure 10.50 shows the branches of this current system, the names of which are given in abbreviated form. The Gulf Stream system is assumed to begin in the Yucatán and Antilles Currents. In the Yucatán Current, a narrow current profile is formed from the broad trade wind driven Caribbean Current. This narrow band is continued and maintained in the Florida Current transporting 32×10^6 m^3 sec^{-1} of water on the average (Schmidt and Richardson, 1966). The section of the current system between the junction of the Florida and Antilles Currents and the area east of the Grand Banks of Newfoundland is called the Gulf Stream. It closely follows the continental slope between the northern exit of the Straits of Florida to Cape Hatteras and then turns to the east in a narrow band like a "river in the ocean," so to speak, only 50 km wide (cf. Fig. 10.42). The current transports an enormous volume of water, averaging about 63×10^6 m^3 sec^{-1} (Richardson and Knauss, 1971), which, in its further course, according to measurements by Warren and Volkmann (1967), reaches 100×10^6 m^3 sec^{-1} (± 20 to 30%) south of New England. Altogether, the water transport by the Gulf Stream from the Straits of Florida over a distance of 2000 km increases by 7% per 100 km (Knauss, 1969). Southeast of the Grand Banks, the Labrador Current invades the Gulf Stream on its left flank (cf. Fig. 5.30). Approximately half of the Gulf Stream turns back to the west (Mann, 1967) while the other half, following the isobaths, turns to the north as far as 55°N (Warren, 1969) and is there called North Atlantic Current. This current then moves eastward (Dietrich, 1964), until it is split into several branches (cf. Fig. 10.50) when crossing the Mid-Atlantic Ridge at about 55°N 30°W. One of these branches with a volume transport of 11×10^6 m^3 sec^{-1} reaches West European waters. One part, 4×10^6 m^3 sec^{-1}, branches off southward in the Portugal Current and thus only 7×10^6 m^3 sec^{-1} cross the Wyville–Thomson Ridge between the Faeroe and Shetland Islands penetrating into the Norwegian Sea as the Norwegian Current. The water transport of 65×10^6 m^3 sec^{-1}, directed poleward in the narrow Gulf Stream, is contrasted by 54×10^6 m^3 sec^{-1} as compensation directed toward the south in near-surface layers and distributed over a width of 4000 km, not concentrated into individual current bands as one might be led to conclude from Fig. 10.50. The remaining 11×10^6 m^3 sec^{-1}, which descend partly west of the Strait of Gibraltar and partly in the Irminger and Labrador Seas, are transported southward in the North Atlantic deep water (cf. Section 10.1.5).

Some idea of the typical hydrographic structure of the J regions is conveyed by the section across the Gulf Stream in Fig. 10.42, based on an extension of earlier representations (Dietrich, 1937a). This cross section shows that the high velocities of the Gulf Stream of more than 150 cm sec^{-1} are associated with co-existing water masses of different densities, as can be expected from the general principles of dynamics (cf. Section 7.3.3). In this case, cold, low-salinity water is found on the left-hand flank, and warm, high-salinity water on the right-hand flank. While the former water mass is of relatively high density, the latter is of relatively low density. At the sea surface, these two water masses manifest themselves in a front, the so-called "Cold Wall."

It should be pointed out that the large-scale picture of the Gulf Stream as part of the North Atlantic circulation (Fig. 10.50) refers to mean conditions, whereas the cross section through the Gulf Stream in Fig. 10.42 represents an individual case. In addition, several remarkable facts have been revealed by recent Gulf Stream research. Some result from processes in the energy sources, that is, in the trade winds, some from the dynamics governing jet streams in stratified water on the rotating earth, and some from the influence of bottom topography. A few of these important facts will be briefly dealt with in the following.

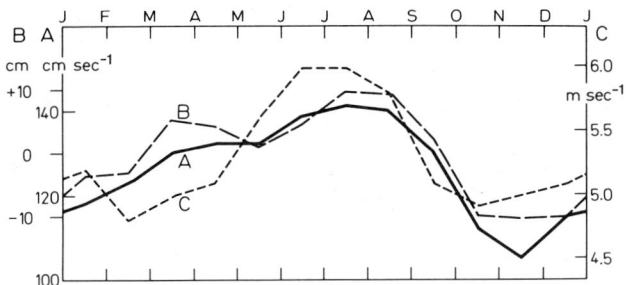

Fig. 10.43. Annual variation of the current velocity at the sea surface (A) and of the water level difference (B) in the Florida Current, as well as of the wind velocity (C) of the trade winds. A: monthly averages of the current velocity according to ships' set and drift (after Fuglister, 1951). B: water level difference, referred to an annual mean of zero between the right- and the left-hand sides of the Florida Current (Cat Key and Miami) (after Stommel, 1953). C: monthly averages of the wind velocity east of the Lesser Antilles.

1. *Periodic variations* of the surface current velocity, and thus of the water transport in the Gulf Stream, predominantly follow the annual rhythm. According to data obtained for the Straits of Florida from ships' set and drift, the mean surface current velocity, averaged over many years, ranges between 105 cm sec^{-1} in November and 140 cm sec^{-1} in July and August. These variations are closely related to the annual variations in the wind velocity of the trade wind zone, as illustrated by curve C in Fig. 10.43 for the area east of the Lesser Antilles. The extremes of the trade winds occur approximately 1 month earlier than the extremes of the current velocity in the Straits of Florida, a phase difference that can be explained by the fact that the Florida Current already lies outside the actual trade wind zone. Besides, curve B indicates the changes in the water level difference across the Florida Current. These data were obtained from monthly averages of the water levels at two gauges, on the left-hand side of the current in Florida and, on its right-hand side, on the Bahama Islands. These water level differences fluctuate by 20 cm in exact accordance with the rhythm of the current. Furthermore if the absolute water level difference is taken into account [which, according to "Oceanic Leveling" by Dietrich (1936), amounts to 45 cm between the two flanks of the Florida Current], the mean water at the coasts of Cuba and the Bahama Islands, in the monthly average, is higher by 35 to 55 cm than at the coast of Florida. The hope that variations and pulsations at the roots of the Gulf Stream might be deduced from water level registrations (cf. also Fig. 10.44) has been fully confirmed by Stommel (1953b). Water level recordings, made since 1938, present a simple way of keeping the Florida Current under surveillance and of detecting any irregularities of its behavior.

2. *Nonperiodic pulsations,* occurring in the Florida Current in addition to periodic variations, have been confirmed by water level records and continuous current registrations, as well as by measurements of the electric potential difference at a submarine cable across the Straits of Florida. This potential difference, caused by electromagnetic induction when seawater moves through the geomagnetic field (cf. Section 3.2.12.1), is approximately proportional to the water transport through the Straits of Florida. According to Stommel

(1954), the water transport can double within 10 days. This, however, is exceptional, as shown by more recent investigations (Wunsch et al., 1969). The water transport varies by about 25%, a quarter of which falls to the annual variation. The causes of such pulsations have not yet been clarified, but it seems possible that, after advancing with the Gulf Stream to the northeast of Cape Hatteras, the pulsations may play a role in the formation of the Gulf Stream meanders (cf. under 4). The importance of the Gulf Stream for climatic conditions in the North Atlantic regions and western Europe has often been the subject of qualitative studies. The variations and pulsations, which today can be measured mechanically as well as electrically in the Florida Current, enable us to investigate any disturbance in the heat budget of ocean and atmosphere in the area of the North Atlantic Ocean at the valve, so to speak.

3. *Horizontal oscillations* of the Gulf Stream axis occur east of Cape Hatteras, where the narrow current band, after leaving the continental slope, is maintained in the Gulf Stream. The total deflection of the Gulf Stream on the cross section from the Chesapeake Bay to the Bermuda Islands (cf. Fig. 10.42) was determined, after repeated measurements, to amount to 150 km with a current width of 50 km (Dietrich, 1937b). Consequently, data of ships' set and drift, supplying mean current values over greater distances, can only give rather a blurred picture of the current. Modern methods of positioning, such as the Loran method (cf. Section 3.1.1.3), in use since 1945, continuous current records from ships underway by means of the GEK (cf. Section 3.2.12.1) since 1949, and quasisynoptic observations of the hydrographic structure of water masses carried out with several vessels during the investigations "Cabot 1950" (Fuglister and Worthington, 1951) and "Gulf Stream '60" (Fuglister, 1963) have provided further insight into the oscillations of the Gulf Stream. Since, especially in winter and spring, the current is characterized by a great horizontal sudden change in surface temperature ("cold wall"), measurements, made from aircraft, of the long-wave temperature radiation of the sea surface during these seasons have yielded information on the position of the meanders in the entire Gulf Stream within 6 hr (Pickett, 1967; Robinson, 1971). Ever since 1965, the radiation temperature of the sea surface measured by weather satellites has been employed with good success as a further important aid for the study of processes in the Gulf Stream that depend on space and time, as well as of current meanders and current eddies (Allison et al., 1967).

4. *Current meanders* develop out of horizontal oscillations of the Gulf Stream, mostly at about 38°N and 67°W, shortly before the current passes over the New England Seamount Chain. The meanders can be recognized in Fig. 10.44 from the temperature distribution at 200 m depth; the distinct thermal front reveals the track of the Gulf Stream. The meandering course was confirmed in 1960 by the trajectory of a drifting buoy, which was traced for 4 months. The current meanders were followed several times, with temperature sensors towed by a vessel at 200 m depth, while the position of the 15°C isotherm was recorded (Fuglister and Voorhis, 1965). Such measuring cruises were repeated every month for more than 1 yr (Hansen, 1967). Thus, the existence of the meanders as well as their deflection of about 400 km were confirmed, as shown in Fig. 10.44. Furthermore, observations have proved that the meanders advance eastward with 7 cm sec^{-1} on the average. The mean wavelength of the meanders approximately amounts to 300 km (Robinson, 1971).

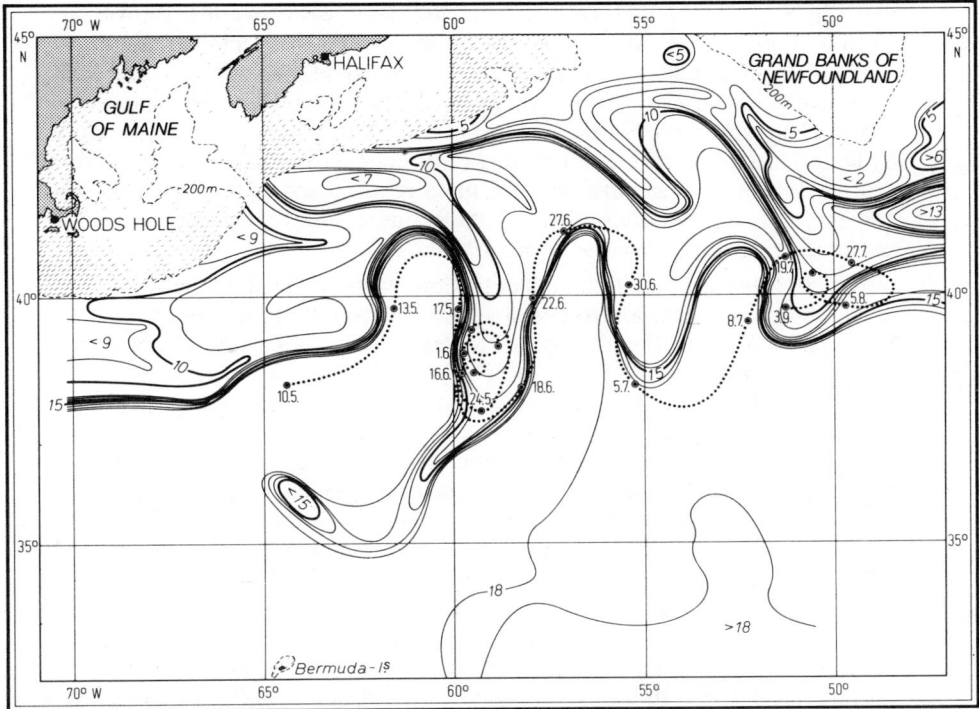

Fig. 10.44. Distribution of water temperature at 200 m depth, based on measurements during the "Gulf Stream '60" investigation (April 2 to June 15, 1960). (After Fuglister, 1963.) In addition, trajectory and dates of positions of a drifting buoy from May 10 to September 3, 1960. (After Woods Hole Oceanogr. Inst., Annual Report for 1960.)

5. *Current eddies* separate from the meanders if the deflection of the latter is particularly large. The eddies are maintained in the subtropical water as independent accumulations of cold water. Such an eddy at 36°N, north of the Bermuda Islands, is shown in Fig. 10.44. Cyclonic eddies of similar sizes could be traced for several months in 1965–1966 (Fuglister, 1967, 1971). They were located between 35 and 39°N and had diameters of about 100 km. They were traveling irregularly in various directions at velocities of approximately 12 cm sec^{-1}. At the sea surface, they had a current speed of around 150 cm sec^{-1}, which remained the same for 6 months. Their lifetime was about 1 yr, which can be concluded by extrapolating the volume decrease observed for 6 months. From five to eight such gigantic eddies are expected to develop every year.

6. *Countercurrents* are associated with the Gulf Stream on both flanks as well as beneath the main current. The former are shown in the example of the geostrophic current in Fig. 10.42. But direct measurements with the GEK, first carried out by von Arx (1950), have also provided relevant evidence. Richardson and Knauss (1971) investigated the countercurrent at Cape Hatteras and found a water transport of 1×10^6 m^3 sec^{-1} on the left flank of the Gulf Stream, while the transport value on the right-hand flank was 10×10^6 m^3 sec^{-1}. The transport in the main current, however, amounted to 63×10^6 m^3 sec^{-1}. The countercurrents in the deep layers can partly be recog-

nized in Fig. 10.09, showing drift measurements with Swallow Buoys. According to measurements by Barret (1965), this holds especially true for the area south of Cape Hatteras, even at measuring depths of only 800 and 1000 m. Wüst (1924) presumed that there must be a countercurrent beneath the Florida Current in the Straits of Florida. Evidence of this has been provided by Düing and Johnson (1971) by way of direct measurements of the current profile. In the further path of the Gulf Stream, the directions of the deep currents vary a great deal (cf. Fig. 10.09), indicating that they must be understood as countercurrents to the meanders.

7. *Cross circulation.* The high concentration of energy in the narrow current band, maintained over a distance of several thousands of kilometers from the Straits of Florida to the Mid-Atlantic Ridge, coincides with the position of the strong horizontal density gradient. With the co-action of friction and mixing, the dynamically enforced inclination of surfaces of equal density (cf. Fig. 10.42) can be maintained only by a cross circulation. On the right-hand flank of the Gulf Stream, this cross circulation causes lateral advection of light (in this case warm and saline) water in the near-surface layer, whereas on the left-hand flank heavy (in this case cold, low salinity) water is brought up from the depth to the main current. Such cross circulation, suggested by Dietrich (1937a) because of the hydrographic structure and represented in Fig. 10.42, has been confirmed by Hansen (1952) on the basis of hydrodynamic considerations. He could prove that such cross circulations are common phenomena in stationary, jetlike currents in a stratified ocean. The cross circulation on the left side of the Gulf Stream brings nutrient enriched water from the depth to subsurface layers. This is associated with an abundance of plankton and reflected in the greenish-blue color of this marginal water, called "slope water" by Americans. Such color contrasts with the deep blue on the right-hand flank of the Gulf Stream.

Another result of the cross circulation is that water masses, moving along with the main current, are continuously discharged laterally and replaced by new water. Hence, the Gulf Stream at Cape Hatteras no longer carries tropical water from the region of the trade wind currents, from which it received its energy, but rather subtropical water from the Sargasso Sea. Besides, the Gulf Stream does not only carry warm water but also cold water, as illustrated by the distribution of temperature and current in Fig. 10.42. Nevertheless, a considerable heat transport is accomplished: subtropical, warm water is continuously absorbed, transported for a while, until it reaches higher latitudes, and then released to the surrounding water. In this way, a large positive temperature difference between water and air develops at the surface of the Gulf Stream area, strongly inducing evaporation. Latent heat of evaporation is one of the main energy sources of the atmosphere over the North Atlantic Ocean. This energy supply is concentrated in the Gulf Stream area between 30 and 40°N, as shown in Fig. 4.11. The Kuroshio exerts a similar influence on the atmosphere over the North Pacific Ocean.

Knowledge about the Gulf Stream has grown considerably in the recent decades. Surprising observational results have not failed to turn up. Some of them have been mentioned in the above items 1 to 7. To date, a general theory of the Gulf Stream, which could give a complete explanation of those items, is not yet available. Parts of certain phenomena are well understood, either through experiments in rotating tanks or with the aid of theoretical models. What is best understood is the Gulf Stream as part of the large subtropical anticyclonic gyre. Stommel (1948) explained the concentration of the Gulf Stream as the result of the increase of the Coriolis acceleration with latitude. By taking the observed wind field into consideration, Munk (1950) computed the main features of surface circulation and mass transport. They are in remarkable agreement with observations. Since then, theoretical models have been studied, with stratification, friction, bottom topography, and time dependence taken into account. The Gulf Stream is a touchstone for studies on the dynamics of ocean currents (Fofonoff, 1962).

Current meanders are particularly difficult to understand. Discrepancies in observations have given rise to different interpretations. While during the "Cabot 1950" investigation meanders were observed to move rapidly eastward, they seemed to be largely stationary when studied during the "Gulf Stream '60" cruises. As to the former case, theoretical studies by Rossby (1951) and others, stated that the meanders must be considered as unstable Rossby waves (cf. Section 7.4.3); whereas, in the latter case, it has been shown that bottom topography is a decisive factor controlling the meanders (Warren, 1963). An extensive interpretation of the observed meanders has been given by Robinson (1971). He explains them by the hydrodynamics of the Gulf Stream, under the assumption of baroclinic instability and some topographic effects. The strange fact that, near Cape Hatteras, the Gulf Stream leaves the continental slope and enters the ocean as a jet stream has also been studied repeatedly (cf. e.g., Veronis, 1966). To some extent, this peculiarity can be understood as topographic influence (Hansen, 1970), since evidence has been found that the Gulf Stream is more than 1000 m thick and extends to near the sea floor at 3000 to 4000 m depth. At Cape Hatteras, these isobaths do not show the same change of direction as the continental slope.

The Gulf Stream equivalent in the North Pacific Ocean is the Kuroshio. The conformity includes many details, as shown in the summarizing studies by Shoji (1963), Uda (1964), and Yoshida (1972). The volume transport of the Kuroshio amounts to 40 to 50 $\times 10^6$ m^3 sec^{-1}, and the meanders have wavelengths of 500 to 800 km with amplitudes of 150 to 400 km. Figure 10.45 shows the position of the current axis in 13 cases for the period from April to November 1964 as derived from GEK measurements and hydrographic cross sections. The 15°C isotherm at 200 m depth approximately coincides with the current axis (Kawai, 1969). Studies of the fluctuations of the Kuroshio are facilitated by monitoring this isotherm. Characteristic features include the narrow current band with current velocities of 2 to 2.5 m sec^{-1}, the meandering of the current axis, and the very stable position of the meanders (Masuzawa, 1969). The latter finding indicates the influence topography has on the position of planetary waves, as confirmed also by a numerical experiment made by Robinson and Taft (1972). The name of Kuroshio is the Japanese word for black current; black because the water looks dark blue, a color indicating nutrient depletion. The Kuroshio, in fact, is poor in fish, like the Gulf Stream. Its left flank is influenced by cold, upwelling water rich in nutrients and fish, partly originating from the Oyashio. Both currents, the Kuroshio and the Oyashio, are subject to long-term fluctuations; the former is under the influence of the North Pacific trade winds, the latter, above all, under the influence of the east Asiatic winter monsoon.

The variability in time and space as well as the fine structure of the jet streams still hold many questions, upon whose solution the understanding of large ocean currents depends. The measuring methods have been developed, in principle (1970): synoptic survey of the processes by means of moored instruments is one way; another one is the repetition of large synoptic surveys like "Gulf Stream '60" together with the monitoring

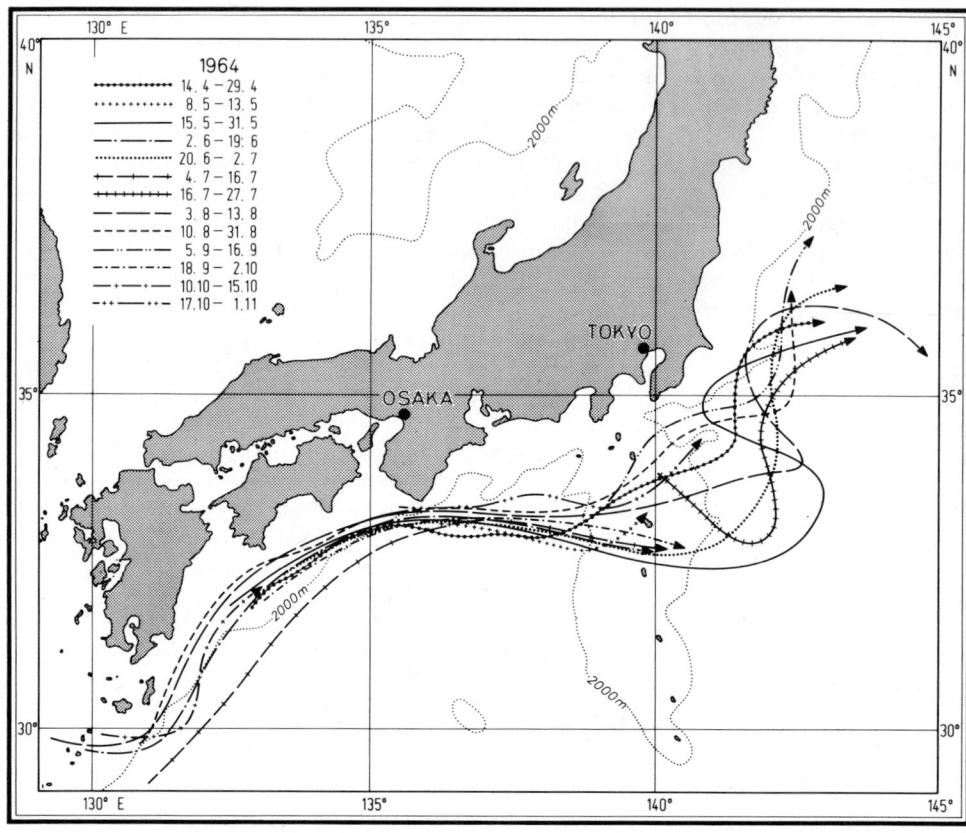

Fig. 10.45. Current axes of the Kuroshio, based on 13 oceanographic surveys from April to November 1964 (GEK measurements and hydrographic cross sections). (After Stommel et al., 1971.)

of sea surface radiation temperature from satellites. Since 1965, the former way has been followed by the Woods Hole Oceanographic Institution in the western North Atlantic Ocean at 70°W, most intensively at "Site D" (39°20′N, 70°00′W) on the left-hand flank of the Gulf Stream (Fofonoff, 1969; Webster, 1969b; Rhines, 1971; Thompson, 1971; Fofonoff and Webster, 1971). Figures 10.14 and 10.15 are based on those measurements.

10.3.7. The regions of west wind drift W_E and W_P

These regions are characterized by currents varying during the entire year and moving predominantly toward the east. Owing to influences of the continents as well as of the systems of the Gulf Stream and the Kuroshio, the conditions in the North Atlantic and North Pacific Oceans are more complicated than in the Southern Hemisphere, where we find only one circumpolar W region. The influence of the jet stream regions J is restricted here to the southwestern Indian Ocean, that is, to the area of the extensions of the Agulhas Current. Only in southern South American waters do continental obstacles exert any appreciable influence on the course of the currents. The Cape Horn and the

Falkland Currents result from these obstacles. As polar boundary of the W region, the ice limit has been chosen, that is, the boundary zone which, in winter, is covered permanently or very frequently with ice of the polar seas (cf. Fig. 5.28). It may be added that the currents of the outer polar region adjoining the W region poleward also move predominantly toward the east. In the Southern Hemisphere, the equatorial boundary of the W region coincides approximately with the subtropical boundary, along which warm subtropical water masses join the cool circumpolar water masses. This manifests itself in the distribution of the surface temperature, showing a narrow zone with strong meridional temperature changes. In the Northern Hemisphere, the equatorial boundary of the W regions cannot be distinguished on the basis of thermal properties.

The atmospheric circulation over the W regions shows a variability similar to that of the surface currents. On the whole, marine meteorological conditions, characterized by great variability, play an important role in these W regions. Direction and velocity of the winds vary in accordance with the cyclones traveling eastward. In winter, when the cyclones are most strongly developed, they increase the frequency of storms within their areas of influence in both hemispheres. These include the "roaring forties" between 40 and 50°S in the Southern Hemisphere, with an average wind force of 6 Beaufort, as well as the storm zones in the North Pacific and North Atlantic Oceans between 40 and 50°N (cf. Fig. 4.16). The latter, where more than 25% of all wind observations show 8 or more Beaufort in winter, covers an area crossed by the main shipping lanes from the western exit of the English Channel to New York, at 40 to 50°N and 35 to 45°W. The swell traveling out of these storm zones is mainly responsible for generating the large surf waves at the west coasts of the continents.

In the W regions, precipitation falls during all seasons; in fall and winter higher and more frequently than in spring and summer. The amounts exceed those of evaporation; thus, surface salinity is decreased, as clearly indicated by the zonal averages for the world ocean in both hemispheres (cf. Fig. 5.14). Further insight into natural conditions is provided by the cartographic representations of oceanographic and climatic data, as compiled in the monthly charts of the North and South Atlantic Oceans, issued by the Deutsche Seewarte and in the *Handbuch des Atlantischen Ozeans* (*Manual of the Atlantic Ocean*), published by the German Hydrographic Institute (1952). In general, the more recent navigational publications for the world ocean, such as the sailing directions or pilots, issued by the hydrographic offices of the seafaring nations, as well as the routing charts in the form of pilot charts, are good sources of information for anyone who intends to acquaint himself in detail with the natural conditions in an oceanic area. For the sub-Arctic Pacific Ocean, there exists a monograph by Uda (1963).

The oceanic polar front lies within the W regions. In the Southern Hemisphere, it coincides with the zone of strongest westerly winds and clearly manifests itself in the hydrographic structure of the near-surface layers as a convergence of wind-generated surface currents. At the sea surface, it separates the cold-water sphere from the warm-water sphere (cf. Section 10.1.1). Furthermore, low-salinity water is accumulated along this front, which, owing to the general descending motion, feeds the subpolar intermediate water of the deep circulation. Along this front, the spreading of the ice of the polar seas also generally finds its equatorial limit (cf. Fig. 5.28). These different observational facts support our notion that it is practical to subdivide the W region into a W_E region (equatorward of the polar front) and a W_P region (poleward of the oceanic polar front).

Figure 10.46 conveys an impression of the structure and circulation of the near-surface layers in these W_E and W_P regions in the Southern Hemisphere, including the polar region P, yet to be mentioned. In a section, the geographic location of which is found in Fig.

538 Regional Oceanography

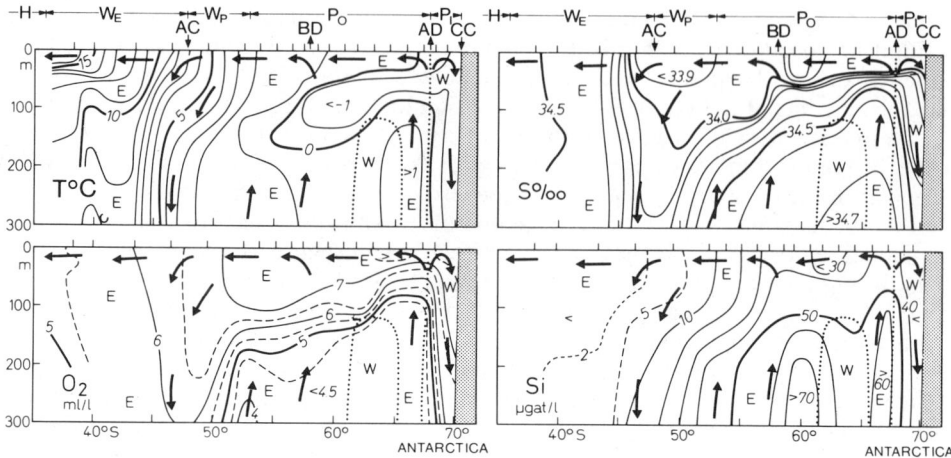

Fig. 10.46. Example of stratification and motion in the regions of the west wind drift W_E and W_P and in the polar regions P_O and P_I. Vertical distribution of temperature, salinity, oxygen, and silicate along a meridional section in the South Atlantic Ocean at 1°E from 35 to 70°S, based on observations of the research vessel *Discovery II* in February to March 1939 (following Koopmann, 1954). E, W: main current towards the east or west, respectively, perpendicular to the section. Dotted lines: current boundaries. Arrows: transverse circulation of the main currents. AC, CC: Antarctic convergence, continental convergence. BD, AD: Bouvet divergence, Antarctic divergence. Marks at the upper edge: oceanographic stations. Vertical exaggeration of the section: 5500 times. Location of the section: cf. Fig. 10.31 (line VII).

10.31, typical summer conditions in the uppermost 300 m are represented, as encountered in a similar form in the Indian and Pacific Oceans, except for the Bouvet divergence and the current setting westward into the Weddell Sea between 62 and 65°S. The areas north of the Bouvet divergence belong to the W region, in particular, the area between the Bouvet divergence (BD) and the Antarctic polar front [or Antarctic convergence (AC)], to the W_P region, the area north of AC to the W_E region. The area south of the Bouvet divergence lies within the polar region P. As shown in Fig. 10.46, the descending motion along the Antarctic polar front carries cold water with low salinity but rich in oxygen into deeper layers. This polar front is the northern boundary of silicate enriched water and also the northern boundary of the diatom distribution, since diatoms require silicate for the formation of their siliceous shells. Therefore, the Antarctic polar front at the sea surface is projected on the ocean bottom as the northern boundary of diatom ooze. In higher latitudes of the W_P region and also in the outer polar region, adjoining the W_P region poleward, divergence phenomena at the sea surface contribute to ascending motion of water from deeper layers, which is reflected in the relatively warm and saline water, rich in silicate but poor in oxygen, as shown in Fig. 10.46. Only close to the Antarctic continent are these processes interrupted by descending water masses. The studies on the relationships of circulation to water masses and biogeographic zones have been extended to the South Pacific Ocean by Knox (1970).

In the Northern Hemisphere, the extensions of the Gulf Stream (cf. Figs. 10.30 and 10.48) and of the Kuroshio [cf. Fig. 10.33(a)] are included in the W_E region. In both cases, however, the North Atlantic and the North Pacific currents (Bulgakov, 1967) have lost their jet stream character, by which these water motions were distinguished in the J region. The ramification of the Gulf Stream was investigated by the two German research vessels *Gauss* and *Anton Dohrn* during the "Polar Front Program" of the International

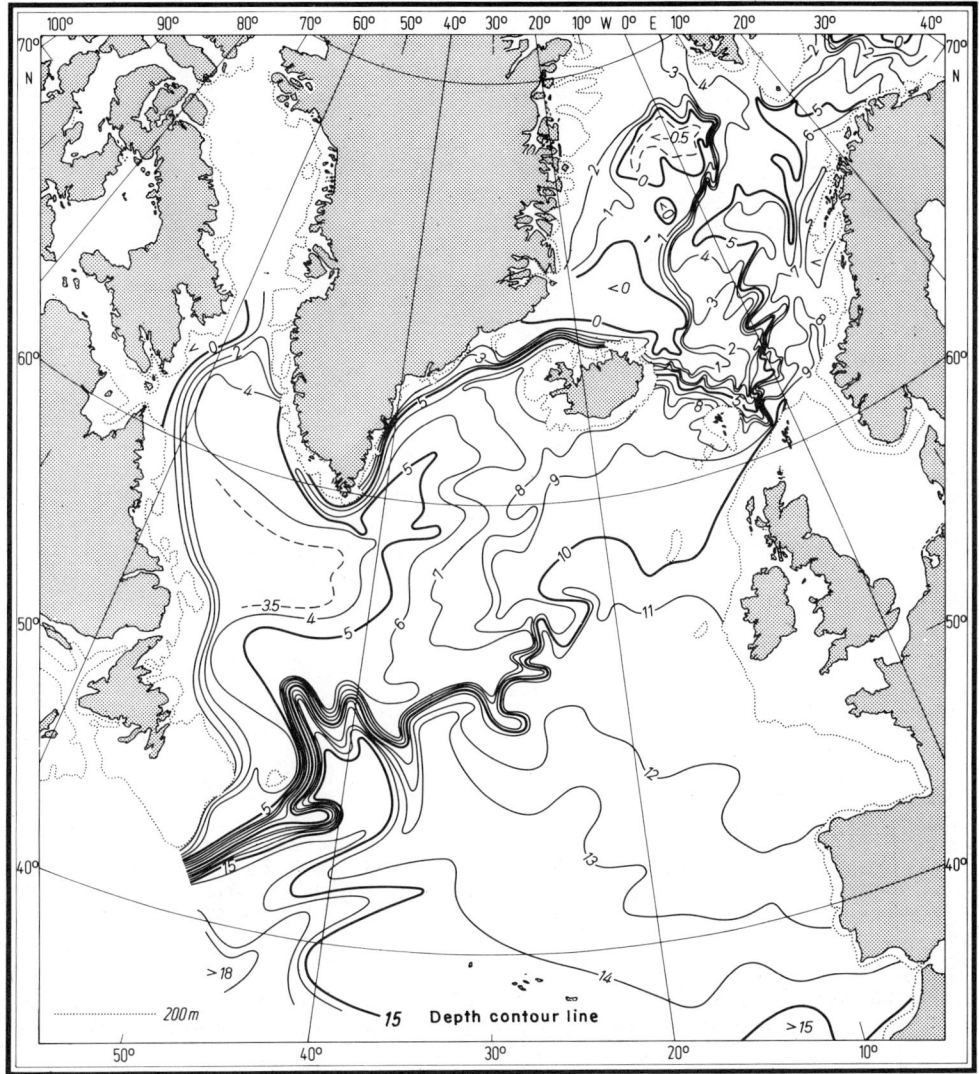

Fig. 10.47. Temperature at 200 m depth in the northern North Atlantic Ocean in the summer of 1958. (After Dietrich, 1969a.) The representation is based on observations of the "Oceanic Polar Front Survey" during the International Geophysical Year.

Geophysical Year (Dietrich, 1957; and Bückmann et al., 1959). A review of the particular conditions in the North Atlantic W_E and W_P regions is given in Figs. 10.47 to 10.50, as well as in Figs. 10.30 and 10.52, all of them based on observations during the Polar Front Program of 1958.

The polar front manifests itself in the temperature distribution at 200 m depth by the strong horizontal temperature gradients in the western Atlantic Ocean (cf. Fig. 10.47). These temperature gradients disappear at approximately 50°N 30°W and appear again

540 Regional Oceanography

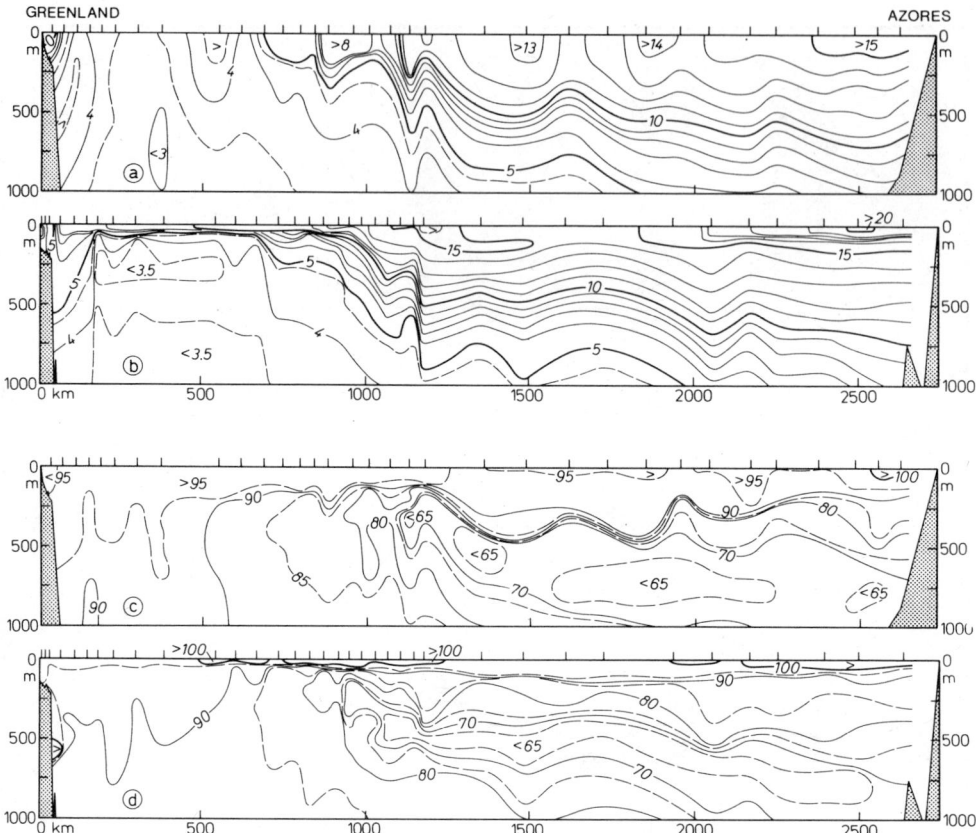

Fig. 10.48. Meridional section for the upper 1000 m across the Labrador Sea between South Greenland and the Azores in the winter and summer of 1958, based on the observations of the research vessels *Gauss, Anton Dohrn,* and *Michael Lomonossov.* (After Dietrich, 1969a.) (a) Temperature in °C in winter 1958, (b) temperature in °C in summer 1958, (c) oxygen (in % saturation) in winter 1958, (d) oxygen (in % saturation) in summer 1958.

in the central Norwegian Sea. The high temperature gradients near Greenland and east of Labrador indicate cold currents of polar origin, the flanks of which form the boundaries against the polar region (cf. Section 10.3.8).

A cross section (Fig. 10.48) between southern Greenland and the Azores demonstrates the change of stratification in the W regions from summer to winter. In summer, the polar front cannot be recognized at the sea surface because of a thin, warm top layer; but at 200 m depth, it is clearly marked, as shown in Fig. 10.47. In winter, however, homothermal conditions are found from the surface down to a depth of 2000 m. Thus, deep-reaching vertical convection can take place, and the Atlantic deep water is renewed, as can be seen in Plate 4. The spreading of the cold, oxygen-enriched water is demonstrated in Fig. 10.08.

The W_E and W_P regions and their boundaries are also to be seen on a cross section between eastern Greenland and middle Norway (cf. Fig. 10.30). Here, too, the warm top layer disappears in winter, and vertical convection can extend down to the bottom of the Greenland Sea. The coldest and heaviest bottom water of the oceans is formed north

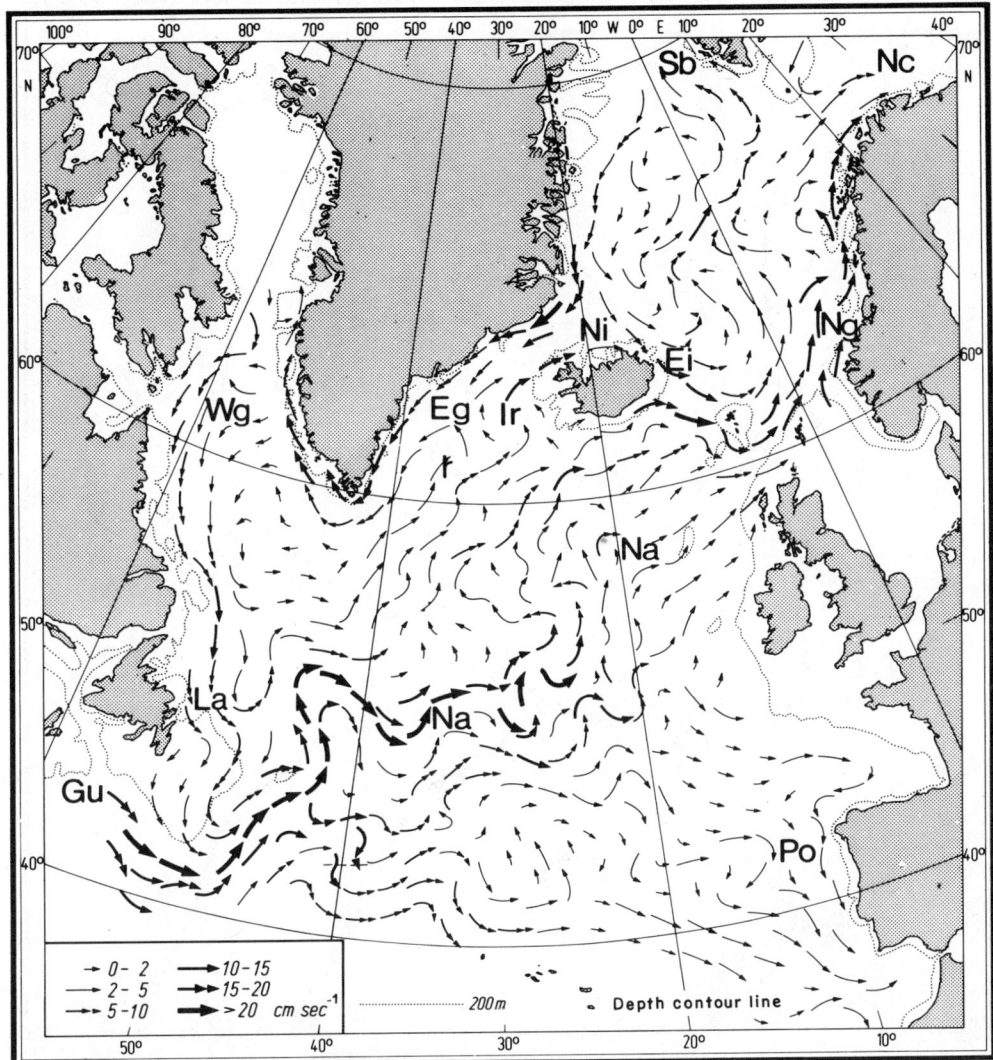

Fig. 10.49. Surface currents in the northern North Atlantic Ocean in the summer of 1958. (After Wegner, 1973.) Derived from the dynamic topography of the sea surface, referred to the 1000 dbar surface. For abbreviations cf. Fig. 10.50.

of Jan Mayen. It flows into the Arctic Ocean (cf. Fig. 10.29) and across the Greenland–Scotland Ridge into the eastern North Atlantic Ocean (cf. Figs. 10.52 and 10.03). The very sharply pronounced boundary between the W region and the polar region is shown in Fig. 10.52, representing conditions in the Denmark Strait between Greenland and Iceland in the summer of 1958. At the sea surface, the contrasts are concealed by a very thin, cold layer of low salinity. At 50 m depth, however, they can be distinguished quite clearly.

542 Regional Oceanography

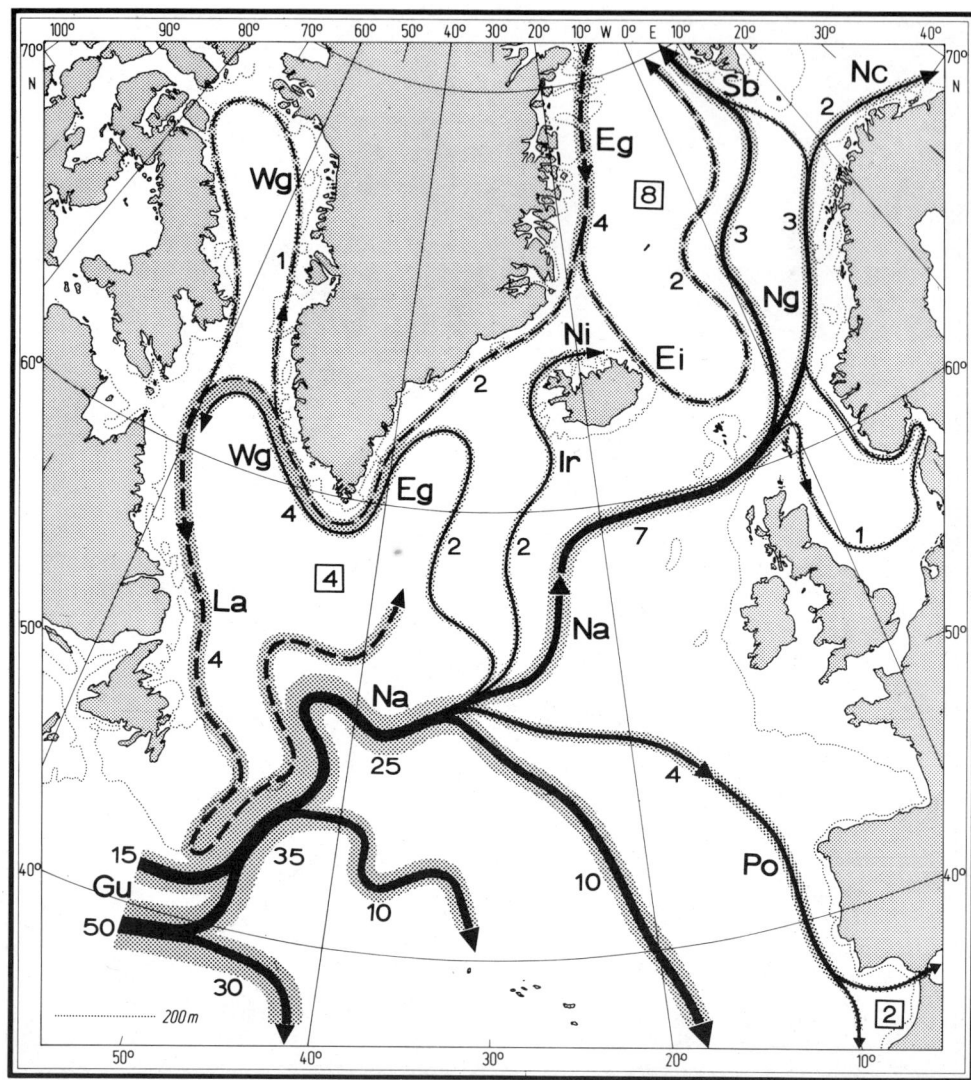

Fig. 10.50. Scheme of water transport (in 10^6 m^3 sec^{-1}) in the layer 0–1000 m in the northern North Atlantic Ocean. (Derived from Wegner, 1973.) Eg: East Greenland Current, Ei: East Iceland Current, Gu: Gulf Stream, Ir: Irminger Current, La: Labrador Current, Na: North Atlantic Current, Nc: North Cape Current, Ng: Norwegian Current, Ni: North Iceland Current, Po: Portugal Current, Sb: Spitsbergen Current, Wg: West Greenland Current. Squares: descending water in 10^6 m^3 sec^{-1}. Solid current axes: relatively warm currents. Broken current axes: relatively cold currents.

The calculations of the geostrophic current in the atlas by Dietrich (1969a) are based on the mass distribution as observed in the summer of 1958. They have been continued by Wegner (1973), who converted them into a chart of surface currents (cf. Fig. 10.49). As a further step, the water transport for the layer of 0 to 100 m has been calculated from the geostrophic current, as schematically summarized in Fig. 10.50. In Figs. 10.49 and

10.50 it is shown that parts of the current branches approach the European continental shelf and consequently lose their prevailing eastward direction because of this obstacle. One branch is sent off northward as the Norwegian Current, another southward as the Portugal Current, and some parts even enter the area of the continental shelf, as, for instance, in the Irish Sea, the English Channel or the North Sea.

Although extended zones of divergence, as found in the Antarctic regions of the oceans, where nutrient enriched deep water is brought to the surface, are missing in the North Atlantic and North Pacific Oceans, deep-reaching mixing processes are present, nevertheless. This is true especially in winter, when, in accordance with the strong annual march, the surface temperature is lowered and therefore deep-reaching thermal convection is made possible, providing the surface water with nutrients. Vertical mixing is particularly effective in the area of the oceanic polar front in the Northern Hemisphere. Cyclonic eddies, which preferentially occur along the polar front, are associated with dynamically enforced upwelling in the center of the eddy, as schematically illustrated in Fig. 7.12. Similar conditions exist near the Grand Banks in the boundary zone between the Labrador Current and the Gulf Stream, or in the Pacific Ocean along the polar front between the Oyashio and the Kuroshio. The high nutrient content provides an opportunity for plankton to develop abundantly, as otherwise found in the open ocean only in higher southern latitudes of the W_P region and the outer polar region P_O.

Plankton provides the nutritional basis for the life of commercial fish. The peculiar mass development of such fish, especially on the shelves, is a characteristic of the W regions. On the shelves, the ocean floor is a source of food or serves as spawning ground. For these reasons, the main catching grounds of deep-sea fishery are found on the shelf areas of North America, north of Cape Hatteras, around Greenland, Iceland, the Faeroe Islands, as well as on the European shelf from the Bay of Biscay to the Barents Sea. On the Patagonian and South African shelves, natural conditions are also favorable for high-sea fishing, which is beginning to develop there more and more. The German landings of saltwater fish originate from the aforesaid areas. In 1968, they amounted to 560,000 tonnes, 20.2% of which came from the North Sea and the waters west of Great Britain, 25% from the waters around Greenland, 21% from Icelandic waters, 20.2% from North American waters, and 1.8% from Norwegian waters, in addition to 5.3% from the Baltic Sea and 0.5% from South African waters. Studies of the oceanographic conditions in those areas also elucidate the problems regarding the environmental conditions of fish. Here lies the great importance of oceanography for fishing, especially for the exploration of new fishing grounds as well as for our understanding of the stock of fish and of its changes in the areas fished commercially. Therefore, the marginal regions of the North Atlantic Ocean have been among the working areas preferred by European and American marine scientists, since oceanography began to flourish toward the end of the last century. European oceanographers are organized in the International Council for the Exploration of the Sea (ICES), with headquarters in Copenhagen, which, since 1902, has fostered international cooperation in oceanography and fishery biology with great success.

But the effort put into research regarding the W regions must be called relatively modest, when one thinks of the extent of these oceanic areas and their extraordinarily complicated nature. Not only do current conditions prove to be intricate, the same is true for the distribution of hydrographic factors. Moreover, there are also violent *fluctuations in hydrographic structure and water motion,* which are, in part, weather-conditioned and, in part, controlled by the annual rhythm as well as by long-term changes, as is known from fairly long series of observations made in the North Atlantic Ocean (cf. Fig. 5.10). Besides, there are irregularities in the large-scale horizontal exchange, taking place in large vortices difficult to comprehend in detail. It is only the recent Gulf Stream research that has shed some light on the laws governing the formation of such vortices.

A further step towards clarifying our notions about the North Atlantic and North Pacific W regions, in particular, has been accomplished by the oceanographic and meteorological observations taken on the 14 *weather ships,* which have been stationed for several years now at various points within these areas to ensure the safety of transoceanic air traffic (cf. Fig. 5.10). The methods of future research in these W regions are thus indicated: the hydrographic structure and water movements must be ascertained synoptically by several research ships, working in conjunction with weather ships and weather satellites. However, this is a task that can apparently be accomplished only in close international cooperation, like it was initiated during the *International Geophysical Year 1957–1958,* when 25 research vessels from 12 nations worked together in the North Atlantic W region for the "Polar Front Survey." In the Pacific Ocean, it was the NORPAC enterprise. In the meantime, several major international programs were carried out in the W regions and adjacent areas: "Overflow 1960," on the Iceland–Faeroe Ridge (Tait, 1967) (cf. Fig. 10.03), "Norwestlant 1963," in the Northwest Atlantic Ocean (ICNAF, 1968), and the "Cooperative Study of the Kuroshio" (CSK 1965–1966) (Japanese Oceanographic Data Center 1967, 1968). In addition, numerous investigations on a national or a multilateral basis were performed, especially in areas of commercial fishery (Redfish Survey in the Irminger Sea, Herring Survey in the North Sea), but also others with a view to special problems of variability (Diffusion Experiment in the North Sea, "Norwegian Sea 1969" in German–Norwegian–Icelandic cooperation, etc.). A summary of the varied literature referring to part of the Northeast Atlantic W region has been given by Lee (1963).

10.3.8. The polar regions P_I and P_O

It seems advisable to depart from the principle of classification based on ocean currents, applied until now, in areas where ice becomes an oceanographic factor of first rank owing to its far-reaching consequences for the atmosphere and the living conditions of marine flora and fauna as well as due to its decisive effect upon navigation and utilization of the ocean by the fishing industry. The region covered with the ice of the polar seas, permanently or at times during the year, has been classified as polar region P. The numerous adjacent seas in higher latitudes of the Northern Hemisphere, which, in fact, are regularly covered with ice in winter (such as the Hudson Bay, the Gulf of St. Lawrence, the Gulf of Bothnia, the White Sea, the Sea of Azov, the northern Sea of Okhotsk, and the Bering Sea) can be excluded from this region for the very good reason that the character of their icing differs significantly from that of the polar seas (cf. Section 5.5.2).

Within the two polar regions we find many differences suggesting further subdivision. In this general survey, we will restrict ourselves to two subdivisions that are significant from the scientific point of view as well as from a practical one regarding navigation and fishery: (1) the inner polar region P_I as the area of permanent ice. This is the area covered with a dark screen in Fig. 5.28. (2) The outer polar region P_O where the ice of the polar seas is encountered with more than 50% probability in the polar winter. In Fig. 5.28, this area is marked by a densely dotted screen.

In the Arctic area, the *inner polar region* P_I includes the deep Arctic basin and the northern passages of the Canadian Archipelago, and in the Antarctic area a narrow belt around the Antarctic ice shelf. The term "permanent ice" refers to the state of the ice coverage only, not to its substance because, although the region is always covered with

Hydrographic Regions of the World Ocean   545

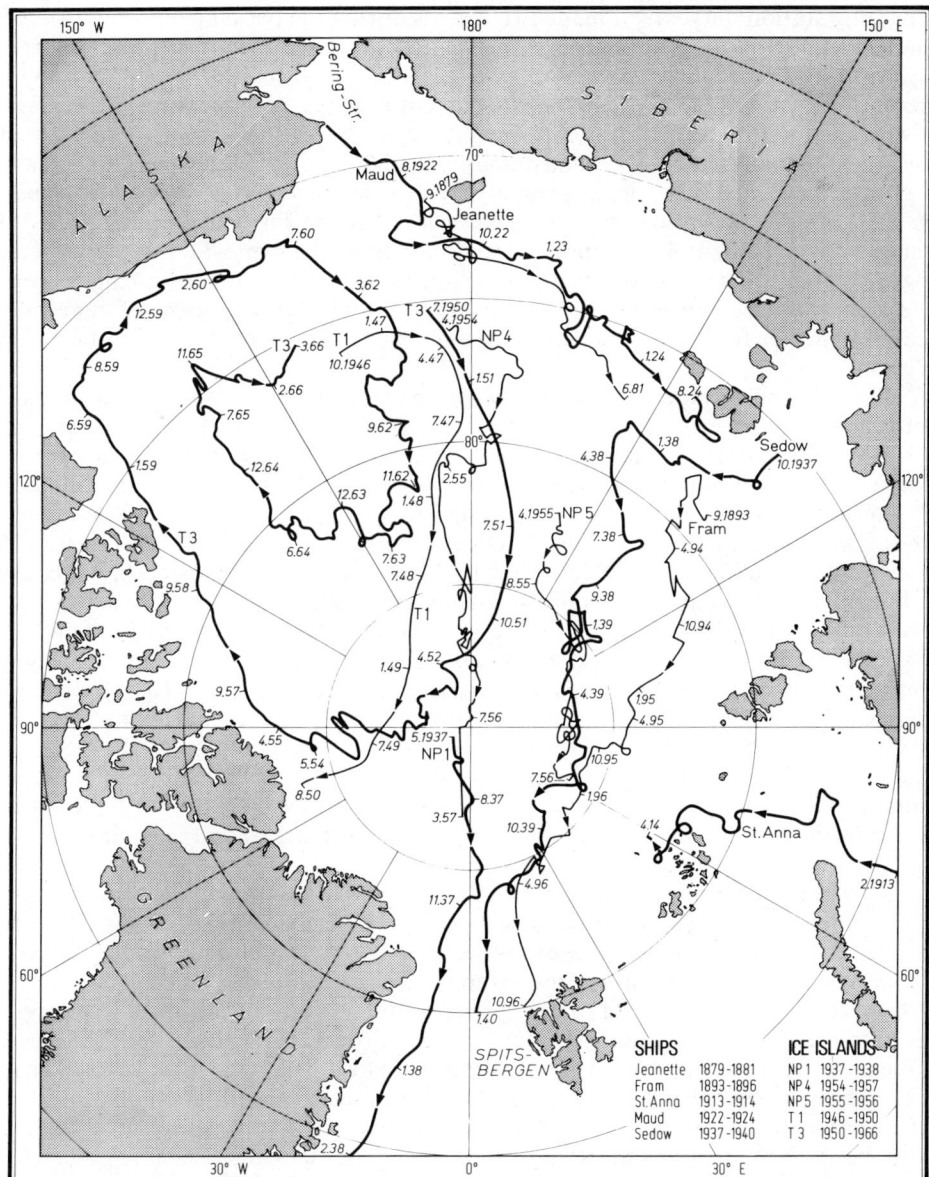

Fig. 10.51. Selected ice drifts in the Arctic Ocean. (After Nansen, 1897; Sverdrup, 1928; Subow, 1940; Papanin, 1947; Koenig, 1952; Schindler, 1968; and U.S. Oceanographic Office, 1968.) The figures give the time of position (month and year).

ice, the ice itself is in continuous motion and leaves the region with oceanic currents, as illustrated by the trajectories of drifting objects in Fig. 10.51, which suggest anticyclonic motion in the Arctic Ocean (cf. Section 7.2.6). In the Canadian part of the Arctic Ocean, the ice stays for a very long time. The ice island T3, also called Fletcher Island, has served as a drifting station for years; it made two rounds during the period from 1950 to 1966. In the Soviet part of the Arctic region, the ice drifts directly towards the exit between Spitsbergen and Greenland. As shown in Fig. 10.51, the 0 to 180° meridian roughly represents the border line between the two current regimes. Newly formed ice replaces that which is continually lost. It is only in winter that the ice sheet is completely sealed; its thickness in the Arctic region then amounts to about 3.5 m. In summer, it decreases to around 2.5 m; in addition, leads and polynyas (wide openings) of open water appear (Fig. 5.23). This ice sheet floats on a thin layer with relatively low salinity (cf. Fig. 10.29), extending down to about 50 m depth and maintained by the melt water of the summer. The great stability of density stratification in this top layer prevents deep-reaching vertical convection; thus, bottom water cannot be formed in the Arctic basin. Its renewal is achieved by water from the Greenland Sea flowing in across the Nansen Sill (cf. Fig. 10.29).

There is a considerable difference between the inner polar regions of the two hemispheres. In the Northern Hemisphere, this area is free of icebergs, because no glaciers touch this region, which is exclusively dominated by pack ice, subjecting the marginal zones to severe ice pressure. Pack ice floes are piled up and form impassable hummocks with intermittent leads and polynyas (cf. Figs. 5.23, 5.25). Any emergency landing of aircraft is impossible; ships are endangered by the pressure of the ice, and sledging is made very difficult, as so impressively described by Nansen (1897) in his book *In Night and Ice*. In contrast, the central parts of the inner polar region are characterized by vast, relatively level pack ice fields, only infrequently interrupted by hummocks. Thus, the heavy-laden planes of the Soviet and American polar drift stations, for example, beginning with the first operation, the Papanin Expedition of 1937 (NP1 in Fig. 10.51), up to the present day, have always been able to land without any special preparations.

In contrast to the Northern Hemisphere, the inner polar region of the Southern Hemisphere is interspersed with icebergs consisting, to a small part, of glacier ice but predominantly of shelf ice that occurs as gigantic tabular bergs (cf. Fig. 5.27). Pack ice and icebergs drift westward with the Polar Current (cf. Plate 6) and eastward with the Weddell Current and the west wind drift.

Only on the western side of Grahamland and in the Ross Sea do the coasts of Antarctica stretch beyond the boundaries of the inner polar region and become ice-free every year during the polar summer. The entrance to the Ross Sea, however, is blocked by a pack ice belt, which first must be penetrated before the open Ross Sea can be reached. Furthermore, it should be noted that only the mean boundary of the inner polar region is given in Fig. 5.28. The actual boundary can vary considerably from year to year. In the Weddell Sea, this boundary may be shifted by up to 300 nautical mi in years of heavy ice formation, as compared to its position in years of little ice formation.

As a result of the cooling of the water and of the salinity increase due to freezing, the density beneath the ice is augmented to such an extent that deep-reaching descending motion is initiated. Such motion is very clearly recognizable at the Antarctic continent in Fig. 10.46, showing a hydrographic cross section. In some areas, the sinking water reaches the ocean floor, where it can be followed as Antarctic bottom water (cf. Section 10.1.2).

The inner polar region is surrounded by the *outer polar region* P_O, which is regularly covered with a pack ice sheet in the winter of the respective hemisphere. Pack ice is encountered also in other seasons, but then, it does not originate within this region but represents alien ice carried there by oceanic currents from the inner polar region. It forms scattered fields of pack ice gradually subjected to melting processes. The considerable

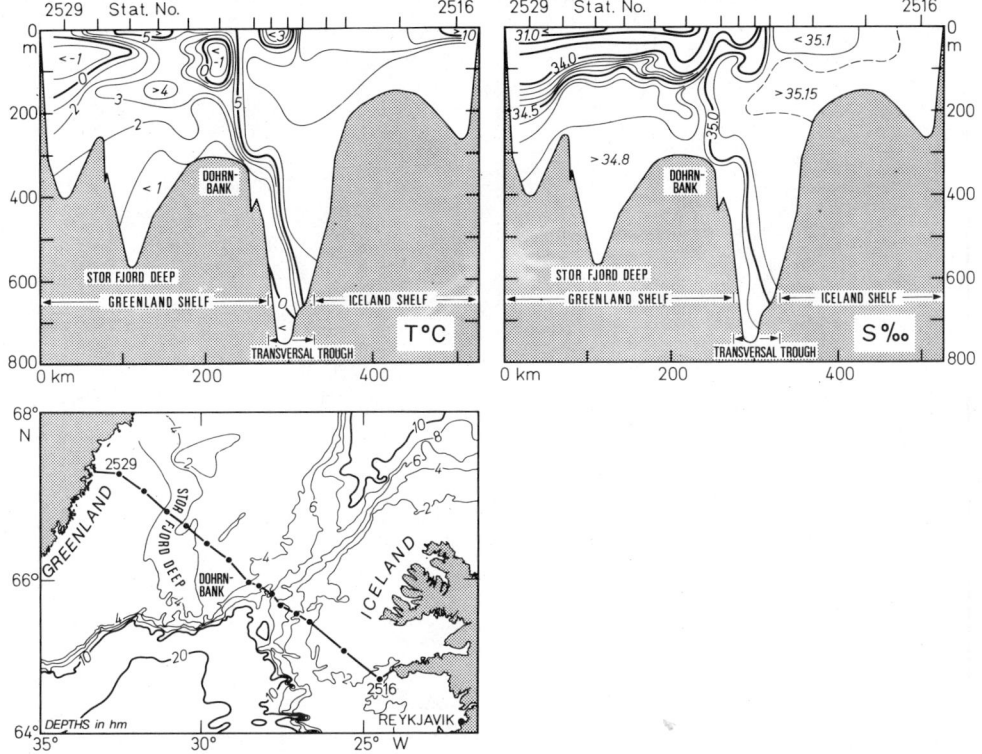

Fig. 10.52. Section across the Denmark Strait between Greenland and Iceland, based on measurements by the research vessel *Anton Dohrn* in the summer of 1958. (After Dietrich, 1969a.) (a) Temperature in °C; (b) salinity in ‰; (c) position of cross section, depth contour lines in hectometers.

influence of oceanic currents is reflected by the boundaries of this region (cf. Figs. 10.30, 10.52, 10.53). Polar pack ice is carried far southward with the cold East Greenland Current and the Labrador Current (cf. Fig. 5.30). At the Grand Banks (46°N), it even reaches the latitude of Venice. Conversely, the eastern part of the Norwegian Sea, up to West Spitsbergen at 79°N, and the southern Barents Sea are kept ice-free during the entire year by the extensions of the warm Gulf Stream system (Figs. 10.49 and 10.50). However, irregularities in the lateral turbulent exchange between cold and warm water masses do not only cause small-scale wriggling of the boundary of that region, which could not be reproduced in Fig. 5.28, but also variations of the boundary with time. This is particularly distinct north of Iceland, where it has been confirmed by observation, as shown in Fig. 10.54 by the ice distribution in March and April for the years from 1960 through 1970. Late in the winter of 1965, as well as from 1967 to 1970, the north coast of Iceland was ice-bound; shipping and fishing were prevented, and heavy damage was inflicted upon Icelandic economy.

Within the outer polar region differences occur that depend, first, on how many icebergs from the continental coasts join the drifting pack ice and, second, to what extent the influence of shallow water becomes effective. Outside the shelf areas, the type of strictly oceanic drifting pack ice is encountered in the Arctic Ocean and in the Norwegian Sea outside the shelf seas. The type of drifting pack ice, interspersed with icebergs

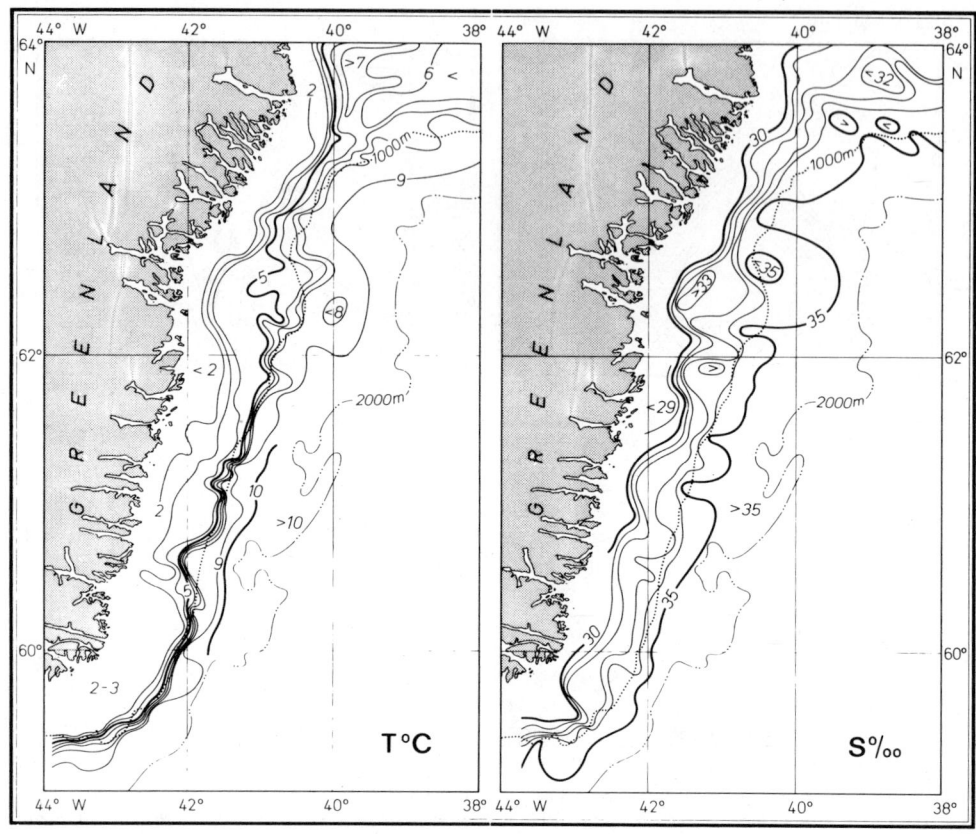

Fig. 10.53. Surface temperature ($T°C$) and surface salinity ($S‰$) off Southeast Greenland, based on the registrations of 90 cross sections taken by the research vessel *Meteor* in August 1966 (from August 12 to September 2). (After Dietrich and Gieskes, 1968.)

originating from valley glaciers is predominant, however, in the area west of Greenland and in the western Labrador Sea as well as in the outer polar region of the Southern Hemisphere. In the latter case, tabular icebergs that have broken off from the Antarctic ice shelf prevail. A third type, which is restricted to the Arctic Ocean, includes the ice in the North Siberian marginal seas and in the passages of the Canadian Archipelago. Here, permanent oceanic currents are masked by tidal currents differing locally. The drifting pack ice is moved back and forth with these tidal currents and is loosened or compressed in accordance with the changing wind conditions. The confining groups of islands produce violent pressure on the ice and, in winter, favor the formation of immovable fast ice. These natural conditions in the marginal seas make navigation extremely dangerous. Once a ship is beset by ice, its chances of getting free under its own power are remote because of the unpredictability of the current drift. On the contrary, the danger of the ship being crushed by the force of the ice pressure grows. This area has been the scene of the great tragedies of polar research, which must be considered lessons in navigation dearly paid for under the particularly hard conditions in the marginal seas of the outer polar region. Here, for example, the British John Franklin Expedition with the two ships *Erebus* and *Terror* perished in the passages of the Canadian Archipelago in 1848;

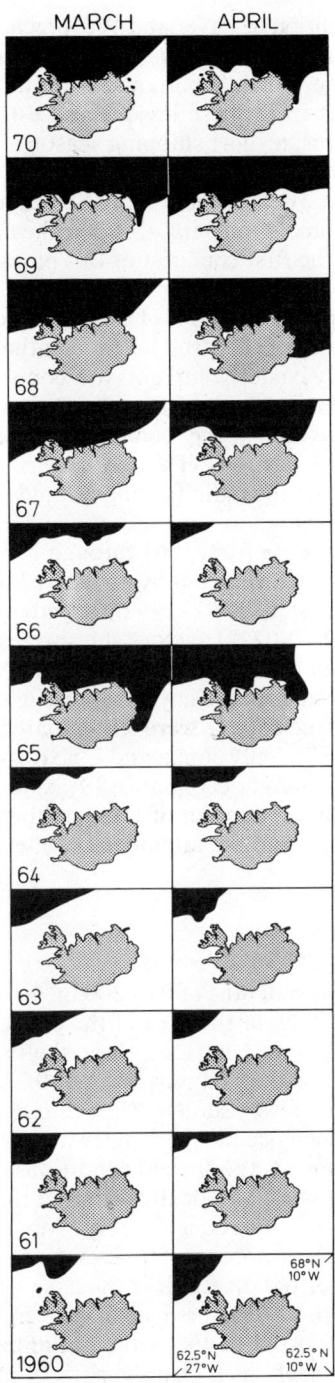

Fig. 10.54. Distribution of ice in Icelandic waters in March and April of each year during the period from 1960 to 1970. (After Sigtryggson, 1969; supplemented until 1970 by the same author.)

so did the Austrian polar expedition of Payer and Weyprecht on the *Tegethoff* near Franz Josef Land in 1874, and the American *Jeanette* Expedition near the New Siberian Islands in 1884. When opening up the North Siberian seaway in the past decades, the Soviet Union drew the consequences from these lessons and established a careful ice patrol service (cf. Section 5.5.3). Thus, a short shipping season in August and September has been made possible in an area stretching from the Barents Sea to the Bering Strait. In 1940, the German auxiliary cruiser *Komet,* using Soviet ice forecasts and the support of ice breakers, succeeded in accomplishing this Northeast Passage within the record time of 14 days; whereas, for the first conquest of this passage, A. E. Nordenskiöld with the *Vega* had needed almost 2 yr, from 1878 to 1880.

In summer, the hydrographic conditions of the P_O region do not differ greatly from those of the W_P region. This is especially true for the Southern Hemisphere (Kort, 1964), where variable, predominantly easterly currents are common in both regions. The two divergence zones, marked in Fig. 10.46 by BD and AD, provide favorable conditions for the development of plankton, which in the food chain forms the basis of krill, on which, in turn, blue whales and fin whales feed. These relationships explain why, in summer, the two divergence zones become the main feeding grounds of whales and, consequently, the main Antarctic whaling grounds.

The decline of Antarctic whaling from 1961 through 1968 was not caused by a change of environmental conditions in the ocean but by excessive killing of whales. While 37,000 whales were landed in the season of 1961–1962, the catch amounted to only 15,000 in 1967-68. In the season of 1961–1962, 21 factory ships were cooperating with 260 catcher boats from five nations. Norway had seven factory ships, Japan seven, USSR four, United Kingdom two, and the Netherlands one. Six years later, in 1967–1968, only eight factory ships with 97 catchers from three nations were left: Japan had four factory ships, USSR three, Norway one. In 1969–1970, only Japan and USSR took part in Antarctic whaling; for the other nations, it was no longer economical. It is hoped that the number of stock can be raised by a considerable reduction of the catching rate and a ban on shooting certain species as the blue whale and the humpback whale. But it will take decades until whaling will pay once more.

10.3.9. Shelf seas

All shallow seas on the shelf with depths of 0 to 200 m, among them also the North Sea and the Baltic Sea, represent 7.8% of the area of the entire world ocean, but only 0.2% of its volume. If, because of their small extent, these shelf seas have not been listed separately in the classification of oceanic regions in Fig. 10.31 but have been included into it, this was done only with some reservation. Although the waters of these marginal seas originate from the adjacent open ocean (their renewal depending, to a large degree, on the size and depth of the entrance), the special conditions of the heat and water budget in the sea near the continents, as well as locally different tidal currents in shallow water, influence the hydrographic conditions considerably. Sometimes, they favor an increase in the stability of density stratification; sometimes a decrease (cf. Fig. 10.55). Stability is increased by a positive water balance; this is the case when the sum of precipitation minus evaporation plus runoff from rivers causes a decrease in surface salinity. Furthermore, stability is also supported by the stronger annual march of sea surface temperature in the vicinity of continents, as is characteristic of moderate and higher latitudes. The consequence of such processes is that all shelf seas in moderate and higher latitudes tend to develop a layer of very high stability, a pycnocline, in spring and summer, more

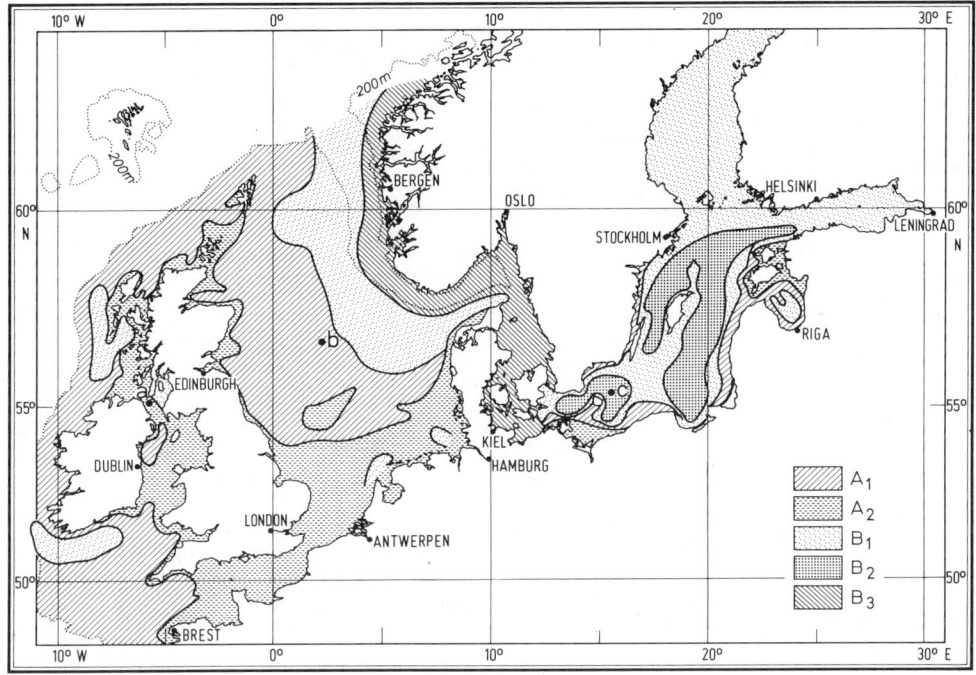

Fig. 10.55. Hydrographic regions of the North Sea and the Baltic Sea. (After Dietrich, 1950.) Points a, b, and c: positions for Fig. 5.08 (mean annual variation of stratification). A_1 and A_2: isohaline during entire year; A_1: thermally stratified at times or during entire year; A_2: isothermal during entire year. B_1, B_2, and B_3: salinity stratification at times or during entire year; B_1: top layer with little, bottom layer with regular annual variation of salinity; B_2: top layer with little, bottom layer without any regular annual variation of salinity; B_3: top layer with strong, bottom layer with regular annual variation of salinity.

so than the open ocean. This layer partly acts as a zone preventing vertical exchange processes. In such a case, the top layer is strongly warmed in spring and summer; it remains well ventilated but is depleted of nutrients. The lower layer, however, which is prevented from participating in the processes of the top layer, remains cool; in fact, it almost maintains winter temperatures. Its oxygen content decreases, and the nutrients are enriched by the oxidation of sinking detritus. Such stratification conditions are characteristics of the North Sea (cf. Fig. 10.56), except in its southern and western parts where high tidal currents occur. The turbulence of these currents, which starts from the sea floor, ceases only under certain conditions depending on the vertical density gradient, the velocity of the tidal current, and the distance from the bottom (Dietrich, 1954). Strong tidal currents of 1.5 m sec^{-1} are sufficient, for example, to prevent the formation of a summer density stratification, even in water more than 100 to 150 m deep. Such a case is given in Fig. 5.08(a) for the North Channel of the Irish Sea. Owing to turbulent mixing, weak tidal currents can still maintain a homogeneous sublayer in summer, as shown in Fig. 5.08(b) for the central North Sea. In areas without any tidal currents, stratification will not be disturbed and may be maintained throughout the year, as illustrated in Fig. 5.08(c) for the Baltic Sea. In particular cases, stratification is much more pronounced, as shown in Fig. 10.57, where the variations of temperature, phosphate, silicate, and ammonia during 24 hr are represented. Isothermal conditions in the top layer and the

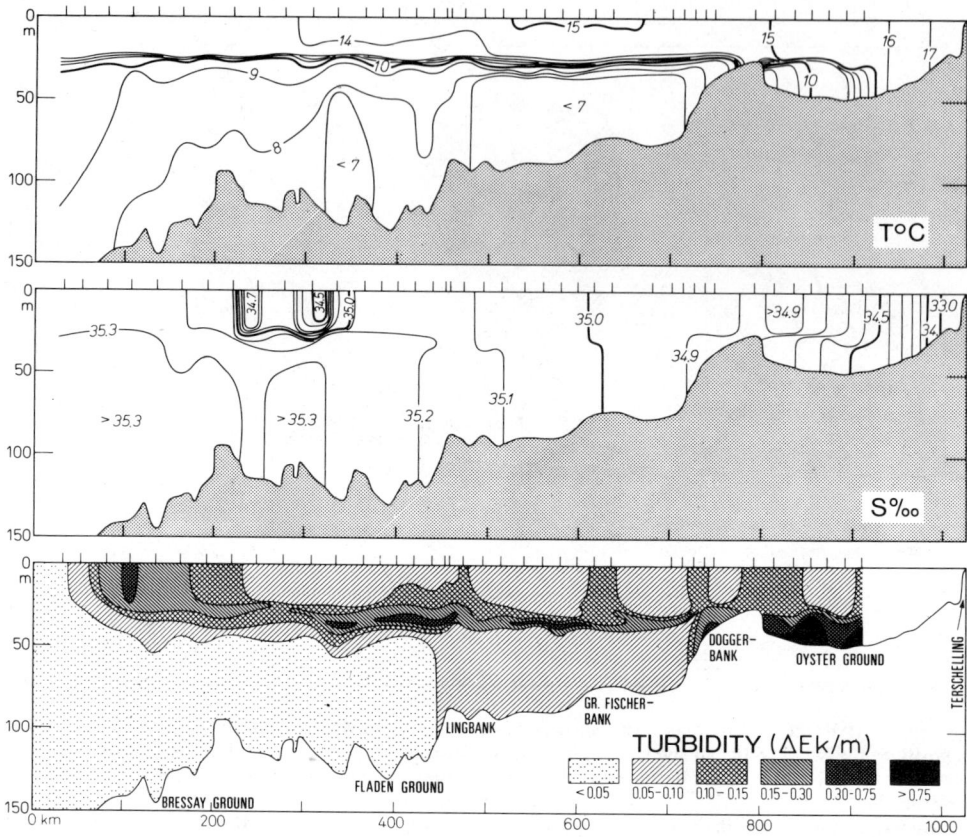

Fig. 10.56. Example of the summer stratification in the North Sea, according to a cross section by the research and survey vessel *Gauss* from August 5–12, 1953. (After Dietrich, 1954 and Joseph, 1957.) Position of cross section: central axis of the North Sea from Terschelling toward NNW up to the transition into the Norwegian Sea. Marks at the upper edge: oceanographic stations. Exaggeration of vertical scale: 2000 times. Turbidity: deviation of extinction m^{-1} from pure water.

sublayer contrast with a top layer practically free of phosphate and silicate and a sublayer very rich in phosphate and silicate. A remarkable fact is this: in contrast to the constant temperature, the chemical factors vary with time, whereby it is demonstrated that the release processes of those substances go so fast that even the turbulence of tidal currents cannot achieve complete mixing. A summarizing description of the hydrography of the North Sea on the basis of recent publications has been given by Lee (1970).

If the shelf seas are considered from the point of view of the stability of stratification and its annual march, one arrives at a classification of more than only formal significance. Natural regions of the shelf seas are obtained that are characterized by fairly uniform behavior with respect to their physical and chemical properties, which does not remain without any influence on the life in those seas. Such a regional classification is given in Fig. 10.55 for the North Sea and the Baltic Sea.

The Baltic Sea, which has the nature of a mediterranean sea (cf. Section 10.2.1), holds a special position among the shelf seas in humid climates. But because of the considerable proportion of fresh water causing strong salinity stratification, it also resembles an estuary

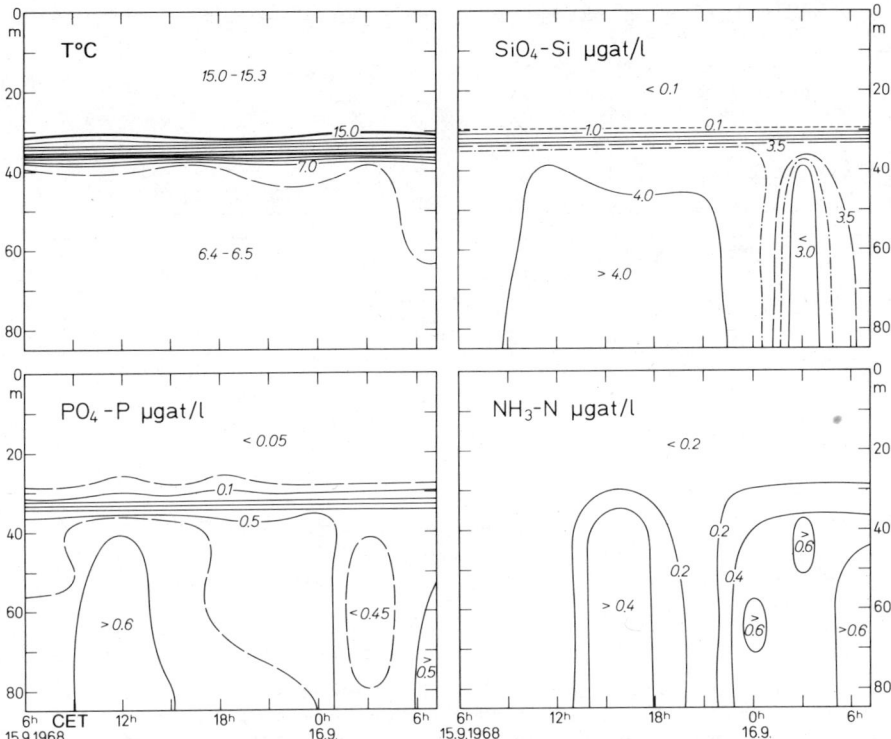

Fig. 10.57. Time series of the stratification of temperature, silicate, phosphate, and ammonia in the central North Sea, based on a 24 hour continuous station ($\varphi = 56°21'$N, $\lambda = 1°0'$E) of the research cutter *Alkor* on September 15–16, 1968. (After Schott and Ehrhardt, 1969.)

(cf. Section 10.3.10). From the oceanographic point of view, it can be considered a large fjord. It is almost free of tides and tidal currents (cf. Section 9.5.2) and, thus, of mixing by tidal currents. Besides, the Baltic Sea is a very young formation (cf. Fig. 1.23). It was 14,000 yr after the last Pleistocene glaciation had receded that it was formed as the Baltic Ice Lake, and from 8000 to 7250 B.C., it was a shallow strait, the Yoldia Sea. Due to the isostatic rising of the continent, it again became a fresh-water lake, the Ancylus Lake, and since 5100 B.C., it began to develop its present nature of a brackish sea under the name of Litorina Sea. It has been investigated very thoroughly. An excellent bibliography has been published by Model (1966). The mean distribution of temperature, salinity, and density in the Baltic Sea has been represented by Lenz (1971) and Bock (1971a and 1971b). Svansson (1972) summarized water level fluctuations and those of water transport, applying numerical methods for the interpretation.

The Baltic Sea consists of a sequence of basins. The decisive sill, upon which the renewal of the water of the Baltic Sea mainly depends, has a saddle depth of only 18 m at the narrows between Gedser and the Darss. Figure 10.58 illustrates the vertical distribution of temperature, salinity, and oxygen, based on synoptic observations at Finnish, Swedish, German, and Danish reference stations in early May 1906 (positions are marked at the upper margin of the figure). This is an example of the peculiar conditions prevailing in the Baltic Sea from the innermost Gulf of Finland to the Skagerrak. The section, in

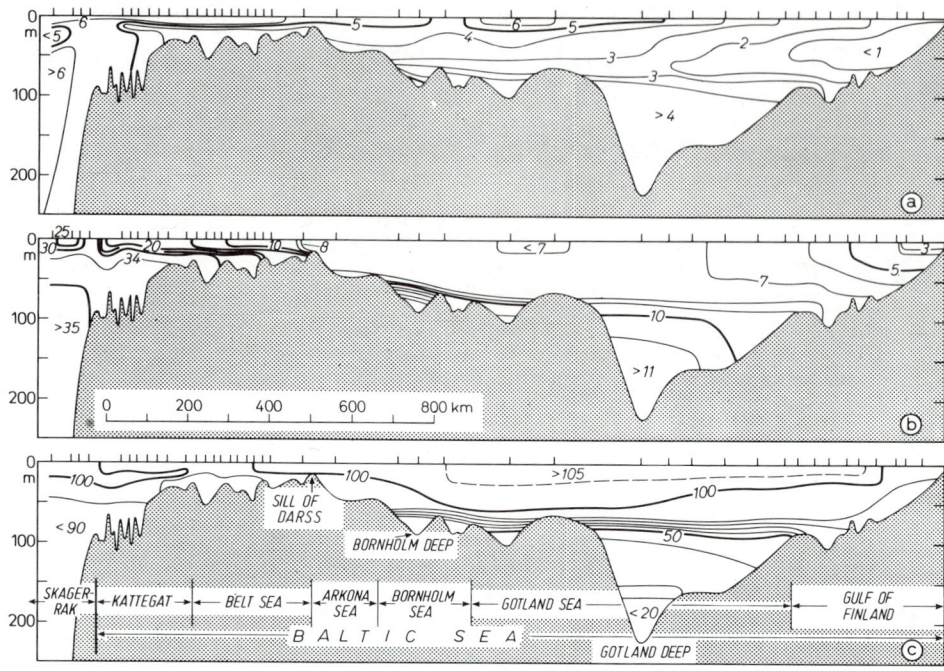

Fig. 10.58. Longitudinal section through the Baltic Sea from the Skagerrak to the eastern end of the Gulf of Finland, based on internationally organized synoptic observations during the period of May 1–7, 1906. (After Dietrich, 1957b.) (a) Temperature (in °C); (b) salinity (in ‰); (c) oxygen content (in % of saturation). Marks at the upper edge: positions of stations. Exaggeration of vertical scale: 2000 times.

detail, runs east of Gotland and north of Bornholm through the Baltic Sea proper, and farther through the Darss Narrows, the Fehmarn Belt, the Great Belt (Store Belt), and the Kattegat into the deep Skagerrak, where it meets with oceanic conditions. Anybody who is surprised to find such an old section of 1906 represented in this book should know that ever since the great time of European cooperation, that is, from 1903 to 1908, no survey of the Baltic Sea as complete as this has been realized. Only in some individual areas have currents and stratification been thoroughly recorded after that time, as especially during the Baltic Year 1969–1970, with research vessels of all countries bordering the Baltic Sea.

Due to the shallow sill depth, the renewal of the deep water in the Bornholm Sea and in the Gotland Sea is restricted to rare cases. Salinity and ventilation of the Baltic Sea basins, both of great importance to living conditions in those regions (Segerstråle, 1969; Jansson, 1972), depend on the volume and the temporal sequence of such water inflow. Reference stations, maintained for decades, provide objective information on the rhythm of the ventilation. Figure 10.59 shows observational results concerning the Gotland Deep, obtained from 1952 through 1970. From these it follows that the renewal, which is associated with an increase of salinity, oxygen, and temperature in the deep water beneath 200 m depth, occurs irregularly. Particularly strong renewals were recorded in the years of 1952, 1961, 1964, 1967, and 1969. In the periods between the large inflows, vertical mixing leads to a decrease of salinity and temperature, as well as to oxygen consumption. As a result, poisonous hydrogen sulfide is present in the deep water beneath 100 m. In

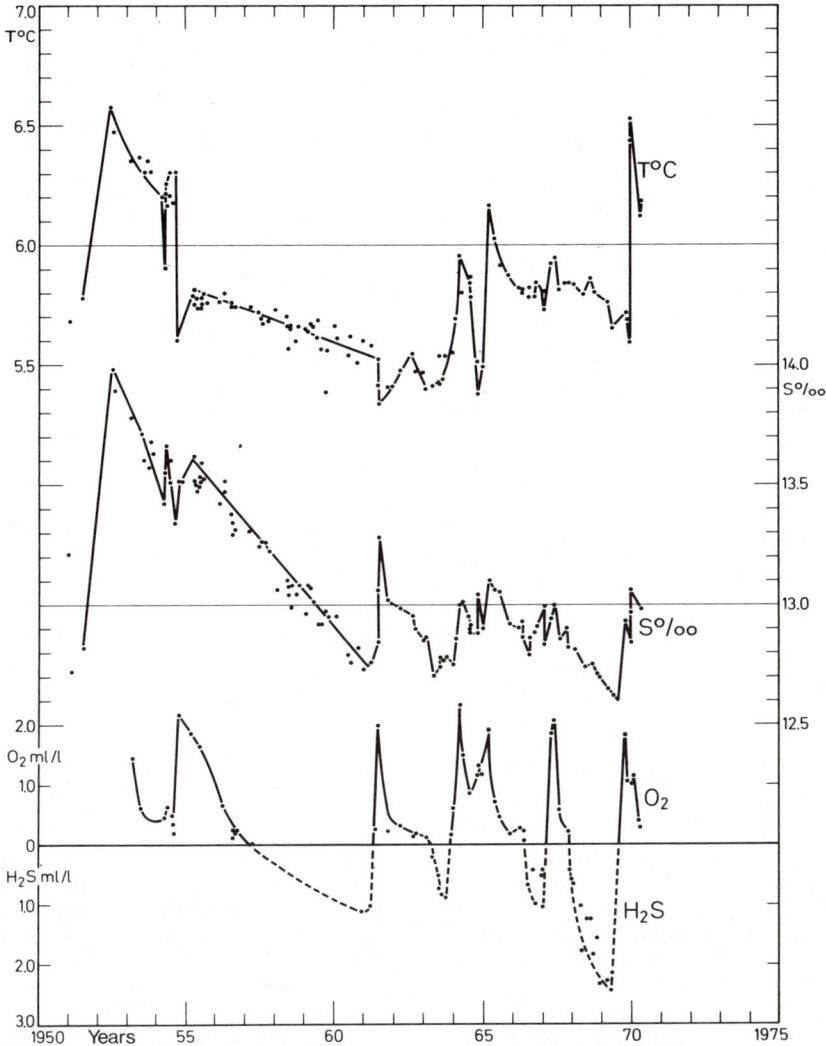

Fig. 10.59. Temperature, salinity, and oxygen content in the Gotland Deep of the Baltic Sea from 1952 to 1970 at the Swedish reference station F 81 ($\varphi = 57°20'N$, $\lambda = 19°49'E$, depth 249 m) below 200 m. (After Fonselius, 1970b.) See also Fig. 10.58.

1969, the process of a thorough renewal of the stagnant deep water in the basins of Bornholm and Gotland could be studied in the research program of the Baltic Year (Fonselius, 1970a and 1970b), in which all countries bordering the Baltic Sea took part.

However, not every peak in the temporal march of temperature, salinity, or oxygen in Fig. 10.59 indicates a renewal of the water in the Gotland Deep. It should be taken into account that strong, short-term variations of the oxygen content may occur in the Gotland Deep, as can be seen in the measurements taken at 3 hr intervals at several depths [Fig. 10.60(a)]. This cannot be explained by vertical displacement, because the oxygen

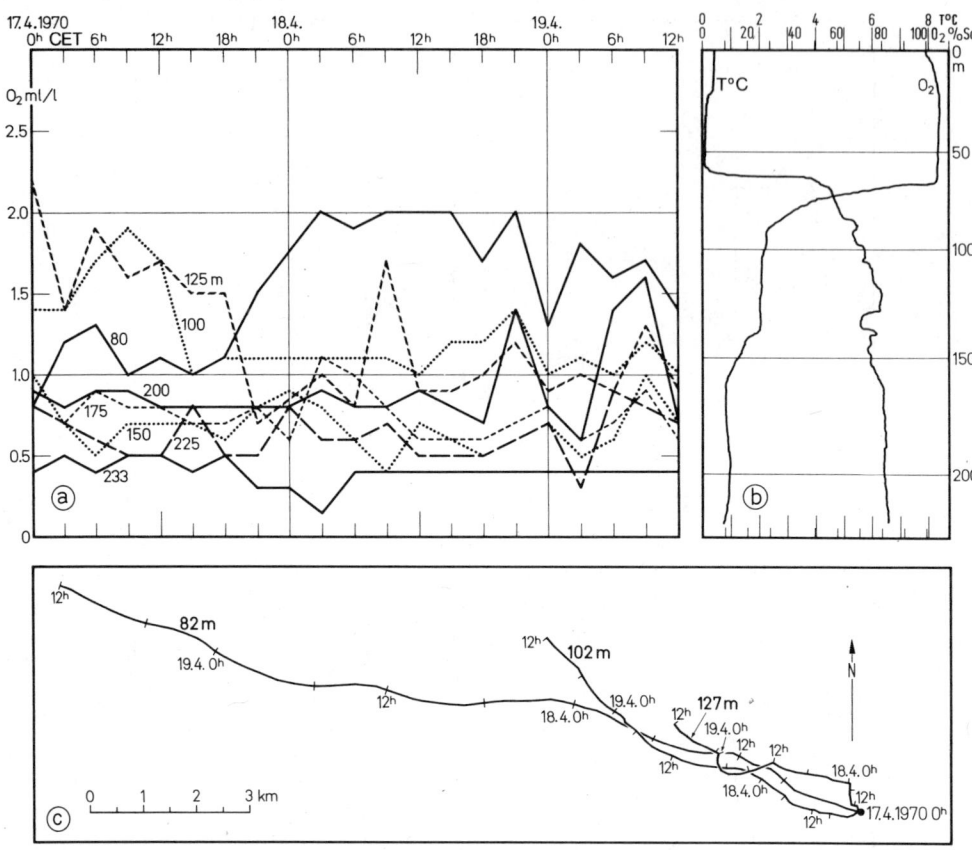

Fig. 10.60. Heterogeneity in the deep water of the Baltic Sea (Gotland Deep) as shown by the example of a continuous station of the research cutter *Alkor* ($\varphi = 57°18'$N, $\lambda = 20°4'$E, depth 237 m, section of April 14 to 19, 1970). (a) Oxygen variation (in ml l^{-1}) at 8 horizons (80 m to 233 m) within 60 hr. (b) Vertical distributions of temperature (°C) and oxygen (% of saturation) on April 17, 1970, 10h 10min. (c) Progressive vector diagrams of the current at 82, 102, and 127 m from April 17, 0h to April 19, 12h, 1970. (a) and (b) after Grasshoff, (c) after Hollan.

values are very uniform in the vertical direction, as demonstrated by the records represented in Fig. 10.60(b). The advective transport, determined with an array of moored current meters, is very small in the depth. At a depth of 127 m, for example, the transport amounts to less than 3 km day^{-1}, that is, less than 3 cm sec^{-1} [cf. Fig. 10.60(c)]. If, nevertheless, the temporal variation in oxygen content is large [cf. Fig. 10.60(a)], small water bodies, with diameters of not more than a few hundred meters, must exist side by side. If changes in the oxygen content of 1 ml l^{-1} occur within 2 hr, as shown in Fig. 10.60(a) for 200 m depth, the representations in Fig. 10.59 must be based on greatly scattered values. Statements on increasing eutrophic conditions in the Baltic Sea should, therefore, be considered with great caution. The heterogeneity in the deep water renders any judgment concerning the degree of pollution in the Baltic Sea very difficult. Repeated continuous measurements would contribute to more thorough knowledge.

The renewal of the deep water in the Baltic Sea is favored when persistent, strong westerly winds press the light, low-salinity top layer of the Baltic water out of the Belt Sea back into the Baltic Sea. The opposite situation is given in Fig. 10.58: easterly winds, blowing for 5 days before hydrographic observations were started, had caused a strong outflow at the surface during this period and pushed the low-salinity surface water as far out as the northern end of the Kattegat. According to Wyrtki (1954), actual wind conditions do not seem to be the only requirements for a large water inflow into the Baltic Sea, as was the case in November and December 1951. In addition, there are several other prerequisites concerning the initial hydrographic conditions before a period of westerly winds that such a large water inflow may occur.

In the North Sea and the Baltic Sea, considerable differences in oceanographic conditions occur even within short distances, which, in addition, are not constant in time. Thus, they permit studies of oceanographic phenomena under different initial conditions. Owing to this diversity, these seas are similar to extended models from which European marine scientists have gained experience and knowledge since 1900. This has not only led to a deeper understanding of these shelf seas but has also given an impetus to entire oceanography.

The Baltic Sea holds a special position among the shelf seas in the areas of humid climate; the same is true for the Persian Gulf in arid climates. The nature of shelf seas will become obvious by a comparison of these two areas. The fundamental differences are restricted not only to stratification and circulation of the water, to tides and tidal currents, but also concern the settlement by marine fauna and flora as well as the bottom sediments. Seibold (1970) has given an informative confrontation of the two cases on the basis of his own investigations in the Baltic Sea (Seibold, 1967) and the results of the *Meteor* Expedition of 1964–1965 in the Indian Ocean (Hartmann et al. 1971). Besides, the information thus obtained was applied to adjacent seas only little investigated so far.

Two rather inconspicuous sills divide the Persian Gulf into three basins (cf. Fig. 10.61). Since there is no closing sill in the Strait of Hormuz, the outer basin passes into the Gulf of Oman of the Arabian Sea, as shown by the new bathymetric chart at the scale of $1:1 \times 10^6$ (Seibold and Vollbrecht, 1969). The Persian Gulf covers an area of 0.226×10^6 km^2. It is a very shallow shelf sea with a mean depth of 35 m; the depth of 100 m is exceeded only in a few places.

In the Persian Gulf, semidiurnal tides prevail, reaching an average tidal range of 1.80 m in the inner part of the Gulf. The tides have two amphidromic systems, one in the western basin, the other in the central basin (cf. Fig. 10.61) with antinodes in the interior, at the central sill, and in the Strait of Hormuz at the exit of the Gulf. Tidal current velocities of 50 cm sec^{-1} are reached in the antinodes (von Trepka, 1968), thus determining also the areas of major turbulence by tidal currents.

In contrast to the Baltic Sea, the surface water of the inner Persian Gulf reaches high salinity values, namely 40 to 41‰, owing to the large excess of evaporation over precipitation (cf. Fig. 4.19). The continental runoff from the Shatt al Arab obviously is too small as to disturb this salt enrichment, as shown by the salinity section in Fig. 10.61. During the cooling of the inner Gulf in winter (beneath 18°C in the example of Fig. 10.61), complete vertical convection is induced, causing the formation of cool and, therefore, rather heavy bottom water of high salinity. This water flows toward the exit of the Gulf in the Strait of Hormuz and, with a still-high salinity of 39.5‰, it enters the Gulf of Oman, where it sinks to a depth of 200 to 300 m, forming high-salinity intermediate water traced by Düing and Schwill (1967) as far as the Arabian Sea. All in all, this is a circulation

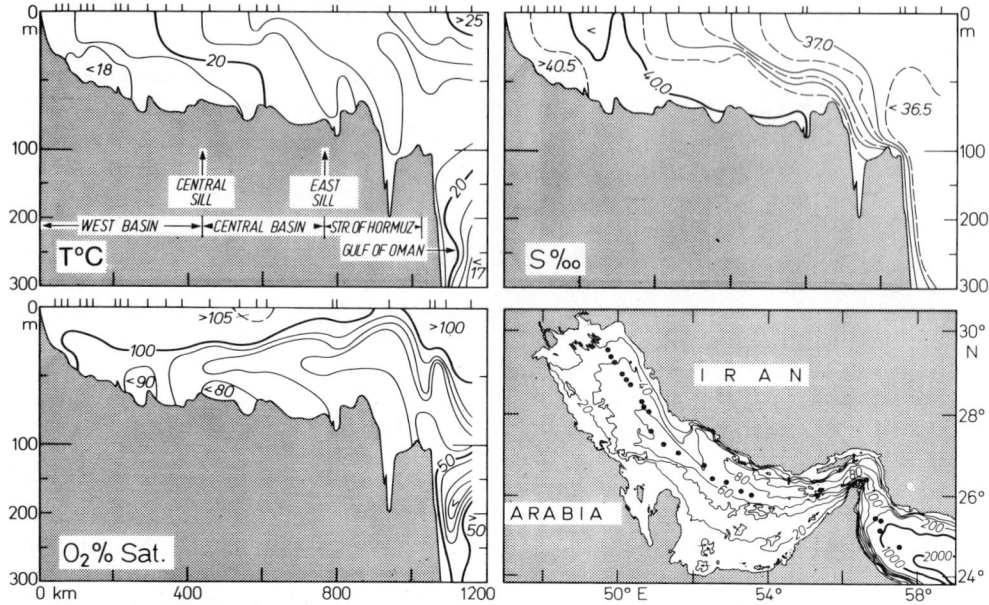

Fig. 10.61. Longitudinal section through the Persian Gulf from the mouth of the Shatt al Arab to the Gulf of Oman, based on observations of the research vessel *Meteor* in March to April 1965. (After Koske and Rabsch, 1972.) Temperature (in °C); salinity (in ‰); oxygen (in % of saturation); position of the stations used in the longitudinal section. Depth contour lines in m. Marks at the upper margin of the section: position of stations. Exaggeration of vertical scale: 0–100 m: 3600 times, 100–300 m: 1800 times.

as schematically shown in Fig. 10.17 under A_{II}. Since the bottom water is very effectively renewed in winter by the full convection in the inner Gulf, the consumption of oxygen in the bottom water is quite small in contrast to the Baltic Sea.

Not only the present water budgets, but also the geological histories of the two shelf seas differ a great deal. Seen within the context of the earth's history, the Baltic Sea is very young, whereas the Persian Gulf is very old. It has existed ever since the Paleozoic period and belongs to a geosyncline, along which the Arabian plate is moved beneath the Eurasiatic plate (cf. Fig. 1.26). The deepening of the Gulf by sinking motion is balanced by sedimentation. The delta of the Shatt al Arab is extended seaward into the Gulf by up to 50 m every year. Consequently, sediments of more than 15 km thickness have been accumulated. Owing to the fact that, during the earth's history, conditions in the area of the primary Persian Gulf often were suitable for production and accumulation of organic matter, large petroleum deposits were formed, which are among the most productive on the globe.

10.3.10. Coastal waters

Coastal waters are parts of the shelf seas. Nevertheless, they are described separately here because, in various respects, their oceanographic conditions differ essentially from those of the open sea in the shelf areas. There is a special coastal oceanography; yet, a border line between coastal waters and the open sea cannot be clearly defined. It varies in space and time and depends on how much importance is attached to various decisive factors, such as bottom topography, continental runoff, tides, tidal currents, and surf. The 20 m depth contour line can be taken as the approximate boundary of coastal waters, because the effects of processes that depend on depth (like continental runoff, surf, and wind induced currents) are generally less important at greater depths. Tidal currents also decrease with growing water depth.

Among other processes, water motion and sediment transport are controlled by bottom topography. To adapt such influence within coastal waters to the requirements of man is the task of coastal engineering.

One of the effects of continental runoff is the decrease of salinity in surface water. It depends on local vertical mixing whether such stratification of salinity will be persistent. Hence, special conditions are dominant in the areas of river mouths, which will be mentioned in the following under estuaries. In humid climates, strong annual variation of the continental runoff is usual, which is reflected in the annual march of salinity, as shown in Fig. 5.12 for the German Bight in the North Sea.

The effect of the tide on coastal waters can be seen, for example, in the amphibious zone, the tidal shoals. They fall dry at low tide, but are covered with seawater at high tide, a process occurring twice a day in areas with semidiurnal tides like the North Sea (cf. Plate 7).

Tidal currents have a twofold effect: (1) they erode and transport bottom sediments, although they move periodically. However, flood current and ebb current are not identical. Hence, lasting morphological changes will finally result. To some extent, they are welcome to man, because channels are kept open in the lower course of tidal rivers, which serve as waterways for shipping, like in the rivers Elbe, Weser, Thames, Gironde, and many others. But such changes can also be unwelcome, if they form and displace sand bars and, thus, endanger shipping. (2) Another effect of tidal currents is the turbulence associated with them and influencing the density stratification of seawater. In extreme cases, this may lead to the destruction of any stratification at all, as shown for the southern North Sea in Fig. 10.56, and for the North Channel of the Irish Sea in Fig. 5.08(a). As a consequence, the annual variation of surface temperature is damped in areas with strong tidal currents. This is reflected in a decrease of the annual amplitude and a delay in the occurrence of the annual maximum of surface temperature, as demonstrated for the western North Sea in Fig. 5.06.

Surf is dominated by strong, turbulent water motion, destroying any stratification in the water and attacking the sea floor. The vertical range of the surf depends on the configuration of the coast. In regions with light winds and in sheltered waters, only a few meters are subjected to surf action; but in areas exposed to the swell from the "roaring forties," a water column of more than 100 m may be strongly affected.

Particularly extreme conditions for coastal waters are found in estuaries. According to Cameron and Pritchard (1963), an estuary, in the oceanographic sense, is a partly enclosed coastal water body that is connected with the open sea and consists of seawater measurably freshened by continental runoff. The concentration and vertical distribution of salinity vary in estuaries. Hence, a classification of estuaries is possible on the basis of salinity stratification, as suggested by Stommel (1953a). It has the advantage that this is also a classification based on circulation. The following four main types of estuaries can be distinguished: (A) estuaries completely mixed in the vertical; (B) estuaries moderately stratified in the vertical; (C) estuaries strongly stratified in the vertical; (D) estuaries with a saltwater wedge. For each of these types, the salinity distribution has been plotted schematically in Fig. 10.62 for four stations (1, 2, 3, and 4), whose positions within the estuary are indicated at the top of the figure. On the left-hand side, the vertical

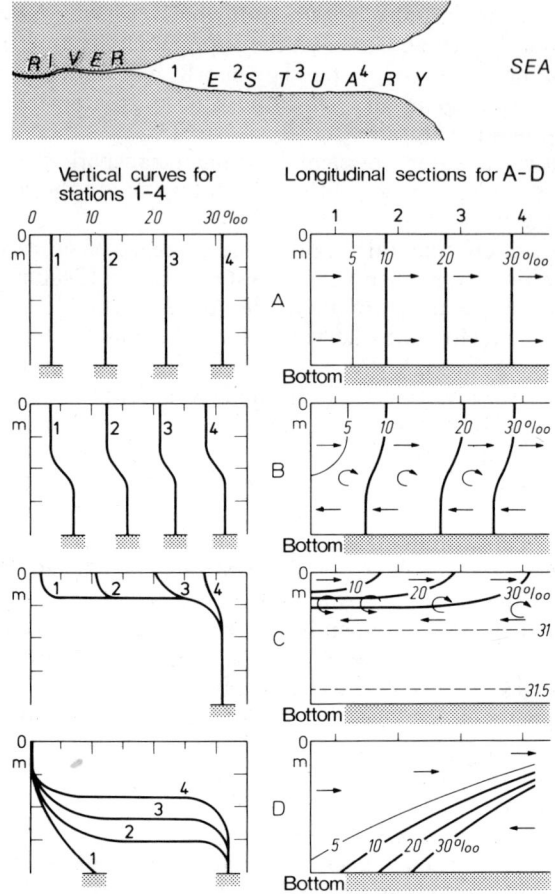

Fig. 10.62. Schematic longitudinal section of salinity stratification in the four main types of estuaries. (After Pickard, 1963.) (A): mixed vertically; (B): slightly stratified; (C): strongly stratified; (D): salt water wedge. Above: positions of stations in the estuary. Arrows: circulation in the estuary.

salinity curves are shown for each of the stations, and on the right-hand side, longitudinal salinity sections are represented, extending from surface to bottom and from one end to the other of the respective type of estuary. In the following, a more detailed characterization of the four types of estuaries will be given, illustrated by examples.

Type A: Not stratified, minor depths, salinity increasing from head to mouth, which means that, with the continental runoff taken into account, the water transport at all depths is seaward on the whole, as indicated by arrows. Due to horizontal exchange, the salt moves inward. An example of this type is given by the outer part of the River Elbe downstream from Scharhörn [cf. Fig. 10.63(a)].

Type B: Two-layered always or at times; on the average, outflow in the upper layer, inflow in the lower layer. Vertical mixing in the transition layer, as

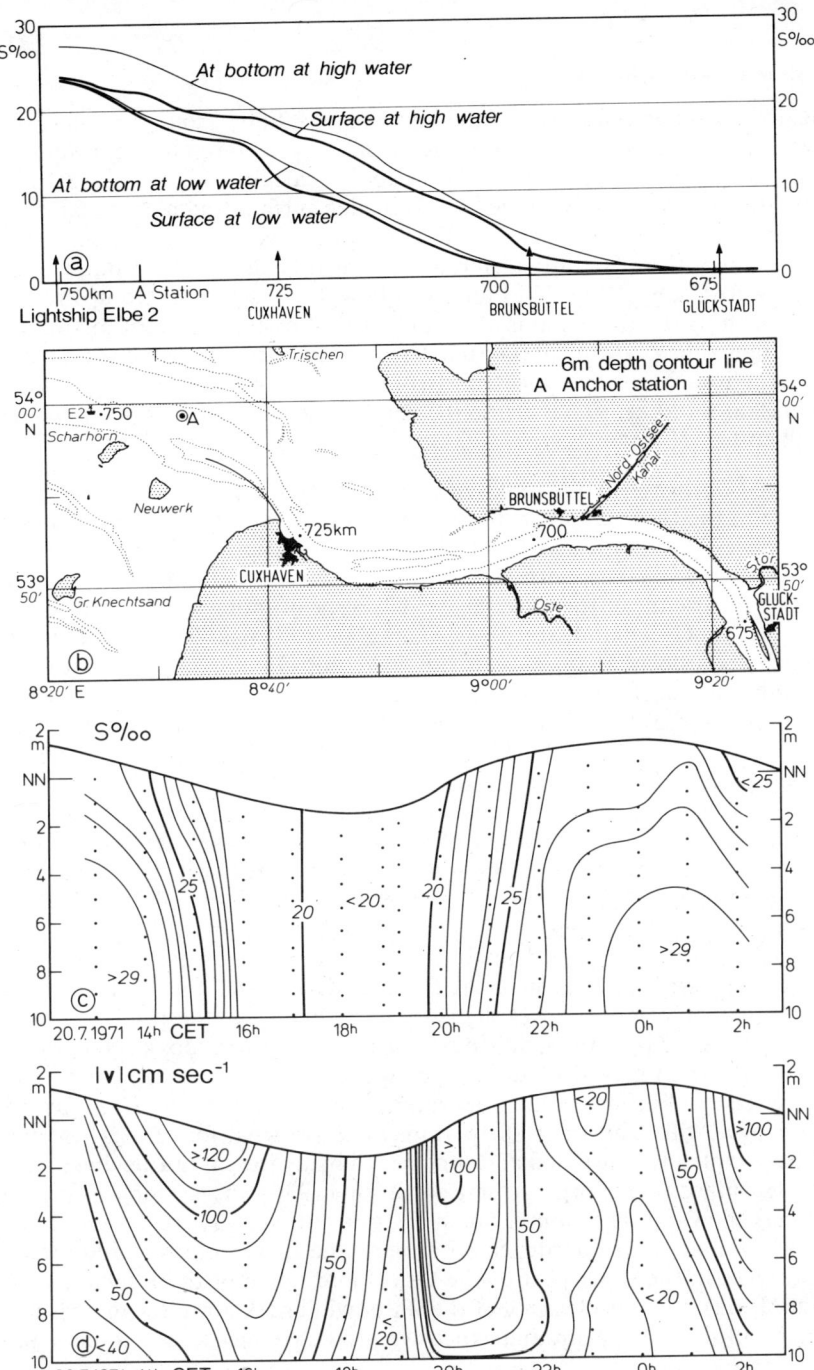

Fig. 10.63. Example of an estuary (type A and B): lower part of the River Elbe between Glückstadt and Scharhörn. (a) Salinity at the surface and at the bottom of the River Elbe at high water (HW) and low water (LW). (After Lucht, 1964.) (b) Positions of (a), (c), and (d). 675–750: Official river distances in kilometers. (c) Vertical distribution of salinity during one tide at anchor station A on July 20 to 21, 1971. (d) Vertical distribution of current velocity during one tide at anchor station A on July 20 to 21, 1971. The tidal current reverses at 20 cm sec^{-1}; ebb current between high water and low water; flood current between low water and high water. The values in (c) and (d) were obtained at anchor station A during a student excursion with the research cutter *Alkor* of the Institut für Meereskunde, Kiel. The water level record originates from the tide gauge of Cuxhaven.

561

indicated by semicircles, causes the increase of salinity from the inner part of the estuary to the mouth. An example of this type is found in the lower course of the River Elbe between Brunsbüttel and Scharhörn. Figure 10.63(b) gives the position of a longitudinal section, for which surface and bottom salinities at high and low water are shown in Fig. 10.63(a) (according to Lucht, 1964). There is a significant salinity difference between surface and bottom, especially at high water, suggesting that high-salinity water has penetrated upstream. At low water, this stratification is almost destroyed by tidal current mixing in the ebb current. The rhythmical fluctuation of stratification within one tidal period can be followed at the position A north of the island of Neuwerk in the two isopleth diagrams of Fig. 10.63(c) and (d), based on measurements taken by the research cutter *Alkor* on July 20 to 21, 1971.

Type C: Always two-layered, mostly at greater depths, fjord type. Salinity increase in the thin top layer, but almost constant salinity in the mighty sublayer between head and mouth of the estuary. Outflow in the upper layer, inflow in the lower layer, with vertical mixing in between, including, however, only the upper edge of the lower layer. This type of estuary comprises all fjords in Norway, in Greenland, and at the Canadian west coast where they are called "inlets." As a peculiarity, most fjords have bottom sills at the mouth. These fjord sills may substantially influence the water circulation in the fjords by preventing the typical full circulation and allowing a partial circulation only. The Oslo Fjord south of the Drøbak Narrows (cf. Fig. 10.64) is an example of full circulation, the part of the fjord north of this passage an example of partial circulation. What is decisive is the saddle depth of the fjord sill: if it remains beneath the level of the light water of the top layer, the heavier water from the open ocean can flow across the sill without any difficulty; the circulation and, thus, the renewal of the lower fjord water will be maintained. This is the case not only in the Oslo Fjord south of Drøbak but also in all large Norwegian fjords, from the Oslo Fjord in southern Norway to the Porsangen Fjord in northern Norway, as well as in the fjords of Greenland and Canada. If, however, the sill of a fjord reaches the level of the lighter water of the top layer, as is nearly the case with the saddle depth of 19.5 m at the Drøbak Sill in the inner Oslo Fjord, the renewal of the deeper water is interrupted and restricted to some occasional inflow of heavier water, when wind conditions are favorable. The renewal will occur irregularly. It may fail to appear for years; then, oxygen will be completely consumed and superseded by hydrogen sulfide due to incomplete decomposition of organic matter. Strøm (1936), in a systematic investigation, has shown quite a number of such fjords in Norway. If those fjords with partial circulation are burdened with more and more domestic and industrial wastes, like in the inner part of the Oslo Fjord north of the Drøbak Narrows (Føyn, 1969), the self-cleaning of the water will break down for some time, hydrogen sulfide will appear, and expensive protective measures will be required.

Type D: Two-layered with wedge-shaped inflow of seawater. Examples can be found in the estuaries of rivers with little vertical mixing and minor tidal currents, as in the Mississippi, but also in shallow, long channels with water inflow at the head, as in the River Trave, downstream from Lübeck (cf. Fig. 10.65). A characteristic of the lower part of the River Trave is the considerable sewage discharge from the town of Lübeck, as well as the small renewal of the

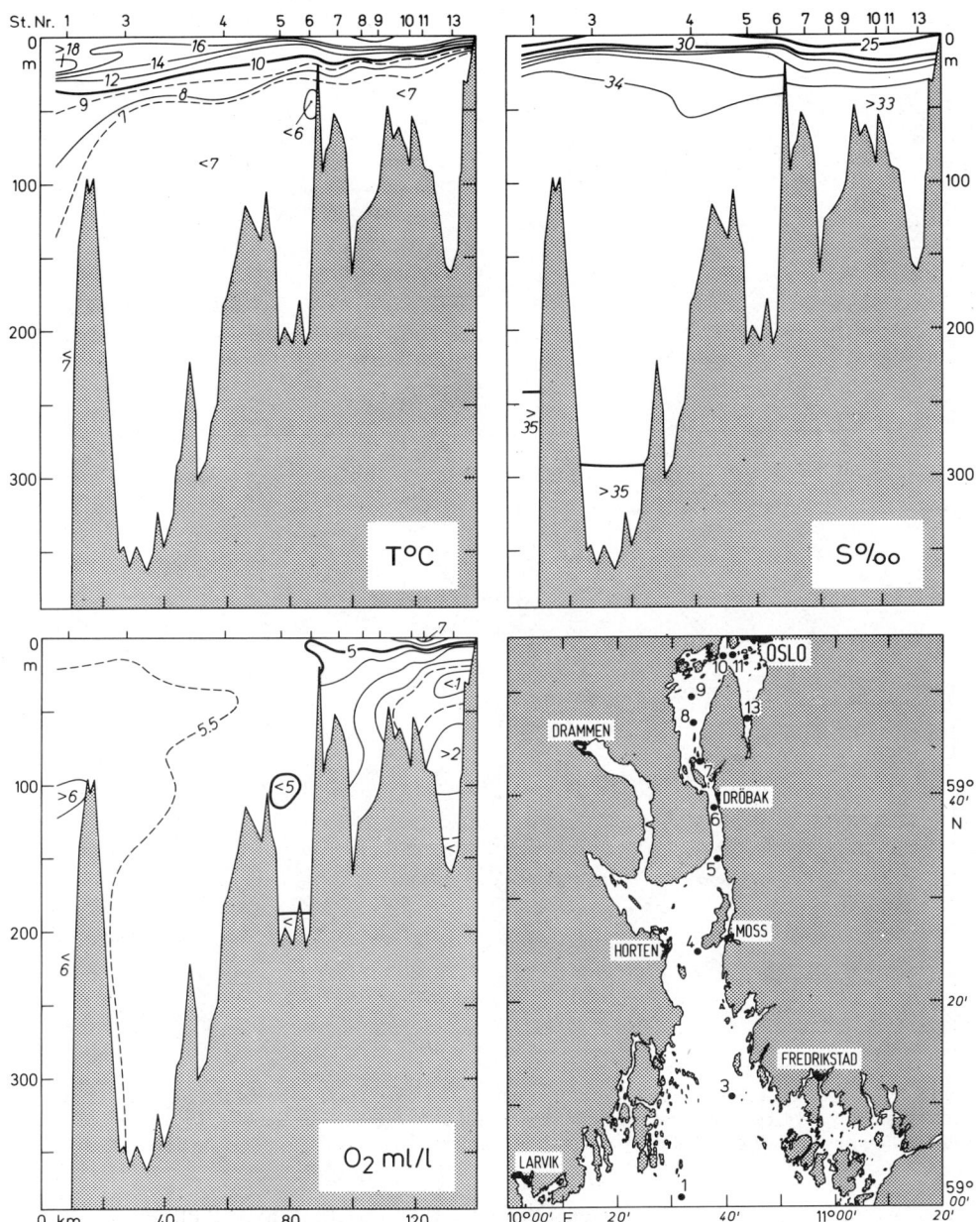

Fig. 10.64. Longitudinal section through the Oslo Fjord with the distributions of temperature, salinity, and oxygen, based on observations taken by the three research vessels *Alkor, Kristine Bonnevie,* and *Gunnar Knudsen* on August 29, 1969. (After Andersen, Beyer and Føyn, 1970.) Exaggeration of vertical scale: 450 times. Positions of stations: see map below on the right-hand side.

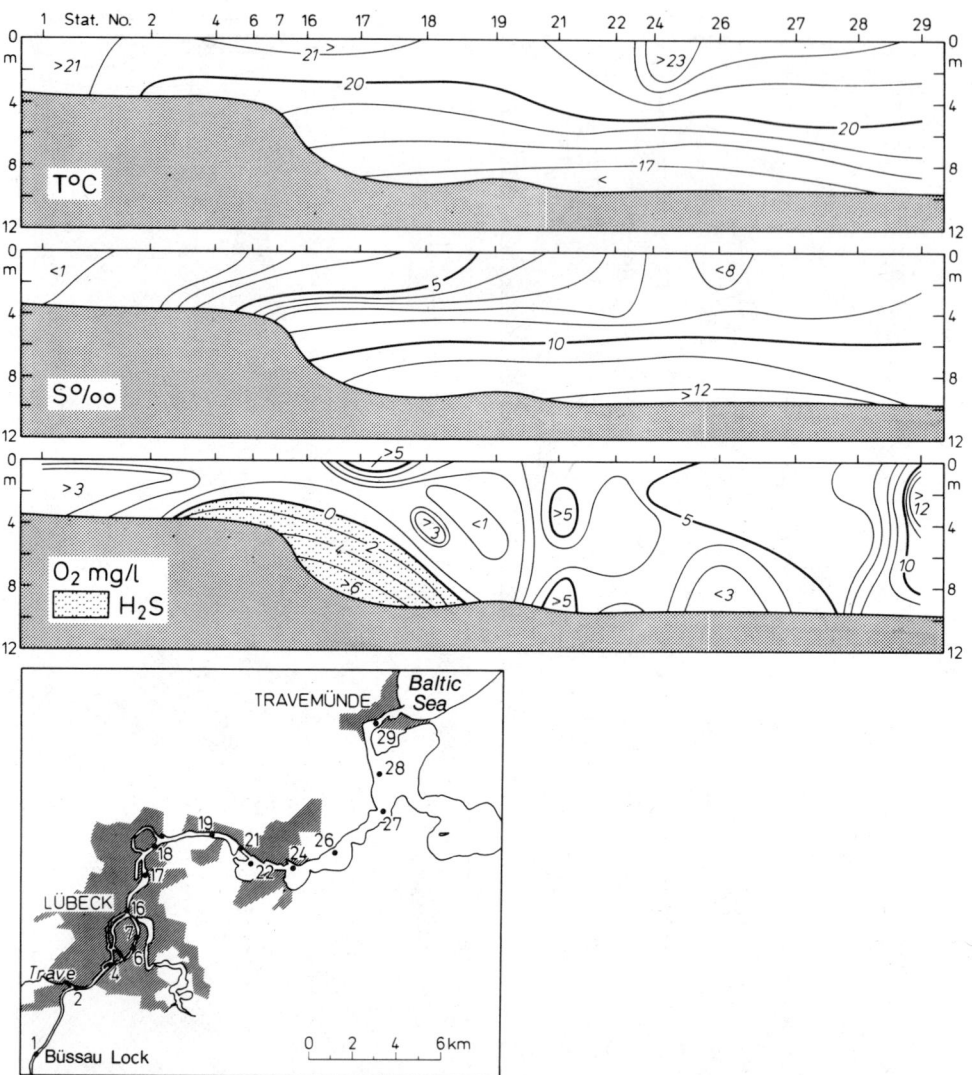

Fig. 10.65. Longitudinal section through the lower part of the River Trave, showing the distributions of temperature, salinity, oxygen, and hydrogen sulfide on August 1–2, 1967. (After Kändler, 1971.) Vertical exaggeration: 700 times. Positions of stations: see map below.

saltwater wedge from the Baltic Sea. The result is that hydrogen sulfide is formed in the wedge, especially in summer. The part of the Baltic Sea east of the Sill of Darss (cf. Fig. 10.58) could be included in this estuary type D, or also in the fjord type C. But according to its dimension it would be more appropriate to give that region a special position among the shelf seas (cf. Section 10.3.9).

That which is schematically classified in Fig. 10.62 and illustrated by the examples given in Figs. 10.63, 10.64, and 10.65, can also be represented by nondimensional parameters concerning stratification and circulation of an estuary, as was attempted by Hansen and Rattray, jr. (1966), but the data required for a reasonable assignment into a diagram are available for few estuaries only.

In ocean straits, circulation and density distribution correspond to the two-layer model, rather like in estuaries. Therefore, the methods used for the investigation of estuaries can also be applied to ocean straits.

11 Appendix

11.1. Salinity as a Function of Electrical Conductivity, Temperature, and Pressure

The salinity of a seawater sample at atmospheric pressure is given as a function of the ratio R_{15} of the conductivity $C(T, S, p)$ of this sample to the conductivity at exactly 35‰ salinity (according to Copenhagen Normal Water) at 15°C [T = temperature (°C), S = salinity (‰), p = pressure (dbar), C = specific electrical conductivity (mS cm^{-1})]:

$$R_{15} = \frac{C(15,S,0)}{C(15,35,0)} \qquad (11.1.01)$$

$$S[‰] = -0.08996 + 28.29720 R_{15} + 12.80832 R_{15}^2 \\ -10.67869 R_{15}^3 + 5.98624 R_{15}^4 - 1.32311 R_{15}^5 \qquad (11.1.02)$$

For the relation between salinity and chloride content, Eq. (2.01) holds. For temperatures deviating from 15°C, R_{15} can be given as a function of the ratio R_S (in the UNESCO tables denoted by R_t):

$$R_S = \frac{C(T,S,0)}{C(T,35,0)} \qquad (11.1.03)$$

$$R_{15} = R_S + 10^{-5} R_S (R_S - 1)(T - 15)[96.7 - 72.0 R_S + 37.3 R_S^2 \\ -(0.63 - 0.21 R_S^2)(T - 15)] \qquad (11.1.04)$$

(example: for $T = 18.94°C$, $R_S = 0.9864$ it follows that $S = 34.466‰$ and $Cl = 19.079‰$).

The temperature dependence has been explicitly given by Brown et al. (1966) as follows:

$$R_T = \frac{C(T,35,0)}{C(15,35,0)}$$

$$R_T = 0.67652453 + 0.20131661 \cdot 10^{-1} T + 0.99886585 \cdot 10^{-4} T^2 \\ -0.19426015 \cdot 10^{-6} T^3 - 0.67249142 \cdot 10^{-8} T^4 \qquad (11.1.05)$$

For the calculation of the salinity from in situ measurements of temperature, conductivity, and pressure, the dependence of the conductivity on pressure is also required. According to Bradshaw et al. (1965), the following relation is valid:

$$R_p = \frac{C(T,S,p)}{C(T,S,0)}$$

$$R_p = 1 + 10^{-2}[g(T)f(p) + h(p)j(T)][1 + l(T)m(S)] \qquad (11.1.06)$$

with

$g(T) = 1.5192 - 4.5302 \cdot 10^{-2}T + 8.3089 \cdot 10^{-4}T^2 - 7.900 \cdot 10^{-6}T^3$
$f(p) = 1.04200 \cdot 10^{-3}p - 3.3913 \cdot 10^{-8}p^2 + 3.300 \cdot 10^{-13}p^3$
$h(p) = 4 \cdot 10^{-4} + 2.577 \cdot 10^{-5}p - 2.492 \cdot 10^{-9}p^2$
$j(T) = 1.000 - 1.535 \cdot 10^{-1}T + 8.276 \cdot 10^{-3}T^2 - 1.657 \cdot 10^{-4}T^3$
$l(T) = 6.950 \cdot 10^{-3} - 7.6 \cdot 10^{-5}T$
$m(S) = 35.00 - S$

If T, C, and p are given, we know R:

$$R = \frac{C(T,S,p)}{C(15,35,0)}$$

According to Reeburgh (1965) one obtains the value $C(15,35,0) = 42.902$ mS cm^{-1}. R_T results directly from Eq. (11.1.05), and R_p is found by iteration from Eq. (11.1.06). Because of $R = R_T \cdot R_S \cdot R_p$, R_S results, and with Eqs. (11.1.04) and (11.1.02) the salinity S is obtained (example: with $T = 2.42°C$, $C = 32.645$ mS/cm, $p = 3907$ dbar, it follows that $S = 34.902‰$). The set of formulas used here has been proven completely by fundamental determinations only within the ranges of 14 to 28°C, 31 to 39‰ and 0 to 10,000 dbar; individual formulas cover larger ranges.

11.2. Density as a Function of Temperature, Salinity, and Pressure

According to Fofonoff et al. (1958), the following formulas are suitable for the computation of σ_0, σ_T, α_{STp} and δ_{STp}. For σ_0,

$$\sigma_0 = \sum_n B_n S^n \qquad (11.2.01)$$

with all $B_n = 0$, except for

$B_0 = -0.0934458632$
$B_1 = +0.814876577$
$B_2 = -4.82496140 \cdot 10^{-4}$
$B_3 = +6.76786136 \cdot 10^{-6}$

For σ_T,

$$\sigma_T = \frac{\sum_i A_i T^i}{T + A_0} + \sum_i \sum_j A_{ij} \sigma_0^i T^j \qquad (11.2.02)$$

with all A_i and $A_{ij} = 0$, except for

$A_0 = +67.26$
$A_1 = +4.53168426$
$A_2 = -0.545939111$
$A_3 = -1.98248399 \cdot 10^{-3}$
$A_4 = -1.43803061 \cdot 10^{-7}$

$A_{10} = +1.0$
$A_{11} = -4.7867 \cdot 10^{-3}$
$A_{12} = +9.8185 \cdot 10^{-5}$
$A_{13} = -1.0843 \cdot 10^{-6}$
$A_{20} = 0$
$A_{21} = +1.8030 \cdot 10^{-5}$
$A_{22} = -8.164 \cdot 10^{-7}$
$A_{23} = -1.667 \cdot 10^{-8}$

For α_{STp},

$$\alpha_{STp} = \alpha_{ST0}\left[1 - \frac{4.886 \cdot 10^{-6}p}{1 + 1.83 \cdot 10^{-5}p} + \sum_i \sum_j \sum_k A_{ijk} p^i \sigma_0^j T^k\right] \quad (11.2.03)$$

with all $A_{ijk} = 0$, except for

$A_{100} = -2.2072 \cdot 10^{-7}$
$A_{101} = +3.6730 \cdot 10^{-8}$
$A_{102} = -6.63 \cdot 10^{-10}$
$A_{103} = +4.00 \cdot 10^{-12}$
$A_{110} = +1.725 \cdot 10^{-8}$
$A_{111} = -3.28 \cdot 10^{-10}$
$A_{112} = +4.00 \cdot 10^{-12}$
$A_{120} = -4.50 \cdot 10^{-11}$
$A_{121} = +1.00 \cdot 10^{-12}$

$A_{200} = -6.68 \cdot 10^{-4}$
$A_{201} = -1.24064 \cdot 10^{-12}$
$A_{202} = +2.14 \cdot 10^{-14}$
$A_{210} = -4.248 \cdot 10^{-13}$
$A_{211} = +1.206 \cdot 10^{-14}$
$A_{212} = -2.000 \cdot 10^{-16}$
$A_{220} = +1.8 \cdot 10^{-15}$
$A_{221} = -6.0 \cdot 10^{-17}$
$A_{301} = +1.5 \cdot 10^{-17}$

For δ_{STp},

$$10^5 \delta_{STp} = \alpha_{STp} - \frac{\sum_n D_n p^n}{1 + 1.83 \cdot 10^{-5}p} \quad (11.2.04)$$

with all $D_n = 0$, except for

$D_0 = +0.97264310$
$D_1 = +1.326963 \cdot 10^{-5}$
$D_2 = -6.227603 \cdot 10^{-12}$
$D_3 = -1.885115 \cdot 10^{-16}$

(example: $S = 34.917‰$, $T = 2.42°C$, $p = 3907$ dbar results in $\sigma_0 = 28.059$, $\sigma_T = 27.892$, $\alpha_{STp} = 0.956459$, and $\delta_{STp} = 43.2 \cdot 10^{-5}$).

11.3. Potential Temperature as a Function of In Situ Temperature, Salinity, and Pressure

For the computation of potential temperature Θ as a function of temperature T (°C), salinity S (‰), and pressure p (dbar), the following formula can be used (Bryden, 1973):

$$\Delta\Theta = \sum_i \sum_j \sum_k A_{ijk} p^i (S - 35)^j T^k \quad (11.3.01)$$

with all $A_{ijk} = 0$, except for:

$A_{100} = +0.36504 \cdot 10^{-4}$
$A_{101} = +0.83198 \cdot 10^{-5}$
$A_{102} = -0.54065 \cdot 10^{-7}$
$A_{103} = +0.40274 \cdot 10^{-9}$
$A_{110} = +0.17439 \cdot 10^{-5}$
$A_{111} = -0.29778 \cdot 10^{-7}$

$A_{120} = -0.41057 \cdot 10^{-10}$
$A_{200} = +0.89309 \cdot 10^{-8}$
$A_{201} = -0.31628 \cdot 10^{-9}$
$A_{202} = +0.21987 \cdot 10^{-11}$
$A_{300} = -0.16056 \cdot 10^{-12}$
$A_{301} = +0.50484 \cdot 10^{-14}$

(example: $S = 34.917‰$, $T = 2.420°C$, $p = 3907$ dbar results in $\Delta\Theta = 0.336°C$ and $\Theta = 2.084°C$).

11.4. Viscosity of Seawater as a Function of Temperature, Salinity, and Pressure

According to Matthäus (1972), the following formula is obtained for the calculation of the viscosity μ (g cm^{-1} sec^{-1}) as a function of temperature T (°C), salinity S (‰), and pressure p (dbar):

$$\begin{aligned}\mu(T,S,p) =\ & 1.7900 \cdot 10^{-2} \\ & -6.1299 \cdot 10^{-4}T + 1.4467 \cdot 10^{-5}T^2 \\ & -1.6826 \cdot 10^{-7}T^3 \\ & -1.8266 \cdot 10^{-7}p + 9.8972 \cdot 10^{-12}p^2 \\ & +2.4727 \cdot 10^{-5}S \\ & +S(4.8429 \cdot 10^{-7}T - 4.7172 \cdot 10^{-8}T^2 \\ & \quad +7.5986 \cdot 10^{-10}T^3) \\ & +p(1.3817 \cdot 10^{-8}T - 2.6363 \cdot 10^{-10}T^2) \\ & -p^2(6.3255 \cdot 10^{-13}T - 1.2116 \cdot 10^{-14}T^2)\end{aligned} \qquad (11.4.01)$$

(example: $S = 34.917$‰, $T = 2.42$°C, $p = 3907$ dbar results in $\mu = 1.6934 \cdot 10^{-2}$ g cm^{-1} sec^{-1}).

11.5. Sound Velocity as a Function of Temperature, Salinity, and Pressure

According to Anderson (1971), the following formula is obtained for computing the sound velocity c [m sec^{-1}] as a function of temperature T (°C), salinity S (‰), and pressure p (dbar):

$$c = \sum_i \sum_j \sum_k A_{ijk} p^i S^j T^k \qquad (11.5.01)$$

with all $A_{ijk} = 0$, except for

$A_{000} = +1.40295 \cdot 10^{+3}$
$A_{001} = +5.04411497177$
$A_{002} = -5.62864935164 \cdot 10^{-2}$
$A_{003} = +2.41590769023 \cdot 10^{-4}$
$A_{010} = +1.24494448604$
$A_{011} = -1.33395409949 \cdot 10^{-2}$
$A_{012} = +1.01470710283 \cdot 10^{-4}$

$A_{020} = +2.29487467399 \cdot 10^{-3}$
$A_{100} = +1.57267431618 \cdot 10^{-1}/g$
$A_{103} = +2.89033197150 \cdot 10^{-7}/g$
$A_{111} = +4.18588753055 \cdot 10^{-6}/g$
$A_{200} = +2.04834941313 \cdot 10^{-5}/g^2$
$A_{201} = -8.35657086395 \cdot 10^{-7}/g^2$
$A_{310} = -2.00539914999 \cdot 10^{-10}/g^3$

(example: $g = 9.80665$ m sec^{-2}, $p = 3907$ dbar, $S = 34.902$‰, $T = 2.42$°C results in $c = 1525.3$ m sec^{-1}).

Bibliography

AAGAARD, K. (1968): Temperature variations in the Greenland Sea deep waters.—Deep-Sea Res. **15**, 281-296.
——(1970): Wind-driven transports in the Greenland and Norwegian Sea.—Deep-Sea Res. **17**, 281-291.
AANDERAA, I. (1964): A recording and telemetering instrument.—NATO Subcomm. Oceanogr. Res., Techn. Rep. **16**, 1-86.
Academy of Sciences (U.S.S.R.) (1950): Sea Atlas (in Russian) I, 1-83. Moscow.
——(1965): Atlas of Antarctica (in Russian) 1-225. Moscow.
ADDISON, J. R. & E. R. POUNDER (1967): The electrical properties of saline ice.—In: H. OURA (Ed.), Physics of snow and ice, **1**, 649-660.
AIRY, G. B. (1842): Tides and waves.—Encycl. Metropol. 241-396. London.
ALBRECHT, F. (1940): Untersuchungen üben den Wärmehaushalt der Erdoberfläche in verschiedenen Klimagebieten.—Wiss. Abh. Reichsamt f. Wetterd. **8**.
——(1949: Die Aktionsgebiete des Wasser- und Wärmehaushaltes der Erdoberfläche.—Z. Meteorol. **1**, 97-109.
——(1951): Monatskarten des Niederschlages und Monatskarten der Verdunstung und des Wasserhaushaltes des Indischen und Stillen Ozeans.—Ber. Dt. Wetterd. US-Zone **29**, 3-39.
——(1960: Jahreskarten des Wärme- und Wasserhaushaltes der Ozeane.—Ber. Dt. Wetterd. **9**, 66.
ALLAN, T. D. & C. MORELLI (1970): The Red Sea.—In: A. E. MAXWELL (Ed.), The sea **4**, part II, 493-542. New York.
ALLISON, L. J., L. L. FOSHEE, G. WARNECKE & J. C. WILKERSON (1967): An analysis of the north wall of the Gulf Stream utilizing Nimbus 2 high resolution infrared measurements.—Trans. Amer. geophys. Un. **48**, 124.
ANDERSEN, T., F. BEYER & E. FOYN (1970): Hydrography of the Oslofjord. Report on the study course in chemical oceanography arranged in 1969 by ICES.—Coop. Res. Results (A) 20, 1-62. Charlottenlund.
ANDERSON, E. W. (1966): The principles of navigation. **1**.—London.
ANDERSON, E. R. (1971): Sound speed in seawater as a function of realistic temperature-salinity-pressure domains.—Naval Undersea Res. and Develop. Center, San Diego, Rep. NUC TP243 (Unpubl. manuscr.).
ANDERSON, F. P. (1967): Time variations in the Agulhas Current off Natal.—I.U.G.G. Abstr. IAPO, 59.
ANDREYEV, E. G., V. S. LAVORKO, A. A. PIVOVAROV & G. G. KHUNDZHUA (1969): On the vertical temperature profile near the air-sea interface.—Oceanology **9**, 287-290.
ARMSTRONG, F. A. J., C. R. STEARNS & J. D. STRICKLAND (1967): The measurements of upwelling and subsequent biological processes by means of the Technicon Autoanalyzer and associated equipment.—Deep-Sea Res. **14**, 381-389.
ARONS, A. B. & C. F. KIENTZLER (1954): Vapor pressure of sea-salt solutions.—Trans Amer. geophys. Un. **35**, 722-728.
ARRHENIUS, G. (1963): Pelagic sediments.—In: M. N. HILL (Ed.), The sea **3**, 655-727. New York.
ARRHENIUS, G., G. KJELLBERG & W. F. LIBBY (1951): Age determination of Pacific chalk ooze.—Tellus **3**, 222-229.
ARUKOUCHINE, W. A. & HSIN-YI LING (1967): Evidence for turbidite accumulation in trenches in the Indo-Pacific region.—Mar. Geol. **5**, 141-154.
ARX, W. S. v. (1950): An electromagnetic method for measuring the velocities of ocean currents from a ship under way.—Pap. phys. oceanogr. meteorol. Woods Hole **11**, 1-62.
Association International d'Océanographie Physique: Bibliography on tides 1665-1939. Publ Scient. **15**, Bergen 1955. Bibliography on tides 1940-1954. Publ. Scient. **17**, Göteborg 1957. Bibliography on generation of currents and changes of surface-level in oceans, seas and lakes by wind and atmospheric pressure 1726-1955. Publ. Scient. **18**, Göteborg 1957. Bibliography on tides 1955-1969. Publ. Scient. **29**, Birkenhead, Engl. 1971. Bibliography on mean sea level 1719-1958. Publ. Scient. **25**, Birkenhead, Engl. 1967. Bibliography on mean sea level 1959-1969. Publ. Scient. **32**, Birkenhead, Engl. 1971.
ASSUR, A. (1958): Composition of sea ice and its tensile strength.—NAS—Nat. Res. Counc., 106-138.
ATKINSON, G. & R. D. CORSARO (1971): Possible chemical explanation for low-frequency absorption in the sea.—J. Acoust. Soc. Amer. **50**, 123.

BACHE, A. D. (1856): Tiefe des Stillen Ozeans.—cit. in Petermanns Mitt., **119**.
BAGNOLD, R. A. (1963): Mechanics of marine sedimentation.—In: M. N. HILL (Ed.), The sea **3**, 507–528. New York.
BAKER, D. J., Jr. (1971): The Harvard deep-sea pressure gauge; progress report November 1971.—Rep. Meteorol. Oceanogr. No. 4. Cambridge.
BANG, N. D. (1971): The southern Benguela Current region in February, 1966. Part II. Bathythermography and air-sea interaction.—Deep-Sea Res. **18**, 209–224.
BARAZANGI, M. & J. DORMAN (1969): World seismicity maps compiled from ESSA, Coast and Geodetic Survey, epicenter data, 1961–67.—Bull. seism. Soc. Amer. **59**, 369–380.
BARKLEY, R. A. (1968): Oceanographic atlas of the Pacific Ocean. 156 figs. Honolulu.
BARNES, H. (1959): Apparatus and methods of oceanography.—Part I, Chemical. 1–341. London.
BARRETT, J. R. (1965): Subsurface currents off Cape Hatteras.—Deep-Sea Res. **12**, 173–184.
BARTELS, J. (1957): Gezeitenkräfte.—In: S. FLÜGGE (Ed.), Handbuch d. Physik **48** (Geophysik II), 734–774. Berlin.
BARTELS, J. & W. HORN (1952): Gezeitenkräfte.—In: LANDOLT-BÖRNSTEIN, Zahlenwerte und Funktionen **3**, 271–283. Berlin.
BASSE (1931): Consolidated pack ice sheet near Novaya Zemyla in summer. (Photograph).
BATH, U. (1959): Seismic surface wave dispersion: A world-wide survey.—Geofis. Pura Appl. **43**, 131–147.
BATHEN, K. H. (1971): Heat storage and advection in the North Pacific Ocean.—J. Geophys. Res. **76**, 676–687.
BEIN, W., H. HIRSEKORN & L. MÖLLER (1935): Konstantenbestimmungen des Meerwassers und Ergebnisse über Wasserkörper.—Veröff. Inst. Meereskde. Berlin. N.F. (A) **28**, 1–240.
BELDING, H. F. & W. C. HOLLAND (1970): Bathymetric maps eastern continental margin, U.S.A. Sheet 1 to 3.—Amer. Assoc. Petrol. Geol. (Ed.), Tulsa, Okla.
BERCKHEMER, H. (1956): Raleigh-wave dispersion and crustal structure in the East Atlantic Ocean.—Bull. seism. Soc. Amer. **46**, 83–86.
BERGERON, T. (1954): The problem of tropical hurricanes.—Quart. J. R. meteorol. Soc. **80**, 131–164.
BERNOULLI, D. (1740): Traité sur le flux et reflux de la mer. Receuil des pièces qui ont remporté le prix de l'Académie Royale des Sciences. 53–191. Paris.
BEZRUKOV, P. L. & A. P. LISITSYN (1967): Soviet research on ocean bottom sediments.—Oceanology **7**, 641–649.
BIALEK, E. L. (1966): Handbook of oceanographic tables. 1–425. U.S. Naval Oceanogr. Office, Washington, D.C.
BJERKNES, J. (1966): Survey of El Niño 1957–58 in its relation to tropical Pacific meteorology. Bull. Inter-Amer. trop. Tuna Comm. **12**, 25–86.
——— (1967): Ocean-atmosphere interaction (macroprocesses).—In: R. W. FAIRBRIDGE (Ed.), Encycl. atmosph. sci. 704–712. New York.
——— (1969): Teleconnections from the equatorial Pacific.—Monthly Weather Rev. **97**, 163–172.
BJERKNES, V. W. (1936): Über thermodynamische Maschinen, die unter der Mitwirkung der Schwerkraft arbeiten.—Abh. Sächs. Akad. Wiss. Leipzig. Math.-Naturw. Kl. **35**, 1.
BJERKNES, V. W. & W. SANDSTRÖM (1910): Dynamic meteorology and hydrography. Part 1: Statics.—Carnegie Inst., Publ. No. 88, 1–146.
BLACKBURN, M. (1965): Oceanography and the ecology of tunas.—In: H. BARNES (Ed.), Oceanogr. Mar. Biol. Ann. Rev. **3**, 299–322. London.
BLACKWELL, D. D. (1971): Heat flow.—Eos **52**, 135–139.
BLANDFORD, R. (1965): Notes on the theory of the thermocline.—J. Mar. Res. **23**, 18–29.
BLOCH, M. R., D. KAPLAN, V. KERTES & J. SCHNERB (1966): Ion separation in bursting air bubbles: An explanation for the irregular ion ratios in atmospheric precipitations.—Nature **209**, 802–803.
BOCK, K.-H. (1971a): Monatskarten des Salzgehaltes der Ostsee dargestellt für verschiedene Tiefenhorizonte.—Dt. hydrogr. Z. Erg.-H. B, 12, 5–146.
——— (1971b): Monatskarten der Dichte des Wassers in der Ostsee dargestellt für verschiedene Horizonte.—Dt. hydrogr. Z. Erg.-H. B, 13, 5–126.
BÖHNECKE, G. (1936): Temperatur, Salzgehalt und Dichte an der Oberfläche des Atlantischen Ozeans.—Wiss. Erg. Deutsch. Atlant. Exp., "Meteor" 1925–1927, **5**, 1–249. Berlin.
BÖHNECKE, G. & G. DIETRICH (1951): Monatskarten der Oberflächentemperatur für die Nord- und Ostsee und die angrenzenden Gewässer.—Dt. Hydrogr. Inst. Nr. 2336. Hamburg.
BOGDANOV, K. T. (1967): Numerical solution to the problem of distribution of semidiurnal tidal waves (M_2 and S_2) in the world ocean. (In Russian)—Dokl. Akad. Nauk. SSSR 172, No. 6, 1315–1317.
BOUGUSLAVSKII, S. G. & YU. M. BELYAKOV (1966): Special features of the dynamics of the water of the

Subantarctic Intermediate Current in the Atlantic.—Izv., Atmosph. Oceanic Phys. **2**, 1082–1088.
BOWEN, J. S. (1926): The ratio of heat losses by conduction and by evaporation from any water surface.—Phys. Rev. **27**.
BRADSHAW, A. & K. E. SCHLEICHER (1965): The effect of pressure on the electrical conductance of sea water.—Deep-Sea Res. **12**, 151–162.
——(1970): Direct measurement of thermal expansion of sea water under pressure.—Deep-Sea Res. **17**, 691–706.
BRANDT, K. & E. RABEN (1919–22): Zur Kenntnis der chemischen Zusammensetzung des Planktons und einiger Bodenorganismen.—Wiss. Meeresunters. Kiel, N.F. **19**, 175–210.
BREKHOVSKIKH, L. M., K. N. FEDOROV, L. M. FOMIN, M. N. KOSHLYAKOV & A. D. YAMPOLSKY (1971): Large-scale multi-buoy experiment in the tropical Atlantic.—Deep-Sea Res. **18**, 1189–1206.
BRENNECKE, W. (1921): Die ozeanographischen Arbeiten der Deutschen Antarktischen Expedition 1911–12.—Arch. Dt. Seewarte **39**, 1–216.
BRESLAU, L. R., D. J. JOHNSON, J. A. MCINTOSH & L. D. FARMER (1970): The development of Arctic Sea transportation.—MTS J. **4**, 19–43.
BRETTSCHNEIDER, G. (1967): Anwendung des hydrodynamisch-numerischen Verfahrens zur Ermittlung der M_2-Mitschwingungsgezeit der Nordsee.—Mitt. Inst. Meereskde. Hamburg H. 7, 1–65.
BREWER, P. G. & J. P. RILEY (1967): A study of some manual automatic procedures for the determination of nitrate.—Deep-Sea Res. **14**, 475–477.
BRILL, R. (1962): Zur Kenntnis der Struktur des Eises.—Angew. Chem. **74**, 895–900.
BROCKS, K. (1955): Wasserdampfschichtung über dem Meer und "Rauhigkeit" der Meeresoberfläche.—Arch. Met. Geophys. Biokl. (A) 8, 354–383.
——(1959): Ein neues Gerät für störungsfreie meteorologische Messungen auf dem Meer.—Arch. Met. Geophys. Biokl. (A) 11, 227–239.
——(1967): Das meteorologisch-aeronomische Programm der Atlantischen Expedition 1965 (IQSY) mit dem Forschungsschiff "Meteor."—"Meteor" Forsch.-Ergebn. (B) No. 1, V–X. Berlin–Stuttgart.
——(1970): Das Atlantische Passat-Experiment 1969 (APEX).—Schiff und Hafen, Sonderh. 1–7.
BROCKS, K. & L. HASSE (1969): Eine neigungsstabilisierte Boje zur Messung der turbulenten Vertikalflüsse über dem Meer.—Arch. Met. Geophys. Biokl. (A) 18, 331–344.
BROCKS, K. & L. KRÜGERMEYER (1972): The hydrodynamic roughness of the sea surface.—In: A. L. GORDON (Ed.), Studies phys. oceanogr. **1**, 75–92. (WÜST Vol.) New York.
BROECKER, W. S. & YUAN-HUI LI (1970): Interchange of water between the major oceans.—J. Geophys. Res. **75**, 3545–3552.
BROGMUS, W. (1952): Eine Revision des Wasserhaushaltes der Ostsee.—Kieler Meeresforsch. **9**, 15–50.
BROSCHE, P. & J. SÜNDERMANN (1969): Gezeitenreibung und Erdrotation.—Naturwiss. **56**, 135.
BROSIN, H.-J. & D. NEHRING (1968): Der äquatoriale Unterstrom im Atlantischen Ozean auf 29°30'W im September und Dezember 1966.—Beitr. Meereskde. **22**, 5–17.
BROWN, N. L. & B. ALLENTOFI (1966): Salinity, conductivity, and temperature relationships of seawater over the range 0–0.50 P.P.T.—Final rep. March 1.—Prep. for U.S. Navy Off. Nav. Res.
BROWN, N. L. & B. V. HAMON (1961): An inductive salinometer.—Deep-Sea Res. **8**, 65–75.
BRUYEVICH, S. V. & V. D. KORZH (1969): Salt exchange between the ocean and the atmosphere.—Oceanology **9**, 465–475.
BRYAN, K. (1963): A numerical investigation of a nonlinear model of a wind-driven ocean.—J. atmos. Sci. **20**, 594–606.
——(1969): A numerical method for the study of the circulation of the world ocean.—J. Comp. Phys. **4**, 3.
BRYAN, K. & D. COX (1967): A numerical investigation of the oceanic general circulation.—Tellus **19**, 55–80.
BRYDEN, H. L. (1972): New polynomials for thermal expansion, adiabatic temperature gradient and potential temperature of sea water.—Deep-Sea Res. **20**, 401–408.
BUCH, K., M. W. HARVEY, H. WATTENBERG & ST. GRIPENBERG (1932): Über das Kohlensäuresystem im Meerwasser.—Rapp. P.-v. Explor. Mer **79**, 1–70.
BUCHANAN, J. Y. (1886): On similarities in the physical geography of the great oceans.—Proc. R. Georg. Soc. **8**, 753–770.
BUDYKO, M. J. (1956): The heat balance of the earth's surface.—Transl. by N. A. Stepanova from Gidromet. Izdat. 1–255. Leningrad.
——(Ed.) (1963). The atlas of the earth's heat balance.—Izdat. Mezhd. Geof. Komit. Leningrad.
——(1967): Energy interactions between the oceans and the atmosphere.—In S. K. RUNCORN (Ed.), Dict. geophys. **1**, 487–497.
BÜCKMANN, A., G. DIETRICH & J. JOSEPH (1959): Die Forschungsfahrten von F.F.S. "Anton Dohrn" und

V.F.S. "Gauß" im nördlichen Nordatlantischen Ozean im Rahmen des Polarfront-Programms des Internationalen Geophysikalischen Jahres 1958.—Dt. hydrogr. Z. Erg.-H. 3, 7–21.
BÜDEL, J. (1950): Atlas der Eisverhältnisse des Nordatlantischen Ozeans und Übersichtskarten der Eisverhältnisse des Nord- und Südpolargebietes.—Dt. Hydrogr. Inst. Nr. 2335. Hamburg.
BULGAKOV, N. P. (1967): Basic features of the structures and position of the subarctic front in the Northwestern Pacific.—Oceanology 7, 680–690.
BULLARD, E. C. (1968): Conference on the history of the earth's crust.—In. R. A. PHINNEY (Ed.), The history of the earth's crust. 231–235. Princeton.
——(1969): The origin of the oceans.—Sci. Amer. 221, 66–75.
BURKOV, V. A. (1969): The bottom circulation of the Pacific Ocean.—Oceanology 9, 179–188.
BURT, W. V. (1953): A note on the reflection of diffuse radiation by the sea surface.—Trans. Amer. geophys. Un. 34, 199–200.
——(1954): Albedo over wind-roughened water.—J. Meteorol. 11, 283–290.
CALDWELL, D. R., F. E. SNODGRASS & M. H. WIMBUSH (1969): Sensors in the deep sea.—Physics Today 22, 34–43.
CALDWELL, D. R. & B. E. TUCKER (1970): Determination of thermal expansion of sea-water by observing onset of convection.—Deep-Sea Res. 17, 707–719.
CAMERON, W. M. & D. W. PRITCHARD (1963): Estuaries.—In M. N. HILL (Ed.), The sea 2, 306–324. New York.
CARLSON, H., K. RICHTER & H. WALDEN (1967): Messungen der statistischen Verteilung der Auslenkung der Meeresoberfläche im Seegang.—Dt. hydrogr. Z. 20, 59–64.
CARRIT, D. E. & J. H. CARPENTER (1958): The composition of sea water and the salinity-chlorinity-density problems. Physical and chemical properties of sea water.—NAS—Nat. Res. Counc., Publ. 600, 67–86.
CARTWRIGHT, E. D. (1963): A unified analysis of tides and surges round North and East Britain.—Philos. Trans R. Soc. (A) 263, 1–55.
CARTWRIGHT, I. D. (1963): The use of directional spectra in studying the output of a wave recorder on a moving ship.—Ocean Wave Spectra. Englewood Cliffs.
CHALUPNIK, J. D. & P. S. GREEN (1962): A Doppler-shift ocean-current meter.—Mar. Sci. Instrum. 1, 194–199, Instrument Soc. Amer. New York.
CHARNEY, J. G. (1955): The generation of oceanic currents by wind.—J. Mar. Res. 14, 477–498.
——(1960): Non-linear theory of a wind-driven homogeneous layer near the Equator.—Deep-Sea Res. 6, 303–310.
CHESTER, R. (1965): Elemental geochemistry of marine sediments.—In: J. P. RILEY & G. SKIRROW (Eds.), Chem. oceanogr. 2, 23–80.
Chief Administration for Geodesy and Cartography of U.S.S.R. (1966): Atlas of Antarctica (in Russian) 1, Map 66, Moscow.
CHRISTENSEN, N. (1971): Observations of the Cromwell Current near the Galapagos Islands.—Deep-Sea Res. 18, 27–33.
CHROMOV, S. P. (1957): Die geographische Verbreitung der Monsune.—Petermanns geogr. Mitt. 101, 234–237.
CHRYSTAL, G. (1905): On the hydrodynamical theory of seiches.—Trans. R. Soc. Edinburgh 41, 599.
CLARKE, G. L. & H. R. JAMES (1939): Laboratory analysis of the selective absorption of light by sea water.—J. Opt. Soc. Amer. 29, 43–55.
CLAUSS, E., H. HINZPETER & P. LOBEMEIER (1969): Zwei Temperaturprofile.—In: H. CLOSS, G. DIETRICH, G. HEMPEL, W. SCHOTT & E. SEIBOLD, "Atlantische Kuppenfahrten 1967" mit dem Forschungsschiff "Meteor." Reisebericht.—"Meteor" Forsch.-Ergebn. (A) No. 5, 1–71. Berlin-Stuttgart.
CLOSS, H., G. DIETRICH, G. HEMPEL, W. SCHOTT & E. SEIBOLD (1969): "Atlantische Kuppenfahrten 1967" mit dem Forschungsschiff "Meteor." Reisebericht.—"Meteor" Forsch.-Ergebn. (A) No. 5, 1–84. Berlin-Stuttgart.
COLLETTE, B. J., J. I. EWING, R. A. LAGAAY & M. TRUCHAN (1969): Sediment distribution in the oceans: The Atlantic between 10° and 19°N.—Mar. Geol. 7, 279–345.
CORKAN, R. H. (1950): The levels in the North Sea associated with the storm disturbance of 8 January 1949.—Phil. Trans. (A) 242, 483–525.
COURTIER, A. (1938): Marées.—Serv. Hydr. de la Marine. Paris.
COX, A. (1969): Geomagnetic reversals.—Science 163, 237–245.
COX, A., G. B. DALRYMPLE & R. R. DOELL (1967): Reversal of the earth's magnetic field.—Sci. Amer. 216, 44–54.
COX, A., R. R. DOELL & G. B. DALRYMPLE (1964): Reversals of the earth's magnetic field.—Science 144, 1537.

Cox, C. S. & W. Munk (1956): Slopes of the sea surface deduced from photographs of sun glitter.—Bull. Scripps Inst. Oceanogr. **6**, 401–488.
Cox, C. S. & H. Sandström (1962): Coupling of internal and surface waves in water of variable depth.—J. Oceanogr. Soc. Japan **20**, 499–513.
Cox, R. A., F. Culkin & J. P. Riley (1967): The electrical conductivity/chlorinity relationship in natural sea water.—Deep-Sea Res. **14**, 203–220.
Cox, R. A., M. J. McCartney & F. Culkin (1970): The specific gravity/salinity/temperature relationship in natural sea water.—Deep-Sea Res. **17**, 679–689.
Cox, R. A. & N. D. Smith (1959): The specific heat of sea water.—Proc. R. Soc. (A) **252**, 51–62.
Crease, J. (1962): The specific volume of sea water under pressure as determined by recent measurements of sound velocity.—Deep-Sea Res. **9**, 209–213.
——(1965): The flow of the Norwegian Sea water through the Faeroe Bank Channel.—Deep-Sea Res. **12**, 143–150.
Cromwell, T., R. B. Montgomery & E. D. Stroup (1954): Equatorial undercurrent in Pacific Ocean revealed by new methods.—Science **119**, 648–649.
Culkin, F. (1965): The major constituents of sea water.—In: J. P. Riley & G. Skirrow (Eds.), Chemical oceanography **1**, 121–158.
Culkin, F. & R. A. Cox (1966): Sodium, potassium, magnesium and strontium in sea water.—Deep-Sea Res. **13**, 789–804.
Curcio, J. A. & C. C. Petty (1951): The near infrared absorption specturm of liquid water.—J. Opt. Soc. Amer. **41**, 302–304.
Cushing, D. H. (1971): Upwelling and the production of fish.—Adv. mar. Biol. **9**, 255–334.
Darbyshire, J. (1964): A hydrological investigation of the Agulhas Current area.—Deep-Sea Res. **11**, 781–815.
Darwin, Ch. (1842): The structure and distribution of coral reefs.—London.
Darwin, G. H. (1911): The tides and kindred phenomena in the solar system. 3rd Ed. 1–251.—London.
Davies, J. T. & E. K. Rideal (1963): Interfacial phenomena. 1–480.—New York.
Deacon, G. E. R. (1937): The hydrology of the southern ocean.—"Discovery" Rep. **15**, 1–124. London.
——(1963): The southern ocean.—In: M. N. Hill (Ed.), The sea **2**, 281–296. New York.
——(1971): A discussion on ocean currents and their dynamics.—Philos. Trans. R. Soc. (A) **270**, 349–465.
Defant, A. (1919): Untersuchungen über die Gezeitenerscheinungen in Mittel- und Randmeeren, in Buchten und Kanälen.—Denkschr. Akad. Wiss. Wien **96**, 57–174, 673–750.
——(1923): Grundlagen einer Theorie der Nordseegezeiten.—Ann. Hydrogr. u. marit. Meteorol. **51**, 57–64.
——(1925): Gezeitenprobleme des Meeres in Landnähe.—In: C. Jensen & A. Schwassmann (Eds.), Probleme der kosmischen Physik **6**, 1–80. Hamburg.
——(1926): Die Austauschgröβe der atmosphärischen und ozeanischen Zirkulation.—Ann. Hydrogr. u. marit. Meteorol. **54**, K.-Supplement.
——(1928): Die systematische Erforschung des Weltmeeres.—Z. Ges. Erdkde. Berlin. Jubiläums-Sonderb., 459–505.
——(1929): Stabile Lagerung ozeanischer Wasserkörper und dazugehörige Stromsysteme.—Veröff. Inst. Meereskde. Berlin, N.F. (A) 19, 1–32.
——(1932): Die Gezeiten und inneren Gezeitenwellen des Atlantischen Ozeans.—Wiss. Ergebn. Deutsch. Atlant. Exp. "Meteor" 1925–27, **7**, 1–318. Berlin.
——(1936a): Schichtung und Zirkulation des Atlantischen Ozeans. Die Troposphäre.—Wiss. Ergebn. Deutsch. Atlant. Exp. "Meteor" 1925–27, **6**, Teil 1, Lfg. 3, 289–411. Berlin.
——(1936b): Ausbreitungs- und Vermischungsvorgänge im Antarktischen Bodenstrom und im Subantarktischen Zwischenwasser.—Wiss. Ergebn. Deutsch. Atlant. Exp. "Meteor" 1925–27, **6**, Teil 2, Lfg. 2, 55–96. Berlin.
——(1938): Aufbau und Zirkulation des Atlantischen Ozeans.—Sitzungsber. Preuss. Akad. Wiss. Phys.-Math. Kl. **15**, 1–29.
——(1940): Scylla und Charybdis und die Gezeitenströmungen in der Straβe von Messina.—Ann. Hydrogr. u. marit. Meteorol. **68**, 145–157.
——(1941): Die absolute Topographie des physikalischen Meeresniveaus und der Druckflächen, sowie die Wasserbewegungen im Atlantischen Ozean.—Wiss. Ergebn. Deutsch. Atlant. Exp. "Meteor" 1925–27, **6**, Teil 1, Lfg. 5, 191–260. Berlin.
——(1961): Physical oceanography **1**, 1–729; **2**, 1–598.—New York.
Degens, E. T., E. H. Reuter & K. N. F. Shaw (1964): Biochemical compounds in offshore sediments and

sea water.—Geochim. Cosmochim. Acta **28**, 45-66.
DEGENS, E. T. & D. A. ROSS (1969): Hot brines and recent heavy metal deposits in the Red Sea. 1-600.—New York.
——— (1970): Oceanographic expedition in the Black Sea. A preliminary report.—Naturwiss. **57**, 349-353.
DEL GROSSO, V. A. (1970): Sound speed in pure water and sea water.—J. Acoust. Soc. Amer. **47**, 947-949.
DEMENITSKAYA, R. M. & K. L. HUNKINS (1970): Shape and structure of the Arctic Ocean.—In: A. E. MAXWELL (Ed.), The sea **4**, part II, 223-249. New York.
Deutsches Hydrographisches Institut (1952): Handbuch des Atlantischen Ozeans.—Hamburg.
——— (1956, 1959): Handbuch für das Mittelmeer, I und III.—Hamburg.
——— (1963): Tägliche Eiskarten.—Hamburg.
——— (1963): Atlas der Gezeitenströme für die Nordsee, den Kanal und die britischen Gewässer.—Hamburg.
——— (1970): Gezeitentafeln für das Jahr 1971.—Hamburg.
——— (1971): Gezeitentafeln für das Jahr 1972.—Hamburg.
DICKSON, R. & A. LEE (1969): Atmospheric and marine climate fluctuations in the North Atlantic region.—In: M. SEARS (Ed.), Progr. oceanogr. **5**, 55-65. Oxford.
DIETRICH, G. (1935): Aufbau und Dynamik des südlichen Agulhasstromgebietes.—Veröff. Inst. Meereskde. Berlin. N.F. (A) 27, 1-79.
——— (1936): Das "ozeanische Nivellement" und seine Anwendung auf die Golfküste und die atlantische Küste der Vereinigten Staaten von Amerika.—Z. Geophys. **12**, 287-298.
——— (1937): "Die dynamische Bezugsfläche," ein Gegenwartsproblem der dynamischen Ozeanographie.—Ann. Hydrogr. u. marit. Meteorol. **65**, 506-619.
——— (1937a): Über Bewegung und Herkunft des Golfstromwassers.—Veröff. Inst. Meereskde. Berlin. N.F. (A) 33, 53-91.
——— (1937b): Die Lage der Meeresoberfläche im Druckfeld von Ozean und Atmosphäre mit besonderer Berücksichtigung des westlichen nordatlantischen Ozeans und des Golfs von Mexiko.—Veröff. Inst. Meereskde. Berlin. N.F. (A) 33, 5-52.
——— (1937c): Fragen der Groβformen und der Herkunft des Tiefenwassers im Amerikanischen Mittelmeer.—Ann. Hydrogr. u. marit. Meteorol. **65**, 345-347.
——— (1939): Das Amerikanische Mittelmeer.—Z. Ges. Erdkde. Berlin, 3-4, 108-130.
——— (1944a): Die Schwingungssysteme der halb- und eintägigen Tiden in den Ozeanen.—Veröff. Inst. Meereskde. Berlin N.F. (A) 41, 1-68.
——— (1944b): Die Gezeiten des Weltmeeres als geographische Erscheinung.—Z. Ges. Erdkde. Berlin, 69-85.
——— (1950a): Über systematische Fehler in den beobachteten Wasser- und Lufttemperaturen auf dem Meere und über ihre Auswirkung auf die Bestimmung des Wärmeumsatzes zwischen Ozean und Atmosphäre.—Dt. hydrogr. Z. **3**, 314-324.
——— (1950b): Die natürlichen Regionen von Nord- und Ostsee auf hydrographischer Grundlage.—Kieler Meeresforsch. **7**, 38-69.
——— (1951): Influences of tidal streams on oceanographic and climatic conditions in the sea as exemplified by the English Channel.—Nature **168**, 6-11.
——— (1952): Physikalische Eigenschaften des Meerwassers.—In: LANDOLT-BÖRNSTEIN, Zahlenwerte und Funktionen. Astr. u. Geophys. 3. Berlin.
——— (1953a): Die Elemente des jährlichen Ganges der Oberflächentemperatur in der Nord- und Ostsee und in den angrenzenden Gewässern.—Dt. hydrogr. Z. **6**, 49-64.
——— (1953b): Verteilung, Ausbreitung und Vermischung der Wasserkörper in der südwestlichen Nordsee auf Grund der Ergebnisse der "Gauβ"-Fahrt im Februar-März 1952.—Ber. Dt. Wiss. Komm. Meeresforsch. **13**, 104-129.
——— (1954a): Einfluβ der Gezeitenstromturbulenz auf die hydrographische Schichtung der Nordsee.—Arch. Met. Geophys. Biokl. (A) 7, 391-405.
——— (1954b): Ozeanographisch-meteorologische Einflüsse auf Wasserstandsänderungen des Meeres am Beispiel der Pegelbeobachtungen von Esbjerg.—Küste **2**, 130-156.
——— (1956a): Beitrag zu einer vergleichenden Ozeanographie des Weltmeeres.—Kieler Meeresforsch. **12**, 3-24.
——— (1956b): Schichtung und Zirkulation der Irminger See im Juni 1955.—Ber. Dt. Wiss. Komm. Meeresforsch. **14**, 255-312.
——— (1956c): Überströmung des Island-Färöer-Rückens in Bodennähe nach Beobachtungen mit dem

Forschungsschiff "Anton Dohrn" 1955-56.—Dt. hydrogr. Z. **9**, 78-89.
——— (1957a): Ozeanographische Probleme der deutschen Forschungsfahrten im Internationalen Geophysikalischen Jahr 1957-58.—Dt. hydrogr. Z. **10**, 39-61.
——— (1957b): Allgemeine Meereskunde. 1-492. 1. Aufl.—Berlin.
——— (1957c): Ergebnisse synoptischer ozeanographischer Arbeiten in der Nordsee.—Verh. Dt. Geographentag Hamburg. 376-383.—Wiesbaden.
——— (1959): Zur Topographie und Morphologie des Meeresbodens im nördlichen Nordatlantischen Ozean.—Dt. hydrogr. Z. Erg.-H. (B) 3, 26-34.
——— (1961a): Some thoughts on the working-up of the observations made during the "Polar Front Survey" in the IGY 1958.—Rapp. P.-v. Explor. Mer. **149**, 103-110.
——— (1961b): Zur Topographie der Anton-Dohrn-Kuppe.—Kieler Meeresforsch. **17**, 3-7.
——— (1962): Mean monthly temperature and salinity of the surface layer of the North Sea and adjacent waters from 1905 to 1954.—Cons. Int. Explor. Mer. 1-150, 25 Karten. Atlas. Charlottenlund.
——— (1964): Oceanic Polar Front Survey in the North Atlantic.—In: H. ODISHAW (Ed.), Res. geophys. **2**, 291-308. Cambridge, Mass.
——— (1965): Die Internationale Indische-Ozean-Expedition und die deutsche Beteiligung mit dem neuen Forschungsschiff "Meteor."—Erde **96**, 5-20.
——— (1966): Veränderlichkeit im Ozean.—Kieler Meeresforsch. **22**, 139-144.
——— (1969a): Atlas of the hydrography of the northern North Atlantic Ocean. 1-140.—Copenhagen.
——— (1969b): Physical variability in the sea and the consequences for fisheries hydrography.—Fisk. Dir. Skr. Ser. HavUnders. **15**, 266-273.
——— (1969c): "Sandbewegung im deutschen Küstenraum" als Schwerpunktprogramm der Deutschen Forschungsgemeinschaft.—Küste, 18, 10-14.
——— (1970): Ozeanographie. 1-166.—Brunswick.
——— (1972): Editorial.—In: G. DIETRICH (Ed.), Upwelling in the ocean and its consequences.—Geoforum 11, 3-8.—Brunswick.
DIETRICH, G., W. DÜING, K. JOHANNSEN & H. OHL (1966): Der neue Forschungskutter "Alkor."—Kieler Meeresforsch. **22**, 145-154.
DIETRICH, G. & J. M. GIESKES (1968): The oceanic polar front in the waters off the east coast of Greenland in August 1966.—Ann. Biol. **23**, 20-22.
DIETRICH, G. & W. HORN (1973): Norwegian Sea Expedition 1969. Report of the Coordinator.—"Meteor" Forsch.-Ergebn. (A) 12, 1-10. Berlin-Stuttgart.
DIETRICH, G. & H. HUNGER (1962): Gezielte Tiefseebeobachtungen: Eine neue Tiefsee-Fernsehkamera mit eingebauter Fotokamera und mit gekoppelten Sammelgeräten.—Dt. hydrogr. Z. **15**, 229-242.
DIETRICH, G. & G. KRAUSE (1969): The observations of the vertical structure of hot salty water by R. V. "Meteor."—In: E. T. DEGENS & D. A. ROSS (Eds.), Hot brines and recent heavy metal deposits in the Red Sea. 10-14. New York.
DIETRICH, G., D. SAHRHAGE & K. SCHUBERT (1959): Locating fish concentrations by thermometric methods.—In: Modern fishing gear of the world. 453-461. London.
DIETRICH, G. & G. SIEDLER (1963): Ein neuer Dauerstrommesser.—Kieler Meeresforsch. **19**, 3-7.
DIETRICH, G. & J. ULRICH (1968): Atlas zur Ozeanographie.—Meyers Gr. Phys. Weltatlas **7**, 1-75, 19 Tables, 114 Figures. Manheim.
DIETRICH, G. & H. WEIDEMANN (1952): Strömungsverhältnisse in der Lübecker Bucht.—Küste **1**, 69-89.
DIETZ, R. S. (1961): Continent and ocean basin evolution by spreading of the sea floor.—Nature **190**, 854-857.
DITTMAR, W. (1884): Report on researches into the composition of ocean-water collected by H.M.S. Challenger during the years 1873-76. Rep. Sci. Res. Voyage "Challenger" 1873-76.—Phys. & Chem. **1**, 1-251.
DODIMEAD, A. J., F. FAVORITE & T. HIRANO (1963): Review of oceanography of the subarctic Pacific Ocean.—Int. North Pacific Fish. Comm. Bull. **13**, 1-195.
DONK, J. v. & G. MATHIEU (1969): Oxygen isotope compositions of foraminifera and water samples from the Arctic Ocean.—J. Geophys. Res. **74**, 3396-3407.
DOODSON, A. T. (1921): The harmonic development of the tide-generating potential.—Proc. R. Soc. (A) **100**, 305-329.
——— (1928): The analysis of tidal observations.—Philos. Trans. R. Soc. (A) **227**, 223-279.
——— (1936): Tides in oceans bound by meridians. II. Ocean bounded by complete meridian: Diurnal tides.—Philos. Trans. R. Soc. (A) **235**, 290-342.
——— (1938): Tides in the ocean bounded by meridians. III. Ocean bounded by complete meridian: Semidi-

urnal tides.—Philos. Trans. R. Soc. (A) **237**, 311–373.
——— (1957): The analysis and prediction of tides in shallow water.—Int. hydrogr. Rev. **34**, 5–46.
DORSEY, N. E. (1940): Properties of ordinary water substances.—New York.
DREVER, R. G. & T. SANFORD (1970): A free-fall electromagnetic current meter-instrumentation.—Proc. Conf. "Electronic Engineering in Ocean Technology," Univ. Coll. Swansea, I.E.R.E.
DÜING, W. (1965): Strömungsverhältnisse im Golf von Neapel.—Pubbl. staz. zool. Napoli **34**, 256–316.
——— (1967): Die Vertikalzirkulation in den küstennahen Gewässern des Arabischen Meeres während der Zeit des Nordostmonsuns. "Meteor" Forsch.-Ergebn. (A) No. 3, 67–83. Berlin–Stuttgart.
——— (1970): The monsoon regime of the currents in the Indian Ocean.—Int. Indian Ocean Exped. Oceanogr. Monogr. **1**, 1–68. Honolulu.
——— (1970): The structure of the sea surface temperatures in monsoonal areas.—Univ. of Miami. Sci. Rep.
——— (1972): The structure of the sea surface temperatures in monsoonal areas.—In: A. L. GORDON (Ed.). Stud. phys. oceanogr. **1**, 1–17. (WÜST Vol.) New York.
DÜING, W., K. GRASSHOFF & G. KRAUSE (1967): Hydrographische Beobachtungen auf einem Äquatorschnitt im Indischen Ozean.—"Meteor" Forsch.-Ergebn. (A) No. 3, 84–92. Berlin–Stuttgart.
DÜING, W. & D. JOHNSON (1971): Southward flow under the Florida Current.—Science **173**, 428–430.
——— (1972): High resolution current profiling in the Straits of Florida.—Deep-Sea Res. **19**, 259–274.
DÜING, W. & W.-D. SCHWILL (1967): Ausbreitung und Vermischung des salzreichen Wassers aus dem Roten Meer und dem Persischen Golf.—"Meteor" Forsch.-Ergebn. (A) No. 3, 44–66. Berlin–Stuttgart.
DÜING, W. & K.-H. SZEKIELDA (1971): Monsoonal Response in the western Indian Ocean.—J. Geophys. Res. **76**, 4181–4187.
DUNCAN, C. P. (1970): The Agulhas Current.—Ph.D. diss. Univ. of Hawaii. Hololulu.
DUNKEL, M. (1967): Eine Apparatur zur Messung des vertikalen Wind-, Temperatur- und Feuchteprofils über dem Ozean.—"Meteor" Forsch.-Ergebn. (B) No. 1, 45–53. Berlin–Stuttgart.
DUUN-CHRISTENSEN, J. T. (1971): Investigation on the practical use of a hydrodynamic numeric method for calculation of sea level variations in the North Sea, the Skagerrak and the Kattegat.—Dt. hydrogr. Z. **24**, 210–227.
DUVANIN, A. I. (1968): A model of the macroscale air-sea interaction processes.—Oceanology **8**, 459–465.
EBER, L. E., J. F. T. SAUR & O. E. SETTE (1968): Monthly mean charts. Sea surface temperature. North Pacific Ocean 1949–1962.—Bur. Comm. Fish. Circ. 258. Washington, D.C.
EDGERTON, H. E. & L. D. HOADLEY (1955): Cameras and lights for underwater use.—J. Soc. Motion Pict. and Telev. Eng. **63**, 345–350.
EHRICKE, K. (1969): Beitrag zur Bestimmung der turbulenten Vertikaldiffusion im geschichteten Meer am Beispiel des Skagerraks.—Kieler Meeresforsch. **25**, 233–244.
EITTREIM, S. & M. EWING (1972): Suspended particulate matter in the deep waters of the North American Basin.—In: A. L. GORDON (Ed.), Stud. oceanogr. **2**, 123–168. (WÜST Vol.) New York.
EKMAN, V. W. (1904): On dead water.—Sci. Res. North Polar Exped. **15**, 1893–1896. Part 20, 1–52.
——— (1905): On the influence of the earth's rotation on ocean-currents.—Ark. Math. Astron. Fys. **2**, 11, 1–53.
——— (1908): Die Zusammendrückbarkeit des Meerwassers nebst einigen Werten für Wasser und Quecksilber.—Cons. Perm. Int. Explor. Mer, Publ. Circonst. **43**, 1–47.
——— (1914): Der adiabatische Temperaturgradient im Meere.—Ann. Hydrogr. u. marit. Meteorol. **42**, 340–344.
——— (1923): Über Horizontalzirkulation bei winderzeugten Meeresströmungen.—Ark. Math. Astron. Fys. **17**, 1–74.
ELMENSORF, C. H. & B. C. HEEZEN (1957): Oceanographic information for engineering submarine cable systems.—Bell. Syst. Tech. J. **36**, 1047–1093.
EL-SABH, M. I. (1969): Seasonal hydrographic variations in the Suez Canal after the completion of the Aswan High Dam.—Kieler Meeresforsch. **25**, 1–18.
EL WAKEEL, S. K. & J. P. RILEY (1961): Chemical and mineralogical studies of deep-sea sediments.—Geochim. cosmochim. Acta **25**, 110–146.
EMERY, K. O. (1965): Geology of the continental margin off eastern United States.—In: W. F. WHITTHARD & R. BRADSHAW (Eds.), Submar. geol. geophys. 1–17. London.
——— (1968): Relict sediments on continental shelves of the world.—Amer. Assoc. Petrol. Geol. Bull. **52**, 445–464.
EMERY, K. O., E. UCHUPI, J. D. PHILLIPS, C. O. BOWIN, E. T. BUNCE & S. T. KNOTT (1970): Continental

rise off eastern North America.—Amer. Assoc. Petrol. Geol. Bull. **54**.
EMILIANI, C. (1966): Paleotemperature analysis of Caribbean cores P 6304-8 and P 6304-9 and generalized temperature curve for the past 425,000 years.—J. Geol. **74**, 109–126.
——(1970): Pleistocene paleotemperatures.—Science **168**, 822–825.
EPSTEIN, S., R. BUCHSBAUM, H. LOWENSTAM & H. C. UREY (1951): Carbonate water isotopic temperature scale.—Geol. Soc. Amer. Bull. **62**, 417–45.
ERMEL, H. (1966): Der deutsche Beitrag zur Neuherstellung der General Bathymetric Chart of the Oceans (GEBCO).—Dt. hydrogr. Z. **19**, 49–57. ·
ESSA (USA) (1969): Northern Europe, photograph taken by an American weather satellite.
EUCKEN, A. (1948): Zur Struktur des flüssigen Wassers.—Angew. Chem. **60**.
EWING, G. C. (Ed.) (1965): Oceanography from space.—Proc. Conf. on the Feasibility of Conducting Oceanographic Explorations from Aircraft, Manned Orbital and Lunar Lab. Woods Hole, Mass., 24–28 Aug. 1964, Ref. No. 65—10, 1–469, Woods Hole.
EWING, J. I., N. T. EDGAR & J. W. ANTOINE (1970): Structure of the Gulf of Mexico and Caribbean Sea.—In: A. E. MAXWELL (Ed.), The sea **4**, part II, 321–358. New York.
EWING, J. & M. EWING (1967): Sediment distribution on the mid-ocean ridges with respect to spreading of the sea floor.—Science **156**, 1590–1592.
EWING, J., M. EWING, T. AITKEN & J. W. LUDWIG (1968): North Pacific sediment layers measured by seismic profiling.—In: L. KNOPOFF et al. (Eds.), The crust and upper mantle of the Pacific area.—Amer. Geophys. Un., Geophys. Monogr. **12**, 147–173.
EWING, J. & M. EWING (1970): Seismic reflection.—In: A. E. MAXWELL (Ed.), The sea **4**, part I, 1–51. New York.
EWING, M., A. VINE & J. WORZEL (1946): Photography of the ocean bottom.—J. Opt. Soc. Amer. **36**, 307–321.
EWING, M. & J. L. WORZEL (1948): Long-range sound transmission.—Geol. Soc. Amer. Mem. **27**.
EYRIES, M. (1968): Maregraphes de grandes profondeurs.—Cah. océanogr. **20**, 355–368.
FARQUHARSON, W. I. (1970): Tides, tidal streams and currents in the Gulf of St. Lawrence. 1–145.—Atlantic Oceanogr. Lab., Bedford Inst.
FEDOROV, K. N. (1971): Formulas for converting the electrical conductivity of sea water into salinity with a digital temperature-salinity probe under average ocean conditions.—Oceanology **11**, 622–626.
FILIPPOV, D. M., S. YE. NAVROTSKAYA & Z. N. MATVEYEVA (1968): The depth of autumn-winter convection in the North Atlantic from mean long-term data.—Oceanology **8**, 19–28.
FILLOUX, J. H. (1969): Bourbon tube deep-sea tide gauges.—Proc. Symp. Tsunami Res., IUGG.
——(1971): Deep-sea tide observations from the northwestern Pacific.—Deep-Sea Res. **18**, 275–284.
FJIELDSTAD, J. E. (1929a): Ein Beitrag zur Theorie der winderzeugten Meeresströmungen.—Gerlands Beitr. Geophys. **23**, 1–237.
——(1929b): Contributions to the dynamics of free progressive tidal waves.—Norw. North Polar Exped. "Maud" 1918–25. Sci. Res. **4**, 1–79. Bergen.
——(1936): Results of the tidal observations.—Norw. North Polar Exp. "Maud" 1918–25. Sci. Res. **4**, 1–88. Bergen.
FLEISCHER, U., O. MEYER & H. SCHAAF (1970): Über den Aufbau der untermeerischen Tafelberge südlich der Azoren anhand eines gravimetrisch-magnetischen Nord-Süd-Profils über die Große Meteorbank.—"Meteor" Forsch.-Ergebn. (C), No. 3, 37–47. Berlin-Stuttgart.
FLEMING, R. H. (1940): The composition of plankton and units for reporting populations and production.—Sixth Pac. Sci. Congr. Calif. Proc. **3**.
FLIEGEL, M. & NOWROOZI, A. A. (1970): Tides and bottom currents off the coast of northern California.—Limnol. & Oceanogr. **15**, 615–624.
FLINT, R. F. (1971): Glacial and quaternary geology. 1–892.—New York.
FÖYN, E. (1969): The composition of sea water and the significance of the chemical components of the marine environment. An introduction.—In: R. LANGE (Ed.) Chem. oceanogr. 11–34. Oslo.
FOFONOFF, N. P. (1956): Some properties of sea water influencing the formation of Antarctic bottom water.—Deep-Sea Res. **4**, 32–35.
——(1962a): Physical properties of sea-water.—In: M. N. HILL (Ed.), The sea **1**, 3–30. New York.
——(1962b): Dynamics of ocean currents.—In: M. N. HILL (Ed.), The sea **1**, 323–395. New York.
——(1969): Role of NDBS in future. Natural variability studies of the North Atlantic.—In: U.S. Coast Guard Acad. New London, Conn. Proc. 1st USCG Nat. Data Buoy Syst. Scient. Advis. Meet., 50–61.
FOFONOFF, N. P. & R. B. MONTGOMERY (1955): The Equatorial Undercurrent in the light of the vorticity equation.—Tellus **7**, 518–521.
FOFONOFF, N. P. & S. TABATA (1958): Programm for oceanographic computations and data processing

on the electronic digital computer ALWAC III-E, DP-1 Oceanogr. Stat. Data Program.—Fish. Res. Board. Can. Ser. No. 25.
——— (1966): Variability of oceanographic conditions between ocean station P and Swiftsure Bank off the Pacific coast of Canada.—J. Fish. Res. Board Can. **23**, 825–869.
FOFONOFF, N. P. & F. WEBSTER (1971): Current measurements in the western Atlantic.—Philos. Trans. R. Soc. (A) **270**, 423–436. No. 1206.
FOLDVIK, A. & T. KVINGE (1969): Remarks on processes pertinent to the formation of bottom water in the Weddell Sea.—Geophys. Inst. Rep. 19. Bergen.
FONSELIUS, S. H. (1970a): On the stagnation and recent turnover of the water in the Baltic.—Tellus **22**, 533–543.
——— (1970b): On the water renewals in the Eastern Gotland Basin after World War II.—ICES C.M./C:8. 1–6. Copenhagen.
Food and Agriculture Organization (1963): Proceedings of the World Scientific Meeting on the biology of tunas and related species.—Fish. Rep. 6, 1–3. Rome.
FOREL, F. A. (1895): Le Léman: monographie limnologique, **1**: Géographie, hydrographie, géologie, climatologie, hydrologie.—In: F. ROUGE (Ed.), Lausanne, 543. **2**: Mécanique, chemie, thermique, optique, acoustique.—In: F. ROUGE (Ed.), Lausanne, 651.
FOX, CH. J. J. (1907, 1909): On the coefficients of absorption of the atmospheric gases in distilled water and sea water.—Cons. Int. Publ. Circonst. **41–44**.
FRANCIS, J. R. D., H. STOMMEL, H. G. FARMER & D. PARSON, Jr. (1953): Observations of turbulent mixing processes in a tidal estuary.—WHOI Techn. Rep. 53–22. (Unpubl. manuscr.)
FREIESLEBEN, H. C. (1968): Position finding at sea.—In: H. BARNES (Ed.), Oceanogr. Mar. Biol. Ann. Rev. **6**, 47–81. London.
FRIEDRICH, H. (1967): Numerical computations of the wind induced mass transport in a stratified ocean.—Mitt. Inst. Meereskde. Univ. Hamburg, **10**, 134–139.
FRIEDRICH, H. J. (1970): Preliminary results from a numerical multilayer model for the circulation in the North Atlantic.—Dt. hydrogr. Z. **23**, 145–164.
FUGLISTER, F. C. (1951): Annual variations in the current speeds in the Gulf Stream system.—J. Mar. Res. **10**, 119–127.
——— (1960): Atlantic Ocean atlas of temperature and salinity profiles and data from the International Geophysical Year of 1957–1958.—Woods Hole, Mass.
——— (1963): Gulf Stream 60.—In: M. SEARS (Ed.), Progr. oceanogr. **1**, 263–373. Oxford.
——— (1967): Cyclonic eddies formed from meanders of the Gulf Stream.—Trans. Amer. geophys. Un. **48**, 123. (Abstr. only.)
——— (1971): Cyclonic rings formed by the Gulf Stream 1965–66.—In: A. L. GORDON (Ed.), Stud. oceanogr. **1**, 137–168. (WÜST Vol.) New York.
FUGLISTER, F. C. & A. D. VOORHIS (1965): A new method of tracking the Gulf Stream. Limnol. & Oceanogr. **10**, Suppl., 115–124. (REDFIELD Vol.).
FUGLISTER, F. C. & L. V. WORTHINGTON (1951): Some results of a multiple ship survey of the Gulf Stream.—Tellus **3**, 1–14.
FUJINO, K. (1967): Electrical properties of sea ice.—In: H. OURA (Ed.), Physics of snow and ice. **1**, 633–648.
FUNNELL, B. M. & W. R. RIEDEL (Eds.) (1971): Micropaleontology of oceans.—Proc. Symp. Cambridge 1967, 52 contributions. Cambridge.
GAST, P. R. (1965): In: S. L. VALLEY (Ed.), Handbook geophys. and space environments. 16.1.–16.10. New York.
GAYE, J. & F. WALTHER (1934): Der "Seebär" vom 19. Aug. 1932 in der Deutschen Bucht der Nordsee.—Ann. Hydrogr. u. marit. Meteorol. **62**, 317–322.
GENTILLI, J. (1967): Tropical cyclones.—In: R. W. FAIRBRIDGE (Ed.), Encycl. atmosph. sci., 1027–1030. New York.
GERSTNER, F. (1802): Theorie der Wellen usw.—Abh. Kgl. Böhmische Ges. Wiss. Prague.
GIERMANN, G. & M. PFANNENSTIEL (1961): Carte topographique 1:100 000.—Mus. Océanogr. de Monaco (Ed.). Annexe Bull. Inst. océanogr. Monaco.
GIESEL, W. & E. SEIBOLD (1968): Sedimentechogramme vom iberomarokkanischen Kontinentalrand.—"Meteor" Forsch.-Ergebn. (C) No. 1, 53–75. Berlin–Stuttgart.
GIESKES, M. T. M. (1968): Some investigations into the sensibility of the membrane salinometer for various ions.—Kieler Meeresforsch. **24**, 18–26.
GIESKES, J. M. & K. GRASSHOFF (1969): A study of the variability in the hydrodynamical factors in the Baltic Sea on the basis of two anchor stations September 1967 and May 1968.—Kieler Meeresforsch. **25**, 105–132.

GILL, A. E. (1971): The equatorial current in a homogeneous ocean.—Deep-Sea Res. 18, 421-431.
GILL, E. E. & K. BRYAN (1971): Effects of geometry on the circulation of a three-dimensional southern-hemisphere ocean model.—Deep-Sea Res. 18, 685-721.
GILLBRICHT, M. (1952): Untersuchungen zur Produktionsbiologie des Planktons in der Kieler Bucht.—Kieler Meeresforsch. 9, 51-61.
GOLDBERG, E. D. (1965): Minor elements in sea water.—In: J. P. RILEY & G. SKIRROW (Eds.), Chem. oceanogr. 1, 163-196. London.
GOLDBERG, E. D. & M. KOIDE (1962): Geochronological studies of deep-sea sediments by the ionium/thorium method.—Geochim. cosmochim. Acta 26, 417-450.
GOLDSBROUGH, G. R. (1915): The dynamical theory of the tides in a zonal basin. Proc. London Math. Soc. 14, 207-229.
——(1928): The tides in oceans on a rotating globe.—Proc. R. Soc. London (A) 117, 692-718.
——(1933): Ocean currents produced by evaporation and precipitation.—Proc. R. Soc. (A) 141, 512-517.
GOLDSCHMIDT, V. M. (1933): Grundlagen der quantitativen Geochemie.—Fortschr. Miner., Kristallogr., Petrogr. 17, 116.
——(1937): The principles of distribution of chemical elements in minerals and rocks.—J. Chem. Soc. London. 1-655.
GORBUNOVA, Z. N. (1966): Clay mineral distribution in Indian Ocean sediments.—Oceanology 6, 215-221.
GORDON, A. L. (1966): Potential temperature, oxygen and circulation of bottom water in the Southern Sea.—Deep-Sea Res. 13, 1125-1138.
——(1967): Circulation of the Caribbean Sea.—J. Geophys. Res. 72, 6207-6223.
——(1971): Antarctic circulation.—Trans. Amer. Geophys. Un. 52, IUGG 230-232.
——(1972a): Oceanography of Antarctic waters.—In: J. L. REID (Ed.), Antarctic oceanology 1, 167-203. 15. Antarctic Res. Ser. Washington.
——(1972b): Spreading of Antarctic bottom water II.—In: A. L. GORDON (Ed.), Stud. oceanogr. 2, 1-17. (WÜST Vol.) New York.
GORDON, A. L. & R. D. GOLDBERG (1970): Antarctic map folio series, folio 13.—Amer. Geograph. Soc. New York.
GRACE, S. F. (1930): The semi-diurnal lunar tidal motion of the Red Sea.—Month. Not. (Geophys. Suppl.) 2. London.
——(1936): Friction in the tidal currents of the Bristol Channel.—Month. Not. (Geophys. Suppl.) 3, 388-395. London.
GRAF, A. (1958): Das Seegravimeter.—Z. Instrumentenkde. 66, 151-161.
GRANT, H. L., A. MOILLET & W. M. VOGEL (1968): Some observations of the occurrence of turbulence in and above the thermocline.—J. Fluid Mech. 34, 443-448.
GRASSHOFF, K. (1965): On the automatic determination of phosphate, silicate and fluoride in sea water.—ICES, C.M. No. 129. Rome.
——(1967): Results and possibilities of automated analysis of nutrients in seawater.—Techn. Symp. Autom. Analyst. Chem. 573-579. New York.
——(1968): Chemische Methoden.—In: C. SCHLIEPER (Ed.), Methoden der meeresbiologischen Forschung. 13-31. Jena.
——(1969a): Über ein Gerät zur gleichzeitigen Bestimmung von sechs chemischen Komponenten aus dem Meerwasser mit analoger und digitaler Ausgabe.—Ber. Dt. Wiss. Komm. Meeresforsch. 20, 155-164.
——(1969b): Zur Chemie des Roten Meeres und des Inneren Golfs von Aden nach Beobachtungen von F.S. "Meteor" während der Indische-Ozean-Expedition 1964-65.—"Meteor" Forsch.-Ergebn. (A) No. 6, 1-76. Berlin-Stuttgart.
——(1971): Hydrographic and chemical data from the Black Sea.—In: P. BREWER (Ed.), Techn. Tep., Ref. No. 71-65. Woods Hole, Mass.
GRASSHOFF, K. & E. HOLLAN (1972): Personal communication.
GREEN, C. K. (1946): Seismic sea wave of April 1, 1946, as recorded on the tide gages.—Trans. Amer. Geophys. Un. 27.
GREENSPAN, M. & C. E. TSCHIEGG (1957): Sing-around ultrasonic velocimeter for liquids.—Rev. Sci. Instrum. 28, 897-901.
GRIFFIN, J. J., H. WINDOM & E. D. GOLDBERG (1968): The distribution of clay minerals in the world ocean.—Deep-Sea Res. 15, 233-459.
GROLL, M. (1912): Tiefenkarten der Ozeane.—Veröff. Inst. Meereskde. Berlin. N.F. (A) 2, 1-78.

GUILCHER, A. (1958): Coastal and submarine geology, 1–274.—London.
GUNTHER, E. R. (1936): A report on oceanographical investigations in the Peru coastal current.—"Discovery" Rep. **13**, 123–298. Cambridge.
GUSTAFSON, T. & B. KULLENBERG (1936): Untersuchungen von Trägheitsströmungen in der Ostee.—Svens. Hydrogr. Biol. Skr. Ny. Ser. Hydrogr. **13**, 1–28.
GUTENBERG, B. (1941): Changes in sea level, postglacial uplift, and mobility on the earth's interior.—Bull. Geol. Soc. Amer. **52**.
GUTENBERG, B. & C. F. RICHTER (1954): The seismicity of earth and associated phenomena. 2nd Ed.— Princeton.
HAGEN, G. (1839): Über die Bewegung des Wassers in engen cylindrischen Rohren.—Poggendorffs Ann. **46**, 423–442.
HAMON, B. V. (1965): The East Australian Current, 1960–1964.—Deep-Sea Res. **12**, 899–921.
——— (1970): Western boundary currents in the South Pacific.—In: W. S. WOOSTER (Ed.), Scient. explor. South Pacific. 51–59. Washington.
HANSEN, D. V. (1967): A sequence of the Gulf Stream meanders observed by thermal front tracking.—Trans. Amer. geophys. Un. **48**, 123.
——— (1970): Gulf Stream meanders between Cape Hatteras and the Grand Banks.—Deep-Sea Res. **17**, 495–511.
HANSEN, D. V. & M. RATTRAY, JR. (1966): New dimensions in estuary classification.—Limnol. & oceanogr. **11**, 319–326.
HANSEN, W. (1948): Die Ermittlung der Gezeiten beliebig gestalteter Meeresgebiete mit Hilfe des Randwertverfahrens.—Dt. hydrogr. Z. **1**, 157–163.
——— (1952): Einige Bemerkungen zum Golfstromproblem.—Dt. hydrogr. Z. **5**, 80–94.
——— (1956): Theorie zur Errechnung des Wasserstandes und der Strömungen in Randmeeren nebst Anwendungen.—Tellus **8**, 287–300.
——— (1966): Die Reproduktion der Bewegungsvorgänge im Meer mit Hilfe hydrodynamisch-numerischer Verfahren.—Mitt. Inst. Meereskde. Univ. Hamburg 5.
HARRIS, R. A. (1904): Manual of tides. Part IVb.—U.S. Coast Geod. Surv. Rep., App. 5, 313–400. Washington.
HART, T. J. & R. I. CURRIE (1960): The Benguela Current.—"Discovery" Rep. **31**, 123–298.
HARTMANN, M., H. LANGE, E. SEIBOLD & E. WALGER (1971): Oberflächensedimente im Persischen Golf und Golf von Oman.—"Meteor" Forsch.-Ergebn. (C) No. 4, 1–76. Berlin–Stuttgart.
HARVEY, H. W. (1927): Biological chemistry and physics of sea water. 1–223.—Cambridge.
——— (1937): The supply of iron to diatoms.—J. Mar. Biol. Assoc. U.K. **22**, 205–219.
HASSE, L. (1968): Zur Bestimmung der vertikalen Transporte von Impuls und fühlbarer Wärme in der wassernahen Luftschicht über See.—Hamburger Geophys. Einzelschr. 11, 1–76. Engl. Transl. Dept. Oceanogr., Oregon State Univ. 1970.
——— (1971): The sea surface temperature deviation and the heat flow at the sea-air interface.—Boundary-Layer Meteorol. **1**, 368–379.
HASSELMANN, K. (1960): Grundgleichungen der Seegangsvoraussage.—Schiffstechnik **7**, 191–195.
——— (1972): The energy balance of wind waves and the remote sensing problem.—Sea surface topography from space **2**, NOAA Techn. Rep. ERL 228—AOML 7-2. Washington, D.C.
HASSELMANN, K., T. P. BARNETT, E. BOUWS, H. CARLSON, D. E. CARTWRIGHT, K. ENKE, J. A. EWING, H. GIENAPP, D. E. HASSELMANN, P. KRUSEMAN, A. MEERBURG, P. MÜLLER, D. J. OLBERS, K. RICHTER, W. SELL & H. WALDEN (1973): Measurements of wind waves growth and swell decay during the joint North Sea Wave Project (JONSWAP).—Dt. hydrogr. Z. Erg.-H. (A) **12**, 1–95.
HAYES, D. E. (1966): A geophysical investigation of the Peru–Chile Trench.—Mar. Geol. **4**, 309–351.
HEAPS, N. S. (1967): Storm surges.—In: H. BARNES (Ed.), Oceanogr. Mar. Biol. Ann. Rev. **5**, 11–47. London.
HEBERLEIN, H. (1968): Historische Entwicklung des Tauchens als Beitrag zur maritimen Forschung und Wissenschaft.—II. Marinemed.-wiss. Symp., 7–42. Kiel.
HECK, N. H. (1947): List of seismic sea waves.—Bull. seism. Soc. Amer. **37**.
HECK, N. H. & J. H. SERVICE (1924): Velocity of sound in sea water. U.S. Coast Geod. Surv. Spec. Publ. 108.
HECKER, O. (1903): Bestimmung der Schwerkraft auf dem Atlantischen Ozean.—Veröff. Königl. Preuβ. Geodät. Inst., No. 11.
HEEZEN, B. C. (1967): Sub-oceanic ridges.—In: S. K. RUNCORN (Ed.), Int. dict. geophys. 1469–1475.
HEEZEN, B. C., D. B. ERICSON & M. EWING (1954): Reconnaissance survey of the abyssal plain south of Newfoundland.—Deep-Sea Res. **2**, 122–133.

HEEZEN, B. C. & M. EWING (1952): Turbidity currents and submarine slumps and the Grand Banks earthquake.—Amer. J. Sci. **250**, 849–873.
HEEZEN, B. C. & C. HOLLISTER (1964): Deep-sea current evidence from abyssal sediments.—Mar. Geol. **1**, 141–174.
HEEZEN, B. C., M. THARP & M. EWING (1959): The floor of the oceans, I. The North Atlantic.—Geol. Soc. Amer. Spec. Pap. **65**, 1–122.
HEEZEN, B. C. & L. WILSON (1968): Submarine geomorphology.—In: R. W. FAIRBRIDGE (Ed.), Encycl. geomorph. **3**, 1079–1097. New York.
HEIRTZLER, J. R. (1970): Magnetic anomalies measured at sea.—In: A. E. MAXWELL (Ed.), The sea **4**, part II. 85–128. New York.
HEIRTZLER, J. R., G. O. DICKSON, E. M. HERRON, W. C. PITMAN & X. LE PICHON (1968): Marine magnetic anomalies, geomagnetic field reversals and motions of the ocean floor and continents.—J. Geophys. Res. **73**, 2119–2136.
HEIRTZLER, J. R., X. LE PICHON & J. G. BARON (1966): Magnetic anomalies over the Reykjanes Ridge.—Deep-Sea Res. **13**, 427–443.
HELA, I. & T. LAEVASTU (1970): Fisheries oceanography. 1–238.—London.
HELLAND-HANSEN, B. (1916): Nogen hydrografiske metoder.—Scand. Naturforsk. Möte. Oslo.
HELLAND-HANSEN, B. & F. NANSEN (1909): The Norwegian Sea.—Rep. Norw. Fish. Mar. Invest. **2**, part 1, 2, 1–390 and suppl.
HELMHOLTZ, H. v. (1868): Über diskontinuierliche Flüssigkeitsbewegungen.—Monatsber. Kgl. Preuss. Akad. Wiss. **23**, 215–228, Berlin.
HENDERSHOTT, M. C. (1966): The numerical integration of Laplace's tidal equations in idealized ocean basins.—Proc. Symp. math.-hydrodynam. invest. of phys. processes in the sea. Moscow.
HERRING, P. J. & M. R. CLARKE (Eds.) (1971): Deep oceans. 1–320.—London.
HERRMANN, G., H. HAUSER & H. J. RIEDEL (1959): Kritische Betrachtungen zur Beurteilung der radioaktiven Verseuchung auf Grund der Gesamt-Aktivitätsmessungen.—Nukleonik **1**.
HERSEY, J. B. (1963): Continuous reflection profiling.—In: M. N. HILL (Ed.), The sea **3**, 47–72. New York.
HERSEY, J. B. (Ed.) (1967): Deep-sea photography. 1–310.—Baltimore.
HERZEN, R. P. v. & W. H. K. LEE (1969): Heat flow in oceanic regions.—In: W. H. HART (Ed.), The earth's crust and upper mantle. 88–95. Washington.
HERZEN, R. P. v. & S. UYEDA (1963): Heat flow through the eastern Pacific Ocean floor.—J. Geophys. Res. **68**, 4219–4250.
HESS, H. H. (1946): Drowned ancient islands of the Pacific basin.—Amer. J. Sci. **244**, 772–791.
——— (1962): History of ocean basins.—In: A. E. J. ENGEL et al. (Eds.), Petrol. stud., 599–620. Geol. Soc. Amer.
——— (1965): Mid-oceanic ridges and tectonics of the sea-floor.—In: W. F. WITTHARD & BRADSHAW (Eds.), Submar. geol. geophys. 317–332. London.
HESSELBERG, T. H. (1918): Über die Stabilitätsverhältnisse bei vertikalen Verschiebungen in der Atmosphäre und im Meere.—Ann. Hydrogr. u. marit. Meteorol. **46**, 118–129.
HESSELBERG, T. H. & H. U. SVERDRUP (1914–1915): Die Stabilitätsverhältnisse des Seewassers bei vertikalen Verschiebungen.—Bergens Mus. Aarb., Nr. 14–15.
HIDAKA, K. (1951): Drift currents in an enclosed ocean. Part III.—Tokyo Univ. Geophys. Notes **4**, 1–19.
——— (1954): A contribution to the theory of upwelling and coastal currents.—Trans. Amer. geophys. Un. **35**, 431–444.
HIGASHI, K., K. NAKAMURA & R. HARA (1931): The specific gravities and the vapor pressure of the concentrated sea water at 0–175°C.—J. Soc. Chem. Ind. Jap. **34**, 166–172.
HINKELMANN, H. (1957): Gerät zur Schnellregistrierung in der Ozeanographie.—Z. angew. Phys. **9**, 500–513.
——— (1958): Ein Verfahren zur elektrodenlosen Messung der elektrischen Leitfähigkeit von Elektrolyten.—Z. angew. Phys. **10**, 500–503.
HINZ, K. (1969): The Great Meteor Seamount. Results of seismic reflection measurements with a pneumatic sound source, and their geological interpretation.—"Meteor" Forsch.-Ergebn. (C) No. 2, 63–77. Berlin–Stuttgart.
HINZ, K., F.-C. KÖGLER, J. RICHTER & E. SEIBOLD (1971): Reflexionsseismische Untersuchungen mit einer pneumatischen Schallquelle und einem Sedimentecholot in der westlichen Ostsee.—Meyniana **21**, 17–24.
HINZPETER, H. (1962): Messungen der Streufunktion und der Polarisation des Meerwassers.—Kieler Meeresforsch. **23**, 36–41.

HINZPETER, H., E. CLAUSS & P. LOBEMEIER (1969): Maritime Meteorologie.—In: H. CLOSS, G. DIETRICH, G. HEMPEL, W. SCHOTT & E. SEIBOLD, "Atlantische Kuppenfahrten 1967" mit dem Forschungsschiff "Meteor." Reisebericht.—"Meteor" Forsch.-Ergebn. (A) No. 5, 1-71. Berlin–Stuttgart.
HISARD, P., J. MERLE & B. VOITURIEZ (1970): The equatorial undercurrent at 170°E in March and April 1967.—J. Mar. Res. **28**, 281-303.
HJULSTRÖM, F. (1935): Studies of the morphological activity of rivers as illustrated by the river Fyris.—Bull. Geol. Inst. Uppsala **25**, 221-527.
HOEBER, H. (1966): Tagesgänge der Luft- und Wassertemperatur im äquatorialen Atlantik.—Naturwiss. **53**, 474-475.
——— (1969): Wind-, Temperatur- und Feuchteprofile in der wassernahen Luftschicht über dem äquatorialen Atlantik.—"Meteor" Forsch.-Ergebn. (B) No. 3, 1-26. Berlin–Stuttgart.
HORN, W. (1948): Über die Darstellung der Gezeiten als Funktion der Zeit.—Dt. hydrogr. Z. **1**, 124-140.
——— (1952): Gezeiten des Meeres.—In: LANDOLT-BÖRNSTEIN **3**, 504-516. Berlin.
——— (1960): Some recent approaches to tidal problems.—Int. Hydrogr. Rev. **37**, 65-84.
HORN, W. F. (1971): Die zeitliche Veränderlichkeit der Temperatur der ozeanischen Deckschicht im Gebiet der Großen Meteorbank.—"Meteor" Forsch.-Ergebn. (A) No. 9, 47-57. Berlin–Stuttgart.
HORN, W. F., W. HUSSELS & J. MEINCKE (1971): Schichtungs- und Strömungsmessungen im Bereich der Großen Meteorbank.—"Meteor" Forsch.-Ergebn. (A) No. 9, 31—46. Berlin–Stuttgart.
HORNE, R. A. (1969): Marine chemistry. 1-568.—New York.
HORNE, R. A. & D. S. JOHNSON (1966): The viscosity of compressed seawater.—J. Geophys. Res. **71**, 5275-5277.
HOUGH, S. S. (1897, 1898): On the application of harmonic analysis on the dynamical theory of tides.—Philos. Trans. R. Soc. (A) **189, 191**.
ICHIYE, T. (1964): An essay on the equatorial current system.—In: K. YOSHIDA (Ed.), Stud. oceanogr., 38–46. Tokyo.
——— (1970): Contributions to dynamics of the Antarctic Circumpolar Current (A.C.C.). (I. Zonal transport.)—J. oceanogr. Soc. Jap. **26**, 340-353.
ICHIYE, T. & H. SUDO (1971): Saline deep water in the Caribbean Sea and in the Gulf of Mexico.—Texas A&M, Oceanogr. Dept., 1-27.
IGY, WDC-A, NODC (1961, 1963, 1966): Oceanographic vessels of the world, 3 vols.—Washington, D.C.
INMAN, D. L. (1963): Physical properties and mechanics of sedimentation.—In: F. P. SHEPARD (Ed.), Submar. geol. 2nd ed., 101-151. New York.
Institut für Meereskunde (1970): Plankton forms in the ocean (Photograph).
Instrument Society of America (1962, 1963, 1965, 1968):—Mar. Sci. Instrument. **1-4.** New York.
Interagency Committee on Oceanography (1965): Undersea vehicles for oceanography.—Fed. Connc. Sci. Techn., Washington, D.C.
International Commission for the Northwest Atlantic Fisheries (1968): Environmental surveys—NORWESTLANT 1-3, 1963, Part I, text (and figs.); part II, atlas.—Spec. Publ. **7**, 1-255. Dartmouth.
International Hydrographic Bureau (1930-1953): Harmonic constants.—Spec. Publ. No. 26. Monaco.
——— (1953): Limits of oceans and seas. 3rd ed.—Spec. Publ. No. 53. Monte Carlo.
——— (1971): Nomenclature of ocean bottom features.—Int. Hydrogr. Rev. **48**, 203-208.
ISELIN, C. O'D. (1936): A study of the circulation of the western North Atlantic.—Pap. phys. oceanogr. meteorol. Woods Hole **4**, 4, 1-101.
IVANOFF, A. (1960): Au sujet des perfectionnements apportés à l'étude de la lumière diffusée par des échantillons d'eau de mer, et des resultats ainsi obtenus au large de Monaco.—Compt. Rend. Acad. Sci. Paris **250**, 736-738.
IVANOFF, A. & A. MOREL (1964): Au sujet de l'indicatrice de diffusion des eaux de mer.—Compt. Rend. Acad. Sci. Paris **258**, 2873-2874.
JACOBS, W. C. (1942): On the energy exchange between sea and atmosphere.—J. Mar. Res. **5**, 37-66.
——— (1951): The energy exchange between the sea and atmosphere and some of its consequences.—Bull. Scripps Inst. Oceanogr. **6**, 27-122.
JACOBSEN, J. P. (1913): Beitrag zur Hydrographie der dänischen Gewässer.—Medd. Komm. Havunders. Ser. Hydr. **2**, 1-94.
——— (1927): Eine graphische Bestimmung des Vermischungskoeffizienten im Meer.—Gerlands Beitr. Geophys. **16**, 404-412.
——— (1929): Contribution to the hydrography of the North Atlantic.—"Dana" Exped. 1921-22. Copenhagen.
JACOBSEN, J. P. & M. KNUDSEN (1940): Urnormal 1937 or primary standard sea water 1937.—Assn.

d'Oceanogr. Phys., Un. Geod. Geophys. Internat., Publ. sci. **7**, 1-38.
JACOBSEN, J. & H. THOMSEN (1934): Periodical variations of temperature and salinity in the Straits of Gibraltar.—John Johnson Mem. **13**. Liverpool.
JANSSON, B. O. (1972): Ecosystem approach to the Baltic problem.—Swed. Nat. Sci. Res. Counc. **16**, 1-82. Stockholm.
Japanese Oceanographic Data Center (1967, 1968): CSK Atlas, summer 1965 and winter 1965-66. 1-31 and 1-44.—Tokyo.
JEFFREYS, H. (1921): Tidal friction in shallow water.—Philos. Trans. R. Soc. (A) **18**, 1-84.
——(1924): On the formation of water waves by wind.—Proc. Soc. (A) **107**, 189-206.
JENSEN, A. S. (1939): Concerning a change of climate during recent decades in the arctic and subarctic regions, from Greenland in the west to Eurasia in the east, and contemporary biological and geographical changes.—Kgl. Danske Vid. Selsk., Biol. Medd. **14**, 8. Copenhagen.
JERLOV, N. G. (1951): Optical studies of ocean water.—Rep. Swed. Deep-Sea Exped. **3**, 1-59.
——(1961): Optical measurements in the eastern North Atlantic.—Medd. Oceanogr. Inst. Göteborg (B) **8**, 1-40.
——(1964): Optical classification of ocean water.—In: Physical aspects of light in sea. 45-49. Honolulu.
——(1968): Optical oceanography. 1-194.—Amsterdam.
JOHANNSEN, K., H. ECKHARDT, H. KOEHLER & H. CHR. PAULSSEN VON BECK (1965): Forschungsschiff "Meteor."—Schiff u. Hafen **17**, 3-23.
"JOIDES" (1969): Initial reports of the deep-sea drilling project **1**, 1-672.—Washington.
JONES, P. G. W. (1971): The southern Benguela Current region in February, 1966. Part I. Chemical observations with particular reference to upwelling.—Deep-Sea Res. **18**, 193-208.
JOSEPH, J. (1948): Meereskundliche Meßgeräte.—Naturforsch. u. Med. in Deutschland 1939-46. **18**, Teil 2. Wiesbaden.
——(1952): Meeresoptik.—In: LANDOLT-BÖRNSTEIN, Zahlenwerte und Funktionen. **3**, 441-459. Berlin.
——(1957): Extinction measurements.—UNESCO and Jap. Soc. Prom. Sci. Tokyo.
——(1959): Über die vertikalen Temperatur- und Trübungsregistrierungen in einer 500 m mächtigen Deckschicht des nördlichen nordatlantischen Ozeans.—Dt. hydrogr. Z. Erg.-H. (B) **3**, 48-55.
——(1962): Der "Delphin," ein Meßgerät zur Untersuchung von oberflächennahen Temperaturschichtungen im Meere.—Dt. hydrogr. Z. **15**, 15-23.
——(1967): Current measurements during the international Iceland-Faeroe Ridge expedition, 30 May to 18 June, 1960.—Rapp. P.-v. Explor. Mer **157**, 157-172.
JUNGE, C. E. (1963): Air chemistry and radioactivity. 1-382.—New York.
KÄNDLER, R. (1971): Untersuchungen über die Abwasserbelastung der Untertrave.—Kieler Meeresforsch. **27**, 20-27.
KALLE, K. (1945): Der Stoffhaushalt des Meeres. 2. Aufl. 1-232.—Leipzig.
——(1950): Der Mechanismus der ozeanischen und des kontinentalen Produktionsvorganges.—Dt. hydrogr. Z. **3**, 62-69.
——(1966): The problem of the Gelbstoff in the sea.—In: H. BARNES (Ed.), Oceanogr. Mar. Biol. Ann. Rev. **4**, 91-104. London.
KANWISHER, J. (1963): On the exchange of gases between the atmosphere and the sea.—Deep-Sea Res. **10**, 195-207.
KATZ, E. J. (1969): Further study of a front in the Sargasso Sea.—Tellus **21**, 259-269.
——(1973): Profile of an isopycnal surface in the main thermocline of the Sargasso Sea.—J. phys. oceanogr. **3**, 448-457.
KAWAI, H. (1969): Statistical estimation of isotherms indicative of the Kuroshio axis.—Deep-Sea Res. Suppl. to **16**, 109-115. (*Fuglister* Vol.)
KELVIN, LORD (1871): Influence of wind at capillarity in waves in water supposed frictionless.—Math. Phys. Pap. **4**, 76-85, London.
KENDALL, T. R. (1970): The Pacific Equatorial Countercurrent. 1-19, 48 figs.—Laguna Beach, Calif.
KEUNECKE, K. H. (1973): On the observation of internal tides at the continental slope off the coast of Norway.—"Meteor" Forsch.-Ergebn. (A) No. **12**, 24-36. Berlin-Stuttgart.
KINNE, O. & H.-P. BULNHEIM (Eds.) (1973): Int. Helgoland Symp. Man in the sea—in-situ studies of life in oceans and coastal waters.—Helgoländ. wiss. Meeresunters. **24**, 1-535.
KING, C. A. M. (1959): Beaches and coasts. 1-403.—London.
KINSMAN, B. (1965): Wind waves; their generation and propagation on the ocean surface. 1-676.—Englewood Cliffs.

KITAIGORODSKII, S. A. (1970): Small-scale interaction between the atmosphere and the ocean. 1–283. (In Russian)—Moscow.
KNAUSS, J. A. (1960): Measurements of the Cromwell Current.—Deep-Sea Res. **6,** 265–286.
——(1961): The structure of the Pacific Equatorial Countercurrent.—J. Geophys. Res. **66,** 143–155.
——(1963): Equatorial current system.—In: M. N. HILL (Ed.), The sea **2,** 235–252. New York.
——(1966): Further measurements and observations of the Cromwell Current.—J. Mar. Res. **24,** 205–240.
——(1969): A note on the transport of the Gulf Stream.—Deep-Sea Res., Suppl. to **16,** 117–123. (*Fuglister* Vol.)
KNOPOFF, L. (1969): The upper mantle of the earth. Inhomogeneities in the earth's interior are related to large motions observed near the surface.—Science **163,** 1277–1287.
KNOX, G. A. (1970): Biological oceanography of the South Pacific.—In: W. S. WOOSTER (Ed.), Scient. explor. South Pacific. 155–182. Washington.
KNUDSEN, M. (Ed.) (1901): Hydrographical tables according to the measurements of CARL FORCH, P. JACOBSEN, MARTIN KNUDSEN and S. P. L. SORENSEN.—G.E.C. Gad. Copenhagen, London.
——(1902): Berichte über die Konstantenbestimmungen zur Aufstellung der hydrographischen Tabellen von CARL FORCH, MARTIN KNUDSEN und S. P. L. SORENSEN.—K. Danske Vidensk. Selsk. Skr., 6. Raekke, Naturvidensk. math. Afd. **12,** 1–151.
——(1903): Gefrierpunkttabelle für Meerwasser.—Publ. Circ. Cons. Explor. Mer. **5,** 11–13.
KOCZY, F. F. (1956): The specific alkalinity.—Deep-Sea Res. **3,** 279–288.
——(1963): Age determination in sediments by natural radioactivity.—In: M. N. HILL (Ed.), The sea **3,** 816–831. New York.
——(1965): Remarks on age determination in deep-sea sediments.—In: M. SEARS (Ed.), Progr. oceanogr. **3,** 155–171. Oxford.
KÖPPEN, V. & A. WEGENER (1924): Die Klimate der geologischen Vorzeit. 1–255.—Berlin.
KÖPPEN, W. (1936): Das geographische System der Klimate.—In: W. KÖPPEN & R. GEIGER (Eds.), Handbuch der Klimatologie **1,** C. Berlin.
KOHL, J. G. (1868): Geschichte des Golfstromes und seiner Erforschung.—Bremen.
KOOPMANN, G. (1953): Entstehung und Verbreitung von Divergenzen in der oberflächennahen Wasserbewegung der antarktischen Gewässer.—Dt. hydrogr. Z. Erg.-H. **2,** 1–36.
KORT, V. G. (1964): Antarctic oceanography.—In: H. ODISHAW (Ed.), Solid earth and interface phenomena **2,** 309–333. Cambridge, Mass.
——(1972): On the structure of the deep currents.—In: A. L. GORDON (Ed.), Stud. phys. oceanogr. **2,** 115–121. (WÜST Vol.) New York.
KORTEWEG, D. J. & G. DE VRIES (1895): On the change of form of long waves advancing in a rectangular canal, and on a new type of long stationary waves.—Phil. Mag. ser. 5, **39,** 422–443.
KOSHLYAKOV, M. N. & V. G. NEYMAN (1965): Some results of measurements and calculations of zonal currents in the Pacific equatorial region.—Oceanology **5,** 37–49.
KOSKE, P. H. (1964): Über ein potentiometrisches Verfahren zur Bestimmung von Chloridkonzentrationen in Meerwasser.—Kieler Meeresforsch. **20,** 138–147.
KOSKE, P. H. & U. RABSCH (1972): Hydrographische Verhältnisse im Persischen Golf auf Grund von Beobachtungen von F. S. "Meteor" im Frühjahr 1965.—"Meteor" Forsch.-Ergebn. (A) No. **11,** 58–73. Berlin-Stuttgart.
KOSLOWSKI, G. (1969): Die WMO-Eisnomenklatur.—Dt. hydrogr. Z. **22,** 256–267.
KOYAMA, T. & T. G. THOMPSON (1959): Organic acids in sea water.—Repr. Int. Oceanogr. **925,** New York.
KRAUS, E. B. (1972): Atmosphere-ocean interaction. 1–275.—Oxford.
KRAUSE, G. (1963): Eine Methode zur Messung optischer Eigenschaften des Meerwassers in großen Meerestiefen.—Kieler Meeresforsch. **19,** 175–181.
——(1968): Struktur und Verteilung des Wassers aus dem Roten Meer im Nordwesten des Indischen Ozeans.—"Meteor" Forsch.-Ergebn. (A) No. **4,** 77–100. Berlin-Stuttgart.
——(1973). Messung von Stromprofilen in Flachwasserwellen.—Küste **22,** 39–645.
KRAUSE, G. & J. ZIEGENBEIN (1966): Struktur des heißen salzreichen Tiefenwassers im zentralen Roten Meer.—"Meteor" Forsch.-Ergebn. (A) No. **1,** 53–58. Berlin-Stuttgart.
KRAUSS, W. (1955): Zum System der Meeresströmungen in den höheren Breiten.—Dt. hydrogr. Z. **8,** 102–111.
——(1958): Untersuchungen über die mittleren hydrographischen Verhältnisse an der Meeresoberfläche des nördlichen Nordatlantischen Ozeans.—Wiss. Ergebn. der Deutsch. Atlant. Exped. "Meteor," 3. Lf. 251–410. Berlin.

—— (1960): Hydrographische Messungen mit einem Beobachtungsmast in der Ostsee.—Kieler Meeresforsch. **16**, 13–27.
—— (1962): Temperatur, Salzgehalt und Dichte an der Oberfläche des Atlantischen Ozeans.—Wiss. Ergebn. Deutsch. Atlant. Exped. "Meteor" 1925–1927, **5**, 4. Lfg., 411–515. Berlin.
—— (1965): Theorie des Triftstromes und der virtuellen Reibung im Meer. Teil 1.—Dt. hydrogr. Z. **18**, 193–210.
—— (1966): Methoden und Ergebnisse der Theoretischen Ozeanographie II, Interne Wellen, 1–248. Berlin–Stuttgart.
—— (1973): Methods and results of theoretical oceanography I—Dynamics of the homogeneous and the quasihomogeneous ocean. 1–302.—Berlin–Stuttgart.
KRAUSS, W., J. KIELMANN & P. H. KOSKE (1973): Observations on scattering layers and thermoclines in the Baltic Sea.—Kieler Meeresforsch. **29**, 85–89.
KRAUSS, W. & L. MAGAARD (1962): Zum System der Eigenschwingungen der Ostsee.—Kieler Meeresforsch. **18**, 184–186.
KREMLING, K. (1970): Untersuchungen über die chemische Zusammensetzung des Meerwassers der Ostsee. II. Frühjahr 1967—Frühjahr 1968.—Kieler Meeresforsch. **26**, 1–20.
—— (1970): New method for measuring density of seawater.—Nature **229**, 109–110.
—— (1972): Comparison of specific gravity in natural sea water from hydrographical tables and measurements by a new density instrument.—Deep-Sea Res. **19**, 377–383.
—— (1973): Voltametrische Messungen über die Verteilung von Zink, Cadmium, Blei und Kupfer in der Ostsee.—Kieler Meereforsch. **29**, 77–84.
KREY, J. (1950): Eine neue Methode zur quantitativen Bestimmung des Planktons.—Kieler Meeresforsch. **7**, 58–75.
KRITSCHEVSKY, I. & A. ILIINSKAYA (1945): Partial molal volumes of gases dissolved in liquids.—Acta Physicochim. (USSR) **20**, 322–348.
KROEBEL, W. (1970): Die absolute Messung der Schallgeschwindigkeit.—In: H. U. ROLL (Ed.): Interozean '70, **2**, 265–268.
—— (1973): Die Kieler Multimeeressonde.—"Meteor"-Forsch.-Ergebn. (A) No. **12**, 53–67. Berlin–Stuttgart.
KRÜMMEL, O. (1901): Neue Beiträge zur Kenntnis des Aräometers.—Wiss. Meeresunters., Abt. Kiel, N.F. **5**, 9–36.
—— (1907): Handbuch der Ozeanographie. **1**, 1–526.—Stuttgart.
—— (1911): Handbuch der Ozeanographie. **2**, 1–766.—Stuttgart.
KRÜMMEL, O. & E. RUPPIN (1905): Über die innere Reibung des Seewassers.—Wiss. Meeresunters., Abt. Kiel, N.F. **9**, 29–36.
KRUPPA, C. & NOWEA (1973): Interozean '73, 2. internationaler Kongreß mit Ausstellung für Meeresforschung und Meeresnutzung, Kongreßberichtswerk, 2 Bde., 1–1170.
KRUSPE, G. (1972): Autocovarianzspektren von Brechungsindex, vertikaler Windgeschwindigkeit, Lufttemperatur und -feuchte, Co-spektren des vertikalen Wärme- und Feuchteflusses über See. Ber. Inst. Radiometeorol. u. marit. Meteorol. 1–95. Hamburg.
KUENEN, PH. H. (1937): Experiments in connection with DALY's hypotheses on the formation of submarine canyons.—Leidse Geol. Med. **8**, 327–335.
—— (1950): Marine Geology.—New York.
KUENEN, PH. H. & F. L. HUMBERT (1964): Bibliography of turbidity, currents and turbidites.—Dev. in sedimentol. **3**, 222–246.
KUHLBRODT, E. (1938): Kritik der Lufttemperatur-Bestimmung auf See; Größe der Temperaturdifferenz Wasser-Luft auf dem Atlantischen Ozean.—Ann. Hydrogr. u. marit. Meteorol. **64**, 259–264.
KUSNETSOVA, L. P. & V. YA. SHAROVA (1964): Physical-geographical World Atlas. Yearly amount of precipitation. p. 42–43. Academy of Sciences, Chief Administration for Geodesy and Cartography, Moscow.
KUWAHARA, S. (1938): The velocity of sound in sea water and calculation of the velocity for use in sonic sounding.—Jap. J. Astr. Geophys. **16**, 1–17.
LACOMBE, M. (1969): Résultats des mésures d'hydrologie et de courants effectuées à bord de la "Calypso."—Cah. océanogr. **21**, Suppl. 1, 1–48.
LACOMBE, H. & P. TCHERNIA (1960): Quelques traits généraux de l'hydrologie Méditerranée d'après diverses campagnes hydrologiques récentes en Méditeranée, dans le proche Atlantique et dans le détroit de Gibraltar.—Cah. océanogr. **12**, 527–548.
LACOMBE, H., P. TCHERNIA, C. RICHEZ & L. GAMBERONI (1964): L'étude du régime du détroit de Gibraltar.—Cah. océanogr. **16**, 283–331.

La Coste, L. J.'B. (1959): Surface ship gravity measurements on the Texas A&M College Ship *The Hidalgo*.—Geophysics **24**, 309–322.
Laevastu, T. (1960): Factors affecting the temperature of the surface layer of the sea.—Soc. Sci. Fenn. Comm. Phys.-Math. **25**, 1–136.
Laevastu, T. & W. E. Hubert (1965): Analysis and prediction of the depth of the thermocline and near-surface thermal structure.—Fleet Numer. Weath. Facil., Techn. Note 10, 1–90. Monterey, Calif.
Lafond, E. C. (1965): The U.S. Navy electronics laboratory's oceanographic research tower.—Dev. Rep. 1342, U.S. Navy Electronics Lab., San Diego. 1–155.
Lake, R. A. & E. L. Lewis (1970): Salt rejection by ice during growth.—J. Geophys. Res. **75**, 583–597.
Lal, D. (1966): Characteristics of large-scale oceanic circulation as derived from the distribution of radioactive elements.—Morn. rev. lect. 2nd Int. Oceanogr. Congr. 29–48, UNESCO, Paris.
Lamb, H. (1931): Lehrbuch der Hydrodynamik, 2. Aufl. 1–738.—Leipzig.
Lamb, H. H. (1969): Climatic fluctuations.—In: H. Flohn (Ed.), General climatology **2**, 173–249.
Langleben, M. P. (1962): Young's modulus for sea ice.—Can. J. Phys. **40**, 1–8.
———(1971): Albedo of melting sea ice in the Southern Beaufort Sea.—J. Glaciol. **10**, 101–104.
Langseth, M. G. & R. P. von Herzen (1970): Heat flow through the floor of the world oceans.—In: A. E. Maxwell (Ed.), The sea **4**, part I, 299–352. New York.
Laplace, P. S. (1775): Recherches sur plusieurs points du systeme du monde. Mém. Acad. Roy. Sci. **88**, Paris. Also in: Méchanique Céleste. Livre **4**, Paris 1799.
Laughton, A. A. (1957): A new deep-sea underwater camera.—Deep-Sea Res. **6**, 193–205.
———(1963): Microtopography.—In: M. N. Hill (Ed.), The sea **3**, 437–472. New York.
———(1967): Abyssal plains.—In: S. K. Runcorn (Ed.), Int. dict. geophys. **1**, 2–4. London.
Laughton, A. S., D. H. Matthews & R. L. Fisher (1970): The structure of the Indian Ocean.—In: A. E. Maxwell (Ed.), The sea **4**, part II, 543–586. New York.
Lauscher, F. (1955): Optik der Gewässer, Sonnen- und Himmelstrahlung im Meer und in Gewässern.—In: F. Linke & F. Möller (Eds.), Handbuch der Geophysik **7**, 723–768. Berlin.
Lee, A. (1963): The hydrography of the European Arctic and Subartic seas.—In: H. Barnes (Ed.), Oceanogr. Mar. Biol. Ann. Rev. **1**, 47–76. London.
———(1970): The currents and water masses of the North Sea.—In: H. Barnes (Ed.), Oceanogr. Mar. Biol. Ann. Rev. **8**, 33–71. London.
———(Ed.) (1972): Physical variability in the North Atlantic.—Rapp. P.-v. Explor. Mer **162**, 1–303. (Proc. Symp. Dublin, 1969.)
Lee, A. & D. Ellett (1965): On the contribution of overflow water from the Norwegian Sea to the hydrographic structure of the North Atlantic Ocean.—Deep-Sea Res. **12**, 129–142.
———(1967): On the water masses of the Northwest Atlantic Ocean.—Deep-Sea Res. **14**, 183–190.
Le Floch, J. (1970): La circulation des eaux d'origine subtropicale dans la partie orientale de l'Atlantique équatorial étudiée en relation avec les mesures faites à bord du N.O. "Jean Charcot" en Mai 1968.—Cah. O.R.S.T.M., sér. océanogr. **8**, 77–113.
Le Grand, Y. (1939): La pénétration de la lumière dans la mer.—Ann. Inst. Oceanogr. **19**, 393–436.
Leinerö, R. (1973): Water masses and current in a section across the Norwegian Shelf off Stad.—"Meteor" Forsch.-Ergebn. (A) No. **12**, 11–23. Berlin–Stuttgart.
Leipper, D. F. (1970): A sequence of current patterns in the Gulf of Mexico.—J. Geophys. Res. **75**, 637–657.
Lenz, W. (1971): Monatskarten der Temperatur der Ostsee dargestellt für verschiedene Tiefenhorizonte.—Dt. hydrogr. Z. Erg.—H. (B) **11**, 5–148.
Leonard, R. W., P. C. Combs & L. R. Skidmore (1949): Attenuation of sound in synthetic sea water.—J. Acoust Soc. Amer. **21**, 1–63.
Le Pichon, X. (1968): Sea-floor spreading and continental drift.—J. Geophys. Res. **73**, 3661–3697.
Lettau, H. (1934): Ausgewählte Probleme bei stehenden Wellen in Seen.—Ann. Hydrogr. u. marit. Meteorol. **62**, 13–20.
Lewis, E. L. (1967): Heat Flow through winter ice.—In: H. Oura (Ed.), Physics of snow and ice, **1**, 611–631.
Liebermann, L. N. (1948): Origin of sound absorption in water and in sea water.—J. Acoust. Soc. Amer. **20**, 868–873.
Lighthill, M. J. (1969): Dynamic response of the Indian Ocean to onset of the southwest monsoon. Philos. Trans. R. Soc. **265**, 45–92.
Lisitzin, A. P. (1970): Sedimentation and geochemical considerations.—In: W. S. Wooster (Ed.), Sci. explor. South Pacific. 89–132. Washington.
———(1972): Sedimentation in the world ocean.—Soc. econ. paleont. miner., spec. publ. **17**, 1–218.

LITOVITZ, T. A. & G. M. DAVIS (1965): Structural and shear relaxation in liquids.—In: W. P. MASON (Ed.), Physical acoustics **2** (A), 282–349.
LONDON, J. (1957): A study of the atmospheric heat balance.—Final rep., Res. Div., New York Univ.
LONGUET-HIGGINS, M. S. (1950): A theory of the origin of microseisms.—Phil. Trans. Soc. (A) **243**, 1–35.
——(1952): On the statistical distribution of the heights of sea waves.—J. Mar. Res. **11**, 245–266.
——(1968): Double Kelvin waves with continuous depth profiles.—J. Fluid Mech. **34**, 49–80.
LONGUET-HIGGINS, M. S., D. E. CARTWRIGHT & N. D. SMITH (1961): Observations of the directional spectrum of sea waves using the motions of a floating buoy.—In: Ocean waves spectra. 111–132. Englewood Cliffs, N.J.
LONGUET-HIGGINS, M. S. & R. W. STEWART (1960): Changes in the form of short gravity waves on long waves and tidal currents.—J. Fluid Mech. **8**, 565–583.
——(1964): Radiation stresses in water waves: a physical discussion with applications.—Deep-Sea Res. **11**, 529–562.
LUBBOCK, J. W. (1831): On the tides in the port of London.—Philos. Trans. R. Soc. **2**, 379–415.
LUCHT, F. (1964): Hydrographie des Elbe-Ästuars.—Arch. Hydrobiol., Suppl. **29**, 1–96.
LUDWIG, W. J., J. E. NAFE & C. L. DRAKE (1970): Seismic refraction.—In: A. E. MAXWELL (Ed.), The sea **4**, part I, 53–84. New York.
LÜNEBURG, H. (1939): Hydrochemische Untersuchungen in der Elbmündung mittels Elektrokolorimeter.—Arch. Dt. Seewarte **59**.
LUMBY, J. R. (1955): The depth of the wind-produced mixed layer in the oceans.—Fish. Invest. (2) **20**, 1–12.
LUVENDYK, B. P. (1970): Dips of downgoing lithospheric plates beneath island areas.—Bull. Geol. Soc. Amer. **81**, 3411–3416.
L'VOV, B. V. (1970): Atomic absorption spectrochemical analysis. 1–324.—London.
LVOVITCH, M. J. (1971): World water balance. 401–415.—In: Sympos. on world water balance. UNESCO-IASH. Publ. 93. Paris.
LYNN, R. L. & J. R. REID (1968): Characteristics and circulation of deep and abyssal waters.—Deep-Sea Res. **15**, 577–598.
MAGAARD, L. (1962): Zur Berechnung interner Wellen in Meeresräumen mit nichtebenen Böden bei einer speziellen Dichteverteilung.—Kieler Meeresforsch. **18**, 161–183.
MAGAARD, L. & W. KRAUSS (1966): Spektren der Wasserstandsschwankungen der Ostsee im Jahre 1958.—Kieler Meeresforsch. **22**, 155–162.
MAILLARD, C. (1971): Etude hydrologique et dynamique de la Mer Rouge en hiver d'après les observations du "Commandant Robert Giraud" (1963).—Thèse, Fac. Sci. Paris. 1–76.
MALKUS, J. S. (1962): Large-scale interactions. Ideas and observations on progress in the study of the seas.—In: M. N. HILL (Ed.), The sea **1**, 88–294. New York.
MALMGREN, F. (1927): On the properties of sea-ice. Norwegian Polar Exped. with the "Maud" 1918–1925.—Sci. Res. **1**, 1–67.
MAMAYEV, O. I. (1969): Generalized T-S diagrams of the water masses of the world ocean.—Oceanology **9**, 49–55.
MANN, C. R. (1967): The termination of the Gulf Stream and the beginning of the North Atlantic Current.—Deep-Sea Res. **14**, 337–359.
MARCINEK, J. (1964): Der Abfluß von den Landflächen der Erde.—Mitt. Inst. Wasserwirtsch. Berlin. **21**, 1–204.
MARGULES, M. (1906): Über Temperaturschichtung in stationärbewegter und ruhender Luft.—Meteorol. Z., 241–244.
Marine Technology Society (1964): Buoy technology.—Trans. 1964 Buoy Techn. Symp., 1–504 and Suppl., 1–127. Washington, D.C.
——(1967): Trans. 2nd Int. Buoy Techn. Symp. Expos. 1–557. Washington, D.C.
MARSILI, L. F. (1681): Osservazioni intorno al Bosforo Tracio overo Canale di Constantinopoli.—Rome.
MASON, R. G. (1958): A magnetic survey off the west coast of the United States between latitudes 32° and 36°N, longitudes 121° and 128°W.—J. Geophys. Res. **1**, 320–329.
——(1967): Results of magnetic surveys.—In: S. K. RUNCORN (Ed.), Int. dict. geophys. **2**, 878–896. London.
MASTERSON, J. E. (1972): Means of acquisition and communication of ocean data.—Proc. WMO Techn. Conf. **1**, 1–19.
MASUZAWA, J. (1964): Flux of water characteristics of the Pacific North Equatorial Current.—In: K. YOSHIDA (Ed.), Stud. oceanogr., 121–128. Tokyo.

———(1969): A short note on the Kuroshio stream axis.—J. Oceanogr. Soc. Jap. **25**, 259–260.
MATTHÄUS, W. (1969): Zur Entdeckungsgeschichte des Äquatorialen Unterstroms im Atlantischen Ozean.—Beitr. Meereskde. **23**, 37–70.
———(1972): Die Viskosität des Meerwassers.—Beitr. Meereskde. **29**, 93–107.
MATTHEWS, D. J. (1939): Tables of the velocity of sound in pure water and sea water for use in echo-sounding and sound-ranging.—Hydrogr. Dept. Admiralty, London.
MAXWELL, A. E. (Ed.), (1970): The sea **4**, parts 1, 1–791 and II, 1–664. New York.
MCALLISTER, E. D. & W. MCLEISH (1969): Heat transfer in the top millimeter of the ocean.—J. Geophys. Res. **74**, 3408–3414.
MCDONALD, W. F. (1938): Atlas of climatic charts of the oceans.—U.S. Weath. Bur. Washington, D.C.
MEINARDUS, W. (1942): Die bathygraphische Kurve des Tiefseebodens und die hypsographische Kurve der Erdkruste.—Ann. Hydrogr. u. marit. Meteorol. **70**, 225–244.
MEINCKE, J. (1967): Die Tiefe der jahreszeitlichen Dichteschwankungen im Nordatlantischen Ozean.—Kieler Meeresforsch. **23**, 1–15.
———(1971): Der Einfluß der Großen Meteorbank auf Schichtung und Zirkulation der ozeanischen Deckschicht.—"Meteor" Forsch.-Ergebn. (A) No. **9**, 67–94. Berlin-Stuttgart.
MEINCKE, J., E. MITTELSTAEDT, K. HUBER & K. P. KOLTERMANN (1975): Strömung und Schichtung im Auftriebsgebiet vor Nordwestafrika.—Meereskundl. Beobachtungen u. Ergebnisse **41**. Deutsch. Hydrogr. Inst., Hamburg.
MENACHÉ, M. (1966): Variation de la masse volumique de l'eau en fonction de sa composition isotopique.—Cah. océanogr. **18**, 477–496.
———(1970): Etude de la variation de la masse volumique de l'eau en fonction de sa composition isotopique.—C. r. Acad. Sci. Paris **270**, 1513–1516.
MENARD, H. W. (1964): Marine geology of the Pacific. 1–271.—New York.
———(1969): Growth of drifting volancoes. J. Geophys. Res. **74**, 4827–4837.
MENARD, H. W. & H. S. LADD (1963): Oceanic islands, seamounts, guyots and atolls.—In: M. N. HILL (Ed.), The sea **3**, 365–387. New York.
MENARD, H. W. & H. C. SHIPEK (1958): Surface concentrations of manganese nodules.—Nature **182**, 1156–1158.
MENARD, H. W. & S. M. SMITH (1966): Hypsometry of ocean basin provinces.—J. Geophys. Res. **71**, 4305–4325.
MENZEL, D. W. (1964): The distribution of dissolved organic carbon in the western Indian Ocean.—Deep-Sea Res. **11**, 757–765.
MERIAN, J. R. (1828): Über die Bewegung tropfbarer Flüssigkeiten in Gefäßen.—Reprinted 1886 in Math. Ann. **27**, 575–600.
MERLE, J., H. ROTSCHI & B. VOITURIEZ (1969): Zonal circulation in the tropical western South Pacific at 170°E.—Bull. Jap. Soc. Fish. Oceanogr., 91–98. (UDA Pap.)
MERO, J. L. (1965): The mineral resources of the sea. 1–312.—Amsterdam.
MERZ, A. (1925): Die Deutsche Atlantische Expedition auf dem Vermessungs- und Forschungsschiff "Meteor."—Sitz.-Ber. Akad. Wiss. Berlin, Math.-Phys. Kl. **31**, 562–586.
METCALF, W. G., A. D. VOORHIS & M. C. STALCUP (1962): The Atlantic equatorial undercurrent.—J. Geophys. Res. **67**, 2499–2508.
MILANKOVITCH, M. (1920): Théorie mathématique des phénomènes thermiques produits par la radiation solaire.—Paris.
———(1930): Mathematische Klimalehre und astronomische Theorie der Klimaschwankungen.—In: W. KÖPPEN & R. GEIGER (Eds.), Handbuch der Klimatologie **1** (A), 1–176. Berlin.
MILES, J. W. (1957): On the generation of surface waves by shear flows. Part I.—J. Fluid Mech. **3**, 185–204.
MILLER, A. R., P. TCHERNIA, H. CHARNOCK & D. A. MCGILL (1970): Mediterranean Sea Atlas. WHOI—Atlas Ser. **3**, 1–190. Woods Hole, Mass.
MILLERO, F. J., G. PERRON & J. E. DESNOYERS (1973): Heat capacity of seawater solutions from 5° to 35°C and 0.5 to 22‰ chlorinity.—J. Geophys. Res. **78**, 4499–4507.
MITTELSTAEDT, E. (1972): Der hydrographische Aufbau und die zeitliche Variabilität der Schichtung und Strömung im nordwestafrikanischen Auftriebsgebiet im Frühjahr 1968.—"Meteor" Forsch.-Ergebn. (A) No. **11**, 1–57. Berlin-Stuttgart.
MIYAKE, M., M. DONELAN, G. MCBEAN, C. PAULSON, F. BADGLEY & E. LEAVITT (1970): Comparison of turbulent fluxes over water determined by profile and eddy correlations techniques.—Quart. J. R. Meteorol. **96**, 132–137.
MIYAKE, Y. (1939a): Chemical Studies of the Western Pacific Ocean. III. Freezing Point, Osmotic Pressure,

Boiling Point and Vapour Pressure of Sea Water.—Bull. Chem. Soc. Jap. **14**, 58–62.
———(1939b): Chemical Studies of the Western Pacific Ocean. IV. The Refractive Index of Sea Water.—Bull. Chem. Soc. Jap. **14**, 239–242.
MIYAKE, Y. & M. KOIZUMI (1948): The measurement of the viscosity coefficient of sea water.—J. Mar. Res. **7**, 63–66.
MODEL, F. (1950): Warmwasserheizung Europas.—Ber. Dt. Wetterd. U.S.-Zone, **12**.
———(1966): Geophysikalische Bibliographie von Nord- und Ostsee.—Dt. hydrogr. Z. Erg.-H. (A) **8**, 3 vol.
MÖLLER, F. (1951): Vierteljahreskarten des Niederschlags für die ganze Erde.—Petermanns geogr. Mitt. **95**, 1–7.
———(1953): Tabellen zur atmosphärischen Strahlung und Optik.—In: LINKE, Meteorologisches Taschenbuch **2**, Leipzig.
MÖLLER, L. (1928): ALFRED MERZ' hydrographische Untersuchungen im Bosporus und Dardanellen.—Veröff. Inst. Meereskde. Berlin. N.F. (A) H. 18.
MOISEYEV, L. K. (1967): Approximate calculation of the depth of the wind-mixed layer in the ocean.—Oceanology **7**, 39–50.
MONTGOMERY, R. B. (1938): Circulation in upper layers of southern North Atlantic deduced with use of isentropic analysis.—Pap. Phys. Oceanogr. Met. **6**, 1–55. Cambridge, Mass.
———(1940): Observations of vertical humidity distribution above the ocean surface and their relation to evaporation.—Pap. phys. oceanogr. and meteorol. Woods Hole **7**.
———(1958): Water characteristics of Atlantic Ocean and of world ocean.—Deep-Sea Res. **5**, 134–148.
MONTGOMERY, R. B. & E. D. STROUP (1962): Equatorial waters and currents at 150°W in July–August 1952. 1–68.—Baltimore.
MORCOS, S. A. (1960): Die Verteilung des Salzgehaltes im Suez-Kanal.—Kieler Meeresforsch. **16**, 133–154.
———(1967): Effect of the Aswan High Dam on the current regime in the Suez Canal.—Nature **214**, 901–902.
———(1970): Physical and chemical oceanography of the Red Sea.—In: H. BARNES (Ed.), Oceanogr. Mar. Biol. Ann. Rev. **8**, 73–202. London.
———(1972a): Sources of the Mediterranean intermediate water in the Levantine Sea.—In: A. L. GORDON (Ed.), Stud. phys. oceanogr. **2**, 185–206. (WÜST Vol.) New York.
———(1972b): Investigations of the Suez Canal waters after its opening to navigation in 1868.—Paper 2nd Int. Congr. History Oceanogr. Edinburgh.
MOREL, A. (1966): Étude expérimentale de la diffusion de la lumière par l'eau, les solutions de chlorure de sodium et l'eau de mer optiquement pures.—J. Chim. Phys. **63**, 1359–1366.
MORGAN, G. W. (1956): On the wind-driven ocean circulation.—Tellus **8**, 301–320.
MOSBY, H. (1936): Verdunstung und Strahlung auf dem Meere.—Ann. Hydrogr. u. marit. Meteorol. **64**, 281–286.
———(1962): Water, salt und heat balance of the North Polar Sea and of the Norwegian Sea.—Geofys. Publ. **24**, 289–313.
———(1966): Bottom water formation.—Symp. Antarctic oceanogr. Santiago. 47–57.
———(1970): Atlantic water in the Norwegian Sea.—Geofys. Publ. **28**, 3–60.
———(1971): South Atlantic bottom water.—In: G. DEACON (Ed.), Symp. Antarctic ice and water masses. 11–20. Tokyo.
MOSKOWITZ, L. (1964): Estimates of the power spectrum for full developed seas for wind of 20 to 40 knots.—J. Geophys. Res. **69**, 161–179.
MÜGGE, H. (1952): Wasser- und Schiffahrttechnisches Handbuch der Wasser- und Schiffahrtsdirektion Hamburg.—Hamburg.
MÜLLER-KRAUSS (1968): Handbuch für die Schiffsführung. 1–7. Aufl., 2 Bde. Berlin.
MUNK, W. H. (1950): On the wind-driven ocean circulation.—J. Meteorol. **7**, 79–83.
MUNK, W. H. & E. R. ANDERSON (1948): Notes on a theory of the thermocline.—J. Mar. Res. **7**, 276–295.
MUNK, W. H. & D. CARTWRIGHT (1966): Tidal spectroscopy and prediction.—Philos. Trans. R. Soc. (A) **259**, 533–581.
MUNK, W. H. & K. HASSELMANN (1964): Super-resolution of tides.—In: K. YOSHIDA (Ed.), Stud. oceanogr., 339–344. (HIDAKA Vol.) Tokyo.
MUNK, W. H. & G. J. F. McDONALD (1960): The rotation of the earth. 1–323.—Cambridge.
MUNK, W. H., F. E. SNODGRASS & F. GILBERT (1964): Long waves on the continental shelf: an experiment to separate trapped and leaky modes.—J. Fluid Mech. **20**, 529–554.

MUNK, W. H., F. E. SNODGRASS & M. J. TUCKER (1959): Spectra of low-frequency ocean waves.—Bull. Scripps Inst. **7**, 283–362.
MUNK, W. H., F. SNODGRASS & M. WIMBUSH (1970): Tides off-shore: Transition from California coastal to deep-sea waters.—Geophys. Fluid Dynamics **1**, 161–235.
MUNK, W. H. & B. ZETLER (1967): Deep-sea tides: a program.—Science **158**, 884–886.
MUNK, W. H., B. ZETLER & G. W. GROVES (1965): Tidal cusps.—Geophys. J., Roy. Abstr. Soc. **10**, 211–219.
MURRAY, J. & A. F. RENARD (1891): Deep-sea deposits.—Scientific results of the exploration voyage of H.M.S. "Challenger," 1872–1876. "Challenger" Rep., 1–525. London.
MYERS, J. J., C. H. HOLM & R. F. MCALLISTER (1969): Handbook of ocean and underwater engineering.—New York.
NACE, R. L. (1970): World Hydrology: Status and Prospects.—Symp. on World Water Balance, Publ. No. 92 IAHS **1**, 1–10.
NAIRN, A. E. M. (Ed.) (1964): Problems in palaeoclimatology. 1–705.—NATO Conf. Proc. London.
NAMIAS, J. (1970): Macroscale variations in sea-surface temperatures in the North Pacific.—J. Geophys. Res. **75**, 565–582.
NANSEN, F. (1897): In Nacht und Eis **1**.—Leipzig.
National Academy of Sciences-National Research Council (1958): Arctic sea ice.—Proc. Conf. Easton, Md., Publ. **598**, 1–271.
National Institute of Oceanography (U.K.): Characteristic tabular iceberg in Antarctic waters. Photograph taken from the British research vessel *Discovery II*.
NELSON, S. B. (1971): Oceanographic ships fore and aft.—Off. Oceanogr. Navy, Washington, D.C.
Netherlands Meteorological Institute (1952): Indian Ocean oceanographic and meteorological data. 2nd ed., No. 135.—De Bilt.
NEUMANN, G. (1940): Die ozeanographischen Verhältnisse an der Meeresoberfläche im Golfstromsektor nördlich und nordwestlich der Azoren.—Ann. Hydrogr. u. marit. Meteorol. **68**, Beih. 1–87.
——— (1941): Eigenschwingungen der Ostsee.—Arch. Dt. Seewarte **61**, Nr. 4, 1–57.
——— (1944): Das Schwarze Meer.—Z. Ges. Erdkde. Berlin.
——— (1946): Stehende zellulare Wellen im Meere.—Naturwiss. **33**.
——— (1947): Über die Entstehung des äquatorialen Gegenstromes.—Forsch. u. Fortschr. Nr. 16–18, 177–179.
——— (1949): Die Entstehung der Wasserwellen durch Wind.—Dt. hydrogr. Z. **2**, 187–199.
——— (1952): Über die komplexe Natur des Seeganges, Teil I und II.—Dt. hydrogr. Z. **5**, 95–110 u. 252–277.
——— (1953): On ocean wave spectra and a new method of forecasting wind-generated sea.—Beach Erosion Board. Tech. Mem. No. 43, Washington, D.C.
——— (1954): Zur Charakteristik des Seeganges.—Arch. Meteorol. Geophys. Bioklimatol. (A) **7**.
——— (1960): Evidence for an equatorial undercurrent in the Atlantic Ocean.—Deep-Sea Res. **6**, 328–334.
——— (1968): Ocean currents. 1–352.—New York.
——— (1969a): Seasonal salinity variations in the upper strata of the western tropical Atlantic Ocean. I. Sea surface salinities.—Deep-Sea Res., Suppl. to **16**, 165–177. (FUGLISTER Vol.)
——— (1969b): The equatorial undercurrent in the Atlantic Ocean.—UNESCO Symp. Proc. Res. ICITA and GTS, 33–44. Paris.
NEUMANN, G. & E. R. WILLIAMS (1965): Observations of the Equatorial Undercurrent in the Atlantic Ocean at 15°W during Equalant I.—J. Geophys. Res. **70**, 297–304.
NEUMANN, H. (1972): Personal communication.
NEWTON, I. (1687): Philosophiae naturalis principia mathematica.—London. A comprehensive treatise of NEWTON's tide theory is given by J. PROUDMAN: NEWTON's work on the theory of the tides.—In: Math. Assoc. (Ed.), Isaak Newton 1642–1729. London 1927.
NEWTON, M. S. & G. C. KENNEDY (1965): An experimental study of the P-V-T-S relations of sea water.—J. Mar. Res. **23**, 88–103.
NOWROOZI, A. A., G. H. SUTTON & B. AULD (1966): Oceanic tides recorded on the sea floor.—Ann. Geophys. **22**, 512–517.
NUSSER, F. (1952): Das Vorkommen von Meereis um den antarktischen Kontinent zur Zeit des südlichen Hochsommers.—Eisübersichtskarte Nr. 19, Dt. Hydrogr. Inst. Hamburg.
ÖPIK, E. J. (1967): Climatic changes.—In: S. K. RUNCORN (Ed.). Int. dict. geophys. **1**, 179–193.
OGURA, S. (1933): The tides in the seas adjacent to Japan.—Hydrogr. Bull. Dep., Imp. Jap. Navy **7**, 1–189.

OKUBO, A. (1971): Oceanic diffusion diagrams.—Deep-Sea Res. **18**, 789–822.
OKUBO, A. & R. V. OZMIDOV (1970): Empirical dependence of the coefficient of horizontal turbulent diffusion in the ocean on the scale of the phenomenon in question.—Izv., Atm. Oc. Phys. **6**, 534–536.
OLSON, B. E. (1968): On the abyssal temperatures of the world oceans. 1–152.—Ph.D. thesis, Oregon State Univ.
OLSON, J. R. (1972): Two-component electromagnetic flowmeter.—Mar. Techn. Soc. J. **6**, 19–24.
OPDYKE, N. D. (1970): Paleomagnetism.—In: A. E. MAXWELL (Ed.), The sea **4**, part I, 157–182. New York.
OPDYKE, N. D., B. GLASS, J. D. HAYS & J. FOSTER (1966): Paleomagnetic study of Antarctic deep-sea cores.—Science **154**, 349–357.
OREN, O. H. (1969): Oceanographic and biological influence of the Suez Canal, the Nile and the Aswan Dam on the Levant Basin.—In: M. SEARS (Ed.), Progr. oceanogr. **5**, 161–167. Oxford.
——— (1970): Seasonal changes in the physical and chemical characteristics and the production in the low trophic level of the Mediterranean waters of Israel.—Ph.D. thesis, Hebrew Univ. Jerusalem.
ORLENOK, V. V. (1968): Crustal structure of the North Atlantic according to seismic data.—Oceanology **8**, 194–202.
OURA, H. (Ed.) (1967): Physics of snow and ice. **1**, part 1, 1–711, part 2, 713–1414.—Sapporo.
OVCHINNIKOV, I. M. (1961): Circulation of water in the northern Indian Ocean during the winter monsoon.—Akad. Nauk. USSR, Moscow **4**, 18–24.
OVCHINNIKOV, I. M. & YE. A. PLAKHIN (1965): Formation of Mediterranean deep water masses.—Oceanology **5**, 40–47.
OVERSTREET, R. & M. RATTRAY, Jr. (1969): On the roles of vertical velocity and eddy conductivity in maintaining a thermocline.—J. Mar. Res. **27**, 172–190.
OWEN, R. W. & B. ZEITZSCHEL (1970): Phytoplankton production: seasonal change in the oceanic eastern tropical Pacific.—Mar. Biol. **7**, 32–36.
PALMEN, E. (1948): On the formation and structure of tropical hurricanes.—Geophysica **3**, 26–38.
PANFILOVA, S. G. (1968): Seasonal variations of sea surface temperature of the Pacific Ocean.—Oceanology **8**, 639–644.
PAPANIN, J. D. (1947): Das Leben auf einer Eisscholle.—Berlin.
PARK, K. (1964): Electrolytic conductance of sea water: Effect of calcium carbonate dissolution.—Science **146**, 56–57.
PARK, K., P. K. WEYL & A. BRADSHAW (1964): Effect of carbon dioxide on the electrical conductance of sea water.—Nature **201**, 1283–1284.
PATZERT, W. C. (1972): Seasonal variations in structure and circulation in the Red Sea. 1–58.—HIG Rep. Univ. of Hawaii. Honolulu.
PAULING, L. (1968): Die Natur der chemischen Bindung. 1–620. 3. Aufl.—Weinheim.
PAYNE, R. E. (1972): Albedo of the sea surface.—J. Atmos. Sci. **29**, 959–970.
PEKERIS, C. L. & Y. ACCAD (1969): Solution of Laplace's equations for the M_2-tide in the world oceans.—Philos. Trans. R. Soc. (A) **265**, 413–436.
PETTERSON, O. (1909): Gezeitenähnliche Bewegungen des Tiefenwassers.—Publ. Circ. Cons. Int. Explor. Mer. **47**, 1–21.
PETTERSSON, H. & K. FREDRIKSSON (1958): Magnetic spherules in deep-sea deposits.—Pacific Sci. **12**, 71–81.
PFANNENSTIEL, M. (1970): Das Meer in der Geschichte der Geologie.—Geol. Rdsch. **60**, 3–72.
PHILLIPS, O. M. (1957): On the generation of waves by turbulent wind.—J. Fluid. Mech. **2**, 417–445.
——— (1958): The equilibrium range in the spectrum of wind-generated waves.—J. Fluid Mech. **4**, 426–434.
——— (1960): On the dynamics of unsteady gravity waves of finite amplitude. Part 1. The elementary interactions.—J. Fluid. Mech. **9**, 193–217.
PICKARD, G. L. (1963): Descriptive physical oceanography. 1–200.—London.
PICKETT, R. L. (1967): Aircraft observations of the Gulf Stream's inner edge.—Trans. Amer. Geophys. Un. **48**, 123.
PIERSON, W. J. & L. MOSKOWITZ (1964): A proposed spectral form for fully developed wind seas based on the similarity theory of S. A. Kitaigorodskii.—J. Geophys. Res. **69**, 181–190.
PIERSON, W. J., G. NEUMANN & R. W. JAMES (1953): Practical methods for observing and forecasting ocean waves by means of wave spectra and statistics.—New York Univ. Coll. Eng. Techn. Rep. No. 1.
PITMAN, W. C. (1971): Sea-floor spreading and plate tectonics.—Eos **52**, 130–135.
PITMAN, W. C., M. TALWANI & J. R. HEIRTZLER (1971): Age of the North Atlantic Ocean from magnetic anomalies.—Earth Planet. Lett. **11**, 195–200.

PLATZMAN, G. W. & D. B. RAO (1964): The free oscillations of Lake Erie.—In: Studies on Oceanography, 359-382.
POINCARE, H. (1910: Leçons de mécanique céleste, Théorie des Marées.—Paris.
POPOVA, A. K., YA. B. SMIRNOV & G. V. UDINTSEV (1969): Deep-water heat flow and its relation to the tectonic structure of the Pacific.—Oceanology **9**, 368-378.
POUNDER, E. R. (1965): Physics of ice. 1-151.—Oxford.
PRATT, R. M. (1968): Atlantic continental shelf and slope of the United States. Physiography and sediment of the deep-sea basin.—WHOI Coll. Repr. No. 1820, B1-B44.
PROUDMAN, J. (1936): Tides in oceans bounded by meridians. I. Ocean bounded by complete meridian. General equations.—Philos. Trans. R. Soc. (A) **235**, 273-289.
PROUDMAN, J. & A. T. DOODSON (1924): The principal constituent of the tides of the North Sea.—Philos. Trans. R. Soc. (A) **224**, 185-219.
PULS (1895): Oberflächentemperaturen und Strömungsverhältnisse des Äquatorialgürtels des Stillen Ozeans.—Arch. Dt. Seewarte **8**, 1.
PUTNAM, J. H., W. H. MUNK & M. A. TAYLOR (1949): The prediction of longshore currents.—Trans. Amer. Geophys. Un. **30**.
RAKESTRAW, N. W., F. P. DE RUDD & M. DOLE (1951): Isotopic composition of oxygen in air dissolved in Pacific ocean water as a function of depth.—J. Amer. Chem. Soc. **73**, 2976.
RAMAGE, C. S. (1970): International Indian Ocean Expedition. Meteorol. Atlas.—Honolulu.
RAO, D. B. (1966): Free gravitational oscillations in rotating rectangular basins.—J. Fluid Mech. **25**, 523-555.
RASOOL, S. J. & J. PRABHARKA (1966): Heat budget of the southern hemisphere.—In: R. V. GARCIA & T. F. MALONE, Problems of atmospheric circulation. 76-92. London.
RAUSCHELBACH, H. (1924): Harmonische Analyse der Gezeiten des Meeres.—Arch. Dt. Seewarte **24**, 1-114.
REEBURGH, W. S. (1965): Measurements of the electrical conductivity of sea water.—J. Mar. Res. **23**, 187-199.
REICHEL, E. (1952): Der Stand des Verdunstungsproblems.—Ber. Dt. Wetterd. US-Zone, **35**, 155-169.
REID, J. L. (1959): Evidence of a South Equatorial Countercurrent in the Pacific Ocean.—Nature **184**, Suppl. 4, 209-210.
——— (1961): On the geostrophic flow at the surface of the Pacific Ocean with respect to the 1000-decibar surface.—Tellus **13**, 489-502.
——— (1962a): Measurements of the California Countercurrent at a depth of 250 meters.—J. Mar. Res. **20**, 134-137.
——— (1962b): On circulation, phosphate-phosphorus content, on zooplankton volumes in the upper part of the Pacific Ocean.—Limnol. & Oceanogr. **7**, 287-306.
——— (1965): Intermediate waters of the Pacific Ocean.—John Hopkins Oceanogr. Stud. **2**, 1-85. Baltimore.
——— (1969): Sea-surface temperature, salinity, and density of the Pacific Ocean in summer and in winter.—Deep-Sea Res., Suppl. to **16**, 215-224. (FUGLISTER Vol.)
REID, J. L. & R. J. LYNN (1971): On the influence of the Norwegian-Greenland and Weddell seas upon the bottom waters of the Indian and Pacific oceans.—Deep-Sea Res. **18**, 1063-1088.
REID, J. L. & W. D. NOWLIN, Jr. (1971): Transport of water through the Drake Passage.—Deep-Sea Res. **18**, 51-64.
REID, R. O. (1958): Effect of the Coriolis force on edge waves. (I) Investigation of the normal modes.—J. Mar. Res. **16**, 109-144.
REVELLE, R. (1944): Marine bottom samples collected in the Pacific Ocean by the "Carnegie" on its seventh cruise.—Carnegie Inst. Publ. 556, Part 1. Washington, D.C.
REVELLE, R. & A. E. MAXWELL (1952): Heat flow through the floor of the Eastern North Pacific Ocean.—Nature **170**, 199-200.
REVELLE, R. & H. S. SUESS (1957): Carbon dioxide exchange between atmosphere and ocean and the question of an increase of atmospheric CO_2 during the past decades.—Tellus **9**, 18-27.
——— (1963): Gases. Interchange of Properties between sea and air.—In: M. N. HILL (Ed.), The sea **1**, 313-321. New York.
REYNOLDS, O. (1895): On the dynamical theory of incompressible viscous fluids and the determination of the criterion.—Phil. Trans. R. Soc. (A) **186**, 123-164.
RHINES, P. (1971): A note on long period motions at Site D.—Deep-Sea Res. **18**, 21-26.
RICHARDS, F. A. (1965): Dissolved gases other than carbon dioxide.—In: J. P. RILEY & G. SKIRROW (Eds.), Chem. oceanogr. **1**, 197-225.
RICHARDSON, L. F. (1920): The supply of energy from and to atmospheric eddies.—Proc. R. Soc. (A) **97**, 686, 354-373.

RICHARDSON, P. L. & J. A. KNAUSS (1971): Gulf Stream and western boundary undercurrent observations at Cape Hatteras.—Deep-Sea Res. **18**, 1089-1109.
RICHARDSON, W. S. & C. J. HUBBARD (1960): The contouring temperature recorder.—Deep-Sea Res. **6**, 239-244.
RICHARDSON, W. S. & W. J. SCHMITZ (1965): A technique for the direct measurement of transport with application to the Straits of Florida.—J. Mar. Res. **23**, 172-185.
RICHARDSON, W. S., P. B. STIMSON & C. H. WILKINS (1963): Current measurements from moored buoys.—Deep-Sea Res. **10**, 369-388.
RIEL, P. M. VAN (1934): The bottom configuration in relation to the flow of the bottom water.—"Snellius" Exped. **2**, part 5. Leiden.
——(1943): The bottom water. Introductory remarks and oxygen content.—"Snellius" Exped. **2**, part 5. Leiden.
RIEL, P. M. VAN, H. C. HAMAKER & L. VAN EYCK (1950): Tables. Serial and bottom observations. Temperature, salinity and density.—"Snellius" Exped. **2**, part 6. Leiden.
RILEY, G. A. (1947): Factors controlling phytoplankton population on Georges Bank.—J. Mar. Res. **6**, 54-103.
——(1963): Organic aggregates in sea water and the dynamics of their formation and utilization.—Limnol. & Oceanogr. **8**, 372-381.
RILEY, J. P. & R. CHESTERY (1971): Introduction to marine chemistry. 1-465.—London.
RILEY, J. P. & G. SKIRROW (Eds.) (1965): Chem. oceanogr. **1**, 1-712; **2**, 1-508. London.
ROBINSON, A. R. (1960): The general thermal circulation in equatorial regions.—Deep-Sea Res. **6**, 311-317.
——(1966): An investigation into the wind as the cause of equatorial undercurrent.—J. Mar. Res. **24**, 179-204.
——(1971): The Gulf Stream.—Philos. Trans. R. Soc. (A) **270**, 351-370.
ROBINSON, A. R. & B. A. TAFT (1972): A numerical experiment for the path of the Kuroshio.—J. Mar. Res. **30**, 65-101.
ROBINSON, A. R. & P. WELANDER (1963): Thermal circulation on a rotating sphere; with application to the oceanic thermocline.—J. Mar. Res. **21**, 25-38.
ROBINSON, M. K. (1969): Theoretical predictions of subtropical countercurrent confirmed by bathythermograph (BT) data.—Bull. Jap. Soc. Fish. Oceanogr., 115-121. (UDA Pap.)
RODEN, G. I. (1970): Aspects of the Mid-Pacific transformation zone.—J. Geophys. Res. **75**, 1097-1109.
RODEWALD, M. (1967): Beiträge zur Klimaschwankung im Meere.—Dt. hydrogr. Z. **20**, 269-275.
——(1972): Einige hydroklimatische Besonderheiten des Jahrzehnts 1961-70 im Nordatlantik und im Nordpolarmeer.—Dt. hydrogr. Z. **23**, 97-117.
ROHDE, J. (1968): Funktionale Zusammenhänge zwischen Größen des Meerwassers und Auflösung formelmäßiger Darstellungen solcher Größen nach unabhängigen Variablen durch Potenzreihenansatz.—Diss., Univ. Kiel (unpublished).
ROLL, H. U. (1952): Über Größenunterschiede der Meereswellen bei Warm- und Kaltluft.—Dt. hydrogr. Z. **5**, 111-114.
——(1958): Zur Niederschlagsmessung auf See: Ergebnisse von Vergleichsmessungen auf Feuerschiffen und benachbarten Inseln.—Dt. Wetterd. Seewetteramt Einzelveröff. **16**, 1-14. Hamburg.
——(1965): Physics of the marine atmosphere. 1-426.—New York.
ROSS, D. A. & E. T. DEGENS (Eds.) (1969): Hot brines and recent heavy metal deposits in the Red Sea. 1-600.—New York.
ROSS, D. B., V. J. CARDONE & J. W. CONAWAY, Jr. (1970): Laser and microwave observations of sea surface conditions for fetch-limited 17- to 25-m/s winds.—IEEE Trans. Geosi. Electr., GE-8 (4), 326-336.
ROSSBY, C. G. (1939): Relations between variations in the intensity of the zonal circulation of the atmosphere and the displacements of the semi-permanent centers of action.—J. Mar. Res. **2**, 38-55.
——(1940): Planetary flow patterns in the atmosphere.—Quart. J. R. Soc. Can. Branch. **66**, 68-87.
——(1951): On the vertical and horizontal concentration of momentum in air and ocean currents.—Tellus **3**, 15-27.
——(1959): Current problems in meteorology. The atmosphere and the sea in motion. 9-50. ROSSBY Mem. Vol.—New York.
ROSSBY, C. G. & R. B. MONTGOMERY (1935): The layer of the frictional influence in wind and ocean current.—Pap. Phys. Oceanogr. Met. **3**, No. 3, 1-101.
ROSSBY, T. & D. WEBB (1971): The four month drift of a Swallow float.—Deep-Sea Res. **18**, 1035-1039.
ROTSCHI, H. & L. LEMASSON (1967): Oceanography of the Coral and Tasman seas.—In: H. BARNES (Ed.),

Oceanogr. Mar. Biol. Ann. Rev. **5,** 49–97. London.
ROUCH, J. (1943): Remarques sur quelques sondages de température et de salinité du "Meteor" et du "Discovery."—Bull. Inst. Océanogr. Monaco, **838,** 1–7.
ROUSE, H. & J. DODU (1955): Turbulent diffusion across the density discontinuity.—La Houille Blanche **10,** 522–532.
RUBNER, M. (1883): Z. Biol. **19**.
RUDOLPH, E. (1887): Über submarine Erdbeben und Eruptionen.—Gerlands Beitr. Geophys. **1,** 133–373.
RUNCORN, S. K. (Ed.) (1962): Continental drift. 1–338.—New York.
——— (Ed.) (1967): Int. Dict. Geophys. **1** and **2,** Oxford.
RUSBY, J. S. M. (1967): Measurements of the refractive index of sea water relative to Copenhagen Standard Sea Water.—Deep-Sea Res. **14,** 427–439.
RUSSEL, Lord (1838): Report of the Committee on waves.—Rep. 7th Meet. Brit. Assoc. for Advancement of Science, 417–496, Liverpool.
——— (1845): Report on waves.—Rep. 14th Meet. Brit. Assoc. for Advancement of Science, York, 1844, 311–390, London.
RYAN, W. B. F., D. J. STANLEY, J. B. HERSEY, D. A. FAHLQUIST & T. D. ALLAN (1970): The tectonics of the Mediterranean Sea.—In: A. E. MAXWELL (Ed.), The sea **4,** part II, 387–492. New York.
SAGER, G. (1964): Die Entwicklung der Methoden zur Bestimmung der Elemente von Tidenhub und Gezeitenströmen.—Beitr. Meereskde. **11,** 1–99. Dt. Akad. Wiss. Berlin.
——— (1967): Die Einrittszeiten des maximalen Tidestroms in der Nordsee, dem Kanal und der Irischen See.—Monatsber. Dt. Akad. Wiss. Berlin **9,** 16–21.
——— (1968a): Zur Inbetriebnahme des Gezeitenkraftwerkes "Rance."—Petermanns geogr. Mitt. **112,** 122–125.
——— (1968b): Zur Mittelwertbildung meereskundlicher Meßergebnisse über geographische Eingradfelder.—Beitr. Meereskde. **22,** 49–63. Dt. Akad. Wiss. Berlin.
SANDSTRÖM, J. W. (1908): Dynamische Versuche mit Meerwasser.—Ann. Hydrogr. u. marit. Meteorol. **36,** 6–23.
SARKISYAN, A. S. (1969): Theory and computation of ocean currents (Transl. from Russ.)—IPST Press, Jerusalem. 1–90.
SAUNDERS, P. M. (1967): Shadowing on the ocean and the existence of the horizon.—J. Geophys. Res. **72,** 4643–4649.
——— (1970): Correlations for airborne thermometry.—J. Geophys. Res. **75,** 7596–7601.
SAUR, J. F. T. (1963): A study of the quality of sea water temperatures reported in logs of ships' weather observations.—J. Appl. Meteorol. **2,** 417–425.
SAURAMO, M. (1958): Die Geschichte der Ostsee.—Ann. Acad. Sci. Fenn. (A) 1–522.
SCHELL, I. I. (1962): On the iceberg severity off Newfoundland and its prediction.—J. Glaciol. **4,** 161–172.
SCHLEICHER, K. E. & A. BRADSHAW (1956): A conductivity bridge for measurement of the salinity of sea water.—J. Cons. **22,** 9–20.
SCHMIDT, W. (1915): Strahlung und Verdunstung an freien Wasserflächen, ein Beitrag zum Wärmehaushalt und zum Wasserhaushalt der Erde.—Ann. Hydrogr. u. marit. Meteorol. **43,** 111–124, 169–178.
SCHMIDT, W. J. & W. S. RICHARDSON (1966): A preliminary report on operation Strait Jacket.—Inst. Mar. Sci. Univ. Miami.
SCHMITZ, H. P. (1964): Modellrechnungen zu winderzeugten Bewegungen in einem Meer mit Sprungschicht.—Dt. hydrogr. Z. **17,** 201–232.
SCHNEIDER, E. D. & P. R. VOGT (1968): Discontinuities in the history of sea-floor spreading.—Nature **217,** 1212–1222.
SCHOTT, F. (1966): Der Oberflächensalzgehalt in der Nordsee.—Dt. hydrogr. Z. Erg.-H. (A) **9,** 1–58.
——— (1972): Personal communication.
SCHOTT, F. & M. EHRHARDT (1969): On fluctuations and mean relations of chemical parameters in the Northwestern North Sea.—Kieler Meeresforsch. **25,** 272–278.
SCHOTT, G. (1935): Geographie des Indischen und Stillen Ozeans. 1–413.—Hamburg.
——— (1942): Geographie des Atlantischen Ozeans. 1–438. 3. Aufl.—Hamburg.
——— (1943): Weltkarte zur Übersicht der Meeresströmungen.—Ann. Hydrogr. u. marit. Meteorol. **71**.
——— (1951): Der Perustrom.—Erdkunde **5**.
SCHOTT, W. (1935): Die Bodenbedeckung des Indischen und Stillen Ozeans.—In: G. SCHOTT (Ed.), Geographie des Stillen und Indischen Ozeans. 109–122. Hamburg.
——— (1942): Die Bodenbedeckung des Atlantischen Ozeans.—In: G. SCHOTT (Ed.), Geographie des Atlantischen Ozeans. 134–143. 3. Aufl. Hamburg.

——(1969): Meeresgeologie.—In: H. CLOSS, G. DIETRICH, G. HEMPEL, W. SCHOTT & E. SEIBOLD: "Atlantische Kuppenfahrten 1967" mit dem Forschungsschiff "Meteor."—"Meteor" Forsch.-Ergebn. (A) No. **5**, 32-37. Berlin-Stuttgart.
SCHUBERT, O. VON (1944): Ergebnisse der Strommessungen und der ozeanographischen Serienmessungen auf den beiden Ankerstationen der zweiten Teilfahrt.—Wiss. Erg. Dt. Atlant. Exp. 1937 und 1938, Ann. Hydrogr. u. marit. Meteorol. **72**, January Suppl.
SCHÜTZ, D. F. & K. K. TUREKIAN (1965): The investigation of the geographical and vertical distribution of several trace elements in seawater using neutron activation analysis.—Geochim. cosmochim. Acta **29**, 259-313.
SCHULKIN, M. & H. W. MARSH (1963): Absorption of sound in sea water.—Radio & Electr. Eng. **26**, 493-500.
——(1962): Sound absorption in sea water.—J. Acoust. Soc. Amer. **34**, 864-865.
SCHULZ, B. (1923): Hydrographische Untersuchungen besonders über den Durchlüftungszustand in der Ostsee im Jahre 1922.-Arch. Dt. Seewarte **41**.
SCHUMACHER, A. (1952): Results of exact wave measurements (by stereo-photogrammetry) with special references to more recent theoretical investigations.—Gravity waves N.B.S. Circ. **521**, 69-71.
SCHUREMAN, P. (1941): Manual of harmonic analysis and prediction of tides.—U.S. Coast Geod. Surv., Spec. Publ. **98**, Washington, D.C.
SCHWARZBACH, M. (1961): Das Klima der Vorzeit. 1-211. 2. Aufl.—Stuttgart.
——(1963): Climates of the past. 1-328.—London.
SCHWEIGGER, E. (1949): Der Perustrom nach zwölfjährigen Beobachtungen.—Erdkunde **3**.
SCHWERDTFEGER, P. (1963): The thermal properties of sea ice.—J. Glaciol. **4**, 789-807.
Scripps Institution of Oceanography (1961-65): Oceanic observations of the Pacific.—Proc. 1949-1959. Los Angeles.
——(1968): Deep-sea drilling vessel *Glomar Challenger*. Photograph.
SEABROKE, J. M., G. L. HUFFORD & R. B. ELDER (1971): Formation of Antarctic bottom water in the Weddel Sea.—J. Geophys. Res. **76**, 2164-2178.
SEARS, M. (Ed.) (1969): Symposium on Variability. Rome 1966.—Progr. oceanogr. **5**, 1-191. Oxford.
SECKEL, G. R. (1968): A time-sequence oceanographic investigation in the North Pacific trade-wind zone.—Trans. Amer. Geophys. Un. **49**, 377-387.
SEGERSTRÅLE, S. G. (1965): On the salinity conditions off the south coast of Finland since 1950, with comments on some remarkable hydrographical and biological phenomena in the Baltic area during this period.—Soc. Sci. Fenn. Comm. Biol. **28**, 3-28.
——(1969): Biological fluctuation in the Baltic Sea.—In: M. SEARS (Ed.), Progr. oceanogr. **5**, 169-184. Oxford.
SEIBOLD, E. (1964): Das Meer.—In: R. BRINKMANN (Ed.), Lehrbuch der Allgemeinen Geologie **1**, 280-500. Stuttgart.
——(1967): La mer baltique prise comme modèle de géologie marine.—Rev. Géogr. phys. et Géol. dynam. **9**, 371-384.
——(1970): Nebenmeere im humiden und ariden Klimabereich.—Geol. Rdsch. **60**, 73-105.
SEIBOLD, E. & K. VOLLBRECHT (1969): Die Bodengestalt des Persischen Golfs.—"Meteor" Forsch.-Ergebn. (C) No. **2**, 29-56. Berlin-Stuttgart.
SELLERS, W. D. (1965): Physical climatology. 1-272.—Chicago.
SEN GUPTA, R. (1971): Oceanography of the Black Sea: inorganic nitrogen compounds.—Deep-Sea Res. **18**, 457-475.
SHEPARD, F. P. (1963): Submarine Geology. 2nd ed., 1-557.—New York.
——(1968): Coastal classification.—In: R. W. FAIRBRIDGE (Ed.), Encycl. geomorph. **3**, 131-133. New York.
SHEPARD, F. P. & R. F. DILL (1966): Submarine canyons and other sea valleys. 1-381.—Chicago.
SHEPARD, F. P. & D. L. INMAN (1951): Near shore circulation.—Proc. 1st Conf. Coast. Eng. Berkeley, California, 50-59.
SHEPARD, F. P., G. A. MACDONALD & C. C. COLX (1950): The tsunami of April 1, 1946.—Bull. Scripps. Inst. Oceanogr. **5**.
SHOJI, D. (1963): Description of the Kuroshio. (Physical aspects.)—Proc. Symp. Kuroshio. 1-11. Tokyo.
SHOR, G. G., Jr., H. W. MENARD & R. W. RAITT (1970): Structure of the Pacific Basin.—In: A. E. MAXWELL (Ed.), The sea **4**, part 11, 3-27. New York.
SIEDLER, G. (1963): On the in-situ measurement of temperature and electrical conductivity of sea-water.—Deep-Sea Res. **10**, 269-277.
——(1968a): Schichtungs- und Bewegungsverhältnisse am Südausgang des Roten Meeres.—"Meteor"

Forsch.-Ergebn. (A) No. **4,** 1-76. Berlin-Stuttgart.

——— (1968b): Physikalische Methoden.—In: C. SCHLIEPER (Ed.), Methoden der meeresbiologischen Forschung. 32-47. Jena.

——— (1970): Über die Fluktuation der Wasserschichtung im Meer.—In: G. DIETRICH (Ed.), Erforschung der Meere. 53-64. Frankfurt.

SIEDLER, G. & G. KRAUSE (1964): An anchored vertically moving instrument and its application as parameter follower.—Trans. 1964 Buoy Techn. Symp., Mar. Techn. Soc. 483-488. Washington, D.C.

SIGTRYGGSSON, H. (1969): Yfirlit um hafis i grennd vio Island.—Hafisinn, 80-94.

SKIRROW, G. (1965): The dissolved gases.—In: J. P. RILEY & G. SKIRROW (Eds.), Chem. oceanogr. **1,** 227-322. London.

SKOPINTSEV, B. A. (1962): Recent work on the hydrochemistry of the Black Sea.—Deep-Sea Res. **9,** 349-357.

SLOWEY, J. F., L. M. JEFFREY & D. W. HOOD (1962): The fatty-acid content of ocean water.—Geochim. cosmochim. Acta **26,** 607-616.

SMED, J. (1943): Annual and seasonal variations in the salinity of the North Atlantic surface water.—Rapp. P.-v. Explor. Mer. **112,** 77-94.

SMITH, E. H. (1931): Arctic ice, with special reference to its distribution to the North Atlantic Ocean.—"Marion" Exped., Sci. Res. Part 3, U.S. Treas. Dep., Bull. 19. Washington, D.C.

SMITH, R. L. (1968): Upwelling.—In: H. BARNES (Ed.), Oceanogr. Mar. Biol. Ann. Rev. **6,** 11-46. London.

SNODGRASS, F. E. (1968): Deep-sea instrument capsule.—Science **162,** 78-87.

SNYDER, R. L. & C. S. COX (1966): A field study of the wind generation of ocean waves.—J. Mar. Res. **24,** 141-178.

SPILHAUS, A. F. (1938): A Bathythermograph.—J. Mar. Res. **1**.

STAHNKE (1943): Compact pack ice with ridges east of Greenland in March. Photograph.

STALCUP, M. C. & W. G. METCALF (1966): Direct measurements of the Atlantic Equatorial Undercurrent.—J. Mar. Res. **24,** 44-55.

STANLEY, E. M. & R. C. BATTEN (1969): Viscosity of sea water at moderate temperatures and pressures.—J. Geophys. Res. **74,** 3415-3420.

STEEMANN-NIELSEN, E. (1952): The use of radioactive carbon (C^{14}) for measuring organic production in the sea.—J. Cons. **18,** 90.

——— (1954): On organic production in the oceans.—J. Cons. **19,** 309-328.

STEEMANN-NIELSEN, E. & E. A. JENSEN (1957): Primary oceanic production.—"Galathea" Rep. **1,** 49-136. Copenhagen.

STEINITZ, H. (1967): A tentative list of immigrants via the Suez Canal.—Israel J. Zool. **16,** 166-169.

STEPANOV, V. N. (1965): Basic types of water structure in the seas and oceans.—Oceanology **5,** 21-28.

——— (1969): General classification of the water masses of the world ocean, their formation and transport.—Oceanology **9,** 613-621.

STERN, M. E. & J. S. TURNER (1969): Salt fingers and convecting layers.—Deep-Sea Res. **16,** 497-511.

STERNECK, R. v. (1919): Die Gezeitenerscheinungen in der Adria, II.—Denkschr. Akad. Wiss. Wien **96,** 277-324.

——— (1920-21): Die Gezeiten der Ozeane.—Sitzungsber. Akad. Wiss. Wien **129,** 131-150, **130,** 363-371.

STEWART, R. W. (1967): Mechanics of the air-sea interface.—In: Boundary layers and turbulence. New York.

STOCKS, TH. (1935): Erkundungen über Art und Schichtung des Meeresbodens mit Hilfe von Hochfrequenz-Echoloten.—Naturwiss **23,** 383-387.

——— (1937, 1938, 1941, 1961): Grundkarte der ozeanischen Lotungen 1: 5 Mill.—Wiss. Ergebn. Dt. Atlant. Exped. "Meteor" 1925-27. **3,** 4 Lfg. (Five sheets.) Berlin.

STOCKS, TH. & G. WÜST (1935): Die Tiefenverhältnisse des offenen Atlantischen Ozeans.—Wiss. Ergebn. Dt. Atlant.-Exped. "Meteor" 1925-27, **3,** 1. Lfg. Berlin.

STOKES, G. G. (1847): On the theory of oscillatory waves.—Trans. Cambridge Phil. Soc. **8,** 441-455.

STOMMEL, H. (1948): The westward intensification of wind-driven ocean currents.—Trans. Amer. Geophys. Un. **29,** 202-206.

——— (1953a): Computation of pollution in a vertically mixed estuary.—Sewage Industry Wastes **25,** 1065-1071.

——— (1953b): Examples of the possible role of inertia and stratification in the dynamics of the Gulf Stream system.—J. Mar. Res. **12,** 184-195.

——— (1954): Circulation in the North Atlantic Ocean.—Nature **173**.

——— (1957): A survey of ocean currents theory.—Deep-Sea Res. **4**, 149–184.
——— (1960): Wind-drift near the equator.—Deep-Sea Res. **6**, 304–308.
——— (1965): The Gulf Stream. 1–202. Berkeley—Los Angeles.
——— (1972): Deep winter-time convection in the western Mediterranean Sea.—In: A. L. GORDON (Ed.), Stud. phys. oceanogr. **2**, 207–218. (WÜST Vol.) New York.
STOMMEL, H. & A. B. ARONS (1971): On the abyssal circulation of the world ocean. Part V. Structure of deep western boundary currents.
STOMMEL, H., B. A. TAFT, T. WINTERFELD, D. SHOJI, J. MASUZAWA & K. YOSHIDA (1971): Axis of the Kuroshio.—In: H. STOMMEL & K. YOSHIDA (Eds.), Kuroshio—its physical aspects. 1–517. Tokyo.
STOMMEL, H. & G. VERONIS (1956): The action of variable wind stresses on a stratified ocean.—J. Mar. Res. **15**, 43–75.
STOMMEL, H. & K. YOSHIDA (Eds.) (1972): The Kuroshio, its physical aspects. 1–517. Tokyo.
STRICKLAND, D. H. (1965): Production of organic matter in the primary stages of the marine food chain.—In: J. P. RILEY & G. SKIRROW (Eds.), Chem. oceanogr. **1**, 476–610. London.
STROM, K. M. (1936): Landlocked waters.—Norske Vid. Akad. Oslo, Math.-Naturwiss. Kl., 7, 1–85.
STRÜBING, K. (1970): Satellitenbild und Meereiserkundung. Ein methodischer Versuch für das Baltische Meer.—Dt. hydrogr. Z. **23**, 193–213.
STURGES, W. (1970): Observations of deep-water renewal in the Caribbean Sea.—J. Geophys. Res. **75**, 7602–7610.
SUBOW, N. N. (1940): Into the center of the Arctic Ocean (in Russian). Moscow.
SUELLENTROP, F. J., A. E. BROWN & E. RULE (1962): An acoustic ocean-current meter.—Mar. Sci. Instrum. **1**, 190–193, Instrument Soc. Amer., New York.
SÜNDERMANN, J. (1966): Ein Vergleich zwischen der analytischen und der numerischen Berechnung windbezeugter Strömungen und Wasserstände in einem Modellmeer mit Anwendungen auf die Nordsee.—Diss., Mitt. Inst. Meereskde. Univ. Hamburg Nr. **4**.
SUGAWARA, K. (1965): Exchange of chemical substances between air and sea. In: H. BARNES (Ed.), Oceanogr. Mar. Biol. Ann. Rev. **3**, 59–77. London.
SUKHOVEY, V. F. & A. P. METALNIKOV (1968): The deep-sea water exchange between the Caribbean Sea and the Atlantic Ocean.—Oceanology **8**, 159–164.
SUNDBORG, Å. (1956): The river Klarälven. A study of fluvial processes.—Geogr. Ann. **38**, 127–316.
SUTTON, G. W., W. G. McDONALD, D. D. PRENTISS & S. N. THANOS (1965): Ocean bottom seismic observatories.—Inst. Elec. Electron. Eng. Proc. **53**, 1909.
SVANSSON, A. (1972): Canal models of sea level and salinity variations in the Baltic and adjacent seas.—Fish. Board Swed., Ser. Hydrogr., Rep. 26, 1–72.
SVERDRUP, H. U. (1926): Dynamic of tides on the North Siberian shelf.—Geophys. Publ. **4**, 1–75.
——— (1928): Norw. North Polar Exped. "Maud" 1918–25, **4**, Bergen.
——— (1932): Arbeiter i luft-og havsforskning.—Chr. Michelsens Inst. Vidensk. of Aandsfrihet **2**, 5. Bergen.
——— (1933): Meteorology. The Norwegian North Polar Exped. "Maud" 1918–25. Sci. Res. **2**, Bergen.
——— (1945): Oceanography for meteorologists. 1–235.—London.
——— (1947): Wind-driven currents in a baroclinic ocean; with application to the equatorial currents of the eastern Pacific.—Proc. Nat. Acad. Sci. Wash. **33**, 318–326.
——— (1951): Evaporation from the oceans.—In: T. F. MALONE (Ed.), Compendium of meteorology 1071–1081. Boston.
SVERDRUP, H. U., M. W. JOHNSON & R. H. FLEMING (1942): The oceans, their physics, chemistry, and general biology. 1–1060.—New York.
SVERDRUP, H. U. & W. H. MUNK (1947): Wind, sea and swell: Theory of relations for forecasting.—U.S. Navy, H.O. Publ. Nr. 601.
SWALLOW, J. C. (1955): A neutral buoyancy float for measuring deep currents.—Deep-Sea Res. **3**, 74–81.
——— (1967): The Equatorial Undercurrent in the western Indian Ocean in 1964.—In: Stud. trop. oceanogr. **5**, 15–36. Miami.
SWINBANK, W. C. (1967): Evaporation.—In: R. W. FAIRBRIDGE (Ed.), Encycl. atmosph. sci., 368–372. New York.
SYKES, L. R., J. OLIVER & B. ISACKS (1970): Earthquakes and tectonics.—In: A. E. MAXWELL (Ed.), The sea **4**, part I, 353–420. New York.
SZEKIELDA, K.-H. (1970): The effect of cyclonic and anticyclonic water movements on the distribution of organic matter.—Goddard Space FLight Center x-622-70-40. Greenbelt, Md.

TAFT, B. A. & J. A. KNAUSS (1967): The Equatorial Undercurrent of the Indian Ocean as observed by the Lusiad Expedition.—Bull. Scripps Inst. Oceanogr. **9,** 1–85. Los Angeles.
TAIT, J. B. (Ed.) (1967): The Iceland-Faroe ridge international (ICES) "Overflow" Expedition May–June, 1960.—Rapp. P.-v. Explor. Mer. **157,** 1–274. Copenhagen.
TAIT, J. B., A. J. LEE, U. STEFANSSON & F. HERMANN (1967): Temperature and salinity distribution and water masses of the region.—Rapp. P.-v. Explor. Mer. **157,** 38–149.
TALWANI, M. (1964): A review of marine geophysics.—Mar. Geol. **2,** 29–80.
——— (1970): Gravity.—In: A. E. MAXWELL (Ed.), The sea **4,** part 4, 251–297. New York.
TALWANI, M., G. H. SUTTON & J. L. WORZEL (1959): A crustal section across the Puerto Rico Trench.—J. Geophys. Res. **64,** 1545–1555.
TAMS, E. (1921): Über Fortpflanzungsgeschwindigkeit der seismischen Oberflächenwellen längs kontinentaler und ozeanischer Wege.—Zbl. Miner. Geol. Paläont., 44–52, 75–83.
TAYLOR, G. I. (1919): Tidal friction in the Irish Sea.—Philos. Trans. R. Soc. (A) **220,** 3–33.
——— (1920): Tidal oscillation in gulfs and rectangular basins.—Proc. London Math. Soc. **20,** 148–181.
——— (1922): Tidal oscillations in gulfs and rectangular basins.—Proc. London Math. Soc. ser **2,** 20, 148–181.
TAYLOR, R. C. (1970): The distribution of rainfall over the tropical Pacific Ocean deduced from island, atoll and coastal stations.—Proc. Symp. Trop. Meteorol., Contr. 347. HIG, J III 1–8. Honolulu.
THIEDE, J. (1971): Planktonische Foraminiferen in Sedimenten vom ibero-marokkanischen Kontinentalrand.—"Meteor" Forsch.-Ergebn. (C) No. **7,** 15–102. Berlin–Stuttgart.
THOMAS, B. D., T. G. THOMPSON & C. L. UTTERBACK (1934): The electrical conductivity of sea water.—J. Cons. **9,** 28–35.
THOMPSON, R. (1971): Topographic Rossby waves at a site north of the Gulf Stream.—Deep-Sea Res. **18,** 1–19.
THOMPSON, T. G. & H. E. WIRTH (1931): The specific gravity of sea water at zero degrees in relation to the chlorinity.—J. Cons. **6,** 232–240.
THOMSON, W. (Lord KELVIN) (1868): Report of committee for the purpose of harmonic analysis of tidal observations.—Brit. Assoc. Adv. Sci. Rep. London.
——— (1879): On gravitational oscillations of rotating water.—Prof. Roy. Soc. Edinburgh **10,** 92–100.
THORADE, H. (1924): Einige Bemerkungen über Amphidromien I und II.—Ann. Hydrogr. u. marit. Meteorol. **52,** 136–140, 184–188.
——— (1928): Gezeitenuntersuchungen in der Deutschen Bucht der Nordsee.—Arch. Dt. Seewarte **46**.
——— (1931): Probleme der Wasserwellen. 1–219.—Hamburg.
THORPE, S. A., E. P. COLLINS & D. I. GAUNT (1973): An electromagnetic current meter to measure turbulent fluctuations near the ocean floor.—Deep-Sea Res. **20,** 933–938.
THOULET, J. & A. CHEVALLIER (1889): Sur le chaleur spécifique de l'eau de mer à divers degrés de dilution et de concentration.—C.R. Acad. Sci. Paris **108,** 794–796.
TOMCZAK, G. (1954): Der Windstau- und Sturmflutwarndienst für die deutsche Nordseeküste beim Deutschen Hydrographischen Institut.—Dt. hydrogr. Z. **7,** 35–41.
——— (1955): Die Sturmfluten vom 20. bis 24. Dezember 1954 bei Cuxhaven.—Dt. hydrogr. Z. **8,** 145–156.
TOMCZAK, M., Jr. (1970a): Schwankungen von Schichtung und Strömung im westafrikanischen Auftriebsgebiet während der "Deutschen Nordatlantischen Expedition" 1937.—"Meteor" Forsch.-Ergebn. (A) No. **7,** 1–109. Berlin–Stuttgart.
——— (1970b): Eine lineare Theorie des stationären Auftriebs im stetig geschichteten Meer.—Dt. hydrogr. Z. **23,** 214–234.
——— (1973): An investigation into the occurrence and development of cold water patches in the upwelling region off NW Africa (Roßbreiten-Expedition 1970).—"Meteor" Forsch.-Ergebn. (A) No. **13,** 1–42. Berlin–Stuttgart.
TREPKA, L. v. (1968): Investigations of the tides in the Persian Gulf by means of hydrodynamic-numerical model.—Mitt. Inst. Meereskde. **10,** 59–63. Hamburg.
TRIPP, R. B. (1971): Personal communication.
TSUCHIYA, M. (1970): Equatorial circulation of the South Pacific.—In: W. S. WOOSTER (Ed.), Scient. explor. South Pacific. 69–74. Washington, D.C.
TUCKER, M. J. (1956): A shipborne wave recorder.—Trans. Inst. Naval Architects **98,** 236–250.
TULLY, J. P. & L. F. GIOVANDO (1963): Seasonal temperature structure in the eastern subarctic Pacific Ocean.—R. Soc. Can., Spec. Publ. **5,** 10–36.
TURNER, J. S. (1967): Salt fingers across a density interface.—Deep-Sea Res. **14,** 599–611.
——— (1973): Buoyancy effects in fluids. 1–367.—Cambridge.

TURNER, J. S. & E. B. KRAUS (1966): A one-dimensional model of the seasonal thermocline, Part I–II.—Tellus **19**, 88–105.
UCHUPI, E. (1968): Atlantic continental shelf and slope of the United States—physiography.—Geol. Surv. Prof. Pap. **529-C**, 1–30. Washington, D.C.
UCHUPI, E., J. D. PHILLIPS & K. E. PRADA (1970): Origin and structure of the New England Seamount Chain.—Deep-Sea Res. **17**, 483–494.
UDA, M. (1955): On the subtropical convergence and the currents in the Northwestern Pacific.—Rec. Oceanogr. Jap. **2**, 141–150.
——— (1963): Oceanography of the subarctic Pacific Ocean.—J. Fish. Res. Can. **20**, 119–179.
——— (1964): On the nature of the Kuroshio, its origin and meanders.—In: K. YOSHIDA (Ed.), Stud. oceanogr. 89–107. Tokyo.
——— (1968): Remarkable recent advances in fisheries oceanography of Japan (from 1966–1967).—Adv. Fish. Oceanogr. **2**, 28–30.
UDA, M. & K. HASUNUMA (1969): The eastward subtropical countercurrent in the western North Pacific Ocean.—J. Oceanogr. Soc. Jap. **25**, 201–210.
ULRICH, J. (1963): Der Formenschatz des Meeresbodens.—Geogr. Rdsch. **15**, 136–148.
——— (1968): Die größten Tiefen der Ozeane und ihrer Nebenmeere.—Geogr. Taschenbuch 1966–1969, 45–48. Wiesbaden.
——— (1969): Topographie und Morphologie.—In: CLOSS, H., G. DIETRICH, G. HEMPEL, W. SCHOTT & E. SEIBOLD: "Atlantische Kuppenfahrten 1967" mit dem Forschungsschiff "Meteor."—"Meteor" Forsch.-Ergebn. (A) No. **5**, 21–23. Berlin–Stuttgart.
——— (1970): Geomorphologische Untersuchungen an Tiefseekuppen im Nordatlantischen Ozean.—In: Tagungsber. u. wiss. Abh.: Dt. Geographentag Kiel, 367–378, Wiesbaden.
UNESCO and National Institute of Great Britain (1966): International Oceanographic Tables.—UNESCO Office Oceanogr., Paris.
UNESCO (1969): Proceedings of the Symposium on tides. 1–264.—Paris.
UNTERSTEINER, N. (1967): Natural desalination and equilibrium salinity profile of old sea ice.—In: H. OURA (Ed.), Physics of snow and ice **1**, 569–577.
UREY, H. C. (1947): The thermodynamic properties of isotopic substances.—J. Chem. Soc., 562–581.
URICK, R. J. (1967): Principles of underwater sound for engineers. 1–342. New York.
U.S. Coast Guard (1950): Iceberg near the Grand Banks in spring. Photograph.
——— (1954): Typical consolidated pack ice sheet in the Arctic Ocean north of Ellesmere Island in September. Photograph.
U.S. Hydrographic Office (1948): World atlas of sea surface temperatures.—Washington, D.C.
U.S. Naval Oceanographic Office (1967): Oceanographic atlas of the North Atlantic Ocean, physical properties. Publ. No. 700.—Washington, D.C.
U.S. Navy, Department of Defense (1954): Typical pancake ice. Initial stage of a pack ice sheet in the open sea. Photograph taken north of Alaska in July 1954.
U.S. Navy (1957): Navy marine climatic atlas of the world. III, Indian Ocean.—NAVAER 50-1C-530. Chief Naval Operat. Washington, D.C.
U.S. Navy Oceanographic Office (1968, 1970): Oceanogr. atlas of the Polar seas, part I and II. Publ. No. 705.—Washington, D.C.
UTTERBACK, C. L., T. G. THOMPSON & B. D. THOMAS (1934): Refractivity-chlorinity-temperature relationships of the ocean waters.—J. Cons. **9**, 35–38.
VACCARO, R. F. (1965): Inorganic nitrogen in sea water.—In: J. P. RILEY & G. SKIRROW (Eds.), Chem. oceanogr. 1, 365–408. London.
VACQUIER, V. (1959): Measurements of horizontal displacement along faults in the ocean floor.—Nature **183**, 452–453.
VÄISÄLÄ, V. (1925): Über die Wirkung der Windschwankungen auf die Pilotbeobachtungen.—Soc. Sci. Fenn., Comm. Phys.—Math. Kl. 2, **19**, 1–46.
VALENTIN, H. (1952): Die Küsten der Erde.—Petermanns geogr. Mitt., Erg.-H. 246. 1–118.
VAUGHAN, T. W. (1937): International aspects of oceanography. Oceanographic data and provisions for oceanographic research.—NAS. Washington, D.C.
VENING-MEINESZ, F. A. (1948): Gravity expeditions at sea 1923–38, **4**.—Delft.
VERONIS, G. (1960): An approximate theoretical analysis of the equatorial undercurrent.—Deep-Sea Res. **6**, 318–327.
——— (1966): Wind-driven ocean circulation, 1 and 2.—Deep-Sea Res. **13**, 17–30, 31–56.
——— (1972): On properties of seawater defined by temperature, salinity and pressure.—J. Mar. Res. **30**, 227–255.

VILLAIN, C. (1952): Les lignes cotidales dans les océans, d'après le Dr. GÜNTER DIETRICH.—Bull. d'Inform. C.O.E.C. **4**. Paris.
VINE, F. J. (1968): Magnetic anomalies associated with mid-oceanic ridges.—In: R. A. PHINNEY (Ed.), The history of the earth's crust. 73–89. Princeton.
VINE, F. J. & H. H. HESS (1970): Sea-floor spreading.—In: A. E. MAXWELL (Ed.), The sea **4**, part II, 587–622. New York.
VINE, F. J. & D. H. MATTHEWS (1963): Magnetic anomalies over oceanic ridges.—Nature **199**, 947–949.
VOIGT, K. (1961): Äquatoriale Unterströmung auch im Atlantik (Ergebnisse von Strömungsmessungen auf einer atlantischen Ankerstation der "Michael Lomonossov" am Äquator im Mai 1959).—Beitr. Meereskde. **1**, 56–60.
VOLLMERS, H. (1969): Carriage of solid matter by currents.—XXII. Int. Navig. Congr., Paris.
VOLLMERS, H. & L. PERNECKER (1967): Beginn des Feststofftransportes für feinkörnige Materialien.—Wasserwirtschaft **6**, 237–241.
VOORHIS, A. D. (1968): Measurements of vertical motion and the partition of energy in the New England slope water.—Deep-Sea Res. **15**, 599–608.
——— (1969): The horizontal extent and persistence of thermal fronts in the Sargasso Sea.—Deep-Sea Res., Suppl. to **16**, 331–337 (FUGLISTER Vol.).
VOORHIS, A. D. & J. B. HERSEY (1964): Oceanic thermal fronts in the Sargasso Sea.—J. Geophys. Res. **69**, 3809–3814.
VOORHIS, A. D. & D. C. WEBB (1970): Large vertical currents observed in a winter sinking region of the northwestern Mediterranean.—Cah. océanogr. **22**, 571–580.
WADATI, K. (1967): Depth of earthquakes.—In: S. K. RUNCORN (Ed.), Int. dict. geophys. **1**, 389–392. London.
WARNECKE, G., L. J. ALLISON, L. M. MCMILLIN & K.-H. SZEKIELDA (1971): Remote sensing of ocean currents and sea surface temperature changes derived from Nimbus II satellite.—J. Phys. Oceanogr. **1**, 45–60.
WARREN, B. A. (1963): Topographic influence on the path of the Gulf Stream.—Tellus **15**, 167–183.
——— (1969): Divergence of isobaths as a cause of current branching.—Deep-Sea Res., Suppl. to **16**, 339–355. (FUGLISTER Vol.).
WARREN, B. A., H. STOMMEL & J. C. SWALLOW (1966): Water masses and patterns of flow in the Somali Basin during the southwest monsoon of 1964.—Deep-Sea Res. **13**, 825–860.
WARREN, B. A. & G. H. VOLKMANN (1967): A measurement of volume transport of the Gulf Stream south of New England.—J. Mar. Res. **26**, 110–126.
WASMUND, E. (1938): Entwicklung der Naturforschung unter Wasser im Tauchgerät. Geol. Meere u. Binnengewässer.—Berlin.
Wasser- und Schiffahrtsamt Cuxhaven (1954–55): Example of tides strongly disturbed by wind set-up. Water level record at the gauge from December 15, 1954 to January 3, 1955.
——— (1955): Example of tides slightly disturbed by wind set-up. Water level record at the gauge from August 16 to September 2, 1955.
WATTENBERG, H. (1929): Die Durchlüftung des Atlantischen Ozeans.—J. Cons. **4**, 68–79.
——— (1933): Kalziumkarbonat- und Kohlensäuregehalt des Meerwassers.—Wiss. Ergebn. Dt. Atl. Exped. "Meteor" 1925–1927, **8**, 1–333. Berlin.
——— (1935): Kalkauflösung und Wasserbewegung am Meeresboden.—Ann. Hydrogr. u. marit. Meteorol. **63**, 387–391.
——— (1939): Die Verteilung des Sauerstoffs im Atlantischen Ozean.—Wiss. Ergebn. Dt. Atlant. Exped. "Meteor" 1925–27, **9**, 1–131. Berlin.
——— (1943): Zur Chemie des Meerwassers.—Z. anorg. u. allg. Chem. **251**.
WATTENBERG, H. & E. TIMMERMANN (1936): Über die Sättigung des Seewassers an $CaCO_3$ und die anorgane Bildung von Kalksedimenten.—Ann. Hydrogr. u. marit. Meteorol. **64**, 23–31.
WEBSTER, T. F. (1968): Observations of inertial-period motions in the deep sea.—Rev. Geophys. **6**, 473–490.
——— (1969a): Vertical profiles of horizontal ocean currents.—Deep-Sea Res. **16**, 85–98.
——— (1969b): On the representativeness of direct deep-sea current measurements.—In: M. SEARS (Ed.), Progr. oceanogr. **5**, 3–15. Oxford.
WEDDERBURN, E. M. (1909): Temperature observations in Lock Garry.—Proc. R. Soc. Edinburgh **29**, 98–135.
WEGNER, G. (1973): Geostrophische Oberflächenströmung im nördlichen Nordatlantischen Ozean im Internationalen Geophysikalischen Jahr 1957–58.—Ber. Dt. Wiss. Komm. Meeresforsch. **22**, 411–426.
WEICHART, G. (1970): Kontinuierliche Registrierung der Temperatur und der Phosphat-Konzentration im

Oberflächenwasser des nordwestafrikanischen Auftriebswasser-Gebietes.—Dt. hydrogr. Z. **23**, 49–60.
WELANDER, P. (1959): On the vertically integrated mass transport in the oceans.—In: B. BOLIN (Ed.), The atmosphere and the sea in motion. (ROSSBY Mem. Vol.). New York.
——(1963): On the generation of wind streaks on the sea surface by action of surface film.—Tellus **15**, 67–71.
——(1969a): An ideal fluid model of the oceanic thermocline.—Göteborgs Univ. Oceanogr. Inst. Rep. No. **1**, 1–27.
——(1969b): Effects of planetary topography on the deep-sea circulation.—Deep-Sea Res., Suppl. to **16**, 360–391. (FUGLISTER Vol.)
WELZ, B. (1972): Atom-Absorptions-Spektroskopie. 1–216.—Weinheim.
WEMELSFELDER, P. J. (1971): Mean sea level as a fact and as an illusion.—Int. hydrogr. Rev. **48**, 115–127.
WEXLER, H. (1959): The Antarctic convergence—or divergence?—In: B. BOLIN (Ed.), The atmosphere and the sea in motion. 107–120. (ROSSBY Mem. Vol.) New York.
WHEWELL, W. (1833): Essay towards a first approximation to a map of cotidal lines.—Philos. Trans. R. Soc. (A) **32**, 147–236.
WILLET, H. C. (1967): Solar-climatic relationships.—In: R. W. FAIRBRIDGE (Ed.), Encycl. atmosph. sci. astro. 869–878. New York.
WILSON, W. D. (1960a): Speed of sound in seawater as a function of temperature, pressure and salinity.—J. Acoust. Soc. Amer. **32**, 641–644.
——(1960b): Equation for the speed of sound in seawater.—J. Acoust. Soc. Amer. **32**, 1357.
WILSON, W. & D. BRADLEY (1968): Specific volume of sea water as a function of temperature, pressure and salinity.—Deep-Sea Res. **15**, 355–363.
WOLFF, P. M., L. P. CARSTENSEN & T. LAEVASTU (1960): Analysis and forecasting of sea-surface temperature (SST).—Fleet Numer. Weath. Facil., Techn. Note 8, 1–30. Monterey, Calif.
WOODCOCK, A. H. (1962): Interchange of properties between sea and air: solubles.—In: M. N. HILL (Ed.), The sea **1**. 305–312. New York.
WOODS, J. D. (1968): Wave-induced shear instability in the summer thermocline.—J. Fluid Mech. **32**, 791–800.
Woods Hole Oceanographic Institution (1967): Deep-sea research submersible *Alvin*. Photograph.
WOOLLARD, G. P. & W. E. STRANGE (1962): Gravity anomalies and the crust of the earth in the Pacific basin.—Geophys. Monogr. **6**, 60–80. The crust of the Pacific basin. Amer. Geophys. Un. Washington.
WOOSTER, W. S. (1970a): Auftrieb am Beispiel des Perustromes.—In: G. DIETRICH (Ed.), Die Erforschung des Meeres. 39–51. Frankfurt.
——(1970b): Eastern boundary currents in the South Pacific.—In: W. S. WOOSTER (Ed.), Scient. explor. South Pacific, 60–68. Washington.
WOOSTER, W. S. & M. GILMARTIN (1961): The Peru-Chile Undercurrent.—J. Mar. Res. **19**, 97–122.
WOOSTER, W. S. & J. H. JONES (1970): California undercurrent off northern Baja California.—J. Mar. Res. **28**, 235–250.
WOOSTER, W. S., A. J. LEE & G. DIETRICH (1969): Redefinition of salinity.—Z. Geophys. **35**, 611–613.
World Meteorological Organization (1970): WMO Sea-ice nomenclature. 1–147.—Geneva.
WORTHINGTON, L. V. (1966): Recent oceanographic measurements in the Caribbean Sea.—Deep-Sea Res. **13**, 731–739.
——(1970): The Norwegian Sea as a mediterranean basin.—Deep-Sea Res. **17**, 77–84.
WRIGHT, W. R. (1970): Northward transport of Antarctic Bottom Water in the Western Atlantic Ocean.—Deep-Sea Res. **17**, 367–371.
WÜST, G. (1920): Die Verdunstung auf dem Meere.—Veröff. Inst. Meereskde. Berlin, N.F. (A) **6**, 1–96.
——(1924): Florida- und Antillenstrom.—Veröff. Inst. Meereskde Berlin, N.F. (A) **12**, 1–2.
——(1935): Schichtung und Zirkulation des Atlantischen Ozeans. Das Bodenwasser und die Stratosphäre. Wiss. Ergebn. Dt. Atlant. Exped. "Meteor" 1925–27, **6**, 1–288. Berlin.
——(1936): Die Vertikalschnitte der Temperatur, des Salzgehaltes und der Dichte.—Wiss. Ergebn. Dt. Atlant. Exped. "Meteor" 1925–27, Teil A des Atlas zu **6**, Beilage II–XLVI. Berlin.
——(1938): Bodentemperatur und Bodenstrom in der atlantischen, indischen und pazifischen Tiefsee.—Gerlands Beitr. Geophys. **54**, 1–8.
——(1939): Die Grenzen der Ozeane.—Ann. Hydrogr. marit. Meteorol., May Supplement.
——(1940): Zur Nomenklatur der Großformen des Ozeanbodens.—Assoc. Oceanogr. Un. Geod. Geophys. Int. Publ. Sci. **8**, 12–124.
——(1949): Über die Zweiteilung der Hydrosphäre.—Dt. hydrogr. Z. **2**, 218–225.
——(1950): Blockdiagramme der atlantischen Zirkulation auf Grund der "Meteor"-Ergebnisse. Kieler Meeresforsch. **7**, 24–34.

———(1952): Der Wasserhaushalt des Mittelländischen Meeres und der Ostsee in vergleichender Betrachtung.—Geofis. pura e appl. **21**.

———(1955): Stromgeschwindigkeiten im Tiefen- und Bodenwasser des Atlantischen Ozeans auf Grund dynamischer Berechnung der "Meteor"-Profile der Deutschen Atlantischen Expedition 1925–27.—Pap. Mar. Biol. Oceanogr.; Deep-Sea Res., Suppl. to **3**, 373–397. (BIGELOW Vol.)

———(1957): Stromgeschwindigkeiten und Strommengen in den Tiefen des Atlantischen Ozeans.—Wiss. Ergebn. Dt. Atlant. Exped. "Meteor" 1925–27, **6**, 2. Teil, 6. Lfg., 261–420. Berlin.

———(1961): Das Bodenwasser und die Vertikalzirkulation des Mittelländischen Meeres. 2. Beitrag zum mittelmeerischen Zirkulationsproblem.—Dt. hydrogr. Z. **14**, 81–92.

———(1963): On the stratification and the circulation in the cold water sphere of the Antillean-Caribbean basins.—Deep-Sea Res. **10**, 165–187.

———(1964a): Stratification and circulation in the Antillean-Caribbean basins.—Vema Res. Ser. **2**, 1–201. New York.

———(1964b): The major deep-sea expeditions and research vessels 1873–1960.—In: M. SEARS (Ed.), Progr. oceanogr. **2**, 1–52. Oxford.

———(1965): Zur Frage stationärer Verhältnisse in der Makrostruktur der Kaltwassersphäre des Atlantischen Ozeans.—Kieler Meeresforsch. **21**, 12–21.

WUNSCH, C., D. V. HANSEN & B. D. ZETLER (1969): Fluctuations of the Florida Current inferred from sea level records.—Deep-Sea Res., Suppl. to **16**, 447–470. (FUGLISTER Vol.)

WYRTKI, K. (1954): Der große Salzeinbruch in die Ostsee im November und Dezember 1951.—Kieler Meeresforsch. **10**, 19–24.

———(1960): The Antarctic Circumpolar Current and the Antarctic Polar Front.—Dt. hydrogr. Z. **13**, 153–174.

———(1961): Physical oceanography of the Southeast Asian waters.—NAGA Rep. **2**, 1–195. La Jolla, Calif.

———(1965a): Surface currents of the eastern tropical Pacific Ocean.—Inter-Amer. Trop. Tuna Comm. **9**, 270–304.

———(1965b): The thermal structure of the eastern Pacific Ocean.—Dt. hydrogr. Z. Erg.-H. A, **6**, 1–84.

———(1965c): The annual and semiannual variation of the sea surface temperature in the North Pacific Ocean.—Limnol. & Oceanogr. **10**, 307–313.

———(1969): The duration of temperature anomalies in the North Pacific Ocean.—Bull. Jap. Soc. Fish. Oceanogr., Spec. Number (UDA Comm. Pap.), 81–86.

———(1971): International Indian Ocean Expedition. Oceanographic Atlas. 1–532.—Washington, D.C.

YOSHIDA, K. (1959): A theory of the Cromwell Current (Equatorial Undercurrent) and of the equatorial upwelling—an interpretation in a similarity to a coastal circulation.—J. Oceanogr. Soc. Jap. **15**, 159–170.

———(1967): Circulation in the eastern tropical oceans with special references to upwelling and undercurrents.—Jap. J. Geophys. **4**, 1–175.

———(1972): Some aspects of theoretical studies on the Kuroshio. A review.—In: H. STOMMEL & K. YOSHIDA (Eds.), Kuroshio—its physical aspects. 433–440. Tokyo.

YOSHIDA, K. & T. KIDOKORO (1967): A subtropical countercurrent in the North Pacific.—J. Oceanogr. Soc. Jap. **23**, 88–91.

ZAHEL, W. (1970): Die Reproduktion gezeitenbedingter Bewegungsvorgänge im Weltozean mittels des hydrodynamisch-numerischen Verfahrens.—Mitt. Inst. Meereskde. Univ. Hamburg **17**, 1–50.

ZEITZSCHEL, B. (1969): Productivity and microbiomass in the tropical Atlantic in relation to the hydrographical conditions (with emphasis on the eastern area).—Proc. symp. oceanogr. fish. resources trop. Atlantic. 69–84. UNESCO, Paris.

ZENK, W. (1971): Zur Schichtung des Mittelmeerwassers westlich von Gibraltar.—"Meteor" Forsch.-Ergebn. (A) No. **9**, 1–30. Berlin-Stuttgart.

ZENKOVITCH, V. P. (1966): Black Sea.—In: R. W. FAIRBRIDGE (Ed.), Encycl. oceanogr., 145–151. New York.

ZENKOVITCH, V. P. & B. A. SHULYAK (1967): Progress in the study of the inshore zone of the sea.—Oceanology **7**, 628–641.

ZETLER, B. D. & R. A. CUMMINGS (1967): A harmonic method for predicting shallow-water tides.—J. Mar. Res. **25**, 103–114.

ZETLER, B. D. & V. D. HANSEN (1970): Tides in the Gulf of Mexico.—A review and proposed program.—Bull. mar. Sci. **20**, 57–69.

ZIMMERMANN, H. B. (1971): Bottom currents on the New England continental rise.—J. Geophys. Res. **76**, 5865–5876.

Author Index

Aagaard, K., 502, 504
Academy of Sciences (USSR), 210
Accad, Y., 436, 437, 439, 440
Addison, J. R., 87
Airy, G. B., 430, 439
Albrecht, F., 178, 181, 182, 196, 200
Allan, T. D., 47
Allison, L. J., 532
Anderson, E. R., 74, 213
Anderson, E. W., 109, 529, 563, 569
Andreyev, E. G., 204
Armstrong, F. A. J., 510
Arons, A. B., 69, 469
Arrhenius, G., 25, 32
Arx, W. S., von, 146, 533
Atkinson, G., 78

Bagnold, R. A., 27
Baker, D. J., Jr., 130
Bang, N. D., 510
Barazangi, M., 34
Barkley, R. A., 525, 526
Barnes, H., 155
Barret, J. R., 534
Bartels, J., 413, 416, 417
Bath, U., 34
Bathen, K. H., 513
Bein, W., 61, 80
Belding, H. F., 9
Belyakov, Yu. M., 470
Berckhemer, H., 34
Bergeron, T., 192
Bernoulli, D., 428
Beyer, F., 563
Bezrukov, P. L., 25

Bialek, E. L., 60, 74
Bjerknes, J., 221, 510, 513
Bjerknes, V. W., 65, 291, 334
Blackwell, D. D., 40
Blanford, R., 337
Bock, K. - H., 553
Bogdanov, K. T., 439
Böhnecke, G., 210, 230
Bouguslawkii, S. G., 470
Bowen, J. S., 180
Bradshaw, A., 62, 65, 566
Brandt, K., 279
Brekhovskikh, L. M., 488
Brennecke, W., 461
Breslau, L. R., 250
Brettschneider, G., 436
Brewer, P. G., 157
Brill, R., 58
Brocks, K., 161, 163, 165, 202
Broecker, W. S., 461
Brogmus, W., 196
Brosche, P., 457
Brosin, H. - J., 517
Brown, N. L., 62, 139, 566
Bruyevich, S. V., 168
Bryan, K., 328, 341, 342, 474, 481
Bryden, H. L., 568
Buchanan, J. Y., 516
Buchsbaum, R., 218
Bückmann, A., 539
Büdel, J., 239, 244, 245, 246
Budyko, M. J., 174, 176, 181, 183, 184
Bulgakov, N. P., 538
Bullard, E. C., 53
Burt, W. V., 80, 170

605

Bush, K., 279, 281

Caldwell, D. R., 65, 113, 130
Cameron, W. M., 559
Carlson, H., 369, 370
Carrit, D. E., 61
Carstensen, L. P., 215
Cartwright, I. D., 372, 421, 424, 425
Chalupnik, J. D., 143
Charney, J. G., 327, 517
Chester, R., 25
Christensen, N., 516
Chromov, S. P., 191
Chrystal, G., 388
Clarke, G. L., 82, 84
Clauss, E., 159
Collette, B. J., 47
Corkan, R. H., 382, 383
Courtier, A., 427
Cox, A., 38, 61
Cox, C. S., 170, 375, 403
Cox, D., 341, 481
Cox, R. A., 65, 66, 142, 491
Crease, J., 65, 465
Cromwell, T., 513
Culkin, F., 88, 491
Cummings, R. A., 425
Curcio, J. A., 82
Currie, R. I., 510
Cushing, D. H., 510

Darbyshire, J., 529
Darwin, Charles, 44
Davies, J. T., 166
Deacon, G. E. R., 461, 476, 483
Defant, A., 146, 188, 216, 226, 234, 311, 313, 315, 318, 389, 396, 403, 405, 413, 428, 431, 433, 434, 435, 442, 464, 469, 470, 473, 477, 480, 483, 520
Degens, E. T., 95, 96, 495, 496
Del Grosso, V. A., 74
Demintskaya, R. M., 47
Dickson, R., 219, 221
Dietrich, G., 5, 8, 10, 11, 12, 17, 25, 29, 37, 48, 49, 60, 76, 84, 99, 111, 180, 182, 185, 203, 206, 208, 210, 211, 214, 215, 219, 221, 224, 226, 228, 237, 284, 299, 377, 403, 411, 423, 424, 439, 442, 458, 459, 463, 465, 466, 480, 483, 484, 496, 498, 501, 502, 503, 504, 505, 506, 509, 510, 514, 523, 527, 529, 530, 531, 532, 534, 539, 540, 542, 547, 548, 551, 552, 554

Dietz, R. S., 39, 47
Dill, R. F., 45
Dittmar, W., 60
Dodimead, A. J., 213
Dodu, J., 340
Donk, J. V., 218
Doodson, A. T., 415, 416, 421, 431, 432, 435
Dorman, J., 34
Dorsey, N. E., 65
Drever, R. G., 146
Düing, W., 146, 185, 228, 517, 522, 524, 534, 557
Duncan, C. P., 529
Dunkel, M., 203
Duun-Christensen, J. T., 454
Duvanin, A. I., 221

Eber, L. E., 210
Edgerton, H. E., 17
Ehrhardt, M., 553
Ehricke, K., 72
Eittreim, S., 465
Ekman, V. W., 64, 66, 146, 320, 321, 322, 323, 329, 330, 332, 518
Ellett, D., 466, 467
Elmendorf, C. H., 31
El-Sabh, M. I., 493
El Wakeel, S. K., 24
Emery, K. O., 10, 42, 44
Emiliani, C., 218
Epstein, S., 218
Ermel, H., 6
Eucken, A., 56
Ewing, G. C., 110
Ewing, J., 33, 34, 36, 47, 48
Ewing, M., 17, 30, 34, 36, 47, 77, 465
Eyries, M., 130, 438

Farquharson, W. I., 442
Fedorov, K. N., 62
Filippov, D. M., 215
Filloux, J. H., 438
Fjeldstad, J. E., 320, 455
Fleischer, U., 37
Fleming, R. H., 279
Fliegel, M., 438
Fofonoff, N. P., 67, 183, 461, 486, 487, 517, 535, 536, 567
Foldvik, A., 461
Fonselius, S. H., 555
Forch, von Carl, 64, 138

Forel, F. A., 30, 388
Fox, Ch. J. J., 93
Føyn, E., 562, 563
Francis, J. R. D., 72
Frederiksson, K., 24
Freiesleben, H. C., 109
Friedrich, H., 341
Friedrich, H. J., 342, 481
Fuglister, F. C., 212, 230, 326, 327, 483, 531, 532, 533
Fujino, K., 87
Funnell, B. M., 21

Gast, P. R., 173
Gaye, J., 382, 384
Gentilli, J., 193
Gerstner, F., 354
Gieser, W., 28, 493
Gieskes, J. M., 157
Gieskes, M. T. M., 157, 548
Gill, E. E., 474, 481, 517
Gillbricht, M., 280
Gilmartin, M., 510
Goldberg, R. D., 464
Goldsbrough, G. R., 336, 337, 341, 431
Goldschmidt, V. M., 251, 255
Gorbunova, Z. N., 25
Gordon, A. L., 461, 462, 463, 464, 474, 499
Grace, S. F., 434, 455
Graf, A., 37
Grant, H. L., 71
Grasshoff, K., 155, 156, 157, 471, 495, 496, 510, 556
Green, C. K., 380, 381
Greenspan, M., 142
Griffin, J. J., 21, 25
Groll, M., 5
Guilcher, A., 44
Gunther, E. R., 510
Gustafson, T., 330
Gutenberg, B., 34, 35, 411

Hagen, G., 296
Hamon, B. V., 528
Hansen, D. V., 523, 535, 565
Hansen, W., 385, 428, 436, 444, 454, 534
Harris, R. A., 391, 439
Hart, T. J., 510
Hartmann, M., 557
Harvey, H. W., 276, 279

Hasse, L., 159, 161, 165, 185, 298
Hasselmann, K., 133, 355, 425
Hasunuma, K., 526
Hayes, D. E., 46
Heaps, N. S., 454
Heberlein, H., 16
Heck, N. H., 74, 379
Hecker, O., 37
Heezen, B. C., 8, 11, 17, 30, 31, 45, 51
Heirtzler, J. R., 34, 38, 48, 52
Hela, I., 222
Helland-Hansen, B., 233, 315, 398, 502
Helmholtz, H. V., 372, 373
Henderschott, M. C., 439
Herring, P. J., 107
Herrmann, G., 260
Hersey, J. B., 17, 36, 526
Herzen, R. P. von, 34, 40, 41
Hess, H. H., 12, 51, 53
Hesselberg, T. H., 304, 305
Hidaka, K., 325, 508
Higashi, K., 69
Hinkelmann, H., 139, 140
Hinz, K., 33, 50
Hinzpeter, H., 82, 159
Hisard, P., 516
Hjulström, F., 27
Hoeber, H., 162, 163, 164, 203, 204
Holland, W. C., 9
Hollister, C., 17
Horn, W., 410, 416, 426, 448, 484, 504
Horn, W. F., 188, 205, 206, 486
Horne, R. A., 57, 58, 73
Hough, S. S., 430
Hubert, W. E., 215
Hurnbert, F. L., 30
Hunger, H., 17
Hunkins, K. L., 47

Ichiye, T., 474, 499, 517
IGY, 103
Inman, D. L., 26, 29, 377
Instrument Society of America, 120
Interagency Committee on Oceanography, 107
International Hydrographic Bureau, 1, 2, 13-14
Ivanoff, A., 82, 152

Jacobs, W. C., 177, 180, 181, 182, 200
Jacobsen, J. P., 61, 72, 234, 404

Jansson, B. O., 554
Jeffreys, H. 372, 375, 457
Jenson, A. S., 222, 509
Jerlov, N. G., 84, 152, 170, 173, 175, 203
  203
Johannsen, K., 100
Johnson, D., 534
Jones, P. G. W., 510
Joseph, J., 60, 84, 137, 151, 465, 466, 467,
  468, 552
Junge, C. E., 168, 259

Kalle, K., 83, 92, 263, 273, 284
Kändler, R., 564
Kanwisher, J., 166
Katz, E. J., 137, 526
Kawai, H., 535
Kelvin, Lord, 372, 387, 421, 425
Kendall, T. R., 518, 519
Keunecke, K. H., 504
Kidokoro, T., 526
King, C. A. M., 44
Kinne, O., 113
Kinsman, B., 344
Kitaigorodskii, S. A., 202
Knauss, J. A., 513, 514, 515, 517, 519, 530,
  533
Knopoff, L., 53
Knox, G. A., 538
Knudsen, M., 60, 64, 71, 128, 142
Koczy, F. F., 89, 259
Kohl, J. G., 529
Koopman, G., 538
Köppen, V., 218, 507
Kort, V. G., 474, 475, 483, 550
Korteweg, D. J., 379
Korzh, V. D., 168
Koshlyakov, M. N., 513
Koske, P. H., 157, 558
Koslowski, G., 239
Koyama, T., 95
Kraus, E. B., 158, 165, 202, 340
Krause, G., 143, 151, 471, 496, 498
Krauss, W., 111, 250, 294, 307, 337,
  338, 339, 344, 388, 390, 401, 404,
  406, 444
Kremling, K., 62, 65, 142, 154, 491
Krey, J., 150
Kritchevsky, I., 65
Kroebel, W., 140, 142, 151, 152
Krügermeyer, L., 163, 165
Krümmel, O., 68, 69, 73, 80, 238, 397, 439

Kruppa, C., 120
Kruspe, G., 165
Keunen, Ph. H., 16, 30, 45
Kuhlbrodt, E., 176
Kullenberg, B., 330
Kusnetsova, L. P., 196, 197
Kuwahara, S., 74
Kvinge, T., 461
Lacombe, M., 146, 492
La Coste, L. J. B., 37
Ladd, H. S., 13
Laevastu, T., 215, 222
La Fond, E. C., 119
Lake, R. A., 85
Lal, D., 461
Lamb, H., 221, 296, 429
Langleben, M. P., 86, 87
Langseth, M. G., 34
Laplace, P. S., 410, 429, 430
Laughton, A. A., 17, 18, 19, 48
Lauscher, F., 81
Lee, A., 219, 221, 466, 467, 483, 502, 544, 552
Lee, W. H. K., 40
LeFloch, J., 517
Le Grand, Y., 82
Leipper, D. F., 529
Lemasson, L., 528
Lenz, W., 553
Le Pichon, X., 51, 52
Lettau, H., 395, 396
Lewis, E. L., 86
Li, Y. - H., 461
Liebermann, L. N., 78
Lighthill, M. J., 522, 523
Lisitzin, A. P., 16, 25, 411
Litovitz, T. A., 78
Longuet-Higgins, M. S., 355, 356, 357, 369,
  370, 387
Lowenstam, H., 218
Lubbock, J. W., 410
Lucht, F., 561, 562
Ludwig, W. J., 34, 35
Lumby, J. R., 213
Luvendyk, B. P., 34, 46
L'Vov, B. V., 153
Lvovitch, M. J., 198, 199
Lynn, R. J., 216, 463, 472

McAlister, E. D., 205
McCartney, M. J., 491
McDonald, W. F., 181, 194, 210, 457
McLeish, W., 205

Magaard, L., 344, 345, 390, 403, 444
Maillard, C., 496
Malkus, J. S., 181, 202
Malmgren, F., 88
Mamayev, O. I., 505
Mann, C. R., 530
Marcinek, J., 197
Marine Technology Society, 120
Marsili, L. F., 495
Mason, R. G., 37, 38
Masterson, J. E., 148, 149
Masuzawa, J., 518, 535
Mathieu, G., 218
Matthäus, W., 517, 569
Matthews, D. H., 39, 49
Matthews, D. J., 74, 76
Maxwell, A. E., 34, 40, 47
Meincke, J., 28, 215, 449, 450, 485
Menaché, M., 65, 491
Menard, H. W., 3, 7, 8, 13, 20, 23, 49
Menzel, D. W., 94
Merle, J., 519, 526
Mero, J. L., 23
Merz, A., 335, 493
Metalnikov, A. P., 599
Metcalf, W. G., 517
Milankovitch, M., 42, 171, 218
Miles, J. W., 374
Miller, A. R., 490
Millero, F. J., 66
Mittelstaedt, E., 510
Miyake, Y., 69, 71, 73, 80
Model, F., 189, 553
Moiseyev, L. K., 213
Möller, F., 176, 196, 493
Montgomery, R. B., 180, 213, 231, 237, 238, 517, 519
Morcos, S. A., 230, 491, 493, 496
Morel, A., 82
Morelli, C., 47
Morgan, G. W., 327
Mosby, H., 174, 461, 462, 502
Moskowitz, L., 371
Mügge, H., 397
Muller-Krauss, 109
Munk, W. H., 72, 170, 213, 325, 360, 363, 367, 370, 376, 385, 386, 421, 424, 425, 438, 441, 535
Murray, J., 16
Myers, J. J., 103

Nace, R. L., 197

Nairn, A. E. M., 217
Namais, J., 221
Nansen, F., 398, 502, 545
National Academy of Sciences, 88
Nehring, D., 517
Nelson, S. B., 103
Netherlands Meteorological Institute, 210
Neumann, G., 226, 232, 357, 360, 363, 364, 365, 389, 390, 396, 398, 479, 481, 493, 516, 517, 519
Newton, I., 411
Neyman, V. G., 513
Nowlin, W. D., Jr., 474
Nowroozi, A. A., 438
Nusser, F., 245, 246

Ogura, S., 445
Okubo, A., 72
Olson, B. E., 463
Olson, J. R., 143
Opdyke, N. D., 34, 38-39
Oren, O. H., 493
Orlenok, V. V., 33
Oura, H., 88
Ovchinnikov, I. M., 490, 517
Overstreet, R., 213
Owen, R. W., 509

Palmen, E., 191
Panfilova, S. G., 206
Papanin, J. D., 545
Park, K., 62, 63
Patzert, W. C., 496
Payne, R. E., 80, 81
Pekeris, C. L., 436, 437, 439, 440
Permecker, L., 27
Pettersson, H., 24, 398
Phillips, O. M., 355, 372, 373, 374
Pickard, G. L., 560
Pickett, R. L., 532
Pierson, W. J., 366, 371
Pitman, W. C., 39, 51
Plakhin, Ye. A., 491
Platzman, G. W., 390
Poincaré, H., 429, 431
Popova, A. K., 41
Pounder, E. R., 88, 239
Prabharka, J., 184
Pratt, R. M., 10
Pritchard, D. W., 559
Proudman, J., 428, 431, 435

Puls, 516, 517
Putman, J. H., 377

Raben, E., 279
Rabsch, U., 558
Rakestraw, N. W., 92
Ramage, C. S., 523
Rao, D. B., 390, 394
Rasool, S. J., 184
Rattray, M., Jr., 213, 565
Rauschelbach, H., 420, 421
Reeburgh, W. S., 62, 567
Reichel, E., 179
Reid, J. L., 210, 216, 463, 472, 474, 510, 511, 519
Renard, A. F., 16
Revelle, R., 24, 40, 257
Reynolds, O., 296
Rhines, P., 536
Richardson, L. F., 297
Richardson, P. L., 530, 533
Richardson, W. S., 145, 146, 530
Richter, C. F., 34, 35
Rideal, E. K., 166
Riedel, W. R., 21
Riel, P. M., van, 499, 500
Riley, G. A., 97, 272
Riley, J. P., 24, 155, 157
Robinson, A. R., 337, 517, 526, 532, 535
Roden, G. I., 467
Rodewald, M., 219, 220
Roll, H. U., 158, 196, 366, 367
Ross, D. A., 60, 495, 496
Rossby, C. G., 184, 213, 332, 333, 485, 487, 535
Rotschi, H., 528
Rouch, J., 222
Rouse, H., 340
Rubner, M., 269
Runcorn, S. K., 51, 109
Rusby, J. S. M., 81
Russel, Lord, 379
Ryan, W. B. F., 47

Sager, G., 442, 443, 451
Sahrage, D., 214, 221
Sandström, W., 291, 315, 334, 403
Saunders, P. M., 79
Saur, J. F. T., 180, 210
Sauramo, M., 43
Schell, I. I., 246

Schleicher, K. E., 65, 139
Schmidt, W., 180
Schmidt, W. J., 530
Schmitz, H. P., 517, 523
Schneider, E. D., 48
Schott, F., 18, 22, 50, 226, 227, 228, 229, 452, 453, 553
Schott, G., 25, 32, 308, 505, 510
Schubert, K., 214, 221, 403
Schulayk, B. A., 44
Schulkin, M., 78
Schulz, B., 279
Schumacher, A., 363
Schureman, P., 417, 421
Schütz, D. F., 153
Schwarzbach, M., 217, 219
Schweigger, E., 510
Schwerdtfeger, P., 85, 86, 87
Scripps Institution of Oceanography, 212
Seabroke, J. M., 461
Sears, M., 483
Seckel, G. R., 183
Segerstråle, S. G., 229, 554
Seibold, E., 16, 28, 44, 493, 557
Sellers, W. D., 171, 172, 176
Sen Gupta, R., 495
Sette, O. E., 210
Sharova, V. Ya., 196, 197
Shepard, F. P., 16, 42, 44, 45, 377, 380
Shoji, D., 535
Siedler, G., 111, 139, 140, 146, 235, 496, 497
Sigtryggson, H., 549
Skirrow, G., 281
Skopintsev, B. A., 493
Smed, J., 226, 228
Smith, E. H., 241, 247
Smith, R. L., 510
Smith, S. M., 3, 7
Snodgrass, F. E., 438
Snyder, R. L., 375
Sørensen, S. P. L., 64, 138
Spilhaus, A. F., 136
Stanley, E. M., 73
Steemann-Nielsen, E., 274, 509, 526
Steinitz, H., 493
Stepanov, V. N., 237, 505
Stern, M. E., 68
Sterneck, R. V., 428, 431
Stewart, R. W., 202, 356, 357
Stocks, Th., 5, 6, 19

Author Index

Stokes, G. G., 386
Stommel, H., 323, 324, 325, 337, 469, 474, 490, 522, 529, 531-532, 535, 536, 559
Strange, W. E., 37
Strickland, D. H., 277
Strøm, K. M., 562
Stroup, E. D., 519
Strübing, K., 248
Sturges, W., 500
Subow, N. N., 545
Sudo, H., 499
Suellentrop, F. J., 143
Suess, H. S., 257
Sugawara, K., 168
Sukhovey, V. F., 599
Sündermann, J., 385, 457
Sutton, G. W., 438
Svansson, A., 553
Sverdup, H. V., 179, 183, 233, 237, 299, 301, 305, 321, 323, 358-360, 363, 367, 445, 455, 456, 505, 520, 545
Swallow, J. C., 149, 517
Swill, W. - D., 557
Swinbank, W. C., 179
Sykes, L. R., 34
Szekielda, K. -H., 185, 186, 523

Tabata, S., 183
Taft, B. A., 517, 535
Talwami, M., 37, 46, 47
Tams, E., 34
Taylor, G. I., 393, 394, 431, 457
Taylor, R. C., 196
Thiede, J., 33
Thomas, B. D., 61
Thompson, R., 536
Thompson, T. G., 65, 485
Thomsen, H., 404
Thorade, H., 392, 398, 428, 435, 455
Thorpe, S. A., 143
Thoulet, J., 66
Timmermann, E., 282
Tomczak, G., 454, 455, 510
Trepka, L. V., 557
Tsuchiya, M., 513
Tucker, M. J., 132
Tully, J. P., 215
Turekian, K. K., 153
Turner, J. S., 235, 340
Turner, J. S., 68

Uchupi, E., 10, 12

Uda, M., 221, 526, 535, 537
Ulrich, J., 3, 5, 6, 8, 10, 11, 12, 13, 37, 38, 48, 49, 501
U.N.E.S.C.O., 61, 566
U. S. Hydrographic Office, 210, 545
U. S. Naval Oceanographic Office, 210, 490
U. S. Navy, 210
Untersteiner, N., 85
Urey, H. C., 32, 218
Urick, R. J., 77, 79
Utterback, C. L., 80

Vaccaro, F., 268
Väisälä, V., 401
Valentin, H., 44
Vaquier, V., 37
Vaughan, T. W., 212
Vening-Meinesz, F. A., 37, 53
Veronis, G., 67, 328, 522, 535
Villian, C., 439
Vine, F. J., 39, 40, 49, 51
Vogt, P. R., 48
Voigt, K., 517
Volkmann, G. H., 530
Vollmers, H., 27
Vollbrecht, K., 557
Voorhis, A. D., 150, 490, 526, 532
Vries, G. de, 379

Wadati, K., 34
Walther, F., 382, 384
Warnecke, G., 185
Warren, B. A., 522, 530, 535
Wasmund, E., 16
Wasser-und Schiffahrtsamt Cuxhaven, 454, 455
Wattenberg, H., 281, 282, 283, 285, 464
Webb, D., 485, 487, 490
Webster, T. F., 329, 487, 536
Wedderburn, E. M., 400
Wegener, A., 218
Wegner, G., 541, 542
Weichart, G., 510
Weidemann, H., 377
Welander, P., 307, 333, 337, 338, 473
Welz, B., 153
Wemelsfelder, P. J., 411
Wexler, H., 476
Whewell, W., 439
Willet, H. C., 220

Williams, E. R., 517
Wilson, L., 8
Wilson, W., 65
Wilson, W. D., 74
Wolff, P. M., 215
Woodcock, A. H., 167
Woods, J. D., 71
Woolard, G. P., 37
Wooster, W. S., 61, 510
World Meteorological Organization, W. M. O., 231
Worthington, L. V., 326, 327, 500, 502, 532
Wright, W. R., 483
Wunsch, C., 532
Wust, G., 5, 11, 98, 179, 180, 193, 196, 199, 212, 216, 222, 231, 233, 235, 322, 335, 340, 465, 469, 471, 472, 473, 481, 482, 483, 490, 491, 499, 500, 501, 505, 534
Wyrtki, K., 182, 183, 206, 207, 216, 221, 476, 498, 519, 521, 523, 557

Yoshida, K., 510, 517, 526, 529, 535

Zahel, W., 439, 440
Zeitzschel, B., 509, 520
Zenk, W., 235, 236, 492, 493, 494
Zenkovitch, V. P., 44, 493
Zetler, B. D., 425, 444
Ziegenbein, J., 496
Zimmermann, H. B., 462

Subject Index

Acoustic properties of seawater, 74-78
Acoustic transponders, 109
Aegean Sea, 434
Agulhas Current, 528-529, 536
Airplanes, for oceanographic observations, 110-111, 133
Algae, phytoplankton, 278
"Alkalinity," 281
Alkor (research vessel), 100-101
Alvin (deep-sea submersible), 16-17, 102, 104-105, 107
American Mediterranean Sea, 3, 498-504
Amphidromic systems, long surface waves, 391-394
Antarctic Convergence, 312
Antarctic Circumpolar Current, 474-475
Antarctic Divergence, 476
Antarctic Ocean, 4
  bottom water, 461-465
  polar regions, 546-547
  subpolar intermediate water, 467-470
  water masses, 474-476
Arabian Sea, 524, 557
Arctic Mediterranean Sea, 3, 501-504
Arctic Ocean, 3, 236
  bottom water, 465-467
  ice drifts, 544-546
  stratification and circulation, 465-467, 501-504
  subpolar intermediate water, 467-470
Assimilation process, 268, 271
Aswân Dam, Egypt, 493
Atlantic Ocean, 3, 7
  deep water, 470-473
  gyres, 481
  North, 312
  salinity, 225
  South, 222
  subpolar intermediate water, 467-470
  warm-water sphere, 480-481
  water masses, 236-237
Atmosphere-ocean interaction, 158-202
  determination of vertical transports in maritime friction layer, 159-162
  large-scale heat budget, 168-189
  large-scale water budget, 181-202
  molecular boundary layer, 159
  parameterization of vertical transports, 164-166
  small-scale transfer processes, 158-168
  transfer of gases and salts, 166-168
  vertical fluxes of momentum, heat, and water vapor, 163-164
Atolls, formation of, 49
Atomic absorption spectroscopy, 153
Attenuation, measurement of, 150-152
Austral-Asiatic Mediterranean Sea, 3
  stratification and circulation, 497-498
  temperature, salinity, and oxygen content, 500
Azores, 528

Baffin Bay, 236
Baltic Sea, 3, 43-44, 247, 284, 550
  hydrographic regions, 551-552
  ice conditions, 240, 246, 248-250
  salinity, 226, 229
  stratification and circulation, 550-558
  tides, 442, 444
  waves, 389-391

Barents Sea, 547, 550
Basins, deep-sea, 8, 11, 48-49
  abyssal hills and plains, 11-12, 48
  formation of, 48-49
  step or fracture zones, 48-49
Bathyscaph (submersible), 16
"Bathysonde," 140
Bay of Fundy, 434, 444
Benguela Current, 270, 508
Bering Sea, 3, 524-525
Bering Strait, 465, 504
Bermuda Islands, 528
Biochemistry, 261-287
  carbon dioxide-calcium carbonate system, 280-284
  energy production of organisms, 266-271
  general fundamentals, 261-264
  mechanism of plankton metabolism, 275-280
  "microbiological" methods, 154
  ocean as source of life, 264-266
  organic production of ocean, 271-275
  regional distribution of nutrients and oxygen, 284-287
Black Sea, 236, 283, 493-495
Blood serum, composition of seawater and, 265
Boiling point, seawater, 70
"Bore," tidal waves in rivers, 397-398
Borha II, French buoy, 112, 114
Bottom topography, 1-15
  aids for interpreting morphology of, 33-41
    geomagnetic measurements, 38-41
    gravimetric measurements, 37
    seismic investigations, 34-41
  boundaries and names of oceans, 1-2, 504-505
  charts, 5-6
  deep-sea basins, 11-12
  deep-sea floor, 17-18
  development of knowledge of, 2-6
  distribution of land and water, 1
  formation of bottom features, 33-53
    continental margins, 41-46
    Mid-Ocean Ridge, 46-48
    oceans, 50-53
    seamounts and oceanic islands, 49-50
  geomorphology of, 1-53
  heat flow measurements, 40-41
  influence on currents, 332-334
    potential vorticity, 332-333
    topographic Rossby waves, 333-334
  large-scale features, 8-13

  manganese nodules, 20, 23
  nature of, 16
  relief of sea floor, 2-6
  sediments, 16-26. *See also* Sediments and sedimentology
  size of oceans, 2
  spreading of ocean floor, 47-52
  statistics of depth distribution, 6-8
  terms and definitions, 13-15
Bottom water:
  Antarctic, 461-465
  Arctic Ocean, 465-467
  coastal waters, 559
  European Mediterranean Sea, 490-491
  polar, 461
  subpolar intermediate water, 467-470
Boundaries of oceans, 1-2
  classification of, 504-505
Bourdon tubes or Well tubes, 127-128, 130-131
Bouvet divergence, 538
Brazil Current, 528
Buoys (measurements), 112, 132-134, 161

Cables, transatlantic, 4
  breaks, 30-32
Calcium carbonate, 25
  carbon dioxide and, 279-284
Canary Current, 408
Canary Islands, 528
Canyons, submarine, 9, 32, 45
Cape Verde Islands, 512
Carbohydrates, 279
Carbon:
  assimilation of, 280
  basic element in living matter, 262
  cycles, 262-264
Carbon dioxide, 25, 262, 279
  -calcium carbonate system, 280-284
  in seawater, 89, 92
Caribbean Sea, 498
Central rift valleys, 10
Charts, 4-6
  bathymetric and nautical, 5-6, 13
  isotherm, 209-212
  sounding depths, 4-6
Chemistry, marine, 153, 251-287
  biochemistry of ocean, 261-287
    carbon dioxide-calcium carbonate system, 280-284
    energy production of organisms, 266-271
    general fundamentals, 261-264

Subject Index **615**

mechanism of plankton metabolism, 275-280
ocean as source of life, 264-266
organic production of ocean, 271-275
regional distribution of nutrients and oxygen, 284-287
chemical budget, 251-287
composition of marine sediments, 24-25
elements in seawater, 88-97, 253-255
marine geochemistry, 251-261
  general fundamentals, 251-253
  primary natural radioactivity, 256-257
  radioactivity generated by cosmic rays, 257-259
  radioactivity produced artifically, 259-261
  regulating mechanisms for chemical elements, 253-255
  sedimentation budget, 24-25, 255-256
China Sea, 524
Chloride content, 61
  measuring, 138
Chromatographic methods, 154
Circulation, oceanic:
  distribution of temperature and salinity, 334-338
  inclination of seawater stratification in currents, 270
  Mediterranean and adjacent seas, 488-490
  meridional, 342, 481-482
  mixing processes and, 337-338
  numerical solutions, 341-342
  and stratification, see Stratification and circulation
  thermal, 334-338
  thermohaline, 334-342
    coastal currents in higher latitudes, 338
    large-scale processes, 334-338
    in ocean straits, 338-340
  vertical convection of water masses, 270
  wind-driven currents, 270, 324-326, 336-338
  See also Currents
Coastal waters, 44-45, 558-565
  bottom topography and, 559
  continental runoff, 559
  estuaries and, 559-564
  tides and tidal currents, 559
Cold-water sphere, 460-461
  water masses, 460-461
Colorimetric methods, 153, 155-157
Color of seawater, 84
Compressibility of seawater, 63-66

Computer analysis, 109-110
Continental Divergence, 476
Continental drift theories, 47, 51
Continental margins, 8-9
  formation of, 41-46
  shelves, 41-46
Continental rises, 10, 45-46
Continental runoff, 197-198, 559
Continental shelf, 9-10
  formation of, 41-45
Continental slopes, 8, 45-46
Continuity, equations of, 292-293
Copenhagen Normal Water, 81, 138
Coral reefs and atolls, 25-26, 44
Coriolis force, 289-290, 329, 331
Cromwell Current, 513-517
Currents, ocean, 288-342
  bottom, 17
    drift measurements, 150
  coastal water, 558-565
  countercurrents, 510
  drift, 319-320
  effect on heat budgets, 183
  equatorial countercurrents, 518-521
  geostrophic, 308-318
    in continuously stratified ocean, 314-318
    in homogeneous ocean, 308-310
    in two-layer ocean, 310-314
  horse latitudes, 190, 525-528
  hydrodynamic equations, 288-299
    boundary conditions, 294-295
    deflecting force of rotation of earth, 289-290
    equations of continuity, 292-293
    equations of motions, 288-289, 291-292
    equations of thermal conduction and diffusion, 294
    field of gravity, 290-291
    friction, turbulence, and mixing, 295-299
  influence of bottom topography, 332-334
    potential vorticity, 332-333
    topographic Rossby waves, 333-334
  jet stream regions, 487, 528-536
  kinetic energy, 484-487
  meandering, 326, 532, 535
  measurement of, 143-157
    current meters, 143-145
    from moored vessels, 146
    sensors for, 143
  monsoons, 190, 521-525
  nonstationary, 328-332, 484
    waves and, 328-329

inertial, 329-331
planetary, 331-332
polar regions, 544-550
shelf seas, 550-558
slope, 528
statics and kinematics, 299-308
  field of mass, 299-301
  field of pressure, 302-304
  representation of field of motion, 307-308
  stability of water stratification, 304-307
stationary, 308-328
  in continuously stratified ocean, 314-318
  drift currents, 319-320
  Ekman's elementary system, 321-322
  linear theory of western boundary currents, 324-327
  nonlinear theory of western boundary currents, 327-328
  Sverdrup regime, 321-323
  in two-layer ocean, 310-314
stratification, *see* Stratification and circulation
thermohaline circulation, 334-342
  coastal currents in higher latitudes, 338
  compensation currents in ocean straits, 338-340
  large-scale processes, 334-338
  numerical solutions regarding, 341-342
  thermoclines, 340-341
tidal, 407-410, 446-453. *See also* Tidal phenomena
trade wind currents, 190, 505-518
  components of motion directed poleward, 518
  components of motion directed equatorward, 508-512
  with strictly westward motion, 512-518
turbidity currents, 29-31, 45-46, 465
  off Grand Banks, 30-31
velocities, 315-317
volume transport, 317-318
west wind drift, 536-544
Cyclones, 191, 193, 379

Dating sediments, 32
Deep water, 470-473
  stratification and circulation, 470-473
  tongue-like spreading, 472-473
Deflecting force of rotation of earth, 289-290
Denmark Strait, 466
Density, 63-66, 230-232
  distribution at sea surface, 230
  distribution with depth, 231

effect of temperature and salinity, 230
equations and formulas, 567-568
horizontal differences, 231
measurement of, 142-143
of seawater, 63-66
vertical distribution of, 231
Detritus, 97, 280
Deuterium oxide, 59
Diatoms, 269, 276
Dissolved substances, measurement of 152-157
Drake Passage, 475
Drift measurements, 148-150
  at arbitrary depth, 149-150
  in bottom current, 150
  in near-surface water, 149
  in surface water, 148-149

Earth, deflecting force of rotation, 289-290
Earthquakes, 34
  epicenters, 52-53
  foci of, 34-35
  off Grand Banks (1929), 30
  submarine, 379
  waves generated by, 345
East Australian Current, 528-529
East China Sea, 3, 442
Echo sounders, 5, 19, 36, 122-126
  for fish finding, 126
Eddies, 373
  cyclonic, 326-327
  stratification in, 312-314
Ekman transport, 321-322, 478
Electrical conductivity of seawater, 61-63
  measurement of, 138-139
Electrical properties of sea ice, 87
Electromotive force (EMF), 146-148
Electronic navigational methods, 107-108
Elements in seawater, 253-255
  distribution of, 254
  necessary for life, 252, 269
  selective adsorption, 255
  solubility properties, 255
"El Niño" Current, 510
Energy, production of organisms, 266-271
Energy and water budgets, 158-202
  large-scale heat budget, 168-169
  large-scale water budget, 189-202
  small-scale transfer processes in atmospheric boundary layer, 158-168
English Channel, 434, 459
Equations:
  boundary conditions, 294-295

of continuity, 292-293
deflecting force of rotation of earth, 288-290
field of gravity, 290-291
friction, turbulence, and mixing, 295-299
hydrodynamic, 288-299
of motion, 288-289, 291-292
of thermal conduction and diffusion, 294
Equatorial Countercurrent, 477-478, 518-521
  hydrographic structure, 518-521
  oceanic regions, 518-521
Equatorial Currents, 312, 476-477, 512
Equatorial Undercurrents, 478, 513-518
Equilibrium theory of tides, 428-429
Erosion, 27-28
Estuaries, 559-564
  types, 560-564
European Mediterranean Sea, 3, 340
  stratification and circulation, 490-495
Evaporation processes, 179-181
  transfers of water and heat, 179-181, 193-196
Expendable Bathythermograph (XBT), 128, 137

Field of mass, 299-301
Field of motion, representation of, 307-308
Field of pressure, 302-303
  representation of, 304
Fish and fishing, 521
  fishing off Peru, 512
  plankton, 275-280, 593
  production rates, 275
  temperature changes, effect of, 222
Flank zones, 10-11
FLIP, American buoy, 113, 114-115
Floor, deep-sea, 17-18
Florida Current, 315, 501, 530-531
Fog, formation of, 459
Food chains, 272, 476
Fracture and step zones, 12, 48-49
Freezing-point depression, seawater, 69-71
Friction:
  coefficient, 163-164
  equations of, 295-299
  influence on long waves, 394-397
  of tidal currents, 455-457

Galapagos Islands, 512
Gases in seawater, 89-93
  saturation values, 92, 93
  transfer between atmosphere and ocean, 166-168

Gauges, 128
  float, 128-129
  high-sea, 129-130
General Bathymetric Chart of the Oceans (GEBCO), 5-6, 13
Geomagnetic electrokinetograph (GEK), 146-148
Geomagnetic measurements, 38-41
  magnetic anomalies, 37-39
  paleomagnetic epochs, 39
  reversals of geomagnetic field, 39, 48
Globigerina ooze, 19, 24, 25-26
Glomar Challenger (research vessel), 16, 18, 23, 33, 53, 101-102
Grand Banks, ice patrol, 246-247
Gravity, field of, 290-291
  measurements of, 37
Greenland Basin, 504
Greenland-Iceland Rise, 466-467
Greenland-Scotland Ridge, 465, 487
Greenland Sea, 215, 504
Gulf of California, 3
Gulf of Finland, 553-554
Gulf of Mexico, 498
Gulf of Oman, 557
Gulf of St. Lawrence, 3, 442
Gulf Stream, 247, 270, 316, 333, 528-529
  branches, 530
  countercurrents, 533-534
  cross circulation, 534
  current eddies, 533
  current meanders, 532, 535
  cyclonic eddies, 326-327
  discovery of, 529
  horizontal oscillations, 532
  nonperiodic pulsations, 531-532
  periodic variations, 531-532
  source of, 501
Guyots, 12-13, 49-50
  See also Seamounts
Gyres of water masses, 476-481
  formation of, 481-483

Hawaiian Islands, 380-381, 528
Heat, 34, 40-41, 168-189
  budget, 168-189
    equation, 168-189
    large-scale, 168-169
    total heat balance, 182-185
    transfer by back radiation, 175-176

618 Subject Index

transfer by direct thermal conduction, 176-178
transfer by evaporation, 179-181
transfer by incoming radiation, 169-175
evaporative, 66, 179-181
flow at sea floor, 34, 40-41
propagation in ocean, 186-189
 dynamic convection, 188
 lateral mixing, 188-189
 molecular thermal conduction, 186-187
 thermohaline convection, 187-188
specific heat of seawater, 66
vertical fluxes from profile measurements, 163-164
High fractured plateau, 10
Horse latitudes regions, 190, 525-528
 color of sea, 528
 hydrographic structure, 525-528
 stratification and circulation, 525-528
Hudson Bay, 3
Humboldt Current, 510
Hurricanes, 191-193, 379
Hydrogen in seawater, 89
 oxygen and hydrogen sulfide, 283
Hydrographic regions, 504-565
 coastal waters, 558-565
 equatorial countercurrents, 518-521
 horse latitudes, 525-528
 jet stream regions, 528-536
 monsoon regions, 521-525
 polar regions, 544-550
 regional classification of oceans, 504-507
 shelf seas, 550-558
 trade wind currents, 505-518
 west wind drift, 536-544
Hydrologic cycle, 198-202

Ice and icebergs, 238-250
 charts of coverage, 244-246
 formation and types of, 238-243
 icebergs, 241-242
  calving of, 241-242
  depth of submerged part, 243
 nomenclature, 239-240
 patrols 246-250, 307-308, 550
 "permanent ice," 544
 physical properties of, 58, 86-88
 polar regions, 544-550
 sea ice, 239
Iceland, 547-550
Iceland-Faeroe Ridge, 465-466, 468

ICES, *see* International Council for the Exploration of the Sea
IGY, *see* International Geophysical Year
Indian Ocean, 3, 7
 deep water, 470-473
 salinity, 225
 water masses, 236-237
Instruments and observational methods, 98-157
 airplanes and satellites, 110-111
 computers, 109-110
 development of, 120
 drifting platforms, 114-116
 fixed observation platforms, 111-114
  moored systems, 111-112
  research towers, 112-114
  underwater laboratories, 112-114
 measurement of depth of, 127-128
 measuring techniques, 116-157, 485-488
  See also Measurements and measuring techniques
 platforms for observations, 98-116
 position fixing, 107-109
 research vehicles, 98-111
 submersibles, 103-107
Instrument Society of America, 120
International Council for the Exploration of the Sea (ICES), 212, 226, 228
International Geophysical Year (IGY), 2, 544
International Hydrographic Bureau in Monaco, 6, 13-14
Irish Sea, 3, 459, 551
Irminger Current, 504
Islands, oceanic, 12
 formation of, 49-50

Jet stream regions, 487, 528-536
 stratification and circulation, 528-536
Joint Oceanographic Institutions for Deep-Earth Sampling (JOIDES), 18

Krakatao eruption (1883), 379
"Krill," 476, 550
Kuril Trench, 46
Kuroshio, 528, 535
 characteristic features, 535-536

Laboratories, underwater, 112-114
Labrador Current, 247, 530, 547
Labrador Sea, 247, 284
Land and water, distribution of, 1
Level of sea surface, 42

Subject Index 619

measurement of variations, 128-131
Life:
　energy production of organisms, 266-271
　ocean as source of life on earth, 264-266
Lime, 24
Lipids, 279
Loran-A navigation method, 108

Madeira, 528
Magma, 49-50
Magnetism, 38-41
　measurement of, 34, 38-41
Manganese nodules, 20, 23
Marginal seas, 2-3
Marine biology, 83, 97
Marine Geology (journal), 16
Marine Technology Society, 120
Mass, field of, 299-301. *See also* Water masses
Measurements and measuring technique, 116-157, 485-488
　basic requirements, 116-120
　bottom topography, 34-41
　　geomagnetic measurements, 38-41
　　gravimetric measurements, 37
　　seismic investigations, 34-41
　content of dissolved substances, 152-157
　　atomic absorption spectroscopy, 153
　　biochemical "microbiological" method, 154
　　chromatographic methods, 154
　　colorimetric methods, 155-157
　　continuous recordings, 157
　　neutron activation analysis, 153-154
　　volumetric analysis, 154-155
　content of suspended material, 150-152
　of currents, 143-157
　　drift measurements, 148-150
　　at fixed position, 143-148
　of density or specific weight, 142-143
　electronic, 120
　fixed observation platforms, 111-114
　of instrument depth, 127-128
　　pressure measurements, 127
　　time measurement at constant sinking velocity, 128
　measured quantities, 116
　optical measuring methods, 150-152
　of salinity, 137-140
　of sound velocity, 140-142
　of surface waves, 131-134
　of temperature, 134-137

of water depth, 122-127
of water level variations, 128-131
water sampling devices, 120-122
See also Instruments and observational methods
Mediterranean seas, 2, 3, 340
　American Mediterranean Sea, 236, 498-501
　Arctic Mediterranean Sea, 501-504
　area, volume, and depths, 3
　Austral-Asiatic Mediterranean Sea, 497-498
　categories of, 488-489
　circulation schemes, 488-490
　European, 236, 490-495. *See also* European Mediterranean Sea
　hydrographic conditions, 488
　Red Sea, 495-497
　stratification and circulation, 488-504
　water balance, 493
Meteor (research vessel), 100-103, 125
Mid-Ocean Ridge, 8, 10
　formation of, 46-48
Minerals, 23
Mixing processes, 295-299, 340
MODE (Mid-Ocean Dynamics Experiment, 488
Mohole drilling, 18, 95
Momentum, vertical fluxes of, 163-164
Monsoon currents, 190-191, 521-525
　effect on Austral-Asiatic Mediterranean Sea, 497-498
　stratification and circulation, 521-525
Morphology of ocean bottom, 1-53
　bottom topography, 1-15
　formation of bottom features, 33-53
　mechanism of sedimentation, 26-33
　sediments at ocean bottom, 16-26
Motion, equations of, 288-289, 291-292
Motion processes, 484
　measuring techniques, 484-486

Nansen water bottles, 121
National Academy of Sciences, 88
Navigation methods, 107-109
　hyperbolic, 108
　position fixing, 107-109
　radio direction finding, 108
　satellites and, 109
Netherlands, storm surge (1953), 381
Neutron activation analysis, 153-154
Nile River, effects of damming, 493
Nitrate, 284

Nitrogen in seawater, 89, 270, 278
  phosphate-phosphorus ratio to nitrate
    nitrogen, 279
Northeast Passage, 550
North Sea, 3, 248, 249, 443, 550
  fishing yield, 275
  hydrography, 552
  salinity, 226-229
  storm surges, 381-385
  tides, 442
Norwegian Current, 338-340
Norwegian fjords, 283, 398, 563-564
Norwegian Sea, 236, 502-503, 548
Nutrients, 509-510
  regional distribution of oxygen and, 284-287
  variation of, 278-279
  See also Plankton organisms

Observational methods, 98-157
  measuring techniques, 116-157
  platforms for, 98-116
  See also Instruments and observational methods
Oceans:
  adjacent seas, 3. See also Mediterranean seas
  area, volume, depths, 3
  bottom, see Bottom topography
  boundaries of, 1-2
  formation of, 50-53
  hydrographic regions, 504-565
  interaction between atmosphere and, 158-202
  islands, 49-50
  level of sea surface, 42-43
  names of, 1-2
  organic production, 271-275
  regional classification, 504-507
  size of, 2
  as source of life on earth, 264-266
  theory of bridge continents, 51
Oil and gas deposits, 22, 558
  explorations, 33
  towers, 113
Optical properties of seawater, 78-84
  measuring methods, 150-152
Organic substances in seawater, 94-96, 271-275
  measurement of dissolved, 152-157
  measurement of suspended, 150-152
Organisms
  energy production of, 266-271

organic production of ocean, 271-275
Oscillations, stability, 405-406
Oslo Fjord, 563-564
Osmosis, seawater, 69
Overflow of ocean margins, 487
Oxygen in seawater, 89
  hydrogen sulfide and, 283
  regional distribution of nutrients and, 284-289
  transformation into carbon dioxide, 279

Pacific Ocean, 3, 7
  deep water, 470-473
  salinity, 225
  water masses, 236-237
Paleoclimatology, 32
Parameterization of vertical transports, 164-166
Persian Gulf, 3, 557-558
pH range in seawater, 281, 287
Philippine Trench, 306
Phosphate, 284
Phosphorus, 270, 278
Photography, underwater, 16
Photosynthesis, 266-267
Phytoplankton, 269
  chemical composition, 276-277
  metabolism, 277
Plankton organisms, 97, 269, 476
  chemical composition, 275-276, 279
  development of, 509-510
  in food chains, 550
  forms of, 286
  mass bloom, 280
  mechanism of metabolism, 275-280
  zooplankton, 269
Plate tectonics, 52-53
Platforms for obervations, 98-116
  drifting, 114-116
  fixed observation, 111-114
    moored systems, 111-112
    research towers, 112-114
    underwater laboratories, 112-114
  research vehicles, 98-111
Polar Current, 546
Polar front, 537-539
  North Atlantic, 312
Polar regions, 544-550
Pollution of sea surface, 74
Portugal Current, 530
Position fixing, 107-109
Potassium in seawater, 255, 270
Power stations, tidal, 451

Precipitation, component of water budget, 196-197
Pressures:
  field of, 302-303
  homogeneous ocean, 291
  measurements, 127
Production of ocean, 271-275
  absolute and effective, 273, 280
  food chains, 272
  growth and decomposition curves, 272-273
  organic, 271-275
Project Mohole, 18, 95
Protein in seawater, 279
Pteropod ooze, 25
Pycnoclines, 311, 320
Pycnometers, 142

Radar, navigation aid, 108
Radiation in seawater, 78-84, 169-175
  albedo, 79
  fundamental laws of, 170-175
  heat source, 169-175
  heat transfer by back radiation, 175-176
  measurements by satellites, 184-185
  shortwave, 78
Radioactive waste material, disposal of, 32
Radioactivity, 256-261
  energy transfer during, 258
  generated by cosmic rays, 257-259
  isotopes, 259-261
  primary natural, 256-257
  produced artificially, 259-261
Red Sea, 3, 22, 84, 231, 236, 495-497
  deep water, 470-471
  hot brines, 496-498
  stratification and circulation, 490-497
  tides, 434
"Red Tide," 22, 280
Reef and atoll formation, 44, 49
  coral, 25, 26, 44
Regional oceanography, 460-565
  hydrographic regions of world ocean, 504-565
    classification of oceans, 504-507
    coastal waters, 558-565
    horse latitudes, 505, 525-528
    jet stream region, 507, 528-536
    monsoon currents, 505, 521-525
    polar regions, 507, 544-550
    trade wind currents, 505-518

  equatorial countercurrents, 505, 518-521
  region of west wind drift, 507, 536-544
  regional classification of oceans, 504-507
  stratification and circulation, 460-504
    in deep layers, 460-488
    in mediterranean seas, 488-504
Research vehicles, 16-19, 98-111
Research vessels, 98-103
  list of, 99
Respiration, 266, 271
"Respiratory coefficient," 279
Reykjanes Ridge, 10-11
Reynolds number, 296
Richardson number, 297
Ridge system, 10-11
  Mid-Ocean Ridge, 46-48
Rift valleys, 10
Rivers, continental runoffs, 197-198, 559

Salinity, 222-230
  distribution of surface, 222-225
  distribution with depth, 225-226
  diurnal and annual variations, 226-228
  electrical conductivity of seawater, 61-63, 566-567
  equations, 566-567
  long-term changes, 228-230
  major constituents, 88-89
  measurement of, 137-140
  sea ice, 85-88
  as thermodynamic variable, 60-61
Salts, transfer processes between atmosphere and ocean, 166-168
Samples:
  collecting devices, 18
  water sampling devices, 120-122
Sandy shores, 377
Sargasso Sea, 528
Sargasso weed, 307
Satellites:
  drift measurement, 148-150
  ice conditions, observation, 248
  navigation systems, 109
  oceanographic observations, 110-111
  radiation measurement, 184-185
  wave data, 133
Sea-air interface
  transfer processes in, 158-168
Sea Floor Spreading, 47-52
Sea of Japan, 3, 236, 467, 524-525
Sea level, 42-43
  measurement of water level variations, 128-130

Seamounts and oceanic islands, 12
  formation of, 49-50
  guyots, 12-13
Sea of Okhotsk, 3, 236, 467, 524-525
Seaquakes, 45
Sea surface, 372-375
  statistical description, 368-372
  waves and sea state, 357-368
Seas:
  adjacent, 3
    stratification and circulation, 488-490
    tidal currents, 449-453
    tides of, 442-445
  marginal, 2-3
  mediterranean, 2, 3, 340, 488-501
  shelf, 550-558
Sea Surface Temperature Analysis and Forecasting (SST), 215
Seawater:
  acoustic properties, 74-78
  chemical composition, 88-97, 253-255
    gases, 89-93
    major constituents of salinity, 88-89
    organic substances, 94-96
    suspended matter, 96-97
    trace elements, 89
  color of, 84
  density, 63-66, 142-143
  electrical conductivity, 61-63
  evaporative heat and adiabatic temperature change, 66-67
  freezing-point depression, 70
  measurement of content of dissolved substances, 152-157
  measurement of content of suspended material, 150-152
  measurement of density or specific weight, 142-143
  measurement of water depth, 122-127
  measuring techniques, 116-157
  molecular thermal conductivity, 68-69
  optical properties, 78-84
  osmosis, 69-70
  pH range, 281, 287
  physical properties of, 54-88
  salinity, 60-61, 88-89. *See also* Salinity
  sea ice, 85-88
  specific heat, 66-67
  surface tension, 72-74
  thermal expansion and compressibility, 63-66
  turbulent exchange coefficients, 71-72

unique characteristics of pure water, 54-60
vapor pressure, 69-71
viscosity and surface tension, 72-74
water sampling devices, 120-122
Sediments and sedimentology, 16-26
  age of core, 32
  calcium content, 283
  chemical composition, 24-25
  geographic distribution, 25-26
  grain size distribution of, 24
  hemipelagic, 25
  investigative methods, 16
    analysis of samples, 16, 19
    direct observations, 16
    measuring thickness and structure, 16, 18-19
    sample-taking, 16, 18
  laboratory methods for analysis, 19
  littoral sediments, 24
  mechanism of sedimentation, 26-33
    erosion, 27-28
    physical defining quantities, 26-27
    sedimentation rates, 32-33
    transport, 28-32
    turbidity currents, 29-30
  nature of, 16
  origin of sediments, 19-24
    biogenous sediments, 21-22
    cosmogenous sediments, 23-24
    halmyrogenous sediments, 23
    terrigenous sediments, 19-21
  pelagic sediments, 24
  red deep-sea clay, 25-26
  sedimentation budget, 255-256
  settling duration, 29
  sinking velocity, 29
  stratification, 25, 33
  turbidity currents and transport of, 28-32
Seiches, 28, 387-391, 431-432
  in Lake of Geneva, 388
Seismic investigations of ocean bottom, 34-41
    reflection investigations, 33, 36
    refraction measurements, 35-36
Sensors, deep sea, 113
Shearing stress, tangential, 295-299
Shelf seas, 550-558
Shelves, continental, 9-10
  formation of, 41-45
Ships, research, 98-103
  list of, 99
Silicate, 284
Silicon, 279

Sills, deep-sea, 12, 48
Skin-divers, 103
SOFAR, see Sound Fixing and Ranging
Somali Current, 522-523
Sound and sounding techniques, 4-5
  acoustic properties of seawater, 74-78
  measurement of sound velocity, 140-142, 569
  profiles of sound velocity, 76
  propagation in ocean, 74-78
    transmission loss, 76
  soundings on bathymetric charts, 5-6
Sound Fixing and Ranging (SOFAR), 76, 150, 485
South Atlantic Ocean, 222
"Spreading of Ocean Floor," 47-52
  development of seamounts and, 49-50
  drifting of crustal plates, 51-52
  hypothesis of, 39-40
SST, see Sea Surface Temperature Analysis and Forecasting
Statics and kinematics, ocean currents, 299-308
  field of mass, 299-301
  field of motion, 307-308
  field of pressure, 302-303
    determination of, 303
    representation of, 304
  stability of water stratification, 304-307
Step or fracture zones, 12, 48-49
Storm surges, 379-385
  warning systems, 454-455
Strait of Gibraltar, 340
  outflowing Mediterranean water, 491-493
Strait of Hormuz, 557-558
Strait of Messina, 404-405
Straits, ocean, 565
  compensation currents in, 338-340
Stratification and circulation:
  Antarctic bottom water, 461-465
  Antarctic water masses, 474-476
  Arctic bottom water, 465-467
  circulation gyres of water masses, 481-483
  in deep layers of three oceans, 460-488
  deep water, 470-473
  in mediterranean seas, 488-504
    American Mediterranean Sea, 498-501
    Arctic Mediterranean Sea, 501-504
    Austral-Asiatic Mediterranean Sea, 497-498
    circulation scheme, 488-490
    European Mediterranean Sea, 490-495
    Red Sea, 495-497
  stability of water stratification, 304-307

subpolar intermediate water, 467-470
variability in ocean, 483-488
water masses in cold-water sphere, 460-461
water masses in warm-water sphere, 476-481
Submersibles, 16, 102, 103-107
Subpolar intermediate water, 467-470
Substances in sea water
  chemical composition, 94-97
  dissolved in sea water, 152-157
  suspended in sea water, 96-97, 150-152
Subtropical Convergence, 336, plate 6
Surf:
  coastal waters, 559
  wave transformation in shallow water, 375-378
Surface tension of seawater, 72-74
Surface waves, see Waves
Surveying techniques, 33
Suspended matter in seawater
  chemical composition, 96-97
  measurement of, 150-157

Tectonic activity, 34, 41, 46, 52-53
Television equipment, 17-18
  deep-sea photocamera, 21
Temperature, 203-222
  adiabatic change, 66-67
  annual variations of surface, 206-209
  changes in ocean temperature, 66-67
  computation of potential, 568
  distribution at depth, 212-217
  distribution at sea surface, 209-211
  diurnal variation at sea surface, 203-206
  isotherm charts, 209-212
  long-term changes, 217-222
  measurement, 134-137
    in deep layers, 134-135
    paleotemperatures, 19
Temperature-salinity relationships, 232-233
  water masses, 236-238
Terminology:
  ice and icebergs, 239
  ocean bottom features, 13-15
Thermal conduction or diffusion
  equations of, 294
  heat transfer by, 176-178
  seawater, 68-69
Thermal expansion of seawater, 63-66
Thermal properties of sea ice, 85
Thermistor chains, 137

Thermoclines, 340-341, 459
  summer, 340, 476
Thermometers, reversing, 135-136
Tidal phenomena, 407-459
  analysis of tide observations, 421-425
  bores, 397-398
  classification of tidal forms, 427
  coastal waters, 559
  definition, 407-411
    reference levels of tides, 410-411
    tides and tidal currents, 407-410
  derivation of, 413-415
    harmonic expansion, 415-417
  friction of tidal currents, 455-457
  inequalities of tides, 408-410
    declination, 409
    diurnal, 409
    parallax or monthly, 409
    semimonthly or fortnightly, 408-409
  "meteormological tides," 421, 454
  ocean tides, 417-428, 438-442
    characteristics, 426-428
    harmonic representation, 417-421
    theory of, 428-438
  power stations, 451
  predicting tides, 407, 409-410, 425-426
    harmonic method of, 425-426
  spring tide, 408
  superposition of astronomic tides, 454-455
  theories of ocean tides, 428-438
    classic hydrodynamic theories, 429-431
    co-oscillating tides, 431-435
    equilibrium theory, 428-429
    numerical integration of tide equations, 435-438
    scope of, 428
  tidal currents, 407-410, 446-449
    of adjacent seas, 449-453
    friction of, 455-457
    harmonic analysis, 446
    harmonic constants, 421-425
    harmonic representation, 417-421
    turbulence of, 457-459
  tidal cycle, 408
  tidal waves, 455-457
  tide-producing forces, 411-417
    derivation of tidal potential, 413-415
    description of system, 411-413
    harmonic expansion of tidal potential, 415-417
  tide tables, 407
  tides of adjacent seas, 442-445
  tides of bays and marginal seas, 434
Titanic, sinking of, 246
Topography of ocean bottom, *see* Bottom topography
Towers, research, 112-114, 119
Trace elements, 270
  measurement of, 152-157
  neutron activation analysis, 153-154
  in seawater, 89
Trade wind currents, 190, 505-518
  with components directed poleward, 518
  with components directed toward equator, 508-512
  with strictly westward motion, 512-518
Transfer processes between ocean and atmosphere, 158-160
  gases and salts, 166-168
  methods for determination of vertical transports, 160-162
  parameterization of vertical transports, 164-166
  small-scale, 158-168
  vertical fluxes of momentum, heat, and water vapor, 163-164
  vertical transports in maritime friction layer, 160-162
Transport of sediment, 28-32
Trenches, deep-sea, 9-10, 45-46
  gravity anomalies, 37
Tritium, 59
Tropics, water masses, 476-481
Tsunamis and storm surges, 28, 45, 345, 379-385
Turbidity of seawater, 32, 96-97. *See also* Currents, turbidity
Turbulence:
  equations of, 295-299
  exchange coefficients, 71-72
  theory of homogeneous, 298-299
  of tidal currents, 457-459
Typhoon, 191, 379

Underwater laboratories, 112-114
UNESCO, 61
Upwelling at ocean margins, 487
U.S. Hydrographic Office, 210

Vapor pressure, seawater, 69-71
Velocities, method of computing, 315-317
Ventilation of world ocean, 284

Vertical transports:
  methods for determination in maritime friction layer, 160-162
  parameterization of, 164-166
Virgin Islands, 498-499
Viscosity and surface tension, 72-74
  equations and formulas, 569
Vitamins, 270
Volcanic eruptions, 13, 50, 379
Volumetric analysis, 154-155
Vorticity, potential, 332-333

Warm-water sphere, 476-481
  Atlantic Ocean, 480-481
  water masses, 476-481
Water:
  budget, 189-202
    continental runoff, 197-198
    evaporation and, 193-196
    hydrologic cycle, 198-202
    precipitation and, 196-197
    regional wind fields, 191-193
    wind systems, 189-190
  depth, 6-8
    density distribution, 231
    determination of, 3-4, 7
    distribution of water temperature, 212-217
    homogenous ocean, 291
    measurement of, 122-127
    statistics of depth distribution, 6-8
    zones in oceans, 7
  "heavy water," 59
  Levantine intermediate water, 491
  measurement of depth, 122-127
  planetary wind system at sea surface, 189-190
  subpolar intermediate water, 467-470
  temperatures in world ocean, 203-222
  unique characteristics of pure water, 54-60
  See also Seawater
Water level data, 410-411
  measurement of variations, 128-131
Water masses, 232-238
  Antarctic, 474-476
  characteristic, 236-238
  circulation gyres, 481-483
  in cold-water sphere, 460-461
  formation of, 233-236
  stratification and circulation, 460-461
  temperature-salinity relationships, 232-233
  warm-water sphere, 476-481

Water sampling devices, 120-122
Water vapor, vertical fluxes from profile measurements, 163-164
Waves, 343-406
  boundary surface, 398-400
  capillary waves, 74, 344-346, 353-354
  classification of, 328, 343-346
    with respect to generating forces, 344-346
    with respect to restoring forces, 343-345
  currents and, 328-329
  "dead water," 398
  duration and fetch, 366
  gravity, 344-346, 352-353
  harmonic, 346
  height of, 366-368
  inertial, 328-331, 402-405
  infinitesimal, 354-355
  influence of bottom topography, 332-333
  internal, 343, 400-402
    tidal waves, 402-405
  Kelvin, 385-387, 393-394
  kinematic properties, 346-352
    damped waves and forced waves, 349-352
    harmonic oscillations and wave fields, 346-349
    standing waves, 349
  long surface, 378-397
    amphidromic systems, 391-394
    edge waves, 385-387
    influence of friction, 394-397
    Kelvin waves, 385-387
    properties of, 378-379
    seiches, 387-391
    tsunamis and storm surges, 379-385
  measurement of surface, 131-134
  nonlinear interactions, 355-356
  planetary, 331-332
  progressive, 343, 378
  Rossby, 332, 333-334, 535
  rotating, 391-393
  sea state and, 357-368
    generation of, 372-375
    statistical description of, 368-372
  seismic surface (Love and Raleigh waves), 34
  shelf, 385-387
  short surface, or deep-water, 352-378
    capillary waves, 353-354
    gravity waves, 352-353
    properties of sea and swell, 357-358
    statistical description of sea state, 368-372
    wave transformation in shallow water, 375-378

waves of finite amplitude, 354-355
stability oscillations, 405-406
standing, 343, 349, 387-388
storm-surge, 28
surf, 375-378, 404-405
surface, 343, 398-400
  generation of, 372-375
  measurement of, 131-134
tidal, 402-405, 455, 457
types of, 328, 343-346
Väisälä period, 401, 405-406
wind velocity and, 357-368
Weddell Current, 546
Weddell Sea, 461-538
West wind drift region, 190
  stratification and circulation, 536-544
Western boundary currents
  linear theory, 324-327
  nonlinear theory, 327-328
Whales and whaling, 476, 550
Wind systems, 190-191
  cyclones, 191-193
  at equator, 478
  hurricanes, 191-193
  monsoons, 191
  oceanic regions and, 504-507
  planetary wind system, 189-190
  special regional wind fields, 191-193
  state of sea surface and, 359-368
  trade winds, 190, 476, 505-518
  in tropical warm-water sphere, 191-192, 479
  west wind drift, 536-544
  See also Currents
Windward Passage, 498-500
Woods Hole Oceanographic Institution, 104, 529

XBT, *see* Expendable Bathythermograph

Yellow Sea, 524
Yucatán Current, 530
Yucatán Sea, 498

Plates 1–8

Names of Oceans and Seas

The numerals 1 to 66 refer to Plate 1

Atlantic Ocean

1. Baltic Sea
2. Kattegat, Sound and Belts
3. Skagerrak
4. North Sea
5. Greenland Sea
6. Norwegian Sea
7. Barents Sea
8. White Sea
9. Kara Sea
10. Laptev Sea
11. East Siberian Sea
12. Chukchi Sea
13. Beaufort Sea
14. Passages of the Canadian Archipelago
14a. Baffin Bay
15. Davis Strait
15a. Labrador Sea
16. Hudson Bay
16a. Hudson Strait
17. Arctic Ocean
17a. Lincoln Sea
18. Waters off the West Coast of Scotland (Sea of the Hebrides and North Minch)
19. Irish Sea with North Channel and St. George's Channel
20. Bristol Channel
21. English Channel
22. Bay of Biscay
23. North Atlantic Ocean
24. Gulf of St. Lawrence
25. Bay of Fundy
26. Gulf of Mexico
27. Caribbean Sea
28. Mediterranean Sea
28a. Adriatic Sea
28b. Aegean Sea
29. Sea of Marmara
30. Black Sea
31. Sea of Azov
32. South Atlantic Ocean
33. River Plate
34. Gulf of Guinea

Indian Ocean

35. Gulf of Suez
36. Gulf of Aqaba
37. Red Sea
38. Gulf of Aden
39. Arabian Sea
40. Gulf of Oman
41. Persian Gulf
42. Laccadive Sea
43. Bay of Bengal
44. Andaman Sea
45. Indian Ocean
45a. Mozambique Channel
45b. Great Australian Bight

Pacific Ocean

46. Strait of Malacca
47. Gulf of Siam
48. Java Sea and Celebes Sea
49. South China Sea
50. East China Sea
51. Yellow Sea
52. Sea of Japan
53. Inland Sea of Japan
54. Sea of Okhotsk
55. Bering Sea
56. Philippine Sea
57. North Pacific Ocean
58. Gulf of Alaska
59. Coastal Waters off Southeast Alaska and British Columbia
60. Gulf of California
61. South Pacific Ocean
62. Bass Strait
63. Tasman Sea
64. Coral Sea
65. Solomon Sea
66. Bismarck Sea

Vertical distribution of temperatu

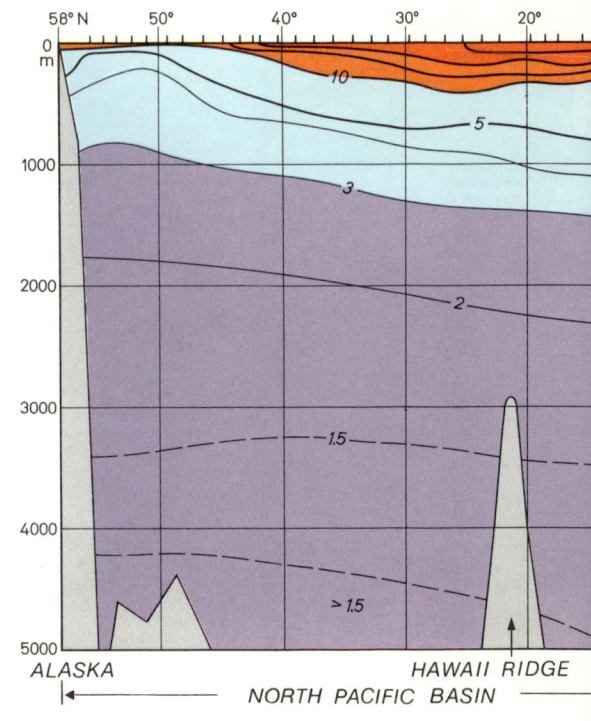

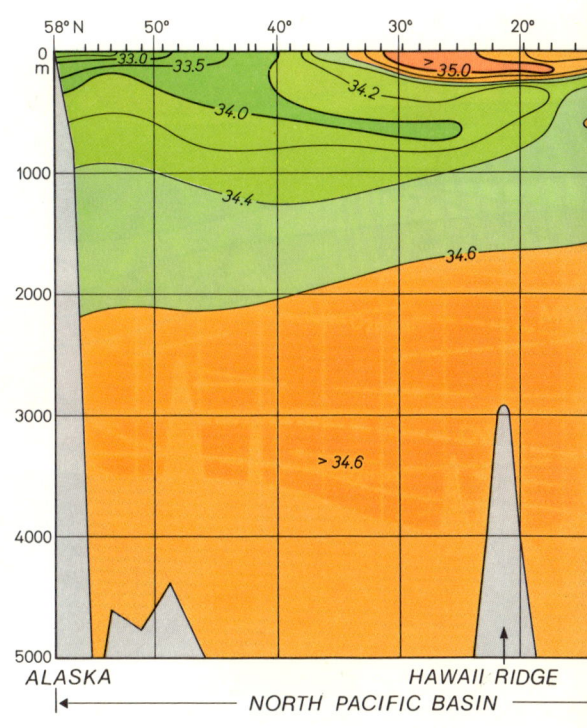

Position see Figure 10.31, sources see C

and characteristic water masses

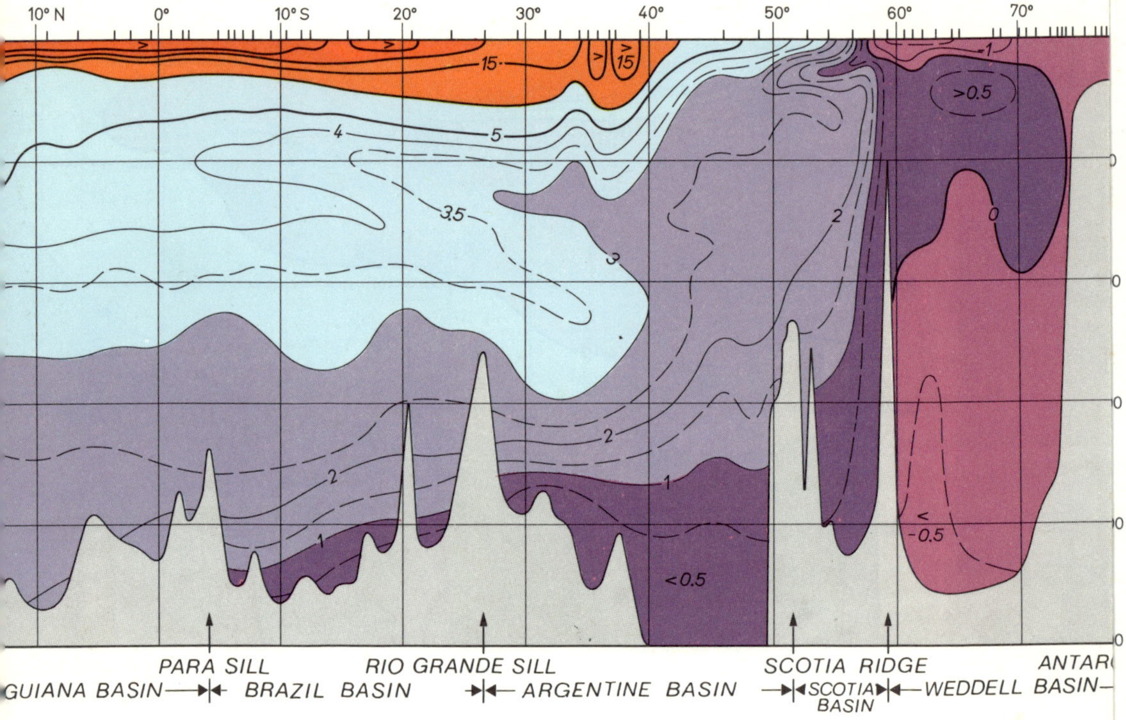

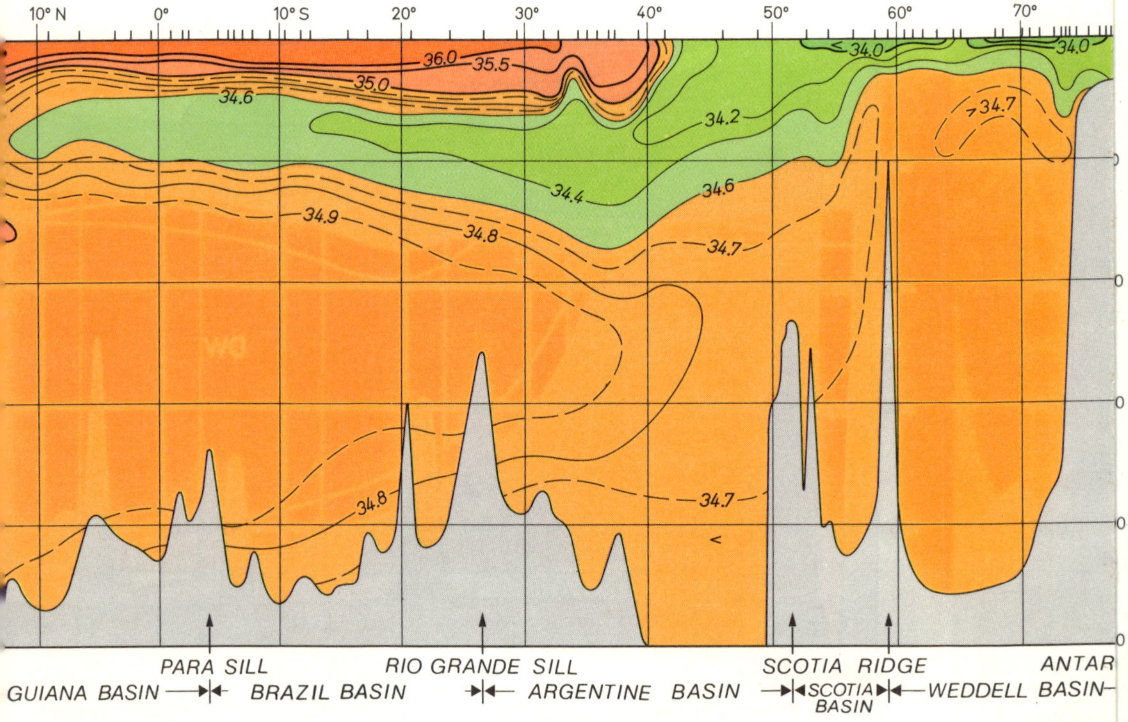

GUIANA BASIN → ← BRAZIL BASIN → ← ARGENTINE BASIN → ← SCOTIA BASIN → ← WEDDELL BASIN →
PARA SILL   RIO GRANDE SILL   SCOTIA RIDGE   ANTAR

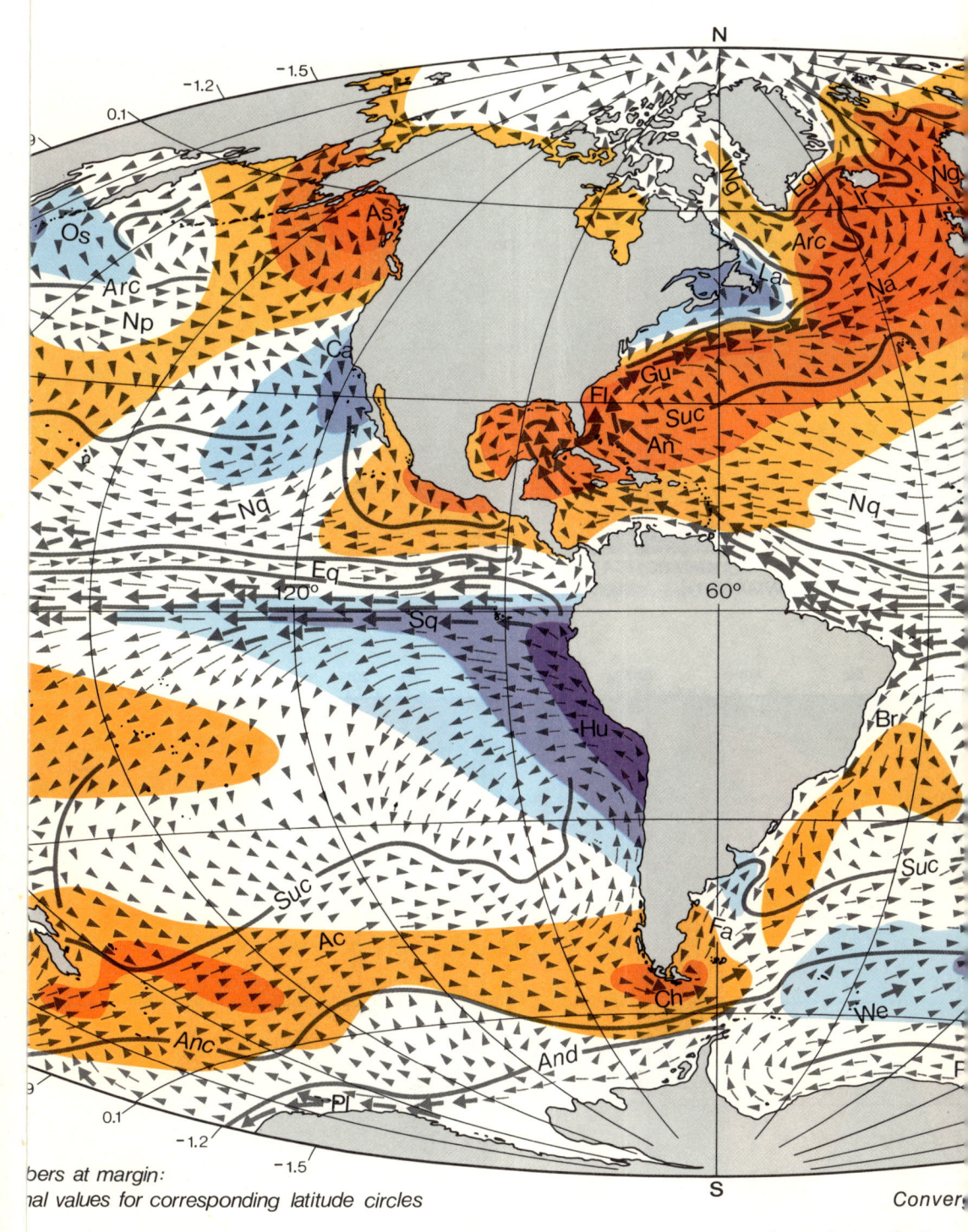

bers at margin:
nal values for corresponding latitude circles

Conver

Anc   Antarctic Conv
And   Antarctic Diver

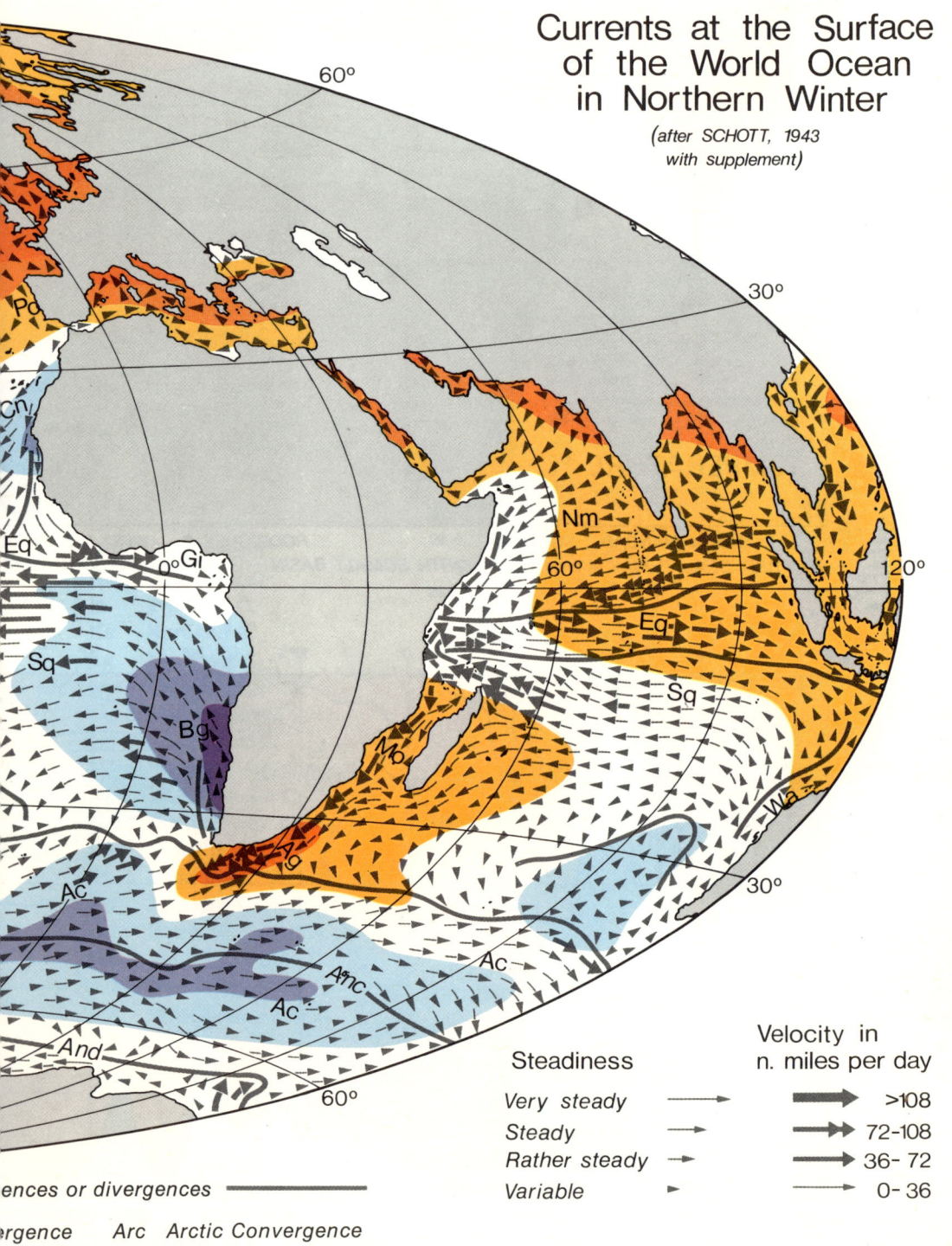